Economic and Business Analysis

Quantitative Methods Using Spreadsheets

Economic and Business Analysis

Quantitative Methods Using Spreadsheets

Frank S. T. Hsiao
University of Colorado, Boulder, USA

NEW JERSEY • LONDON • SINGAPORE • BEIJING • SHANGHAI • HONG KONG • TAIPEI • CHENNAI

Published by

World Scientific Publishing Co. Pte. Ltd.
5 Toh Tuck Link, Singapore 596224
USA office: 27 Warren Street, Suite 401-402, Hackensack, NJ 07601
UK office: 57 Shelton Street, Covent Garden, London WC2H 9HE

British Library Cataloguing-in-Publication Data
A catalogue record for this book is available from the British Library.

ECONOMIC AND BUSINESS ANALYSIS
Quantitative Methods Using Spreadsheets

Copyright © 2011 by World Scientific Publishing Co. Pte. Ltd.
All rights reserved. This book, or parts thereof, may not be reproduced in any form or by any means, electronic or mechanical, including photocopying, recording or any information storage and retrieval system now known or to be invented, without written permission from the Publisher.

For photocopying of material in this volume, please pay a copying fee through the Copyright Clearance Center, Inc., 222 Rosewood Drive, Danvers, MA 01923, USA. In this case permission to photocopy is not required from the publisher.

ISBN-13 978-981-283-492-8
ISBN-10 981-283-492-3

Typeset by Stallion Press
Email: enquiries@stallionpress.com

Printed in Singapore.

In the memory of my parents

Preface

Most 21st century students are familiar with microcomputers. They are adept in visually-oriented learning and playing, whether through video games, music videos, DVDs, iPod, or the internet. This book appeals to the computer skills of modern day undergraduate and graduate students through innovative uses of spreadsheets, including built-in spreadsheets, equations, and formulas.

This computer skill-intensive book covers major topics in both economic and business analysis. Students will learn how to build complex spreadsheet layouts and perform high-level calculation and analysis using the popular spreadsheet program Microsoft Excel. The exposition of the basic concepts, models, and interpretations are presented intuitively and graphically without compromising the rigor of analysis. We study Economics, Business, Mathematics, and Statistics systematically and with hands-on practice while learning Excel.

To encourage students' active learning and critical thinking, they will be given hands-on experiences by creating tables and graphs presented in the text and practice questions. They will be able to change parameters within spreadsheets and see the effects of change instantly. At the same time, by acquainting themselves with Microsoft Excel, students will directly acquire practical and advanced job skills.

The book is intended for third or fourth year (upper-level undergraduate) economics or business majors in colleges and universities, and first-year graduate students in MA programs in economics and business. The book is a stand-alone full textbook for one semester, but, depending on the pace of instruction (if Excel commands have to be covered extensively), it can be extended to two semesters. It can also serve as a supplement to textbooks on principles of economics, intermediate economics, and introductory statistics.

The Features, Tools, Structure, and Advantages of this Book

There are many unique and innovative features in this book, which have been developed by the author over the past twenty five years through teaching a course using the manuscript of this textbook and from his research on computer assisted learning (CAL). Our new systematic approach of unifying economics, business, mathematics, statistics, and spreadsheet programs is probably the first of its kind in the area of quantitative methods. The following is a summary and explanation of the unique features, tools, structure, and advantages of our approach.

A. The features of this book

(1) The book contains major topics **in economics**, such as the derivation of the demand and supply curves and utility and production optimization models in microeconomics, national income models, economic policy analysis, and dynamic models in macroeconomics. These topics are also shared by managerial economics. On the other hand, the book also covers some of the major topics in **business economics**, including future and present value problems with or without annuity, probability, statistics, regression analysis, flow charts, and PowerPoint slide presentations. Due to recent emphasis on integrated and interdisciplinary studies, more and more economics majors are also interested in these business topics, and more business majors are also interested in microeconomics and macroeconomics.

(2) For both students and instructors, instead of spending time in programming, using software like C+, we feel that it is much easier to spend time on **spreadsheet construction** and learn economic modeling first-hand, using spreadsheet commands directly. If the proprietary software programs are used in teaching, the students download the program as add-ins, they will miss hands-on practice opportunities. This is a black box we would like to avoid. While it is true that proprietary software programs may be more powerful and may introduce more advanced or sophisticated methods or topics, we submit that such advanced or sophisticated methods or topics may not be appropriate at the undergraduate level.

(3) We emphasize the economic interpretation and **policy implications** of the models and computer output results, and the **applicability** of the models to real world issues in economics and business. For example, in teaching linear policy models, we systematically present the effects of fiscal, monetary, and consumption policies on equilibrium income and consumption, by simply changing the parameters in the model. We present it as the prototype of the more sophisticated federal government's economic policy, and all analyses are carried out easily by finding an inverse matrix. If a model does not make economic sense, we ask students to change the model (like parameter values) so that the model is plausible.

(4) The book combines major concepts and tools in **economics and related topics in statistics, mathematics, and spreadsheet programs**. As an upper level undergraduate textbook, it can also serve as a review of the materials for students who have previously studied introductory economics and business courses, and give students a different, unified perspective on previous courses. For example, we emphasize the similarity of demand and supply relations in microeconomic market models and in the aggregate demand/supply macroeconomic income determination models, the relationships between static and dynamic microeconomic and macroeconomic models, the relationship between numerical demand and supply equations, the method of estimating these equations, and other topics.

(5) The book appeals to the **intuitive and visual understanding** of complex economics, business, statistics, and mathematics topics, such as unconstrained and constrained optimization problems, through the use of spreadsheets. In fact, our method of using 3D

and 2D contour maps makes it easy for students to visualize optimization problems on spreadsheets, before going through the first and second order conditions or LaGrange function method for optimization using calculus.

(6) "**Computer assisted teaching**" also enables the students to take the "**computer assisted testing**". Unlike traditional written tests, this book allows students to use computers to perform complicated calculations, such as accurate comparative static analysis (Chapter 4), estimating multiple regression coefficients, hypothesis testing (Chapter 7), finding the solution to a large system of simultaneous equations (Chapter 10), and giving economic interpretations to the results during the examinations. This eliminates the need for take-home examinations.

(7) Since the calculations are performed by computer, the instructor has more time to concentrate on teaching the procedures of computing, economic and business concepts, and interpretation or explanation of results. Our basic philosophy is reflected in the questions presented in the **homework assignments** for each chapter. They are mostly collections of past mid-term or final examinations that the author has used in this course. Note that few homework questions are purely mathematical or statistical; yet, due to the complete integration with microcomputer techniques, the level of these homework questions is quite advanced and challenging, and at least comparable to, if not exceeding, the difficulty of course materials in intermediate microeconomics or macroeconomics.

(8) More importantly, the tables and charts in the book are not only used to illustrate explanation of definitions, theories, and applications of topics, as in most textbooks, but are also presented to that students can **reproduce or reconstruct the tables and charts** using Microsoft Excel, to gain better understanding of the topics and deeper insight into the concepts and methods.

(9) The textbook is **self-contained**. Through innovative uses of spreadsheets and built-in spreadsheet equations and formulas, we are able to present all these topics in one book. We do not use homemade add in spreadsheet macro programs, proprietary software programs, or additional data sets by attaching a CD to the book. We emphasize **hands-on practice**; all models are built with the students, step-by-step, on site in the classroom. All data sets, except in Chapter 14, where students practice how to download on-line data, are also generated by the students in the classroom through the use of a random number generator.

In general, unlike the practice of workbooks or study guides of other textbooks, the basic concepts are explained concisely and rigorously in each chapter. We provide key terms in economics and business and key terms in Excel commands at the end of each chapter for review, and the basic topics are illustrated by a flowchart at the end of chapters.

B. On analytical tools

(1) **We use only algebraic operations. Calculus is not required**. The exposition of the basic concepts, models, and interpretations are presented intuitively and graphically without compromising the rigor of analysis. We study Economics, Business, Mathematics,

and Statistics systematically and with hands-on practice while learning Excel. To maintain rigor and clarity of exposition, we introduce the concept of difference to explain the change of variables, which is then used extensively in the last chapter on difference equations.

(2) We place emphasis on teaching the class to understand **intermediate and some advanced concepts in economics and business** without being inhibited by the technicalities of mathematics and computer programming. For example, we introduce and illustrate the income and substitution effects of price change in demand, total factor productivity, and the golden rule of capital accumulation without getting into complicated mathematics or computer programming.

(3) In this book, learning spreadsheet commands advances with the progress of each chapter, along with improvement in statistical and mathematical modeling skills. From statics to comparative statics, from dynamics to comparative dynamics, the sequence of commands builds up naturally, and the **spreadsheet commands are a joy for students to learn and instructors to teach**. For example, when we study comparative statics in Chapter 4, we introduce the naming method and picture copy command, so that students will know immediately the uses and advantages of these commands. Thus, the spreadsheet commands are introduced in the context of economic and business applications. These commands are also used in other chapters, and students have opportunities to reinforce their previous learning.

(4) In addition to simulation methods and constructing data sets in the classroom setting, we also introduce how and where to **download government and international data sources**, and we practice the **research methods** of extracting useful information through Excel's data analysis tools: sorting, subtotaling, auto-filtering, and pivotal tables and graphs.

C. Structure of the book

(1) Each chapter has **six parts**: chapter outline, objectives, basics (theoretical background), programming (on how to use Excel to learn, verify, and understand the basic theories), appendices (explanation of the text and additional Excel techniques), and homework. These are integral parts of systematic learning.

(2) At the end of each chapter we include a **summary**, list of **key economic concepts**, a list of **new spreadsheet commands**, a **review of mathematical and statistical formulas**, **problems, and applications**. Problems and applications are covered extensively in each chapter of the book.

(3) **Homework** includes a collection of the past mid-term and final examination questions from my upper level undergraduate course, *Microcomputer Applications in Economics*. It presents useful practical problems and is an integral part of learning. Some homework problems are taken from various existing textbooks on mathematical economics or economic statistics, to show the linkages between our approach (using spreadsheets) and the traditional approach (without using spreadsheets).

(4) We emphasize the **link between basic theories and hands-on programming**. Students learn abstract basic theories and then verify or understand the theory by hands-on programming through Excel. They also learn the meanings and sources of formulas and equations through hands-on programming. Unlike some textbooks or workbooks in mathematical economics or statistics, this book tries to present theory and practice as two sides of one coin.

D. Advantages of using spreadsheets

(1) By using the **random number generating device** in the Microsoft Excel program and what we call the "naming method", students can generate their own simulated and yet realistic data sets, and change the parameters to find out how they change the economic solutions and the graphic images by their hands-on practice. The ease of generating a data set, as compared to entering external data by typing or downloading the data from a website, encourages students to engage in active learning that they initiate themselves. In addition, the random numbers are "live" and students really enjoy "**dancing**" **numbers and curves** by simply pressing the recalc (F9) key.

(2) We have devised a powerful and versatile "naming method" to prepare sensitivity tables to illustrate the functional relations in **three-dimensional graphs** instantly. For example, this method can illustrate easily any two-independent-variable function $z = f(x, y)$. Our method avoids the problems of relative and absolute references in copying a formulas or the need for either prepacked commercial or homemade add-in programs. We have also solved the classroom problem of quickly implementing a complicated equation into tables without importing add-ins.

(3) We apply the "**naming method**" and the "**picture copy command** (PCC)" to illustrate and learn comparative statics and comparative dynamics in economics. Many current computer-assisted instruction (CAI) programs do not use the naming method or the picture copy command, making teaching and learning comparative statics and comparative dynamics quite tedious and time consuming, as the students have to change formulas and equations each time, instead of just changing the parameters.

(4) The naming method, the picture copy command, and the range copy command (RCC) used extensively in this book enable students to reproduce different **comparative statics graphs** and arrange them in one sheet in the same workbook beautifully. All the above RCC and PCC can be performed within 15 minutes or less, even during course examinations. In this sense, we submit that our methods "revolutionize" the presentation of economics in Excel, and can be applied in many other fields in business and sciences.

(5) By combining the **built-in Excel Solver** for the optimization program and the graphic methods, students can find and illustrate solutions to optimization problems as accurately as possible. By changing the parameters, they can experiment with the impacts on solutions as many times as they like. They even can invent their own optimization problems.

(6) We emphasize the one-to-one **correspondence between data table and chart**. Students are not only able to draw the charts, they are required to find the correspondence between lines and curves in charts and numbers in the table. For example, they must find the equilibrium price and quantity of the demand and supply curves in the chart as well as in the table. They also will be asked to change the parameters of the model and trace out how the correspondence has been changed.

(7) Since the charts can show different widths for horizontal and vertical grid lines, using them is the same as using plotting paper in the pre-computer era several decades ago. Instead of the students drawing charts manually, the computer can **draw the charts** accurately and neatly. Students can change the scales of drawings, zoom in and zoom out, and get a feeling for the charts.

(8) Since charts and tables can be colored or boxed, we encourage students to present the tables and charts in **as colorful and pleasant forms** as they can. For example, students enter the formula of the CES function themselves and set up the three-dimensional surface. Then we encourage them to color the table and charts in presenting their results. They learn and have a lot of fun painting. This experience will not be possible in traditional economics, business, and quantitative textbooks.

(9) We have devised a method of **constructing a sample with large data set instantly**, such as randomly generated uniform, normal, or student-t distributions, or large sets of records with many fields (like region, month, year, gender, political affiliation, etc.), using random numbers and the IF **function**, for practice in the classroom. This cannot be done in traditional textbooks.

E. Preparing computer skill for job markets

(1) This is **a skill intensive book** in which students learn the skills of using spreadsheets for research and presentations on the job after they graduate. For example, we include the methods of constructing flow charts and presentations by Microsoft PowerPoint. We also include some aspects of business mathematics, like future and present value problems, with and without annuity calculations, which are then tied to the exponential functions and logarithmic functions used in the theory of economic growth. We also build loan amortization tables and methods of constructing index numbers. Students as well as the general readers may use the techniques to work on their personal financial planning.

(2) One of the difficulties in using Excel is that it requires students to learn "idiosyncratic" spreadsheet commands. In this book, **the spreadsheet commands are introduced systematically from simple to complex commands in a natural way as we cover more chapters**. In recent years, most students already have used or are even familiar with some Excel commands before taking the class, and thus learning Excel commands is not a big problem. They can quickly get into economics and business applications in this book. Furthermore, the spreadsheet skills and applications learned in this book are useful when the students graduate and take jobs in banks, business, or government,

or international organizations. In fact, the spreadsheet skills they learn from this book may be among the **most useful tools** that they learn in their college years.

As an example of what we are going to present in this book, a summary chart of the book is shown below. The readers can see immediately the relations among various methods of analysis contained in the book through the flowchart; at the same time, in Chapter 14, they also learn the skill and fun of constructing such a chart (see Fig. 14.2).

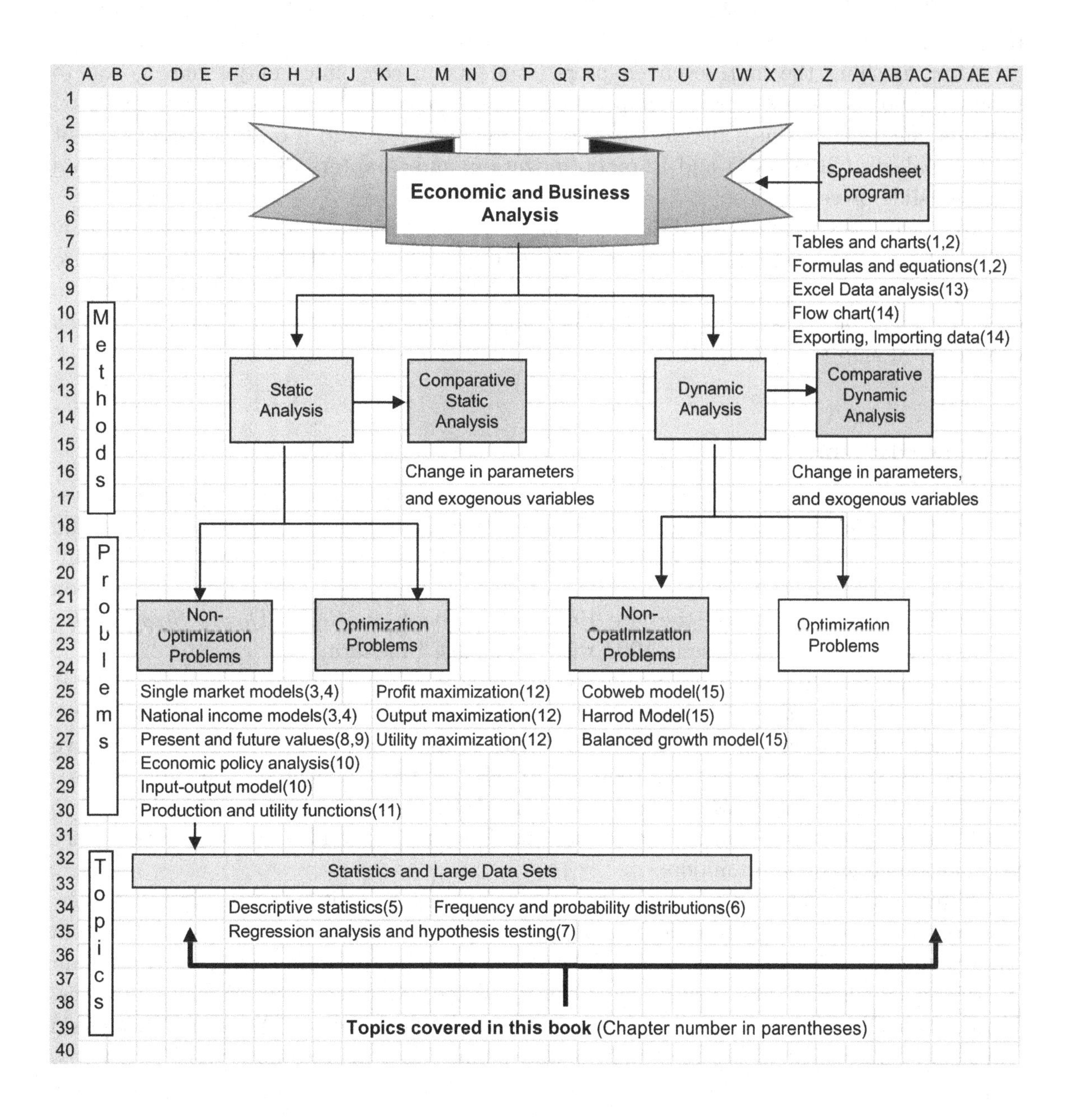

Some Suggested Uses of This Book

The contents of this book can cover more than one semester. Thus, the instructor has flexibility in choosing the topics of the chapters or sections. A common ground might include Part I and Part II. Part I includes the first four chapters, which are introduction to basic Excel spreadsheet commands as well as basic economic and business analysis (microeconomics and macroeconomics). The subsequent three chapters (Chapters 5 to 7 in Part II) cover the basic statistics, which are also common topics for both economic and business majors.

After finishing the first seven chapters, students in economics major may go on to Chapter 10 on economic policy, Chapters 11 and 12 on optimization, and Chapter 15 on dynamic analysis. Business majors may go on to Chapters 8 and 9 on future and present value problems, Chapters 11 and 12 on optimization, and Chapters 13 and 14 on large data sets and slide presentations.

We have tried to build up Excel commands and mathematical techniques gradually as the chapter proceeds. Hence, if some chapters are skipped, the readers may want to review the commands and techniques explained in the skipped chapters. The key terms in economics and business, along with Excel commands and terms, listed at the end of each chapter, will help them locate the missed commands or techniques.

Acknowledgments

When personal computers became popular in the 1980s, I was fascinated by their usefulness in Computer Assisted Teaching (CAT) and Learning (CAL). I created ECON 4838: *Microcomputer Applications in Economics* in 1985 with a new course development grant from the University of Colorado at Boulder. I then published 13 academic papers on computer assisted learning in the late 1980s, and a book on *Game Theory Step by Step Using Spreadsheets* in Japanese in 1997 with Yoshisuke Umehara.

The original course name for this book was *Microcomputer Applications in Economics.* It can be *Economic (and/or Business) Analysis — Quantitative Methods using Spreadsheets.* In our department, since the late 1990s, this course has been one of any three quantitative courses that economics majors are required to take for the BA degree. The three courses are this course, namely, *Micro computer Applications in Economics, Introduction to Mathematical Economics*, and *Introduction to Econometrics.* This class has been very popular among Economics majors.

Numerous students and colleagues, in and out of the University of Colorado at Boulder, have contributed to this book project over the past two decades. I have benefited from the comments and suggestions of (in alphabetical order) Professors Changmo Ahn, Nam T. Hoang, Hong-Van Thi Hoang, Mei-Chu W. Hsiao, Xiaodong Liu, William Mertens, Keun Yeob Oh, Changsuh Park, Katsuhiro Saito, Myles Wallace, Yongkul Won, Akio Yamashita. They helped me read part of the manuscript. In addition to the artist Ms. Corinne Wang, many students also helped me. Henry Yen Heng Chen, Robert Howard,

Shao Yan Lee, Ying Lin, Jonathan D. Mathiew, Maung H. Myat, Ha Manh Nguyen, and others, made many valuable suggestions. Professor Mary Coberly assisted me in making this book more readable during the various stages of writing, and Professors Charles R. Goeldner, Kishore G. Kulkarni, James Markusen, Henry Wan, Jr., and William Wei advised me on publication. Many current and previous staff members of the Editorial Office of World Scientific Publishing spent a long time transforming the rough manuscript into a finished book, and Ms. Alisha Nguyen, the Social Science Editor, pressed forward at the final stage to see this book published. To all of them, I am grateful. My sincere appreciation goes to my wife, Mei-Chu, and our children, Edward and Victoria, for their encouragement and unflagging support during my academic career and the writing of this book.

About the Author

Frank S.T. Hsiao received his B.A. and M.A. in Economics from the National Taiwan University, Taipei, Taiwan, and M.A. and Ph.D. in Economics from the University of Rochester, Rochester, NY. From 1966 to 2007, he was a full-time professor in the Department of Economics at the University of Colorado at Boulder. He taught undergraduate and graduate courses on *Mathematical Economics (Statics and Dynamics)*, *Economic Statistics*, and *Microcomputer Applications in Economics*, using many of the techniques and models described in this book.

Professor Hsiao's expertise has been recognized internationally through many lectures at universities and institutes in Japan, the Netherlands, Mexico, Taiwan, Korea, Malaysia, Thailand, Hong Kong, and China. He has also been a visiting scholar at the Fairbank Center for East Asian Research of Harvard University; the Hoover Institute of Stanford University; Hitotsubashi University, Nagoya University, the International Centre for the Study of East Asian Development, etc., in Japan.

Professor Hsiao's research interests include economic growth and development and quantitative methods in economics. He has published extensively in leading professional journals, including *American Economic Review, Journal of Political Economy, Journal of Finance, Southern Economic Journal, Metroeconomica.* He has also coauthored papers in many journals such as *World Development*, *Review of Development Economics*, *Journal of Asian Economics*, *Journal of the Asia Pacific Economy, Journal of Korean Economy*, and *Journal of Productivity Analysis.*

Currently, Dr. Hsiao is Professor Emeritus of Economics in the Department of Economics, University of Colorado at Boulder, and is an Associate Editor for the *Journal of Asian Economics.*

Brief Contents

Contents

Part I

Basic Economic and Business Analysis

Chapter 1

The Excel Worksheets

Chapter Outline

Objectives of this Chapter

This chapter acquaints the readers with the appearance of the Excel spreadsheets by moving around the four corners of the spreadsheet window (Sec. 1.2). We explain the basic features of the tabs, ribbons, groups, toolbars, and logos. Since the "Home" tab and mini toolbar are the basis of all Excel operations, their contents are explained in detail (Sec. 1.3). We also introduce the use of arrows and the text box step by step (Sec. 1.4). To add some fun to learning the otherwise rather boring Excel commands and features, we have added WordArt in Sec. 4. In learning the use of WordArt, the readers also learn the major objectives of this class. The appendix explains how to customize the quick access toolbar (Appendix 1A).

Some Conventions Used in this Book

Before we start, here are some of the notation conventions that will be used throughout this book. They are listed in Table 1.1. In cases 1 and 2 of Table 1.1, we may encounter a sequence of commands like

<"Insert"><Text, WordArt!>(Choose a character)<"Drawing Tools, Format">
<Shape Styles, Shape Fill!><Gradient>, etc.

This will occur when we want to show a drawing object (see Example 1.4 WordArt and Fig. 1.14). In this case, the first <.> tells you to click "Insert" tab button in the menu, and then, click the down arrow ▼ next to WordArt in the "Text" group of the ribbon. Then, the instruction in (.) is that you choose a character. The third <.> consists of two

Table 1.1 Some conventions used in this book

Case	Symbol	Example	Explanation
1	< >	<F2>	Select and press the F2 (edit) key, or name of key F2.
2	<x, y>	<Font, Bold>	In Font group (x), click the Bold button (y).
3	<"x">	<"Home">	Select the "Home" (x) tab in the menu bar.
4	[]	[2]	Enter optional number 2. It can be any number.
5	( )	(A dialogue box opens)	Explanation of results or procedure.
6	!	Arial!	Click the down arrow ▼ at the right of the Arial box.
7	@	@All	Select the bullet eye next to "All".
8	x	xPrint to file	Click the blank square left of "Print to file".
9	^ (Cat)	^C = control+C	Hold the control key and then press the C key.
10	# (music Sharp)	#C = shift+C	Hold the shift key and then press the C key.
11	RM (or LM)	<RM> (or <LM>)	Click the right (or left) button of the mouse. We call this procedure **Right Mouse** or **Left Mouse**.

parts, "Drawing Tools", under which there is a new tab "Format". The first entry in this <.> is for location: it lets you locate the "Format" tab under the "Drawing Tools". The second entry in <.> is for clicking, you click the second tab to open the "Format" ribbon. The fourth <.> indicates that, after you open the "Format" ribbon, you locate the "Shape Styles" group among the other groups (see Table 1.2) and click the down arrow (▼) next to the Shape Fill button. A drop-down list appears. The last command <.> has a single entry. It asks you to select and click the down arrow of Gradient button.

In the text, we also give examples. At the end of an example, we add □ to indicate its end. Sometimes, for clarity, <x> is also used to show the name of button x rather than requesting the reader to click x. <RM> or <LM> is also used to denote right mouse or left mouse.

1.1 General Worksheet Properties

A **spreadsheet** is a software program for entering numerical data for calculation, manipulation, and analysis. It usually comes with graphic and drawing facilities with limited word processing capacity.

When you start Excel, the monitor screen will be filled with a **worksheet** (1) or **spreadsheet** (see Fig. 1.1), which is embedded in a **document window**, that is, the **screen**. It displays command menus on the top and bottom parts, with blank grids of rows and columns, and lets you enter data in the blank spaces. When the mouse is moved inside the document window, an empty fat cross sign called the (mouse) **pointer** or **curser** (2) appears. When the pointer is pressed, a **cell** is **selected**. The selected cell is enclosed with

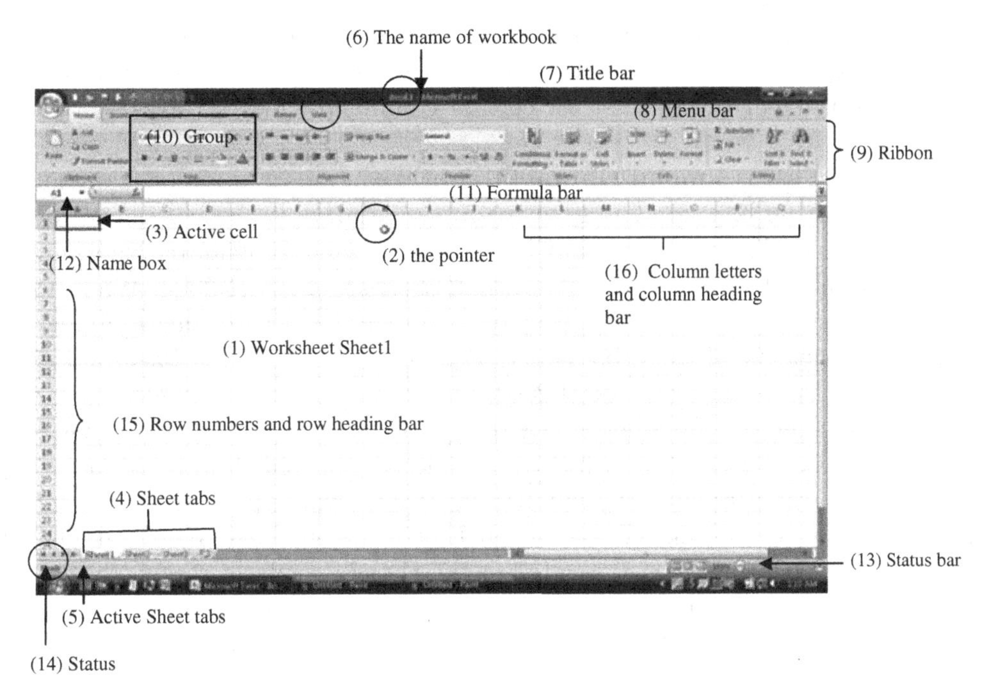

Fig. 1.1 Worksheets and a workbook

dark borders, and the pointer still remains as a fat cross. That cell is called the **active cell** (3). Similarly, a column or a group of columns, or a row or a group of rows, may be selected by clicking the **column letters** (16) or the **row numbers** (15), and the selected columns or rows will be shaded.

You change a document window to another window by clicking the **sheet tab** (4) at the bottom part of the left-hand side. Now, type your name or any word anywhere in Sheet1, and click the Sheet2 tab. A new worksheet, Sheet2, will open. Click the Sheet1 tab to return to Sheet1. In the illustration, there are three **sheets** or **worksheets**: Sheet1, Sheet2, and Sheet3. The **active sheet** (5) is indicated by a white tab. A **workbook** is a collection of all the worksheets; in the current example, the **name of the workbook** (6) is in the **title bar** (7), the first row of the worksheet, it is Book1, as shown in the first row of the worksheet. The names of worksheets and workbooks can be changed, as noted below.

Use the mouse to move the pointer (2) to point to the "Sheet1" tab (5). The fat cross pointer changes to a fat arrow pointer. Click the "Sheet1" tab twice to invert the black and white background of the tab. The tab now is in the **editing mode**. Enter "Question1" and click anywhere in the worksheet. This changes the sheet name. Similarly, change "Sheet2" and "Sheet3" to "Question2" and "Question3". Save the changed workbook as file name "Test1". The name of the workbook (6) is changed from "Book1" to "Test1", which will be shown in the first row of the worksheet.

The second row of the worksheet is called the **menu bar** (8): it gives the menu of available commands. The menu bar has a **tab list**. Its default list has seven tabs: "Home",

"Insert", "Page Layout", "Formulas", "Data", "Review", and "View". Some more tabs called "tools" (like "Table Tools" or "Chart Tools" consisting of "Design", "Layout") may be added to the right of the tab list for certain commands, like <Tables> or <Charts> under the "Insert" tab.

Below the menu bar is a **menu ribbon** (9), which contains **groups** of submenus related to the tab in the menu bar (8). The **group name** is shown at the bottom of the ribbon, like Clipboard, Font, Alignment, etc. under the "Home" tab. For example, the **Font group** (10) of the "Home" tab contains the **standard toolbar** of Excel 2003. When we click a tab in the menu bar (8), the ribbon associated with the tab also changes, giving different groups. The arrangement of the tabs and commands in the ribbon is fixed and cannot be changed or moved, as it can be in Excel 2003.

The row below the group names is called the **formula bar** (11), which will show the formula or the entry of the cell at which the pointer is located. Enter 123 or =rand( ) in any cell, when the cell is selected the number or formula will appear in the formula bar.

The last row at the bottom of the spreadsheet is called the **status bar** (13). In the illustration, the worksheet is in the "ready" mode, that is, ready to receive data.

The major part of the worksheet consists of a grid of **rows** and **columns**. The rows are denoted by Arabic numbers, called the **row numbers** or **headings** (15), shown in the **row-heading bar** (15), and the columns by alphabets, called the **column letters** or **headings** (16), shown in **column heading bar** (16). In Excel 2007, there are 1,048,576 rows and 16,384 columns (A,..., Z, AA,..., AZ, ..., ZA,..., ZZ, AAA,..., AZZ, BAA,..., BZZ,..., XFA,..., XFD). The intersection of a row and a column is a **cell**. There are 17,179,869,184 cells in a worksheet. It is a huge worksheet. Usually, the amount of data you can enter on a worksheet is limited only by the memory of the computer.

A cell has a **cell address** (or **cell reference**) like B5, which means that the cell is located[1] at the intersection of column B and row 5. A rectangular collection of cells is called a **range**. A range may be part of a column or row, or may include parts of rows and columns. It is defined by choosing the cell address at the northwest corner and the cell address at the southeast corner. In the last spreadsheet of Fig. 1.2, for example, the range is denoted as C5:F10, with the colon between the two cell addresses.

Pointer Movements

There are several ways of moving around the spreadsheet.

- The up, down, left, or right arrow keys move the active cell one cell up, down, left, or right from the current active cell.
- If the pointer has selected any cell in a given data set, Ctrl + up, Ctrl + down, Ctrl + left, or Ctrl + right arrow keys move the active cell to the upper, lower, left, or right edge of the data set if the active cell is inside the data set.

[1]This is different from matrix algebra notation. In matrix algebra, a_{ij} denotes the element at the ith row and the jth column, while on the spreadsheet, it denotes the entry in column i and row j.

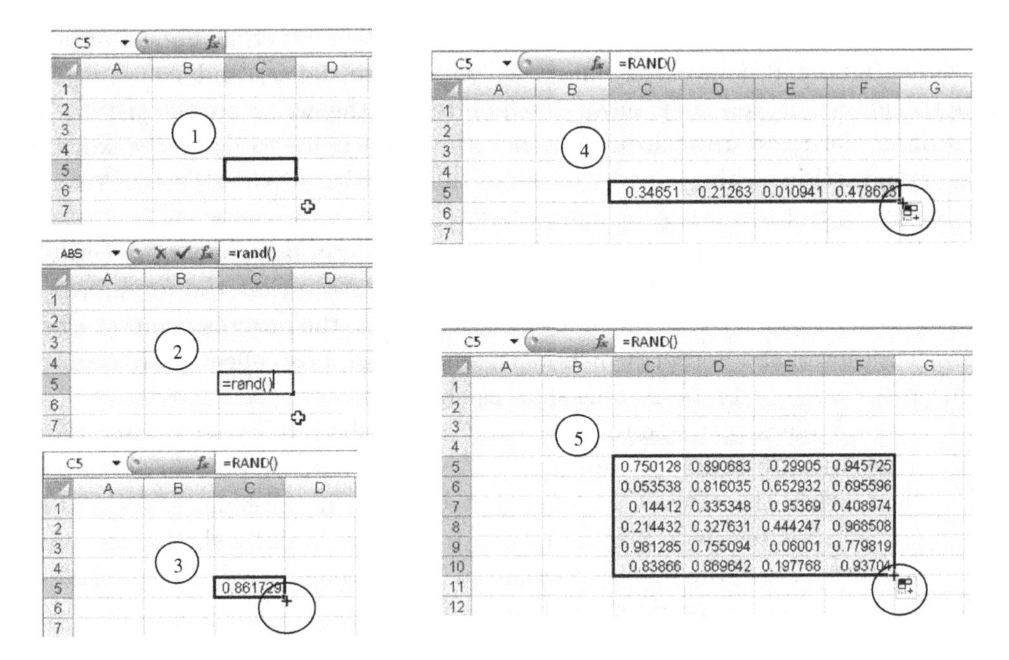

Fig. 1.2 Generating numbers and creating a range

- The PageUp and PageDown keys move the active cell one screen up or down from the current active cell position.
- The Alt + PageUp and Alt + PageDown keys move the active cell one screen left or right from the current active cell position.
- The Ctrl + PageUp and Ctrl + PageDown keys move the current active sheet one sheet left or right from the current sheet.
- The Home key moves the active cell to the leftmost cell of the same row.
- The Ctrl + Home keys move the active cell to cell A1.
- The Ctrl + End keys move the active cell to the lower right corner of the range including all the data sets (not including the charts).
- If the Shift key is combined with the up, down, left, or right arrow key, or the PageUp or PageDown key, the pointer will select the range defined by the original key's movement.

Example 1.1 We want to fill the range C5:F10 with random numbers.

- Step 1. Move the pointer to C5, as shown in step 1 of Fig. 1.2, and click the cell. The pointer changes to a fat cross. Note that the name box denotes the active cell C5, and the cell itself is enclosed with heavy borders. The column heading and the row heading in which the cell is located are shaded.
- Step 2. Enter the **random number generating function.** It is defined as

$$=\text{rand}(). \tag{1.1}$$

Enter it in C5. Note that when =r is typed, a list of function names appears. You may either ignore it (since you know what you are doing) and continue typing to finish the entry; or you may move the pointer to the word on the list to read the definition of the **rand function**, and click the word twice to enter the word into the cell as =rand(. (Finish the word by entering the right-hand parenthesis). A vertical bar is flashing next to ( ), waiting for more entry.

- Step 3. When the return key is entered, the random number appears in cell C5. Each time the <F9> key is pressed, the number changes. Note that the formula bar shows the formula in capital letters (which are default). Move the fat cross pointer close to the southeast corner of the border, which has a small black box called the **fill-handle**, and the fat cross changes to a black-plus sign, as shown.
- Step 4. Drag the black-plus sign to cell F5 and release the LM. This will copy the formula and its value from C5 to F5, as shown.
- Step 5. Next, while the range C5:F5 is selected, drag the fill-handle down to F10 and the range C5:F10 is selected. Release LM and the range will be shaded and filled with random numbers. Move the pointer away from the fill-handle and click the LM, and the shade disappears but the random numbers remain. □

In step 4, if you stop at F5, move the pointer away and click anywhere, and the shade and the black-plus sign disappear. In this case, to resume filling the range C5:F10, you have to repeat step 3 by selecting and shading C5:F5, moving the pointer to the southeast corner of F5, and dragging the range down to F10.

For a given range, we **select a range** by clicking a cell, say C5, pressing LM, and dragging the pointer with the fat cross to the desired range, say to F10, and then releasing the LM. The selected range is shaded. Another method is to select the northwest corner of cell C5, hold the shift key, and then click the southeast corner of cell F10. This second method is particularly useful when the range is large, say several pages long.

When the range is selected, use the **tab key** to move the pointer inside the shaded range, C5:F10. The pointer will move along a row within the shaded area only, and will bounce back to the beginning of the next row at the end of a row. This feature is useful when we enter data in the range. You may press the **F9 key** (which is called the **recalculation key**) several times to change the numbers.

1.2 The Four Corners of the Worksheet

To be acquainted with a person, we first note the person's face, dress, and manners. Similarly, to acquaint ourselves with a worksheet, we have to note and study the appearance of the worksheet and its functions. This way we will not feel strange facing a worksheet.

1.2.1 *Upper left corner and the formula bar*

The first corner of the worksheet is almost the same as other Microsoft Office programs, like Microsoft Word 2007. We explain it in two parts.

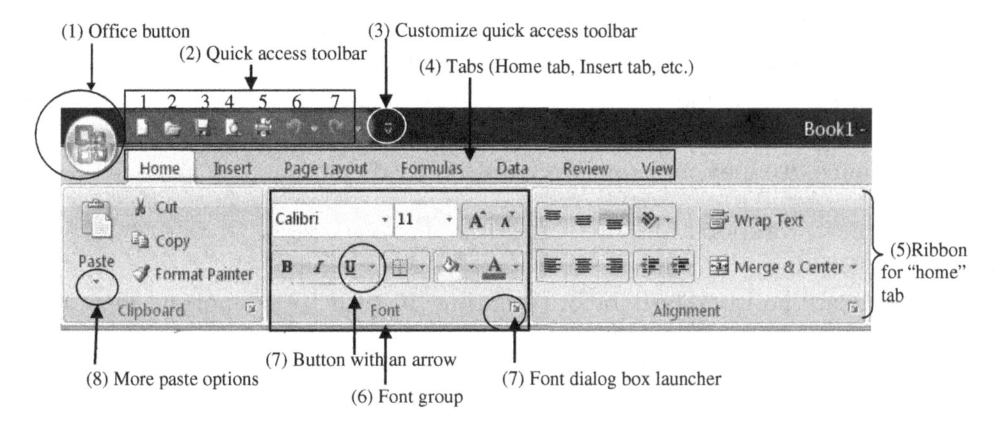

Fig. 1.3 The northwest corner of the worksheet

The first part of the upper left corner

The first part of the upper left corner, see Fig. 1.3, consists of the menus of commands. The first circle with a Microsoft Office logo is called the **office button** (1). Clicking it will open a file management menu. It consists of New (starting a new file), Open (open an old file), Save (save a currently working file under the same file name), Save As (change the filename or location to save), Print (chooses from print, quick print, or print preview), Prepare, Send, Publish, and Close. Probably the first five commands are the most frequently used.

The row of logos on the right of the office button is the **quick access toolbar** (2). As its name indicates, you can activate the command by simply clicking the logo on the menu without going through the steps. The original Vista Operating System comes with only three logos:

3Save (quick save of the file). It saves the current workbook in the current directory automatically without asking what to do. If you want to save in a different file name or in a different disk, you have to use the <Save As> command in the Office button.
6Undo Undo the last action. Type 123 in cell B2. Delete the number. Click <"Undo">. 123 reappears. When you make an error by erasing some entry, you can always use this to restore it.
7Redo Redo the action that you previously undid. In B2, you have the number 123, the result of the "Undo" command. If you click "Redo", the number disappears; that is, you have redone what you undid. Click Undo and Redo several times to get the feeling of the commands.

Saving a file

To save a file for the first time, click <Office><Save as>. You can save in two formats: if you use Excel 2007, then click "**Excel Workbook**", but if you use Excel 1997–2003, you should click "**Excel 97–2003 Workbook**". After you decide which format you are using, select your disk drive, give the new file name as [Ch1RandomNumber] and <Save>. If you choose the former, the file will be saved with extension .xlsx; if the latter, it will be saved with extension .xls. Note that if you save in Excel Workbook (that is in Excel 2007), then you cannot open the saved file using your Excel 97–2003, but if you save the file in Excel 97–2003, you can open it in Excel 2007 (backward compatibility).

For convenience, we may add some more often-used commands:

1**New** Open a new workbook file.
2**File** Open an existing saved file.
4**Preview** Preview the worksheet before printing. Return to the normal view by clicking <Close>.
5**Print** Quick print of the current worksheet. If you want to change the printing specifications, like double-sided printing, multiple copies, etc., you have to use <"Office"><Print> command.

These four menu buttons are taken from the menu inside the office button (1) by using the "**customize quick access toolbar**" (3), as explained in Appendix 1A.

The second row is the tab row (4). Each tab is associated with a **Ribbon** (5). A ribbon consists of several **groups**, listed at the last row of a ribbon. Each group, like the **Font group** (6) in Fig. 1.3, has different buttons with different shapes and sizes. Some buttons come with a selection arrow, like the **Underline** button U (7) or the **More Paste Options button** (8). In both cases, clicking the arrow will give you a drop-down list with a menu. If the button is on the right-hand side of the logo, like **Underline** (7), after you choose an item from the menu, Excel will remember that item, and the item will replace the old item and will be shown on the button. If the arrow is below the logo, like (8), Excel will not remember the item and the logo and the function (in this case, the plain "**Paste**") stay the same. Most buttons come without an arrow, like the **bold face button** B. In this case, you simply select the cell and click B. The text in the cell will be bold faced.

The second part of the upper left corner

In Fig. 1.4(a), the intersection of the **row-heading bar** (1) and the **column heading bar** (2) is called **the select-all button** (3). It will select the whole worksheet by shading the worksheet. Click the select-all button. Move the pointer close to a **column separator** (4) (the short vertical line between any consecutive columns in the column heading bar), say between A and B, as in Fig. 1.4(b). The fat cross changes to a black cross with left and right arrows, indicating that the column separator is ready to move in the direction shown by the arrows. Click LM and an indicator appears, such as "Width: 8.11(80 pixels)", or a similar indicator. See the first part of Fig. 1.4(b). The number changes when you move the separator. Move the separator toward the left to 30 pixels. Click anywhere in the window to **deselect the select-all button**. A "plotting paper" with squared cells is created. Select button (4) again and manually move the column separator back to 80 pixels. Similarly, you may change the height of the rows by clicking and dragging the **row separator** (5) up and down, as in the second part of Fig. 1.4(b).

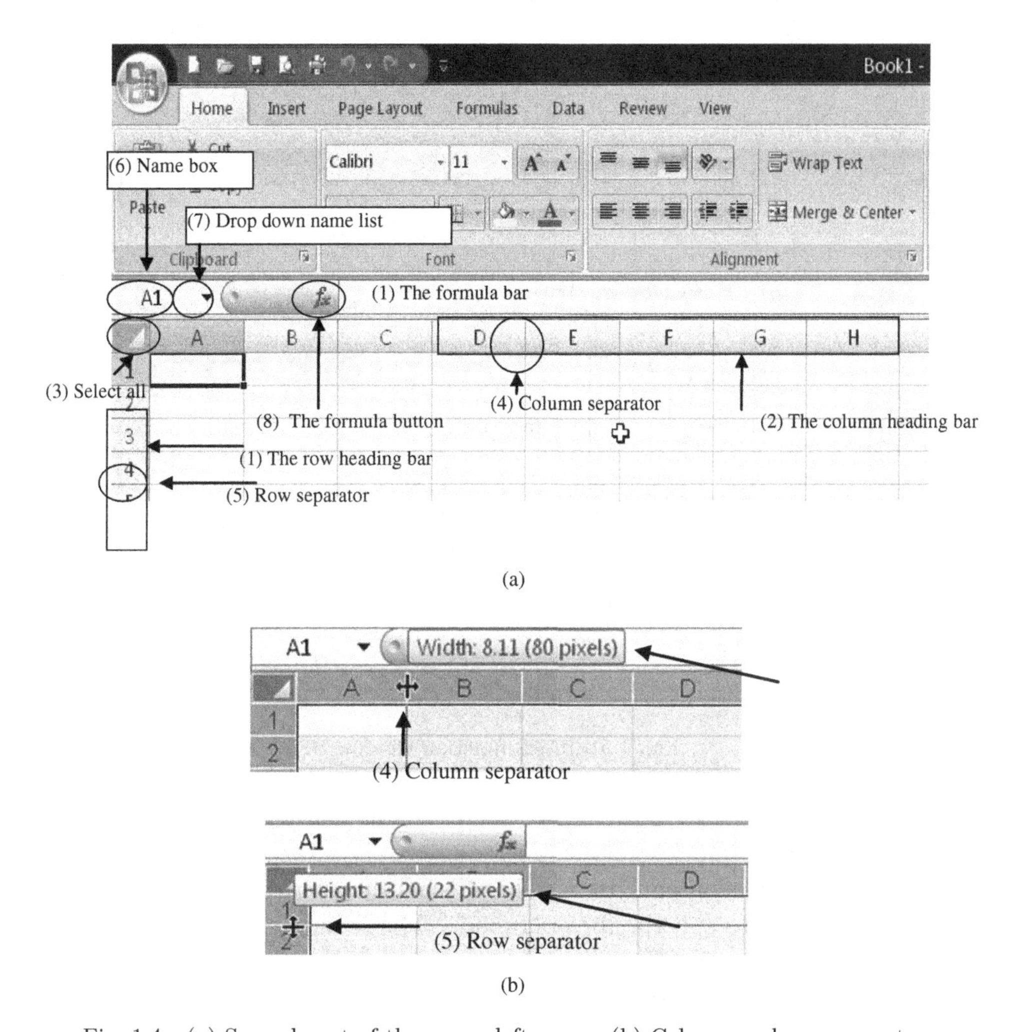

Fig. 1.4 (a) Second part of the upper-left corner (b) Column and row separators

The **Name box** (6) in Fig. 1.4(a) contains the cell address or the named cell address (see Chap. 3 for how to name a range) depending on whether the cell or range is named. In the illustration above, the A1 cell is selected, so the name box shows A1. If a range on the worksheet is given a name, say, "sample", then the name range will show "sample" when the range is selected. If there are other named ranges, **the drop-down name list** (7) contains all other defined names of the ranges on all the worksheets in the workbook.

If you move the pointer to a blank cell, say B5, and click **the formula button** (f_x) (8), an **Insert Function** dialogue box appears (see Fig. 1.5). In the dialogue box, you may click the down arrow on the right of "Select a category" box, and select "Math & Trig" from the drop-down list. Then select RAND from the "Select a function" box. This is a random function we introduced in Fig. 1.2. Click <ok>, then **Function Arguments** box appears.

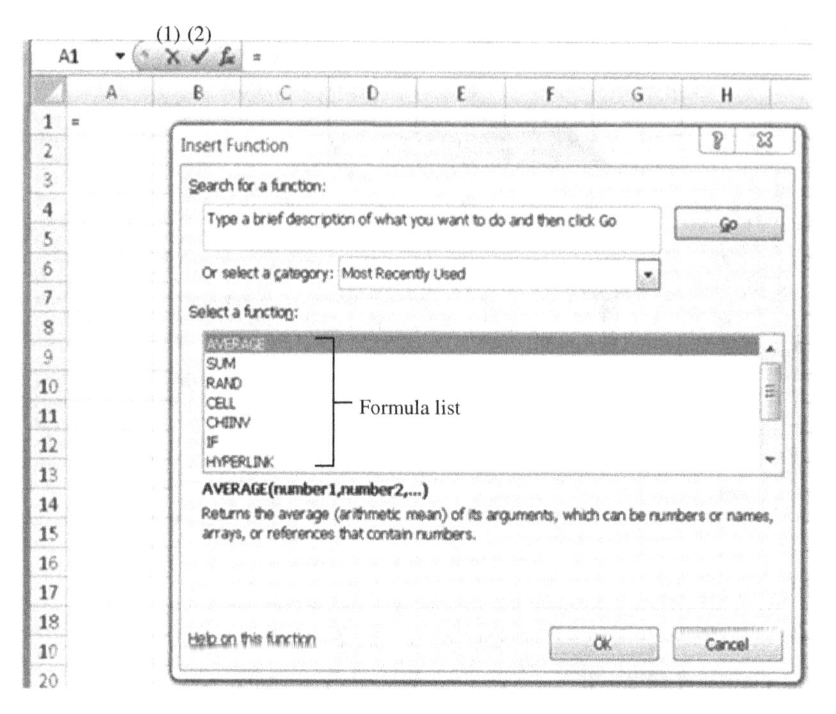

Fig. 1.5 Insert function window

It says: "This function takes no arguments", <OK>. The cell will be filled with a fraction, as in step 3 of Fig. 1.2.

The value or formula contents of a cell will be shown in the **formula bar** (1) in Fig. 1.4(a). Enter <Esc> to exit the formula mode. The X button (1) in Fig. 1.5 is "**cancel the editing**" without changing the original formula (it is the same as <Esc>). The check button (2) is "**enter the formula buttons**", which is the same as <Enter> without the pointer moving one cell down.

In general, compared with Fig. 1.2, the formula button (8) is so time consuming and redundant that it is seldom useful. If you know the formula, as you usually do in most cases, you simply enter =rand( ) manually and get a random number. Thus, we do not use this method of entering formulas in this book. Other methods of entering a formula are explained in Chap. 2.

1.2.2 *Upper right corner*

The upper right corner, Fig. 1.6, consists of a group of basic workbook and worksheet operations: the upper three buttons are to **minimize** (1) the whole workbook to the bottom of the window, **reduce** (2) the sheet to less than full screen, or **close** (3) the Excel workbook; the lower three buttons, 4, 5, and 6, do the same for worksheet. Button (7) is the **Help** button.

Use the mouse to click **the horizontal split box** (9) and drag it to the middle of the worksheet, then enter 1 in the upper part of the sheet and 2 in the lower part of the sheet.

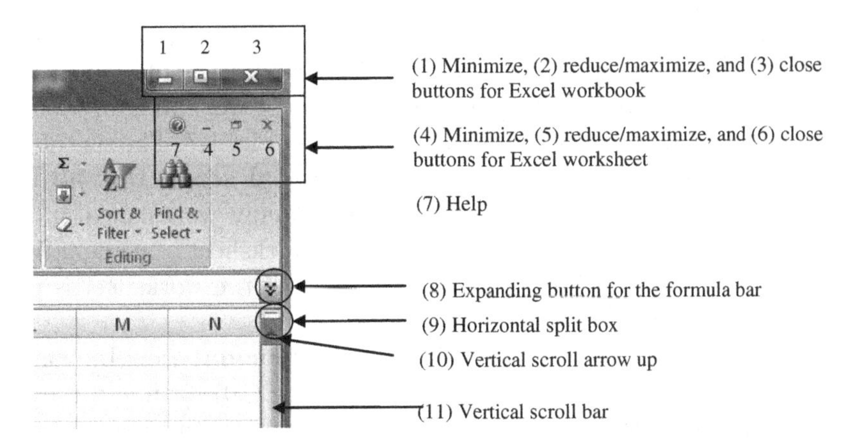

Fig. 1.6 The upper right corner of the worksheet

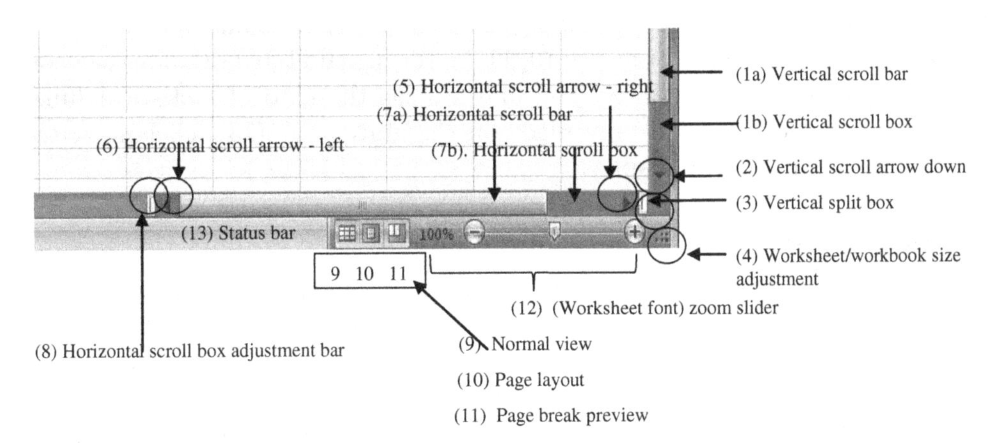

Fig. 1.7 The lower right corner of the worksheet

Move the **vertical scroll bar** (11) in the **vertical scroll box** up and down to see how the numbers move, or move the **vertical scroll-up arrow** (10). To eliminate the split, double click the horizontal split line, or use the mouse to select and drag the horizontal split line back to the original place in (9). Both numbers are now on the same sheet.

1.2.3 *Lower right corner*

In Fig. 1.7, using the mouse pointer, press and drag the **vertical split box** (3) to the center of the worksheet. Enter number 3 anywhere on the left-hand side of the split, enter number 4 anywhere on the right-hand side, and move the **horizontal scroll bar** (7a) in the **horizontal scroll box** (7b) left and right, or click the **horizontal scroll-right** (5) **or scroll-left** (6) arrow. Watch how the numbers move. Click the vertical split box (3) twice or drag the divided line to the right to cancel the split. The numbers are now on one sheet.

You may also split the screen into four parts simultaneously by using the horizontal and vertical split boxes ((9) in Fig. 1.6 and (3) in Fig. 1.7). You may move the vertical or horizontal scroll bars ((11) in Fig. 1.6 and (7a) in Fig. 1.7) to see how the four numbers move. In addition, there are several ways to make an adjustment. When the workbook or worksheet is reduced ((2) or (5) in Fig. 1.6), the size of workbook or worksheet can be adjusted by dragging the **size adjustment button** (4) in Fig. 1.7. The size of the horizontal scroll box (7b) can be adjusted by moving the **horizontal scroll box adjustment bar** (8) to the right or left. This is convenient if you have many worksheets and you want to show the worksheet tabs. The size of the fonts on the worksheets can be adjusted to large or small by moving the **zoom slider** (12). This is useful if the worksheet is projected on a whiteboard for presentation.

The three buttons in the **status bar** (13) are the **Normal** view button (9), which is the default view of the worksheet, the **Page Layout** view (10) for fine-tuning pages before printing, and the **Page Break Preview** (11) for adjusting the page break line before printing. These three commands are also available under <"**View**"> tab.

When you enter numbers in a range, say 1 to 6 in B2:D3, and select a range of numbers, then the results of some statistical functions, such as average, count, and sum, show up on the left-hand side of the view buttons (9, 10, 11) in the **status bar** (13). This is a convenient way to do simple statistics instantly. You may choose the kind of statistical functions to display by moving the pointer to any place on the status bar (13), clicking <RM>, and selecting the item you want to display.

1.2.4 *Lower left corner*

The lower left corner, Fig. 1.8, shows the sheets management. The default number of the worksheet in a workbook is 3, but you can create many more new worksheets by clicking **insert sheet tab button** (3). Click the button 12 times to increase the number of sheet tabs from 3 to 15. The sheet tabs now overflow to the left of the horizontal scroll box adjustment bar (8) of Fig. 1.7. Select Sheet3 as the active sheet. Using the mouse, click the **last left sheet arrow** (4) to jump back to Sheet1, click the **next right sheet arrow** (6) to scroll the sheets to the right one by one, click the **last right sheet arrow** (7) to jump

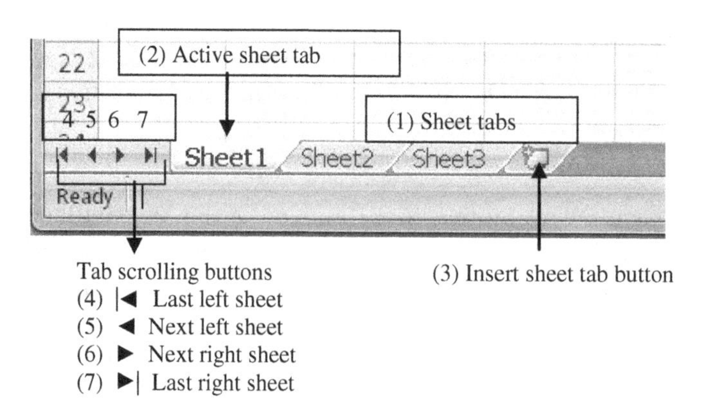

Fig. 1.8 Lower left corner of the worksheet

to Sheet15, and then click **next left sheet arrow** (5) to scroll the sheets to the left one by one. Watch how the active sheet tab, Sheet3, moves.

Multiple worksheets can be selected from the sheet tabs (1). To do so, in A1, enter 1 anywhere in Sheet1, click and open Sheet2, and enter 2. Lastly, click and open Sheet3, and enter 3. This will identify each sheet. Click Sheet1, and then, pressing and holding the control key, click both the Sheet2 and Sheet3 tabs. The tab background changes to white, and (**Group**) indicator appears after the workbook name in the title bar (see Fig. 1.1 (7)). Enter your name in cell B1 in the current worksheet. Click any one of the sheet tabs in the group to move inside the grouped sheets; your name should be shown in cell B1 in all three sheets in the group. Deselect the group tabs by clicking any sheet tab that is not in the group.

1.3 The Structure of Ribbons

The structure of ribbons is quite complicated. It is almost impossible to comprehend and is very confusing for beginners. We first summarize the general structure of the ribbon and the group names under each ribbon. Since the "Home" ribbon contains the basic formatting commands, we discuss this ribbon in some detail. Some of the more useful or popular formatting commands are also available in the mini toolbar (Fig. 1.11). They are explained in the following subsections.

1.3.1 *A summary of ribbon and group names*

Table 1.2 presents the group names associated with each of seven basic tabs and five additional tabs. In this book, we do not cover all the groups and topics, as many are simply cosmetic arrangements related to the appearance of tables and charts, not directly related to understanding economics and business. The important groups and topics for this book are shown in bold face. As you can see, the most useful groups are located under <"Insert"> and <"Formulas">, and the most useful tools are the <"Chart Tools">. The Chart Tools will be introduced in detail in Chapter 3.

1.3.2 *The "Home" tab*

Most of the basic commands and formatting tools are located under the "Home" tab. Some of these commands and tools apply to the whole worksheet or workbook (like Open or Save), and some apply only locally to a cell or a range of cells (like Sum or Copy). Most of the ribbon names are self-explanatory. We explain the contents of the home tab ribbon in two parts, as in Figs. 1.9(a)–1.9(c).

The left-hand side of the Home ribbon — basic formatting

For hands-on practice, we may construct random numbers in the range C5:F10, as in Fig. 1.2, and try the following commands. Unless specified otherwise, use <Undo> (in (2) of Fig. 1.3) to return to the original range, C5:F10.

Table 1.2 Tab names and group names

	Tab name	Group name						
1	**Home**	Clipboard	Font	Alignment	**Number**	Styles	**Cells**	**Editing**
2	**Insert**	**Tables**	**Illustrations**	**Charts**	Links	Text		
3	**Page Layout**	Themes	Page Setup	Scale to Fit	Sheet Options	Arrange		
4	**Formulas**	**Function Library**	**Defined Names**	Formula Auditing	Calculation			
5	**Data**	Get External Data	Connections	**Sort and Filter**	Data Tools	**Outline**		
6	**Review**	Proofing	Comments	Changes				
7	**View**	**Workbook Views**	Show/Hide	Zoom	Window	Macros		
8	**Table Tools**							
	Design	Properties	Tools	External Table Data	Table Style Options	Table Styles		
9	**Chart Tools**							
	Design	**Type**	**Data**	**Chart Layouts**	**Chart Styles**	Location		
	Layout	**Current Selection**	**Labels**	**Axes**	**Background**	**Analysis**	Properties	
	Format	Current Selection	**Shape Styles**	WordArt Styles	Arrange	Size		
10	**Textbox Drawing Tools**							
	Format	Insert Shapes	Shape Styles	WordArt Styles	Arrange	Size		
11	**Header and Footer Tools**							
	Design	Header and Footer	Header and Footer Elements	Navigation	Options			
12	**SmartArt Tools**							
	Design	Create Graphic	Layouts	SmartArt Styles	Reset			
	Format	Shapes	Shape Styles	WordArt Styles	Arrange			

Note: Some contents of tabs 8–12 will be introduced in later chapters.

1**Cut**. Cut the selected range. (If you want to move a range, "cut" it to the clipboard and "paste" it back from the clipboard.) Select the range C5:F10, "cut" the range, and "paste" it to H5. <Undo> to go back to the original range, C5:F10.

2**Copy**. Copy the selected range. (If you want to copy a range, "copy" it and "paste" it.) Select the range C5:F10, "Copy" the range, and "paste" it to H2.

3**Paste**. Paste (from the clipboard) the range that is cut or copied in 1 or 2 above to a designated cell that is selected by the pointer.

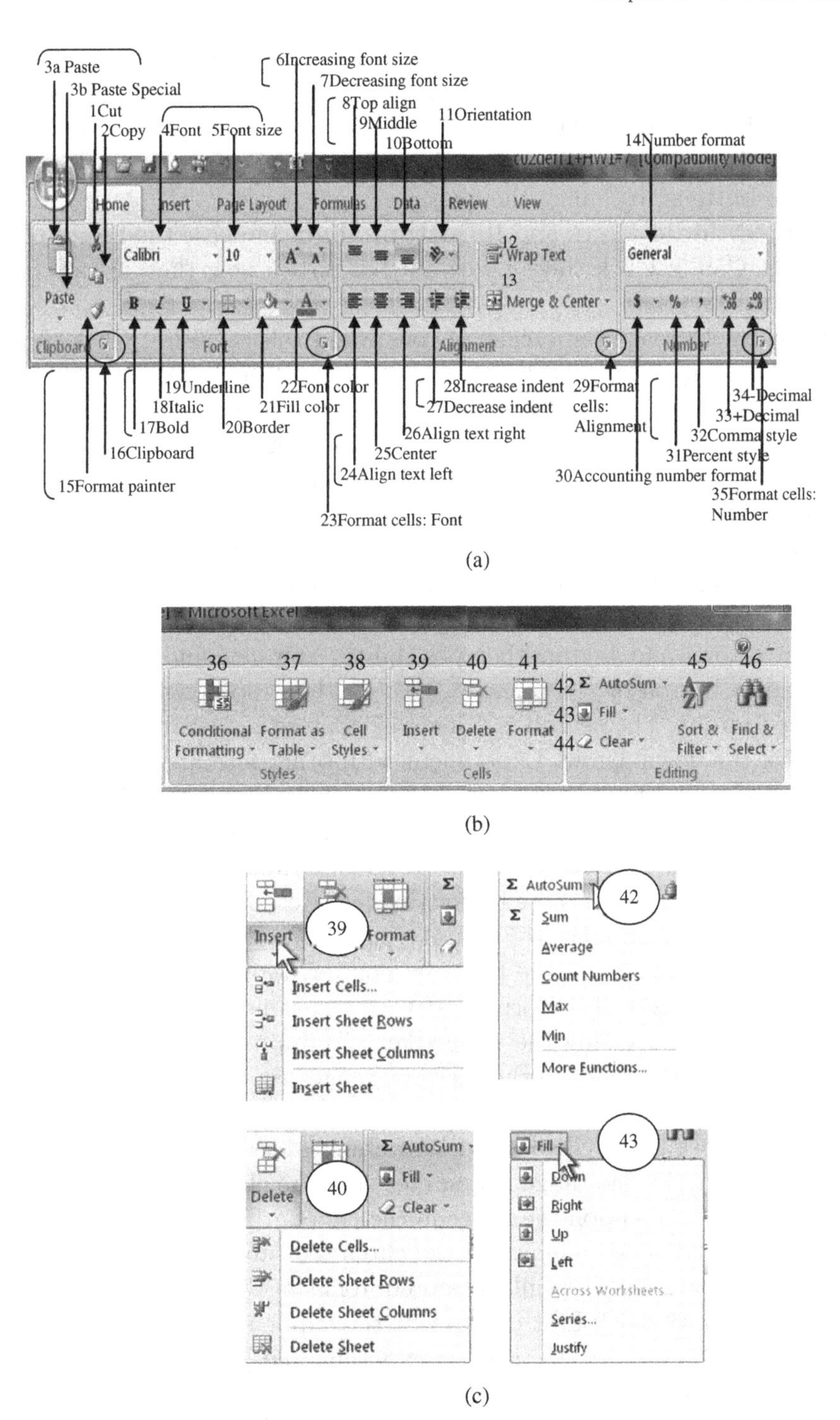

Fig. 1.9 The ribbon for the “Home” tab. (a) the left-hand side, (b), (c) the right-hand side

4**Font**. The default font for Excel 2007 is Calibri. You may change it to Arial (the default font in Excel 2003) by clicking the down arrow and choosing Arial.

5**Font Size**. The default is 11.

6**Increasing font size;** 7**Decreasing font size.** Shade a range, say, C5:D10, and click either one of the buttons to change the font size.

8**Top align**; 9**Middle align**; and 10**Bottom align**. Increase the height of row 5 to 100 pixels, and select C5:F5. Click these three buttons to see how the position of the number changes.

11**Orientation**. Increase the height of row 5 and select C5:F5. Click this button and rotate the numbers by clicking each command in the drop-down list. Also click the last command, **Format Cell Alignment**, and study all the possible arrangements of the contents of the cells. Each time click <Undo> to go back to the original range.

12**Wrap Text**. In C3, say, enter National Income. The words overflow to D3. If the 12Wrap Text button is clicked, the height of row 3 expands and the two words will be contained in one cell C3.

13**Merge & Center**. In C3, enter National Income and select C3 and D3. When (13)Merge and Center is clicked, C3 and D3 merge to contain and center the two words. Click the down arrows and try another possibility. You can undo the effects by clicking either the <Undo> button or <Unmerge Cells> in the drop-down list.

14**Number format**. The default is "General". Enter random numbers in C5:F10. Select C5:F10 and click the down arrow. Experiment with other formats.

15**Format painter**. "Formatting" means you change how the worksheet range/table or chart looks. All the buttons in Fig. 1.9(a) are formatting tools. As an example, select a cell, say C5, and Click 17**Bold**, 18**Italic**, 19**Underline**!, 21**Fill color**!, and 25**Center**. Then, cell C5 contains all five of these formatting features. To paint all these to another cell, place the pointer on C5 and click 15**Format painter** once upon which a paintbrush logo appears at the right-hand side of the fat cross pointer. Move the pointer to select C5:F10. The numbers in the range will be formatted with the same formatting features. Close the format painter mode by pressing the <esc> key. To apply the format painter repeatedly, click this logo twice, and apply to other cells or ranges repeatedly by selecting the cells or ranges. Use <Esc> to undo the **repeated formatting**.

16**Clipboard**. Click the 16**Clipboard launcher**. Observe that when C5:F10 is selected and Copied (2) or Cut (1), the range is first stored in the clipboard. Select a blank cell, say B12, and **Paste** (3a), whereupon the range in the clipboard will be reproduced in B12:E17. Nevertheless, the stored range remains in the clipboard, unless it is cleared ("Clear All" in the <Clipboard>). You may paste all the stored items on the worksheet by clicking <Paste All> from the clipboard. Click Exit (X) to exit the clipboard.

Other features, from 17 to 35, can be experimented similarly by using the random numbers in C5:F10. We especially note the following buttons.

20**Border**. Add borders of a range. Choose a border from the drop-down list. The last border used will stay and show in the menu until another border is chosen.

27**Decreasing indent** and 28**Increasing indent** apply to the text.

33**Increasing decimal**. The rounded decimal will return to the original number.
34**Decreasing decimal**. The remaining decimal will be rounded.

The right-hand side of the Home ribbon — advanced formatting

The right-hand side of the Home ribbon is shown in Fig. 1.9(b). 36**Conditional Formatting**, will be explained in Chap. 4 and 37**Format as Table will be explained** in Chap. 13, but 38**Cell Style** and 41**Format** are seldom used for our purpose. 45**Sort & Filter**, 46**Find & Select** will be explained when we come to use them. Other buttons are explained below in Fig. 1.9(c). Again, we construct the random numbers in the range C5:F10 as in Fig. 1.2, and try the following commands.

Unless specified otherwise, use <Undo> (in (2) of Fig. 1.3) to return to the original range, C5:F10, and start the new experiment.

39**Insert**. In Fig. 1.9(c), move the pointer to D6 and experiment with **Insert Cells**, **Insert Sheet Rows**, and **Insert Sheet Columns**. These commands are useful in editing/adjusting a data table. The last one, **Insert Sheet**, is the same as the **Insert sheet tab** button (3) in Fig. 1.8. This button is convenient when the sheet tabs overflow and the **Insert sheet tab** button hides under the **Horizontal scroll box adjustment bar** (8) (see Fig. 1.7). Another way of inserting a new sheet is to select a tab, click the RM, and choose <Insert> <Worksheet> <Ok> from the opened drop-down list.

40**Delete**. This key has the opposite function from the 39**Insert button**. "**Delete sheet**" will delete the current sheet. Another way of deleting a sheet is to select the tab to be deleted, click the RM, and select <delete> <delete> (twice).

42**AutoSum**. This is a useful button. Select C5:G11 of the range in Fig. 1.2, leaving the last row and the last column blank. Click AutoSum (or one of the functions in the drop-down list), the row and column sums (or the results of the function) will be filled in the last row and the last column. If you click **More Function** in the drop-down list, it will give you the same window as in Fig. 1.5.

43**Fill**. Clicking this button will give the drop-down list shown in Fig. 1.9(c). In Fig. 1.2, select F5:H10, and click **Right** in the drop-down menu. The blank area G5:H10 will be filled with the formula (or value) of F5:F10. Thus, this command is the same as dragging the fill-handle in Fig. 1.2.

The last command in the drop-down menu, "**Justify**", is useful in justifying the text when we have a statistical table and want to add the data sources. For example, if C5:F10 is the table, as in Fig. 1.10, then below it, in cell C11, we enter "Data Sources: Economic Report of the President, transmitted to the Congress, February 2008". The words overflow to the right of the table. In this case, select C11:F14 (enough space for the words), and

<"Home"> <Editing, Fill> <Justify>.

The words will be contained in C11:F13, as in the lower part of Fig. 1.10.

1.3.3 *The mini toolbar*

When you click a cell, say F10, and click RM, a double-layered mini toolbar appears next to the cell, as in Fig. 1.11. It lists some of the most popular buttons, most of them from the "Home tab" ribbon. The numbers in Fig. 1.11 locate the corresponding buttons in the "Home" tab ribbon, or in the ribbons of other tabs. This mini toolbar is convenient if you are using non-Home tabs and need to use some buttons in the "Home" tab ribbon.

	A	B	C	D	E	F	G	H	I	J	K
10			0.325801	0.673366	0.507676	0.453636					
11			Sources: Economic Report of the Presiden, transmitted to the Congress. February 2008..								
12											
13											
14											

	A	B	C	D	E	F	G
10			0.228234	0.444629	0.666545	0.98365	
11			Sources: Economic Report of the				
12			Presiden, transmitted to the Congress.				
13			February 2008..				
14							

Fig. 1.10 The <"Home"> <Editing, Fill> <Justify> operation

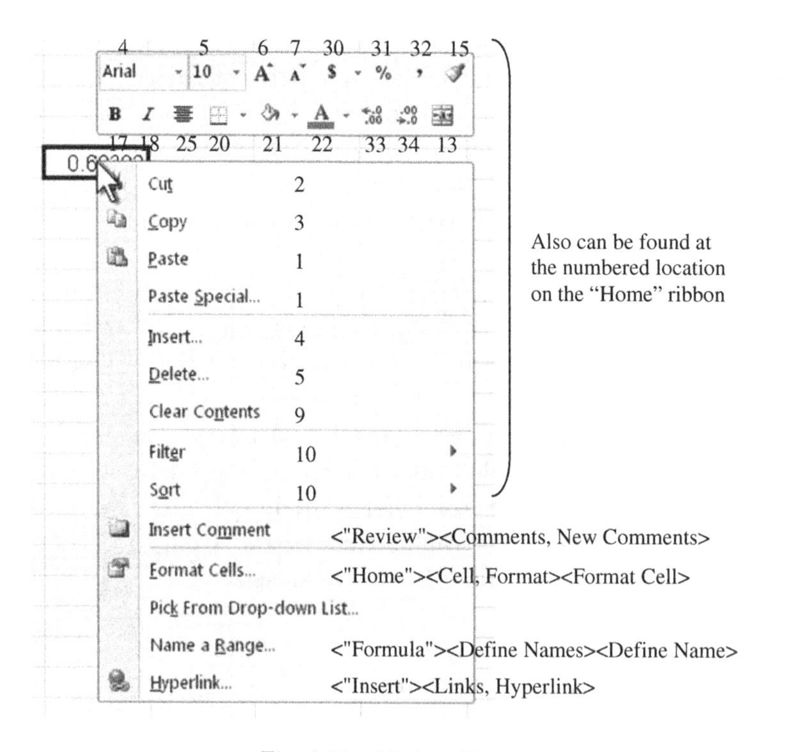

Fig. 1.11 Mini toolbar

1.4 Drawing Toolbar

Compared with most of the standard and formatting tools explained in the previous sections, the drawing tools under the "Insert" tab are an important part of economic and business analyses. Most buttons are self-explanatory. The most useful button combinations are <"Insert"> <Illustrations, Shapes>, <"Insert"> <Charts, Column>, <"Insert"> <Charts, Line>, and <"Insert"> <Charts, Other Charts> <Surface>. We will introduce 2D Charts in Chap. 2 and 3D charts in Chap. 9.

1.4.1 *Use of arrows*

Among the different **Shapes**, the most useful one is **Arrow**.

(1) When

<"Insert"> <Illustrations, Shapes!>

is selected, a drop-down list appears. Click to select the Arrow under <Lines>, whereupon the list disappears and the pointer changes to a lean + sign.

(2) When LM is pressed, the + sign becomes larger (like part c in Fig. 1.12). Drag it to make a line and release the LM. An arrow is drawn.

(3) When the pointer clicks the arrow, two circles at both ends of the arrow appear (see part a). At the same time, the <Drawing Tools> tab appears inside the title bar, and below it is a new <Format> tab (see part e) next to the <"View"> tab.

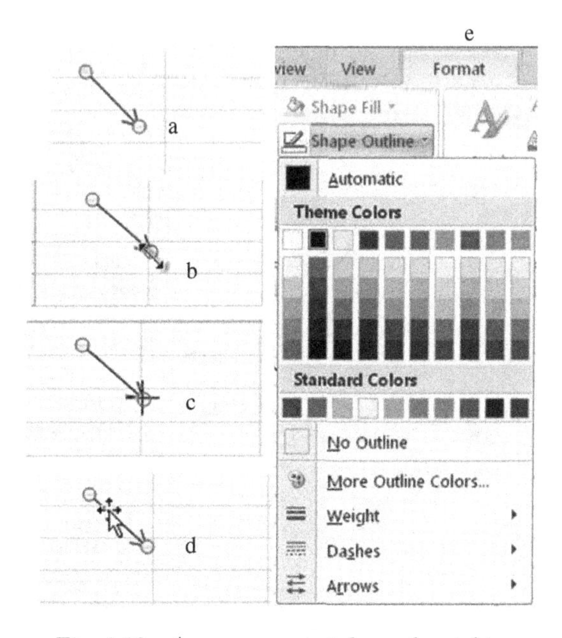

Fig. 1.12 Arrow, arrow style, and weights

(4) When the fat cross pointer gets closer to the tip of the arrow, the pointer changes to a double-headed arrow (see part b), and when LM is pressed, it changes again to a lean plus sign (see part c). This means you can stretch or shorten the arrow, or rotate the arrow with center at the other circle.

(5) If you want to move the whole arrow, move the pointer to the body of the arrow. The pointer changes to an arrow-headed cross (see part d), implying that you can move the arrow in all four directions.

(6) When the arrow is selected and

<"Drawing Tools, Format"><Shape Styles, Shape Outline>

is clicked, a drop-down color palette appears, from which you can change the color, style, weight, and dashes of the arrow.

1.4.2 *The text box*

We may have some fun with WordArt and learn the use of the Text Box at the same time. Follow the steps below using Fig. 1.13.

(1) Choose <"Insert"> <Text, Text Box> and move the pointer inside the spreadsheet. The pointer changes to an **upside-down cross**. Click anywhere in the blank space and release LM. A bordered square with double-headed arrow and flushing vertical line

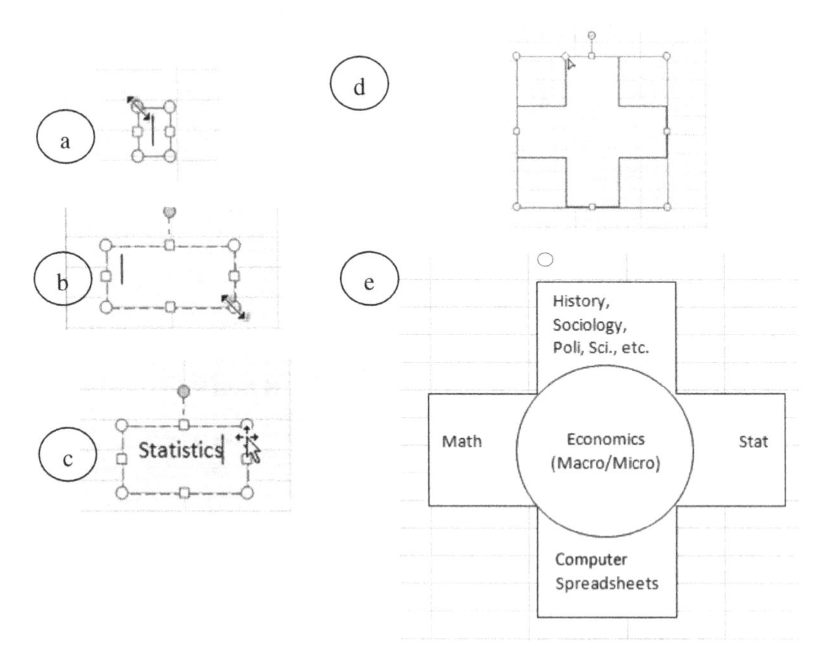

Fig. 1.13 Text box operations

appears, like part a of Fig. 1.13. In this case, if you enter words directly without moving the square, then you can **write a text without borders**.

(2) If you **drag the pointer** and make a square, the square will look like part b, and then you can write a text **with a border**. There are four white circles and four white squares. When you move the pointer close to a circle or a square, a double-headed arrow appears, indicating that you can drag the corner to **expand or reduce** the chart in both directions shown.

(3) You may also move the pointer close to a **border**. A four-arrow-head cross appears (see part c). You then **move the whole box** anywhere on the worksheet. Move the chart to the middle of the worksheet.

(4) Click anywhere inside the text box. The flushing vertical line indicates that the text box is in the **editing mode** and is ready to receive text entries. Type the word "Statistics", as in part c.

(5) Note that there is a green circle on top of the box. When you move the pointer close to the green circle, a black clockwise arrow-headed circle appears, indicating that you can **rotate** the chart. Press and hold LM and rotate the chart 360°.

(6) Move the pointer to the border of the box, the pointer changes to a four-arrow-head cross, as shown in part c of Fig. 1.13. Click the box border, and it changes from dotted borders to solid borders. This means that the pointer has **selected the whole text box** and you can now **edit the whole box**, as by adding color, changing font color, changing font size, or deleting the box (by entering <Delete> key), etc. Change the font size to 9 points.

(7) Notice that the dotted border with circles and squares disappears when you click a cell outside of the border, and appears again if you click anywhere inside the box.

(8) When you have selected the box, the <"Drawing Tools, Format"> tab appears next to the "View" tab in the Title bar, and it disappears when the box is deselected.

(9) If the text box is bordered, we can **remove its borders**. Click the box, make sure the whole box is selected (see (6) above). Enter

<"Drawing Tools, Format"> <Shape Styles, Shape Outline> <No Outline>.

Click anywhere outside the text box and the border disappears.

Example 1.2 Use of Text Box and Shapes. We would like to construct an illustration like part e in Fig. 1.13. It illustrates economics in relation to other fields of science. This requires some practice and patience.

(1) Click <"Insert"> <Illustrations, Shapes> <Basic Shapes, Cross>. From the Basic Shapes dialog box, select the **fat cross** in the second row.

(2) Click any cell on the worksheet and a shape like part d in Fig. 1.13 is created, with a green circle and a yellow diamond. As usual, you can **rotate** the shape by selecting and moving the green circle (see (5) under Sec. 1.4.2 and Fig. 1.13).

(3) If the pointer moves close to the diamond, it changes to a **fat triangle** as shown in part d. Move the diamond to change the size of the shape and move the white circles to enlarge its shape as in part e.
(4) Add a **circle** in the middle part of the cross as in part e of Fig. 1.13. Adjust the circle and make sure that it touches the four corners of the shape as shown.
(5) Type "Economics (Macro/Micro) as shown inside the circle. **Text** can be added to any shape by just typing the letters. Choose the **font size 9**.
(6) Construct a text box like part c, with the word "Stat" The font size should be 9 points. After removing the border, we also **make the Text Box transparent** by entering

<"Drawing Tools, Format"> <Shape Styles, Shape Fill> <No Fill>.

(7) **Copy** the Stat text box by clicking the border (see part c) and, at the same time, holding the control key and clicking the box borders, drag the box to a blank place and release the mouse. The Stat box is reproduced. Change the word to Math. Similarly, reproduce two more Stat boxes, and one enters the words in the upper square, one in the lower square, of part e in Fig. 1.13.
(8) **Move** the four text boxes created above one by one to overlap with the protruded squares as shown in part e. The size of the text box may be larger than the protruded squares, and adjustment may be needed.
(9) You may **group all the components** together by holding the control key and selecting the two shades and four text boxes, and enter

<"Drawing Tools, Format"> <Arrange, Group> <Group>.

You may now click the completed cross (part e), and copy or move the whole picture to the middle of the screen. This completes the construction. □

1.4.3 *WordArt*

The use of WordArt can be viewed as an application of the text box.

Example 1.3 Use of WordArt

(1) Choose <"Insert"> <Text, WordArt> and click the first character A. The screen shows "Your Text Here" with large 54-point letters, shaded, and surrounded by four white circles and four white squares.
(2) The shading disappears when you click a blank cell outside the shade, and appears again if you **select** any letter inside the chart. Notice that, when you have selected the chart, the <"Drawing Tools, Format"> tab appears next to the "View" tab in the Title bar, and it disappears when the chart is deselected.
(3) You can expand the chart like the text box, or move the chart to the middle of the worksheet.

(4) When you click any letter, say, T, a vertical flushing line appears to the right or left of the letter, indicating it is in the **editing mode**, and you can edit the letter and words. Type "Message", and see the chart expand automatically to the right.

(5) **Reduce** the box so that "Your Text" is on the first line and the other two words on the second line.

(6) Select the word "Message" and **paint** the word in red (<"Home"> <Font, Font Color!>).

(7) If you move the pointer to a border until an arrow-headed cross appears, and click LM, the dotted vertical line disappears, but the words are still bordered. This means that you have **selected the whole chart**. Reduce the entire text to font size 32 ((<"Home"> <Font, Font Size!> [32]), and paint the background color in green (<"Home"> <Font, Fill Color!>). You may delete the WordArt by moving the pointer to any border, selecting the box, and entering the <delete> key. □

Example 1.4 WordArt. Continuing from the above example, erase the four words (undo the box if you already erased it), and type the message about the class coverage as in Fig. 1.14. We now want to add the rainbow colors to the box. Moving the pointer in the box, enter

<"Drawing Tools, Format"> <Shape Styles, Shape Fill!> <Gradient>

<More Gradient> <Format Shape, Fill, xGradient fill> <Preset color!> <Rainbow>.

The whole chart is then painted with rainbow colors. If you reselect the chart, and

<"Drawing Tools, Format"> <Shape Styles, Shape Fill!> <Gradient>

again, then, Excel will present 13 variants of rainbow shape. You may click <More Gradient> for more choices or changes as shown in Fig. 1.14. Print it out by clicking the print tab. □

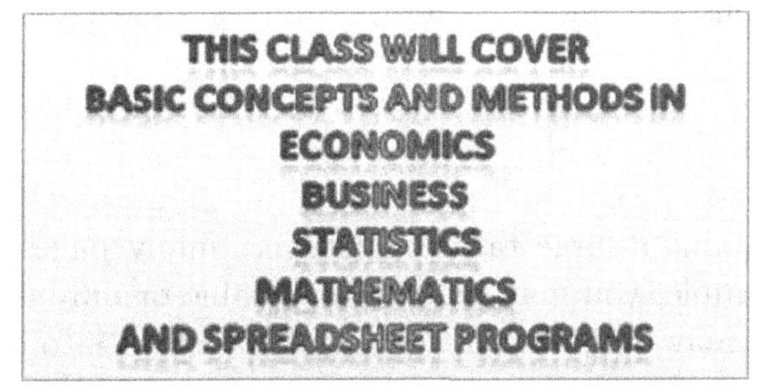

Fig. 1.14 WordArt

1.5 Printer Basics

If you answer homework assignments on a same worksheet, then printing will take two or more pages. To print in one page, or to adjust printing in general, the following procedure may be taken.

1.5.1 *The contents of <"Office"> <Print> and <Print Preview>*

If you are printing the current worksheet as it is, you may simply click the 5**Quick Print** button in the Quick Access Toolbar ((2) of Fig. 1.3). You may also open <"Office"> <Print> <Quick Print>. In both cases, the printer will start printing immediately.

If you write homework (or test) questions separately in several worksheets, or you want to print only part of the worksheet, or you want to print more than one copy, you may open the Print dialogue box (<"Office"> <Print>) and click <**Entire workbook**> or <**Selection**>, or just change the <**Number of copies**>. The Print default is to print **All** the contents of the "Active sheet(s)" only and also print one copy only. You may also open the **Print Preview** dialogue box indirectly by clicking the last button on the bottom of the Print dialogue box.

If you want to print a large worksheet on one regular page, or print it in landscape, then you have to open <"Office"> <Print, Print Preview> to change the default. The upper part of Fig. 1.15 shows the menu of **Print Preview**. All buttons are self-explanatory except **Page Setup**. When you click the <Page Setup> button, the lower part of Fig. 1.15 appears. It has four tabs. In the <Page> tab, you can choose the following four items.

(1) **Orientation** (see part a): **Portrait** (vertical) **or Landscape** (horizontal).
(2) **Scaling** (see part b): you may manually **adjust the size** of the printing output, larger or smaller, or let Excel adjust automatically to fit in a 1-page-wide by 1-(or 2)-pages-tall space. This will **print the large worksheet in one page**, and font size will be automatically reduced. You may go back to the normal size by <Scaling, @Adjust to> [100%] normal size.

1.5.2 *Print area and Print Titles*

Other often-used buttons are under <Sheet> tab, as shown on the right-hand side of Fig. 1.15.

Print area

When a worksheet contains a large table that covers many pages, you may like to print only part of it. For example, you may select only a table or only a chart to print, but not both. In this case, you may designate it as the <Print Area> to print it. Print Area can be activated or deactivated only through <"Page Layout">. First, select the area to be

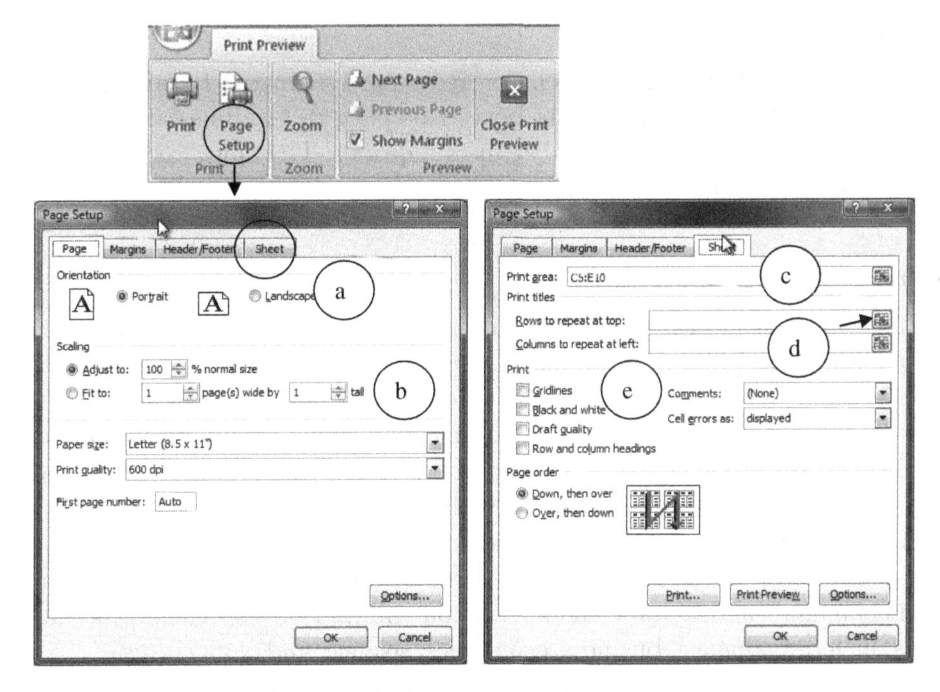

Fig. 1.15 Print preview and page setup

printed, then enter

<"Page Layout"> <Page Setup, Print Area!><Set Print Area>

The print area will be bordered with dashed lines and can be printed as usual from the Print Setup dialog box.

Note that <Set Print Area> only sets one print area, and clears all the previous print areas. If you want to select multiple print areas, then select the additional print area first, and click <Add to Print Area>, instead of <Set Print Area>, as in the last procedure above. (If the additional area is not selected, the <Add to Print Area> item will not show.

Print titles

Another useful item is **Print titles** (see part d of Fig. 1.15). They also can be activated only through <"Page Layout">. Enter

<"Page Layout"> <Page Setup, Print Titles>

<Page Setup, Sheet, Print titles, **Rows to repeat at top**>

or <**Columns to repeat at left**>. When Print Titles is clicked, the same Page Setup dialog box as the RHS of Fig. 1.15 comes up. Just click the box at the end of the space (see part d with arrow) and select the rows and/or columns to be repeated at the top or left of each page. Print out from the Page Setup dialog box.

Gridlines and headings

You may also "Print" **Gridlines** of cells or **Row and column headings** (see (15) and (16) of Fig. 1.1) by checking appropriate boxes (see part e). Gridlines and Print Titles may also be added from <"Page Layout"> <Sheet Options, Gridlines, xPrint> and <"Page Layout"> <Page Setup, Print Titles>.

1.6 Summary

The command buttons in new Excel 2007 are many and very complicated, if not confusing. Based on our long experience of introducing spreadsheets to economics students, we have tried our best to introduce the above commands. It is suggested that the most useful features of the spreadsheets are moving, copying, pasting, adding, and deleting cells and ranges, columns and rows, text boxes, tables, and charts, worksheets and workbooks. We especially recommend the use of the control key method to move and copy a range. Other commands can be learned as the need for them arises in the process of using spreadsheets in economic and business applications in later chapters. Thus, the more you use the spreadsheets, the faster you will become familiar with Excel commands and become an expert. You will find that spreadsheets can be used in many areas, including finance, business, economics, science, and engineering.

Appendix 1A. Customize Quick Access Toolbar

Using **Customize Quick Access Toolbar** (3) in Fig. 1.3, double click the down arrow, and a drop-down list appears. Click the item you want to put in the Quick Access Toolbar (2). When you click the item, the Customize Quick Access Toolbar will close and the item appears in the Quick Access ToolBar (2). To delete the item on the Quick Access Toolbar (2), point to the item and double click the right mouse, whereupon a menu appears. Click the first item "Remove from Quick Access Toolbar". The item will be removed.

If you cannot find the item in the menu when you click (3), go to the lower part of the drop-down list, click "More Commands", and choose the item that you are looking for.

Key terms: Economics and Business

Key terms: Excel

Homework Chapter 1

Try your best to present all the tables below. Make them colorful, with different fonts, colors, or sizes, etc. Experiment with different column widths, background colors, italics, bolding, underlining, etc. Use print preview before you print.

For Section 1.4 Drawing Toolbar

1-1 Illustration of the theory of supply The relation between costs and revenues is illustrated in Fig. HW1-1 below (revised from Fisher and Dornbusch, 1983, p. 157). Reproduce Fig. HW1-1. (Hints: Create a plotting paper with squared cells, and then use various shapes).

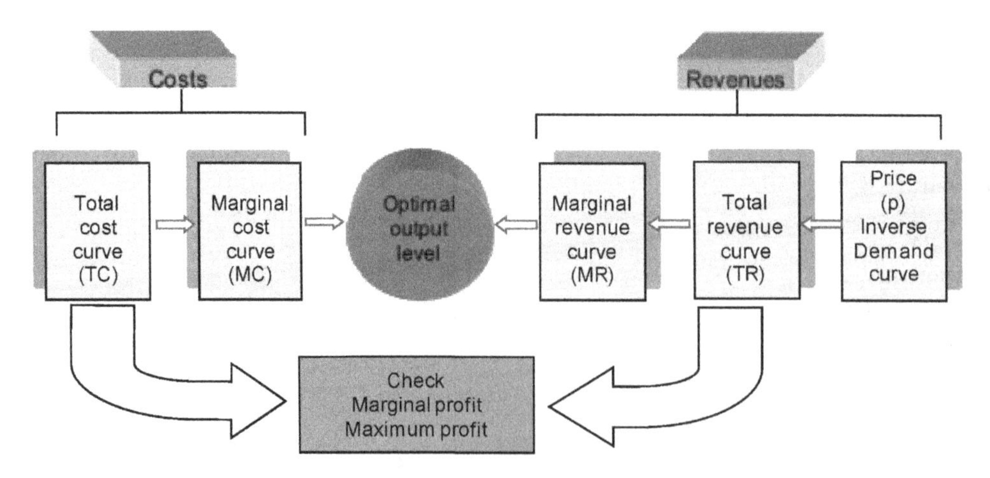

Fig. HW1-1 The theory of supply (the firm's output decision)

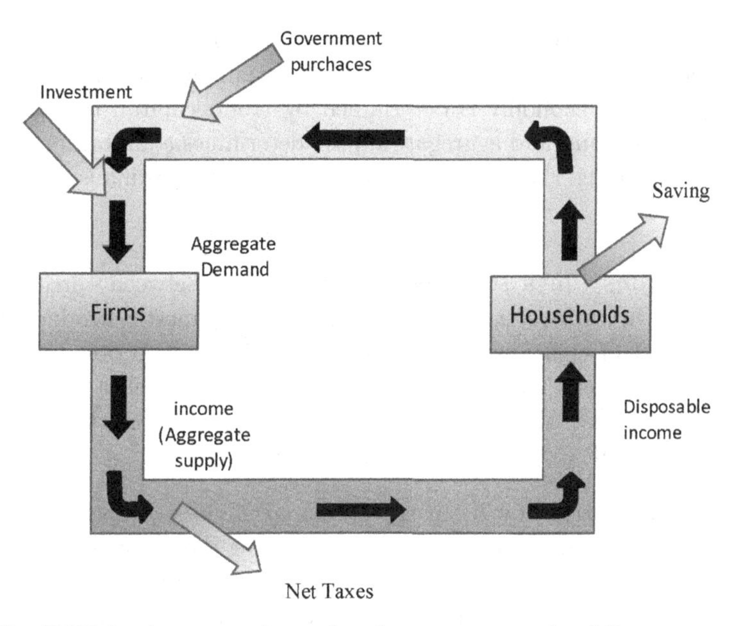

Fig. HW1-2 Aggregate demand and aggregate supply of the economy

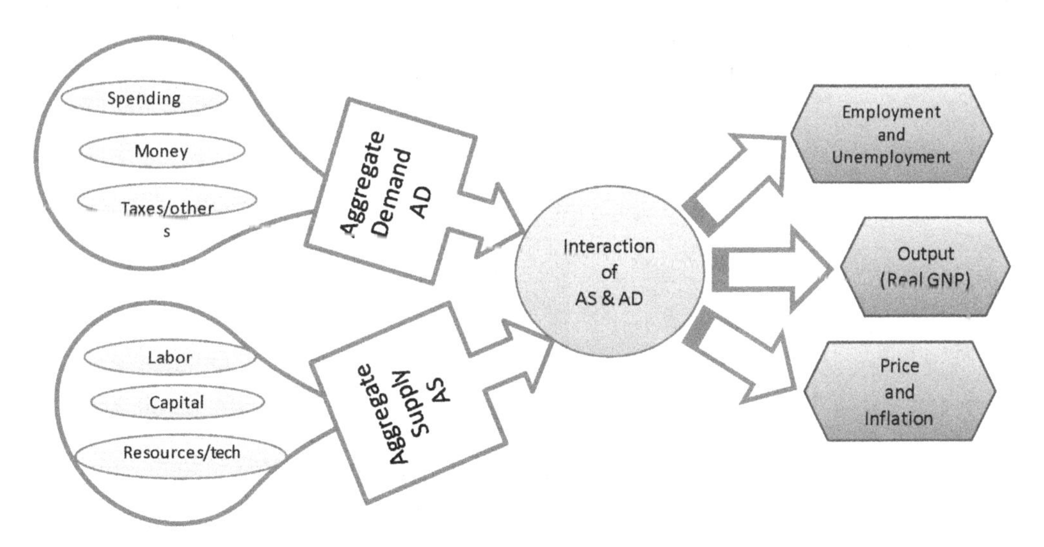

Fig. HW1-3 The macroeconomic system

1-2 Illustration of the equilibrium circular flow of the economy (Fig. HW1-2) In equilibrium, aggregate demand is equal to aggregate supply, which in this chart is income. Saving and net taxes are leakages from the circular flow, while investment and government purchases are injections to the flow. In economic equilibrium, the leakages must be equal to the injections, that is, saving plus net taxes is equal to the government purchases plus investment ($\mathrm{I}+\mathrm{G} = \mathrm{S}+\mathrm{T}$). (see Fisher and Dornbusch, 1983, p. 605). Reproduce Fig. HW1-2.

1-3 Illustration of the macroeconomic system The aggregate demand is determined by the spending of households, governments, and firms, money demand and supply, taxes, and other forces. The aggregate supply of the economy is determined by labor, capital, resources, and technology. Interaction of aggregate demand and aggregate supply determines equilibrium output, employment, and prices. This is shown in Fig. HW1-3 (based on Samuelson and Nordhaus, 1985, p. 90). Reproduce Fig. HW1-3.

1-4 Causality directions In a study of interdependence between three economic variables, Exports (EX), Foreign Direct Investment (FDI), and Gross Domestic Product (GDP), it is found that, for a panel data set of eight Asian economies, there is a bidirectional causality between EX and GDP: EX (Granger) causes GDP (10%, that is, at the 10% level of significance, same below) and GDP causes EX (15%). On the other hand, there is a unidirectional causality from FDI to EX (10%), and from FDI to GDP (5%) (Hsiao and Hsiao, 2006).

(a) Illustrate the above relations by using three colored circles to represent the three variables, with the variable names inside the circles, and connecting the three circles with differently sized directional arrows representing the levels of significance. For example, 1% uses 3 points solid arrow, 5% uses 2 1/4 points solid arrow, 10% uses 1 point solid arrow, and 15% uses 1 point dotted line.

(b) The following bidirectional causality is found in Thailand: GDP causes EX (15%), EX causes GDP (5%), and also GDP causes FDI (15%), and FDI causes GDP (10%). There is one unidirectional causality: FDI causes EX (15%). Draw the diagram.

Chapter 2

Total Revenue, Total Cost, and Profits — Excel Tables

Chapter Outline

Objectives of this Chapter

In the previous chapter, we acquainted the readers with the appearance of the Excel spreadsheets by moving around the four corners of the spreadsheet window. We also introduced the use of arrows and the text box step by step.

This chapter applies Excel commands and techniques to construct simple tables of economic and business applications. First using the random number generator function to generate numbers to construct a firm's financial table (Sec. 2.1), we then introduce mathematical equations of total revenue and total cost and show how to implement equations in a table (Sec. 2.2).

The last two sections are devoted to explaining some more basic features of the spreadsheet lifelines: moving, copying, pasting a range, a worksheet, columns, and rows (Sec. 2.3). The three appendixes explain how to use a simple macro program for adding and deleting columns and rows (Appendix 2A), constructing AutoFill lists (Appendix 2B), and getting rid of the copyright logo (Appendix 2C). Appendixes 2B and 2C are used in the homework section.

2.1 Generating Random Numbers

In this section, we will first discuss the basic mathematical operations in the spreadsheet environment, and then explain in detail the three methods of entering a formula, four

Table 2.1 Basic mathematical operations

	A	B	C	D	E	F	G
1		2	4				
2							
3	Table 2.1 Basic Mathematical Operations						
4		Operator	Name		Example	Result	Precedence
5	1	+	Addition		=B1+C1	6	3
6	2	-	Subtraction		=B1-C1	-2	3
7	3	*	Multiplication		=B1*C1	8	2
8	4	/	Division		=B1/C1	0.5	2
9	5	^	Exponential		=B1^2	4	1
10		Logical comparison					
11	6	=, >, <	equal, greater than, smaller than				5
12	7	>=	Greater than or equal to		=B1>=C1	FALSE	5
13	8	<=	Smaller than or equal to		=B1<=C1	TRUE	5
14	9	<>	Not equal to		=B1<>C1	TRUE	5

methods of editing a formula, and how to construct a table, using a financial statement as an example.

2.1.1 *Entering and editing formulas*

The mathematical operators in Excel are listed in Table 2.1. For example, we enter number 2 in B1 and 4 in C1. **Enter the formulas** of E5:E14 into F5:F14 as formula (see the process of entering a formula in Fig. 1.2), producing the results shown in column F. All the formulas should be preceded by an equality sign, so as to let Excel know that it is a formula, not a text. On the other hand, if you want to **enter a formula as a text**, instead of as a formula, you should make a space before the equality sign, as in E5:E14. When you enter a formula, you may enter the number directly, like $= 2 + 4$ in cell E5. In this case, the numbers are fixed. You may use cell names, like E5:E14, to represent the numbers in the cells. In this case, if the numbers in B1 and C1 change, the results also change. Change the number in B1 to 6 and that in C1 to 4, and get the feeling of the use of formula.

When entering the formula, say =B1+C1 in cell E5, there are three ways to enter the formula:

- The **direct method**: You may type the cell addresses B1 and C1 directly. In this case, you have to identify the cell address each time.
- An intuitive method is using the **pointing method**, that is, you enter = sign first, move the pointer to click B1, enter + sign, and then move the pointer to click C1. When you finish, press the enter (return) key. In this method, you do not need to identify the cell address each time.
- You can also use the **formula palette**, that is, use Fig. 1.5, but we do not recommend this.

The lower part of Table 2.1 shows **logical comparisons**. The equality and inequality equations, formulated as in rows 11–14, give results as either False or True. The last column shows the **operator precedence**. Exponentials are operated first, then the multiplication and division, then addition and subtraction, and lastly, the logical comparison. To avoid confusion, it is recommended that the parentheses should be used. For example, in case =B1/C1+C1, Excel calculates B1/C1 first and then adds the result to C1. However, it is clearer if the formula is written as =(B1/C1)+C1.

If you need to **edit (or correct) a formula** in the cell, there are four ways to do so. Move the pointer to the cell, which has a formula, say F5 in Table 2.1, and select the cell.

- **In-cell editing (I)**: Double click the cell and edit directly in the cell. <enter> when finish.
- **In-cell editing (II)**: Press <F2>, which is the "edit" key, and edit the formula directly in the cell. <Enter> when finish.
- Click the **formula bar**, and edit the formula in the formula bar. <enter> when finish.
- Click the **formula button** (f_x) in the formula bar, and edit the formula in the Formula Palette.

We recommend the first method, which edits the formula directly in the cell.

2.1.2 *Construction of a table*

The random function is shown in step 2 of Fig. 1.2 and its value is shown in step 3. It shows six digits after the decimal point in the cell. In fact, the variable =**Rand**( ) can yield a 15-digit random number ranging from 0 to 1. This can be seen by selecting the random number cell and clicking <"Home"><Number, Increase Decimal> buttons. Click the **Increase Decimal** button until zeros appear at the end. The number will end at 15 digits. Let k be a constant, then

$$=\text{rand}()*\text{k}. \tag{2.1}$$

If k = 100, then, =rand()*100 in B2, and we have a random number ranging from 0 to 100. The number of digits in a cell depends on the width of the cell. In general, we have an eight-digit random number. In addition, we may click the "**decreasing decimal**" button to reduce the number of decimals.

We now use the random function introduced in (2.1) to construct a table of Total Revenue, Total Cost, and Profits for the GiGo Company from 1980 to 2000. Formulas may be entered in upper case letters or lower case letters; in the latter case, Excel will change the letters to upper case automatically.

Box 2.1: In this lecture, the tables are bordered by row and column heading bars. They are for you to locate the cells and for your reference only. Do not reproduce them when you construct your own table.

Table 2.2 Financial statement of the GiGo company

	A	B	C	D	E	F	G
1	**Table 2.2 Total Revenue, Total Cost, and Profit, *GiGo Company***						
2	Formula	=RAND()*20	=rand()*5	=2.00	=VC+FC	=TR-TC	
3	Year	TR	VC	FC	TC	Profit	
4	1980						
5	1982						
16	2004						
17	2006						
18	2008						
19	average						

(a)

	A	B	C	D	E	F
1	**Table 2.2 Total Revenue, Total Cost, and Profit**					
2		***GiGo Company***				
3	Formula	=RAND()*20	=rand()*5		=VC+FC	=TR-TC
4	Year	TR	VC	FC	TC	Profit
5	1980	2.63	0.26	2.00	2.26	$ 0.36
6	1982	14.64	2.85	2.00	4.85	$ 9.80
17	2004	2.51	2.85	2.00	4.85	$ (2.34)
18	2006	1.92	2.85	2.00	4.85	$ (2.93)
19	2008	10.82	2.85	2.00	4.85	$ 5.97
20	average	5.08	2.85	2.00	4.85	$ 0.23
21	Table compiled by ***Your Name***, Department of					
22	Accounting and Finance, GiGo Company					

(b)

Entering the titles

(1) As in Table 2.2(a), enter **the table title** (row 1) and **the column title** (rows 2 and 3) as in rows 1 to 3. Note that =rand()*20 in row 2 should be entered as a **text, not formula**. Hence, when you click the cell, enter <Space> and = sign (that is, make a space before the = sign) and start typing. Another way of entering a formula as text is to enter ' =rand()*10, where " ' " is an apostrophe. In this case, no space is needed before the = sign.

(2) Row 2 is aligned to the middle of each cell (<"Home"><Alignment, Center>). You may change **the size of font** in B2 to 8 points (<"Home"><Font, Font Size!> [8]) or 7 points (the default is 11 points) to fit the column width.

(3) Select A1 and press <F2> to change the cell to the <Edit> mode. Then, use the mouse to select "GiGo Company" and put it in **italics** and a **smaller font** (GiGo = Garbage In Garbage Out).

(4) Enter the row title "Year" in A3. To enter the **row titles** (A4:A18), enter 1980 and 1982 first, then select A4:A5, and use the **fill-handle box** (step 3 in Fig. 1.2) to drag down the fill-handle to fill the numbers from 1980 to 2008. Note that, you have to select

two cells, A4 and A5, together so that Excel will know the interval between the two years; in this case, the interval is 2. In Table 2.2(a), in order to save space, we cut the years between 1982 and 2004. In this book, we use an arrow to show where the interim data are hidden.

(5) Enter the column titles in row 3: TR (Total Revenue), VC (Variable Cost), FC (Fixed Cost), TC (Total Cost), and Profit.

(6) Before we go to the next step, we use the "Justify" command (see Fig. 1.10) to correct the table title, by moving "GiGo Company" down to row 2, as shown in Table 2.2(b). To use the "Justify" command, insert a row below row 1 in Table 2.2(a), as shown in Table 2.2(b), select A1:F2, and use <"Home"><Editing, Fill><Justify!>. Move "GiGo Company" is moved from A2 to B2.

This completes the table title and the column and row titles. We continue to complete the data part of the table as in Table 2.2(b), using the following steps, steps (7)–(18).

Creating the data table

(7) Enter the **formulas** in row 5 of Table 2.2(b) as follows. This time, no space is allowed at the beginning of "=". At each cell, follow the steps as below:

B5: =rand()*20<Right> C5: =rand()*5<Right> D5: 2<Right>
E5: =C5+D5<Right> F5: =B5–E5<Enter>

where <Right> means press the right arrow to move to the right cell after typing the formula. (Do not enter it in the cell.) In the above, B5:, C5:, etc. are cell names listed just for your reference, and should not be entered in the cell. Use the pointing method to **enter the formula** in E5 and F5. As row 3 shows, formula can be entered in capital or small letters.

Box 2.2: **Entering formulas.** Formula can be entered in capital letters or small letters. In the latter, Excel will change them to capital letters when you press the return key. No space is allowed in a formula.

(8) After filling row 5, select B5:F19 and use the fill-handle to fill the range (see Fig. 1.2). To save time, always remember to enter the formula in **the first row of the data table first and then copy the row down to fill the rest of the rows.** Do not waste time by copying/typing the formula column by column.

(9) Note that row 20 is currently blank. The formula for average is

$$=\text{average(range)}, \quad (2.2)$$

where range is the range of numbers to be averaged. Thus, using the pointing method, enter

=average(B5:B19)

in cell B20, and use fill-handle to copy it to C20:F20. This completes the data part.

(10) Select B5:F19, and use **decreasing decimal** to show two decimal places. Select F5:F20 and click <"Home"><Number, **$**> to add the $ sign to the profit column. Note that **negative numbers** are enclosed in parentheses.

(11) **Color** TR column in yellow. Use <**Format Painter**> brush to color the TC and Profits columns in the same light blue color (see item 15 of Fig. 1.9(a)).

(12) Enclose the data part, B5:F19, and the average row, B20:F20, with a **heavy bordered box** (enter <"Home"><Font, Border!><Thick Box Border> (Not shown on the table)).

(13) Press the **F9 key(the "recal" or recalculation key**) several times to enjoy the changes in numbers. The number changes because we entered the formula of the random numbers.

Finishing the table and saving the file

(14) We preserve the original table by copying the table from Sheet1 to Sheet2 (by selecting whole Sheet1: <^A><^C> and click <Sheet2> tab, then, <A1><^V>). We use the old table in Sheet1 (Sheet2 is used for backup).

In Sheet1, to **change the random numbers to fixed values**, select B5:C19, and enter <^C>. and

<"Home"><Clipboard, Paste><Paste Values>.

Now, when <F9> is pressed, the numbers will not change. To make sure we have values in B5:C19, click any data cell in B5:C19. There should be no formula in the formula box.

(15) In row 21, write your name as shown. Click A21 again and select your name. Click <"Home"><Font, B> to add **bold (B)**, and also **italics (I)** and **underline (U)** to your name (the best practice is to use ^B, ^I, and ^U).

(16) Use "Justify" (see Fig. 1.10) to justify the long note in row 21.

(17) Click <Office><Print><Print Preview> to preview the table. Click **Print** to print out the table.

(18) Save the file (see Sec. 1.2.1).

This completes the construction of Table 2.2.

2.2 Entering Equations in a Table

In Table 2.2, we enter the data as random variables. In this section, we would like to enter data as formulas and use the definition of mathematical operators in Table 2.1.

2.2.1 *The profit table*

As in any Microeconomics textbook, we define the following equations for a business firm. Let the **inverse demand**[1] **function** (which is the same as the **average demand function**) be $p = 40 - Q$. Then the total revenue is, by definition, price (p) multiplied by quantity (Q); thus,

$$TR \equiv pQ = (40 - Q)Q. \tag{2.3}$$

Let the **total cost function** be

$$TC = Q^3 - 12Q^2 + 60Q + 20. \tag{2.4}$$

In these two equations, output Q, the variable on the right-hand side, is the independent variable, and the variables on the left-hand side, TR and TC, are the dependent variables. The definition of profits is $\pi = TR - TC$. As shown in Table 2.3, output Q ranges from 0.0 to 0.5, ..., 10.0, which we label as output level i in column A, for example, $Q_2 = 0.5$, $Q_5 = 2.0$, etc. Marginal profit ($m\pi_i$) is defined as the profit per unit of output produced at the output level i, that is,

$$m\pi_i = \frac{\pi_i - \pi_{i-1}}{Q_i - Q_{i-1}}, \tag{2.5}$$

where Q_i indicates the ith output level. Q_{i-1} is the $(i-1)$th output level, or output at the previous level. Thus, $Q_i - Q_{i-1}$ is the difference of outputs between the consecutive levels. Similarly, π_i is the profit at the ith output level.

We would like to implement these equations as in Table 2.3. Enter the table title, formulas, and column and row headings of Table 2.3, which is similar to Table 2.2(b), except that we have the independent variable Q in column B5:B25, with an equal output interval of 0.5 unit. Using the mathematical operators defined in Table 2.1, in cell C5 we enter the TR formula (2.3),

C5: =(40−B5)*B5.

To do so, select C5 and type =(40−B5)*B5. While still in C5, with the pointer blinking immediately after the *, use the mouse to click on cell B5. B5 should appear after the * just as if you had typed it. This **B5 in the formula represents the value of the variable Q at output level i = 1**. That is, the columns B5 to B25 represent the variable Q, where $i = 1, \ldots, 21$. In particular, the value in B5 is a particular value of variable Q. This method of representation is very important and will be used repeatedly. Press the return key. 0.0 should appear in C5.

Next, in cell D5, enter (2.4) as

D5: =B5^3−12*B5^2+60*B5+20

[1]It is called inverse demand function since the demand function is generally written as $Q = 40 - p$, in which p is an independent variable, and Q is a dependent variable. See Chap. 3.

Table 2.3 Entering equations

	A	B	C	D	E	F
1	Table 2.3. Profits of the GiGo Company					
2			TR=Q(40-Q)	TC=Q^3-12Q^2+60Q+20		
3		Q				
4	i		TR	TC	Profit	mProfit
5	1	0.0	0.0	20.0	-20.0	
6	2	0.5	19.8	47.1	-27.4	-14.8
7	3	1.0	39.0	69.0	-30.0	-5.3
8	4	1.5	57.8	86.4	-28.6	2.8
9	5	2.0	76.0	100.0	-24.0	9.3
15	11	5.0	175.0	145.0	30.0	16.8
16	12	5.5	189.8	153.4	36.4	12.8
17	13	6.0	204.0	164.0	40.0	7.3
18	14	6.5	217.8	177.6	40.1	0.3
19	15	7.0	231.0	195.0	36.0	-8.3
25	21	10.0	300.0	420.0	-120.0	-90.8
26		C5	=(40-B5)*B5			
27		D5		=B5^3-12*B5^2+60*B5+20		
28		E5			=+C5-D5	
29		F6			=(E6-E5)/(B6-B5)	

Cell formulas (rows 26–29)

Then, entering profit =C5−D5 in cell E5, and using the fill-handle, copy row C5:E5 down to C25:E25. Lastly, leaving F5 blank, enter the marginal profit (2.5) in F6,

F6: =(E6−E5)/(B6−B5),

and copy F6 down to F25. This completes the table.

After setting up the table, we observe that the marginal profit changes from negative to positive between rows 7 and 8, and between rows 18 and 19, as in Table 2.3. We paint these four rows with yellow.

2.2.2 *Cell formulas of the table*

Although row 2 of Table 2.3 shows the formula for TR and TC, it is not clear how these formulas are actually entered. Rows 26 to 29 reproduce the formulas entered in the first row, row 5, of each column of the data part of the table. We call them **cell formulas**.

To enter the cell formula of C5 in C26:

(a) Move the pointer to C5, and <F2(edit)><Home><Space><Enter>. C5 will show the formula, instead of a number.
(b) Moving the pointer to C5 again, and copy (^C) the formulas and paste it to C26 (^V), we have the formula in C26, as shown in Table 2.3.
(c) Moving the pointer back to C5 and closing the space before the equality sign and <Enter>, the cell returns to the number.

Do the same to obtain cell formulas in D27, E28, and E29.

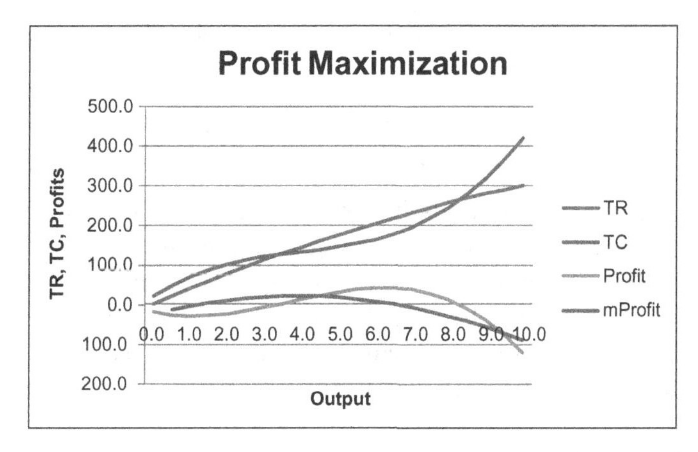

Fig. 2.1 Profit maximization

2.2.3 *A simple chart*

While we will discuss the method of Excel graphics in detail in Chap. 3, a simple use of chart is helpful in visualizing the variables in Table 2.3. The chart is shown in Fig. 2.1. To draw the chart, we select the data first, namely, select B4:F25, including the column labels in row 4, and click

<"Insert"><Charts, Lines!><Line(the first line box)>.

The four lines will appear automatically as in Fig. 2.1. To add the chart and axis titles, click the chart, and in the Menu Bar, click

<"Chart Tools, Design"><Chart Layouts, Chart!><Layout 10>.

The "Chart Title" and "Axis Title" appear, showing the place to put the titles, along with vertical lines at each output. Click one of the vertical lines and enter <Delete> key to delete the lines. The chart and axis titles are in text boxes. Click each title and add the appropriate title as shown in Fig. 2.1.

From the chart, it is clear that minimum profit (largest negative profit) occurs between Q = 1.0 and 2.0, and the maximum profit occurs between Q = 6.0 and 7.0.

2.2.4 *Economic interpretation*

What is the economic interpretation of the changes? It is easy to construct the Excel table, but to interpret the results, we need economic knowledge and common sense. We first note that when output increases from 0.5 to 1.0 to 1.5, marginal profit *increases* from −14.8 to −5.3 and to 2.8, so it is profitable for the firm to continue increasing output. But when output increases from 6.5 to 7.0 and to 7.5, marginal profit *decreases* from 0.3 to the negative value of −8.3 and −18.3. So, instead of invoking negative marginal profit, the firm should stop producing at Q = 6.5, or, if the output is divisible, at some output level between 6.5 and 7.0, at which marginal profit is zero. At this point, the firm's profit is maximized.

Thus, from the table, we can easily find the level of output at which the firm's profit is at maximum.

2.3 Some Notes on Copy, Move, and Paste Commands

Two of the most important commands in Excel are **copy** (**Copy** and **Paste**) and **move (Cut** and **Paste)**. You may copy or move a range, a column, a row, or a sheet. There are four methods of copying and moving a range. To see this, as in the previous chapter, enter random number, =rand(), in C5:F10 on a new sheet (see Fig. 1.2). To perform the copy or move operations, we always **select the range first**. After selecting C5:F10, practice the following four methods of moving and copying.

2.3.1 *Moving and copying a range*

The four methods are summarized in Table 2.4.

- **The clicking button method.** Selecting the range and clicking the **Cut** or **Copy** button in the "Home" ribbon (in the Clipboard group), move the pointer to the desired cell and click <Paste>. This method has been discussed in buttons 1, 2, and 3 of Fig. 1.9(a) in the previous chapter.
- **The right mouse method.** Select the range and <RM> (click the right-hand mouse button), and a mini toolbar appears. Select <Cut> or <Copy> and then move the pointer to the desired cell. Clicking the cell and <RM> again, select <paste>. This method uses the right wrist too often.[2]

Table 2.4 Four methods of coping and moving range

Method	Action from			to
1. Icon	Click Click		→	Click Click
2. Right Mouse	<Cut> <Copy>		→	<Paste>
3. Control Key	^X ^C		→	^V
4. Click and Drag Move	0.823876 0.882137	0.823876 0.882137	LM →	0.823876 0.882137
Copy	0.823876 0.882137	0.823876 0.882137		

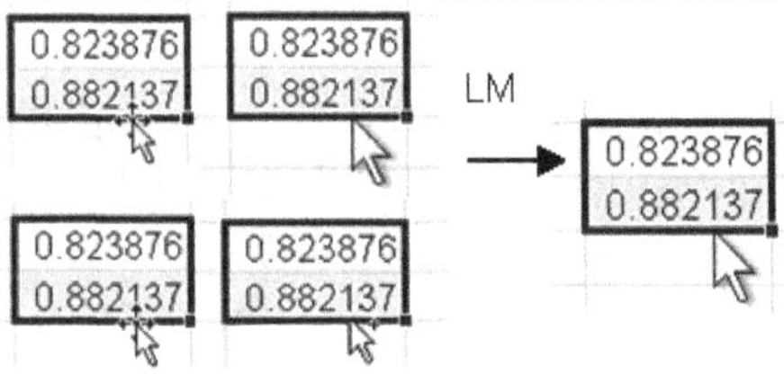

[2]In general, since we use the right wrist quite often using the mouse, it is suggested that we use the control key method to move or copy the range. This may prevent wrist tendonitis in the future.

- **The control key method.** Using the left hand, press ^X (for Cut/Move) or ^C (for Copy), move the pointer to the desired cell, and press[3] ^V. This method is the easiest, and reduces the use of the right wrist.
- **The click and drag method.** For moving or copying a range to the adjacent cells or range, you can use the fill-handle to click and drag, as explained in Example 1.1 of Chap. 1. If the destination is separate from the original range, you do the following steps, as shown in the last row of Table 2.4:
  (a) Move the pointer near the border of the selected range. The fat cross changes to a fat arrow, and a four-arrow-head cross appears at the tip of the fat arrow.
  (b) For **moving**, press the LM. Only the fat arrow remains. Drag it to the desired place and release the mouse.
  (c) For **copying**, use the left index finger to press the control key, and the four-arrow-head cross changes to a **small plus at the tip arrow**. You then copy by continuing to hold the control key, at the same time pressing LM and dragging the plus-tipped arrow to the desired place and releasing the LM. This method is intuitively clear, but limited only to making a copy inside the current worksheet.

Box 2.3: **Moving and copying a range**. We recommend that you use the control key method, namely, ^X (move) and ^C (copy) and then ^V (paste).

2.3.2 *Moving and copying a worksheet*

(1) **To move an entire sheet to the left or right of other sheets**, like the "Move" sheet in part a of Fig. 2.2, select the sheet tab and press the LM. A fat arrow appears with a sheet logo at the tip, and a black down arrow ▼ at the left end of the tab (see part a). Drag the down arrow to the desired position and release the LM.

(2) **To copy a sheet to another sheet**, say the "Copy" sheet in part b of Fig. 2.2, move the pointer to the sheet tab. A fat arrow appears. Pressing the control key by the left hand, and at the same time, use the right hand to press the "Copy" sheet tab until a sheet logo with + sign appears next to a down arrow ▼ (see part b). Drag the down arrow to the **right or left** and release the LM. The "Copy" sheet will be copied to "Copy (2)" sheet.

Fig. 2.2 Moving and copying a sheet

[3]X for "X out (to erase or delete)." "V" looks like an arrow pointing downward and showing "insert" the wedge.

(3) To **delete** "Copy (2)" sheet, move the pointer to the "Copy (2)" sheet tab, and press RM. A menu window appears. Select <Delete>.

2.3.3 *Column and row operations*

Columns and rows may be added or deleted easily by selecting the desired column or row headings. Create a random number table in C5:F10 again to practice.

1. To **add columns** (or **rows**), use the mouse to select column headings, say C and D (or rows 8 and 9). This is called "**selecting columns** (or **rows**)."
   (a) The two selected columns (or rows) will be shaded. Click <"Home"><Cells, Insert>. Two columns (or rows) will be added to the left of (or above) the shaded columns (or rows). Note that after you have selected columns (or rows), Excel can detect whether you want to insert columns or rows. Hence, you only need to click the upper half of the <Cells, Insert> button.
   (b) You may click the <Insert!> (with the down arrow), and a drop-down list appears (see 39 of Fig. 1.9(c)). Click <Insert Sheet Columns (or Rows)>.
   (c) You may also click RM to invoke the mini toolbar (Fig. 1.11), and click <Insert> from the mini toolbar. An Insert dialog box appears. Select <Insert, @Entire column><OK>.
2. To **delete columns** (**rows**), the procedure is the same as in above three methods, except that instead of <Insert>, you select <Delete>. Note that if you enter the Delete key from the keyboard, it erases the contents of the columns, not the columns themselves.

Box 2.4: **Adding or deleting columns or rows.** It may be more convenient to add or delete columns by Macro commands (^#C and ^#X) and rows by Macro commands (^#R and ^#E). See Appendix 2A.

3. **Columns and rows can be moved or copied** by the same four methods used to move or copy a range, except that at the destination the pointer must be placed at row 1 for moving columns (or column A for moving rows). This is so since you are moving whole columns (or rows), and the number of rows in the column (or number of columns in the rows) contained in the original columns (or rows) must be the same.
4. **Resizing and optimizing column (or row) size.** Enter random numbers in C5:F10. Reduce the default 6 digits to 2 digits. The column width looks "oversized". Select columns from C to F, and double click any column separator (see Fig. 1.4(a)). Columns in the range are **optimized** in the sense that each column has just the right size to contain the contents of the columns. When a range of columns is selected, an adjustment of the size of any one column affects the columns in the whole range.

   Similarly, select the row headings from 5 to 10. Click any row separator and drag to "height: 30.00", which is shown by an indicator just above the pointer when a row separator is clicked (see Fig. 1.4(b)). Click anywhere to deselect the row range. The height of rows in the range is enlarged. To optimize the height of the data part, select the row headings from 5 to 10, and click any row separator twice.

2.4 Summary

Tables 2.2 and 2.3 introduced in this chapter are different. One uses random numbers to simulate the actual situation and generate a financial data table, while the other uses equations to generate the data. As we have seen in the previous chapter, the random numbers are useful in the sense that you do not need to spend a long time entering the actual numbers, and yet you still can get the feeling of constructing a table. The Excel program has an extensive set of methods for formatting tables for presentation purposes. Since the purpose of this book is not to construct models for presentation in a seminar or a conference, we do not explore methods of formatting tables.

As mentioned in Box 2.4 of Sec. 2.3.3, editing rows and columns will be more fun and efficient if we use macro commands to add and delete entire columns and rows (Appendix 2A). The fill-handle box is another very useful feature (see Example 1.1 of Chap. 1 and Part (4) of Sec. 2.1.2). The macro method introduced in Appendix 2A can also be applied to other applications.

In general, the use of spreadsheets just for the sake of using spreadsheets is counter-productive. It is tedious and dry, and many spreadsheet commands and steps are strange and idiosyncratic. Even if you learn the commands, you will forget them soon. Thus, the best way to learn and remember the spreadsheet commands is to learn them along with a specific subject, in our case, economic and business applications. Thus, in this chapter, in addition to acquainting ourselves with the spreadsheet commands and buttons, we have introduced and explained how to construct a basic economic/business table of total revenue, variable cost, total cost, and profits, taking advantage of the random-number-generating function (Sec. 2.1) and the spreadsheet implementation of equations (Sec. 2.2). We will use either one of these two methods to generate tables, and most of them will be used as the basis of drawing charts, in the rest of this book.

We will introduce other commands and features of spreadsheets as we introduce economic and business applications in the following chapters, and the readers will have more opportunity to practice the commands and buttons they have learned in this chapter. The key terms at the end of this and following chapters are provided for the readers to review the basic concepts in Excel, Economics and Business, Mathematics and Statistics.

Appendix 2A. Macro Program for Adding and Deleting Entire Columns and Rows

The following Macro program is very helpful in avoiding right-hand tendonitis. We define the following keyboard macros:

$$\text{^\#R (Control + Shift + R, same below) as "rowadd" (add entire } Rows\text{),}$$

and its neighboring key,

^#E, as "rowdelete" (delete or *E*liminate entire rows)

We also define

^#C, as "columnadd" (add entire *C*olumns),

and its neighboring key,

^#X, as columndelete (delete, or *X*-out, the entire columns).

For practical purposes, enter random numbers in C5:F10 as before. Click D7. To define ^#R, follow the five steps below:

(1) Enter

<"View"><Macros, Macros><Use Relative References>. (2A.1)

If you click <"View"><Macros, Macros> again, you will see that the logo on the left of <Use Relative References> is shaded/boxed, indicating it is active. "**Use Relative References**" will make the macro work at any cell (otherwise, the macro will work only at the cell, say D7, where the macro is defined). You only do this once.

(2) Next, enter

<"View"><Macros, Macros!><Record Macro>. (2A.2)

The worksheet is ready to receive macro commands.

(3) The "Record Macro" dialogue box appears, as in the LHS of Fig. 2A.1. In part a, enter **rowadd**, in part b, press the Shift+R key, and in part c, click the downward arrow and select **Personal Macro Workbook** to store in the Excel template (rather than the current workbook). <OK> to exit the dialogue box.

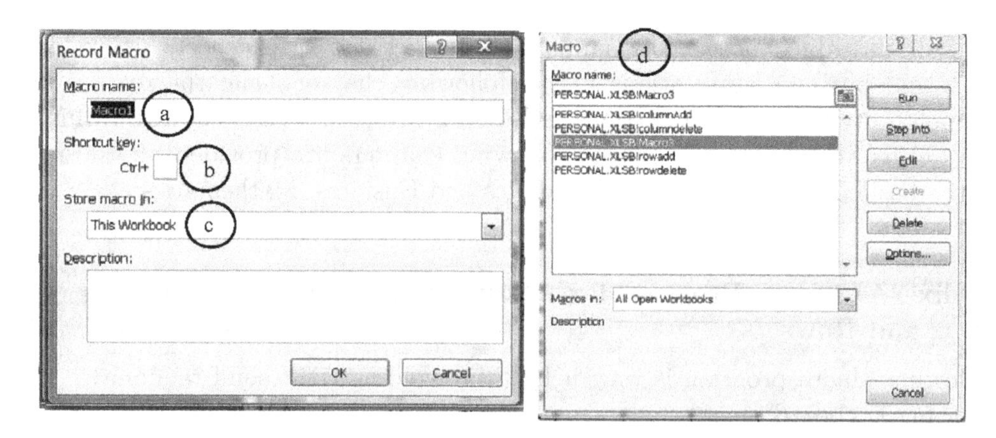

Fig. 2A.1 Recording and viewing macros

(4) Enter

<"Home"><Cells, Insert><Insert Sheet Rows>. (2A.3)

A row will be inserted.

(5) To stop the macro recording,

<"View"><Macros, Macros><Stop Recording>. (2A.4)

(You may also click the Stop Recording logo, which appears on the left of the status bar). This completes the macro assignment.

You may continue to define ^#C by replacing <Insert Sheet Rows> in (2A.3) by <Insert Sheet Columns>.

For deleting rows (^#E), enter in (2A.3) that

<"Home"><Cells, Delete!><Delete Sheet Rows>,

and for columns (^#X), replace <Delete Sheet Rows> by <Delete Sheet Columns>.

If you made an **error in assigning a macro**, enter

<"View"><Macros, Macros!><View Macros>.

A Macro window appears, as in part d of Fig. 2A.1. You may select a macro to **Run**, or **Delete** a macro, like Macro 3 in the figure. If you are familiar with programming, you may make a correction directly by clicking the **Edit** button. The **Option** button will show you the key you have defined. Sometimes you may have to **unhide the Personal File** in order to delete a macro. In this case, enter <"View"><Window, Unhide> and then click <Personal>.

Appendix 2B. Consecutive Months Custom AutoFill Lists

In Homework 2.2, months can be entered by typing Jan in a cell and copying Jan down by fill-handle. If the fill-handle does not work, do the following. Open

<"Office"><Excel Options><Popular><Edit Custom Lists>.

Then select the New List under Custom lists dialog box. Enter Jan, Feb, ..., Dec in "List entries". Click <Add> to close the small window (you may create a Custom AutoFill List on a spreadsheet first, select it, and <Import> the list).

Appendix 2C. The symbol (c) and the Copyright Logo

In Homework 2.7, when (c) is entered in a worksheet, it may change into a copyright logo ©. To avoid the change, enter

<"Office"><Excel Options><Proofing><AutoCorrect Options>.

Select the copyright logo and <Delete>. Click <OK> to return to the original cell and enter (c) again. A temporary method is to enter "c)" first, and then move the pointer to the left and enter "(", and click any other cell. This will fool the AutoCorrect command.

Review of Basic Equations and Formulas

(2.1) =rand()*k (2.3) Total revenue $TR \equiv pQ = (40 - Q)Q$ (2.4) Total cost $TC = Q^3 - 12Q^2 + 60Q + 20$ Profit $\pi = TR - TC$ (2.5) Marginal profit $m\pi_i = \dfrac{\pi_i - \pi_{i-1}}{Q_i - Q_{i-1}}$	(HW2-9) $MR = \dfrac{TR_i - TR_{i-1}}{Q_i - Q_{i-1}}$ $VC = Q^3 - 12Q^2 + 60Q$ $FC = 20$ $ATC = TC/Q$, $AVC = VC/Q$, $AFC = FC/Q$, $AFC = FC/Q$ $MC = \dfrac{TC_i - TC_{i-1}}{Q_i - Q_{i-1}}$,

Key terms: Economics and Business

Key terms: Excel

Homework Chapter 2

Try your best to present all the tables below. Make them colorful, different font, colors, or sizes, etc. Experiment with different column width, background color, italic, bold, underline, etc. Use print preview before you print.

For Section 2.1 Generating Random Numbers

2-1 A simple table construction If the GiGo Company has the data of TR ranging from 0 to 10, and TC ranging from 0 to 5, and the year ranging from 1970 to 2010, construct GiGo Company's financial statement as in Table 2.2.

(a) Please show the formulas above the row titles, and find the average TR, TC, and profits of the data.
Questions:
(b) Calculate the average of TR, TC, and profit over the years;
(c) Enclose the data part with borders;
(d) Color the average row in blue, and change the font color to white;
(e) Change to two decimals.
(f) Show your cell formulas of TR, TC, and profits for 1970 above the column titles, and show the cell formulas of the average TR, TC, and profits of the data a row below the table.
(g) Print out only the data from 1970 to 1975, and 2005 to 2010 (hide the other rows), with the average and formula rows of the original table.

(Hints: (f) See Sec. 2.2.2. (g) Selecting the rows from 1976 to 2004, use <"Home"><Cells, Format!><Visibility, Hide & Unhide!><Hide, Rows>. Then print out.)

2-2 Generating a large set of monthly data In Table 2.2, instead of yearly data, we have only monthly data from 2000 to 2006. Supposing that TR ranges from 0 to 100 and TC ranges from 0 to 50, construct the financial statement of the firm for TR, TC, and Profit as follows. (a) The row titles range from Jan, Feb, ... up to Dec for each year. (b) Change the data to two decimal places

(after the decimal point). (c) Find the monthly average of TR, TC, and Profit of each year, and of the seven years from 2000 to 2006.
(Hints: (a) Sometimes, depending on the setting of the Excel program, when Jan is entered, you may use the fill-handle button to drag Jan down to Dec. Do the same for other years. If this is not the case, enter Jan to Dec for 2000, and copy Jan to Dec for 2001, and up to 2006, and then insert a row for year. You may use the custom autofill method in Excel to enter months. See Appendix 2B. (b) Enter the first row of Jan 2000 for TR, TC, and Profit, and then copy the first row up to the last row, namely, Dec 2006. Use the "copy command". Do not enter the numbers one row by one row, or even one year by one year. Then, since the Year rows, 2001, 2002, etc. should be blank, delete the row entries corresponding to each year. (c) After entering the data, find the monthly average of the seven years first, then insert a row below December of each year to calculate the monthly average of each year).

2-3 Constructing a table The table of TR, TC, and profit of the GiGo Company from 1990 to 2005 is given as Table 2.2 in the text: The TR and VC are random numbers. TR ranges from 0 to 50 and VC ranges from 0 to 25. FC is 10 and TC = VC + FC.

	TR	VC	FC	TC	Profit
1990					
1991					
1992					
⋮					
2005					

(a) Find the total revenue, variable cost, fixed cost, total cost, and profit of the firm from 1990 to 2005. All entries should have two decimal places.
(b) Copy the data to sheet2. Name Sheet2 as Q1. Change the data to values so that the data will not change when the "return" key is pressed. Optimize the columns.
(c) Enclose the data part of the table with a heavy box AFTER you have completed the table. The row and column titles should be in bold face. The column title should be centered.
(d) Use light yellow paint to highlight profits, and light blue to highlight TR.
(e) Find the sum (=sum()), average (=average()), max (=max()), min (=min()) of each column, and list them at the bottom of the table.

2-4 Entering formulas In Table HW2-4, we enter the sales information of 60 regions. In addition to the average function, other formulas of some basic descriptive statistics are listed and defined below. In each formula, the range must be specified. Do not reproduce the colored borders or the hints. In these lecture notes, colored borders are for your reference only.

Table HW2-4 Sales of GiGo company

	A	B	C	D	E	F	G
1	**Table HW2-1 Sales of GiGo Company**						
2		rand()*100	rand()*30		Earning Rate	Profit ratio	
3	Region	TR	TC	Profits	TR/TC	Profits/TR	Profits/TC
4	1						
5	2						
	..						
65	60						
66	max						
67	min						
68	sum						
69	average						
70	var						
71	varp						
72	stdev						
73	stdevp						
74	CV (%)						
75	CVp (%)						

Explanation of the formulas (see Chap. 5 for details)

=max(range): the maximum number of the specified range.
=min(range): the minimum number of the specified range.
=sum(range): the sum of the specified range.
=average(range): the average value of the specified range.
=var(range): sample variance (the **sum of squared deviation (SSD)**) from the mean divided by n − 1).
=varp(range): population variance (the SSD from the mean divided by n).
=stdev(range): sample standard deviation (the positive square root of var).
=stdevp(range): population standard deviation (the positive square root of varp).
=CV: sample coefficient of variation =stdev*100/average, in %.
=CVp: population coefficient of variation =stdevp*100/average, in %.

Questions:

(a) Enter the formulas in the shaded part of the TR column (B66:B75), and then copy to the TC column.(Hint: Use the fill-handle to fill in the region).
(b) Enter the first row, B4:G4, and then copy the first row down. Do not copy column by column.
(c) Formulas can be entered by first entering =max(B$4:B$65) in cell B66, and copying B66 down to B73. Change the formula name. Enter CV and CVp using the formula given. Then copy B66:B75 to G66:G75. Here $ indicates that rows 4 and 65 are fixed. See the explanation in Chap. 5.
(d) Write the cell formulas in B66:B75 on the right-hand side of the table, H66:H75.
(e) Enclose the table with boxes as shown in Table HW2-4.

(Hints: (a) Use the fill-handle to fill in the "Region". (d) Copy B66:B75 to H64:H73, and edit the formulas).

2-5 Use of the help button The help button of Excel 2007 is at the upper right corner "?" in the menu bar. You may also enter <F1> to invoke the help menu. Using the help button <F1>, find the definition of var, varp, stdev, and stdevp. Write the formula for each. You may simply copy the formula from the Help, print out the explanation from the help section (Table HW2-2).

> Box 2.5: To find the definition of, say, VAR, press <F1>, the Excel Help drop-down window appears. Enter "VAR" or "VAR worksheet function" and click <Search>.

2-6 Formulas in the rows and columns Table HW2-6 shows the 2009 sales by 50 regions. Fill in the table and do your best to present your table in color.
(Hints: (a) You may enter =rand()*40 in Q1, select Q1 cell and copy them to the right up to Q4. Then enter <F2> in each cell to make an appropriate correction. (b) Enter all the formulas in the first row (Region 1), and copy them down to Region 50).

2-7 Construction of a Large Table Reproduce Table HW2-7, the Balance of Payments Table below, including the explanation column, the braces, and the small box of Accounting Rules at the bottom of the table. You may choose your formatting design to make the table easier to read and appealing. Note that the subtotal column should use the sum formula: Do not type in the numbers. The whole table should fit into one page. In the debits and credits columns, item a in the debits column corresponds to item a in the credits column, and so on with b, c, etc. This is to show the **principle of double entry in accounting**. (In some cases, when you type (c), the computer gives a copyright symbol ©. See Appendix 2C.)

2-8 TR and TC TR and TC are random numbers ranging from 0 to 100 for TR and 0 to 50 for TC, respectively. Profits = TR−TC. The year ranges from 1970 to 2000 (see Table HW2-8).

(a) Construct the table exactly as shown and fill in the table. All entries should have two digits after the decimal point. Center all the entries of the columns. Optimize the column.
(b) Enclose the data part of the table with a heavy box AFTER you have completed the table. The row and column titles should be in bold face. Both column and row titles should be centered.

Table HW2-6 2002 Sales of 50 regions

	rand*40	rand*60	rand*80	rand*100							
Region	Q1	Q2	Q3	Q4	sum	average	varp	stdev	stdevp	CV	CVp
1											
2											
⋮											
50											
sum											
average											

Table HW2-7 The balance of payment table

The Balance of Payments Table (IMF format)

The Republic of GiGo, December 31, 2000

Compiled by ***Your Name*** (Millions of dollars)

Use formula

		Debits	Credits	Cr - De	Explanations
1		Debits	Credits	Cr - De	Net Debit (-) or Credit (+)
2	**A. Current Account**				
3	1. Goods (Merchandise trade)	600	500		Imports/Exports
4			25		
5	*Net Balance on Trade*	*600 a*	*525 b*	-75	Imports/exports
6	2. Services				Surplus (+)/deficit (-)
7	(a) Transportation	100	50		*National income item*
8	(b) Travel	75	60		
9	(c) Investment income	50	100		interests, dividends, fees
10	(d) Other	25	50		
11		*250 c*	*260 d*	10	
12	*Net Balance on Goods and Services*	850	785	-65	(5)+(11)
13	3. Unilateral Transfers	75	25	-50	grant, remitance
14	*Net Balance on Current Account*	*925*	*810*	*-115*	Subtotal Surplus (+)/deficit (-)
15	**B. Capital Account**				
16	1. Long-term Capital				over one year *Basic balance*
17	(a) Direct Investment	100	50		Outward/Inward = -140
18	(b) Portfolio Investment	25	50		Outward/Inward
19		*125 f*	*100 e*	*-25*	net long-term capital flow
20	2. Short-term Capital	525 b	600 a		less than a year
21		260 d	250 c		
22		100 e	125 f		
23			150		
24		*885*	*1125*	*240*	net short term capital flow
25	*Net Balance on Capital Account*	*1010*	*1225*	*215*	Subtotal (19+24). Net capital flow
26	**C. Official Reserves Account**				Surplus (+)/deficit (-)
27	1. Monetary Gold (net)	5			
28	2. Other Reserve Assets (net)	100			Mostly in foreign exchange (money)
29	3. Liabilities to Foreign Central Banks (net)		15		
30	*Net Balance on Official Reserves Account*	*105*	*15*	*-90*	Subtotal
31	*Net Errors and Omissions*	*10*		*-10*	
32	*Total Debits and Credits*	2050 h	2050 h	0	sum of the shaded parts. The net is identically 0.

Sources: Based on Franklin R. Root: Internaitnal Trade and Investment, 6th ed., South-Western Publishing Co. 1990, p. 351. The letters a, b, ... g show+B9 the double accounting entries.

Use 8 points font

Accounting rules	
Debits	Credits
Purchase	Sale
Payment of fund	Receipt of fund
Assets +	Assets -
Liability -	Liability +
Money Departed	*Money Created*
Money is departed due to	**Money is created due to**
Imports (M)	Exports (X)
Lending (L)	Borrowing (B)
Outward FDI	Inward FDI

$ $

(c) Use light yellow paint to highlight profits, and light blue to highlight TR. Add a $ sign to the value of the profits column.

(d) Find the average (=average()), the standard deviation (=stdev()), and the coefficient of variation (=stdev*100/average) for all five columns. Place them below the table as shown in the table.

(e) Show the Excel cell formula you entered in the 1970 row below the table.

Table HW2-8 Sales of GiGo Company

	rand*100	rand*50		Profit ratio	
Year	TR	TC	Profits	Profits/TR	Profits/TC
1970					
1971					
⋮					
⋮					
1999					
2000					
average					
stdev					
CV (%)					

(Hint: Copy the 1970 row to the last (formula) row and use the edit key (F2)). Draw the chart as in Fig. 2.1.

For Sec. 2.2 Entering Equations in a Table

2-9 The total revenue (TR) function The demand function of firm ABC is given as $p = 30 - Q$. Find the firm's total revenue (TR) and the marginal revenue (MR), which is defined below. Draw the demand schedule and TR and MR for $Q = 0, 0.5, \ldots, 10$.

$$MR = \frac{TR_i - TR_{i-1}}{Q_i - Q_{i-1}}.$$

2-10 The total cost (TC) function (TC) function of firm ABC is given as

$$TC = Q^3 - 12Q^2 + 60Q + 20.$$

The variable total cost (VC), fixed cost (FC), average total cost (ATC), average variable cost (AVC), average fixed cost (AFC), and marginal cost (MC) are defined as follows:

$$VC = Q^3 - 12Q^2 + 60Q, \quad FC = 20,$$
$$ATC = TC/Q, \quad AVC = VC/Q, \quad AFC = FC/Q,$$
$$MC = \frac{TC_i - TC_{i-1}}{Q_i - Q_{i-1}}.$$

Construct the table for TC, VC, ATC, AVC, AFC, and MC for $Q = 0, 0.5, \ldots, 10$. Draw the chart as in Fig. 2.1.

2-11 The profit function Construct the profit function as $\prod = TR - TC$, where TR is given in HW2-9 and TC is given in HW2-10 above, for $Q = 0, 0.5, \ldots, 10$. Thus, you have four columns: Q, TR, TC, and $\prod$ on the table.

(a) Find the marginal profits as defined in the text (Equation (2.5)), and enter it as the fifth column.

(b) Enter MR, MC, and the difference MR-MC next to the marginal profit column (thus, you have a total of eight columns including the column for the independent variable Q).
(c) Draw the chart of mProfit, MR, and MC like Fig. 2.1.
(d) Do you find any relationship between the marginal profits and the difference MR-MC? Give an economic interpretation.
(e) Are the profit level in this example reasonable? What is the economic interpretation?

(Hints: (a) $Q^{\#}$ is between Q = 5.5 and 6.5, or $Q^{\#} = 5.75$ $\pi^{\#}$ is between −18.63 and −20.00)

2-12 Profit maximization The total cost, total revenue and the profit, of a firm is given as follows:

$$TR = Q(500 - 5Q)$$
$$TC = Q^3 - 10Q^2 + 100Q + 50$$

where $Q = 1, 2, \ldots, 20$. As usual, Profit = TR − TC, and marginal profit is

$$\text{mprofit} = \frac{\text{profit}_i - \text{profit}_{i-1}}{Q_i - Q_{i-1}}$$

(a) Construct the table of output, TC, TR, profit, and marginal profits. All data and formula entries should have two digits after the decimal point.
(b) What is the average (=average()) of TC, TR, and profits? Place the averages below the table.
(c) Enclose the Profit column of the table with a heavy box AFTER you have completed the table. The row and column titles should be in bold face. The column headings should be centered, and in bold face.
(d) Use light yellow paint to highlight the maximum profit row(s).
(e) Draw a chart of TC, TR, Profits, and marginal profits without markers. Enter the chart title, and horizontal and vertical axis titles.
(f) Using a text box to denote the name of each of the four curves. Make sure the text boxes are embedded in the chart.
(g) Denote the point of maximum profit with an empty circle. Draw a vertical dotted line through the point of maximum profit.
(h) Using right or left braces, show that, at the maximum point of the profit curve, the height of the maximum profit must be the same as the difference between the TR and TC curves at that point.
(i) In the chart, how is the marginal profit curve related to the profit curve at the point of the maximum profit?
(j) Show the cell formula for Q = 20 below the table.

Excel operations

2-13 Blank cells in a range This problem concerns with the Excel average and variance formulas when their ranges contain blanks, letters, and zeros. In the following table, explain why the average and variance in the last two rows are different in each case. Reproduce the table and fill in the blanks:

(a) In cases 2 and 3, average and variance are divided by ------ (give a number), since the formulas in the last two rows do or do not (choose one) count the blank cells and letters.
(b) In case 4, 0 is counted as a -------- (value, number). Hence the average and the variance are divided by -------- (give a number), as in Case 1.

	Case 1	Case 2	Case 3	Case 4
	1	1	1	1
	2		letters	0
	3	3	3	3
	4	4	4	4
	5	5	5	5
sum	15	13	13	13
sum/5	3	2.6	2.6	2.6
average	3	3.25	3.25	2.6
variance	2.5	2.92	2.92	4.3

For Sec. 2.3 and appendixes: Moving and copying, and macro programs

2-14 Rows and columns Generate 30 × 10 random numbers, and practice four methods of moving and copying a range, worksheet, columns, and rows, as we have studied in Sec. 2.3. (Hints: Enter the frame of the table first).

2-15 Macro program Generate any set of 10 × 10 random numbers.

(a) Construct a keyboard macro program for ^#V such that, after a range is selected, the macro will change the random numbers into value (eliminate the formulas).
(b) Construct a keyboard macro program for ^#J such that, after a range is selected, the macro will justify the extra long text range into a smaller specific range.

(Hints: (a) and (b) To make the section of the range flexible, do not include the selection of the range in the macro program.)

Chapter 3

Static Analysis in Economics and Business — Excel Graphics

Chapter Outline

Objectives of this Chapter

The purpose of this chapter is to introduce the basic structure of simple economic models. There are two kinds of models. One is the market equilibrium model, which is a basic model in microeconomics. The other is the national income model, which is a fundamental model in macroeconomics. Although these two kinds of models are treated in different fields of economic theory, from the mathematical point of view, the basic idea underlying the analysis is essentially the same. They come under the topic of solving systems of simultaneous equations. In this chapter, we first deal with the market equilibrium model. The national income model will be introduced in the second part of the chapter.

In the process, we learn Excel graphics in detail. We show how to present demand and supply equations in microeconomics graphically and how to find equilibrium values. This method is also applied to aggregate demand and aggregate supply equations in macroeconomics. The last section is devoted to the introduction of comparative static analysis in economics. We show the saving and investment analysis. This leads to the concept of Keynesian investment multiplier theory.

3.1 Single Market Models

Economic models consist of equations or functions, variables, constants, and the coefficients of variables. Other fields of science share the definitions of these terms. The set of total revenue, total cost, and profit functions that we introduced in Sec. 2.2 is an economic model of a firm.[1] In this chapter, we introduce other simple economic models.

We start with the following example of a microeconomic model.

Numeric model	Parametric model[2]	Name of equation	
$D = 18 - 2P$	$D = a - bP$, $a > 0$, $b > 0$	Demand function	(3.1)
$S = -6 + 6P$	$S = -c + dP$, $c > 0$, $d > 0$	Supply function	(3.2)
$D = S$	$D = S$	Equilibrium condition	(3.3)

The first column is a system of equations called **a numeric model**, since the constant terms and the coefficients of variables are numbers. The second column is a system of equations called **a parametric model**, since the constant terms and the coefficients of variables are parameters. These two models are the same. The difference is only that one is given in numbers and the other is given in unknown **parameters**, a, b, c, and d, which can represent real numbers, including the corresponding numbers in the numeric model.[3] As we will show in the next chapter, the parametric model can be illustrated using the **range name method** in Excel.

In both numeric and parametric models, the first equation shows that the buyer's demand D for a certain commodity,[4] say Q, depends on Q's price P. The second equation shows that the seller's supply S of commodity Q also depends on Q's price P. D, S, and P are **variables**. The expressions (3.1) and (3.2) indicate that demand D or supply S depends on price P.

In both equations, the variable on the right side of the equality sign is called the **independent variable**, and the variable on the left is called the **dependent variable**. Here, for simplicity, D and S denote the same[5] commodity, say, wheat, apples, etc.

[1]On the other hand, the TR and TC defined as random numbers in Chap. 1 are not economic models. They are economic data sets.

[2]For simplicity, we assume all parameters are positive. We may relax these conditions as: $a \geq 0$, $b \geq 0$, $c \geq 0$, $d \geq 0$, and $b + d \neq 0$. The last condition is to assure that the equilibrium values exist. See Eqs. (4.6) and (4.7).

[3]Theoretically, the numbers are estimated by using econometric methods, such as the ordinary least squares (OLS) estimation method. We will discuss the OLS method in Chap. 7.

[4]Note that the letter Q has two meanings. One is that it denotes the name of commodity, another is that it denotes the quantity of commodity demanded or supplied. In general, the difference can be seen from the context of the writing. We follow the convention and do not distinguish the two uses rigorously.

[5]Sometimes, to show that they represent the same commodity, D is denoted as Q_d and S as Q_s. To avoid cluttering the notations, we simply use D and S.

The third equation (3.3) shows the *a priori* condition that the demand is equal to the supply when the market is **in equilibrium** (that is, when the market is "cleared"). It is called the **equilibrium condition**. The first two equations are called **behavioral equations**, because they explain the behavior of the buyer and the seller. These three equations complete a **simple economic model of single market**. It is the prototype of all other economic equilibrium models. The model is also called **a microeconomic model**, since it depicts the determination of the price of an individual commodity in a market.

Since the higher the price, the lower the quantity demanded, the demand curve is generally sloped downward; that is, **the Law of Demand** holds. Hence, the coefficient of price in the demand equation is negative. The positive constant term means that when the price is zero, the buyer's demand has a maximum of 18. In this simple model, the constant 18 (or the constant term "a") represents some other factors, other than Q's price, that may influence the buyer's decision, such as prices of other commodities, income of the consumer, assets held by consumer, etc. These factors are held constant in this simple model.

Similarly, since the higher the price, the larger the quantity supplied, the supply curve is generally sloped upward; that is, **the Law of Supply** holds. Hence, the coefficient of price in the supply equation is positive. The negative constant term here means that when the price is zero, the supplier will not supply any commodity Q, but will hold 6 units of the commodity. In general, the constant 6 (or c) represents some other factors, other than the price of commodity Q, which may influence the seller's decision, such as prices of other commodities, rainfall, wages, other cost of production, etc. These factors are held constant in this simple model.

In both cases, we have three variables (D, S, P) in three equations. a, b, c, and d are called **parameters**, since they are constant but not assigned with numbers. The variables (D, S, P) are called **endogenous variables**, since their equilibrium values are determined by the parameters in the model or numbers, see Sec. 3.5, and (4.6) and (4.7) in the next chapter.

3.2 The Nature of Static Analysis

The problem now is to find the **equilibrium values**[6] ($D^{\#}$, $S^{\#}$, $P^{\#}$) such that the values ($D^{\#}$, $S^{\#}$, $P^{\#}$) satisfy the model (or the system) **identically** and **simultaneously** when these values are substituted into the variables in the model.

Note that mathematically, $D^{\#}$, $S^{\#}$, and $P^{\#}$ are the **solutions** of the system of simultaneous equations, while at the same time, economically, they are the **equilibrium values** (or **equilibrium solutions**) of the model, since they satisfy the equilibrium condition that $D = S$. The word equilibrium is distinctively an economic concept.

The study of the existence and uniqueness of the equilibrium values of the variables for given parameters is called **static analysis**. Mathematically speaking, it is equivalent to the

[6]We use $D^{\#}$, instead of conventional D^{*}, to denote the equilibrium value, since * is used to denote multiplication in Excel.

theory of a system of simultaneous equations. The solution of the system is the **equilibrium values** of the economic model.

The concept of **equilibrium** is important. At the equilibrium price, the quantity the buyer wants and is able to buy equals the quantity the seller wants and is able to sell, and both sides are satisfied and the **market is cleared**: no more buyers looking for sellers and no sellers looking for buyers. We generally do not experience turmoil in the market and economy. Millions and millions of transactions are conducted and concluded, which means that what is demanded and what is supplied reach the same amount, and many transactions are completed every moment of time.[7] Thus, economists are interested in finding the equilibrium price, the factors influencing the equilibrium price, and how the price is determined.

For this purpose, we first set up an economic model based on our daily experience, observations, or theory, and then the problem becomes that of finding a mathematical solution of a system of simultaneous equations. Here we see that economics and mathematics can be two sides of a coin. They are closely related. This is also the reason that economists have to learn mathematics.

There are at least four kinds of mathematical tools that can be applied to solve the problem of **Economic Statics**. They are

1. The graphic method,
2. The algebraic method,
3. The matrix method, and
4. The Cramer's rule method.

In this and the next chapters, we concentrate on the algebraic and graphic methods. The matrix method will be discussed in Chap. 10. The Cramer's Rule does not come with Excel, and will not be introduced in this book.

3.3 The Graphic Method of Solution — The Market Model

Before we introduce how to draw demand and supply curves using Excel graphics, in Sec. 3.4, some theoretical background on curve drawing will be helpful. In this section, we first discuss whether price, the independent variable, should be on the horizontal or vertical axis, and what the slope of a line is.

3.3.1 *The mathematical and Marshallian conventions*

Before getting into the details of the graphic method, we have to discuss which variable is on the horizontal axis and which is on the vertical axis. This is very confusing for the beginners.

[7]When you go to the supermarket and buy three pounds of orange at $1.00 per pound, this means that the oranges are supplied by the supermarket at $1.00 per pound, and your demand for the oranges is three pounds at that price. When you pay the price at the counter, the transaction is complete. The price $1.00 per pound is the equilibrium price, and the quantity you bought is the equilibrium quantity.

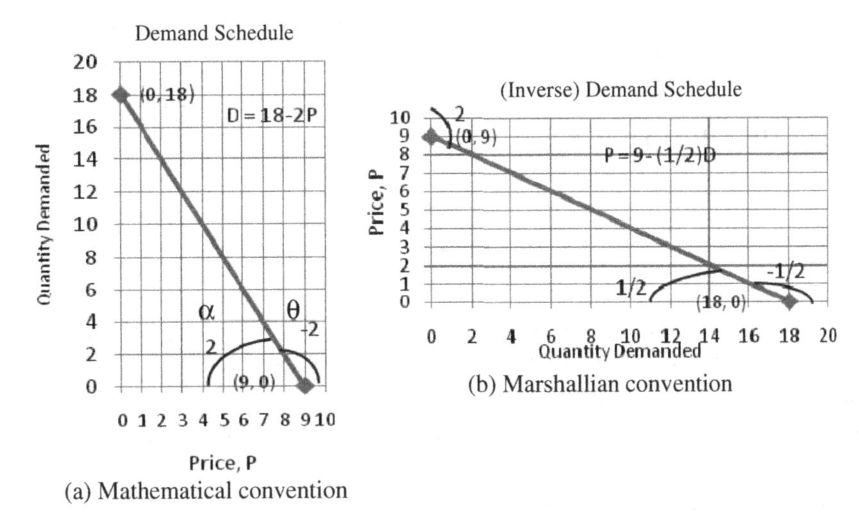

(a) Mathematical convention

(b) Marshallian convention

Fig. 3.1 The mathematical and Marshallian conventions

The **mathematical convention** is that the horizontal axis is always taken to show the independent variable, and the vertical axis, the dependent variable. Hence, mathematically, the horizontal axis should be taken as price P, the independent variable, and the vertical axis should be quantity Q, the dependent variable, namely, the corresponding demand D or supply S for a given price P. This is illustrated in Fig. 3.1(a).

In many economics textbooks, however, the horizontal axis is taken to show quantity Q, and the vertical axis its price, as illustrated in Fig. 3.1(b). This practice is called the **Marshallian convention**. It started with Alfred Marshall (1843–1924), one of the great economists of the turn of the 20th century. In a **Marshallian diagram**, it is intuitively easy to see how the price converges, moving up or moving down, to the equilibrium price, but it is rather hard to visualize the slope of a curve. In this book, we follow the mathematical convention.

3.3.2 *The intercept method*

With simple linear models like (3.1)–(3.3), there are two ways to find the equilibrium solution. The first one is the **intercept method** and the second one is the **schedule method**.

We first find the intercepts of the demand and supply equations, that is, we find the point of the line when one variable is zero.

For the **demand curve**,

$$\text{when } P = 0, \text{ we have } D = 18;$$
$$\text{when } D = 0, \text{ we have } P = 9.$$

Hence, the two intercepts are $(P, D) = (0, 18)$ and $(P, D) = (9, 0)$, where the first number denotes the coordinate of P on the horizontal axis, and the second number, the coordinate of D on the vertical axis. Fig. 3.1(a) illustrates the demand curve with these two intercepts.

Similarly, for the **supply curve**,

$$\text{when } P = 0, \text{ we have } S = -6;$$
$$\text{when } S = 0, \text{ we have } P = 1.$$

Hence, the two intercepts are $(P, S) = (0, -6)$ and $(P, S) = (1, 0)$, and connecting the two points will give the supply curve. Extend the line to point (4, 18) from the supply curve. This is illustrated in Fig. 3.2 with these two intercepts.

Imposing Fig. 3.1(a) on Fig. 3.2, we have Fig. 3.3. The intersection of the two lines gives the equilibrium quantity $Q^{\#} = D^{\#} = S^{\#}$ and the equilibrium price $P^{\#} = 3$. Substituting $P^{\#} = 3$ into either the demand function or supply function, we have $Q^{\#} = 12$.

Example 3.1 Write the demand curve (3.1) in Marshallian form, and find the intercepts of the two axes. See Fig. 3.1(b). This demand function is called the **inverse demand function** (or average demand function), which we encountered in Chap. 1.

The inverse demand function for (3.1) is

$$P = 9 - 0.5D.$$

Hence, when $D = 0$, we have $P = 9$, and when $P = 0$, we have $D = 18$. Connecting $(D, P) = (0, 9)$ and $(D, P) = (18, 0)$, as in Fig. 3.1(b), we have the graph for the inverse demand function. □

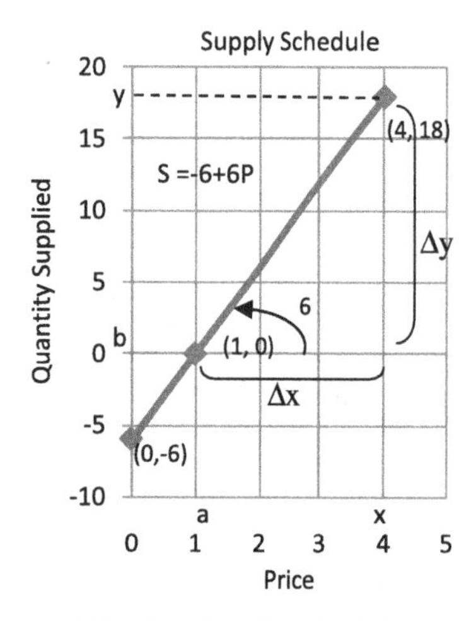

Fig. 3.2 Supply schedule

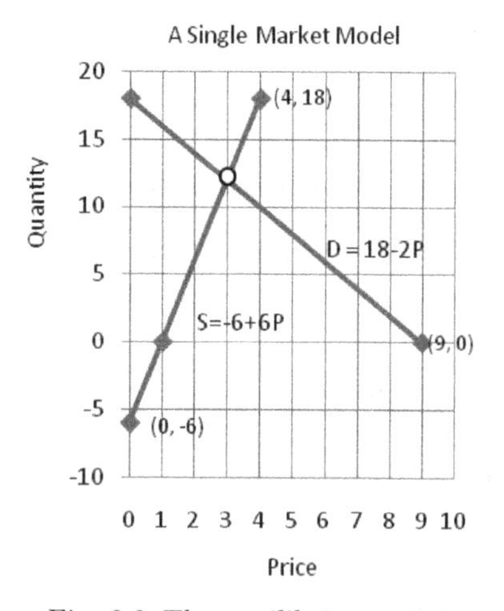

Fig. 3.3 The equilibrium point

3.3.3 *Slope of a line*

We now introduce an important concept, the slope of a line. The **slope** of a straight line is a measure of the steepness of the line. It is the **tangent of the (smallest) positive angle**[8] that the line makes with the horizontal axis. Using the supply schedule Fig. 3.2 as an example, take any two points, a fixed point say, $(\mathrm{a}, \mathrm{b}) = (1, 0)$ and another point $(\mathrm{x}, \mathrm{y}) = (4, 18)$ on the line. Then define

$$\Delta \mathrm{x} = \mathrm{x} - \mathrm{a} = 4 - 1 = 3 \text{ as the "run"},$$
$$\text{the difference from the independent variable}$$

and

$$\Delta \mathrm{y} = \mathrm{y} - \mathrm{b} = 18 - 0 = 18 \text{ as the "rise"},$$
$$\text{the difference from the dependent variable.}$$

See Fig. 3.2. The **difference quotient** $\Delta \mathrm{y}/\Delta \mathrm{x}$ is defined as the **slope** of the line at the fixed point (a, b). If the positive angle between the horizontal axis and the supply curve is defined as θ, then the tangent of the line is the same as the tangent (tan) of the angle θ:

$$\text{Slope} \equiv \frac{\mathrm{y} - \mathrm{b}}{\mathrm{x} - \mathrm{a}} = \frac{\Delta \mathrm{y}}{\Delta \mathrm{x}} = \frac{\text{rise}}{\text{run}} = \frac{18 - 0}{4 - 1} = \tan \theta = 6. \tag{3.4}$$

That is, the supply curve has the slope of 6, as seen in Fig. 3.2. This also holds if the points are reversed, that is, if (a, b) is taken as (4, 18), and (x, y) as (1, 0), because, as in this case, the negative signs in the numerator and the denominator cancel out. Note that in Fig. 3.2, we have placed the name of the angle, θ, and the value of the tangent of the angle, tan $\theta = 6$, together. They should be distinguished clearly.

If we denote the slope as $\mathrm{m} = 6$, then, from Eq. (3.4), we can solve for y and write $\mathrm{y} - (\mathrm{b} - \mathrm{ma}) + \mathrm{mx}$, or more generally, letting the constant $(\mathrm{b} - \mathrm{ma}) \equiv \mathrm{c}$, we may write the equation of a line as

$$\mathrm{y} = \mathrm{c} + \mathrm{mx}, \tag{3.5}$$

which is called **the slope–intercept form of the line** (since c is the intercept of the vertical axis when $\mathrm{x} = 0$). Thus, one way to find a slope of a linear equation is to write, or convert, the equation in this form, and then we can read the tangent of the line instantly as the coefficient m of the independent variable x. For example, since the demand equation is already in the slope–intercept form, we can read immediately that the slope of this demand curve is -2. To verify this, we take any two points on the demand curve in Fig. 3.1(a), as $(\mathrm{x}, \mathrm{y}) = (9, 0)$ and $(\mathrm{a}, \mathrm{b}) = (0, 18)$. Substituting into Eq. (3.4), we have $\mathrm{m} = -2$. Any angle

---

[8] An angle is positive if it is measured counterclockwise from its base line and negative if it is measured clockwise. Thus, in Fig. 3.2, the base line is the line through point 0. Rotating the line counterclockwise, we measure the positive angle as θ.

that is greater than 90° has a negative slope. Similarly, since the supply equation is also in the slope–intercept form, we can see that its slope is +6.

The angle at the other side of the line is called the **supplementary angle** (that is, the two angles add to $\pi = 180°$). If we denote the supplementary angle as α, then, from the trigonometric formula, we have

$$\tan(\alpha) = \tan(\pi - \theta) = -\tan(\theta) = -\mathrm{m}.$$

Thus, if the demand curve has the slope $\mathrm{m} = -2$, then the tangent of the supplementary angle α, as shown in of Fig. 3.1(a), is the absolute value of m, that is, $|\mathrm{m}| = |-2| = 2$.

Example 3.2 Slope of the inverse demand function. Write the demand equation in (3.1) in the Marshallian convention form, and find its slope by (3.5), and also by using the definition of slope in (3.4). What is the slope of its supplementary angle? See Fig. 3.1(b).

Answer: From Example 3.1 and Eq. (3.5), the slope of the inverse demand function is −0.5. To verify this by the definition of slope, take any two points on the inverse demand curve, say, $(\mathrm{D}, \mathrm{P}) = (\mathrm{a}, \mathrm{b}) = (0, 9)$ and $(\mathrm{D}, \mathrm{P}) = (\mathrm{a}, \mathrm{b}) = (18, 0)$, from Example 3.1. Substituting them into definition (3.4), we have the slope $= (9-0)/(0-18) = -0.5$. Hence, the slope of its supplementary angle is $|-0.5| = 0.5$. □

3.4 A Quick Introduction to Excel Graphics

We now show how the demand curve (Fig. 3.1(a)), the supply curve (Fig. 3.2), and the market model (Fig. 3.3) are drawn. The two lines can be drawn in two ways, one by using the intercept method and the other by using the demand and supply schedule method.

3.4.1 *Drawing of the demand curve*

We first construct a simple table, Table 3.1. For the time being, ignore the boxes and the pointer shown in the table. Row 3 has the column label, Q, and column A has the row labels. Rows 4 and 5 contain the two intercepts of the demand curve derived in Sec. 3.3.2, and rows 7 and 8 contain the two intercepts of the supply curve. Cell B3 should be blank (to let Excel know that P is an independent variable). Row 6 should also be blank (so that demand and supply are separate).

To draw the demand curve, select range B3:C5 (including the labels). Enter

<“Insert”><Charts, Scatter>
<Scatter with Straight Lines (the last chart)>.

A chart like Fig. 3.4(a) will be created.

The default chart gives the **chart title** Q, and the legend is on the right-hand side of the chart. Click the legend and press the <Delete> key. (We do not need the legend in this example because there is only one line).

When the pointer moves inside the chart, the fat cross pointer changes to fat arrow with a black arrow-headed cross at the tip, as shown in the upper right corner of Fig. 3.4(a),

Table 3.1 The intercept method of drawing (a) Demand curve and (b) Supply curve

	A	B	C
1	**Table 3.1 The Intercept Method of**		
2	**Drawing Lines**		
3			Q
4	Quantity Demanded	0	18
5	(P, D)	9	0
6			
7	Quantity Supplied	0	-6
8	(P, S)	1	0
9	S = -6 + 6P		
10			
11			

(a)

	A	B	C
1	**Table 3.1 The Intercept Method of**		
2	**Drawing Lines**		
3			Q
4	Quantity Demanded	0	18
5	(P, D)	9	0
6			
7	Quantity Supplied	0	-6
8	(P, S)	1	0
9	S = -6 + 6P	2	6
10		3	12
11		4	18

(b)

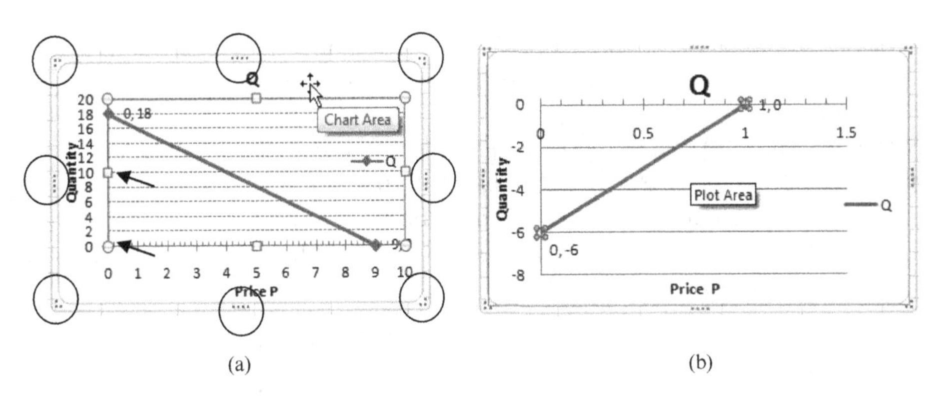

(a) (b)

Fig. 3.4 (a) The demand schedule (b) Part of the supply curve

indicating that you can click and **move the chart**. Press LM to drag the chart to the middle of the screen. When you click anywhere inside the chart, four things will happen.

(a) The chart will be enclosed by **light borders**, the **four corners** of the borders showing three dots in a triangle and the **four sides** showing four dots in line, as enclosed by eight circles in Fig. 3.4(a), indicating that you can click these dots to resize the chart.

(b) When you move the pointer inside the chart, and move it close to a place or an item in the chart, a small box with the name of the place or item appears below the pointer. In Fig. 3.4(a), the pointer is in the "Chart Area" as indicated by the small box.

(c) When the pointer clicks anywhere in the plot area, the plot area is enclosed by four circles and four squares, as indicated by the two arrows, and when the pointer moves close to them, or close to a border of the plot area, it will change to either a double-headed arrow or an arrow-headed cross, indicating that you can resize or reposition the plot area.

(d) The "Chart Tools" tab appears in the Title Bar of the spreadsheet with three new tabs, "Design", "Layout", and "Format" in the Menu Bar to the right of the "View" tab. (see Appendix 3A for details.)

We now want to **add the axis titles**. Select

<"Chart Tools, Design"><Chart Layouts!>
<Layout 5 (the fifth chart)>.

Layout 5 will show the places of Axis Titles, Chart Title, and the data labels for you to write. Click the horizontal Axis Title and type "Price P" and click the vertical Axis Title and enter "Quantity". Note the Layout 5 chart comes with the data labels. Thus, in Fig. 3.4(a), points (0,18) and (9,0) are labeled automatically.

3.4.2 *Drawing of the supply curve*

Three methods of copying a chart

After we have spent time drawing the demand curve, we do not want to repeat the same drawing procedure to draw the supply curve. The best way is to copy the chart of the demand curve and revise it.

A way to copy the chart is as follows. When the pointer clicks inside the newly created chart, Fig. 3.4(a), the fat cross pointer changes to a four-arrow-head cross at the tip, indicating that you can click and **move the chart**. Holding down LM, drag the chart to the middle of the screen. If you click the pointer inside the chart area and press <RM>, one of two things will occur:

(a) If you tap the chart heavily, the chart will dim and will be enclosed with a dark border. A message window will appear: "Move Here, Copy Here, and Cancel." Click "Copy Here" and a copy of the chart will lie overlapping slightly with the original chart. Click and drag the chart to a place below the original chart. Or
(b) If you tap the chart area (not the plot area) of the chart lightly, a mini toolbar will appear — you may have to do this several times to get the mini toolbar. In the mini toolbar, select <Copy> and move the pointer to a blank place, clicking <RM> again. The mini toolbar will reappear and you then select <Paste>.

In either case, Fig. 3.4(a) will be copied to Fig. 3.4(b). The copied chart still shows the demand curve. We also find the simple ^C or ^X and ^V sequence works in copying and moving the chart.

Connection between the table and the chart

We now want to change the demand curve in copied Fig. 3.4(b) to the supply curve without redrawing. Note that when the demand curve in Fig. 3.4(b) is selected, the data area of Table 3.1(a) of the chart will be enclosed by **three boxes** (if not, try to click the line. This also happens if the demand curve in Fig. 3.4(a) is selected). One box on the left, with light blue borders, shows the location of the independent variable, the second box on the right, with deep blue borders, shows the dependent variable(s), and the box on the top with red borders shows the column label of the dependent variable(s). The three boxes show the connection between data range B3:C5 in Table 3.1(a) and the chart Fig. 3.4(b).

Using fill-handle to extend the table data range

When you select the demand curve in Fig. 3.4(b), the three boxes appear. If you move the pointer close to a border of the blue box at the right, the fat cross pointer changes to a fat arrow with a four-arrow-head cross at the tip, as shown on the right-hand side of Table 3.1(a), implying that you can move the blue box. Click the border and drag the box down to cover the range B7:C8, as shown in Table 3.1(b). The original demand line in copied Fig. 3.4(b) changes to a shortened supply curve as shown in Fig. 3.4(b): The supply curve only ranges from $(0, -6)$ to $(1, 0)$. Note that the horizontal axis labels ranges from 0 to 1.5 only and is located close to the axis. Click anywhere outside the chart to get out of the chart.

To extend the supply curve, select range B7:C8 (without clicking the chart), and drag the black fill-handle down to row 11. The range B9:C11 will be filled with numbers, as shown in B9:C11 of Table 3.1(b). Going back to click the short supply curve in Fig. 3.4(b), we find the range B7:C8 will be enclosed with two purple/blue boxes (as shown in Table 3.1(b)). The four corners of the blue box have a small fill-handle square box. If we click and hold any of the lower fill-handle, as shown by an arrow in Table 3.1(b), and drag it down to row 11, the supply curve will be extended to point (4, 18), as shown in Fig. 3.2.

Note that unlike Fig. 3.2, the extended supply curve has three extra coordinates along the supply curve. They can be eliminated by deleting rows 8, 9, and 10 in Table 3.1(b). Then, the entries in B8 and C8 would be 4 and 18, respectively. We now have reproduced Fig. 3.2 in the text, except that the horizontal axis labels are not at the bottom, and they show only the even numbers. We will correct both problems in part (4) of Sec. 3.7.4.

3.4.3 *The equilibrium point*

We reproduce Fig. 3.2 below the original Fig. 3.2 and change the chart title of the copied chart to "Fig. 3.3 The Equilibrium Point". (This title will be changed later.) When the extended supply curve in Fig. 3.4(b) is selected, three boxes appear in Table 3.1(b), among which two boxes enclose B7:C8. Note that there is a blue fill-handle box at each corner of the two boxes. Clicking and dragging one of the upper fill-handles and dragging upward to row 4, we have both the demand and supply curves in one chart like Fig. 3.3. Select the chart and use

<"Insert"><Illustrations, Shapes><Basic Shape, Oval>.

Then select the Oval from the drop-down window. When the pointer changes to a plus sign, draw **a small circle** at the intersection of the two lines to show the equilibrium point, as in

Fig. 3.3. This intersection shows D = S, and the equilibrium price $P^{\#}$ can be read from the horizontal axis as 3, while the equilibrium quantity $Q^{\#}$can be read from the vertical axis as about 13. They can be verified by using the algebraic method explained below in Sec. 3.5.

Note that each of titles, equations, and coordinate labels are in a text box and so can be edited like a usual text box. To practice, you may click a coordinate label, say (4,18), and enclose the pair of numbers in parentheses. Change the chart title as "A Single Market Model" and the vertical Axis Title to "Supply and Demand". This completes the graphing.

Box 3.1: **Embedding a text box and a shape in a chart.** The chart must be selected first before you embed the text box and shapes in the chart. The text box in the chart will not be bordered. If text box borders appear in the chart, it implies that you have not embedded the text box in the chart. Select the chart and try again.

Box 3.2: The method of creating new chart from an old chart, as we practiced in this section, applies to any chart, and is a very convenient feature of Excel.

One step method of finding the equilibrium

We have discussed the method of finding the demand and supply curves separately. As a matter of fact, after setting up Table 3.1(a), selecting B3:C11, we can draw the demand and supply curves of Fig. 3.3 simultaneously by using the scatter diagram as we did in Sec. 3.4.1.

3.5 The Algebraic Method of Solution

Using the intercept method, we draw the demand and supply curves and find the intersection of the two curves, which is the equilibrium point. The problem with this method is that it is rather difficult to see the exact values of the equilibrium price and quantity from the graph. The algebraic method will give the exact equilibrium values.

The **algebraic method** is also called the **substitution method** or **elimination method**. Substituting the demand and the supply equations into the equilibrium condition, we eliminate D and S, and the three equations reduce to one equation, as shown below. We then solve the equation for P to obtain the equilibrium price $P^{\#}$:

$$18 - 2P = -6 + 6P,$$
$$8P = 24, \text{ then, } P^{\#} = 3.$$

Substituting $P^{\#}$ into either the demand equation or the supply equation, we obtain the equilibrium quantity demanded and supplied, denoted by $D^{\#}$ and $S^{\#}$:

$$D^{\#} = 18 - 2(3) = 12,$$
$$S^{\#} = -6 + 6(3) = 12.$$

We now check that the solutions obtained above do satisfy the condition stated in the Nature of Static Analysis introduced in Sec. 3.2 above. Obviously, the following equality holds identically: Substituting $P^{\#} = 3$, $D^{\#} = S^{\#} = 12$ into the numeric model, we see that

$$12 \equiv 18 - 2(3), \tag{3.6}$$

$$12 \equiv -6 + 6(3), \tag{3.7}$$

$$12 \equiv 12. \tag{3.8}$$

Hence, the equilibrium values ($D^{\#}$, $S^{\#}$, $P^{\#}$) satisfy the model **identically** and **simultaneously**.

Another way of characterizing equilibrium values $P^{\#}$, $D^{\#}$, and $S^{\#}$ is that at the value of $P^{\#}$, we have $D^{\#} = S^{\#}$; that is, quantity demanded = quantity supplied, at which $P^{\#}$ does not increase or decrease.

3.6 Excess Demand and Stability of the Equilibrium Price

The difference $E = D - S$ is called the **excess demand function**, which is a function of price. In our numerical model, it is

$$E = D - S = 24 - 8P.$$

It has a negative slope — How do we know? — and is zero at the equilibrium price. This can be shown by substituting $P^{\#}$ in the equation. In Fig. 3.3, the excess demand is shown by the vertical distance between the demand curve and the supply curve.

In Fig. 3.3, when E is positive, $D > S$ and the price is on the left-hand side of the equilibrium price, $P^{\#} = 3$, but when E is negative, $D < S$ and the price is on the right-hand side of the equilibrium price. In the latter case, we also say that the market has an **excess supply**.

When there is an **excess demand**, the quantity of commodity Q that is demanded is larger than the quantity supplied. Hence, the actual price tends to rise. On the other hand, if there is an excess supply, then the quantity of commodity Q that is supplied is larger than the quantity demanded. Thus, the actual price tends to fall. Only at price $P^{\#}$ does the quantity demanded equal the quantity supplied, so that the price does not change. Since the price does not change at $P^{\#}$, this price is called the **equilibrium price**. It is the price that satisfies the equilibrium condition.

Thus, the excess demand function has two uses: One, it can be used to find the direction of price change: If it is positive, price will increase, if negative, price will decrease. Two, it can locate the equilibrium price, and so the equilibrium quantity, since we can find the equilibrium price and quantity when $E = 0$. The analysis of price stability belongs to the field of **dynamic analysis**, in which changes of variables over time will be explicitly considered. Dynamic Analysis is introduced in Part VI of this book. This chapter will concentrate on **static analysis**.

Example 3.3 Excess demand and equilibrium values. Find the equilibrium price and quantity algebraically by using the excess demand function.

Answer: Since excess demand equals zero at equilibrium price, setting equation E above to zero, we have $24 - 8P = 0$, and so $P^{\#} = 3$. Substituting into either demand curve or supply curve, we have the equilibrium quantity $Q^{\#} = 12$. □

3.7 On Excel Graphics — The Schedule Method

To use Excel's graphic method, we have to construct the calculation table first. The graph is always based on the table.

3.7.1 *Construction of the table*

Using the single market model (3.1)–(3.3) introduced in Sec. 3.1, namely,

$$D = 18 - 2P, \quad S = -6 + 6P, \quad \text{and} \quad D = S,$$

we observe that there are three variables, (D, S, P), in three equations. The excess demand function is

$$E = D - S = 24 - 8P.$$

In constructing Table 3.2, we use the format shown in Table 3.1(b). Enter the labels as shown in rows 1–3 in Table 3.2. We will worry about the borders and lines of the table only after we finish the table. Fill cells A4:A14 with values from 0 to 10 by using the fill-handle. Select cell B4 and type = 18 – 2*A4. Press the return key. The value 18 should now be presented in cell B4. Perform similar operations to fill cells C4 and D4. Select B4:D4 and then click on the fill-handle, dragging down to cell D14. The appropriate values for quantity demanded and supplied, and also the excess demand, should appear in B4:D14. Column B shows the **demand schedule**, in the sense that it shows the quantity demanded at each price, and column C shows the **supply schedule**, as it shows the quantity supplied at each price.

Table 3.2 shows that the excess demand is zero ($E = 0$) in row 7, and hence the **equilibrium price, at which the quantity demanded equals the quantity supplied,** is $P^{\#} = 3$. In other words, the equilibrium price is located on the cell in which the excess demand changes the sign from positive to negative. The corresponding **equilibrium quantity** is $D^{\#} = S^{\#} = 12$, as you can read from row 7. We draw the bottom border in row 7 for clarity. Save the table as Ch3DeSu. Print out the table.

> Box 3.3: The **equilibrium price** and **quantity** is located at the point where the excess demand is zero or where the sign changes from positive to negative or negative to positive.

Table 3.2 A single market model

	A	B	C	D
1				
2	P	D=18-2P	S=-6+6P	E=D-S
3		D	S	E
4	0	18	-6	24
5	1	16	0	16
6	2	14	6	8
7	3	12	12	0
8	4	10	18	-8
9	5	8	24	-16
10	6	6	30	-24
11	7	4	36	-32
12	8	2	42	-40
13	9	0	48	-48
14	10	-2	54	-56

3.7.2 *Graphing procedure*

Table 3.2 shows the demand and supply schedules. To draw the demand and supply curves from Table 3.2, we take the following six steps:

Step 1. Select the data range, including **the column and the row titles**: A3:D14.

Box 3.4: In defining the range of a graph, row and column titles must be included in the range. The intersection of the row and column titles (**the Northwest corner**) **must be blank**, as shown in cell A3 of Table 3.2. The blank cell will let Excel know the location of the independent variable.

Step 2. Click <“Insert”><Charts, Line>. A window with seven line figures appears, as shown in Fig. 3.5. Each line figure has its own name, as listed on the right-hand side of Fig. 3.5. Click to select #4, **Line with Markers**. The diagram appears instantly as Fig. 3.6.

In this case, the demand curve is marked with diamond markers, the supply curve with square markers, and the excess demand curve with triangle markers. Notice that the horizontal axis shows the independent variable, namely, the price, and the vertical axis shows the quantity demanded and supplied and the excess demand.

Box 3.5: When you draw the chart, make sure that the first **axis label** corresponds to the first number of the independent variable. In Fig. 3.6, the horizontal axis labels should start from 0, not 1. If they start from 1, that means you forgot to define the independent variable by leaving Cell A3 blank.

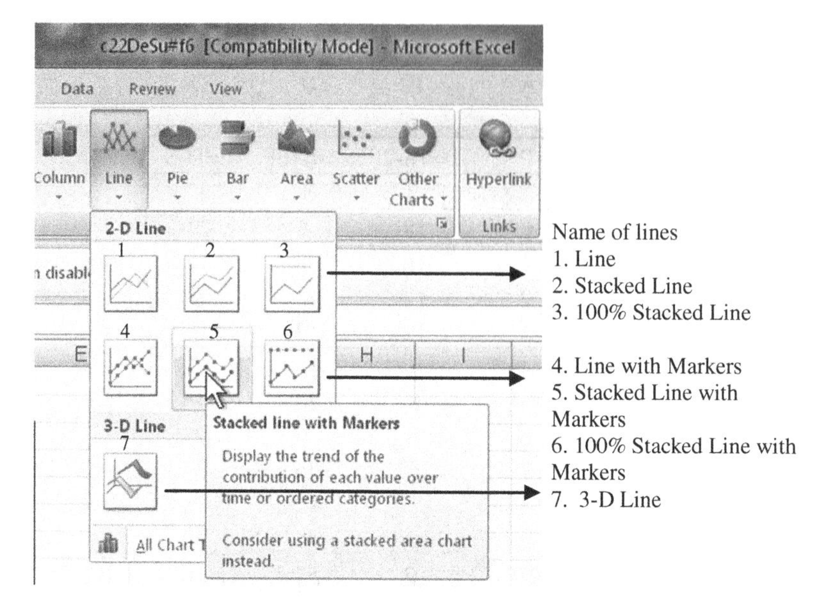

Fig. 3.5 A 3-D line

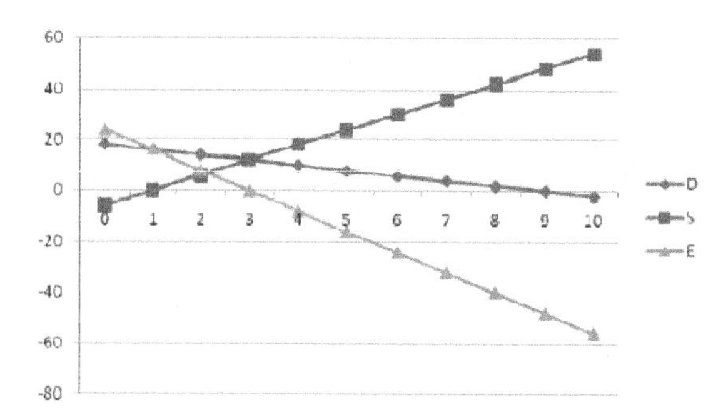

Fig. 3.6 A single market model

Graphically, the **excess demand curve** crosses the horizontal axis at price = 3, at which excess demand is equal to zero, indicating that the quantity demanded equals the quantity supplied at 12 units (read from the table), and the market is cleared. It indicates the location of the equilibrium price.

Step 3. We now add some basic ingredients of the chart. They include the chart title, the horizontal axis title, and the vertical axis title. First, we notice the change in the menu bar in the ribbon when the mouse is clicked in or out of the chart. When Fig. 3.6 is selected, the menu bar adds three more tabs under new Chart Tools, namely, Design, Layout, and Format. The group name and its major contents are listed in Appendix 3A. Note that some contents, when selected, will open a drop-down list for an explanation or to give more choices.

To avoid cluttering, click the excess demand curve and press the <Delete> key to delete the excess demand curve. We have Fig. 3.6 without E.

Step 4. Add chart title, axis titles. Clicking the chart, enter

<"Chart Tools, Design"><Chart Layouts, Quick Layout!>;

that is, click the down arrow in the Chart Layout group. The drop-down list with 12 layout designs appears, as you can see in Fig. 3.7. Since we are looking for all three titles, choose <Layout 10>, and the chart changes to Fig. 3.8.

Step 5. The **vertical lines** between the demand and supply curves in Fig. 3.8 show the distance between the two curves, that is, the absolute value of the excess demand curve ($|E| = |D-S|$). The distance is zero at the intersection of the two curves. To eliminate these lines, simply click any line and press the <Delete> key.

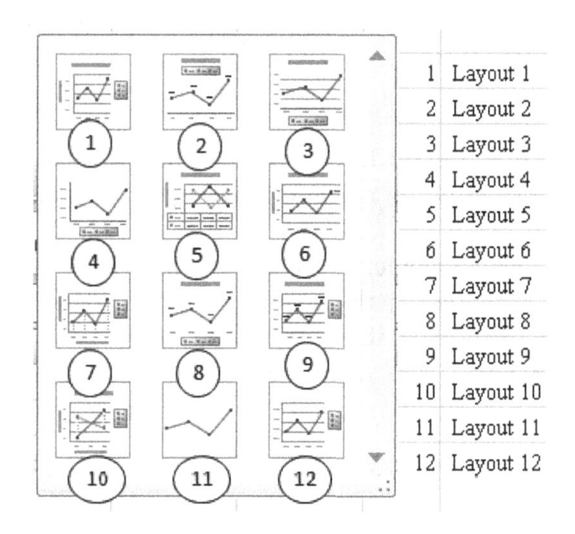

Fig. 3.7 The 12 chart layout designs for line

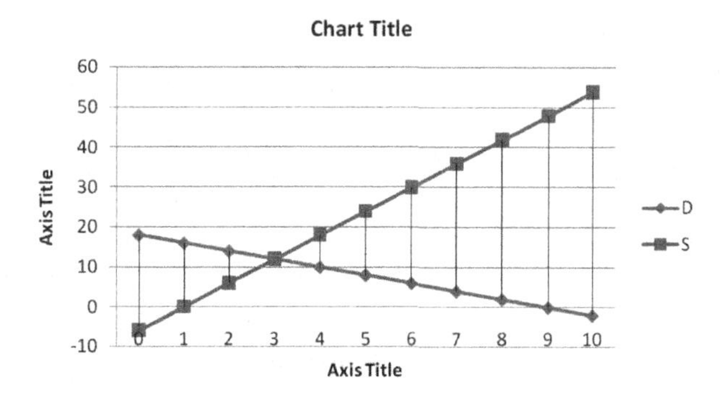

Fig. 3.8 The chart for layout 10

Step 6. Change the Chart Title to "A Single Market Model", the horizontal Axis Title to Price, and the vertical Axis Title to "Quantity", as shown in Fig. 3.12.

3.7.3 *Definitions of some graph elements*

For detailed editing of the chart, we need to know some definitions of **graph elements**. This is shown in Fig. 3.9. They are mostly self-explanatory. The **3Plot area** is the area of the inner box, and the **1Chart area** is the area outside the plot area, which is enclosed by the **2Chart area border**. The chart **has** four axes: the **10Primary horizontal category (X) axis** on the bottom, the **secondary** X-**axis** on the top, the **16Primary vertical value (Y) axis** on the left, and the **secondary Y-axis** on the right.

3.7.4 *Editing a graph*

We start from Fig. 3.6 with three lines in the chart, and proceed to obtain Fig. 3.8. Our objective is to change the appearance of Fig. 3.8 to that of Fig. 3.12. There are three ways to edit or reformat the chart elements marked from 1 to 16 in Fig. 3.9. In the following, we assume that the chart is selected.

- **The simple formatting method.** The chart elements, especially the chart area and the plot area, can be formatted by using formatting buttons in the <"Home"> tab (see Sec. 1.3.2).
- **The RM method.** Click an element (except the chart area) and press <RM>, and an editing menu appears. The item at the bottom serves to <Format> that element. If you click it, Excel will show you more items to choose from.
- **The chart tools method.** Click anywhere in the chart to invoke the "Chart Tools" tab in the title bar (see (7) of Fig. 1.1). Three additional tabs appear in the menu bar

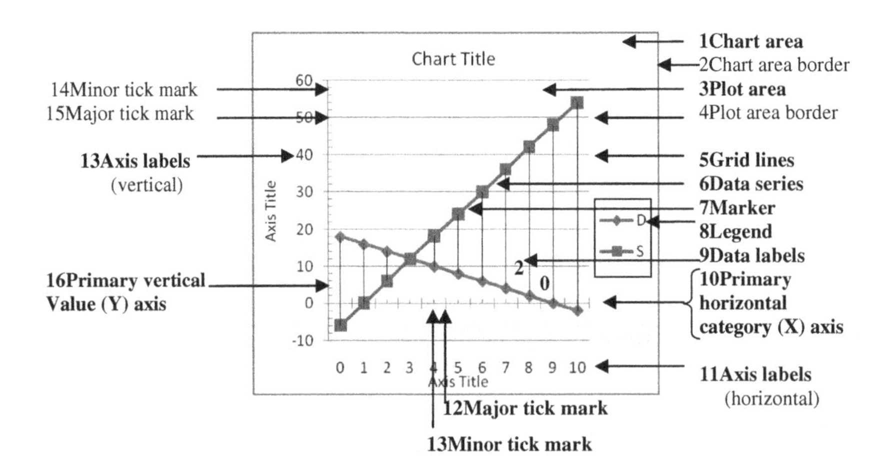

Fig. 3.9 Definition of some chart elements

(see (8) of Fig. 1.1) next to the View tab: "Design", "Layout", and "Format". Click any of the three new tabs and choose a chart element to format. For some buttons, a drop-down list will appear for further selection. This method requires that you are familiar with the location of the chart elements in these three tabs. Click the chart and use the drop-down list under

<"Chart Tools, Layout"><Current Selection, Format Selection>,

located at the upper left corner. The same button can also be found under <"Chart Tools, Format">. Then a formatting menu of the selected element appears.

We find that **the RM method** is probably the most convenient way to edit the chart, although the chart tools method has visual appeal, coming with a help window and more choices. In the RM method, when any one of the elements in the boldface font in Fig. 3.9 is selected, say **6data series**, and the <RM> is pressed, a drop-down list like Fig. 3.10(a) appears. The item at the end is always the formatting button for that element. Click it, and a formatting dialogue box for that element opens, in this case, **Format Data Series,** as shown in Fig. 3.10(b). You can select the items on the left column to edit the element. You may move to other graph element without clicking the <Close> button and the mini menu changes automatically to the formatting window of the new element so that you can **continue editing**. This is one of the new convenient features of Excel 2007.

In Fig. 3.8, we simply enter the **chart title** and **axis title** as in Fig. 3.12. They can be formatted by using the <"Home"> ribbon, or by clicking them and using RM to open <Format Chart Title> or <Format Axis Title>.

To edit other graphic elements, we proceed as follows. We start from editing the data series, and use the number before each element in Fig. 3.9.

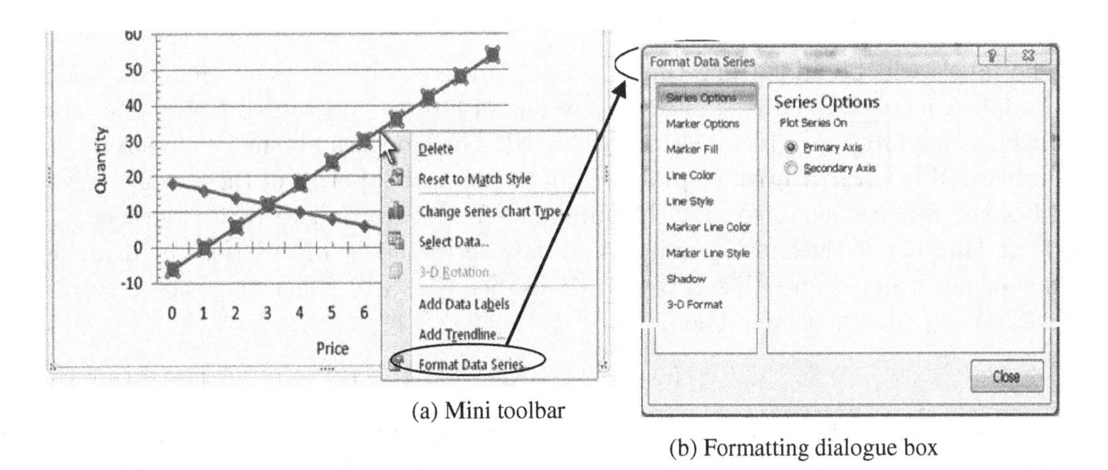

(a) Mini toolbar

(b) Formatting dialogue box

Fig. 3.10 Format data series

> Box 3.6: If the formatting dialog box, like Fig. 3.10(b) or Fig. 3.11(a) is already open, you can continue editing other elements by simply clicking the chart element without closing the dialog box. Otherwise, you have to click the element and <RM>, and click the formatting button at the end of the list to open the formatting dialog box.

(1) **6Data Series** This is explained in the above paragraph. After you open the formatting dialog box of "**Format Data Series**" (as in Fig. 3.10(b)), you can choose and edit the line and marker of the curve on the left side of the dialogue box. To change the supply curve to the same format as that of Fig. 3.12, we proceed as follows:

<Marker Options><@Build-in, Type!><Square><Size:, 7>
<Marker Fill><@Solid fill><Color!><White>
<Line Color><@Solid line><Color!><Black>
<Marker Line Color><@Solid line><Color!><Black>.

Whenever you move the pointer over a specification, the element in the chart will change automatically to the specification so that you can make your choice.

Similarly, without exiting the menu window, click the demand curve, following the above procedure except that under the <Marker Options>, you choose <Triangle>.

Note that you may edit all the **7markers** at once by clicking any one of the markers once. If you want to edit only one **7marker**, you click that marker again after all markers are activated.

(2) **9Data Labels** If the data labels are not present, they can be added when you start editing the data series by clicking <Add Data Labels> in the mini toolbar. See Fig. 3.10(a). A menu window of <Format Data Labels> appears. You may also add data labels by choosing

<"Chart Tools, Layout"><Labels, Data Labels><below>.

The labels will be added like those in Fig. 3.12. As with markers, you can select all the labels at once to edit by clicking any one of labels. You can edit only one label by clicking that label after all labels are activated. The labels can be moved, resized, edited, and erased like a text box. To practice, move the data label 18 of the supply curve and 14 of the demand curve so that the numbers will not overlap with the markers.

(3) **Text Box** Inside the chart, you may add data series labels, or any text, by using a text box or any other shapes, like arrows or lines. See Fig. 3.12. Click the chart area or plot area to activate the chart. Then,

<"Insert"><Text, Text Box>

and draw a text box. Write S in the text box for the supply curve. Edit the text box as usual (see Sec. 1.4.2) such that S is in the center of the box (that is, use

<"Home"><Alignment, Center> and also <Alignment, Middle Align>) with a green background color (<"Home"> <Fonts, Fill Color><Olive Green>).

Box 3.7: **Embedding a Text Box into a chart.** You should always activate the chart before adding the text box or any other shapes to embed them into the chart.

Box 3.8: **Creating a second or other text boxes by copying.** To save time and effort, you should complete the formatting of the first text box, then copy the text box to create other text boxes in the chart. Change and edit the contents of the copied text boxes.

Box 3.9: **Moving a text box in a chart.** It is not easy to select the text box in a chart, because often the pointer will select the whole chart area. Make sure that your pointer is **exactly on a border** and that when the box is selected, the text box will be enclosed by a darker border, indicating that you have indeed selected the text box.

In Fig. 3.12, you may create the text boxes for line D and point E by copying the S text box to the appropriate place and changing the letters, and you may eliminate the background color by using Drawing Tools. Also, you may add the vertical dotted line through point E by using the line in <"Insert"><Illustration, Shapes><Line>. To change it to a dotted line, select the line and use

<"Drawing Tools, Format"><Shape Styles>
<Shape Outline><Dashes><Square Dot>.

Box 3.10: **Drawing vertical or horizontal lines in Excel 2007.** They may be slanted. After a vertical line is drawn from <"Insert"><Illustrations, Shapes>, select the line and <"Drawing Tools, Format"><Size, height!>, and enter 0. If a horizontal line is drawn, then select <"Drawing Tools, Format"><Size, width>, and enter 0.

(4) **10The Primary Horizontal Axis** Click any one of the horizontal **11Axis labels**, and the labels and axis will be enclosed by a box. The formatting dialogue box changes to **Format Axis** dialog box (as in Fig. 3.11, assuming that you still have the previous dialogue box open; if not, you will have to use the RM method to get it out). Figure 3.11(a) shows the menu window for the horizontal axis (which is different from that of the vertical axis). Under the <Axis Options>, we accept the first part

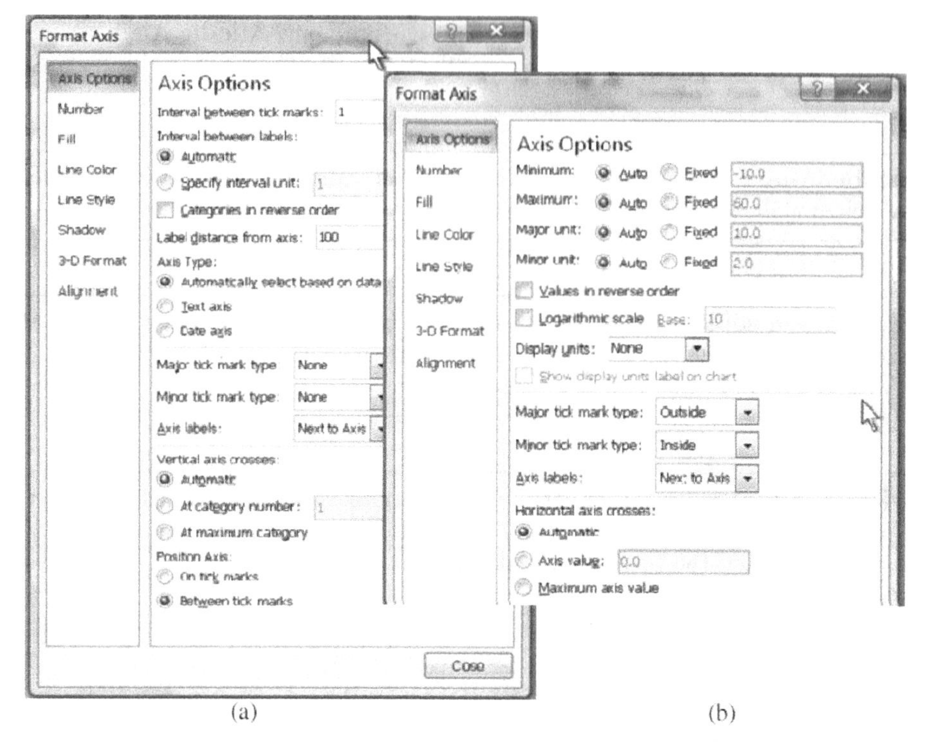

(a) (b)

Fig. 3.11 Format axis (a) horizontal (b) vertical

and go on to change the following:

<Major tick mark type!><Cross>	(cross the axis line)
<Minor tick mark type!><Inside>	(inside the plot area)
<Axis labels!><Low>	(below the plot area)
<Position Axis><@Between tick marks>	(centered at the minor tick mark).

If zero is not placed at the bottom of the vertical axis, as in Fig. 3.12, the chart may look crowded if the 11Axis labels are placed next to axis(10). Thus, we recommend putting them to "low", that is, below the plot area, as shown in Fig. 3.9 or 3.12. Note that, in Fig. 3.12, the markers(7) and labels(11) are placed between the major tick marks(12). This is default. You may click <@On tick marks> (at the bottom of Fig. 3.11(a)) to place them on the major tick marks.

(5) **16The Primary Vertical Axis** In Fig. 3.12, the vertical axis has the minimum value −10 and the maximum value 60. They are automatically set and shown in the boxes on the right of Fig. 3.11(b). You may change them manually by entering numbers in the boxes on the right by clicking the "Fixed" radio button: say, change the maximum to 50 and minimum to 0. The interval between the numbers along the vertical axis has 10 major units, and the minor unit is 2. You may change them by entering numbers,

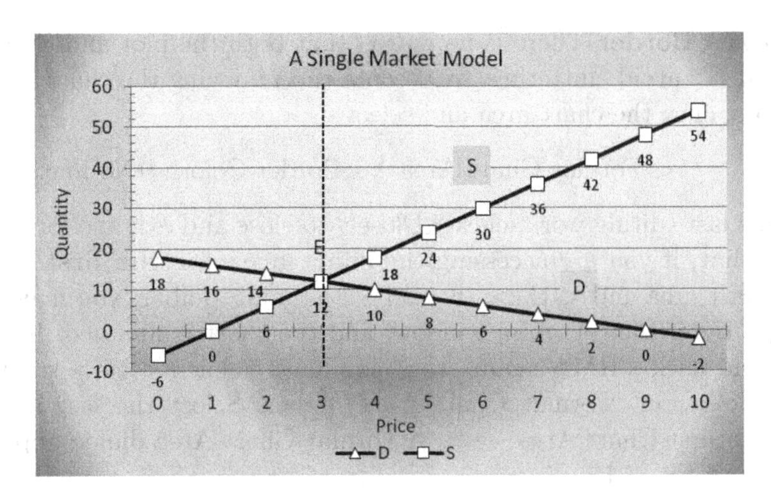

Fig. 3.12 A single market model

such as changing the major unit to 5 and minor unit to 1 in the boxes on the right. Change back to the Auto setting by clicking the radio button on the left.

(6) **8Legend** Clicking Legend will open "Format Legend" dialogue box, you may edit the format of legend, or place it anywhere on the side of the plot area. In Fig. 3.12, we enter

<Format Legend><Legend Options, Legend Position>@Bottom.

You may also use the pointer to move it inside the plot area.

(7) **5Grid lines** The default grid lines are solid lines, as in Fig. 3.12. To change them to dotted lines, click a grid line and select

<RM><Format Grid lines···><"Format Major Grid lines",
Line Style><Dash type!><Square Dot>.

The solid lines change to dotted lines.

(8) **3Plot Area** To reproduce the color shade of the plot area in Fig. 3.12, click anywhere inside the plot area (but away from any grid lines) and choose

<"Format Plot Area", Fill><@Gradient fill><Preset Colors!>
<Parchment(#14 box)><Type!><Linear>
<Direction!><Linear Down (#2 box)>.

(9) **1Chart Area** To reproduce the color shade of the chart area in Fig. 3.12, click anywhere inside the chart area, then choose

<"Format Chart Area", Fill><@Gradient fill>
<Preset Colors!><Fog(#10 box)><Type!>
<Radial><Direction!><From Center(#3 box)>.

(10) **2Chart Area Border** There is no reason that both the plot and chart areas should be bordered. To avoid cluttering, we recommend removing the chart area borders. For this purpose, click the chart area and select

<"Format Chart Area"><Border Color>@No line.

If this is the last editing work, click <Close> to close and exit the formatting dialogue box. Note that, if you are accessing the **chart area for the first time**, when you click the chart area and <RM>, instead of the mini toolbar, you may get the "Move Here, Copy Here, Cancel" mini window, and the chart is dimmed. In this case, click "Cancel" and click <RM> again. You may have to click <RM> several times until the mini toolbar of "Format Chart Area" opens. Select the last item of the mini toolbar: <Format Chart Area ⋯ >. A Format Chart Area dialog appears, similar to Fig. 3.10(b) or Fig. 3.11.

The final version of Fig. 3.8 should look like Fig. 3.12. The demand and supply curves show that the intersection of these two curves at point E gives the **equilibrium price** $P^{\#} = 3$, reading from the horizontal axis, and **equilibrium quantities** $D^{\#} = S^{\#} = 12$, reading from the vertical axis or data labels. Note that the tick label is positioned between the tick marks. At $P = 3$, tick label 3 is located between the tick marks 2.5 and 3.5.

To **print** the chart, click the chart area, and either clicking the "Print" logo in the Quick Access Toolbar or <Office><Print, Print Preview>. In the Print Preview mode, you may still adjust the chart for printing according to the menu. Click <Print> to print out.

3.8 A National Income Model

After we have introduced the method of graphing in details in the construction of the market model, we may draw charts for any economic applications. For the rest of this book, we will take the graphing procedures explained above for granted and will explain the graphing procedures whenever we encounter new features.

3.8.1 *A national income model*

The market model expounded above can be extended to the national income model, which is also called the **Keynesian theory of income determination**.

The simplest numerical national income model may be specified as follows:

$AD = C + I + G$	Aggregate demand function	(3.9)
$C = 0.3 + 0.6Y$	Consumption function	(3.10)
$I = 0.7$	Investment function	(3.11)
$G = 0.8$	Government expenditure.	(3.12)
$AS = Y$	Aggregate supply function	(3.13)
$AD = AS$	Equilibrium condition	(3.14)

The corresponding parametric model is

$$
\begin{aligned}
AD &= C + I + G && \text{Aggregate demand function} && (3.9')\\
C &= a + bY && \text{Consumption function} && (3.10')\\
I &= I_0 && \text{Investment function} && (3.11')\\
G &= G_0 && \text{Government expenditure} && (3.12')\\
AS &= Y && \text{Aggregate supply function} && (3.13')\\
AD &= AS && \text{Equilibrium condition,} && (3.14')
\end{aligned}
$$

where AD is aggregate demand (consisting of consumption, C, investment, I, and government expenditure, G), Y is gross national product, a and b are parameters, and I_0 and G_0 are given constants (we use subscript 0 to show that they are fixed numbers). AS is the aggregate supply of output in the economy.

The first Eq. (3.9) and Eq. (3.13) are **definitional equations**, defining what the aggregate demand and supply functions are. Equations (3.10)–(3.12) are **behavioral equations** of consumption, investment, and government expenditure. The last equation is the **equilibrium condition of the economy**: like the single market model, the aggregate demand must be equal to the aggregate supply in the aggregate commodity market.

We have six equations and six variables (Y, C, I, G, AD, AS). I and G in this model are called **exogenous variables** since they are determined outside of the model by the given I_0 and G_0. In the above case, $I_0 = 0.7$ and $G_0 = 0.8$. The other four variables, Y, C, AD, and AS, are **endogenous variables**, since their values are determined either by numbers or by the parameters a and b and the exogenous variables, I_0 and G_0. See, for example, the equilibrium variables of Y in (3.17) and C in (3.18) below.[9] This model is also called a **macroeconomic model**, since it depicts the determination of the national income in the aggregate commodity market.

Clearly, I, G, and C can be substituted into the first equation (3.9) or (3.9′), and the model reduced to three equations. In the parametric model, they are

$$
\begin{aligned}
&AD = a + bY + I_0 + G_0 && \text{Aggregate demand function}\\
&AS = Y && \text{Aggregate supply function}\\
&AD = AS && \text{Equilibrium condition.}
\end{aligned}
$$

Thus, we have five variables in three equations: three endogenous variables (AD, AS, Y) and two exogenous variables (I_0, G_0). As a general principle, the number of equations and the number of endogenous variables usually match so that the system of equations has a solution.

[9]When the numeric values of these parameters and exogenous variables are inserted into the equilibrium values, $Y^{\#}$ and $C^{\#}$, the equilibrium values become numbers, which represent the parameters and exogenous variables given outside the model.

In this national income model, it is apparent that national income Y plays the role of price P in the market model. Both the national income (macroeconomic) model and the single market (microeconomic) model have an equilibrium condition and the behavioral equations that explain the elements in equilibrium conditions.

Substituting AD and AS into the equilibrium condition and retaining the consumption function C as a separate equation, we then have the usual simple Keynesian model of two equations and two endogenous variables: Y and C. From the parametric model, we obtain

$$Y = C + I_0 + G_0 \qquad \text{Equilibrium condition} \qquad (3.15)$$

$$C = a + bY \qquad \text{Consumption function} \qquad (3.16)$$

The original six equations in six variables are now reduced to two equations and two endogenous variables. Equation (3.15) shows directly the equilibrium condition that aggregate supply (**national income produced**) must be equal to aggregate demand (**national income spent**).[10] The second equation (3.16) shows the aggregate behavior of the consumers. $a = 0.3$ is called the **subsistence** or **basic level of consumption**, as it is the amount of consumption when income falls to zero. $b = 0.6$ is called the **marginal propensity to consume**. It shows the change in consumption when national income increases by one unit. The empirical law shows that b is greater than 0 but less than 1, as consumers, in general, do not spend all extra income when income increases one extra unit.

Our problem is that, given the values of parameters a and b, we want to use the graphic method to find the equilibrium national income. First, we need some more definitions. We solve the two equations, (3.15) and (3.16), algebraically. Substituting (3.16) into (3.15), the above model reduces to

$$Y^{\#} = k(a + I_0 + G_0), \qquad (3.17)$$

where $k = 1/(1 - b)$. k is called a **multiplier**. # indicates the **equilibrium value**,[11] in the sense that $Y^{\#}$ satisfies the equilibrium condition (3.14). In our numerical example, $k = 2.5$ and $Y^{\#} = 4.5$. If Y, C, I, and G are measured in trillions, this means that the equilibrium income is 4.5 trillion. An increase in a unit of a, or I_0, or G_0, other things being equal, will increase $Y^{\#}$ by a magnitude of a multiplier $k = 2.5$. This also means that, if investment increases by 1 trillion, other things being equal, income will increase by 2.5 trillion. Thus, k is also called the **investment multiplier**.

$Y^{\#}$ is the **equilibrium national income**. It is in equilibrium since it satisfies the equilibrium condition that aggregate demand equals to aggregate supply. At $Y^{\#}$, the

---

[10]Strictly speaking, the consumption function should depend on the **national income distributed**, ND. In this case, we have added one variable to the model. The equilibrium condition then is ND = AS, and we have seven variables and seven equations. The equilibrium condition that AS=AD=ND is called the **Law of three-sided equivalence of national income**.

[11]Note the difference in notations. $Y^{\#}$(Y sharp) means the equilibrium value of Y, while #Y means entering "shift key +Y". (See the conventions used in this book, Sec. 1.1)

equilibrium consumption, $C^{\#}$, can be derived as

$$C^{\#} = a + bY^{\#} = k(a + b(I_0 + G_0)). \tag{3.18}$$

Both equilibrium values are expressed in terms of the parameters in the model. In our numerical model, $C^{\#} = 3$, that is, 3 trillion. Mathematically, the set $(Y^{\#}, C^{\#})$ is the solution of the simultaneous equations (3.15) and (3.16), or equivalently, (3.9)–(3.14).

The difference between the aggregate demand and the aggregate supply is the **excess aggregate demand**, ED = AD − AS. If ED is positive, too many people are chasing too few goods, the prices tend to increase, the economy is inflationary, and as the firms produce more, the national income will increase; if ED is negative, too few people buy goods, the prices tend to decrease, the economy is deflationary, the firms produce less, and national income will decrease. Only at the equilibrium national income, aggregate demand equals aggregate supply and the aggregate market is cleared.

3.8.2 *Graphic illustration of the macroeconomic model*

The method of graphing the national income model is the same as that for the single market model. From the numeric national income model, we can write

$$\begin{aligned} AD &= 0.3 + 0.6Y + 0.7 + 0.8 && \text{Aggregate demand function} \\ AS &= Y && \text{Aggregate supply function} \\ AD &= AS && \text{Equilibrium condition.} \end{aligned}$$

Since Y plays the same role as P in the single market model, we can merely substitute Y for P in Table 3.2. Here, the values of Y (in trillions) are assumed to run from 0, 0.5, up to 8. We may construct a new table as Table 3.3(a), in which we enter equations in the following cells:

	Formula	Equation
B4:	=0.3+0.6 * A4+0.7+0.8	AD = a + bY + I + G
C4:	=A4	AS = Y
D4:	=B4−C4	ED = AD − AS.

Using the graphing method expounded in the previous sections, we may draw the AD and AS curves in Table 3.3(a) like the single market model in Fig. 3.12 (see Sec. 3.7). In Table 3.3(a), both AD and AS curves must be sloped upward. See Fig. 3.13. The intersection of the aggregate demand line and aggregate supply line is called **the Keynesian cross**. It is marked by a circle and a dotted vertical line.

To complete the chart, in addition to the basic six steps in Sec. 3.7.2, you add major and minor tick marks to the horizontal axis, add text boxes, move the legend to the Northwest corner of the plot area, and add a vertical dotted line and a circle at the Keynesian cross.

Table 3.3 A national income model

(a)

	A	B	C	D
1	A National Income Model			
2	Y	=1.8+0.6*Y	=Y	=AD-AS
3		AD	AS	ED
4	0.0	1.8	0.0	1.8
5	0.5	2.1	0.5	1.6
...	...			
12	4.0	4.2	4.0	0.2
13	4.5	4.5	4.5	0.0
14	5.0	4.8	5.0	-0.2
15	5.5	5.1	5.5	-0.4
16	6.0	5.4	6.0	-0.6
17	6.5	5.7	6.5	-0.8
18	7.0	6.0	7.0	-1.0
19	7.5	6.3	7.5	-1.2
20	8.0	6.6	8.0	-1.4

(b)

	A	B	C	D	E	F
1	**A National Income model - with C**					
2	Y	.3+0.6*Y	=C+I0+G0			=AD-AS
3		C	C+I0	AD	AS	ED
4	0.0	0.3	1.0	1.8	0.0	1.8
5	0.5	0.6	1.3	2.1	0.5	1.6
...	...					
12	4.0	2.7	3.4	4.2	4.0	0.2
13	4.5	3.0	3.7	4.5	4.5	0.0
14	5.0	3.3	4.0	4.8	5.0	-0.2
15	5.5	3.6	4.3	5.1	5.5	-0.4
16	6.0	3.9	4.6	5.4	6.0	-0.6
17	6.5	4.2	4.9	5.7	6.5	-0.8
18	7.0	4.5	5.2	6.0	7.0	-1.0
19	7.5	4.8	5.5	6.3	7.5	-1.2
20	8.0	5.1	5.8	6.6	8.0	-1.4

AD components

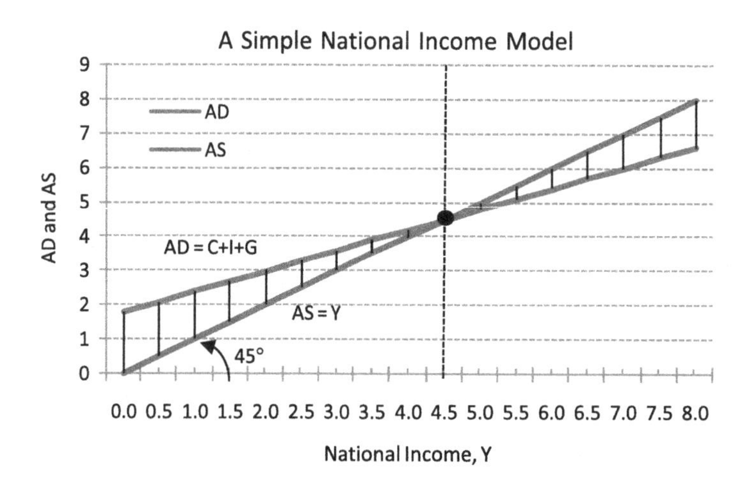

Fig. 3.13 A simple national income model and the Keynesian cross

To add a circle at the Keynesian cross, click the chart. Choose the oval logo in the <“Insert”><Illustrations, Shapes> and click again at the intersection, the circle is now embedded in the chart at the Keynesian cross. Since the circle is small, you may have difficulty in moving the circle. Similarly, you may find the arc from the Basic Shapes collection under <Shapes>. Writing “45°” is a little bit tricky. It is obtained by inserting circle to the chart, and moving the text box of “45” closer to the circle, as shown.

The Keynesian cross shows that the equilibrium income is \$4.5 trillion, which is the same as the income at zero excess demand in Table 3.3(a). Depending on the values of parameters and the exogenous variables, this equilibrium income can be anywhere on the 45° line.

3.8.3 *Another method of illustration*

On many occasions, information on consumption C is also important. Thus, in drawing the chart, we may separate the aggregate expenditure into its components: C, I, and G. These components are drawn in a cumulative way. We first copy the sheet with Table 3.3(a) and Fig. 3.13 to another sheet, and call the copied sheet Sheet3. We now revise the copied Table 3.3(a) to become Table 3.3(b). We first erase B3:D20, then enter the new column titles in rows 2 and 3, as shown in Table 3.3(b). We now implement the first five equations (3.9) to (3.13) of the original equations. In Table 3.3(b), enter the equations in the following cells:

	Formula	Equation
B4:	=0.3+0.6*A4	$C = a + bY$
C4:	=B4+0.7	$C + I$
D4:	=C4+0.8	$AD = C + I + G$
E4:	=A4	$AS = Y$
F4:	=D4–E4	$AD = AS$ at equilibrium,

and copy the range B4:F4 down to B20:F20. Note the differences between the entries of row 4 of Table 3.3(a) and Table 3.3(b).

Another method of entering Table 3.3(b) is as follows. Select columns B and C and insert two blank columns to the left of the AD column. Enter the formula for consumption C in B4 and $C + I_0$ in C4, and copy B4 and C4 down to row 20, as shown in Table 3.3(b).

From the tables, we find that the **equilibrium national income** is $Y^{\#} = 4.5$ (trillion), and **the equilibrium consumption** is $C^{\#} = 3.0$ (trillion).

Table 3.3(b) is illustrated in Fig. 3.14. The vertical lines have been eliminated so that the chart looks more like the ordinary macroeconomic diagram. Equilibrium income and its corresponding equilibrium consumption are shown by a circle and a dotted line through 4.5 on the X-axis. Note that the C + I line is now changed to a dashed line to show the differences among the lines. Three right braces are added, and explained by text boxes. The legend is now enclosed by borders and the background is white (not transparent as in Fig. 3.13). The vertical axis title reflects the changes.

Example 3.4 Table 3.3(a) is a particular table of Table 3.3(b). Using Table 3.3(b), draw the AD/AS curves as in Fig. 3.13. There are at least two ways to draw Fig. 3.13 from Table 3.3(b) without reconstructing the table. What are they?
Answer: One is to select only columns Y, AD, and AS in Table 3.3(b) and draw a new chart; the other is to delete C and C + I lines from Fig. 3.14, and edit the chart. □

Example 3.5 Excess aggregate demand equation. Using the excess aggregate demand equation, find the equilibrium income and consumption.
Answer: From $ED = AD - AS$ in D4 of Table 3.3(a), we have $ED = 1.8 - 0.4Y = 0$. Solving, we have $Y^{\#} = 4.5$. □

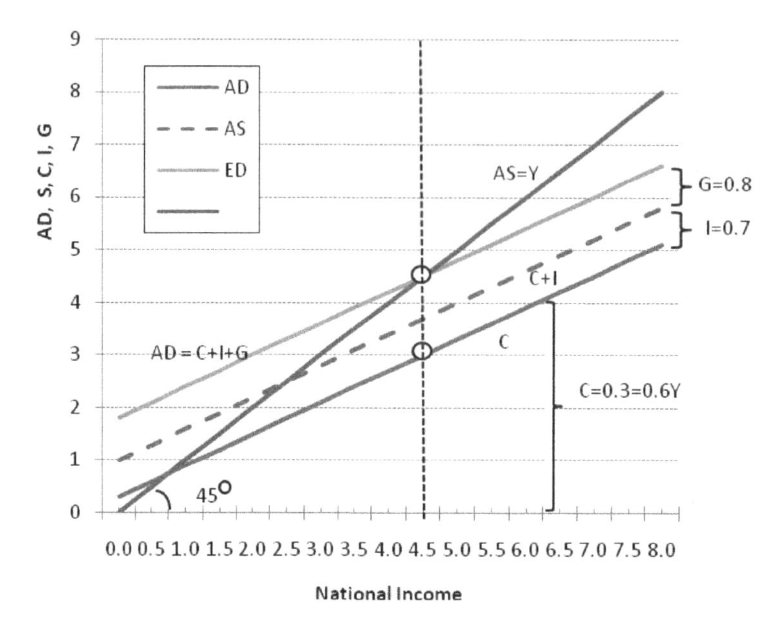

Fig. 3.14 A national income model with consumption, investment, and government expenditure

3.9 Saving–Investment Analysis — Prelude to Comparative Static Analysis

The saving–investment analysis is equivalent to the AD/AS analysis. The saving and investment schedule can be drawn easily using the same AD/AS table as Table 3.4.

3.9.1 *The data table*

For simplicity, we eliminate the separate government expenditure $G_0 = 0.8$ in (3.12) and add it to the basic consumption level a, and thus, Eq. (3.10) becomes $C = 1.1 + 0.6Y$. The equilibrium income is still at \$4.5 trillion, as shown in Table 3.4, which is similar to Table 3.3(b) except that now $AD = C + I$. We also insert saving $S = Y - C$ in column G, as shown in Table 3.4. Since consumption and saving are opposite concepts, the aggregate demand and aggregate supply analysis can be studied using the saving–investment relation. They are shown simultaneously in Fig. 3.15. When AD = AS at $Y^{\#} = 4.5$, we also have the equilibrium condition that saving equals investment, $S = I$, at $Y^{\#} = 4.5$, and vice versa. Hence, both analyses are equivalent in the Keynesian model. The double-headed arrows show the excess demand or excess supply. Table 3.4 shows that when $Y = 1.5$, we have $S = -0.5$, and when $Y = 7.5$, $S = 1.9$.

Note that the slope of the consumption function is $b = 0.6$, which is the **marginal propensity to consume** (MPC). The slope of the saving function, $S = Y - C = -a + (1 - b)Y$, is $(1 - b) = 0.4$, which is the **marginal propensity to save** (MPS). MPC and MPS must add to 1 (i.e., 45°).

Table 3.4 The AD–AS analysis and the I–S analysis

	A	B	C	D	E	G	F
1							
2	Y	=1.1+0.6*Y		=C+I	=Y	Y-C	=AD-AS
3		C	I	AD	AS	S	ED
4	0.0	1.1	0.7	1.8	0.0	-1.1	1.8
5	0.5	1.4	0.7	2.1	0.5	-0.9	1.6
6	1.0	1.7	0.7	2.4	1.0	-0.7	1.4
7	1.5	2.0	0.7	2.7	1.5	-0.5	1.2
8	2.0	2.3	0.7	3.0	2.0	-0.3	1.0
9	2.5	2.6	0.7	3.3	2.5	-0.1	0.8
10	3.0	2.9	0.7	3.6	3.0	0.1	0.6
11	3.5	3.2	0.7	3.9	3.5	0.3	0.4
12	4.0	3.5	0.7	4.2	4.0	0.5	0.2
13	4.5	3.8	0.7	4.5	4.5	0.7	0.0
14	5.0	4.1	0.7	4.8	5.0	0.9	-0.2
15	5.5	4.4	0.7	5.1	5.5	1.1	-0.4
16	6.0	4.7	0.7	5.4	6.0	1.3	-0.6
17	6.5	5.0	0.7	5.7	6.5	1.5	-0.8
18	7.0	5.3	0.7	6.0	7.0	1.7	-1.0
19	7.5	5.6	0.7	6.3	7.5	1.9	-1.2
20	8.0	5.9	0.7	6.6	8.0	2.1	-1.4

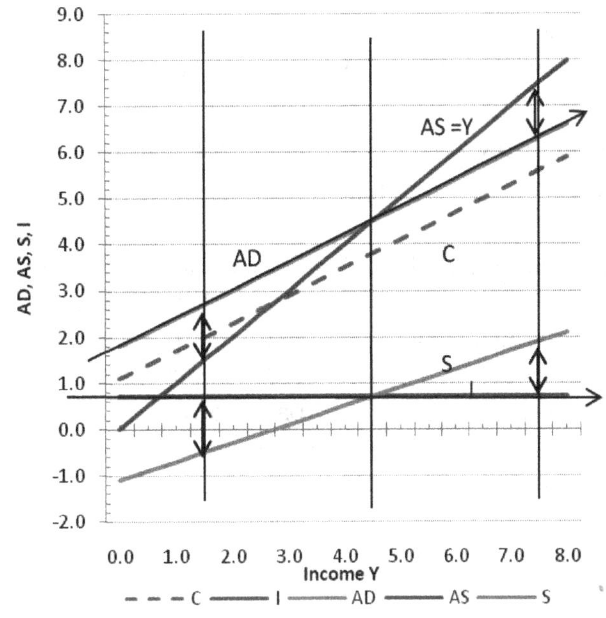

Fig. 3.15 The aggregate D–S analysis and the I–S analysis

3.10 Summary

In this chapter, we have defined the **equilibrium values** of the variables in an economic model as those values that satisfy the equations of the model identically and simultaneously. The equilibrium values can be solved by using the intercept method, as in Sec. 3.3.2, or the algebraic method, as in Sec. 3.5, and the graphic method, as in Sec. 3.7. Another method, the matrix method, will be explained in detail in Chap. 10. We have used simple examples from the microeconomic market model and the macroeconomic national income model to explain these methods. More elaborate models and applications to other area of economics are given in the homework.

After static analysis, the next step is to find the effects of change in a parameter or an exogenous variable on the equilibrium values of the model. This kind of analysis in Economics is called **comparative static analysis**. In order to compare the model or chart before and after the change, we may manually draw artificial lines to denote the original position of the aggregate demand and the investment, and then examine the new and old equilibrium points. Since a parameter or an exogenous variable can change in infinite ways, it is very cumbersome to conduct comparative static analysis in this way. Excel provides a better method, called the **naming method**, to do comparative static analysis. We will explore the naming method in the next chapter.

Appendix 3A. Excel Menu for Chart Tools

"Chart Tools" appear when a chart is selected. It consists of three subtabs or ribbons: Design, Layout, and Current selection. The group in each ribbon and the major contents are shown in Table 3A.1.

Appendix 3B. Reconnecting Table and Chart

When we select a chart or a data series drawn from a table, three boxes appear in the data part, as described in step (1) of Sec. 3.4.2. If not, we have to use the following procedure to reconnect table and chart. Suppose Table 3.3(b) is disconnected from Fig. 3.14.

Select Fig. 3.14, and click <RM>, repeatedly if necessary, until a mini toolbar appears. Select <Select Data> in the middle part of the toolbar, and a "Select Data Source" dialogue box, Fig. 3B.1 below, appears. The first row is "Chart data range:" (part a). At the end of the fill-in box is the **collapse dialog** button (part b). Click the button, and the dialog box collapses to part c, letting you see more worksheet area. Select A3:F20 in Table 3.3(b), not including the ED column, as shown in part d, and click the **expand dialog** button (part e) to return to Select Data Source dialog box; then click <OK>. The data range is selected. When you click Fig. 3.14, Table 3.3(b) will be enclosed by three boxes: an upper green box

Table 3A.1 The major contents of "Chart Tools"

Tab	Group	Major contents
Design	Type	Chart, Save as Template
	Data	Switch Axes, Change the Data Range
	Chart Layouts	**Titles, Legend, Grid**
	Chart Styles	Color, Plot area
	Location	Move to
Layout	Current Selection	Click and Edit the Selection
	Insert	Picture, Shapes, Text box
	Labels	Titles, Legend, Data Labels and Tables
	Axes	Horizontal, Vertical, Grid lines
	Background	Plot Area, Chart/Wall/Floor, 3D
	Analysis	Trendlines, Lines, Updown bars, Error bars
	Properties	Current Chart Name
Format	Current Selection	Click and Edit the Selection
	Shape Styles	Color/Width/Height, Fill, Outlines, 3D effects
	WorldArt Styles	Fill, Outlines, Effects
	Arrange	Print to Front/Back, Align, Group, Rotate
	Size	Scale, Crop, Positioning

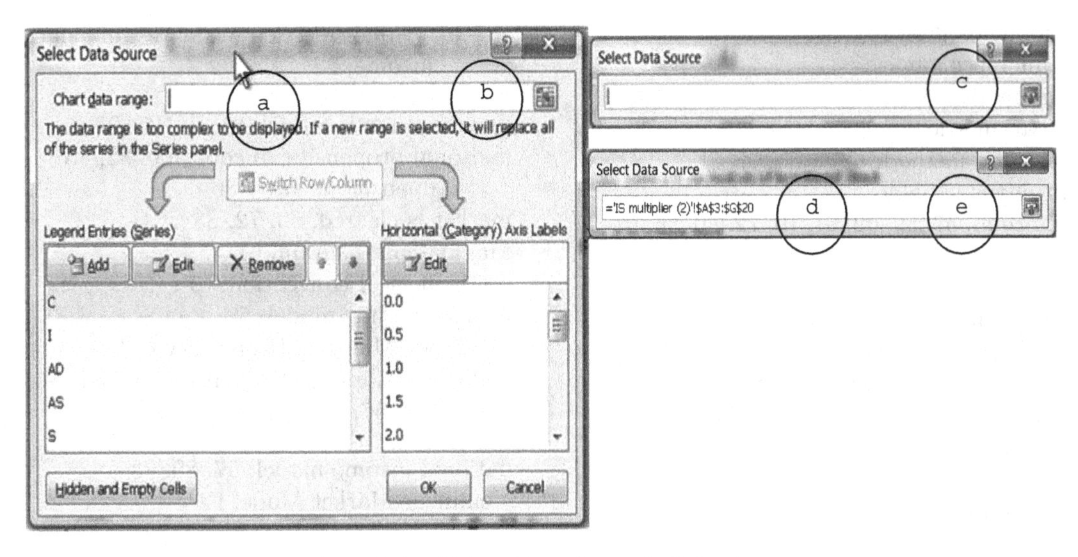

Fig. 3B.1 "Select Data Source" dialog box

for column titles, a purple long box on the left for the independent variable, and a blue box on the right for the dependent variables.

Review of Basic Equations and Formulas

A single market model

(3.1) $D = a - bP,\ a > 0,\ b > 0$

(3.2) $S = -c + dP,\ c > 0,\ d > 0$

(3.3) $D = S$

(3.4) Slope $\equiv \tan\theta$

$$= \frac{y - b}{x - a} = \frac{\Delta y}{\Delta x} = \frac{\text{rise}}{\text{run}} = \frac{18 - 0}{4 - 1}$$

(3.5) The slope − intercept form of the line

$$y = c + mx$$

Let $\alpha + \theta = \pi$,

$$\tan(\alpha) = \tan(\pi - \theta) = -\tan(\theta) = -m$$

Excess demand

$$E = D - S$$

$E = D - S$

A national income model

(3.9) $AD = C + I + G$

(3.10) $C = a + bY$

(3.11) $I = I_0$

(3.12) $G = G_0$

(3.13) $AS = Y$

(3.14) $AD = AS$

(3.17) Equilibrium income

$$Y^{\#} = k(a + I_0 + G_0)$$

(3.18) Equilibrium consumption

$$C^{\#} = k(a + b(I_0 + G_0)).$$

Key terms: Economics and Business

Key terms: Excel

Homework Chapter 3

Microeconomic models

3-1 A Simple market model Suppose the demand function is given by $D = 30 - 2P$ in Eq. (3.1), and the supply function is the same, $S = -6 + 6P$. Quantity is measured in millions.

(a) What are the slopes of the demand and supply curves?
(b) Using the intercept method, draw the simple market model as in Fig. 3.3. Give the title of the chart, and label the axis. Using the text box, give the name of the curves. The horizontal axis labels should be placed below the plot area. Major unit of the horizontal axis should be 1, and the major tick mark type should be "cross." Draw a vertical line through the equilibrium point.
(c) Using the schedule method, construct the table as Table 3.2, and draw the diagram of the single market like Fig. 3.12. Let $P = 0, 1, \ldots, 10$. Format the chart as in (b).
(d) Using the algebraic method, find the equilibrium price and quantity
(e) Are the three methods of solution, (b), (c), (d), give the same solution? What are the differences among these three methods?
(f) What is the economic meaning of the fact that the constant term in the demand function has changed from 18 to 30?
(g) This model, as originally given, has three equations in three variables, D, S, and P. Can you reduce the model to two equations in two variables? Show your new model, what are the variables? From the two equation model, can you reduce to a one-equation model? How? (this process is called the **elimination method of solving system of simultaneous equations**).

(Hints: (d) $(S^{\#}, Q^{\#}, P^{\#}) = (21, 21, 4.5)$)

3-2 A market model Draw the chart for the following market model. What are the equilibrium price and quantity? Use a white circle and a dotted line to show the equilibrium values.

$$D = 24 - 2P, \quad S = -5 + 7P, \quad D = S,$$
$$P = 0, 1, \ldots, 10.$$

Answer the same questions as HW3-1 above.
(Hints: (d) $(S^{\#}, Q^{\#}, P^{\#}) = (17.6, 17.6, 3.2)$)

3-3 A nonlinear market model Set up a table for the following market model, draw the demand and supply chart, and find the equilibrium price and quantities.

$$D = 4 - P^2, \quad S = 4P - 1, \quad D = S$$
$$P = 0, 0.2, \ldots, 3.$$

Chart formatting instructions:

(a) Enter the proper chart title, and X- and Y-axes titles. Label the curves as D and S.
(b) The Y-axis and the X-axis labels should have one decimal place, and the font size should be 9.
(c) The grid lines are black dotted lines, and the background of the plot area is white.
(d) Identify demand and supply curves by using text boxes, as shown in Fig. 3.12. The text boxes should be embedded into the chart.
(f) The demand curve is a black medium-size (2.25 pt) line with empty delta (triangle) markers and also with solid black marker line color.
(g) The supply curve is a red medium-size (2.25 pt) line with empty square markers and also with solid red marker line color.
(h) The axis labels should correspond to the independent variable (starting from 0, not 1).
(i) The tick labels should be placed below the plot area, not next to the X-axis.
(j) The major tick mark type is cross, and minor tick mark type is inside.
(k) The legend should be placed at the bottom of the plot area.
(l) The chart has no border.
(m) Denote the intersection of the demand and supply curves with a circle, and run a light dotted vertical line through the intersection.

Answer the following questions:

(a) What are the slope of the demand and supply curves?
(b) Can you use the intercept method or the algebraic method to find the equilibrium values? If you can, use the method to find the equilibrium values.
(c) How many variables are in this nonlinear market model? What are they?
(d) Using algebraic method, find the equilibrium price and quantity.
(e) Why are economists interested in finding equilibrium price and quantity?
(f) Why are they called "equilibrium" price and quantity?
(g) What is the use of the excess demand function?
(h) Locate the equilibrium price in the table and paint the corresponding row in the table with yellow. Do the equilibrium price and quantity obtained in the table correspond to those in the chart? Why?

(Hints: (d) $(D^{\#}, S^{\#}, P^{\#}) = (3, 3, 1)$).

3-4 A nonlinear market model Set up the table to be similar to Table 3.2 and draw the nonlinear demand and supply curves.

$$\mathrm{D} = 10 + 100/\mathrm{P}^2, \quad \mathrm{S} = -5 + \mathrm{P}^2, \quad \mathrm{D} = \mathrm{S}$$
$$\mathrm{P} = 0, 1, \ldots, 10.$$

Using the algebraic method to find the equilibrium price and quantity, and then snswer the same questions as in HW3-3.
(Hints: the entry in B4 should be = 10+100/(A4^2), etc. Since P starts from 0, the entry of B4 happens to divide by zero, shown as #DIV/0!. When drawn in a chart, the demand curve will connect 0 in B4 to 110 in B5. Since the connection does not make sense, we erase the connecting line simply by deleting the entry in B4).

3-5 The Marshallian market model In the nonlinear model in Question 5, the price is on the horizontal axis and the quantity is on the vertical axis. Change the axis so that the price is on the vertical axis and the quantity is on the horizontal axis (see Fig. 3.1(b)). Draw the demand and supply curves.

Note that, in this case, D and S must be solved for P. Let the price obtained from the demand equation be $\mathrm{P_D}$, and call it the **demand price**. Similarly, solve P from the supply equation and call it $\mathrm{P_S}$, the **supply price**. Let D = S = Q. Then we have[12]

$$\mathrm{P_D} = (4 - \mathrm{Q})^{1/2}, \quad \mathrm{P_S} = \mathrm{Q}/4 + 1/4, \quad \mathrm{P_D} = \mathrm{P_S}.$$

Draw the diagram with Q on the horizontal axis. What are the equilibrium demand and supply prices? The equilibrium quantity?

3-6 An exchange market model If, in a usual demand and supply model, the commodity is replaced by a foreign currency, say, the Japanese yen (JPY), then the model is a foreign exchange model. It shows the demand and supply of foreign exchange in the foreign exchange market. Suppose that the United States is the domestic country, and let e be the (home currency) price of foreign currency in terms of dollars (USD or US$, the home currency), then e is called the **foreign exchange (FX) rate in direct term** or **in American term**, or **direct quotation for exchange rate** (or **direct exchange rate**). It is the price of yen expressed in U.S. dollars, like US$ 0.00833 per yen, that is, 0.83 cent per yen, or 83 cents (= 0.00833 × 100 = 0.83) per 100 yen.[13]

Let the U.S./Japan foreign exchange model be given as follows:

$\mathrm{D} = 8 - 350\mathrm{e}$	Demand function for yen
$\mathrm{S} = -1 + 550\mathrm{e}$	Supply function for yen
$\mathrm{D} = \mathrm{S}$	Equilibrium condition of the Exchange Market.

[12]In Excel, $(4 - \mathrm{Q})^{1/2}$ is written as = sqrt(4 − Q), the square root of 4 − Q.
[13]Another way of quoting an exchange rate is the (foreign currency) price of a US dollar in terms of foreign exchange. This is an **exchange rate in indirect**, or **foreign, terms**, and is denoted by e*. For example, 120 yen per one US dollar. These two exchange rates are inverse to each other: e = 1/e*. Thus, e = US$0.00833 per yen is equivalent to e* = 1/0.00833 = 120 yen per dollar.

The quantity is measured in trillions of yen, and e ranges from 0.005, 0.006 to 0.015 in dollar. The model states that when the direct FX rate increases, demand for Yen decreases and supply of yen increases in the Yen market in the United States (Yen here is the same as goods, and e is the price of Yen).

(a) Using the graphic (schedule) method, draw the demand and supply curves of the exchange market for yen. What are the equilibrium exchange rate and the equilibrium amount of yen in this model?
(b) Use a callout and a vertical dotted line to show the location of the equilibrium exchange rate $e^{\#}$ at the intersection of the two curves (in the callout, enter $(e^{\#}, Q^{\#}) = (0.01, 4.5)$).
(c) Who demands yen mostly in the US domestic market?
(d) Who supplies yen mostly in the US domestic market?
(e) What are the factors that influence exchange rates?
(f) At this exchange rate $e^{\#}$, what is the equilibrium exchange rate of dollar in terms of yen?
(g) What is the relation between the FX in American terms and FX in foreign terms (or direct and indirect quotation of exchange rates)? Draw the relation in the demand and suppy chart of (a).

3-7 A nonlinear exchange market model Let the foreign exchange model be

$$D = 10 - 70e^{1/2}$$
$$S = -3 + 50e^{1/2}$$
$$D = S.$$

Answer the same questions as in HW3-6.
(Hints: In this exercise, the equilibrium exchange rate and quantity lie in an interval of the independent variable.)

Macroeconomic models

3-8 A Keynesian macroeconomic model It is given as follows:

$$AD = C + I + G, \quad C = 30 + 0.75Y, \quad I = 60, \quad G = 40$$
$$AS = Y, \quad AD = AS.$$

Let $Y = 50, 100, \ldots, 850$. Set up the data table, draw the chart, and find the equilibrium national income and consumption. The chart should be exactly the same as Fig. HW3-8.

(a) The circles must be at the exact location.
(b) Use a text box to indicate the lines for the C, C + I, AD, and AS curves.
(c) Use a curled brace to indicate the amount of investment in the figure.
(d) What are the equilibrium income and consumption for this model? If you cannot find the exact value of the equilibrium income and consumption from the table or chart, why this is the case? Can you make a guess about what are they?
(e) Find the exact levels of the equilibrium income and consumption using the aggregate excess demand function.

Table HW3.8 A national income model					
Y	=30+0.75Y		=C+I+G	=Y	=AD−AS
	C	C+I	AD	AS	ED

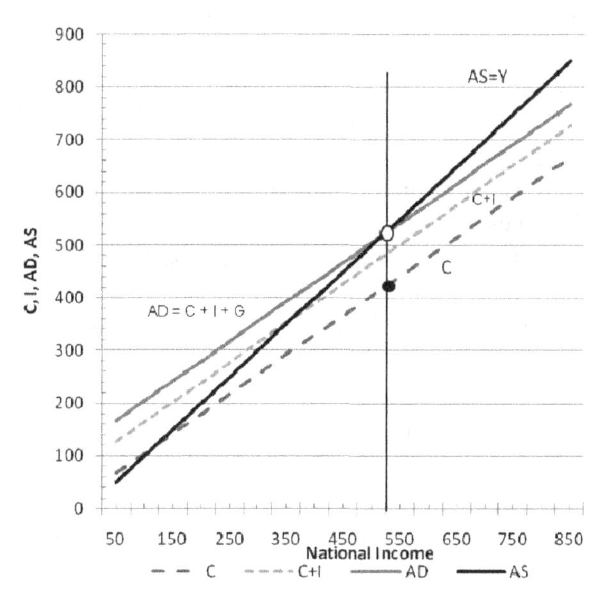

Fig. HW3-8 A keynesian macroeconomic model

(f) Use yellow paint to paint the row (or rows) to locate the equilibrium income and consumption in the table.

(g) What do you call b = 0.75 in Economics? What will happen to the AD curve in Fig. HW3-8 if b increases, say, from 0.75 to 0.9? Calculate the new equilibrium income.

3-9 AD–AS analysis and I–S Analysis Let the simple macroeconomic model be given as

$$Y = C + I, \quad C = 0.3 + 0.6Y,$$

where I = 1.5, and Y ranges from $0, 0.5, \ldots, 8$. Draw the AD and the AS curves. Derive the IS model and show that the AD–AS analysis is equivalent to the I–S analysis, as illustrated in Fig. 3.15 (show the AS, AD, I, and S lines). Use a circle to indicates the equilibrium points, and use a yellow pen to color the row of equilibrium values.
(Hints: See Table 3.4 and Fig. 3.15 of the text.)

3-10 A Keynesian model Let a Keynesian model be

$$AD = C + I + G, \quad C = 130 + 0.6Y, \quad I = 70, \quad G = 30,$$
$$AS = Y, \quad AD = AS.$$

Draw the chart of the decomposed AD model (similar to Table 3.3(b), with columns consisting of Y, C, C + I, AD = C + I + G, AS = Y, and E = AD – AS). Calibrate Y as $0, 50, \ldots, 800$.

(a) How many equations and variables are in this model? What are they?
(b) How many exogenous variables and how many endogenous variables are in this model? What are they?
(c) In what sense are they called "exogenous" variables? "Endogenous" variables?
(d) What are the equilibrium income and consumption? Find these values algebraically.
(e) Why are they called "equilibrium" income and consumption?
(f) Why are economists interested in finding equilibrium income and consumption?
(g) What is the use of the excess aggregate demand function?
(h) Construct the data table with AD components consisting of C, C + I, and AD, AS, and ED.
(i) Find the levels of equilibrium income and consumption. Denote the levels with white circles with a dotted line through the equilibrium income, and paint the corresponding row in the data table with yellow.
(j) Use text box to denote each of four lines.
(k) Using the aggregate excess demand function, derive the exact equilibrium income and consumption. Are they the same as those you have derived from the algebraic method, the table, and the chart?
(l) Reduce the equations to three equations, consisting of the AD function, the AS function, and the equilibrium condition. How many variables are in this three-equations model? Are they endogenous or exogenous variables?
(m) From the data table, draw a separate chart for the AD, AS curves as in Fig. 3.13, and reproduce the format of the chart exactly as in Fig. 3.13.

3-11 AD with the trade sector Let a Keynesian model be

$$\mathrm{AD} = \mathrm{C} + \mathrm{I} + \mathrm{G} + \mathrm{F}, \quad \mathrm{C} = 60 + 0.6\mathrm{Y}, \quad \mathrm{I} = 90,$$
$$\mathrm{G} = 50, \quad \mathrm{F} = 40, \quad \mathrm{AS} = \mathrm{Y}, \quad \mathrm{AD} = \mathrm{AS},$$

where F = X – M, the net exports (exports X subtracting imports M). Calibrate Y as $0, 50, \ldots, 800$. Answer the questions a to m in HW3-10 above.
(Hints: In the data table, the decomposed AD consists of four columns: C, C + I, C + I + G, and AD = C + I + G + F; see Table 3.3(b))

3-12 AD with taxation Let a Keynesian model be

$$\mathrm{AD} = \mathrm{C} + \mathrm{I} + \mathrm{G}, \quad \mathrm{C} = 150 + 0.5\mathrm{D}, \quad \mathrm{I} = 30, \quad \mathrm{G} = 60,$$
$$\mathrm{D} = \mathrm{Y} - \mathrm{T}, \quad \mathrm{T} = 40, \quad \mathrm{AS} = \mathrm{Y}, \quad \mathrm{AD} = \mathrm{AS},$$

where D = Y – T, the disposable income. Calibrate Y as $0, 50, 100, \ldots, 850$. Answer the questions a to m in HW3-10 above.
(Hints: In the data table, the columns should consist of Y, D, C, C + I, AD = C + I + G, AS = Y, E = AD – AS).

3-13 The simple IS/LM macroeconomic model In this model, we introduce the monetary sector into the simple Keynesian model:

$$Y = C + I + G, \qquad C = 300 + 0.9Y,$$
$$I = 2000 - 2000r, \quad M = -1000r + 0.1Y,$$

where M is real money balances and r is interest rate. The endogenous variables are Y, C, I, and r. The levels of government purchases G and money supply M are exogenous, and are given as $G = 3700$, $M = 5850$. The first three equations are called the **IS equations**, and the first equation is the **equilibrium condition of the aggregated commodity market**. The fourth equation is called the **LM equation**, it is the **equilibrium condition of the money market**. We have four equations and six variables, (Y, C, I, M, G, r). G and M are given numbers, hence, they are exogenous variables. Thus, we have four endogenous variables: (Y, C, I, r). Substituting C and I equations into the equilibrium condition in the commodity market, and solving the money market equation in terms of Y, we have

$$Y_{IS} = 60000 - 20000r \quad \text{IS curve,}$$
$$Y_{LM} = 58500 + 10000r \quad \text{LM curve,}$$

where Y_{IS} is income determined by the commodity market, and Y_{LM} is income determined by the money market. The equilibrium condition is that the income determined by the commodity market must be the same as that determined by the money market. The usual practice is to take r as the independent variable, as in the single market model in Sec. 3.7 and Fig. 3.12.

(a) Derive the IS and LM curves step by step from the original four equations.
(b) Set up a data table, using r as independent variable, where $r = 0, 0.01, \ldots, 0.1$.
(c) Draw the IS–LM chart similar to Fig. 3.12.
(d) How many variables are there in the original system of equations? How many endogenous variables and how many exogenous variables? What are they?
(e) Answer the same equation as (d) for the second system of equations.
(f) Why the fourth equation is called the equilibrium condition of the money market? What does it represent?

(Hints: Let $E = Y_{IS} - Y_{LM}$. $r^{\#} = 0.05$, and $Y^{\#} = 59,000$)

3-14 Fancy colors and fonts Make the best colorful presentation of the chart you obtained in any two of the questions in HW3-3, HW3-4, HW3-6, HW3-9, HW3-10, HW3-11, HW3-12, H 3-13. (Your imagination is the limit. Later in the class, you will have an opportunity to show off your creation in the slide presentation. This is a field in industrial arts, but, unfortunately, there is not much use of it in economics *per se*, except that you can enjoy your creation.)

Chapter 4

Comparative Static Analysis — Name that Range!

Chapter Outline

Objectives of this Chapter

The purpose of static analysis is to find the equilibrium values of the variables in an economic model. Mathematically, this is the problem of finding the solution of a system of simultaneous equations. When the equilibrium values are found, there are two possible cases. In a numeric economic model, the equilibrium values are expressed in real numbers. In a more general parametric model, the equilibrium values consist of parameters and/or exogenous variables, which are given from outside of the model. Note that if the values of parameters and exogenous variables are given numerically, then, after the values are substituted into the parameters and exogenous variables, the equilibrium variables will become real numbers.

Thus, it is natural to ask what will happen to the equilibrium values of the model if there is any change in the values of parameters and/or exogenous variables. Will the equilibrium values increase or decrease, or not change? And, if so, how large will the effect be? These are the questions in comparative static analysis studies and they are the subjects of this chapter.

The mathematical tool we use to answer these questions is the concept of difference of dependent and independent variables, and the new Excel tools we will use are range names, conditional formatting, and the picture copy command. They are explained in detail in this chapter. As in Chapter 3, we explain the single market model in detail, and then apply the method to the national income model. The emphasis here is on the similarity in the methods of analysis.

4.1 Parametric Market Models

In Chapter 3, we introduced parametric models of a macroeconomic national income model in detail. In this chapter, we begin by reintroducing parametric models of a single market model.

4.1.1 *Parametric expression of equilibrium values*

For convenience, the model of a market price determination in Chapter 3 is reproduced as follows.

A numeric model	A parametric model	Name of the equation	
$D = 18 - 2P$	$D = a - bP, \quad a > 0, b > 0$	Demand function	(4.1)
$S = -6 + 6P$	$S = -c + dP, c > 0, d > 0$	Supply function	(4.2)
$D = S$	$D = S$	Equilibrium condition	(4.3)

where a, b, c, and d, are positive **parameters**, and D, S, and P are endogenous variables.[1] The major difference between the numeric model and the parametric model is that the equilibrium values are now given in terms of the parameters and/or exogenous variables. But they are essentially the same models.

Using the intercept method of Sec. 3.3.2, from **the demand equation**, we have

$$\begin{aligned} &P = 0 \text{ implies } D = a > 0, \\ &D = 0 \text{ implies } P = a/b > 0. \end{aligned} \tag{4.4}$$

The intercepts are expressed in parameters. Connecting the two intercepts, (P, D) = (0, a) and (P, D) = (a/b, 0), we have the demand curve.

Similarly, from **the supply equation**, we find

$$\begin{aligned} &P = 0 \text{ implies } S = -c < 0, \\ &S = 0 \text{ implies } P = c/d > 0. \end{aligned} \tag{4.5}$$

Connecting the two intercepts, (P, S) = (0, −c) and (P, S) = (c/d, 0), we have the supply curve. Corresponding to the numeric model of Fig. 3.3, the intercepts in terms of the parameters are shown in Fig. 4.1. The intersection of the curves gives the equilibrium price and quantity. Unless the parameters are given in numerical values, Excel cannot draw the chart for the parametric model like those in Sec. 3.3.

The algebraic method of solution corresponding to Sec. 3.5 is to substitute the demand and the supply equations into the equilibrium condition. In this way, we eliminate D and S,

[1]Sometimes, the parameters and exogenous variables are difficult to distinguish. Generally, parameters are the constant terms or the coefficient of endogenous variables, and denoted by a, b, c, etc., while the exogenous variables are denoted by the variable names with a zero subscript, like G_0, I_0, etc.

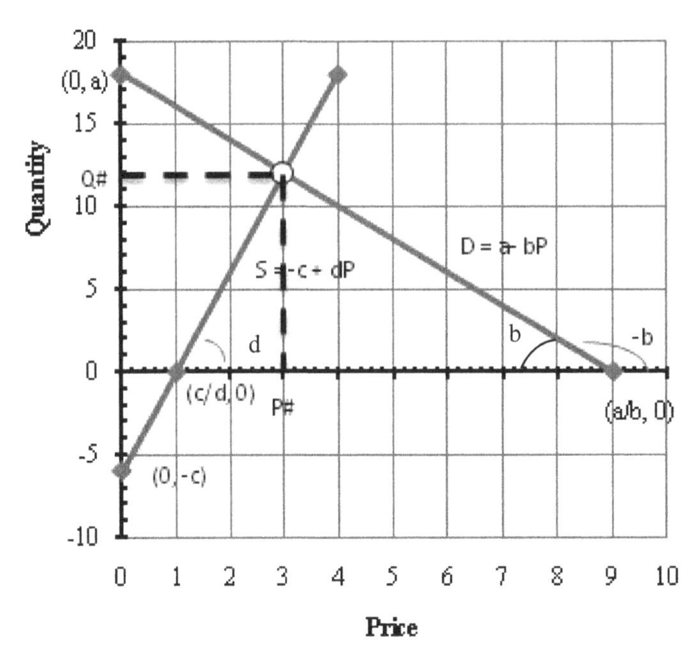

Fig. 4.1 Intercepts of a parametric model

and the three equations reduce to one equation:

$$\mathrm{a - bP = -c + dP}.$$

Solving P from the equation, we have

$$\boxed{\mathrm{P}^{\#} = \frac{\mathrm{a+c}}{\mathrm{b+d}} > 0} \tag{4.6}$$

where $\mathrm{b + d} \neq 0$. Substituting $\mathrm{P}^{\#}$ in the demand or the supply equation, we have

$$\boxed{\mathrm{D}^{\#} = \frac{\mathrm{ad - bc}}{\mathrm{b+d}} = \mathrm{S}^{\#} = \mathrm{Q}^{\#}} \tag{4.7}$$

which is positive if $\mathrm{ad{-}bc} > 0$. This condition is called the **nonnegativity condition**. It guarantees that the solution makes economic sense. In (4.6) and (4.7), $\mathrm{P}^{\#}$, $\mathrm{D}^{\#}$, and $\mathrm{S}^{\#}$ depend on the parameters of the model, that is, a, b, c, and d. This is the reason that these three variables are called **"endogenous" variables**, in the sense that they are determined by the parameters in the model. This is not clear in the numerical model, since the parameters in the model collapse and become numbers.[2]

[2]In other cases, a variable may be given from outside of the model, like income in the demand function or rainfall in the supply function, which are not determined by parameters of the model. They are then called **exogenous variables**.

From Sec. 3.4.1, we see that the slope of the demand curve is negative, −b, and the slope of the supply curve is positive, d, as shown in Fig. 4.1. The tangent of the complementary angle of the demand curve is b, which is positive.

The advantage of using the parametric model is that it is a more general formulation of an economic model. Comparing the numeric model and parametric model in Eqs. (4.1) and (4.2), we see that in the numeric model, we have assigned the following values:

$$a = 18, \quad b = 2, \quad c = 6, \quad d = 6.$$

Substituting into the equilibrium values, we have

$$P^{\#} = 3 \quad \text{and} \quad Q^{\#} = 12$$

as obtained in Chapter 3.

However, there is no particular reason that the parameters should take these specific values. In fact, we can assign other values so long as the nonnegative solutions exist. In economics, these values should be estimated using real world data and econometric methods, such as regression analysis (see Chapter 7). Unless there is a special reason, economic theory generally tries to avoid such specific parameters or a specific functional form. At present, we assume linear demand and supply functions, and formulate economic models as generally as possible.

For the parametric model, the excess demand function is

$$E = D - S = (a + c) - (b + d)P. \tag{4.8}$$

As we learned in Chapter 3, the equilibrium price $P^{\#}$ can also be obtained by equating the excess demand equation to zero and solving for P.

Example 4.1 Using the definition (3.5) of slope in Sec. 3.3.3, show that the excess demand function (4.8) has a negative slope. Show that E is zero at the equilibrium price.
Answer: From $y = c + mx$, we know $m = -(b + d) < 0$ since $b > 0$ and $d > 0$. Substituting $P^{\#}$ into (4.8) we can show $E = 0$ at $P^{\#}$. □

Since the equilibrium values (that is, the **solution** of the system of simultaneous equations) are functions of parameters and exogenous variables, after the equilibrium values ($P^{\#}$, $Q^{\#}$) are derived, we may then study the effects of changes in parameters and/or exogenous variables on $P^{\#}$ and $Q^{\#}$. This study is called **comparative static analysis** (CSA).

These effects may be of two kinds:

(1) **Qualitative effects**, in cases in which we are concerned with the direction of changes, either positive, negative, or no changes, from the original equilibrium values to the new equilibrium values;
(2) **Quantitative effects**, in cases in which we are concerned with the magnitude of changes from the original equilibrium values to the new equilibrium values.

Thus, it is a general practice in economics that after we perform static analysis of a model, we go further to conduct comparative static analysis. There are two methods of studying a comparative static analysis: the discrete method and the continuous method. We explain the discrete method in this book (as the continuous method uses calculus).

4.1.2 *The discrete method of comparative static analysis*

In this section, we discuss the theoretical foundation of the comparative static analysis, and relate the theory with tables and Excel charts. In the next sections, we will show how to make comparative static analysis using the naming method.

As in the definition of slope in Eq. (3.4), we use Δ to denote the change of a variable.

Case a: Effect of a change in a on equilibrium values

In (4.6), if $P^{\#}$ changes by $\Delta P^{\#}$ from a point, say, $P_0^{\#}$, to $P_1^{\#}$, then we write, $\Delta P^{\#} \equiv P_1^{\#} - P_0^{\#}$, which is the **difference of $P^{\#}$**. We may take $P_0^{\#}$ as the equilibrium price before the change, and $P_1^{\#}$ as the equilibrium price after the change. Similarly, let $\Delta a \equiv a_1 - a_0$, which is the **difference of parameter a** in demand function. Other things being equal (or, ceteris paribus), we assume that only parameter a changes, holding other parameters the same before and after the change. Since

$$P_0^{\#} = \frac{a_0 + c}{b + d}, \quad P_1^{\#} = \frac{a_1 + c}{b + d}, \tag{4.9}$$

where $b + d \neq 0$, then subtracting, we have

$$\boxed{\Delta P^{\#} \equiv P_1^{\#} - P_0^{\#} = \frac{a_1 - a_0}{b + d} = \frac{\Delta a}{b + d} >, =, < 0, \quad \text{as } \Delta a >, =, < 0} \tag{4.10}$$

Thus, the **qualitative effect** of a change in parameter a on $P^{\#}$ depends on the sign of Δa; that is, if Δa is positive, then $\Delta P^{\#}$ is also positive, or, in other words, when "a" increases, $P^{\#}$ also increases. The **quantitative effect** is the value of the change, which is given by $\Delta a/(b + d)$ on the right-hand side of (4.10).

Comparing (4.10) with (4.6), we see immediately that if the changed parameter is only in the numerator (the parameters in the denominator are all constant in this case), and it is linear (no squares), then the difference (4.10) can be derived immediately by treating other parameters as constant when we subtract. This is an easy way of deriving the difference. From (4.7), we can derive $\Delta Q^{\#}$ as

$$\boxed{\Delta Q^{\#} \equiv Q_1^{\#} - Q_0^{\#} = \frac{d\Delta a}{b + d} >, =, < 0, \quad \text{as } \Delta a >, =, < 0} \tag{4.11}$$

Example 4.2 A numerical example of case a. In Table 4.1(a) below, we enter the labels in rows 1–4, the parameter names in row 5, and the current parameter values in row 6. We also enter the default parameter values in row 7, which are taken from the numeric values in Eqs. (4.1)–(4.3). The current values in row 6 are, at the beginning, the same as the default values in row 7, but will be changed in the comparative static analysis later. The default equilibrium values are at ($P^{\#}$, $Q^{\#}$) = (3, 12) as shown in F7 and G7. For the details of constructing Table 4.1(a), see Sec. 4.3.1. We now ask if, other things being equal (ceteris paribus), parameter a changes from $a_0 = 18$ to $a_1 = 26$, as in (4.9), and suggested in row 10 of Table 4.1(a), what is the effect of this change on the equilibrium values $P^{\#}$ and $Q^{\#}$? Substituting $a_0 = 18$ and $a_1 = 26$ in (4.9), we have $P^{\#}_0 = 3$ and $P^{\#}_1 = 4$. Similarly, we have $Q^{\#}_0 = 12$ and $Q^{\#}_1 = 18$. Hence, from (4.10), we can calculate $\Delta P^{\#} = 4 - 3 = 1 > 0$, and from (4.11), we have $\Delta Q^{\#} = 18 - 12 = 6 > 0$. Thus, we can see that, when the intercept a increases from 18 to 26,

> the **qualitative effect** is that both equilibrium price and quantity increase,
> the **quantitative effect** is that the equilibrium price increases from 3 to 4, and the equilibrium quantity increases from 12 to 18 units.

In short, both qualitative and quantitative effects can be summarized as

$$(a; P^{\#}, Q^{\#}) = (18; 3, 12) \rightarrow (a; P^{\#}, Q^{\#})' = (26; 4, 18), \tag{4.12}$$

where the prime "′" denotes the values after the change. This relation is shown in rows 6 and 7 of Table 4.1(a). Equivalently, row 8 of Table 4.1(a) shows the change of current values from the default values (for example, B8 = B6 – B7). The direction of arrows shows the qualitative effect: the up arrow shows an increase, the down arrow a decrease, and the horizontal arrow shows no change. Row 8 of change shows that, if a increases by 8,

> the **qualitative effect** is that both equilibrium values increases (up arrows),
> the **quantitative effect** is that the equilibrium price changes by 1, and the equilibrium quantity changes by 6 units.

We may also denote both effects simply as

$$(\Delta a; \Delta P^{\#}, \Delta Q^{\#}) = (8; 1, 6). \tag{4.13}$$

The qualitative effect is shown by the sign of the numbers, and the quantitative effect is shown by the numerical values. □

Example 4.3 The chart of case a. Rows 5–8 are reproduced in Table 4.2(a) in Sec. 4.4 (ignore the method of constructing the chart for the time being), and illustrated in Fig. 4.8(a). The original demand and supply curves are shown as the dotted line as those in Fig. 4.7, the original equilibrium point (3, 12) is denoted as a white circle (O for original). After a increases from 18 to 26, the whole demand line shifts upward. The new line is shown as the solid line in the chart. The new demand line intersects with the supply line at ($P^{\#}$, $Q^{\#}$)′ = (4, 18), which is denoted by a filled dark circle. The movement of the equilibrium values from the white circle to the dark circle represents the following relation,

$$(a; P^{\#}, Q^{\#}) = (18; 3, 12) \rightarrow (a; P^{\#}, Q^{\#})' = (26; 4, 18)$$

This is the **impact vector in levels**. The three short dark arrows then represent the changes in a parameter and equilibrium values:

$$(\Delta a; \Delta P^{\#}, \Delta Q^{\#}) = (8; 1, 6)$$

This is the **impact vector in differences**. In general, the direction of the arrows shows the qualitative effect, and the numerical values of change shows the quantitative effect in the chart. □

Case b: Effect of a change in b on equilibrium values

If a parameter or an exogenous variable is in the denominator, like b and d in $P^{\#}$ or $Q^{\#}$, the calculation will become a little bit complicated. Starting from b_0, if b changes to b_1, and $\Delta b = b_1 - b_0$, then,

$$P_0^{\#} = \frac{a+c}{b_0+d}, \quad P_1^{\#} = \frac{a+c}{b_1+d} \tag{4.14}$$

Thus, simplifying the denominator, we have

$$\boxed{\Delta P^{\#} = P_1^{\#} - P_0^{\#} = -\frac{(a+c)\Delta b}{(b_0+d)(b_1+d)} <,=,> 0, \quad \text{as } \Delta b >,=,< 0} \tag{4.15}$$

Table 4.1 (a) Four cases of comparative static analysis

	A	B	C	D	E	F	G	H
1	**Comparative Static Analysis**							
2			**of a Single Market Model**					
3	**A. The Parameter Table**							
4		Parameters				Equi Values		
5		a	b	c	d	P#	Q#	
6	Curr	26	2	6	6	4	18	
7	Deflt	18	2	6	6	3	12	
8	Chg	⇧ 8	⇨ 0	⇨ 0	⇨ 0	⇧ 1	⇧ 6	
9	Chg	chg a	chg b	chg c	chg d	chg P#	chg Q#	
10	Sugg	⇧ 26	→	→	→	⇧ 4	⇧ 18	Case a
11	value		⇧ 6	→	→	⇩ 2	⇩ 6	Case b
12				⇧ 30	→	⇧ 6	⇩ 6	Case c
13					⇧ 10	⇨ 2	⇧ 14	Case d

(a)

(Continued)

Table 4.1 (b) Comparative static analysis of a single market model

	A	B	C	D	E	F	G
1	**Comparative Static Analysis**						
2			**of a Single Market Model**				
3	**A. The Parameter Table**						
4		Parameters				Equi Values	
5		a	b	c	d	P#	Q#
6	Curr	26	2	30	6	7	12
7	Deflt	18	2	6	6	3	12
8	Chg	⇧ 8	⇨ 0	⇧ 24	⇨ 0	⇧ 4	⇨ 0
9	Chg	chg a	chg b	chg c	chg d	chg P#	chg Q#
10	Sugg 1	26	6	30	10		
11	Sugg 2	15	1	4	2		
12	Non-negativity condition:Q# >0						TRUE
13							
14	**B. The Data Table**						
15			=-6+6P		D=a-bP		E=D-S
16	P	=18-2P		=D0-S0		S=-c+dP	
17		D0	S0	E0	DD	S	E
18	0	18	-6	⇧ 24	26	-30	⇧ 56
19	1	16	0	⇧ 16	24	-24	⇧ 48
20	2	14	6	⇧ 8	22	-18	⇧ 40
21	3	12	12	⇨ 0	20	-12	⇧ 32
22	4	10	18	⇩ -8	18	-6	⇧ 24
23	5	8	24	⇩ -16	16	0	⇧ 16
24	6	6	30	⇩ -24	14	6	⇧ 8
25	7	4	36	⇩ -32	12	12	⇨ 0
26	8	2	42	⇩ -40	10	18	⇩ -8
27	9	0	48	⇩ -48	8	24	⇩ -16
28	10	-2	54	⇩ -56	6	30	⇩ -24
29							
30		Numeric			Parametric		
31		Model			Model		

This shows that Δb and $\Delta P^{\#}$ have opposite signs. Similarly, denoting

$$Q_0^{\#} = \frac{ad - bc}{b_0 + d} \quad \text{and} \quad Q_1^{\#} = \frac{ad - bc}{b_1 + d},$$

$$\boxed{\Delta Q^{\#} \equiv Q_1^{\#} - Q_0^{\#} = \frac{-(a+c)d\Delta b}{(b_0 + d)(b_1 + d)} <, =, > 0, \quad \text{as } \Delta b >, =, < 0} \tag{4.16}$$

Hence, Δb and $\Delta Q^{\#}$ have opposite signs for parameter b.

Example 4.4 Numerical example of Case b. Given the default values as in row 7 of Table 4.1(a), when b increases from 2 to 6, as shown in row 11, the new equilibrium values are ($P^{\#}$, $Q^{\#}$) = (2, 6)′, as in F11:G11 in Table 4.1(a). This relation is also shown in the upper part of Table 4.2(b). Hence, we may write the impact vector in levels as

$$(b; P^{\#}, Q^{\#}) = (2; 3, 12) \rightarrow (b; P^{\#}, Q^{\#})' = (6; 2, 6)$$

This is also shown by the movement of the equilibrium values from the white circle to the dark circle in the southwest direction in Fig. 4.8(b).

Equivalently, in terms of changes, from the result in the "change" row of Table 4.2(b), we can write the impact vector in differences as

$$(\Delta b; \Delta P^{\#}, \Delta Q^{\#}) = (4; -1, -6)$$

This is shown by the direction of the three black arrows in Fig. 4.8(b). □

Case c: Effect of a change in c on equilibrium values

Using the same technique, we may derive the comparative static result when c changes:

$$\Delta P^{\#} = \frac{\Delta c}{b+d} >, =, < 0, \quad \text{as } \Delta c >, =, < 0 \tag{4.17}$$

$$\Delta Q^{\#} = \frac{-b\Delta c}{b+d} <, =, > 0, \quad \text{as } \Delta c >, =< 0 \tag{4.18}$$

The above results show that an increase in parameter c, that is, Δc, will increase $P^{\#}$ but will decrease $Q^{\#}$. In other words, the qualitative effects are: When c changes, $P^{\#}$ changes in the same direction as c but $Q^{\#}$ changes in the opposite direction from c.

Example 4.5 Numerical example of Case c. Given the default values as in row 7 of Table 4.1(a), when c increases from 6 to 30 (since c is entered as –c in the supply equation, increase in c means the supply function shifts to the right, that is, supply decreases at each price), as suggested in row 12 of Table 4.1(a), the new equilibrium values are ($P^{\#}$, $Q^{\#}$) = (6, 6)′ as shown in row 12 of Table 4.1(a). This relation is reproduced in the upper part of Table 4.2(c). Hence, we may write the impact vector as

$$(c; P^{\#}, Q^{\#}) = (6; 3, 12) \rightarrow (c; P^{\#}, Q^{\#})' = (30; 6, 6)$$

This is also shown by the movement of the equilibrium values from the white circle to the dark circle in the southeast direction in Fig. 4.8(c).

Equivalently, in terms of changes, from the result in the "Change" row in Table 4.2(c), we can write impact vector as

$$(\Delta c; \Delta P^{\#}, \Delta Q^{\#}) = (24; 3, -6).$$

This is shown by the direction of the three arrows in Fig. 4.8(c). □

Case d: Effect of a change in d on equilibrium values

Since

$$P_0^{\#} = \frac{a+c}{b+d_0} \quad \text{and} \quad P_1^{\#} = \frac{a+c}{b+d_1},$$

we have

$$\boxed{\Delta P^{\#} = \frac{-(a+c)\Delta d}{(b+d_0)(b+d_1)} <,=,> 0, \quad \text{as } \Delta d >,=,< 0.} \tag{4.19}$$

Since

$$Q_0^{\#} = \frac{ad-bc}{b+d_0} \quad \text{and} \quad Q_1^{\#} = \frac{ad-bc}{b+d_1},$$

we have

$$\boxed{\Delta Q^{\#} = \frac{(a+c)b\Delta d}{(b+d_0)(b+d_1)} >,=,< 0, \quad \text{as } \Delta d >,=,< 0.} \tag{4.20}$$

Hence, Δd and $\Delta P^{\#}$ have opposite sign, while Δd and $\Delta Q^{\#}$ have the same signs.

Example 4.6 Numerical example of Case d. Given the default values as in row 7 of Table 4.1(a), when d increases from 6 to 10, as suggested in row 13 of Table 4.1(a), the new equilibrium values are $(P^{\#}, Q^{\#}) = (2, 14)'$ as shown in F13:G13 of Table 4.1(a). This relation is also reproduced in the upper part of Table 4.2(d). Hence, we may write

$$(d; P^{\#}, Q^{\#}) = (6; 3, 12) \rightarrow (d; P^{\#}, Q^{\#})' = (10; 2, 14).$$

This is also shown by the movement of the equilibrium values from the white circle to the dark circle in the northwest direction in Fig. 4.8(d).

Equivalently, in terms of changes, from the result in the "change" row, we can write

$$(\Delta d; \Delta P^{\#}, \Delta Q^{\#}) = (4; -1, 2).$$

This is shown by the direction of the three black arrows in Fig. 4.8(d). □

4.2 Name that Range!

Comparative static analysis can be best studied by using the Excel spreadsheet method of naming a cell or a range of cells by a parameter, endogenous variable, word, or string of characters as the name. There are two ways to name a range.

4.2.1 *The name box method and the menu method*

The simplest method is to use the **name** box when only one range is defined at any one time. For example, in a worksheet, construct a table as shown in the range B5:E6 in Table 4.1(b).

We may name the whole range "parameters". The procedure is as follows:

Step 1. Select the **range**, that is, B5:E6.
Step 2. Click **the name box**. The cell reference in the name box moves to the left and is shaded. Range B5:E6 is also shaded.
Step 3. Enter **a name**, in this case, "parameters", or your name (without using the quotation marks) and press <Enter>.

The name "parameters" is created. Clicking a cell anywhere on the spreadsheet, clicking the down arrow next to the name box, and clicking "parameters" in the drop-down box, you select and shade B5:E6. On the other hand, the name "parameters" should appear in the name box whenever B5:E6 is selected.

You may also use the "Define Name" dialog box. In this case, in step 2 above, click

<"Formulas"><Defined Names, Define Name!>.

The "New Name" dialog box appears (see Fig. 4.2(a)). In this dialog box, in the "Name:" box, enter "parameters", and in the "**Scope**!" box, select workbook or Sheet1, or the name of the current sheet on the sheet tab. In the "Refers to:" box, the selected range and the current sheet name in the sheet tab will appear automatically. Click <OK> to complete the naming.

Note that in Fig. 4.2(a), when the down arrow of **Scope** is clicked, we have workbook and all the names of the sheets in the workbook, as shown in the dialog box with a bracket. Click a specific sheet if you want the name to be applicable only in a specific sheet (that is, **locally**), in this case, in Sheet1. This means that the name appears only in the name box of Sheet1. It will not appear, and is not defined, in other sheets.

If you want the name "parameters" to be applicable/defined in other sheets (**globally** in the current workbook), then click "workbook" (always on the first item) under the "New Name" dialog box as in Fig. 4.2(a). Note that, when the name box is used, the defined name is applicable to the workbook (global).

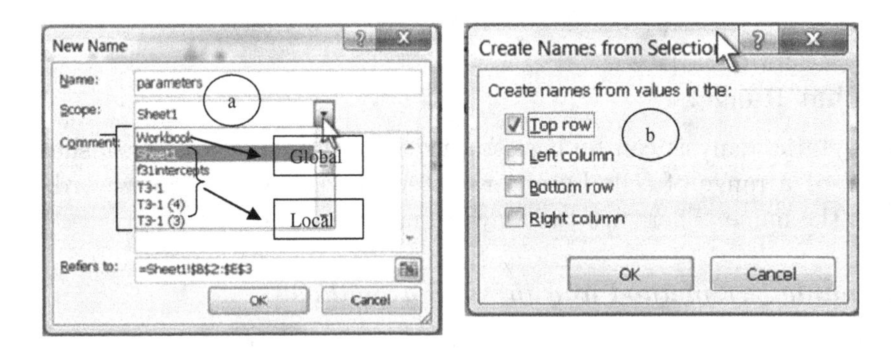

Fig. 4.2 Naming methods

If you want to define multiple names, for example, to name B6 as a, C6 as b, D6 as c, and E6 as d, follow these steps:

Step 1. Select a **range**, which includes the names B5:E6.
Step 2. Enter

<"Formulas"><Defined Names, Create from Selection> (a dialog box Fig. 4.2(b) appears)<"Create Names from Selection", Top row><OK>.

Thus, the four cells are named a, b, c, and d, respectively, at once.[3] The named range, B6:E6, includes only the data part, no range names. Thus, in order to know the name of the data cell or the range, the name of the cell or the range must be posted around the cell or the range.

Click the name box down arrow to confirm that a, b, c, and d are listed. Click each name to confirm that the range is selected correctly in Sheet1 and other sheets. The **scope** of the range selected by the above method is the worksheet (global). This means that if you are in Sheet2 or in a sheet other than Sheet1, and you click the name box down arrow in the other sheet, a, b, c, and d will appear, and if you click a, b, c, or d in the name box of other sheets, the pointer will jump back to the named cells in Sheet1.

Example 4.7 Global and local range names. Now, we may add or multiply the cells by names. Since a, b, c, and d are defined globally, in a blank space of Sheet1 or any other sheet, enter =a+b, =a−b, =a*b, and =a/b, and make sure that the correct number appears in the cell. Next, in Sheet1, define 6 as e and 7 as f locally (see Fig. 4.2(a)). Then, we have $e + f = 13$ in Sheet1 but e + f = #NAME? in other sheets. □

4.2.2 *Editing and deleting range names*

Note that if the range name is defined globally (for the entire worksheet), **the range name is sheet specific. It does not move with the sheet.** If a table in Sheet1, which contains a range name, is copied to Sheet1 (2), the range name stays with the original table in Sheet1. Clicking the range name in the name box of sheet2 will move the pointer back to the named range in Sheet1.

Change or deleting a range name
You cannot change or delete the range name using the name box. To change or delete the range name, click

<"Formulas"><Defined Names, Name Manager>.

A dialog box appears, as in Fig. 4.3. It includes all the names we have defined in this section. Note that the scope column indicates clearly whether the scope is global (worksheet) or local (Sheet1), and the parameter value includes all the contents of the cells defined in the range. Click the item you want to edit or delete, say, "parameters", and the edit and delete button

[3]Note that in Excel, c is named c_ (c underscore). See Box 4.2 for explanation.

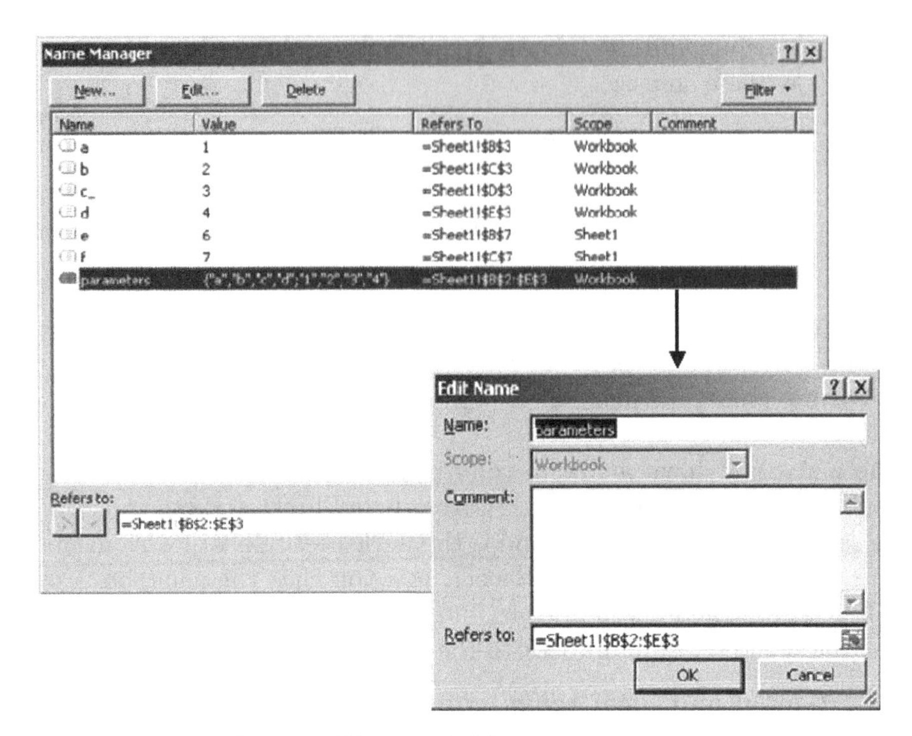

Fig. 4.3 Editing and deleting range names

appears. For editing a range name, the dialog box of "Edit Name" appears (see the lower part of Fig. 4.3). Change "parameters" to "para" and select <OK>. The name in "Name Manager" changes to "para". From Name Manager, you may also add new names by clicking the "New..." button.

> Box 4.1: **Overlapping names.** If parameters a and b in equation =a+b are defined globally in Sheet1, it is effective in other sheets, and the calculation depends on the numbers given in Sheet1. If =a+b is also to be defined in Sheet2 and globally, then we suggest that parameter a be named as aa (or aaa) and b as bb (or bbb), so the equation becomes =aa+bb (or =aaa+bbb) to avoid the conflict with a and b in Sheet1.
>
> If Sheet1 is **copied** to another sheet, say, Sheet1(2), then the parameters a and b in Sheet1(2) become **locally defined** in Sheet1(2). Local names override the global names in that specific sheet. Thus, =a+b in Sheet(2) takes the value from that sheet, not from globally defined a and b in Sheet1.

4.3 The Table for Comparative Static Analysis of a Market Model

We continue using the data table used in Chapter 3. Table 4.1(b) illustrates and solves the single market model of (4.1)–(4.3). The illustration has three steps: the construction of the parameter table, the construction of the data table, and the drawing of the figures.

4.3.1 *The parameter table*

The major difference between statics and comparative statics is the use of the parameter table. We first construct the parameter table for the market model as shown in Table 4.1(b).

Entering the parameters

Table 4.1(b) is basically the same as Table 4.1(a), except that rows 10 to 13 of Table 4.1(a) are combined into row 10 of Table 4.1(b). In Table 4.1(b), as is shown, the current value of a is changed from 18 to 26, and that of c is changed from 6 to 30. $P^{\#}$ and $Q^{\#}$ columns will be entered after the cells in the current row are named as follows.

Naming the parameters

As we have explained in Sec. 4.2.1, select range B5:E6, which includes the parameters and the values, and, using the menu method (Fig. 4.2(b)), **name** each cell in B6:E6 as the corresponding name in B5:E5. Each cell in row B6:E6 is now named a, b, c_, and d, respectively. Note that Excel names D6 as c_, (c underscore), not straight c. This is so since Excel reserves c and r to denote column and row.[4] In order to avoid conflict, Excel automatically names c as c_ and r as r_.

Box 4.2: **c and r in the name box.** When you name a parameter c or r, Excel will change it automatically to c_ and r_, since c and r are reserved by Excel to denote column and row.

In F6 and G6, enter the formulas (4.6) and (4.7) for $P^{\#}$ and $Q^{\#}$ as:

F6: =(a+c_)/(b+d)

G6: =(a*d−b*c_)/(b+d)

Then, the current equilibrium values 7 and 12 should appear in F6 and G6. In F7 and G7, enter manually the default equilibrium values 3 and 12, respectively. These values were derived previously in Chapter 3, either from the algebraic method (Sec. 3.5) or from the graphic method (Table 3.2 and Fig. 3.12). The current equilibrium values will be compared with the default equilibrium values when the current parameters change.

Enter the row of direction and magnitude of changes

Row 8 is the difference between the current value in row 6 and the default value in row 7 for each column. This row will show the qualitative effect and the quantitative effect of the

[4]Placing the pointer anywhere on the spreadsheet, typing c (or r) in the name box, and pressing <Return> key, you will select the whole column (or row) of the cell in which the pointer is located. Thus, c means column and r means row in Excel. If c is a parameter, c should have an underscore, c_ to make it different from column c.

change of the equilibrium value(s). Row 9 is merely the labels of entries in row 8 above. The effect of change in the parameter(s) in row 8 on the equilibrium values $P^{\#}$ and $Q^{\#}$ will be shown directly in F8:G8 as positive, negative, or zero magnitude. A positive number indicates increase, a negative number decrease, and zero indicates no change.

Nonnegativity condition

If the parameters are positive, from (4.6), the equilibrium price is always positive. But, mathematically, the equilibrium quantities (4.7) may be negative, depending on the magnitude of ad and bc. Negative equilibrium quantity does not make sense in economics. Thus, for this economic model to be meaningful, we have added the **nonnegative condition** $ad - bc > 0$ for $Q^{\#}$ in row 12. In G12, enter

=a*d−b*c_>=0   or   =G6>=0,

where G6 contains formula (4.7). The current value in G6 gives positive $Q^{\#}$, and G12 shows "TRUE". However, if we change c to say, 100, in D6, we will have negative $Q^{\#}$ and G12 will show "FALSE" (see Sec. 2.1.1, Table 2.1). If $Q^{\#}$ is negative, we say that the **model is in error** or is **economically incompatible**, and the model should be reconstructed.

Suggested values

Rows 10 and 11 show the suggested values for the parameters. Although any value can be used, we have chosen the values shown in row 10 to present clearer changes in the positions of the lines in Figs. 4.7 and 4.8. When a value in row 10 is substituted in the "Current" row (row 6), other things being equal, the range B8:E8 ("change") will show the direction and the magnitude of the changes of the parameter, and F8:G8 will show the direction and the magnitude of the corresponding change in the equilibrium values $P^{\#}$ and $Q^{\#}$. If there is no change in parameters, all the changes will be zero.

For example, change a in B6 from 18 to 26, and observe how the equilibrium values change. Then return a to 18, and change b in C6 from 2 to 6, and observe how the equilibrium values change. We change the parameters one at a time and see the changes in the equilibrium values. This is the basic idea of **comparative static analysis** expounded in Sec. 4.1.2.

4.3.2 *The data table*

The data table is the same as the data table we have constructed in Chapter 3, except that now the numbers in the equations and formulas are substituted by the named parameters.

Construction of the data table

The data table in Table 4.1(b) is similar to Table 3.2. We first enter the frame of the table by entering the labels in rows 14–17 and the independent variable P in range A18:A28. The underlining equation for each column is listed in rows 15 and 16 for convenience. The equations are entered in two rows to avoid cluttering. D0, S0, and E0 denote the original (default) equations, that is, the numeric model in (4.1)–(4.3), and D, S, and E denote the current equations, which are the parametric model of (4.1)–(4.3).

Name the independent variable P

Using the name box, give the name P (single name) to the column A18:A28, and also create a name for each column in the range B18:G28 for the name in the top row, B17:G17 (multiple names). Thus, we name the column in range B18:B28 as D0, C18:C28 as S0,..., E18:E28 as DD,..., etc.

Entering the data

This will be a continuation of Table 3.2 in Chapter 3. The first four columns are the same as in Table 3.2. In columns E–G, instead of a numeric model, we enter the **parametric equations** exactly like the parametric model (4.1)–(4.3). The equations in row 18 are as follows:

B18: =18−2*P
C18: =−6 + 6*P — the numeric model
D18: =D0−S0,

E18: =a−b*P,
F18: =−c_+d*P — the parametric model
G18: =DD−S

The computer can recognize that P in row 18 means that P = A18 = 0, and P in row 19 means that P = A19 = 1, etc. Copy B18:G18 to B19:G28; now we have the data table.

4.3.3 *Conditional formatting*

The changes in row 8 may be positive or negative. We may use Excel's conditional formatting to show the sign, and also locate the zeros of the excess demand before and after the changes in the data table. To color the cells and add the direction of changes — an up arrow for increase, a down arrow for decrease, a horizontal arrow for no change or zero — we take the following steps.

Adding the color of changes

Select the "change" row, B8:G8, and holding the control key, continue to select the E0 column, D18:D28, and the E column, G18:G28.[5] Then select

<"Home"><Styles, Conditional Formatting!><Highlight Cells rules>
<Greater Than···>.

[5]Note that you can select different rows and columns by holding the control key and selecting the desired rows or columns.

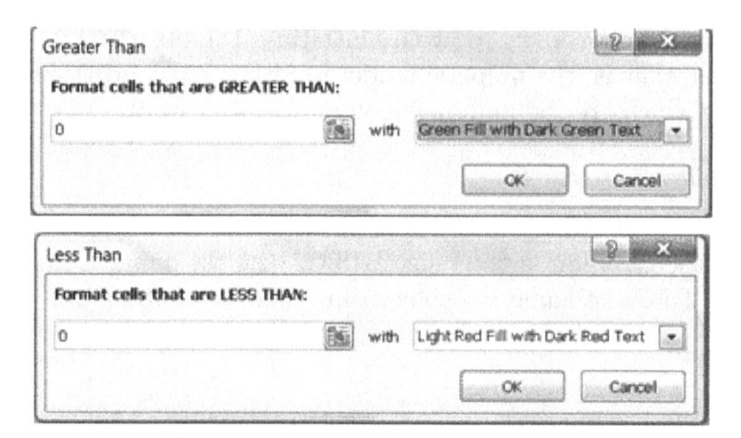

Fig. 4.4 Conditional formatting — highlight cell rules

The "Greater Than" dialog appears (Fig. 4.4, upper part). Enter 0 and select the down arrow (!) <Green fill with Dark Green Text> and <OK>. Repeat the process and select <Less Than···>. The "Less Than" dialog box appears (Fig. 4.4, lower part). Enter 0 and select the down arrow (!) <Light Red Fill with Dark Red Text> and <OK>. The selected three different ranges will be colored, except the cells with 0's.

Adding directional arrows

To add arrows, select the three ranges as above, and then select

<"Home"><Styles, Conditional Formatting!><Icon Sets><More Rules···>.

A dialog box appears (Fig. 4.5). Under <Icon Style:!>, select <3 Arrows (Colored)>. Under "icon", choose the following conditions:

For the up arrow, when the value is "> 0", click <Type!> and select "Number";
For the horizontal arrow, when < = 0 and "> = 0", click <type!> and select "Number";
For the down arrow, when < 0, select <OK>.

Clearing the conditional formatting

You can clear conditional formatting by selecting

<"Home"> <Styles, Conditional Formatting!> <Clear Rules>
<Clear Rules from Selected Cells>

(or from Entire Sheet). You may also change or erase the conditional formatting by choosing <Managing Rules···> instead of <Clear Rules>.

4.3.4 *The chart*

Now we are ready to draw the chart based on the data table.

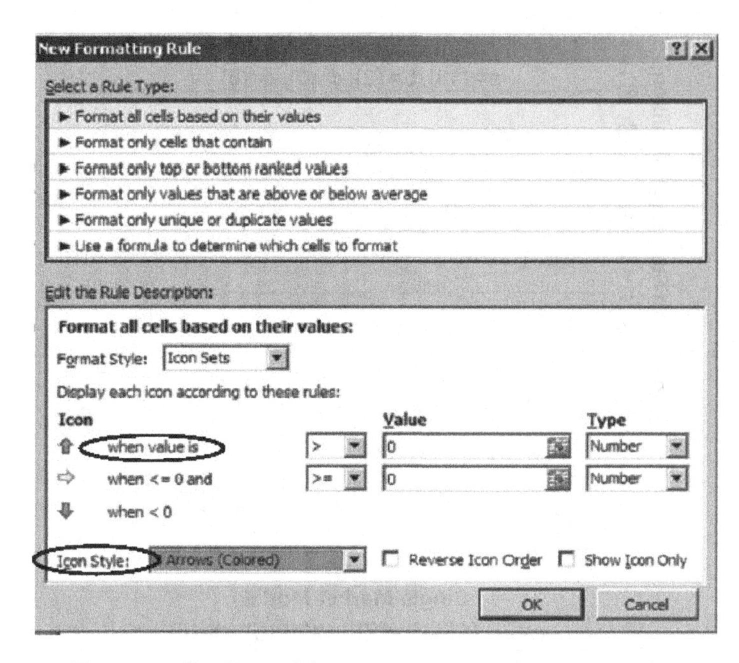

Fig. 4.5 Conditional formatting — new formatting rules

Chart

We would like to draw the diagram as in Fig. 3.12 in the previous chapter. This can be done in several ways. The simplest method is to select the range A17:C28 and E17:F28 (to avoid cluttering, we exclude the excess demand) in Table 4.1(b), and draw the chart. If the current values (row 6) are exactly the same as the default values (row 7), we should have two lines as in Fig. 4.6, which is the same as Fig. 3.12.

At the beginning, in Table 4.1(b), D0 and D, S0 and S have the same values. Therefore, each of the demand curve and the supply curves has two lines overlapping. In the chart title, we enter the values of the parameters in parenthesis to show that they are default values. Eliminate the legend for simplicity. We should have a chart, Fig. 4.6, after editing. This is the prototype or original chart.

Adding data labels

To distinguish the four lines, we want to add data labels to the curves. For this purpose, we have to **separate four lines** and work on each line in turn. Table 4.1(b) has already separated the lines by changing a from 18 to 26 in cell B6, and c from 6 to 30 in D6, and the chart is shown automatically as Fig. 4.7. In this chart, we use a heavy dashed line for the default demand curve D_0, and a heavy dotted line for the supply curve S_0, both without markers. The current D and S use solid lines with markers, triangles (that is, Greek *D*elta Δ) for the *d*emand curve and *s*quares for the *s*upply curve.

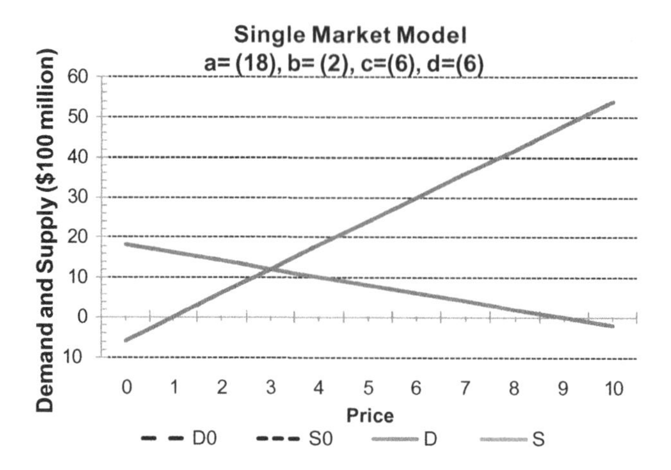

Fig. 4.6 The default market model

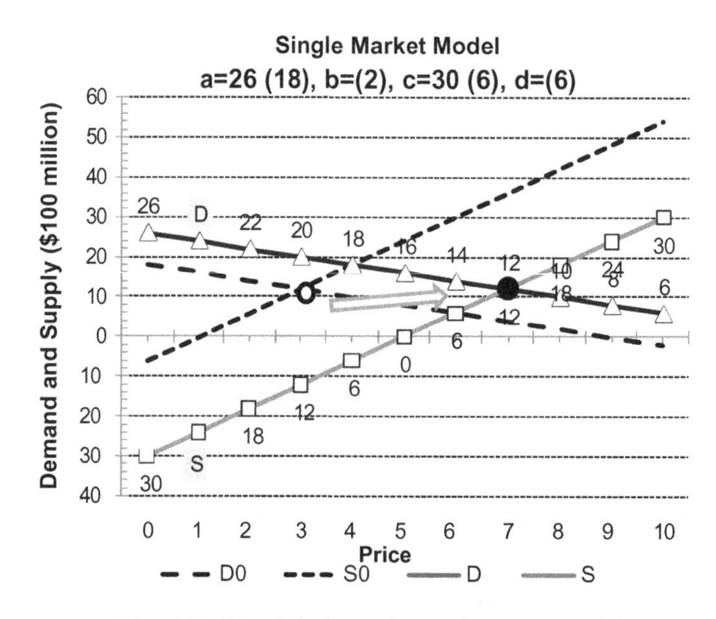

Fig. 4.7 The default market and current model

We also **add data labels** to the current demand and supply curves (see (2) of Sec. 3.7.4 for data labels). The data label position for the demand curve is aligned "above" and that for the supply curve is aligned "below". Then we **edit the data labels** as follows: The data label for P = 1 is changed to D and S with yellow background, respectively, so that when the curves move, the letters also move with the curves. Their label boxes are painted in yellow for easy identification. The final chart should look like Fig. 4.7.

Add a white circle to the original equilibrium point (O for original), a black circle to the new equilibrium point, and add a white arrow to show the direction of change.

4.3.5 *Copying a chart*

Copying and moving a chart is similar to copying and moving a range, which we discussed in Chap. 2, Sec. 2.3.1. Click the chart area and ^C (or ^X) to copy (or move), and then ^V to paste (see Sec. 2.3.1). Copy Fig. 4.7 and paste it below the original chart. In Table 4.1(b), change parameter a back to 18 and c back to 6, we see that both original Fig. 4.7 and copied Fig. 4.7 return to two lines as in Fig. 4.6. This is very inconvenient for comparative static analysis since we want to preserve the different charts for different parameter values, without reconstructing the chart.

The picture-copy method

The method we want to use is called the **picture copy method**. Click <Undo> twice to return to Fig. 4.7 (the first <Undo> returns c to 30, and the second <Undo> returns a to 26) for both charts.

The method is to select the original chart,[6] say, the four-line picture Fig. 4.7, and then <^C>, moving the pointer to a blank space, and

<"Home"><Clipboard, Paste!><As Picture><Paste Special>.

<@Paste: As: Picture(Enhanced Metafile)><OK>

The chart is now pasted as a **picture** for Fig. 4.7. When we change a from 26 back to 18 and c from 30 back to 6 in Table 4.1(b), we have Fig. 4.6, but the picture does not change (since it is a "picture"), it remains as Fig. 4.7.

Editing a picture

When the picture is clicked, there are four small empty circles around the corners and four small empty squares on the four borders, with a green circle above the picture. This will signal you that the figure is a picture. Since it is a picture, its contents cannot be edited, and will not change when the entries in the table change. You may still **move or resize the picture** as usual by dragging the borders or one of the empty squares. The green circle can be used to rotate the picture. When the picture is selected, <"Picture Tools", Format> appears in the menu bar of the spreadsheet for further editing. You can also format the picture by clicking RM to open a drop-down menu. Choose <Format Picture>, which is at the bottom of the menu. Note that color can only be applied to the frame of the picture, not to the inside of the picture. If you want to paint the picture, do so on the original chart before you convert it to a picture.

[6]Either the original chart or copied chart will work. This picture copy method is also applicable to any range of the worksheet, including tables.

4.4 Method of Graphing Comparative Static Changes

Use the horizontal split box to divide the screen into two parts: the upper part shows the parameter table A1:G12, and the lower part shows the data chart, Fig. 4.7.

As shown in Sec. 4.1.2, we have four cases of comparative static analysis. They are

Case (a) Parameter a changes from 18 to 26, denoted as a = 26(18)
Case (b) Parameter b changes from 2 to 6, denoted as b = 6(2)
Case (c) Parameter c changes from 6 to 30, denoted as c = 30(6)
Case (d) Parameter d changes from 6 to 10, denoted as d = 10(6)

The number in the parentheses is the default value. These four cases are derived in Tables 4.2(a)–4.2(d) as special cases of Table 4.1(b).

Case (a) Only parameter a changes

Change the value of parameter a from 18 to 26 in the parameter table. Change the title of the chart to Fig. 4.8(a), and add a = 26(18) to the chart title. Other parameters are the same: b = (2), c = (6), d = (6). In this case, since the slope of the demand curve is

Table 4.2 (a) Comparative static analysis: Case a. (b) The comparative static analysis: Case b. (c) The comparative static analysis: Case c. (d) The comparative static analysis: Case d.

	a	b	c	d	P#	Q#
Curr	26	2	6	6	4	18
Deflt	18	2	6	6	3	12
chg	⇧ 8	⇨ 0	⇨ 0	⇨ 0	⇧ 1	⇧ 6
	Δa				$\Delta P^{\#}$	$\Delta Q^{\#}$

	D0	S0	E0	D	S	E
0	18	-6	⇧ 24	26	-6	⇧ 32
1	16	0	⇧ 16	24	0	⇧ 24
2	14	6	⇧ 8	22	6	⇧ 16
3	12	12	⇨ 0	20	12	⇧ 8
4	10	18	⇩ -8	18	18	⇨ 0

(a)

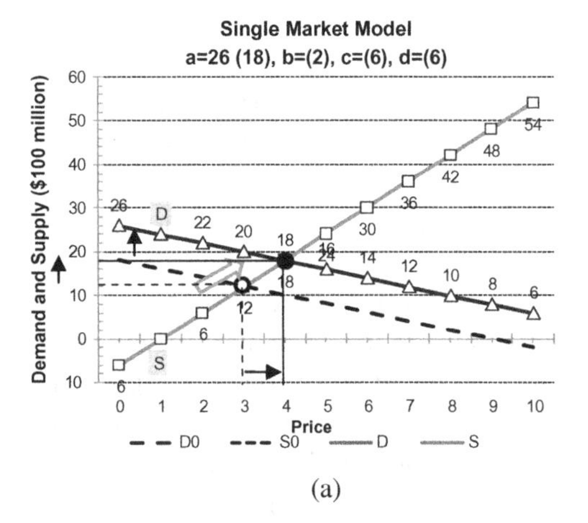

(a)

Fig 4.8 Case a: Comparative static analysis of change in a. (b) Case b: The comparative static analysis of change in b. (c) Case c: The comparative static analysis of change in c. (d) Case d: The comparative static analysis of change in d

Table 4.2 (*Continued*)

	Parameters				Equi Values	
	a	b	c	d	P#	Q#
Curr	18	6	6	6	2	6
Deflt	18	2	6	6	3	12
chg	→ 0	↑ 4	→ 0	→ 0	↓ -1	↓ -6

Δb → $\Delta P^{\#}$ $\Delta Q^{\#}$

	D0	S0	E0	D	S	E
0	18	-6	↑ 24	18	-6	↑ 24
1	16	0	↑ 16	12	0	↑ 12
2	14	6	↑ 8	6	6	→ 0
3	12	12	→ 0	0	12	↓ -12

(b)

	Parameters				Equi Values	
	a	b	c	d	P#	Q#
Curr	18	2	30	6	6	6
Deflt	18	2	6	6	3	12
chg	→ 0	→ 0	↑ 24	→ 0	↑ 3	↓ -6

Δc → $\Delta P^{\#}$ $\Delta Q^{\#}$

	D0	S0	E0	D	S	E
0	18	-6	↑ 24	18	-30	↑ 48
1	16	0	↑ 16	16	-24	↑ 40
2	14	6	↑ 8	14	-18	↑ 32
3	12	12	→ 0	12	-12	↑ 24
4	10	18	↓ -8	10	-6	↑ 16
5	8	24	↓ -16	8	0	↑ 8
6	6	30	↓ -24	6	6	→ 0

(c)

	Parameters				Equi Values	
	a	b	c	d	P#	Q#
Curr	18	2	6	10	2	14
Deflt	18	2	6	6	3	12
chg	→ 0	→ 0	→ 0	↑ 4	↓ -1	↑ 2

Δd → $\Delta P^{\#}$ $\Delta Q^{\#}$

	D0	S0	E0	D	S	E
0	18	-6	↑ 24	18	-6	↑ 24
1	16	0	↑ 16	16	4	↑ 12
2	14	6	↑ 8	14	14	→ 0
3	12	12	→ 0	12	24	↓ -12

(d)

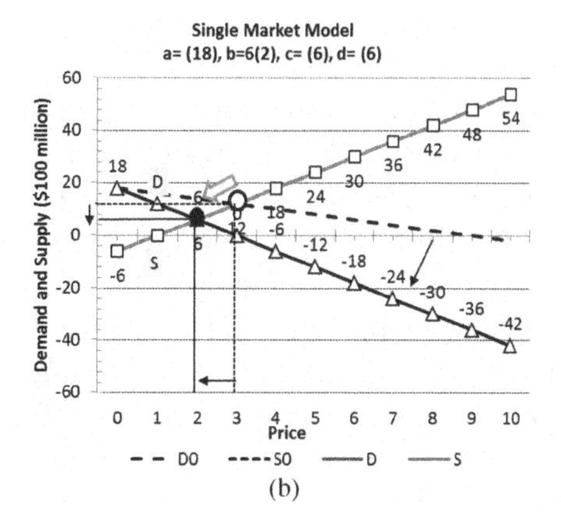

(b)

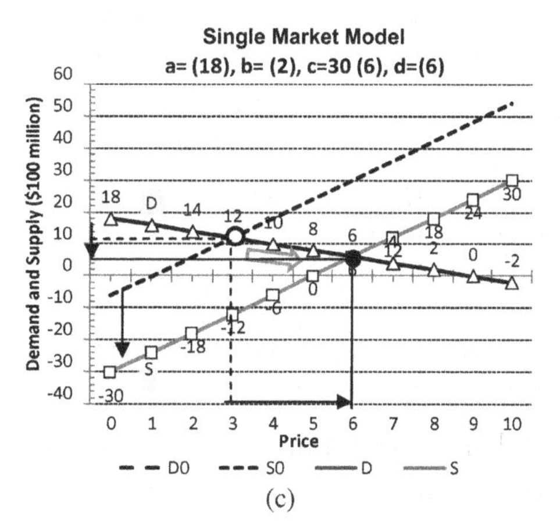

(c)

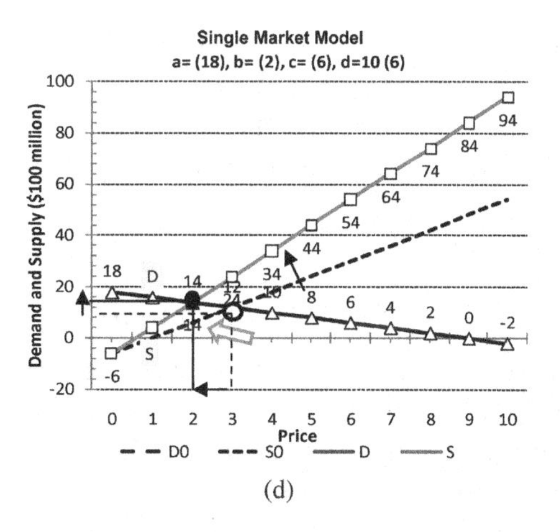

(d)

Figure 4.8 (*Continued*)

held constant, the whole demand curve shifts upward. This is the case if consumer income increases (for normal goods), consumer taste for Q increases, or the price of related goods increases (for substitutes like coffee and tea). Note that this is a shift of the whole demand curve (due to a change in intercept a), and is different from moving along the demand curve when the price changes. See the D curve in Fig. 4.8(a).

The algebraic effect of the changes in parameter table for case (a) has been given in Example 4.3. In addition to the parameter table, the impact can also be seen from the **data table** in Table 4.2(a). $E_0 = 0$ shows the row of the original equilibrium, and $E = 0$ shows the row of the new equilibrium. The up arrows and down arrows show how the values of the function change at each price, and the changes can be traced in the graph. **Graphically**, the fat white arrow indicates the direction of a change: In this case, both equilibrium price and quantity change upward (the qualitative effect). To preserve the results, picture copy Fig. 4.8(a) to Sheet2.

Four methods of showing comparative statics

In general, we have four ways to show comparative statics: the parameter table (most informative) either in levels or in differences; impact vector in levels or differences; the data table (only the equilibrium results are shown); and the chart (visually appealing). Each has its own characteristics.

Case (b) Only parameter b changes

Suppose b increases from 2 to 6. This means that, originally, when price increased by 1 unit (say, one dollar), other things being equal, the demand would decrease by 2 units. But now, when price increases by one unit, the demand will decrease by 6 units along the demand curve, holding a constant. This is due to change in the consumer's evaluation or taste of the value of the commodity for any given price. The commodity is now less valuable or likable than before. The new demand curve will tilt downward at intercept a, and will be steeper, as shown in Fig. 4.8(b).

The algebraic effect of the changes in parameter table for case (b) has been given in Example 4.4. The new equilibrium values can be found in the data table at $E = 0$ in Table 4.2(b).

Case (c) Only parameter c changes

In this case, parameter c increases from 6 to 30 and the whole supply curve shifts to the right. This is the case if costs of production, like wages, increase, or government taxes increase, or safety regulations or environmental regulations increase, so that the supply decreases at each price level. This is a shift of the whole supply curve to the right, due to an absolute increase (or algebraic decrease) in intercept c, see Fig. 4.8(c), and is different from a change in quantity along the supply curve when the price changes.

The algebraic effect of the changes in parameter table for case (c) has been given in Example 4.5. The results are also shown in the data table at $E_0 = 0$ and $E = 0$. The old and new equilibrium points can be compared in Fig. 4.8(c). Picture copy it to Sheet2, and rearrange the three pictures so they do not overlap each other.

Case (d) Only parameter d changes

In this case, parameter d changes from 6 to 10. This means that originally, when price increased by 1 unit (say, 1 dollar), other things being equal, supply increased by 6 units, but now, supply will increase by 10 units. Since c is held constant, the supply curve will tilt upward at intercept c, see Fig. 4.8(d). This may be the case because of new technology or increased efficiency, which enables the producer to supply more at any given price.

The algebraic effect of the changes in parameter table for case (d) has been given in Example 4.6. The new equilibrium values can be found in the data table at $E = 0$ in Table 4.2(d). The results are also shown in the data table and the original and new equilibrium points can be compared in Fig. 4.8(d). Copy Fig. 4.8(d) to Sheet2, rearrange the four pictures in one page of the sheet, and print out the page.

Example 4.8 Other suggested values of a. If a changes from 18 to 24 (instead of 26 as in the text above), the new equilibrium values are

$$(\mathrm{a}; \mathrm{P}^{\#}, \mathrm{Q}^{\#}) = (18; 3, 12) \rightarrow (\mathrm{a}; \mathrm{P}^{\#}, \mathrm{Q}^{\#})' = (24; 3.75, 16.5);$$

or, in terms of increments,

$$(\Delta \mathrm{a}; \Delta \mathrm{P}^{\#}, \Delta \mathrm{Q}^{\#}) = (6; 0.75, 4.5).$$

In this case, zero will not be shown in the E column of Table 4.1(b), only the color changes from green to red between $P = 3$ and 4, indicating that the equilibrium price lies between these two values. However, we still can find the intersection of the new demand and supply curves in the chart. To find the exact value of $P^{\#}$ algebraically, use the excess demand function (4.8), or substitute the four parameters in (4.6) and (4.7). □

Example 4.9 Other suggested values. Experiment with other possible values of parameters. In Table 4.1(b), row 11 lists the other suggested values, which are less than the default values (that is, the parameter values decrease). Construct the four comparative static analysis diagrams and state the qualitative and quantitative effects for each case. Do the effects make economic sense? Why and why not. □

Example 4.10 Change two parameters at a time. In suggested values of parameters in row 10 (or row 11), change two parameters at the same time (like Fig. 4.7) and see how the four curves interact. There are 6 possibilities. Try to change a and b, a and c, a and d, b and c, b and d, and c and d. Give an economic interpretation of the changes in the equilibrium values in each case. □

Example 4.11 Economic interpretation of parameter changes. In this section and Sec. 4.1, we have seen the effect of a change in a parameter on the equilibrium values. This is the definition of comparative static analysis. Why do the parameters change? More specifically, what is the economic meaning of a change in parameters a, b, c, and d?
Answer: The economic meaning of a change in parameters a, b, c, and d is already explained at the beginning of the discussion of each case in this section. □

4.5 Comparative Static Analysis of A National Income Model

The comparative static analysis of the national income model may proceed in the same way as the market model.

4.5.1 *A national income model*

For convenience, we will continue to use the national income model presented in Sec. 3.8.1. Renumbering the equations, we have

Numeric Model	Parametric Model		
$AD = C + I + G,$	$AD = C + I + G,$		(4.21)
$C = 0.3 + 0.6Y,$	$C = a + bY$	$a > 0,\ 1 > b > 0$	(4.22)
$I = 0.7$	$I = I_0$	$I_0 \geq 0$	(4.23)
$G = 0.8$	$G = G_0$	$G_0 \geq 0$	(4.24)
$AS = Y$	$AS = Y$		(4.25)
$AD = AS$	$AD = AS$		(4.26)

The numeric model will be taken as the default model, and the parametric model as the current model. As in the market model, the comparative static analysis is to change a parameter in the parametric model one by one and study the effect of the change on the equilibrium values of the model. The procedure is the same as in the market model, and is described briefly as follows.

Comparative static analysis

As we have seen in Chap. 3, the six variables (AS, AS, Y, C, I, G) in six equations can be reduced to two variables (Y, C) in two equations, with two exogenous variables (I_0, G_0) and two parameters (a, b). The equilibrium income and consumption are, from Sec. 3.8.1,

$$Y^{\#} = k(a + I_0 + G_0), \tag{4.27}$$

$$C^{\#} = k(a + b(I_0 + G_0)). \tag{4.28}$$

where $k = 1/(1 - b)$, the investment multiplier. They are expressed in terms of the two exogenous variables and two parameters. Thus, we should have eight ($= 2 \times 4$) comparative statics cases for this national income model. Each equilibrium value can be changed in four ways by changing one of a, b, I_0, and G_0. They are, using similar calculations to those in (4.9) to (4.20):

$$\Delta Y^{\#} = k\Delta x > 0, \quad \text{for } x = a, I_0, G_0, \quad \text{for } \Delta x > 0 \tag{4.29}$$

$$\Delta Y^{\#} = k_0 k_1 (a + I_0 + G_0)\Delta b \cong kY^{\#} > 0, \quad \text{for } \Delta b > 0 \tag{4.30}$$

$$\Delta C^{\#} = bk\Delta x > 0, \quad \text{for } x = I_0, G_0, \quad \text{for } \Delta x > 0 \tag{4.31}$$

$$\Delta C^{\#} = k\Delta a > 0, \quad \text{for } \Delta a > 0 \tag{4.32}$$

$$\Delta C^{\#} = k_0 k_1 (a + b(I_0 + G_0))\Delta b \cong kY^{\#} > 0, \quad \text{for } \Delta b > 0 \tag{4.33}$$

where, for simplicity, x represents the variables indicated, $k_0 = 1/(1-b_0)$ and $k_1 = 1/(1-b_1)$. Equations (4.30) and (4.33) can be derived by using a method similar to that for deriving (4.14) and (4.15). Note that $k_0 \cong k_1$ if $b_0 \cong b_1$. In this case, k_1 can be approximated by $k_0 = k$.

Equation (4.29) states that qualitatively, an increase in fixed investment I_0 (or basic level of consumption a, or government expenditure G_0) will increase $Y^{\#}$, and quantitatively, (4.29) states that one unit increase in fixed investment I_0 (or basic consumption a, or government expenditure G_0) will increase $Y^{\#}$ by a magnitude of the multiplier k, say $k = 2.5$ if $b = 0.6$. As shown in Fig. 4.9, Case (a), the change in $Y^{\#}$ is manifested in the diagram as the result of a shift of the AD line.

Equation (4.30) or (4.33) shows that a 1% increase in marginal propensity to consume (MPC) (say, b changes from 0.6 to 0.606), will increase the equilibrium income or equilibrium consumption approximately by the amount of $kY^{\#}(0.01)$ or $kC^{\#}(0.01)$ units, respectively.

All the changes in this case are positive. This means that an increase in a parameter (a or b) or an exogenous variable (I_0 or G_0) will increase the equilibrium values.

4.5.2 *The parameter table*

Starting from a new workbook, you may copy and revise Table 3.3(b) of the static analysis to the new workbook,[7] or construct a new table, as in Table 4.3. Construct the parameter table as in rows 2–7 in Table 4.3. Enter the values of parameters in B4:E4. Name B4:E4 and H4 using the names in row 3. Enter the equilibrium income (4.27) in cell F4 and the equilibrium consumption (4.28) in cell G4. In cell H4, enter $= 1/(1 - b)$. The "change" in row 6 is the difference between the current value and the default value. The default values are taken from Table 3.3 in the previous chapter. The changes show the direction

[7]Note again that, in Excel, **every workbook can have only one range name among the worksheets.** Thus, if a and b are already defined in the previous worksheet which contains the market model, then a and b in row 2 must be changed to another name, say, aa and bb, to distinguish them from a and b in the market model. We prefer to use aa or aaa to make the entry of the name easier and recognizable.

Table 4.3 Comparative static analysis — national income model

	A	B	C	D	E	F	G	H
1	**A National Income model**							
2	**A. Parameter Table**							
3		a	b	I0	G0	Y#	C#	k
4	Current	0.3	0.6	1.1	0.8	5.5	3.6	2.5
5	Default	0.3	0.6	0.7	0.8	4.5	3	2.5
6	Chg	⇨0.0	⇨0.0	↑0.4	⇨0.0	↑1.0	↑0.6	⇨0.0
7	Sugg	1.5	0.7	1.1	1.6			
8								
9	**B. The Data Table**							
10	Y	=0.3+0.6Y		=AD0-AS	=a+bY	=C+I+G	=Y	=AD-AS
11		C0	AD0	ED0	C	AD	AS	ED
12	0.0	0.3	1.8	↑1.8	0.3	2.2	0.0	↑2.2
13	0.5	0.6	2.1	↑1.6	0.6	2.5	0.5	↑2.0
14	1.0	0.9	2.4	↑1.4	0.9	2.8	1.0	↑1.8
15	1.5	1.2	2.7	↑1.2	1.2	3.1	1.5	↑1.6
16	2.0	1.5	3.0	↑1.0	1.5	3.4	2.0	↑1.4
17	2.5	1.8	3.3	↑0.8	1.8	3.7	2.5	↑1.2
18	3.0	2.1	3.6	↑0.6	2.1	4.0	3.0	↑1.0
19	3.5	2.4	3.9	↑0.4	2.4	4.3	3.5	↑0.8
20	4.0	2.7	4.2	↑0.2	2.7	4.6	4.0	↑0.6
21	4.5	3	4.5	⇨0.0	3.0	4.9	4.5	↑0.4
22	5.0	3.3	4.8	↓-0.2	3.3	5.2	5.0	↑0.2
23	5.5	3.6	5.1	↓-0.4	3.6	5.5	5.5	⇨0.0
24	6.0	3.9	5.4	↓-0.6	3.9	5.8	6.0	↓-0.2
25	6.5	4.2	5.7	↓-0.8	4.2	6.1	6.5	↓-0.4
26	7.0	4.5	6.0	↓-1.0	4.5	6.4	7.0	↓-0.6
27	7.5	4.8	6.3	↓-1.2	4.8	6.7	7.5	↓-0.8
28	8.0	5.1	6.6	↓-1.4	5.1	7.0	8.0	↓-1.0
29								
30		Numeric model			Parametric model			

and the magnitude of changes in parameters and the equilibrium values numerically. As in Table 4.1(b), each number in row 7 of "Suggested values" is to be entered one by one into the "Current row", row 4.

For clarity, models (4.21)–(4.26) are listed in row 10 of Table 4.3. They may be listed in a column next to the table. It is a good practice to show the contents and purpose of the table.

4.5.3 *The data table*

In Table 4.3, enter the value of Y in A12:A28 using the fill-handle box. Here Y plays the role of the independent variable P in the market model. Name the range A12:A28 as Y and

E12:E28 as C. Enter the formula in row B12:H12 and copy the row to B13:H28. We have Table 4.3. The numbers are entered with one decimal.

In Table 4.3, the current value for I_0 is the suggested value, 1.1. Change it to the default value of 0.7, and find that the equilibrium income and consumption are 4.5 and 3, respectively, and that k = 2.5. These are the default equilibrium values. Enter these numbers in F5:H5. Make sure that the change in the equilibrium values are zero.

Since we want to hold constant the original AD and AS curves, we need to change the contents in the column of AD_0 to values (no formula underneath the numbers), so that AD_0 will not change when I_0 and G_0 change. **To change a range to values**, select C12:C28, then select

<"Home"><Clipboard, Copy><Paste Special><Values><OK>.

To check that the formulas are erased, click the cells in the column. A formula should not appear in the formula box.

Conditional formatting

As in Table 4.1(b), select the "change" row, B6:H6, the ED0 column, D12:D28, and the ED column, H12:H28, and apply conditional formatting to these three ranges as described in Sec. 4.3.2, Figs. 4.4 and 4.5.

4.5.4 *Drawing the charts*

To draw the chart, let a = 1.5, say, in B4 of Table 4.3, so that all the lines are distinguishable. Select A11:H28 in Table 4.3. Make the chart like case a in Fig. 4.9. Reformat the lines so

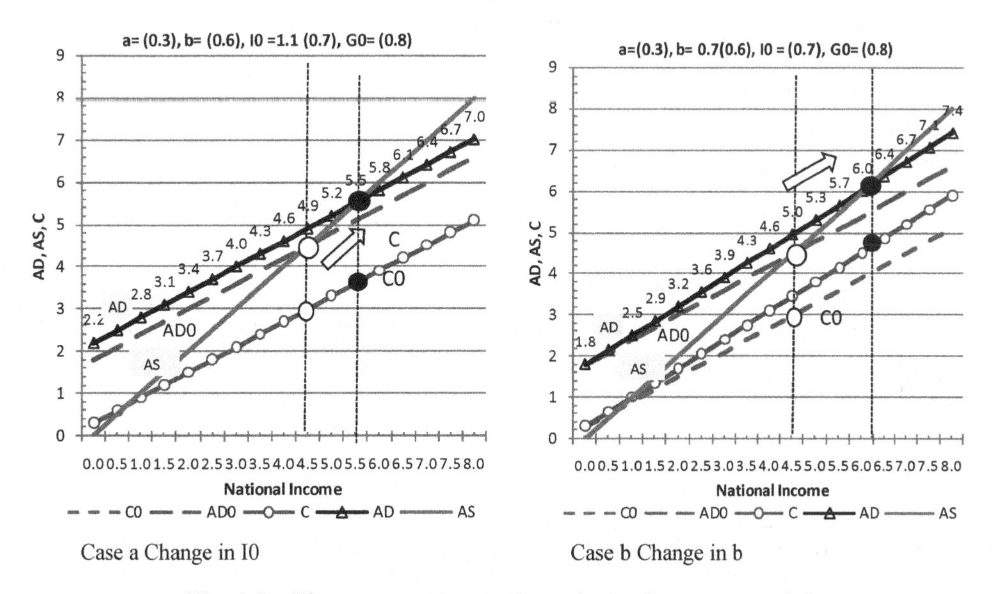

Fig. 4.9 The comparative static analysis of a macro model

that the default consumption C_0 is a dotted line, the default aggregate demand AD_0 is a dashed line, the current aggregate demand AD is a solid line with triangle markers, and the current consumption C is a solid line with circle markers. The aggregate supply curve AS is a solid line without markers. Add small data labels to the current AD line and change the series label at 0.5 to AD and AS with yellow backgrounds. Draw dotted lines[8] through the default and current Keynesian crosses; you should draw the dotted lines before drawing circles. Add empty circles O (O for *O*riginal point) to the default Keynesian cross and the default equilibrium consumption, and filled black circles to the current Keynesian cross and the current equilibrium consumption. Place the legend box at the bottom.

4.5.5 *Performing comparative static analysis*

Starting from the four default values (in row 5 of Table 4.3) as the current values in row 4 of Table 4.3, we now experiment with the suggested values (in row 7) of the parameters one by one (other things being equal). We have four cases of comparative static analysis. They are

Case (a) Fix investment I_0 changes from 0.7 to 1.1 denoted as $I_0 = 1.1(0.7)$
Case (b) MPC b changes from 0.6 to 0.7, denoted as $b = 0.7(0.6)$
Case (c) Parameter a changes from 0.3 to 1.5, denoted as $a = 1.5(0.3)$
Case (d) Government expenditure G_0 changes from 0.8 to 1.6, denoted as $G_0 = 1.6(0.8)$

We examine how the equilibrium point changes in the table and chart, compared with the default equilibrium point. Note that, as usual, the excess demand (ED) column in Table 4.3 is used to locate the equilibrium income. It is either exactly at the income level, or between the two levels of income, where the excess aggregate demand changes to the excess aggregate supply (shown as negative ED). Since the variables are continuous, but we can only show the values of the variable in A12–A28 as an interval, the equilibrium income often falls between the two income levels.

Case (a) Impact of change in investment, I_0

The change in investment probably is the most important topic of comparative static analysis in this model. Investment may change due to changes in income, consumption, interest rates, expectation of the future economy, business conditions, etc.

Now, from the parameter table of Table 4.3, we see the effect of a change in the investment I_0 from \$0.7 trillion to \$1.1 trillion, as recommended in the "suggest" row in Table 4.3. The equilibrium income $Y^{\#}$ increases from \$4.5 trillion to \$5.5 trillion, and equilibrium consumption changes from \$3 trillion to \$3.6 trillion, which are shown in range F4:G4 of the parameter table. As in Sec. 4.4, the impact can be expressed in the **impact**

[8]In the new Excel 2007, it is extremely difficult to draw a vertical line through the Keynesian cross accurately, because it has lost the ability of "Nudging" that Excel 2003 had. The readers may skip drawing the vertical lines.

vector in levels as

$$(I_0; Y^{\#}, C^{\#}) = (0.7; 4.5, 3) \rightarrow (I_0; Y^{\#}, C^{\#})' = (1.1; 5.5, 3.6).$$

Equivalently, in Table 4.3, from the "change" row, row 6 of the parameter table, we see that an extra \$0.4 trillion of fixed investment induces an extra \$1.0 trillion of equilibrium income (F6), and also an extra \$0.6 trillion of equilibrium consumption (G6). The results are predicted by comparative static analysis in equations (4.29) and (4.31) as

$$\Delta Y^{\#} = k\Delta I_0 = (2.5)(0.4) = 1.0,$$
$$\Delta C^{\#} = bk\Delta I_0 = (0.6)(2.5)(0.4) = 0.6.$$

To express in the **impact vector in differences**, we can write

$$(\Delta I_0; \Delta Y^{\#}, \Delta C^{\#}) = (0.4; 1.0, 0.6).$$

In the **data table** of Table 4.3, the original equilibrium values are located at $ED_0 = 0$ at $Y^{\#} = 4.5$. After I_0 changes from 0.7 to 1.1, AD changes, and the new equilibrium income and consumption are located at ED = 0 at $Y^{\#} = 5.5$. Graphically, we observe in Fig. 4.9(a) the change in the position of the AD line as compared with the original line. The original equilibrium income and consumption are marked by white circles. We locate the new equilibrium income and consumption, and denote them by black-filled circles. We then compare the changes in income and consumption before and after the changes.

Case (b) Impact of change in marginal propensity to consume, b

Returning to the default value of $I_0 = 0.7$ in cell D4, we now change the MPC b from 0.6 to 0.7. MPC changes means that the consumer will spend higher proportion of extra income due to change in taste, increases in wealth (bullish stock markets), improved financial institution (like availability of credit cards), better welfare system (like consumers do not need to save for old age), etc.

The income multiplier increases from 2.5 to 3.33, the equilibrium income, $Y^{\#}$, increases from \$4.5 trillion to \$6 trillion, and equilibrium consumption, $C^{\#}$, changes from \$3 trillion to \$4.5 trillion, which should be shown in cells F4:G4. The **impact vector in levels** is

$$(b; Y^{\#}, C^{\#}) = (0.6; 4.5, 3) \rightarrow (b; Y^{\#}, C^{\#})' = (0.7; 6.0, 4.5).$$

Thus, an increase of MPC by an extra 0.1 will induce an extra \$1.5 trillion of equilibrium income and also an extra \$1.5 trillion of equilibrium consumption, which may also be calculated from (4.30) and (4.33) as

$$\Delta Y^{\#} = \Delta C^{\#} = kY^{\#}\Delta b = (2.5)(6.0)(0.1) = 1.5.$$

The results are again predicted by comparative static analysis. The **impact vector in differences** is:

$$(\Delta b; \Delta Y^{\#}, \Delta C^{\#}) = (0.1; 1.5, 1.5).$$

In Fig. 4.9(b), the magnitude of change is small, the upward shift of the AD line is barely visible. Change b to 0.75 or 0.8 and observe the change in the slope of the AD line, and compare it with the original line. Locate the new equilibrium income and consumption and indicate them by black-filled circles.

Case (c) Impact of change in a

The parameter a can change also. Since $C = a + bY$, the parameter a is the amount of consumption when Y is zero. In economics, a is called the **basic consumption** or **subsistence level of consumption**. Thus, "a" may increase due to increase in the subsistence level of consumption due to inflation, increase in older population, etc.

We can change a from 0.3 to 1.5, other things being equal, and we obtain a graph similar to Fig. 4.9(a). The new parameter table will show the **impact vector of levels:**

$$(a; Y^{\#}, C^{\#}) = (0.3; 4.5, 3) \to (a; Y^{\#}, C^{\#})' = (1.5; 7.5, 6.0).$$

Cases (d) Impact of change in G_0

Lastly, the government expenditure G_0 may change due to inflation, fighting recession, increased social welfare expenditure, economic construction, etc. The chart will show the effects of doubling G_0 from 0.8 to 1.6 on $Y^{\#}$ and $C^{\#}$. From the parameter table (not shown), the impact vector of levels is (see HW4–14)

$$(G_0; Y^{\#}, C^{\#}) = (0.8; 4.5, 3) \to (G_0; Y^{\#}, C^{\#})' = (1.6; 6.5, 4.2),$$

A concluding observation

From the charts in Fig. 4.9, the policy implication of the above analysis is obvious. If there is a deflationary gap (recession) in the economy, the government should increase public expenditure G_0, encourage investment I_0, increase the basic consumption a, or marginal propensity to consume b. If there is an inflationary gap, the policy should be reversed.

Example 4.12 Other cases. In Table 4.3, we have chosen the current values of parameters and exogenous variables such that the values of equilibrium income and consumption fall exactly on the value of Y shown in the Y column, so that $ED_0 = 0$ and $ED = 0$ will show on the table. Do the comparative static analysis when the current parameters and exogenous variables decrease relative to the default values in Table 4.3. Choose the current values of the parameters and exogenous variables such that the equilibrium value of Y falls in the range between the two values of Y defined in Table 4.3.

For each experiment, you may use <Undo> to return to the original values.

Answer: Use Table 4.3 as a template and experiment with different values of parameters and exogenous variables. For example, let a = 0.15, b = 0.5, $I_0 = 0.2$, and $G_0 = 0.5$. □

4.6 Summary

This chapter introduces parametric models of single market price and quantity determination and the Keynesian theory of income determination. In Sec. 4.1, we first use a parametric market model as an example to show the intercepts of demand and supply curves, the excess demand equation, and the equilibrium price and quantity in terms of parameters. We then define **comparative static analysis** as the study of the **effects of changes** in a parameter (or an exogenous variable if it is present) on the equilibrium variables which are expressed in parameters (and exogenous variables). The effects have two parts, **qualitative and quantitative**. These two effects are illustrated by using the **discrete** method. In the case of the continuous method, we use partial derivatives, but the knowledge of calculus is not required in this book and the method is not explained.

The changes of parameters and exogenous variables and their effects on equilibrium variables can be dealt best with by using the **naming method of Excel**. Thus, in Sec. 4.2, we introduce Excel's naming method. There are slightly different methods of naming a range, depending on whether we are defining a **single name** or **multiple names**.

The Excel tables for comparative static analysis consist of two parts: the **parameter table**, which defines the parameter and equilibrium values, the **data table**, which consists of a numerical model which is fixed, and the parametric model, the entries of which change when parameters change. To trace the changes, we introduce the method of **conditional formatting** in Excel in Sec. 4.3.3. The **charts** are then based on the data table, using the Excel method of graphs and graph formatting techniques. After a chart is drawn, it is **picture-copied** and put aside or moved to a different sheet. This is explained in detail using the market model in Secs. 4.3.5 and 4.4, and is also applied to a national income model in Sec. 4.5.5.

Review of Basic Equations and Formulas

A market model

(4.1) $D = a - bP$, $a > 0$, $b > 0$
(4.2) $S = -c + dP$, $c > 0$, $d > 0$
(4.3) $D = S$

Static Analysis

(4.6) (4.7) Equilibrium price and quantity

$$P^{\#} = \frac{a+c}{b+d} \quad Q^{\#} = \frac{ad-bc}{b+d}$$

Comparative Static Analysis

(a) Effect of change in a (4.10), (4.11)

$$\Delta P^{\#} = \frac{\Delta a}{b+d}, \ \Delta Q^{\#} = \frac{\Delta a}{b+d}$$

Quantitative and qualitative effects

(4.12) $(a; P^{\#}, Q^{\#}) \rightarrow (a; P^{\#}, Q^{\#})'$

(4.13) $(\Delta a; \Delta P^{\#}, \Delta Q^{\#})$

(b) Effect of change in b (4.15), (4.16)

$$\Delta P^{\#} = -\frac{(a+c)\Delta b}{(b_0+d)(b_1+d)},$$

$$\Delta Q^{\#} = \frac{-(a+c)d\Delta b}{(b_0+d)(b_1+d)}$$

Quantitative and qualitative effects (similar to Case (a))

(c) Effect of change in c (4.17), (4.18)

$$\Delta P^{\#} = \frac{\Delta c}{b+d}, \quad \Delta Q^{\#} = \frac{-b\Delta c}{b+d}$$

Quantitative and qualitative effects

(d) Effect of change in d (4.19), (4.20)

$$\Delta P^{\#} = \frac{-(a+c)\Delta d}{(b+d_0)(b+d_1)},$$

$$\Delta Q^{\#} = \frac{(a+c)b\Delta d}{(b+d_0)(b+d_1)}$$

Quantitative and qualitative effects

Non-negativity condition

$ad - bc \geq 0$

A national income model

(4.21) $AD = C + I + G$

(4.22) $C = a + bY$

(4,23) $I = I_0$

(4.24) $G = G_0$

(4.25) $AS = Y$

(4.26) $AS = AD$

Static Analysis

Equilibrium income and consumption

(4.27) $Y^{\#} = k(a + I_0 + G_0)$,

(4.28) $C^{\#} = k(a + b(I_0 + G_0))$,

where $k = 1/(1-b)$

Comparative Static Analysis

(a) Effect of change in a, I_0, G_0 on $Y^{\#}$

(4.29) $\Delta Y^{\#} = k\Delta x$, for $x = a, I_0, G_0$

$(I_0; Y^{\#}, C^{\#}) \rightarrow (I_0; Y^{\#}, C^{\#})'$

$(\Delta I_0; \Delta Y^{\#}, \Delta C^{\#})$

Effect of change in b on $Y^{\#}$

(4.30) $\Delta Y^{\#} = k_0 k_1 (a + I_0 + G_0)\Delta b \cong kY^{\#}$

where

$k_0 = 1/(1-b_0)$, $k_1 = 1/(1-b_1)$

(b) Effect of change in I_0, G_0, a, b on $C^{\#}$

(4.31) $\Delta C^{\#} = bk\Delta x$, for $x = I_0, G_0$

(4.32) $\Delta C^{\#} = k\Delta a$

(4.33) $\Delta C^{\#} = k_0 k_1 (a + b(I_0 + G_0))\Delta b \cong kY^{\#}$

Key terms: Economics

Key terms: Excel

Homework Chapter 4

4-1. Static analysis and comparative static analysis

(a) *Static Analysis*
Draw the chart of the following numeric market model.

$$D = 24 - 2P, \quad S = -3 + 7P, \quad D = S,$$

where price P ranges from 0.1, ..., 10. From the chart, what is the equilibrium price and quantity? (Answer in a text box.)
(Hint: You may use Table 4.1(b)).

(b) *Comparative Static Analysis*

The parametric model for comparative static analysis of the above market model can be written as

$$D = a - bP, \quad S = -c + dP, \quad D = S,$$

where a, b, c, and d are parameters representing the corresponding numbers in the numeric model in part (a). Using the numbers in the numeric model as the default values, draw the chart of the comparative static analysis for the following two cases.

The parameter table should consist of four rows, parameter, current, default, and difference. (Hint: Use the following tables.)

Parameter Table

Parameter	a	b	c	d	P#	Q#
Current						
Default						
Difference						
Data Table	D0	S0	E0	D	S	E

You first draw four lines differently, by giving different values to a and c, say 40 and 20, respectively, as shown in Fig. 4.7. (To receive full credit, you should reproduce all the chart elements of Fig. 4.7 exactly.)

Case 1. Other things being equal, parameter a changes from 24 to 30. From the parameter table what is the equilibrium price and quantity? Express them in impact vector in levels and in differences:

$$(\mathrm{a}; \mathrm{P}^{\#}, \mathrm{Q}^{\#}) = (?; ?, ?) \rightarrow (\mathrm{a}; \mathrm{P}^{\#}, \mathrm{Q}^{\#})' = (?; ?, ?),$$
$$(\Delta\mathrm{a}; \Delta\mathrm{P}^{\#}, \Delta\mathrm{Q}^{\#}) = (?; ?, ?).$$

(Answer in a text box.)

Case 2. Other things being equal, parameter b changes from 2 to 5, and parameter d changes from 7 to 5. From the parameter table what is the equilibrium price and quantity? Express them in impact vector in levels and differences:

$$(\mathrm{b}, \mathrm{d}; \mathrm{P}^{\#}, \mathrm{Q}^{\#}) = (?; ?, ?) \rightarrow (\mathrm{b}, \mathrm{d}; \mathrm{P}^{\#}, \mathrm{Q}^{\#})' = (?; ?, ?),$$
$$(\Delta\mathrm{b}, \Delta\mathrm{d}; \Delta\mathrm{P}^{\#}, \Delta\mathrm{Q}^{\#}) = (?; ?, ?).$$

(Answer in a text box below each question)

(c) *Conditional formatting*

Use the three arrows as in Table 4.1(b), with a green cell background for up and a red cell background for down, to show the differences.

(d) *Charting*

You should have one table only (that is, you should use the naming method) and at least two comparative statics charts (use the picture copy command), in addition to the original chart. The charts should be similar to Fig. 4.8 in their formatting. The charts must have a white background, light dotted grid lines, and the legend at the bottom of the chart.

Note: In the table of part (a), you need not use the range name. If you do, you have to give the new name of the current values in question (b), like aa, bb, etc.

The charts should contain the following chart elements.
D0, in a heavy dash line,
S0, in a heavy dotted line,
D, in a light red solid line with triangle data labels below the line.

4-2 In the **nonlinear numeric model** of HW3-3:

$$\mathrm{D} = 4 - \mathrm{P}^2,$$
$$\mathrm{S} = 4\mathrm{P} - 1,$$
$$\mathrm{D} = \mathrm{S}, \quad \text{and } \mathrm{P} = 0, 0.2, \ldots, 3.$$

convert the model to a parametric model. Note that the coefficient of P^2 is 1, and should be denoted by a parameter, say b.

(a) Increase the constant term a of the demand equation by one unit and find the new equilibrium values of the variables. Express the impact in vector form in levels and differences. Give an economic interpretation of the change. What factors change a?
(b) Increase the coefficient b of the price in the demand equation by one unit and find the new equilibrium values of the variables. What factors will change b?

4-3 The demand and supply curves (static analysis) Begin with a market model of apples:

$$\begin{aligned} D &= a - bP & D0 &= 10 - 2P \\ S &= -c + dP & S0 &= -5 + 3P \\ D &= S & D0 &= S0, \end{aligned}$$

where the default value of the parameters are $a = 10$, $b = 2$, $c = 5$, and $d = 3$. $P = 0, 0.5, \ldots, 10$.

(a) Using the naming method, construct the parameter and data tables, as we have done in the class. The table should be properly labeled. Make sure that you enter the rows of current values and default values.
(b) Draw the demand curve and the supply curve for the default values. What are the equilibrium price, the equilibrium supply, and the equilibrium demand? Use a text box to answer on the graph. The text box should not have borders, and its font should be size 8, in italics. Reproduce the chart format exactly as shown in Figure 4.7. Note that the borders of the chart area have been removed.
(c) Answer the following questions in the text box:
   i. How many variables are in this model? What are they?
   ii. How many endogenous variables are in this model? What are they?
   iii. Why are they called endogenous variables? (In what sense they are "endogenous?")
   iv. How many parameters are in this model? What are they?
   v. Why are they called "parameters"?
   vi. Why are D#, S#, and P# called the equilibrium values? (In what sense they are "equilibrium"?)
   vii. Why is the coefficient of P in the demand equation negative (what is the economic law)?
   viii. Why is the coefficient of P in the supply function positive (what is the economic law)?
(d) In the table, color the equilibrium price and quantities with light blue, and in the chart, denote the equilibrium price and quantity with a vertical dotted line. Make sure that the dotted line moves with the chart.

4-4 Comparative static analysis (This part continues from HW4-3.) In the above question,

(a) Change a to 20, and c to 10. What are the new equilibrium values of price and quantities? Show the impact using the impact vectors in levels and differences.
(b) Answer the following questions. Write the answer below each question.
   i. What makes "a" increase? (Give an economic example under which "a" may increase.)

ii. What are the qualitative effects of an increase in a on the equilibrium price and quantity? Do they make economic sense? Why?
iii. What are the quantitative effects of an increase in a on the equilibrium price and quantity?

(c) Reproduce the chart right below the original chart. Make sure that the dotted line moves with the chart. Then name the new chart Figure 2. Transform Figure 2 into a picture (copy and paste as a picture).
(d) Change a and c in Fig. 1 back to their original values (You should have two different figures and one table.)
(e) What is the economic meaning of an increase in parameters a, b, c, and d, respectively?

4-5 The demand and supply curves in health economics (static analysis) Let the demand D and supply S of the quantity of physician services be

$$\begin{aligned} \mathrm{D} &= \mathrm{a} - \mathrm{bP} & \mathrm{D0} &= 24 - 2\mathrm{P} \\ \mathrm{S} &= -\mathrm{c} + \mathrm{dP} & \mathrm{S0} &= -6 + 6\mathrm{P} \\ \mathrm{D} &= \mathrm{S} & \mathrm{D0} &= \mathrm{S0} \end{aligned}$$

where the default values of the parameters are a = 24, b = 2, c = 6, and d = 6. The dollars per unit of the services are denoted as $\mathrm{P} = 0, 0.5, \ldots, 10$.

(a) Using the naming method, construct the parameter and data tables, as we have done in class. Call the table Table 1. The table should be properly labeled, as shown in Table 4.1(b). Make sure that you enter the rows of current values, default values, and the differences. Enter the formulas as in rows 15 and 16. The formulas for $\mathrm{P}^\#$ and $\mathrm{Q}^\#$ are given as Eqs. (4.6) and (4.7).
Given the value of parameters in the model, what are the initial equilibrium price and quantity of physician services?
(b) Draw the demand curve and the supply curve for the default values. Label the curves. The text box should not have borders, its font should be size 8 and in italics. Reproduce the chart format exactly as shown in Fig. 4.7. Note that the borders of the chart area have been removed.
(c) Answer the same eight questions as in part (c) of HW4-3 above.
(d) In Table 1, color the row(s) showing equilibrium price and quantities with yellow, and in the diagram, denote the equilibrium price and quantity with a vertical dotted line. Make sure that the dotted line moves with the chart.

4-6 Comparative static analysis — supply induced demand theory (continued from the above HW4-5) We are given $(\mathrm{P}^\#, \mathrm{Q}^\#) = (3.75, 16.5)$.

(a) If the number of practicing physicians increases and thus also their services. This means that the supply curve shifts to the left, and that c changes from 6 to 2. What are the new equilibrium price and quantity? What are the qualitative and quantitative effects of increase in physician's services?

(b) Facing the decrease in their income, physicians may induce[9] the patients to increase demand for their services, as for instance by requiring more office visits, additional tests, and unnecessary surgery. This means that a increases from 24 to, say, 35. What are the new equilibrium values of price and quantities?

(c) After four lines are shown in the chart, draw the default (original) demand and supply lines as heavy solid lines without markers, the current demand line as a solid light line with triangular markers, and the current supply line in solid light line with square markers.

(d) Add the values of demand and supply as the data labels, as shown in Fig. 4.7.

(e) Case 1. Other things being equal, change parameter a from 24 to 35. Change the chart title, name it Figure 1. Then, picture copy it and place it below Table 1 of HW4-5. In Table 1, color the row(s) showing the new equilibrium price and quantities with light gray, and in the diagram, denote the equilibrium point with a white circle.

(f) From the table, what is the new equilibrium price and quantity? Show the changes by the impact vectors in levels and differences.

(g) Case 2. Other things being equal, change parameter b from 2 to 6. Change the chart title: name it Figure 2. Then, picture copy it and place it next to Figure 1. In Table 1, color the row(s) showing the new equilibrium price and quantities with light blue, and in the diagram, denote the equilibrium point with a white circle.

(h) From the table, what is the new equilibrium price and quantities? Show the changes by impact vectors in levels and differences.

(i) Change all the current values to the default values in Table 1. (Thus, you now have three charts, the original chart has only two lines). Print out the charts and Table 1 in one page.

4-7 The exchange market model (static analysis) Let the demand D and supply S of the amount of Japanese yen be

$$\begin{aligned} D &= a - be & D0 &= 8 - 350e \\ S &= -c + de & S0 &= -1 + 550e \\ D &= S & D0 &= S0, \end{aligned}$$

where the default value of the parameters are a = 8, b = 350, c = 1, and d = 550. The exchange rate of yen in terms of dollars is $e = 0.005, 0.006, \ldots, 0.015$. Thus, e is the domestic price of yen. The vertical axis is measured in millions of yen.

(a) Using the naming method, construct the parameter and data tables, as we have done in class, and name it Table HW4-7, as shown below. The table should be properly labeled. Make sure that you enter the rows of current values, default values, and the differences. The formulas for

[9]The inducement is most likely to succeed due to asymmetry of information, as the patients have to rely on the physicians' decisions. The physicians may exploit (or abuse) the patients by manipulating their demand curve. In Health Economics, this model is called the **supplier-induced demand (SID) model**. It can be described by the principal (patients)–agent (physicians) theory. See Santerre and Neun (1996), pp. 226–227.

$e^{\#}$ and $Q^{\#}$ are

$$e^{\#} = (a + c)/(b + d) \quad \text{and}$$
$$Q^{\#} = (ad - bc)/(b + d).$$

(b) Given the values of the parameters in the model, what are the initial equilibrium exchange rate and equilibrium quantity of yen?

(c) Construct the chart for comparative static analysis, and call it Figure 1. Make sure the format of Figure 1 is exactly the same as that of Fig. 4.7 in the text. Label the curves. Note that the borders of the chart area have been removed.

(d) Draw the default (original) demand and supply lines as dotted lines without markers, the current demand line as a solid light line with triangular markers, and the current supply line as a solid line with square markers.

(e) Answer the same eight questions in part (c) of HW4-3 for e and Q. Write all the answers below each question.

4-8 Comparative static analysis of the exchange market model (continued from HW 4-7)

(a) *Other things being equal*, change parameter c from 1 to 4. Change the chart title, name it Figure 2. Reproduce Figure 2 same as shown in Figure 4.8 of the text. Use the picture copy method and place it next to Table HW4-7. In Table HW4-7, color the row(s) showing the new equilibrium price and quantities with light gray, and in the diagram, denote the equilibrium point with a white circle.

(b) From the *table*, what are the qualitative and quantitative effects of the change in c from 1 to 4? What economic force(s) or factor(s) will make c change?

(c) Change all the current values, including c, to the default values. (Thus, you now have two charts, Figures 1 and 2. Figure 1 is the original chart from HW4-7, and has only two lines.) Print out the two charts and the revised Table HW4-7 in one page.

4-9 A macroeconomic model

(a) Set up the two models in Table HW4-9 similar to Table 4.3. Both are same models, but one is numerical model, one is the parametric model. In the parametric model, use the following table form of Table HW4-9(b) for the parameters. Use the numbers in the numeric model as the default values of the parametric model. The data table should be the same as that in Table 4.3.

Table HW4-7 The demand and supply of yen in the US market

e	D0	S0	E0	D	S	E
0.005	6.25	1.75	4.5	6.25	1.75	4.5
...						

Table HW4-9 A macroeconomic model

(a)

The numeric model	The parametric model
AD = C + I	AD = C + I
C = 0.4 + 0.6Y	C = a + bY
I = 2	$I = I_0$
AS = Y	AS = Y
AD = AS	AD = AS

(b)
The parameter table

	a	b	I0	Y#	C#	k
Current						
Default						
Change						

(b) How many variables are in this model? How many endogenous variables and exogenous variables? What are they?
(c) What are the equilibrium values of the variables?
(d) Combine the two tables and find the comparative static results when the current investment reduces by 1 unit. Draw the diagram. Express the impact in vector form in levels and in differences. What are the new equilibrium values? Does the change in equilibrium value make economic sense? Why?
(e) Do and answer the same question when b decreases from 0.6 to 0.4. Illustrate the change and call the new chart Figure 2.
(f) Do and answer the same question when a increases from 0.4 to 0.8. Illustrate the change and call the new chart Figure 3.

4-10 Static analysis Draw the chart of the following equivalent parametric and numeric national income models:

$AD = C + I_0$	AD = C + 1000
C = a + bY	C = 3000 + (2/3)Y
AS = Y	AS = Y
AD = AS	AD = AS,

where income Y ranges from 5000, 6000, up to 22,000 billion dollars ($G_0 = 0$ in the model).

(a) Set up a parameter table(using the naming method), and construct the aggregate demand (AD) and aggregate supply (AS) table, as we have done in class for Table 4.3. The table should be properly labeled. Make sure that you enter the rows of current values and default values. (to draw four lines differently, it will be easier if you give different values to a and I_0, say, a = 5000 and $I_0 = 2000$, respectively.)
(b) We first draw the AD curve and the AS curve for the default values similar to Fig. 4.9, call your chart Figure 1. What is the equilibrium income and the equilibrium consumption? Note that the borders of the chart area have been removed.
(c) Paint the row of the equilibrium income and consumption of the default values in the table in yellow.
(d) Use a white circle to denote the Keynesian cross in the chart.

(e) Answer the following questions. Write the answer below each question.

i. How many variables are in this model and what are they? Which ones are endogenous variables and which ones are exogenous variables?
ii. Why are they called endogenous variables? (In what sense are they "endogenous"?)
iii. Why are $Y^{\#}$ and $C^{\#}$ called the equilibrium values? (In what sense are they "equilibrium"?)
iv. Indicate the row(s) which show $Y^{\#}$ and $C^{\#}$ in the table.
v. What do economists call the coefficient (2/3) of Y? What is the economic meaning of 2/3?
vi. In the above model, we let I = 1000 in the aggregate demand function. In this case, is the variable "I" endogenous or exogenous? Why?

4-11 Comparative statics (continued from HW4-10)

(a) Other things being equal, parameter a changes from 3000 to 5000 billion

i. Picture copy the new chart and place it under Figure 1; call it Figure 2.
ii. Paint the row of the equilibrium income and consumption in the table in light gray.
iii. Use conditional formatting to add directional arrows to indicate the impact of changes.
iv. Use a dotted line to denote the new Keynesian cross; use a white empty circle to denote the original Keynesian cross.

(b) Answer the following questions. Write the answer below each question in a text box.

i. What are the new equilibrium income and consumption?
ii. What is the economic meaning of an increase in a? (Give an example under which a may increase.)
iii. What is the qualitative effect of an increase in a on the equilibrium income and consumption? Do they make economic sense? Why?
iv. What is the quantitative effect of an increase in a on the equilibrium income and consumption?

4-12 Statics and comparative statics Suppose the current US economy can be represented by the model in Table HW4-12(a).

Part I. Statics

All the numbers are in $US billion. We take the numeric model as the default model, and the parametric model as the current model whose parameters can be changed to perform comparative static analysis.

(a) Set up a parametric table as shown in Table HW4-12(b), where $Y^{\#} = k(a + I_0 + G_0)$, $C^{\#} = a + bY^{\#}$. $k = 1/(1 - b)$.
(b) Construct the data table, which is the same as Table HW4-12(c), where Y is calibrated as $100, 150, \ldots, 900$.
(c) Color the row in which equilibrium income and consumption of the default model are located.
(d) Draw the chart of the decomposed AD model. The chart should consist of Y, C, and AD.
(e) Give a name to the chart, the horizontal axis, and the vertical axis.
(f) Make sure the major and minor ticks are shown on the horizontal axis.

Table HW4-12 A macroeconomic model

(a)

The Numeric model	The Parametric model
$AD = C + I + G$	$AD = C + I + G$
$C = 130 + 0.6Y$	$C = a + bY$
$I = 70$	$I = I_0$
$G = 30$	$G = G_0$
$AS = Y$	$AS = Y$
$AD = AS$	$AD = AS$

(b)

Parameters	a	b	I_0	G_0	Y#	C#	k
Current							
Default							
Difference							

(c)

Y	C0	AD0	ED0	C	AD	AS	ED
	...						

(g) Identify each line by using a text box.
(h) Add a dotted vertical line through the Keynesian cross.
(i) Add a white circle with black border at the equilibrium income and equilibrium consumption.

Part II. Comparative Statics

Suppose the economy is in a dire recession, and the government has a rescue plan to increase consumption by deducing taxes so that the basic consumption expenditure (parameter a) changes from the default value of 130 to 200. Indicate by color the row(s) of the new equilibrium income and consumption levels in the table, and by a vertical line and filled black circles in the circle. You must perform the comparative static analysis by using the naming method, and you should retain the original chart (before parameter a changes) by applying the picture copy command to the current (after changes of a) chart.

Thus, you should have two tables, one is the parametric table and another is the data table (the format of the data table is given in Table HW4-12(c)). You also have two charts, one is the chart after the parameter a changes. The other chart is before the parameter a changes (you should restore to the original chart of the default parameters).

Answer the following questions:

(a) How many equations and variables are in this model? What are the variables?
(b) How many exogenous variables and how many endogenous variables are in this model? What are they?
(c) In what sense are they called "exogenous" variables? "endogenous" variables?
(d) What are equilibrium income and consumption for the default model?
(e) Why are they called "equilibrium" income and consumption?
(f) What is the use of the excess aggregate demand function?

(g) What is the economic meaning of increase in a, b, I_0, and G_0, respectively?
(h) From this model, can you tell us what are the qualitative and quantitative effects of government economic stimulus plan on consumption?
(i) From this model, are there any other economic stimulus policy the government can take to fight recession?

4-13 Economic meaning of a change in a, b, I_0, and G_0 In Sec. 4.5.5, we have discussed the effects of a change in parameters a and b, and the exogenous variables I_0 and G_0, on the equilibrium values $Y^\#$ and $C^\#$.

(a) Why do these parameters and exogenous variables change? What is the economic meaning of the change?
(b) Using the difference method in Sec. 4.1.2, taking the difference on the equations $D = a - bP$, $S = -c + dP$, $C = \alpha + \beta Y$, with respect to a change in the independent variable, and give an economic interpretation of its coefficients, b, d, and β.
(c) Give some numerical example of changes.

(Hints: (a) See the discussion at the beginning of the four cases in Sec. 4.5. (b) Similar to (4.10) and (4.11), $\Delta D = -b\Delta P$. This means that when P increases by ΔP, D decrease by $b\Delta P$. In other words, $b = -\Delta D/\Delta P$, that is, if P increases by one unit, D will decrease by b. (c) If $\Delta P = 3$ and $b = 2$, then $\Delta D = -b\Delta P = -2 * 3 = -6$, that is, if price increases by 3 dollars, demand will decreases by 6 units. In other words, whenever price increases by one dollars, demand will decrease by 2 units, $\Delta D/\Delta P = -6/3 = -2 = -b$. Hence, b is the change of demand per unit change of P. Similarly for d and β).

4-14 CSA using parameter and data tables In comparative static analysis of macroeconomic model discussed in Fig. 4.9, we have not shown the parameter tables and the data tables.

(a) Reproduce the diagrams with the relevant parameter tables and data tables as we have done in the four cases of Table 4.2.
(b) Complete the comparative static analysis for cases c and d by setting up parameter and data tables and drawing the chart for cases c and d.
(c) Write the impact vector of levels and differences for cases c and d.

Part II

Basic Statistics

Chapter 5

Some Useful Statistic Functions — Equations and Formulas

Chapter Outline

Objectives of this Chapter

In Chaps. 2–4, we have introduced economic equations and models. Some of them are numeric, like the demand equation, $D = 18 - 2P$, some are parametric, like $D = a - bP$. In Chap. 4, we use numeric models as default, and examine the effects of change in a parameter, like a or b, on the equilibrium values of endogenous variables, like $P^{\#}$ and $Q^{\#}$. The basic assumption here is that the parameters take certain values as default, like $a = 18$ and $b = 2$. However, the assignment of these default values is arbitrary. How do we know $a = 18$ and $b = 2$? Economics as an empirical science solves this question by using statistics and econometrics. The demand equation indicates that for each price P of a certain commodity, say, apple, wheat, or gasoline, consumers will decide the quantity demanded Q of that commodity. Thus, economists can observe the relations by taking a sample of prices and quantities of a commodity in a certain region or among certain consumers. Statisticians and econometricians then use this sample to estimate the numeric value of a and b using real world data. In general, the specific values of a and b depend on the sample period, place, society, and the method used to estimate a and b.

The purpose of this chapter is to introduce the basic concepts of some statistical terms. Since Excel comes with many statistical functions, we first show how to implement mathematical and statistical formulas and equations on a spreadsheet, and relate their numerical results with the results obtained directly from the Excel equations. An Excel equation will provide a quick numerical result without going through the details of calculation, but the user should know how the equation or formula is defined, derived, and interpreted, and especially what the meaning of the equation is and how it is applied. Fortunately, spreadsheets provide both features through hands-on practice. After we understand how the equations and formulas are constructed, we may use the equations directly with a sense of understanding and confidence.

This chapter introduces some important and often-used descriptive statistics: the central tendency of a statistical sample: the mean; the spread of a distribution: the variance and the standard deviation. We then introduce other useful statistical measures and the simple least squares method of estimating coefficients of a linear equation using the "trend line". The theory of ordinary regression analysis will be introduced in Chapter 7 after we study probability distributions. Since this is not a statistics textbook, readers should consult with some statistics textbooks for more topics.

Statistics and econometrics are very important to the economic and business professions. All major universities and colleges require these courses for undergraduate economics and business majors. Thus, this chapter serves as a review of the concepts underlying some commonly used statistical measurements, and at the same time introduces some Excel statistical equations and the connection between statistical definitions and Excel equations.

5.1 Basic Definitions and Tools

Statistics is a branch of science that deals with collection, classification, and evaluation of numerical data, called a **sample**, obtained from a group or an aggregate, called a **population**. The data are often used as a basis of statistical inference for drawing conclusions about the numerical properties of that population. A **population** is the totality of the group elements about which information is desired. The name statistics is derived from "state affairs". In the mid-eighteenth century, politicians were concerned with the conditions of the state, and in the 19th century, they started compiling numerical data about the current condition of the state. Recently, the word "statistics" is also used as the plural of "statistic", denoting the estimates or the results of statistical calculation, like the mean, variance, and correlation coefficient, in contrast to "parameters".

A **sample** is a subset of a population. We say "**a sample of size n**". It is a collective noun. The individual element in the sample is called a **sample point**, or a **sample element**. For example, to find the average amount of pocket money carried by students in a college, we may draw a random sample of 100 students on the campus to estimate the average amount the pocket money carried by the students in that college.

There are two basic categories of statistics: descriptive statistics and statistical inference. In this book, we mostly deal with the first category to show the uses of Excel in statistics. Descriptive statistics deals with collection, classification, and evaluation of data, including methods of data collection and presentation, the measures of central tendency, dispersion, relation, position, and the distribution of the data.

The most important measure of central tendency is the **mean**; the most used measures of dispersion are the **variance**, the **standard deviation**, and the **coefficient of variation**; and the most popular measure of relation is the **correlation coefficient**. In recent years,

growth rates and **elasticity** have also become popular among journalists and scholars. We derive formulas of these statistics, show how the formulas are implemented on spreadsheets, and compare the manual results with the results obtained directly from Excel equations. However, before we introduce these concepts, we first introduce the rules of summation, which are the basis of statistical definitions.

5.1.1 *Rules of summation*

We have introduced the sum $\sum$ in Chapters 1 and 2. It is one of the most useful functions that spreadsheets are good at. Suppose we have a sample of five values for a variable X, denoted as X_1, X_2, X_3, X_4, and X_5. Mathematically, the sum of the five values is expressed as

$$\sum_{i=1}^{i=5} X_i = X_1 + X_2 + X_3 + X_4 + X_5, \tag{5.1}$$

where $\sum$ is the Greek capital letter for s, and its subscript or superscript indicates the range of integer i, which is called the **index**. In this case, the range means that the sum is taken for each i consecutively from $i = 1$ (lower i) to $i = 5$ (upper i). Index i can be positive, zero, or negative, for example, from -4 to 0 or from -2 to $+2$. If the beginning value or the ending value is known, or if both are known from the context, we may not write them explicitly and denote the sum simply as,

$$\sum_{1}^{5} X_i, \sum_{i}^{5} X_i, \sum X_i, \sum X_i, \quad \text{or} \sum X. \tag{5.2}$$

In addition, the letter i can be any letter, like j or t, hence it is just a "dummy" notation. Using mathematical operations, the following rules of summation apply:

$$\sum aX_i = a \sum X_i; \tag{5.3}$$

$$\sum^{n} a = na; \tag{5.4}$$

$$a \sum (X_i + Y_i) = a \sum X_i + a \sum Y_i; \tag{5.5}$$

$$\sum (aX_i + bY_i) = a \sum X_i + b \sum Y_i. \tag{5.6}$$

Note that (5.5) and (5.6) still hold if sum (+) is replaced by difference (−).

In general, (5.3) states that constant a can be factored out, and (5.5) and (5.6) state that the summation sign is like a constant, it can be "distributed" inside the parentheses.[1]

[1]Mathematically, we say that the summation sign is a linear operator in the sense that you can take out the parentheses on the left-hand side of equation and apply the sum directly to each variable in the parentheses, as shown on the right-hand side.

Table 5.1 Rules of summation

	A	B	C	D	E	F	G	H	I	J	K	L	M	N	O
	Rules of Summation														
1	i	Xi	aXi	Xi	5Xi	Yi	bYi	Yi	10Yi	Xi+Yi	aXi+aYi	aXi+bYi	Xi+Yi	5(Xi+Yi)	5Xi+10Yi
2	1	X1	aX1	8	40	Y1	bY1	2	20	X1+Y1	a(X1+Y1)	aX1+bY1	10	50	60
3	2	X2	aX2	3	15	Y2	bY2	3	30	X2+Y2	a(X2+Y2)	aX2+bY2	6	30	45
4	3	X3	aX3	1	5	Y3	bY3	3	30	X3+Y3	a(X3+Y3)	aX3+bY3	4	20	35
5	4	X4	aX4	3	15	Y4	bY4	0	0	X4+Y4	a(X4+Y4)	aX4+bY4	3	15	15
6	5	X5	aX5	8	40	Y5	bY5	3	30	X5+Y5	a(X5+Y5)	aX5+bY5	11	55	70
7	Sum			23	115			11	110				34	170	225
8	LHS	ΣXi	ΣaXi	ΣXi	ΣaXi	ΣYi	ΣbYi	ΣYi	ΣbYi	Σ(Xi+Yi)	aΣ(Xi+Yi)	Σ(aXi+bYi)	Σ(Xi+Yi)	aΣ(Xi+Yi)	Σ(aXi+bYi)
9					a=5				b=10					a=5	b=10
10	RHS				aΣXi				bΣYi					aΣXi+aΣYi	aΣXi+bΣYi
11	Equation #				(5.3)				(5.3)					(5.5)	(5.6)

These rules are rather obvious and can be proved algebraically. In fact, (5.4), (5.5), and (5.6) can be derived from (5.3) in a straightforward manner (exercise).

Spreadsheet entries

The value X_i can be entered in spreadsheet as a column. Table 5.1 shows the spreadsheet implication for the above rules. Column A shows the range of index i, $i = 1, \ldots, 5$; column B enters the value of X_i; and column C is X_i multiplied by constant a. Numerical examples are shown in columns D and E, and similarly in columns F through I when $b = 10$. Row 8 gives the left-hand side of the equations, and row 10 gives the right-hand side, and the corresponding equation numbers are listed in row 11. The numerical examples show that the left-hand side is identically equal to the right-hand side. For example, (5.3) states that if each X_i is multiplied by $a = 5$ and then summed up to the sum of 115 (cells E7 and E8), it is then the same as the sum of X_i (cell D7) multiplied by $a = 5$ (cells E9 and E10), as shown by an arrow. Similar interpretation holds for columns J to O.

5.1.2 *Some important inequalities*

The following inequalities hold:

$$\sum X_i^2 \neq \left(\sum X_i\right)^2, \tag{5.7}$$

$$\sum X_i Y_i \neq \left(\sum X_i\right)\left(\sum Y_i\right). \tag{5.8}$$

That is, the sum of squares is not the same as the square of the sum, and the sum of the products is not the same as the product of sums.

The method of proving these inequalities is similar to (5.3)–(5.6): we prove that the RHS is equal or not equal to the LHS, or vice versa. To show (5.7), let $i = 1, 2$, that is, X takes values (X_1, X_2). The LHS is $X_1^2 + X_2^2$, but the RHS of (5.7) is $X_1^2 + 2X_1X_2 + X_2^2$, which is not the same as the LHS.

Similarly, let $i = 1, 2$ for Y_i. Then the LHS of (5.8) is

$$\sum X_iY_i = X_1Y_1 + X_2Y_2.$$

The RHS is

$$\left(\sum X_i\right)\left(\sum Y_i\right) = (X_1 + X_2)(Y_1 + Y_2) = X_1Y_1 + X_1Y_2 + X_2Y_1 + X_2Y_2.$$

The RHS has extra cross-product terms. Thus, both sides are not equal. Hence, (5.8) holds.

Similarly, we may prove for the case when $i = 1, 2, 3$ (exercise).

5.2 Basic Descriptive Statistics

We now introduce some basic concepts of descriptive statistics. We are interested in the measurements of centrality and dispersion of a sample.

5.2.1 *The mean*

If a sample of size n is given by $X_1, X_2, \ldots, X_n$, then the **average** or the **mean**[2] of this set of data is **defined** as

$$\boxed{\bar{X} = \frac{X_1 + X_2 + \cdots + X_n}{n} = \frac{\sum_{t=1}^{n} X_i}{n} = \frac{\text{sum}}{n},} \tag{5.9}$$

where the sign $\sum$ means summation. The subscript and the superscript show that the range of the sum is from $i = 1$ to n for all X_i in the sample.[3] As we have seen in Chapter 1, the **Excel equation for the mean** is **=average(X)**, where X is the range of the data on the worksheet.

There are several interesting interpretations of the average. Table 5.2 shows that for a random sample of five students A, B, C, D, and E (**a sample of 5**), we have observed the following amount of pocket money, denoted as variable X, where X takes the values of \$8.00, \$3.00, \$2.00, \$10.00, and \$7.00, respectively. We sum them up and divide the sum by 5,

$$\bar{X} = \frac{8 + 3 + 2 + 10 + 7}{5} = 6,$$

which is the average of the pocket money of the five students.

[2]Also called the **arithmetic mean** to distinguish this mean from other means like **geometric mean** or **harmonic mean**.

[3]Excel calls (5.9) an **average**, but statisticians call it a mean to distinguish it from **median** and **mode**. In this book, we use "mean" and "average" interchangeably.

Table 5.2 Calculation of mean and variance

	A	B	C	D	E
1	The meaning of average and variance				
2				=X-mX	
3		X	mX	x	x^2
4	A	8	6	2	4
5	B	3	6	-3	9
6	C	2	6	-4	16
7	D	10	6	4	16
8	E	7	6	1	1
9	avg	6	6	0	9.2
10		=Avg(X)		↑	=Var(X)
11				Mean deviation	

We may also use the **Excel formula for the mean**. The arguments of the mean function can be **delineated individually** as follows:

=average(8,3,2,10,7),

which counts each sample point and returns 6. Note that no space is allowed between the numbers.

If the data are entered on a spreadsheet like B4:B8 in Table 5.2, the formula can be based on the **data range** as

=average(B4:B8)=6,

which is entered in B9.

Frequency table

Another way of entering the data in this example is to use a **frequency table**, as shown in Table 5.3. The amount of money is an independent variable X. The range of X is taken from 0 to 10. Clearly, X = \$2.00 occurred once, that is, the frequency f_2 of \$2.00 is 1; \$3.00 also occurred once, that is, the frequency f_3 is 1; and so on. The frequencies are listed in column B. This column shows the **sample frequency distribution** of the sample at each value of X. It records the number of times a different X_i is observed. The five observations (or occurrences) are **realized values of X** between 0 and 10, and are listed in the product f * X column. Each element in the f * X column shows that X_i is repeated f_i times, and $f_i * X_i$ is the total number of X_i at i. Thus, the mean in the frequency table is defined as

$$\bar{X} = \frac{\sum_{i=1}^{i=n} f_i X_i}{\sum_{i=1}^{i=n} f_i} = \frac{\sum_{i=1}^{i=n} f_i X_i}{n} = \sum_{i=1}^{i=n} \left(\frac{f_i}{n}\right) X_i.$$

Table 5.3 Calculation of mean and variance — using frequency table

	A	B	C	D	E	F	G
1	Table 5.3 The Mean and the Variance						
2	X	freq f	amt	mean	x=mdev	fixi	dev^2
3		fi	fi*Xi	m	Xi-m	fi(Xi-m)	fixi^2
4	0	0		6	-6	0	
5	1	0		6	-5	0	
6	2	1	2	6	-4	-4	16
7	3	1	3	6	-3	-3	9
8	4	0		6	-2	0	
9	5	0		6	-1	0	
10	6	0		6	0	0	
11	7	1	7	6	1	1	1
12	8	1	8	6	2	2	4
13	9	0		6	3	0	
14	10	1	10	6	4	4	16
15	sum	5	30		-11	0	46
16	sum/5		6				9.2
17	average(C4:C14)		6.0				9.2
18	varp=avg(G4:G14) =sn^2		9.2				
19	stdevp = sqrt(varp) =sn		3.03				
20	var=varp*n/(n-1) =s^2		11.5			n=5	11.5
21	stdev=sqrt(var) =s		3.39				
22	CVp = stdevp*100/avg =sn*100/m						50.55
23	CV = stdev*100/avg = s*100/m						56.52

In this example, each entry has only one frequency. Hence, $f_i = 1$ for all i. The sum is entered in C15, and the average is in C16. Note that Excel calculates the average for the cells with numbers and ignores the blank cells.[4] The result is illustrated in Fig. 5.1(a). In Fig. 5.1(a), the independent variable (the amount of money) is shown on the horizontal axis (the X-axis), the frequency is shown on the vertical axis (the Y-axis), and the average is indicated by a triangle below the X-axis and also by the line along $X = 6$. Using the charts, there are three ways to interpret the mean, as explained below.

The average as the center of gravity

Figure 5.1(a) shows that if we take the X-axis as a lever and place unit weights at the five points defined by the amounts of pocket money (letting the down arrows show the force of gravitational attraction at each point), then the mean is the point where the total gravitational forces concentrate, as if all the weights were moved to a single point at the mean. Thus, the mean is the **center of gravity**. It is the place at the top of a fulcrum supporting all the weights in Fig. 5.1(a).

[4]But if you enter 0 in a blank cell, then 0 is counted as a number and the sum will be divided by 6 instead of 5. As shown in HW2-12, if the 6 blank cells in C4:C14 are filled with 0, then the average in C17 will show 2.7, and other calculations in the lower part of Table 5.3 will be wrong.

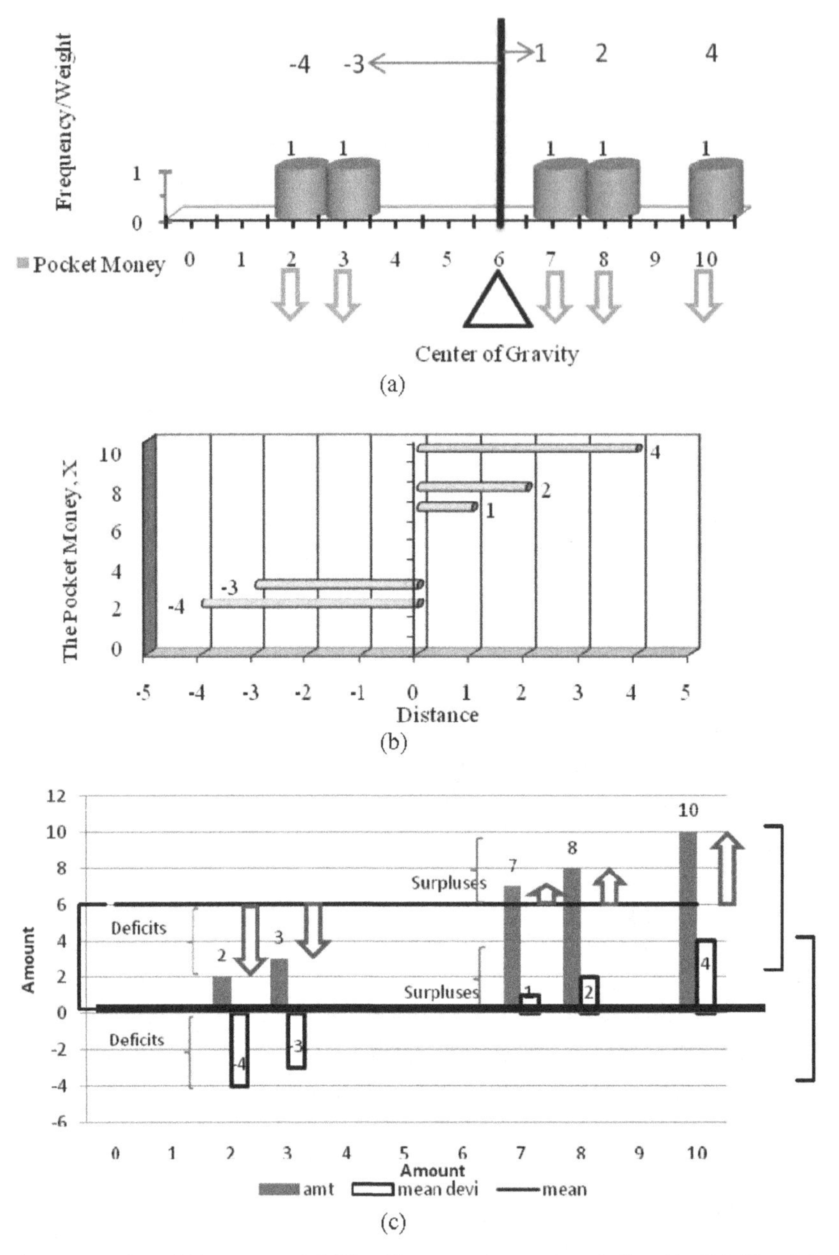

Fig. 5.1 The three meaning of the mean. (a) The Mean as the center of Gravity (b) The Distance from the mean (c) Average as the "Fair" amount

The average as the offsetting forces (or mean deviations)

Figure 5.1(b) shows the distance of sample point from the mean, denoted as $x_i = X_i - \bar{X}$, which is called the **mean deviation** at i, and is usually denoted by small case x. The positive distance can be regarded as the upward force exerted at each point on the right of the mean, and the negative distance as the downward force exerted at each point on the left of the mean. At the mean, the upward and downward forces balance, and there will be no motion. It is in a state of "equilibrium", that is,

$$\boxed{\sum_{i=1}^{5}(X_i - \bar{X}) = \sum_{1}^{5} x_i \equiv 0,} \tag{5.10}$$

as shown at point 0 in Fig. 5.1(b).

In Table 5.2, the mean is entered manually in C4:C8, and the mean deviation is entered in D4:D8. We use m to denote the mean of X, as Excel cannot write $\bar{X}$ (or subscripts, or superscripts). In the frequency table, Table 5.3, the mean is entered manually in D4:D14, and the mean deviations (mdev), $x_i = f_i(X_i - m)$, are entered in F4:F14, which sums to zero, as shown in F15.

The average as the "Fair" amount

The mean has a special characteristic as **the "fair amount"**. In Fig. 5.1(c), the solid filled columns show the amounts of pocket money X, and the upper thick line shows the average at Y = 6. If we subtract X from the average, there is a **surplus**, shown by an up arrow along the mean line at 6, and there is a **deficit**, shown by a down arrow along the mean line. The surpluses and the deficits of the five students will cancel each other and sum to zero, as shown by the white columns along the lower thick line along the X axis at Y = 0. The white columns and the arrows above them have the same length. In other words, if you "rob" the "richer" students who have more than average pocket money, and, to make up the deficits, distribute them to the "poorer" students who have less than average pocket money, all students will have the same amount of the pocket money. Hence, the average is the "fair" amount of money.[5]

Drawing

Figure 5.1(a) is based on columns A and B of Table 5.3, using the first **3D column** with data labels, and shortening the Y-axis and changing its interval from 0 to 1. Other parts are manually drawn, using the text box and shapes.

Figure 5.1(b) is based on columns A and F using the first **bar chart** with data labels.

Figure 5.1(c) is based on columns A, C, D, and F, using the column chart, and changing the mean column, D4:D14, to the line chart, with data labels for columns C and F. Others are drawn manually (exercise: it is recommended that the readers draw the three Excel charts).

---

[5]This was the basic idea of "communism".

5.2.2 *The population variance*

If a series of data is given as $X_1, X_2, \ldots, X_n$, then the population variance of this set of data is **defined**[6] as the average of the **squared deviations from the mean**, that is, the **Sum of Squared deviations from the mean** (SSx) divided by the sample size:

$$\mathrm{varp}(X) = \frac{(X_1 - \bar{X})^2 + (X_2 - \bar{X})^2 + \cdots + (X_n - \bar{X})^2}{n}. \tag{5.11}$$

Or, using the summation sign,

$$\boxed{\mathrm{varp}(X) = \frac{\sum_{i=1}^{n}(X_i - \bar{X})^2}{n} = \frac{\sum_i^n x_i^2}{n} = \frac{SSx}{n} \equiv s_n^2,} \tag{5.12}$$

which is divided by the sample size n. $SSx \equiv \sum x_i^2$. Note that the mean deviations are squared first, and then summed up, since the sum of mean deviations is identically zero. This measure of dispersion is commonly used in theoretical analysis, called the **population variance**, and is denoted as s_n^2 (sn squared). However, sometimes this formula is also applied to a sample, especially when the sample size is greater than 30, since it has an intuitive appeal as the average of the squares of mean deviations. In this case, to illustrate the use of Eq (5.11), we use the sample (8, 3, 2, 10, 7) of the pocket money in the previous example. If we calculate the variance of this sample by using the equation for the population variance, we have, by (5.11),

$$\begin{aligned}\mathrm{varp}(X) &= \frac{(8-6)^2 + (3-6)^2 + (2-6)^2 + (10-6)^2 + (7-6)^2}{5} \\ &= \frac{(2)^2 + (-3)^2 + (-4)^2 + (4)^2 + (1)^2}{5} = \frac{46}{5} = 9.2.\end{aligned}$$

As in the case of the mean, this calculation can be done most conveniently in a table like Table 5.2 or 5.3. In Table 5.2, since we have already calculated mean deviations in D4:D8, their squares are calculated in E4:E8, and the average is in E9 as

=average(E4:E8)=9.2,

which is the population variance of the pocket money of the five students. Using the frequency table, in column D of Table 5.3, we enter the mean; and in column F, we calculate the mean deviations, which sum to zero in F15. The squares of mean deviations multiplied by their frequency are calculated in column G, their sum is 46. When it is divided by 5, the average is 9.2, which is entered in cell G16. This is the "population" variance (varp).

---

[6]The definition in Excel has a different form, but can be derived from (5.12). The numerator of (5.12) can be written as $\sum(X_i^2 - 2X_i\bar{X} + \bar{X}^2) = \sum X_i^2 - 2\sum X_i\bar{X} + n\bar{X}^2 = \sum X_i^2 - n\bar{X}^2$. Using the definition of mean in (5.9), we have $\mathrm{varp}(X) = (n\sum X_i^2 - (\sum X_i)^2)/n^2$, which is the formula defined in Excel's Help document. The formula of varp for grouped data is $\sum f_i(X_i - m)^2/n$, where $n = \sum f_i$. This formula is used in G16, G17, C18 in Table 5.3. The formula of varp in C20 is $\sum f_i(X_i - m)^2/(n-1)$.

The **Excel function for the population variance** is

=varp(X),

where X is the range of the data, and p denotes population. For sample (8, 3, 2, 10, 7), we obtain

=varp(8,3,2,10,7) =9.2.

Alternatively, if the data is entered in range B4:B8 on the spreadsheet as in Table 5.2, then we can simply use the range,

=varp(B4:B8) =9.2,

Note that, like the calculation of the mean, Excel does not count the blank space as zero in the range. In general, the formula is not applicable to grouped data.

The variance as average areas

Figure 5.2 illustrates the meaning of the variance. The average of the squared deviations from the mean is denoted by the height of a horizontal line through 9.2. The average of the heights of the light colored columns is the variance of X, which is $9.2, and is indicated by the horizontal black solid line.

Another way of looking at the variance is that it is the average area of all the squares (x^2) of the surpluses and deficits (x) in column E of Table 5.3. The five right rectangles are shown in Fig. 5.3. The mean deviations are measured from 0 along the X-axis, and we build a square manually above a mean deviation. The area of each square is shown at the upper corner of the square. The variance is the average area of all the areas of these five squares.

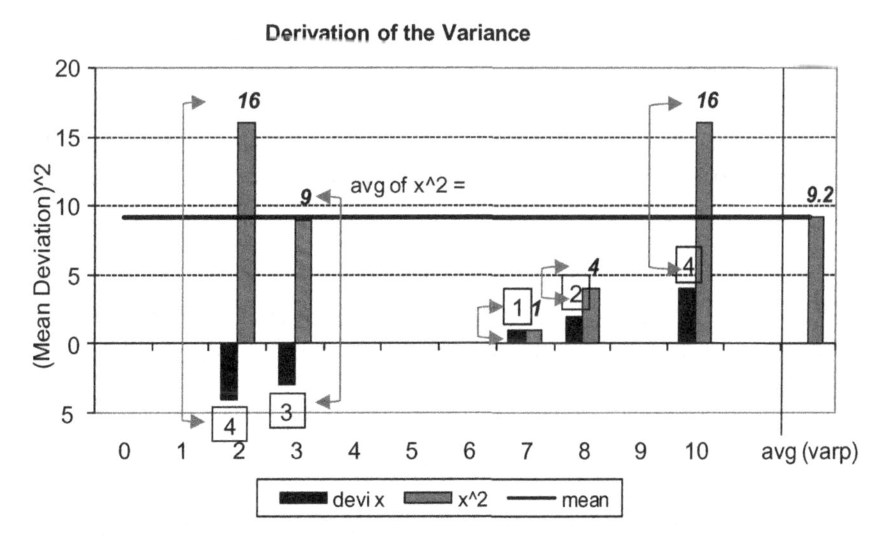

Fig. 5.2 Derivation of the variance

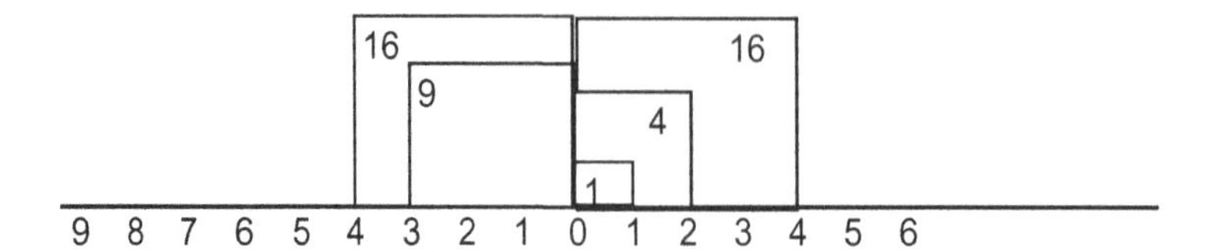

Fig. 5.3 An interpretation variance

The unit of measurement

In Fig. 5.2, the deviations are shown as black columns. They are measured in dollars \$, say \$4.00 or −\$3.00. However, when the deviations are squared, the unit of measurement is also squared, $\$^2$, and the sum of the deviations and its average are also in the $\$^2$ units, that is, $\$^2$ 9.2. The unit is in a different dimension than that of the sample.

5.2.3 *The population standard deviation*

The population standard deviation is **defined** as

$$\text{stdevp(X)} = \sqrt{\frac{\sum_i^n (X_i - \bar{X})^2}{n}} \equiv \sqrt{\frac{\sum_i^n x^2}{n}} = \sqrt{\frac{SSx}{n}} = \sqrt{\text{varp(X)}} \equiv s_n, \qquad (5.13)$$

which is the positive square root of the population variance. Thus, unlike the variance, the stdevp has the same dimension as the data, and so is comparable with other statistical measurement like mean. This is the most widely used measure of dispersion of the data (the "standard" among statisticians) and is denoted as s_n. Like population variance, this measurement of dispersion is also applied to sample, especially when the sample size is large (greater than 30). In the example of Table 5.3, the variance is $\$^2$ 9.2, as shown in cell G16, and so the population standard deviation of the pocket money is \$3.03 in C19.

The **Excel function for the population standard deviation** is

$$= \text{stdevp(X)},$$

where X is the range of the data. In Table 5.2 or 5.3, the range is B4:B8 or C4:C14.

If we already have calculated varp, then stdevp can be obtained directly by taking the **square root** of varp:

$$= \text{sqrt(varp(X))},$$

as shown in (5.13). Otherwise, we need to set up the whole table to calculate the square root of the variance. We will discuss this method in the next chapter.

Table 5.4 Mean variance and standard deviation

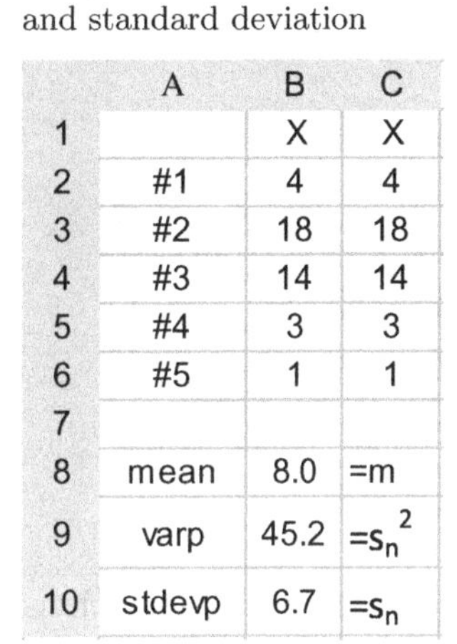

	A	B	C
1		X	X
2	#1	4	4
3	#2	18	18
4	#3	14	14
5	#4	3	3
6	#5	1	1
7			
8	mean	8.0	=m
9	varp	45.2	$=s_n^2$
10	stdevp	6.7	$=s_n$

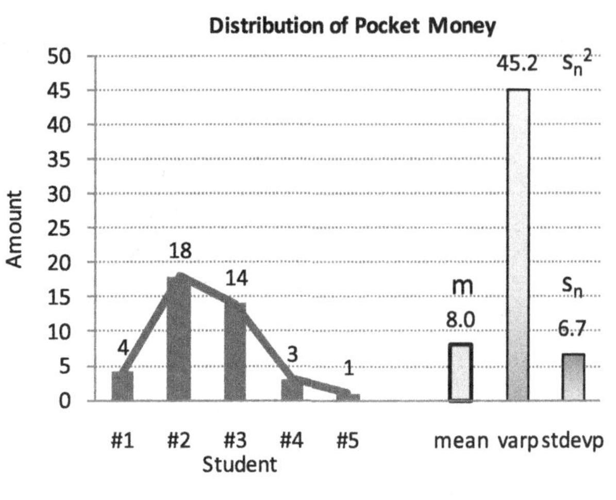

Fig. 5.4 Mean, variance, and standard deviation

Example 5.1 The mean, variance, and standard deviation. If the amount of pocket money carried by the five students is a random number from 0 to 20, as shown in Table 5.4, find the mean (m), variance (s_n^2), and standard deviation (s_n) of their pocket money.

In the above table, in B2, we enter =RAND()*20, and in C2, we enter =B2, and then copy B2:C2 down to B6:C6. Enter the equation for mean, variance, and standard deviation in B8:B10, making sure that row 7 is blank. Draw the column for the range A1:C10, which will give a chart with two columns (not shown). Click the second column and change the chart type from column to line. Change the color of the "mean" column to yellow and the color of the "varp" column and the "stdevp" columns to gradient fill, which has a lighter color at the upper part of the column. Use text box to enter m, s_n^2, and s_n as shown. □

Enter <F9> several times and see how the columns and the three statistics change. This means that we can calculate a different mean, variance, and standard deviation for each random sample of size 5. In other words, if we take a sample 1000 times, that is, enter <F9> 1000 times, and calculate these three statistics each time, we will obtain 1000 statistics of m, s_n^2, and s_n.

Naturally, for such a small sample, (sample) var and (sample) stdev defined in the next section are more appropriate. However, we use varp and stdev for their intuitive appeal of being the average of the mean squared deviations.

Example 5.2 Mean and standard deviation of normal curve. A normal curve is a bell shaped symmetric curve of a large sample, as shown in Fig. 5.5 (we will discuss this curve in Chapter 7).

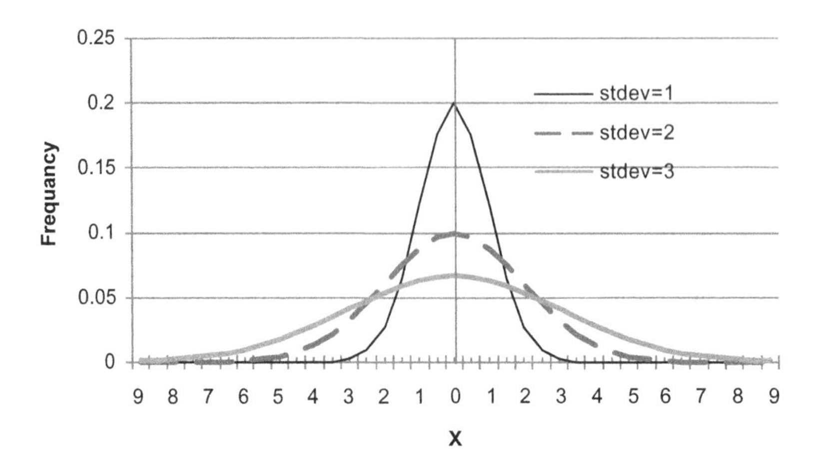

Fig. 5.5 The standard deviation and the shape of the curve

The three normal curves in Fig. 5.5 all have means at 0, but the standard deviations differ. The light curve has stdevp = 1, the dashed curve has stdevp = 2, and the heavy curve has stdevp = 3. This shows that the larger the standard deviation, the more spread the curve will be from the mean. □

5.2.4 *The variance*

In Excel, the variance of the data is **defined** as

$$\mathrm{var}(X) = \frac{\sum_i^n (X_i - \bar{X})^2}{n-1} = \frac{\sum_i^n x^2}{n-1} = \frac{SSx}{n-1} = s^2, \tag{5.14}$$

which is similar to the population variance, except that **the sum of squared deviations** (SSx) from the mean is divided by sample size minus one. As it is, (5.14) is also generally called the **sample variance** of X and is denoted as s^2. The number (n − 1) here is called the **degrees of freedom** of SSx. It means that when a sample of size n is taken, while the value of each sample point can vary freely, when the variance is calculated, the sample mean has to be calculated first. Since the mean has to be given before calculating the variance, one of the sample points cannot be taken freely.[7] Thus, we lose one degree of freedom in the sample, and the sample only has n − 1 degrees of freedom.

In Table 5.3, the variance is the sum of x^2 column (G15) divided by 5 − 1 = 4, that is, 11.5, as shown in cell C20. Statistically, this will give an **unbiased estimator** of the population variance.[8] For a large sample (say 30 or larger), the sample variance is almost the

[7]For example, if $(X_1 + X_2 + X_3)/3 = 10$, then one of the sample points, say X_1, is constrained by $X_1 = 30 - X_2 - X_3$, while X_2 and X_3 can take any values.

[8]That is, if we take a large sample repeatedly (say thousands of times) and calculate the variance of X in (5.14) each time, then the sample variance will approach the population variance when the sample size goes

same as the "population" variance, thus, we may use either one. The "population" variance and the variance are related by the following formula:

$$\mathrm{var(X)} = \frac{\mathrm{n}}{\mathrm{n}-1}\mathrm{varp(X)}. \tag{5.15}$$

Thus, given varp(X), var(X) can be derived from varp(X) multiplied by sample size n and divided by $n - 1$. In Table 5.3, value in C20 can be obtained from G17 by $(5/4)(9.2) = 11.5$, as expected. Excel equation for the variance (5.14) is =var(X), where X is the range of the data.

5.2.5 *The standard deviation*

In Excel, the standard deviation of X is **defined** as

$$\boxed{\mathrm{stdev(X)} = \sqrt{\frac{\sum_{\mathrm{i}}^{\mathrm{n}}(\mathrm{X_i} - \bar{\mathrm{X}})^2}{\mathrm{n}-1}} = \sqrt{\frac{\sum_{\mathrm{i}}^{\mathrm{n}}\mathrm{x}^2}{\mathrm{n}-1}} = \sqrt{\frac{\mathrm{SSx}}{\mathrm{n}-1}} = \sqrt{\mathrm{var(X)}} = \mathrm{s},} \tag{5.16}$$

which is the positive square root of the variance. It is also called simply the **sample standard deviation** of X and is generally denoted by s. The **Excel equation for the standard deviation** is =stdev(X), where X is the range of the data. The standard deviation in Table 5.3 is $\sqrt{11.5} = 3.39$, as shown in cell C21.

5.2.6 *The coefficient of variation*

The mean and the standard deviation depend on the units of measurement. To facilitate comparison among standard deviations, statisticians divide the standard deviation by the mean to cancel the effects of different measurement units of dispersion of the data. This is the **coefficient of variation** and is defined formally as[9]

$$\boxed{\mathrm{CVp} \equiv \frac{\mathrm{stdevp}}{\mathrm{mean}}100 = \frac{\mathrm{s_n}}{\bar{\mathrm{X}}}100,} \tag{5.16$'$}$$

to infinity. If the sample variance is divided by n instead of $n - 1$, the sample variance will be denoted by varp, which, for a large n, will approach a constant different from population variance. Thus, the formula varp, calculated for a sample, is a biased estimator of the "population" variance. Excel's use of the name "population variance" is very confusing. As a method of calculation, the "population" variance varp can also be applied to sample. Thus, the "sample variance" can be calculated by either the var or the varp method. In this book, whenever the "population" variance varp(X) is calculated for a sample data, we use it with quotation marks.

[9]To avoid further confusion, we define CV as the sample standard deviation stdev(X)=s times 100 divided by mean. If we use the "population" standard deviation, stdvp(X)=s_n then it will be denoted as CVp. Most recent textbooks use the definition of CV (instead of CVp) to measure variability of sample distribution.

and

$$CV \equiv \frac{\text{stdev}}{\text{mean}} 100 = \frac{s}{\bar{X}} 100. \tag{5.17}$$

CV is a pure number and in general is shown in percentage.

Excel does not provide the formula for the coefficient of variation. Thus, it has to be constructed from the stdev and the mean from the table. In the pocket money example of Table 5.3, from (5.17), the CVp (or CV) is calculated by using stdevp (or stdev) in C19 (or C21), depending on whether stdevp or stdev is used: In G22, we have $(3.03*100)/6 = 50.55$, and in G23, we have $(3.391 * 100)/6 = 56.52$. The amount of the pocket money varies by 50.55% (or 56.52%) among the five students. Note that CV converges to CVp as n increases: the difference between them will be close if n is larger than 30. It measures the volatility of a variable, like stock prices, growth rates, foreign direct investment inflow, etc. Other things being equal, the smaller the CV, the better, because a smaller CV has less variation or volatility.

Example 5.3 Comparing the variability of stock prices. Suppose, during the past 52 weeks, the average stock price of Company A is $100 and its standard deviation is $25, while the average stock price of Company B is $50 and its standard deviation is $15. Other things being equal (that is, same rating of the companies, same dividends, etc.), which company shows less volatility in its stock price?
Answer: Since Company A's stock prices have $CV = (25/100) * 100 = 25$, and those of Company B have $CV = (15/50) * 100 = 30$, Company A's stock has a 25% CV while Company B's has 30%. Thus, Company A's stock price has less volatility or less risk, and appears to be more desirable than that of Company B. □

5.3 Measurements of Relations

5.3.1 *Covariance*

The **covariance** between two variables, say X and Y, is defined as the average of the products of the mean deviations of the two variables:

$$\begin{aligned} \text{Cov}(X, Y) &= \frac{(X_1 - \bar{X})(Y_1 - \bar{Y}) + (X_2 - \bar{X})(Y_2 - \bar{Y}) + \cdots + (X_n - \bar{X})(Y_n - \bar{Y})}{n} \\ &= \frac{x_1 y_1 + x_2 y_2 + \cdots + x_n y_n}{n}, \end{aligned} \tag{5.18}$$

where we denote $x_i \equiv X_i - \bar{X}$ the **mean deviation of X**, and $y_i \equiv Y_i - \bar{Y}$ the **mean deviation of Y**. Using the summation sign, covariance can be expressed as

$$\boxed{\text{cov}(X, Y) = \frac{\sum_i (X_i - \bar{X})(Y_i - \bar{Y})}{n} = \frac{\sum_i x_i y_i}{n} = \frac{SSxy}{n},} \tag{5.19}$$

where SSxy $\equiv \sum x_i y_i$ is the sum of the products of the mean deviations of X and mean deviations of Y. Equation (5.19) is similar to the population variance (5.12), except that the numerator shows the interaction terms between the mean deviations of both variables.

As an example, in addition to the pocket money X of the five students in Table 5.2 or 5.3, we also survey the students' lunch money Y. The pairs of data turn out to be $(X, Y) = (2, 4)$, $(3, 3)$, $(7, 6)$, $(8, 4)$, and $(10, 8)$. The second number shows the lunch money in dollars (we assume that the students can borrow some money). Since $\bar{X} = 6$ and $\bar{Y} = 5$, the covariance between X and Y can be calculated from (5.18) as

$$\begin{aligned}\text{Cov}(X, Y) \\ &= \frac{(2-6)(4-5)+(3-6)(3-5)+(7-6)(6-5)+(8-6)(4-5)+(10-6)(8-5)}{5} \\ &= \frac{4+6+1-2+12}{5} = \frac{21}{5} = 4.2.\end{aligned}$$

If we calculate by using a spreadsheet tablc, the five pairs of sample points are entered in the X and Y columns of Table 5.5. The mean deviations of X and Y are calculated in the columns of $x \equiv X - \bar{X}$ and $y \equiv Y - \bar{Y}$. The products xy are entered in column F. The average of the sum of these products is 4.2 in cell F9, which is the covariance.

The **Excel equation for the covariance** is

$$=\text{covar}(X,Y), \tag{5.20}$$

where X is the range of X, and Y is the range of Y. For a sample of $X = (2, 3, 7, 8, 10)$ and $Y = (4, 3, 6, 4, 8)$, we have to enter X and Y as arrays, that is, as column vectors or row vectors. In Table 5.5, X and Y are entered as columns. At any blank cell, say F12, enter

Table 5.5 Covariance and correlation coefficient

	A	B	C	D	E	F	G	H
1	**Calculation of Covariance**							
2		**and Correlation Coefficient**						
3		X	Y	x	y	xy	x^2	y^2
4	C	2	4	-4	-1	4	16	1
5	B	3	3	-3	-2	6	9	4
6	E	7	6	1	1	1	1	1
7	A	8	4	2	-1	-2	4	1
8	D	10	8	4	3	12	16	9
9	avg	6	5	0	0	4.2	9.2	3.2
10	stdev						3.0	1.8
11	correl					0.78		

the covariance (not shown)

=covar(B4:B8,C4:C8),

which will show in F12. Note that the Excel formula gives covar(X, Y) directly by skipping the calculation of x, y, and xy.

Interpretations

Figure 5.6 shows an interpretation of covariance. We first locate the means of X and Y at point[10] (6, 5) and draw a horizontal line and a vertical line to divide the chart into four quadrants. Use each pair of mean deviations (x, y) as the sides to draw a rectangle with each point in the chart. Thus, we have five rectangles. If both sides have the same sign, the rectangle has a positive area; this is the case when the rectangles are located in the first or the third quadrant. If one side is negative and the other side positive, the rectangle has a negative area; this is the case when the rectangles are located in the second or the fourth quadrant. The covariance is the average of the sum of these positive and negative areas.

Thus, if the sum of the areas is positive, we know that most of the rectangles are located in the first and third quadrants, the relation between the two variables is positive, and the general trend of the sample points is upward sloping. Generally speaking, this means that the larger the pocket money, the larger the lunch money, and vice versa. On the other hand,

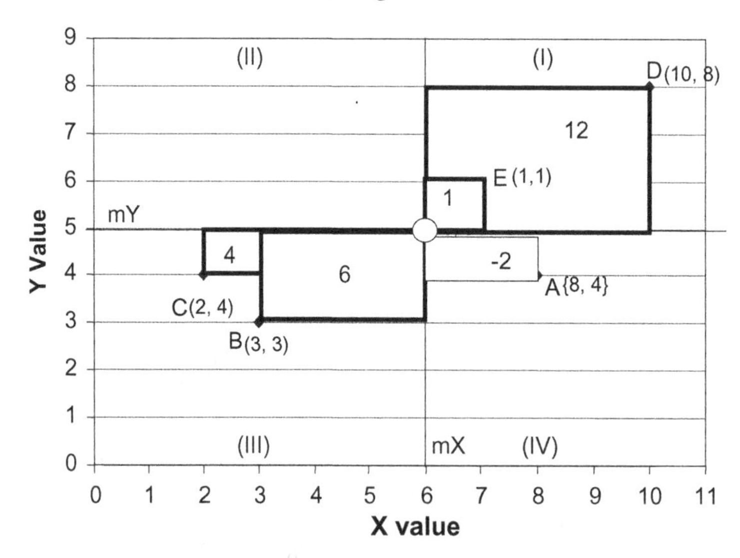

Fig. 5.6 An interpretation of covariance

[10]As usual, in the expression (X, Y) = (6, 5), the first number, 6, shows the coordinate of the X-axis, and the second number, 5, the coordinate of the Y-axis.

if the sum is negative, then most of the rectangles are located in the second and the fourth quadrants, the relation between the two variables is negative, and the sample points are downward sloping. Generally speaking, this would mean that the larger the pocket money, the smaller the lunch money, and vice versa. Finally, if the sum is zero, or tends to be very small, then there is a lack of **linear relationship** between the two variables, and we say there is no linear relationship between them.

5.3.2 *Correlation coefficient*

As a measure of relationship, the covariance has a major problem in the units of measurement in X and Y. In our example, both X and Y are measured in dollar units. However, on many occasions, the units may be different: one variable may be measured in thousands and one in millions, or one variable may be in dollars and one in yen. Thus, it is impossible to compare the variances among samples with different units of measurement. To avoid this problem, statisticians convert covariance into pure numbers by dividing the mean deviations of each variable by the population standard deviation[11]:

$$\frac{X_i - \bar{X}}{\text{stdevp}(X)}, \quad \frac{Y_i - \bar{Y}}{\text{stdevp}(Y)},$$

which are called the **standardized variable** of X and Y, respectively. This means that the sides of the mean deviation of the rectangles in Fig. 5.6 are divided by the standard deviation and so the basic relations of the sum of squares do not change. Since both the denominator and the numerator are in the same units, they cancel out and give a pure number. The **correlation coefficient** r between X and Y is then **defined** as the average of the sum of the products of the standardized variables,

$$\begin{aligned} r(X,Y) &\equiv \sum_{i=1}^{n} \left(\frac{X_i - \bar{X}}{\text{stdevp}(X)} \frac{Y_i - \bar{Y}}{\text{stdevp}(Y)} \right) \frac{1}{n} = \frac{\sum_i^n (X_i - \bar{X})(Y_i - \bar{Y})}{\text{stdevp}(X) * \text{stdevp}(Y)} \frac{1}{n} \\ &= \frac{\text{cov}(X,Y)}{\text{stdevp}(X) * \text{stdevp}(Y)} = \frac{\text{SSxy}}{\sqrt{\text{SSx SSy}}}. \end{aligned} \tag{5.21}$$

[11]Note that, since in Excel, covariance is defined as the sum of cross products of mean deviations divided by n (see (5.19) and Table 5.5), the correlation coefficient in (5.21) should be divided by stdevp(X) and stdevp(Y). In some textbooks, covariance is defined as the sum of cross products of mean deviation divided by n − 1, then, the denominator of (5.21) must be stdev(X) and stdev(Y). To avoid confusion, it is better to define r(XY) by using the last expression: SSxy/sqrt(SSx*SSy). Note that, as explained in footnote 8, p as "population" in stdevp(X) does not make sense sine we are dealing with sample only.

In general, we may say that the correlation coefficient is the covariance[12] of two standardized variables, and is a pure number independent of units of measurement. It is a measure of linear interdependence between two variables, and is a fraction ranging from -1 to 1. If there is a perfect positive (or negative) correlation, then $r = 1$ (or $r = -1$). If there is no correlation, then $r = 0$.

The last two columns of Table 5.5 calculate the standard deviations of X and Y as 3.0 (in G10) and 1.8 (in H10). Since the covariance is 4.2 from the previous section, we have

$$r = \frac{4.2}{(3.0)(1.8)} = 0.78$$

in cell F11. Thus, the correlation coefficient between the pocket money and lunch money is highly positively correlated: It is highly possible that an increase in pocket money will increase the lunch money, and vice versa.

The **Excel equation for the correlation coefficient** is

$$=\text{correl}(X,Y), \tag{5.21$'$}$$

where X and Y are the ranges of the X and Y data. Note again that the covariance and the correlation coefficient in Excel use "population" covariance and "population" standard deviations. In terms of the example of Table 5.5, we have to enter X and Y as either column ranges or row ranges. From Table 5.5, we may derive the correlation coefficient directly as

r=correl(B4:B8,C4:C8)=0.78.

Scatter diagrams and the correlation coefficient

Figure 5.7 presents various cases of correlation coefficients for a sample with 10 pairs of X and Y, where both variables are entered as rand()*10 in columns B and C in Table 5.5. To construct the chart, draw a scatter diagram of X and Y like Fig. 5.7. Keep the first chart as the original chart. The correlation coefficient is shown in F11 of Table 5.5. Enter <F9> repeatedly until the desired coefficient, say $r = 0$, shows up. Picture Copy the $r = 0$ chart to a blank space, as we did in Chapter 3, and go back to the original chart. Repeat the process for the six charts in Fig. 5.7.

For $r = 1$ or 0.9, and $r = -1$, the points show clearly a positive or negative linear trend. However, for $r = 0$ or 0.2, no clear trend can be discerned. For $r = 0.5$, we can see a slightly positive trend. Note that, for the diagram of $r = 0$, the points tend to be nonlinear. Hence, $r = 0$ indicates either that the points are random or that there is a nonlinear trend.

---

[12]Let the standardized variable of X_i be u_i and that of Y_i be v_i. Then, by (5.10), the means of u and v are zero. Thus, from (5.21), $r(X,Y) = \sum u_i v_i/n = \sum(u_i - 0)(v_i - 0)/n = \text{cov}(u,v)$. Note that both the covariance and the correlation coefficient are divided by n.

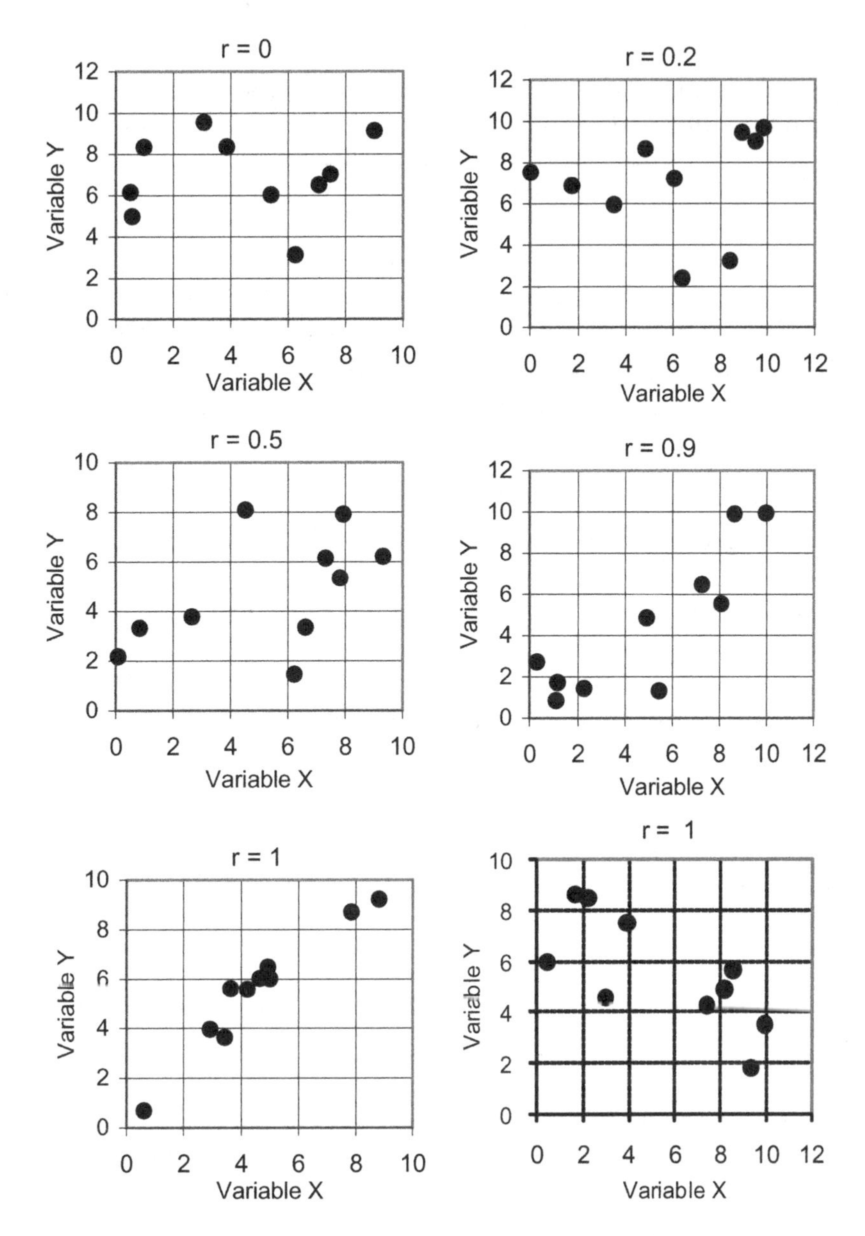

Fig. 5.7 Various cases of the correlation coefficient

5.4 Some Other Statistical Measurements

5.4.1 *Growth rates*

There are two kinds of growth rates: discrete and continuous. Here we discuss the discrete growth rate. We will discuss the continuous growth rate in Chapter 8.

Definition

Let X_t be an economic variable, say GDP or consumption, at period t. The **growth rate**, g_t, of X_t at period t is **defined** as

$$\boxed{g_t = \frac{X_t - X_{t-1}}{X_{t-1}} = \frac{\Delta X_t}{X_{t-1}} = \frac{\text{Change}}{\text{Total}},} \tag{5.22}$$

where X_{t-1} denotes GDP or consumption at the previous time period $t-1$, and Δ denotes a change of X between period $t-1$ and period t, as we defined in Chapters 3 and 4. Equation (5.22) generally is a fraction, and is multiplied by 100 to show the growth rate in percentage. More specifically, if GDP grows from 100 in time $t-1$ to 105 in time t, the growth rate is

$$g_t = \frac{105 - 100}{100} = \frac{5}{100} = 0.05,$$

and we say that the **growth rate** of GDP is 0.05, or in percentage term, $5\% (= 0.05 * 100)$ at period t.

This definition of growth rate is the same as the **percentage change of a variable**. In the above example, we may also state that at period t, GDP increases 5% over the last period, or the **percentage change** of X from period $t-1$ to period t is 5%. In general, when the data consist of time series, we call g_t the **growth rate series**.

Illustration

The definition of growth rate is illustrated in Fig. 5.8. There you are standing at time t and are comparing the current year's X_t (say GDP) with the previous year's X_{t-1}.

Growth equation form

Rewriting Eq. (5.22), we have the **growth equation form** for the discrete time period:

$$X_t = (1 + g_t)X_{t-1}. \tag{5.23}$$

According to the theory of compounded interest, this equation means that GDP is compounded once in each period and increases suddenly at the end of the period by the amount of g_tX_{t-1} (see Chapter 8 for the compounding interest rate).

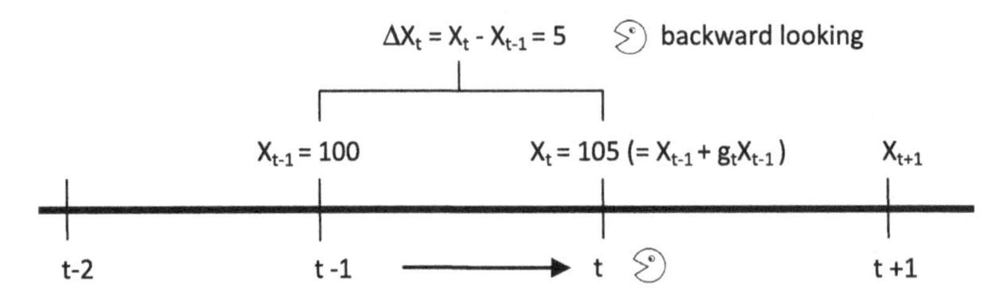

Fig. 5.8 Derivation of the growth rate of X

Table 5.6 Growth rate and elasticities

	A	B	C	D	E	F
1	Percentage changes and price elasticities of gasoline					
2	Months	$	m gallons	% change		
3		price	Gasoline	Price	Gasoline	elasticity
4	Jan	2.69	25.5			
5	Feb	2.49	27.9	-7.4	9.4	-1.27
6	March	2.99	24.6	20.1	-11.8	-0.59
7	April	3.29	23.3	10.0	-5.3	-0.53
8	May	3.39	22.7	3.0	-2.6	-0.85
9	avg	2.97	24.8	6.4	-2.6	-0.81

Difference equation form

Rewriting (5.23) as

$$X_t = X_{t-1} + g_t X_{t-1}, \tag{5.24}$$

which shows that GDP (X_t) at period t is calculated as the GDP (X_{t-1}) of the previous period, $t-1$, plus the additional GDP ($g_t X_{t-1}$) grown over one period. Equation (5.24) is called the **difference equation form** of the growth equation.

Example 5.4 Percentage changes of price and quantity demanded. Table 5.6 shows five-month average prices of gasoline per gallon measured in dollars (column B) and monthly quantity demanded of gasoline in millions of gallons (column C) in a certain city. We enter the formula (5.22) as follows:

D5: =(B5−B4)∗100/B4 and E5: =(C5−C4)∗100/C4,

and copy D5:E5 to D6:E8. Enter the average of each column in row 9. The data show that in February, price decreased by 7.4%, but after February, price increased by 20.1% in March, 10% in April, and 3% in May. We can say that the **average price increase**[13] from January to May is 6.4%. On the other hand, as the price increases, the quantity of gasoline demanded has decreased in general. It increased in February by 9.4%, but it decreased by 11.8% in March, −5.3% in April, and −2.6% in May. The average decrease of demand from January to May is 2.6%. □

5.4.2 *Elasticity*

Elasticity is an abstract concept, which relates two variables by asking if one variable (like a price) changes one percent, what the response is of another variable (like quantity

[13]If the price is measured as annual price index of all commodities, then the annual percentage change of price, commonly measured by the consumer price index (CPI), is called the **inflation rate** (if the index increases) or **deflation rate** (if the index decreases). In our example, we are dealing with monthly percentage changes of price of a single commodity, gasoline.

demanded) in terms of the percentage change. For macroeconomics, we may use two series, national income, X_t and consumption, C_t. **Income elasticity of consumption** is defined as the growth rate of consumption C, g_{ct}, at time t, divided by the growth rate of income X, g_{Xt}, at time t. This is the percentage change of consumption divided by the percentage change of income, and is a pure number, independent of the unit of measurement. It is **defined** as,

$$e_{CX} = \frac{g_C}{g_X} = \left(\frac{C_t - C_{t-1}}{C_{t-1}}\right) \Big/ \left(\frac{X_t - X_{t-1}}{X_{t-1}}\right) = \frac{\Delta C_t}{C_{t-1}} \Big/ \frac{\Delta X_t}{X_{t-1}} = \frac{\Delta C_t}{\Delta X_t}\frac{X_{t-1}}{C_{t-1}}, \quad (5.25)$$

which is

$$\begin{aligned} e_{CX} &= \frac{\text{\% change of consumption}}{\text{\% change of income}} \\ &= \text{\% change of consumption due to the 1\% change of income.} \end{aligned}$$

Note the various expressions of the definition of elasticity in (5.25). For example, referring to the example in Table 5.7, if in 1981 consumption changes by 10.28% and income by 11.91% (see columns I, J, and K of Table 5.7), then the income elasticity[14] of consumption is 0.86. It means that in 1981, if income changes by 1%, consumption will change by 0.86% in 1981 from the base year 1980.

By definition, elasticity is **inelastic** if it is less than 1, **elastic** if it is larger than 1, and **unitary** if it is 1. In the case of the income–consumption relations shown in Table 5.7, from 1980 to 1997, the income elasticity of consumption is generally elastic.

Example 5.5 Price elasticity of demand for gasoline. Continuing from Table 5.6, we can find the price elasticity of demand for gasoline. Enter E5/D5 in cell F5 and copy it to F5:F8. In February, the price elasticity is −1.27, meaning that as the gasoline price increases by 1%, the quantity of gasoline demanded decreases by 1.27%. In March, as the price of gasoline increases by 1%, the quantity of gasoline demanded decreases by 0.59%, and so on. On average, during January through May, a 1% increase in gasoline price decreases gasoline demand by 0.81%. Thus, the price elasticity of gasoline is negative and generally inelastic. □

5.5 The Computation Table

Table 5.7 presents a computational table for the definitions we have discussed in the previous sections. We want to show the methods of calculation systematically, and also to show that calculation using definitions and calculation using Excel equations yield the same results. In this book, we emphasize that the reader should know the definitions and the process

[14]Naturally, we can invert (5.25) and write e_{XC} = (% change of income)/(% change of consumption). In this case, we have consumption elasticity of income. In general, we consider income as the independent variable and consumption as the dependent variable, hence, definition (5.25) is generally accepted.

Table 5.7 National income and consumption

	A	B	C	D	E	F	G	H	I	J	K
1	**Calculation of some descriptive statistics**										
2		US$ bil	US$ bil	=C-avg(C)	=X-avg(X)				%	%	
3		dep var	ind var	(5.10)	(5.10)	(5.12)	(5.12)	(5.19)	(5.22)	(5.22)	(5.25)
4	Year	C	X	c	x	c^2	x^2	c*x	gC	gX	Elast
5	1980	1760.4	2784.2	-1743.2	-2482.2	3038669	6161179	4326867			
6	1981	1941.3	3115.9	-1562.3	-2150.5	2440712	4624531	3359635	10.28	11.91	0.86
7	1982	2076.8	3242.1	-1426.8	-2024.3	2035695	4097678	2888187	6.98	4.05	1.72
8	1983	2283.4	3514.5	-1220.2	-1751.9	1488834	3069056	2137596	9.95	8.40	1.18
9	1984	2492.3	3902.4	-1011.3	-1364.0	1022683	1860420	1379355	9.15	11.04	0.83
10	1985	2704.8	4180.7	-798.8	-1085.7	638046	1178684	867211	8.53	7.13	1.20
11	1986	2892.7	4422.2	-610.9	-844.2	373172	712627	515686	6.95	5.78	1.20
12	1987	3094.5	4692.3	-409.1	-574.1	167345	329559	234840	6.98	6.11	1.14
13	1988	3349.7	5049.6	-153.9	-216.8	23678	46990	33356	8.25	7.61	1.08
14	1989	3594.8	5438.7	91.2	172.3	8321	29697	15720	7.32	7.71	0.95
15	1990	3839.3	5743.8	335.7	477.4	112709	227937	160283	6.80	5.61	1.21
16	1991	3975.1	5916.7	471.5	650.3	222333	422926	306644	3.54	3.01	1.18
17	1992	4219.8	6244.4	716.2	978.0	512974	956538	700485	6.16	5.54	1.11
18	1993	4459.2	6558.1	955.6	1291.7	913214	1668561	1234404	5.67	5.02	1.13
19	1994	4717.0	6947.0	1213.4	1680.6	1472393	2824510	2039311	5.78	5.93	0.97
20	1995	4953.9	7269.6	1450.3	2003.2	2103435	4012922	2905326	5.02	4.64	1.08
21	1996	5215.7	7661.6	1712.1	2395.2	2931363	5737116	4100923	5.28	5.39	0.98
22	1997	5493.7	8110.9	1990.1	2844.5	3960586	8091338	5660958	5.33	5.86	0.91
23	(5.9)	3503.6	5266.4	0.0	(0.0)	1,303,676	2,558,459	1,825,933	6.9	6.5	1.1
24	avg					↓	↓	↓			
25	Equ #		**Manual calculation**								
26	(5.11) varp(C), varp(X);				sn^2(C)	1303676	2558459				
27	(5.13) stdevp(C), stdevp(X);				sn(C)	1141.8	1599.5				
28	(5.14) var(C), var(X);				s^2(C)	1380362	2708957				
29	(5.16) stdev(C), stdev(X);				s(C)	1174.9	1645.9				
30	(5.16)' CVp(C), CVp(X)					32.6	30.4				
31	(5.17) CV(C), CV(X)					33.5	31.3				
32	(5.19) cov(C,X)							1825933			
33	(5.21) r = cov(C,X)/(stdevp(C)*stdevp(X))							0.999794			
34			**Calculation using Excel formula**								
35	(5.11) varp(C), varp(X)				sn^2(X)	1303676	2558459				
36	(5.13) stdevp(C), stdevp(X)				sn(X)	1141.8	1599.5				
37	(5.14) var(C), var(X)				s^2(X)	1380362	2708957				
38	(5.16) stdev(C), stdev(X)				s(X)	1174.9	1645.9				
39	(5.20) covar(C,X)							1825933			
40	(5.21) correl(C, X)							0.999794			
41	Sources of data: *Economic Report of the President* , U.S. Government Printing Office, Jan. 1999.										

of calculation of Excel equations and formulas, rather than applying them without such knowledge. Instead of using traditional pen and paper, we set up an Excel table and let Excel do the detailed calculation. As we have pointed out in the introduction, this is one of the advantages of using spreadsheets.

The data in Table 5.7 consist of the levels of consumption and GDP in $US billions, and are listed in columns B and C. We first calculate the basic statistical measures (columns D–H) in Sec. 5.5.1, and then apply the corresponding Excel equations (rows 34–40) in Sec. 5.5.2. The growth rate and elasticity (columns I–K) will be calculated separately in Sec. 5.5.3.

5.5.1 *Calculation using definitions*

As the nine statistics (means, varp, stdevp, var, stdev, CVp, CV, covar, and r) are related each other, we can conveniently set up one table and calculate the values of these nine statistics. The table will show clearly how they are related. We proceed by the following steps.

(1) **Data** First use the fill-handle to enter the years in column A. Enter the **data** of consumption C and GDP X in columns B and C as shown.
(2) **Labels** Columns D to H present calculation of the basic statistics. Enter the title of the table and the column labels in rows 1–4 as shown. The labels are aligned to the right. The **auto-correction of (c)** is ©. To avoid auto-correction, see Appendix 2C.
(3) **Average** Row 23 calculates the **average** of the corresponding column.

$$\text{B23: =average(B5:B22).}$$

Copy B23 to B23:K23. For the time being, the rows, except B23 and C23, will show #DIV0! (since n = 0). They will be filled automatically when data in other columns are entered.

(4) Calculate the **mean deviations of C and X** in columns D and E, denoted by lowercase letters c and x. They are squared in columns F and G. The product of c and x is entered in column H. Enter

$$\begin{aligned}
&\text{D5: =B5–\$B\$23,} && \text{Mean deviation of C = c}\\
&\text{E5: =C5–\$C\$23,} && \text{Mean deviation of X = x}\\
&\text{F5: =D5\^{}2,} && \text{Square of c}\\
&\text{G5: =E5\^{}2,} && \text{Square of x}\\
&\text{H5: =D5*E5,} && \text{Product of c and x}
\end{aligned}$$

Copy D5:H5 down to row 22. Note that in D5 and E5, a $ sign has been added to the cell address of the means. This is the Excel method to fix or to fasten (that is, to make an **absolute reference**) the cells B23 and C23 when D5 and E5 are copied down to row 22. We will discuss absolute reference in the next chapter.

The "population" variances (varp) of C and X (see (5.12)) are the averages of the sum of squares of deviations (SSx) from the mean, (that is, the averages of the numbers in F5:F22 and G5:G22), and are calculated in F23 and G23, respectively. The covariance (cov in (5.19)) between C and X is calculated as the average of values in column cx, H5:H22, and entered in H23. Note that **the mean (or the sum) of the mean**

deviations c and x must be zero, as shown in D23 and E23 (see Eq. (5.10)). This property can be used to check the accuracy of calculation of mean deviations.[15]

(5) **Equation numbers** for this chapter are entered at the left side of the table A23:A40 and row 3. We use arrows to show that varp and cov are obtained directly from row 23.

(6) Row 27 finds the **"population" standard deviations** (stdevp) of C and X by taking the square roots of the variances of C and X, respectively. The formula for taking the square root is **=sqrt(cell)**, hence,

F27: =sqrt(F26) and G27: =sqrt(G26).

(7) Rows 28 and 29 calculate the variance (var in (5.14)), and the standard deviation (stdev in (5.16)) for C in

F28: =(18/(18–1))*F23 and F29: =sqrt(F28).

Similarly, we calculate for X in G28 and G29 by simply copying F28:F29 to G28:G29.

(8) Rows 30 and 31 calculate the **"population" coefficient of variation** CVp (5.16)′ and the coefficient of variation CV (5.17) for C in

F30: =F27*100/B23 and F31: =F29*100/B23.

Similarly, we calculate for X in G30 and G31 by copying F30:F31 to G30:G31.

(9) Rows 32 and 33 calculate the relations between C and X: the **covariance** (5.19) and the **correlation coefficient** r (5.21) between C and X. The covariance between C and X is already calculated manually in H23. Thus, copy H23 down to H32, and in H33, enter

H33: =H32/((F27)*(G27)).

The correlation coefficient r in cell H33 is almost equal to 1. This means that consumption and GDP have an almost perfect positive correlation.

The last three columns, I, J, and K, will be explained in Sec. 5.5.3.

5.5.2 *Calculation using Excel equations*

Rows 35–40 of Table 5.7 calculate statistics by applying Excel equations directly to the data without considering the details of calculation. In the following cells, we enter the

[15]Note that Excel encloses the mean in cell E23 by parentheses, (0.0). It means that zero is obtained by rounding a negative fraction.

corresponding Excel equations:

F35: = varp(B5:B22),	"Population" variance of C
F36: = stdevp(B5:B22),	"Population" standard deviation of C
F37: = var(B5:B22),	(Sample) Variance of C
F38: = stdev(B5:B22),	(Sample) standard deviation of C
H39: = covar(B5:B22,C5:C22),	Covariance of C and X
H40: = correl(B5:B22,C5:C22),	Correlation coefficient between C and X.

We have shaded the cells to show the corresponding calculation-by-definition cells and calculation-by-equation cells. They yield, as expected, exactly the same results. Table 5.7 shows the advantages of using a spreadsheet. We compare the calculations using the definitions and calculations using the Excel equations, and gain the understanding and feeling of the Excel equations.

5.5.3 *Calculation of growth rates and elasticities*

There is no Excel equation for calculating growth rates and elasticities. Columns I, J, and K of Table 5.7 calculate the growth rate of C and X and the income elasticity of consumption for each year. The calculations must start from row 6, as we do not have data for 1979. Calculate the **growth rate** of C in I6, that of X in J6, and the **income elasticities of consumption** in column K6 as follows:

I6: = (B6−B5)*100/B5,	Growth rate of C, in %, from (5.22)
J6: = (C6−C5)*100/C5,	Growth rate of X, in %, from (5.22)
K6: = I6/J6,	Income elasticity of consumption, from (5.25)

Copy I6:K6 down to row 22. In K6, the %Δ of dependent variable C is in the numerator and the %Δ of independent variable X is in the denominator. As usual, we change the independent variable and see what its effects are on the dependent variable.

The range I23:K23 shows that the average growth rate of consumption from 1981 to 1997 in the United States is 6.9%, that of the national income is 6.5%, and the average elasticity is 1.1. Hence, the income elasticity of consumption is elastic; that is, on average, when income increases by 1%, consumption will increase by 1.1%, indicating over-consumption in the United States, and this is one of the reasons that the US Government is hoping to reduce taxes to stimulate consumption and thereby pump up the economy during the economic recession.

5.6 Simple Least Squares Regression — The Graphic Method

The correlation coefficient only shows a very rough trend of linearity, as shown by Fig. 5.7. To be more definite, statisticians try to draw a trend line among the sample points on the scatter diagram, as will be explained in Fig. 5.11. The Excel program provides six different methods of "fitting" a trend line. The most commonly used one is the linear (least squares) method.

5.6.1 *The scatter diagram of consumption and income*

Drawing the scatter diagram

The relationship between consumption and income can be visualized diagrammatically in a **scatter diagram**. To draw the scatter diagram between C and X, we first note that C is the dependent variable and X is the independent variable. The data are entered in Table 5.7 so that C is on the left side and X is on the right side. But, in previous chapters, we have seen that in order to draw a chart correctly, **the mathematical convention** is that the independent variable must be to the left of the dependent variable, so that when the table is charted, the independent variable will be on the horizontal X-axis, and the dependent variable on the vertical Y-axis. This is what we have taken for granted in previous chapters. In this example, we want income X be on the horizontal (X) axis and consumption C on the vertical (Y) axis. Therefore, we should copy columns B and C of Table 5.7 to another sheet and rearrange the columns so that the X column is on the left and the C column is on the right, as shown in Table 5.8. Then, select the range including the column title, and click.

<"Insert"><Charts, Scatter><Scatter with only markers>.

Excel will automatically recognize that X is the independent variable.

Changing the axes of a chart

For a large set of data, this switching of columns may not be convenient or practical. Another method is to switch the axes of the chart. Note that unlike previous charts, both column titles must be given (that is, cell B4 in Table 5.7 must be filled with title C) as follows:

1. Select the data set B4:C22 in Table 5.7, including the column titles. Then draw the column chart:

<"Insert"><Charts, Column><Clustered column (the first chart)>.

Table 5.8 Rearrangement of columns

	A	B	C
1	**coeff**	a	b
2		US$ bil	US$ bil
3		ind var	dep var
4	Year	X	C
5	1980	2784.2	1760.4
6	1981	3115.9	1941.3
...	...	...	...
22	1997	8110.9	5493.7

The chart shows that both C and X are represented by columns. The horizontal axis labels range from 1 to 18.

2. Moving the pointer in the chart and click <RM> until a drop-down menu appears. Choose <Select data···> from the menu. A "Select Data Source" dialog box like Fig. 5.10(a) appears. The "Legend Entries (Series)" (that is, dependent variable) has two variables, C and X. Since X should be the independent variable, select X and <Remove>.
3. To enter X in "Horizontal (Category) Axis Labels", in order to make X the independent variable, click <Edit> under "Horizontal (Category) axis Labels". An "Axis Labels" dialog box like Fig. 5.10(b) appears. There is a flushing vertical line in the box. Select C5:C22, the value (the data part) of independent variable X from Table 5.7 (do not include the column title) and clicking <OK>. The dialog box returns to Fig. 5.10(c), and the right-side window is filled with the values of X. Click <OK>.
4. **To draw the scatter diagram**, select the chart, press <RM>, and click

<Change Chart Type><XY (Scatter)>,
<Scatter with only markers (the first scatter chart)>, and <OK>.

Coordinates of data points

We have the scatter diagram with marker as shown in Fig. 5.9.

Make sure that the horizontal axis represents income (by noting that the maximum values of X and C are different). When the sample points (markers) are selected and the pointer tip is placed on one of the sample points, a small dialog box appears, like the one shown by an arrow in Fig. 5.9, indicating the coordinates of the sample point. For example,

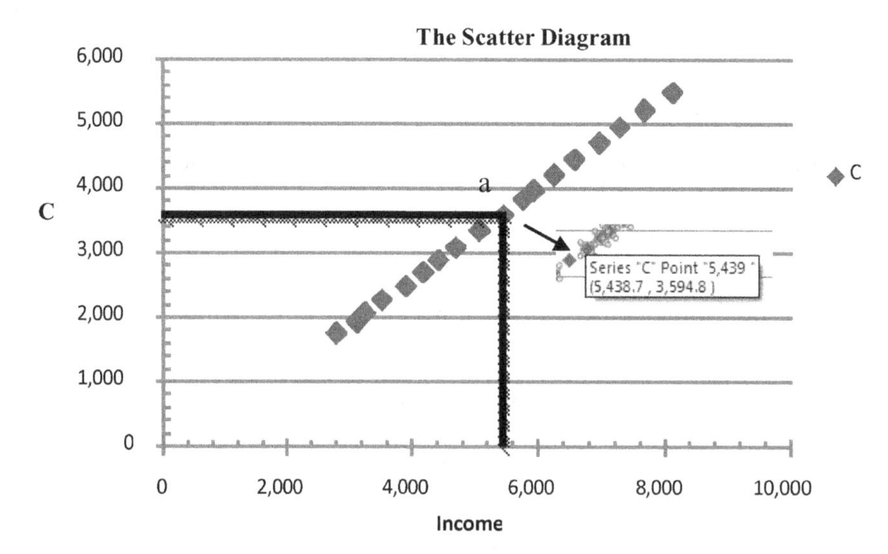

Fig. 5.9 The scatter diagram for income and consumption

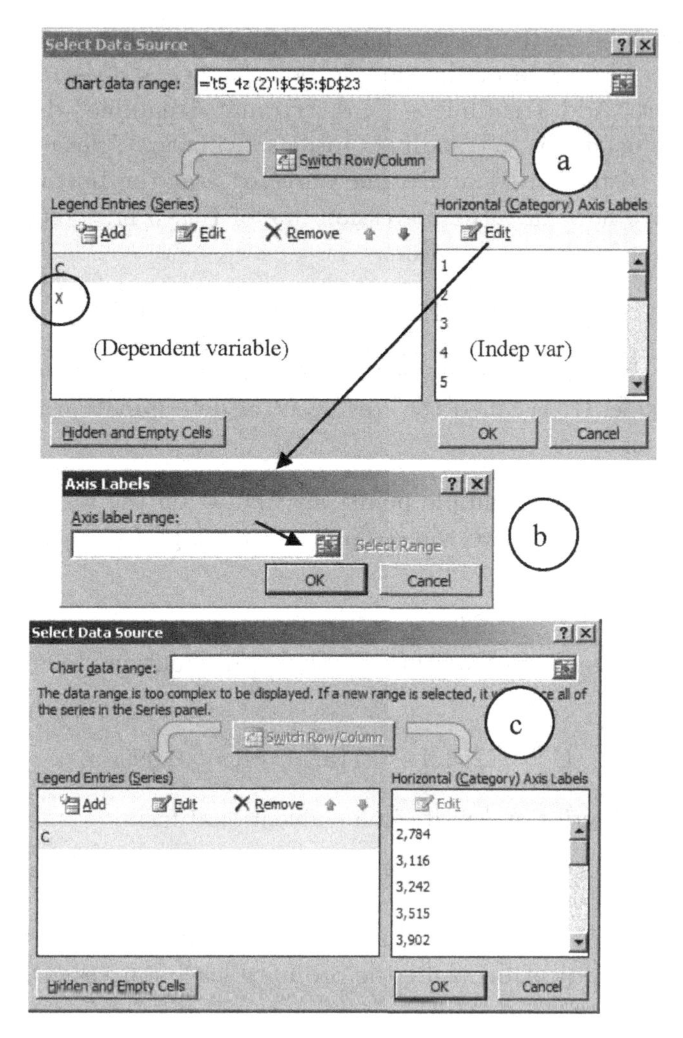

Fig. 5.10 Change of independent and dependent variables

in Fig. 5.9, the dialog box shows that sample point a, which the pointer locates on Series C, is the amount of consumption at income 5438.7 (the X-axis), and the coordinate is at $(X, C) = (5438.7, 3594.8)$, which is the pair in 1989 in Table 5.7. The dialog box changes if you move to another sample point.

5.6.2 *Fitting a linear regression line*

The scatter diagram shows the **sample points** of the observed pairs of X and C, denoted as (X, C). In our example, we have 18 sample points taken from 1980 to 1997 in the United States. Figure 5.11 shows that the relation between consumption and income is almost linear and positively sloped. We may add a linear regression trend line to represent the trend of the sample points as follows.

1. Click any sample point in the scatter diagram. All the scatter points are selected.
2. While the pointer is on one of the sample points, click <RM> and a drop-down list will appear. Select "Add Trendline ⋯". A "Format Trendline" dialog box appears, as in Fig. 5.12. The default is <Trendline Options, @Linear>. Select <Display Equation on chart> and <Display R-squared value on chart> at the bottom of the dialog box. Click <Close>. We now have the regression line of Fig. 5.11, along with the estimated regression equation in generic notation,

$$\mathrm{y} = 0.713\mathrm{x} - 254.9, \quad \mathrm{R}^2 = 0.999, \tag{5.26}$$

where, in Excel notation, y is the generic dependent variable, and x is the generic independent variable. R^2 is called the coefficient of determination and will be explained in detail in Chapter 7.

Note that, in Fig. 5.11, the sample points are almost on the regression line. When the height of the chart is expanded, we will see more scattering of the sample points.

Conventional way to write a regression equation

Following the general practice of writing a regression equation, (5.26) may be rewritten using our notations as

$$\hat{\mathrm{C}}_\mathrm{i} = -254.9 + 0.713\mathrm{X}_\mathrm{i}, \quad \mathrm{R}^2 = 0.999, \tag{5.27}$$

or we may omit the subscript and write in a conventional form

$$\hat{\mathrm{C}} = -254.9 + 0.713\mathrm{X}.$$

The line drawn by (5.27) is called **a linear regression line**, and the equation is **the estimated regression equation** using the ordinary least squares (OLS) method.

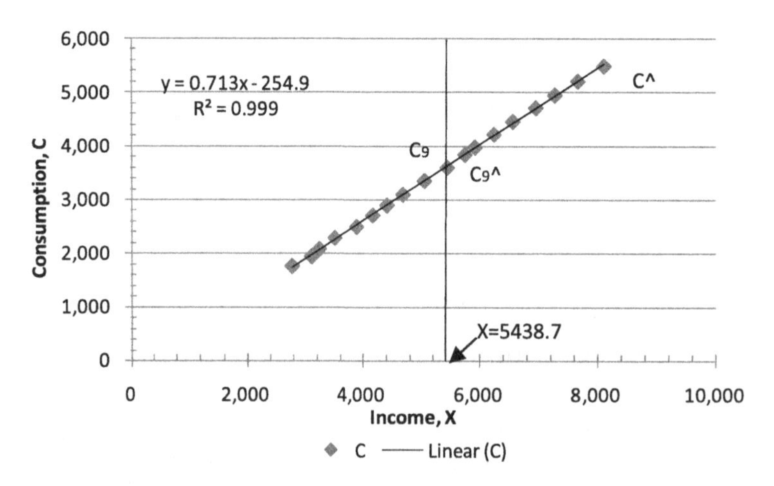

Fig. 5.11 The trendline and least squares equation

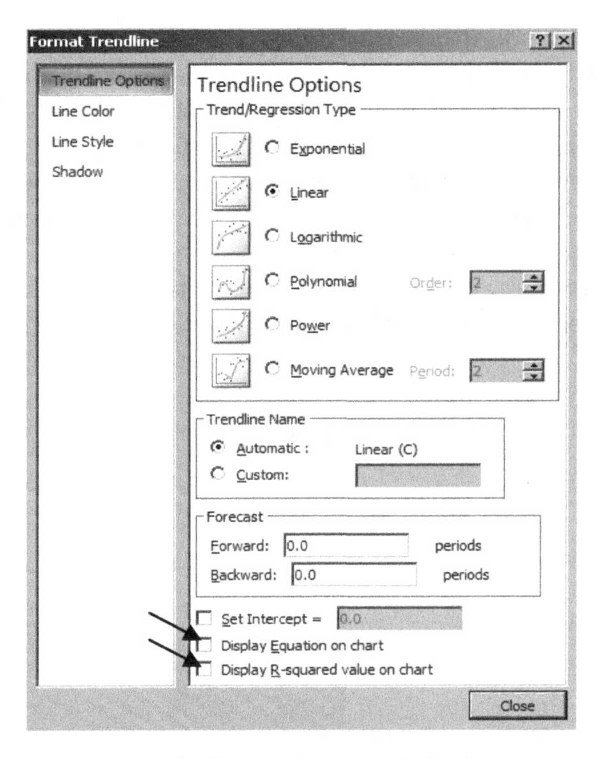

Fig. 5.12 Format trendline dialog box

Mathematically, the simple linear regression equation is written as

$$\boxed{\hat{Y}_i = a + bX_i} \tag{5.28}$$

where a and b are unknown parameter values which are estimated by the least squares method using the data. Figure 5.11 shows that **the intercept** of the line with the vertical consumption axis is estimated as $a = -254.95$. This means that when the income is zero, there is a negative consumption, 254.95 (billion dollars), and the economy needs positive income to have positive consumption. **The slope** of the line is estimated as $b = 0.714$. It means that when income increases by \$1 (billion), consumption will increase by \$0.714 (billion), and the rest will be saved. The values $a = 254.9$ and $b = 0.713$ are called **the least squares estimates** of the above parametric equation. $\hat{Y}_i$ (or $\hat{C}_i$ in (5.27)) is the estimated value of Y_i (or C_i) when the estimated values of a and b are substituted in (5.28). It is different from the actual consumption C_i.

5.7 Summary

In this chapter, we have discussed derivation and calculation of some basic descriptive statistics, using spreadsheet tables and charts. We first discussed the rules of summation, and then defined the mean, the population variance and standard deviation, and the (sample) variance and standard deviation. These are defined for a single variable. If two variables are considered, then the relations between the two variables are measured by the covariance and the correlation coefficient. We emphasized intuitive interpretations of these statistical measurements, and illustrated and simulated by using spreadsheet tables. We also introduced the concepts of growth rate and elasticity, which is defined as the ratio of two growth rates. These two concepts are used extensively in Economics and Statistics, and also in recent newspapers and popular magazines.

Lastly, we show how to draw a scatter diagram for two variables and find the regression equation using the Excel trend-line method. The theory of linear regression will be discussed in Chapter 7.

Review of Statistical Formulas used in this Chapter

$$\bar{X} = \frac{X_1 + X_2 + \cdots + X_n}{n} = \frac{\sum_{t=1}^{n} X_i}{n} = \frac{\text{sum}}{n} \tag{5.9}$$

$$\sum (X_i - \bar{X}) = \sum x_i \equiv 0 \tag{5.10}$$

$$\text{varp}(X) = \frac{\sum_i^n (X_i - \bar{X})^2}{n} \equiv \frac{\sum_i^n x^2}{n} = \frac{\text{SSx}}{n} \equiv s_n^2 \tag{5.12}$$

$$\text{SSx} = \sum_i^n (X_i - \bar{X})^2$$

$$\text{stdevp}(X) = \sqrt{\frac{\sum_i^n (X_i - \bar{X})^2}{n}} \equiv \sqrt{\frac{\sum_i^n x^2}{n}} = \sqrt{\frac{\text{SSx}}{n}} = \sqrt{\text{varp}(X)} = s_n \tag{5.13}$$

$$\text{var}(X) = \frac{\sum_i^n (X_i - \bar{X})^2}{n-1} = \frac{\sum_i^n x^2}{n-1} = \frac{\text{SSx}}{n-1} = s^2 \tag{5.14}$$

$$\text{var}(X) = \frac{n}{n-1} \text{varp}(X) \tag{5.15}$$

$$\text{stdev}(X) = \sqrt{\frac{\sum_i^n (X_i - \bar{X})^2}{n-1}} = \sqrt{\frac{\sum_i^n x^2}{n-1}} = \sqrt{\frac{\text{SSx}}{n-1}} = \sqrt{\text{var}(X)} = s \tag{5.16}$$

$$\text{CVp} = \frac{\text{stdevp}}{\text{mean}} 100 = \frac{s_n}{\bar{X}} 100, \quad \text{CV} = \frac{\text{stdev}}{\text{mean}} 100 = \frac{s}{\bar{X}} 100 \tag{5.17}$$

$$\text{cov}(X, Y) = \frac{\sum_i (X_i - \bar{X})(Y_i - \bar{Y})}{n} = \frac{\sum_i x_i y_i}{n} = \frac{\text{SSxy}}{n} \tag{5.19}$$

$$\text{SSxy} = \sum_i (X_i - \bar{X})(Y_i - \bar{Y})$$

$$r(X, Y) \equiv \frac{\text{cov}(X, Y)}{\text{stdevp}(X) * \text{stdevp}(Y)} = \frac{\text{SSxy}}{\sqrt{\text{SSx SSy}}} \tag{5.20}$$

$$g_t = \frac{X_t - X_{t-1}}{X_{t-1}} = \frac{\Delta X_t}{X_{t-1}} \tag{5.21}$$

$$X_t = (1 + g_t) X_{t-1} \tag{5.22}$$

$$X_t = X_{t-1} + g_t X_{t-1} \tag{5.23}$$

$$e_{CX} = \frac{g_C}{g_X} = \left(\frac{C_t - C_{t-1}}{C_{t-1}}\right) \Big/ \left(\frac{X_t - X_{t-1}}{X_{t-1}}\right) = \frac{\Delta C_t}{C_{t-1}} \Big/ \frac{\Delta X_t}{X_{t-1}} = \frac{\Delta C_t}{\Delta X_t} \frac{X_{t-1}}{C_{t-1}} \tag{5.24}$$

$$\hat{Y}_i = a + b X_i$$

Key terms: Economics and Business

Key terms: Excel

Homework Chapter 5

Descriptive statistics

5-1 Interpretation of the mean As in Fig. 5.1(a), we redraw the mean as the center of gravity as follows (Fig. HW5–1):

(a) Using the x and f columns in Table 5.3, reproduce the chart.
(b) Use a ruler, five pennies or dimes, and a pencil, placing the pencil beneath the ruler at 6 inches distance, and placing each coin at the distance shown in the diagram. Make sure that 6 is indeed the center of gravity in the sense that the ruler is balanced at 6 inches.

(Hints: (a) Select columns A and B, and <"Insert"><Charts, Column!><Cylinder>. Then, click the cylinders, <RM> <3-D Rotation ...><3-D Rotation><Y:,[30]><Close>. To reduce the gap between cylinders, click the cylinders, <RM><Format Data Series><Series Options, Gap Width, [0]><Close> and also change the color of the cylinders. Click any corner of the base and rotate until the desired shape comes out. Draw the other lines using the drawing tool.)

5-2 The frequency table In a survey of 12 students, the amount of pocket money is distributed as follows:

$$0, 1, 1, 2, 2, 2, 5, 5, 5, 5, 10, 10.$$

(a) Find the mean and variance by constructing a table similar to Table 5.3. What is the difference between this example and the example in Table 5.3? Call the new table Table HW5-2a, where $\mathrm{mX} = (\sum \mathrm{f_i X_i})/\mathrm{n}$, the weighted average of $\mathrm{X_i}$, and $\mathrm{n} = \sum \mathrm{f_i}$.
(b) Find the varp, var, stdevp, stdev, CVp, CV for this data set.
(c) Prove algebraically that $\sum \mathrm{f_i}(\mathrm{X_i} - \mathrm{mX}) = 0$. What is the intuitive explanation of this result?
(d) Without using the frequency table like Table 5.3, assume that the sequence of data above is drawn from 12 students A, B, ..., L, like the five students in Table 5.2. Call the new table Table HW5-2b Using the original raw data directly as Table 5.2, construct the mean deviations $\mathrm{X} - \mathrm{mX}$ like column D of Table 5.2, and show that the sum of the mean deviation is zero. Thus, the above formula is an identity.
(e) Show that the "population" variance (varp) and variance (var) calculated from the frequency table Table HW5-2a is the same as those calculated from non-classified data table Table HW5-2b.
(f) Using the frequency table, Table HW5-2a, reproduce the three figures in Fig. 5.1. Give the three interpretations of the mean.

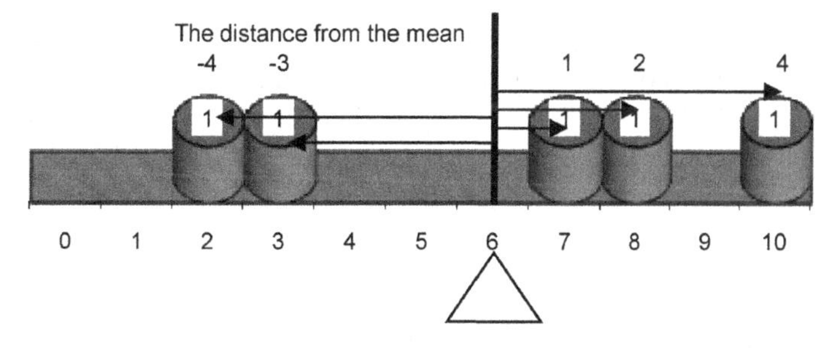

Fig. HW5-1 Interpretation of the mean

(Hints: (a) (b) Table HW5-2a should have 7 columns: X_i, f_i, $f_i * X_i$, mX, Xi-mX, $f_i(X_i - mX)$, and $f_i(X_i - mX)/2$. (c) Use the definition of mX. (d) Use Table 5-2.)

5-3 Correlation coefficient Using the format of Table 5.5, (a) enter 10 integer random numbers from 0 to 10 in columns X and Y, and the correlation coefficient r. (b) Reproduce Fig. 5.7 for r = 0, 0.3, 0.5, 0.7, and 1 by entering <F9> repeatedly until the desired r comes out. When it does, stop and picture copy the desired chart to a blank space. Repeat for all the r's. (c) Rearrange the six pictures as in Fig. 5.7. (d) Can you find r = 1 easily? If not, devise a way to produce r = 1, using restricted random numbers for X or Y.

(Hints: (d) You may artificially restrict the entries of Y to produce an r = 1 chart. Enter X as =rand() in B5, say, then in the next cell, C5, enter Y as B5+rand()*4 to restrict the range of Y, instead of using the pure random number.)

Descriptive statistics and regression

5-4 Consumption and income Given income and consumption data (C_t, X_t), as shown in the enclosed box in Table 5.7:

(a) Reproduce the whole table.
(b) Reproduce the scatter diagram of income and consumption, Fig. 5.11.
(c) Reproduce the simple regression equation (5.27).

5-5 A simulation model We may generate a dataset as follows. Let consumption C and income X be generated by the following formula:

$$C(t) = 400 + C(t-1) + 2000^{*}\text{rand}(),$$

$$X(t) = 500 + X(t-1) + 3000^{*}\text{rand}(),$$

where $C(t-1)$ and $X(t-1)$ are the values of C(t) and X(t) in the previous period, starting from $t = 0, 1, \ldots$ to $t = 20$. Let $C(0) = 1500$, $Y(0) = 2000$. (C(0) and X(0) are called the **initial conditions.**) Construct a table similar to Table 5.7 for C and X. C and X are in billions of dollars and are integers. Note that the formulas should start from t = 1 in Table HW5-5. Before you calculate the descriptive statistics, move the original table to sheet2 and then **change columns**

Table HW5-5 The dataset of the simulation model

Period	C(t)	Y(t)
0	C(0)	Y(0)
1	C(1)	Y(1)
2		
⋮		
20	C(20)	Y(20)

C and X to values. To change the active cells to values, select the dataset of C and X and enter[16]

<^C> <"Edit"> <Clipboard, Paste!> <Paste Values>.

Move the point to any data cell and make sure that you have a number, not formulas, in the formula bar. Answer the three questions in HW5-4.

Regression analysis

5-6 Demand analysis The **demand for Asian pears** in Denver area from 1990, 1991, ..., 2004 is given in Table HW5-6. D is pears measured in 1000 dozens, and P is the price per dozen. D and P are random variables, D ranges from 50 to 100, and P from 5 to 10. All are in integers. (Note that a random number from a to b is given by y=round(rand()*(b−a)+a, n), where $a \le b$, and n is the number of digits after the decimal point).

Let the OLS regression equation be

$$D = a + bP.$$

(a) Draw the scatter diagram for (D, P). Make sure that P is the independent variable, and should be on the horizontal X-axis.
(b) Fit a regression line to the above scatter diagram using the trend-line command of Excel.
(c) Find the OLS estimate of a and b, and write the regression equation from the trend line.
(d) What do you call a and b mathematically? What is the economic meaning of a and b?

Table HW5-6 A demand analysis in Denver area

Variable	1000 dozen	per dozen, US$
	Pears D	Price P
1990	45.56	9.23
...		...
2004	40.02	8.54

5-7 Estimation of supply function Table HW5-7 shows the data on Jamaican coffee supply (S) from 1995 to 2010. The output of coffee is measured in pounds and the price of coffee (P) is in

Table HW5-7 A supply analysis for Jamaican coffee

Variable	Coffee (Pound) S	Price P
1995	29213	1.41
...	...	...
2010	45242	1.39

[16]You may also do as <^C > <"Edit"> <Clipboard, Paste!> <Paste Special> <Paste, @Values> <OK>. In this case, a Paste Special dialog box appears and you have a choice of many options.

US$. S and P are random variables: S ranges from 20,000 to 50,000, in integers, and P from 1.00 to 1.50, which has two digits after the decimal point.
(Note that a random number from a to b is given by y=round(rand()*(b−a)+a,n), where $a \leq b$, and n is the number of digits after the decimal point).

Let the OLS regression equation be

$$S = a + bP.$$

(a) Draw the scatter diagram for (S, P). Make sure that P is the independent variable, which should be on the horizontal X-axis.
(b) Fit a regression line to the above scatter diagram using the trend-line command of Excel.
(c) Find the estimated (or predicted) regression equation (that is replace a and b by their estimated values).
(d) What do you call a and b mathematically? What is the economic meaning of a and b in this example?

5-8 SAT scores Table HW5-8 shows the SAT scores of college-bound seniors, where V_f, V_m, and V_t denote the verbal scores for female, male, and male and female, respectively (and similarly for the mathematics scores: M_f, M_m, and M_t). Copy the table to a new sheet and answer the following questions (You should delete irrelevant data columns).

(a) Using the manual method shown in Table 5.7, find the mean, variance, standard deviation, correlation coefficient, and CV for V_m and M_m
(b) Using the Excel formula as shown in the lower part of Table 5.7, find the mean, variance, standard deviation, correlation coefficient, and CV for V_m and M_m
(c) Draw a scatter diagram for V_m and M_m. Make sure that M_m is the independent variable and V_m is the dependent variable. What general observation can you make from the diagram?
(d) What is the relationship between male verbal scores (V_m) and male math scores (M_m)? Answer this question by finding the OLS estimation of the following linear form:

$$V_m = a + bM_m.$$

Table HW5-8 Mean SAT scores for college-bound seniors 1985–1990

	Verbal			Math		
	Males	Females	Total	Males	Females	Total
Var	V_m	V_f	V_t	M_m	M_f	M_t
1985	437	425	431	499	452	475
1986	437	426	431	501	451	475
1987	435	425	430	500	453	476
1988	435	422	428	498	455	476
1989	434	421	427	500	454	476
1990	429	419	424	499	455	476

Source: The College Board. The New York Times, August 28, 1990, August 28, 1990, p. B-5 as listed in Gujarati (1992), 136.

Rewrite the model in an estimated (numeric) equation. What are the economic meanings of the values of a and b in this case? Do they make sense?

(e) Will it make sense to regress V_m on M_f, or M_f on V_m? Why?

5-9 SAT scores Copy Table HW5-8 to a new sheet and answer the following questions (You should delete irrelevant columns).

(a) Draw a scatter diagram for V_f and M_f. What general observation can you make from the diagram?

(b) What is the relationship between female verbal scores (V_f) and female math scores (M_f):

$$V_f = a + bM_f.$$

Rewrite the model in an estimated (numeric) equation. What are the economic meanings of a and b in this case? Do they make sense?

(c) Will it make sense to regress M_f on V_f? Why?

5-10 US patents and R&D expenditure Table HW5-10 provides the total number of US patents (P, in thousands) issued annually from 1983 to 1989, and the R&D expenditures (R, in billions of 1982 dollars).

(a) Draw a scatter diagram for (R, P), using R as the independent variable.

(b) Find the OLS estimation of the following equation:

$$P = a + bR.$$

(c) Do the estimated values of a and b make sense? Give an economic interpretation of the estimated values of a and b.

(d) Does the equation as whole make sense? Why?

Table HW5-10 US patents and R&D expenditures

Var	US$ R&D	1000 Patents
	RD	Pa
1983	85.7	112.1
1984	93.8	120.3
1985	102.4	126.8
1986	105.2	132.6
1987	108.3	139.5
1988	111.4	151.5
1989	112.1	165.8

Source: Ramanathan (1995), p. 127.

Chapter 6

Random Numbers and Frequency Distributions — Organizing a Large Data Base

Chapter Outline

Objectives of this Chapter

Using the provision of the random number generator in Excel, this chapter relates the random numbers with the standard uniform distribution, and transforms it into more general uniform distribution. Since the value of a random number depends on chance, random numbers and probability are a basic part of statistics. We therefore introduce the different definitions of probability, and show how frequency theory and the axiomatic approach to probability can be explained using the Excel spreadsheets. In fact, using a spreadsheet, a sample of tossing a die 1000 times can be generated instantly, and using the range copy command (RCC), the 1000 sample points can be counted and conveniently classified into class intervals called bins. We then generate the relative frequency distribution of the sample and compare it with the theoretical probability distribution of the population. Excel will show how the former "dances" around the latter.

This chapter shows the method of calculating the mean, the variance, and the standard deviation of grouped data. The statistical characteristics calculated from a sample of a large data base are then compared with those of the population. The ability of spreadsheets to split the window into two parts makes copying and managing the data range a breeze. The method of working on a discrete variable distribution can be applied directly to a continuous normal distribution. In the last part of the chapter, we show how the methods are applied.

6.1 Random Variables

By now we are familiar with the use of random numbers on spreadsheets. The best real life example of random numbers is probably the weekly or monthly **lottery** drawings. In a lottery drawing, ten balls are bouncing around inside a box. Each ball has a number from 0 to 9. One ball is drawn "randomly". After the number is recorded, the ball is returned to the box, and mixed with other balls again. Thus, we can say that we have an infinite **population**. In each drawing, each ball has an **equal chance** of being chosen. Generally, 10 digits are drawn consecutively. Each digit is a sample point or unit. The number of sample points in the sample, in this case, the 10 digits, is the **sample size**, and each **sample** is an **observation** or "**realization**" of the **random variable**[1] X of the lottery drawing.

The word **random** usually refers to a selection process in which each object in a finite set has an equal chance of being chosen. More specifically, the number is **random** in the sense that

(a) Each number is equally likely to be chosen (**uniformity**);
(b) The choice of each number does not depend on the appearance of the previous numbers (**independence**);
(c) The consecutive numbers have no discernable pattern or repetition (**no periodicity**).

6.1.1 *Discrete and continuous random variables*

A random variable can be discrete or continuous. In the previous chapters, we have used the random number generating function =rand(), which gives a 15-digit random number from 0 to 1, like

$$0.609913568710908$$

Clearly, this number is continuous, since there are infinite digits behind the decimal point. Thus, we have seen two kinds of random variables: one is a **discrete random variable**, like the 10 digits in a lottery drawing, and the numbers obtained in tossing a die; the other kind is a **continuous random variable**, like the whole (not truncated) random number generated from the random number generating function above, the height or weight of children at ages, say, 5–10, or the measurement of errors in product weights.

---

[1]Mathematically, a random variable is a numerical-valued function defined on a sample space of events. We leave the details of the theory of probability and statistics to statistics textbooks; see, for example, Hoel (1956), Chapter 2.

Random variates

Since a random variable is a variable whose value depends on chance, we can assign values to a random variable with a certain probability. In this sense, a random variable is different from the usual mathematical variable we have encountered in Chapters 1–4. To distinguish it from the usual mathematical variable, it is also called a **random variate**. In general practice, we use the terms random variable, random number, and random variate interchangeably.

In the lottery drawing, if each digit, $0, 1, \ldots, 9$, is equally likely to be drawn, then we can assign probability of $1/10$ to each digit. This leads to the concept of a **frequency function.**[2] "A function $\mathrm{p(x)}$ that yields the probability that the random variate x will assume any particular value in its range is called the **frequency function** of the random variate x". It is also called the **probability mass function** (pmf), or simply, the **mass function**. Hence, in our example, we can write the mass function of the lottery drawing as

$$\mathrm{p(x)} = 1/10 \quad \text{for } \mathrm{x} = 0, 1, \ldots, 9. \tag{6.1}$$

Example 6.1 Tossing a die or a coin. In tossing a fair die, let the random variable x be the number on the face of the die, $\mathrm{x} = 1, 2, \ldots, 6$. Then, the mass function is

$$\mathrm{p(x)} = 1/6 \quad \text{for } \mathrm{x} = 1, 2, \ldots, 6. \tag{6.2}$$

In tossing a fair coin, let x be either a tail $(\mathrm{x} = 0)$ or a head $(\mathrm{x} = 1)$. Then, pmf is $\mathrm{p}(0) = 1/2$, and $\mathrm{p}(1) = 1/2$. □

6.2 Definitions of Probability

In the above examples, the random variable x is associated with probability. There are several ways to define probability.

6.2.1 *The classical theory of probability*

In general, if there are n numbers or chips, then the probability of the appearance of each number or chip is $1/\mathrm{n}$. This definition of probability is called the **classical probability**, or **a priori probability**. This is what we reasoned in (6.1): that the probability of the occurrence of any digit in drawing a ball is $1/10$, and the probability of a particular number coming up in tossing a fair die in (6.2) is $1/6$. Since there is no solid reason to believe that the probability is indeed $1/\mathrm{n}$, this definition of probability is also called **the probability based on insufficient reason.**

6.2.2 *The frequency theory of probability*

Another definition of probability is based on experiments. The **frequency theory of probability** says that as the sample size n increases (that is, as we toss the die more times), the

---

[2]See Holt (1956), p. 16. There are some confusion in terminologies. When referring to sample data, we use (relative) frequency function, and for the theoretical data, we use mass function or probability function.

sample relative frequency of the ith event (say, of an ace — one dot — coming up) is defined as the number of times the ace comes up, f_1, divided by the number of tosses, f_1/n (assuming the die is fair, of course). Thus, f_1/n will approach the population frequency of the occurrence of ace, that is, the probability $p_1 = 1/6$. In mathematical notation, it is the limit expressed as

$$\lim_{n \to \infty} \frac{f_1}{n} = p_1. \tag{6.3}$$

This holds for the occurrences of other events. The probability defined in this way is a theoretical or conceptual counterpart of the empirically observed sample relative frequency.

Example 6.2 The occurrence of the ace in tossing a die. Table 6.3 shows tossing a fair die 1000 times. We will explain later how the table in columns B–K is constructed. Here we only need to note that the die we use is a fair die and it is tossed 1000 times. Using an Excel command, which we will explain later, we have counted that the ace appeared $f_1 = 172$ times, 2 dots appeared $f_2 = 179$ times, and 6 dots appeared $f_6 = 165$ times. These are the **frequencies** $f_i, i = 1, 2, \ldots, 6$. of the occurrence of the i dots corresponding to the ith value of the random variable X. In Table 6.3, the random variable is listed in L4:L9, and the corresponding frequencies are listed in M4:N9. The dataset grouped in a frequency table like L4:M9 is called **grouped data**. The **relative frequency**, which is the frequency f_i divided by the total number of tosses, $n = 1000$, is listed in N4:N9. Thus, the relative frequency of the occurrence of ace in this experiment is 0.172, and that of the occurrence of two dots is 0.179, etc. The probability of each occurrence, 1/6, as defined in (6.2), is listed in P4:P9.

What Table 6.3 implies is that the relative frequency of the ace, which is currently 0.1720, will converge to $1/6 = 0.16666667$ if we continue tossing a die without limit, that is, to $n = 10{,}000$, $100{,}000$, $1{,}000{,}000$, and up to infinity. The limiting value is the probability of the occurrence of the ace. In other words, the relative frequency f_i/n of the ace, $i = 1$, and for that matter, the other numbers, $i = 2, 3, \ldots, 6$, will approach the theoretical probability $p = 1/6 \approx 0.1667$ as n goes to ∞. □

6.2.3 *Subjective probability*

In many cases, experiments like tossing a die cannot be done. For example, we say that the chances (probabilities) are 50–50 that the recent strike of automobile workers will end before Christmas, or our university football team will probably win its next game. These are **subjective probabilities** which measure or reflect individuals' degrees of confidence. They are not based on repeated experiments.

6.2.4 *Axiomatic theory of probability*

All the above theories of probability satisfy certain fundamental relations which may be postulated as **axioms**. Let $p(x)$ be a function of (discrete) random variable x, which takes values x_i. p(x) is a **probability mass function (pmf)** if it satisfies the following properties:

$$p(x_i) \geq 0, \tag{6.4}$$

$$\sum_{i=-\infty}^{i=\infty} p(x_i) = 1, \tag{6.5}$$

$$\sum_{i=a}^{i=b} p(x_i) = \text{Prob}(a \leq x_i \leq b), \tag{6.6}$$

where $\text{Prob}(a \leq x_i \leq b)$ is the probability of x_i in any interval between arbitrary numbers a and b of the random variable. Equation (6.4) says that the mass function must be nonnegative, (6.5) states that the function must sum to one for all values of the random variable, and (6.6) says that the function must represent the probability of the occurrence between any arbitrary interval of a and b, $a < b$.

The listing of all the possible values of x and the probabilities associated with x is called the **discrete probability distribution**, comparable to the relative frequency distribution from a sample.

Example 6.3 Lottery drawing. In the case of lottery drawing (6.1), condition (6.4) is clearly satisfied for $i = 0, 1, \ldots, 9$. Condition (6.5) holds since

$$\begin{aligned}\sum_{i=-\infty}^{i=+\infty} p(x_i) &= 1 = p(0) + p(1) + \cdots + p(9) \\ &= 1/10 + 1/10 + \cdots + 1/10 = 1.\end{aligned}$$

For any other point in the interval between $-\infty$ and $+\infty$, say, $x_i = -5$ or $x_i = 11$, we have $p(-5) = 0$ or $p(11) = 0$, etc. Hence condition (6.5) is satisfied. Lastly, condition (6.6) is satisfied since the probability of the joint occurrence of any interval, say, $a = 3$ to $b = 5$, is

$$p(3) + p(4) + p(5) = 1/10 + 1/10 + 1/10 = 3/10,$$

which is the probability of the joint occurrence of $x_i = 3, 4$, and 5. In fact, any fixed interval of three numbers, as shown along the X-axis in Fig. 6.1, gives the same probability. □

Example 6.4 Illustration. Example 6.3 can also be seen from the frequency table of the last chapter. Table 6.1 illustrates the three conditions of (6.4)–(6.6). The 10 possible values of the random variable x in lottery drawing are listed in column A, and its mass function is listed in column B. They are illustrated in Fig. 6.1. Note that the mass function (6.4) gives the height of the bars. Clearly, from the entries in B3:B12, the nonnegativity condition (6.4) is satisfied. Cell B13 shows that all the probabilities add to one. Hence, condition (6.5) is satisfied. Any fixed combination of the values of the random variable in column A yields the same probability of the occurrence of the same combinations in column B. For example,

Table 6.1 Lottery drawing

	A	B	C	D
1	x			
2		p	px	p(x-mu)^2
3	0	0.1	0	2.025
4	1	0.1	0.1	1.225
5	2	0.1	0.2	0.625
6	3	0.1	0.3	0.225
7	4	0.1	0.4	0.025
8	5	0.1	0.5	0.025
9	6	0.1	0.6	0.225
10	7	0.1	0.7	0.625
11	8	0.1	0.8	1.225
12	9	0.1	0.9	2.025
13	sum	1.000	4.5	
14	mean (mu)		4.5	
15	variance (sig2)			8.25

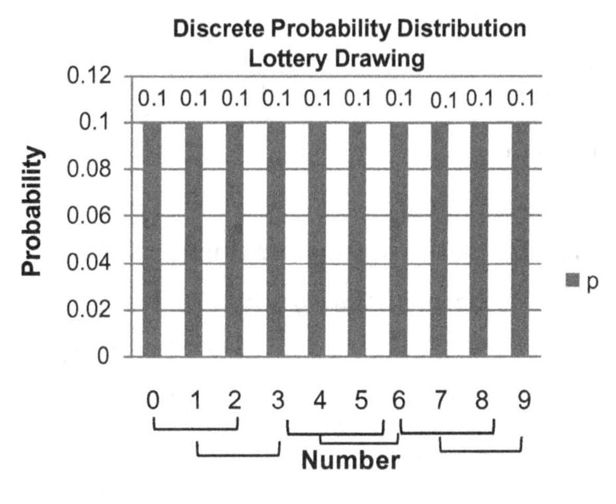

Fig. 6.1 Discrete uniform distribution

the combinations of any three values of x, (0, 1, 2), (1, 2, 3), (4, 5, 6), (7, 8, 9), etc., all yield the same total probability of 3/10. □

Discrete uniform distribution

In general, if x is a discrete random variable ranging from a to b, where a and b are any integers $(\ldots, -2, -1, 0, 1, 2, \ldots)$ and $a < b$, then its probability mass function is

$$\boxed{P(x) = \frac{1}{b - a + 1} = \frac{1}{n},} \tag{6.7a}$$

where $n = b - a + 1$. The ordered list of this frequency function is called the **discrete uniform distribution**. The distribution is "uniform" since, as shown in Examples 6.3 and 6.4, any interval of this random variable will give the same (uniform) probability. It can be shown that (6.7a) is distributed with

$$\text{Mean } \mu = \frac{a + b}{2} \text{ and variance } \sigma^2 = \frac{n^2 - 1}{12}, \tag{6.7b}$$

where, for the theoretical (population) distribution, we use μ (mu) to denote the mean and σ^2 (sigma square) to denote the population variance. The population standard deviation is then the positive square root of σ^2 and is denoted as σ (sigma). (See Sec. 6.6 below).

Example 6.5 Mean and variance of a discrete uniform distribution. In the case of lottery drawing, there are 10 digits ranging from 0 to 9. Hence, $a = 0$, $b = 9$, and $n = 9 - 0 + 1 = 10$. Substituting in the mass function, we have $p(x) = 1/10$, and from (6.7b), $\mu = (0 + 9)/2 = 4.5$, $\sigma^2 = [(10^2) - 1]/12 = 8.25$. The direct calculation is shown in

Table 6.1. The p(x) = 1/10 is entered in column B, px in column C, the values of which sum to mean 4.5 in C13. The squared mean deviation is calculated in column D and the column sum is the variance, which is 8.25 in D15.

Since the value of x can be arbitrary, let x ranges from 4 to 8. Then, n = 5, p(x) = 1/5, and $\mu = 12/2 = 6$, $\sigma^2 = 24/12 = 2$. These results can be calculated directly using a table like Table 6.1. □

6.3 The Continuous Uniform Distribution

6.3.1 *The probability density function (pdf)*

In the previous sections, for a discrete random variable, the unit of width can be taken as 1, and we calculate the probability associated with each value of the random variable, as we have done in Sec. 6.2. For a continuous random variable, there are infinite numbers of values (or points) in any interval, and thus we cannot list all the possible numbers to assign as probabilities. Even if we could, the probability would be zero for all values of a random variable, as can be seen from (6.7a), when n goes to infinity. In the continuous case, the frequency function has height but no width, and it is called the **probability density function** (pdf), or simply the **density function**, and is denoted f(x). In this case, probability can only be defined as an area under the curve of the pdf on an interval, say between a and b, where the interval may be infinitesimally small, and may be denoted dx. Hence, for the continuous random variable, the probability is defined as f(x)dx. The product, f(x)dx is called the **probability element**.

For a continuous random variable, the summation signs in the definition of probability (6.4)–(6.6) should be replaced by integral signs, and the subscript i should be dropped (since we cannot count x), as shown in Appendix 6A. In this case, condition (6.4) holds for the pdf, $f(x) \geq 0$. Condition (6.5) is the integral of f(x) from $-\infty$ to $+\infty$, and must be 1. Condition (6.6) also holds, that is, the area of an arbitrary interval between a and b represents probability.

6.3.2 *The continuous standard uniform distribution*

Let x be a continuous random variable between 0 and 1. This means that the random variable =rand() is a point in the interval [0, 1]. From condition (6.5), the total area under the curve in this interval must be 1. Since the area is width*height, the height (f(x)) of the curve must be 1, that is, the density function for the random numbers from 0 to 1 is

$$\boxed{f(x) = \begin{bmatrix} 1 & \text{for } 0 \leq x \leq 1 \\ 0 & \text{for } x < 0 \text{ and } x > 1 \end{bmatrix}} \tag{6.8a}$$

Furthermore, any interval of the same length, say, 0.05, has the same probability (that is the area under the horizontal line through f(x) = 1), whether it is between 0 and 0.05, or between 0.10 and 0.15, or 0.70 and 0.75. Thus the continuous uniform distribution has a shape like that in Fig. 6.2(a). This distribution is called the **standard uniform distribution**, and denoted $U(0, 1)$.

It can be shown that this uniform distribution has the following distribution characteristics:

$$\text{mean}\ \mu = 1/2, \quad \text{variance}\ \sigma^2 = 1/12, \tag{6.8b}$$

which is a special case of (6.10b) below.

6.3.3 *Transformations of random numbers*

In Chap. 1, we have shown that in order to find a random number from 0 to k, say $k = 30$, we multiply a random number =rand() by k, say, =rand()*30. Thus, in this case, we take $a = 0$ as the initial value and $b = 30$ as the ending value. How can we generate a random number, say from $a = 50$ to $b = 70$? From the explanation above, we know that if we take $k = b-a = 70-50$, then, =rand() $*$ $(70-50)$ will generate a random number from 0 to 20. Since we are interested in a random number starting from 50, a natural way to generate a random number from 50 to 70 is to add the initial value $a = 50$ to the random number generated between 0 and 20. Therefore, a random number y from 50 to 70 can be derived from the formula

$$y = \text{rand}()^*(70-50)+50 = \text{rand}()^*20+50.$$

Formally, we have the following proposition.

Proposition 6.1. *Let* x *be a random number*[3] *from* 0 *to* 1, *and let* a *and* b *be two fixed numbers such that* $a < b$. *Then the random number* y *between* a *and* b *is given by*

$$\boxed{y = (b - a)x + a} \tag{6.9}$$

Illustration

Equation (6.9) is illustrated in Fig. 6.2. Since $(b - a)$ is the distance between a and b, and since x, say 0.55, is a fraction given in Fig. 6.2(a), when x is multiplied to the interval $(b - a)$, the product $(b - a)x$ is the proportion of x (in this case, 55%) in interval $(b - a)$ measured from a, as shown in Fig. 6.2(b). Adding a to $(b - a)x$, we have the distance of x from the origin, which is y in (6.9). Column (2) of Table 6.2 gives the random number x_i, and column (3) gives the transformation of x_i to y_i in the interval between $y = a = 1$ and $y = b = 6$.

Example 6.6 The density function of a continuous random variable from a to b. Since y is a function of a random variable x, it is also a random variable. What is the frequency function of y?

---

[3]Here the words random number may be substituted by random variate or random variable. Thus, the proposition is about transforming a standard uniform random variate to a general uniform random variate.

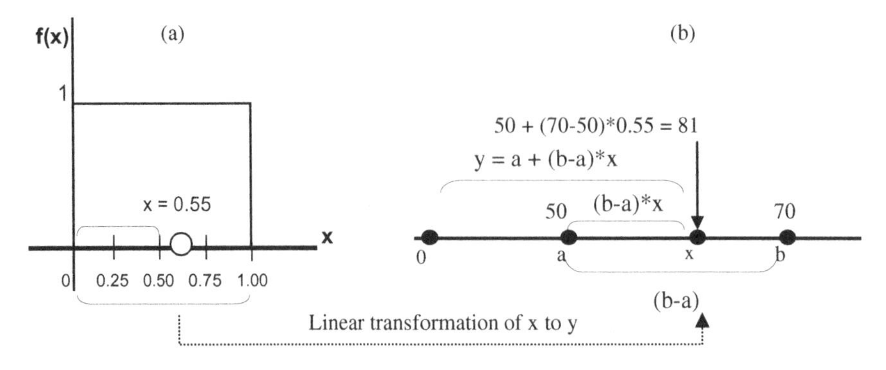

Fig. 6.2 The standard uniform distribution (a) and its transformation (b)

Answer: We use condition (6.5): the area of the probability distribution between a and b must be 1, since area = width $*$ height = 1. But we already know that the width is b − a, hence, the height = 1/(b − a). The density function is

$$f(y) = \begin{cases} \dfrac{1}{b-a} & \text{for } a \le y \le b, \\ 0 & \text{for } y < a \text{ and } y > b. \end{cases} \tag{6.10a}$$

□

Any density function which has this functional form is called the **continuous uniform distribution**, and denoted $U(a, b)$, indicating that the position and the shape of the distribution depend only on the parameters a and b. Note that when $a = 0, b = 1$, (6.10a) reduces to the standard uniform distribution[4] $U(0, 1)$ in (6.8).

It can be proved that the uniform density function (6.10a) is distributed with

$$\text{Mean } \mu = \frac{a+b}{2}, \quad \text{variance } \sigma^2 = \frac{(b-a)^2}{12}. \qquad ((6.10b))$$

6.4 Tossing a Fair Die

Since the random variable =rand() is continuous, and the occurrence of the numbers in tossing a die is expressed in integers (discrete random variable), if =rand() is used to simulate the occurrence of integers, we have to convert the former to the latter.

[4]What we have done so far is to transform standard uniform random variate x in (6.8) to general uniform random variate y in (6.10). Conversely, we may ask how to transform the general uniform random variate y in (6.10) to the standard uniform random variate x in (6.8). Or more generally, given the frequency function of certain distribution, like (6.10) or the normal distribution, how to derive the random variate with that distribution (like uniform or normal distribution). This is the problem known as **generating random variate with a particular distribution**, and used in the simulation analysis. One of the methods is the **inverse function method**. See Meier *et al.* (1969), Chapter 8.

6.4.1 *Transformation of a continuous random variable to a discrete random variable*

For tossing a die, we take a = 1, b = 6 in (6.9). Since x is =rand() in Excel, we have

$$y = (6-1)^{*}\text{rand}()+1 \tag{6.11}$$

This formula will give a 15-digit continuous random number between 1 and 6.

Since the numbers on a die are integers, we have to round them to convert the decimal numbers into **integers.** The **rounding formula** in Excel is:

$$=\text{round}(x, n), \tag{6.12}$$

where x is a number or the cell name in which a number is located, and n is the number of digits after the decimal point. For integers, we should take n = 0.

Combining (6.9) and (6.12), a random number of integers from 1 to 6 is

$$=\text{round}((6-1)^{*}\text{rand}()+1,0). \tag{6.13}$$

The following example explains the meaning of this transformation.

Example 6.7 Illustration. For tossing a die 10 times, (6.12) and (6.13) are illustrated in Table 6.2. Column (1) shows the order of 10 experiments, column (2) shows 10 random numbers x in $[0, 1]$, which are transformed into other random numbers between 1 and 6 in column (3). Using (6.12), they are then rounded to integers in column (4). Note the process of converting a **continuous random number** to a **discrete random number** in column (4) by the use of Excel equation, =round(y, 0). □

Table 6.2 The difference between a crooked die and a fair

		A crooked die	Sample	A fair die	Sample
	x	y		y	= round(y, 0)
Tossing No.	rand()	=rand()∗(6 − 1)+1	=round(y,0)	=rand()∗(6.5−0.5)+(0.5)	
(1)	(2)	(3)	(4)	(5)	(6)
1	0.20137	2.00685	2	6.23089	6
2	0.92743	5.63715	6	6.30846	6
3	0.89898	5.49492	5	5.78419	6
4	0.94130	5.70648	6	5.13928	5
5	0.80839	5.04197	5	0.79796	1
6	0.89516	5.47581	5	3.56174	4
7	0.92423	5.62117	6	0.68498	1
8	0.72313	4.61566	5	5.74960	6
9	0.36423	2.82114	3	1.68946	2
10	0.29753	2.48766	2	6.43198	6

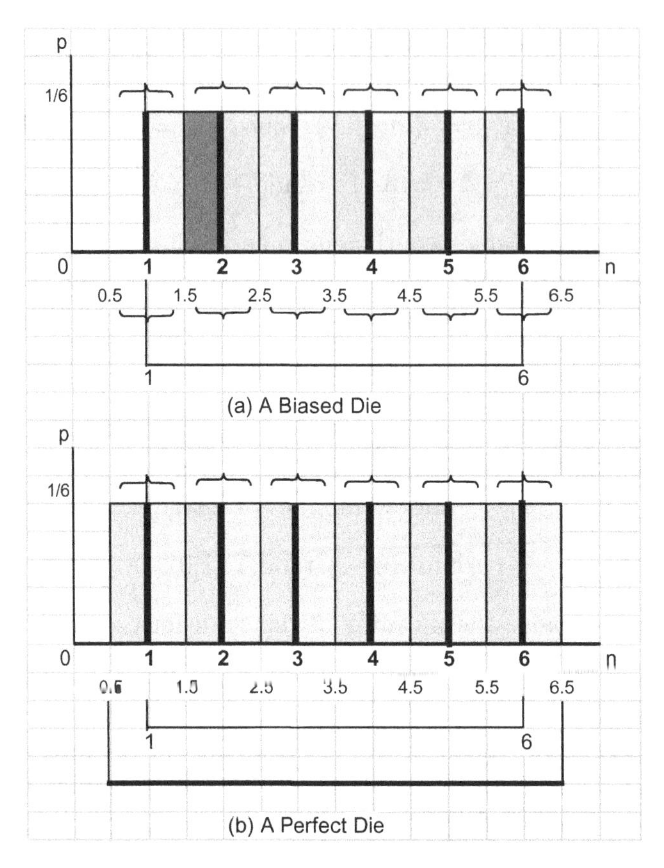

Fig. 6.3 Finite sample corrections

6.4.2 *Finite sample correction*

As shown in Fig. 6.3(a), the frequency at number 2, for example, is counted by the random numbers falling in the range from 1.5 to 2 and also from range 2 to 2.5 and rounded to 2. However, if a random number falls between 0.5 and 1, or between 6 and 6.5, that number will be discarded as these numbers are not included in the defined range. Thus, the probability of the appearance of 1 and 6 will be halved, and the die is biased. Hence, to insure fairness of the die, the range of the formula should be from a = 0.5 to b = 6.5, as shown in Fig. 6.3(b). This is called the **finite sample correction**. The formula for a fair die should be

$$=\text{round}((6.5-0.5)*\text{rand}()+0.5,0) \tag{6.14}$$

The difference between a biased die and a fair die is also illustrated in columns (3)–(6) in Table 6.2. Random numbers for each tossing are generated in column (2), and are transformed into y without finite sample correction in column (3). In column (4), the random numbers from 1 to 6 are rounded to integers. We can see that the random numbers in column (3) do not include any number from 0.5 to 1 nor from 6.0 to 6.5. However, in column (5), five tosses show these numbers: tossing #1, #2, and #10 are rounded to 6, and tossing #5 and #7 to 1, and these are counted in column (6), but would not appear in column (3), and so will not be counted in column (4).

6.4.3 *The Excel command for a discrete random variable*

Excel has a command for a discrete random variable,

$$=\text{randbetween(a,b)}, \quad a < b, \tag{6.15}$$

which will give integers between a and b, where a and b can be fractions. If we are only generating random integers, then (6.15) can be used. In drawing a number from 1 to 6 by tossing a die, we can define a = 1 and b = 6 in (6.15). However, since Excel does not explain how the equation is constructed, it is not clear whether (6.15) has made the finite sample corrections.

6.5 Constructing a Sample of Size 1000

6.5.1 *The table*

We first prepare the frame for entering data in B3:K103. In Table 6.3, enter the numbers from 1 to 100 in column A, starting from A4, and enter numbers from 1 to 10 in row 3, starting from B3 (please use **fill-handle**). This will define the **range of the data**. You may overshoot the numbers for more than 100 or 10. In this case, select the overshot part and enter <delete> to erase the extra numbers.

Table 6.3 Relative frequency and probability of tossing a fair die 1000 times

	A	B	C	D	E	F	G	H	I	J	K	L	M	N	O	P	Q
1	**Sample of size 1000**											The frequency table					
2	Tossing a die 1000 times											Bin	Freq	Relative freq.	Cum.rel.freq	Probab.	Cum Prob
3		1	2	3	4	5	6	7	8	9	#		f	rf	crf	p	cp
4	1	3	1	1	5	4	2	6	2	5	1	1	172	0.172	0.172	0.16667	0.16667
5	2	5	3	5	2	6	5	3	1	6	6	2	179	0.179	0.351	0.16667	0.33333
6	3	2	3	2	3	6	2	5	4	4	5	3	171	0.171	0.522	0.16667	0.50000
7	4	3	6	6	3	6	5	2	5	4	6	4	161	0.161	0.683	0.16667	0.66667
8	5	6	2	1	1	5	5	6	6	3	6	5	152	0.152	0.835	0.16667	0.83333
9	6	5	5	1	4	1	2	2	1	1	6	6	165	0.165	1.000	0.16667	1.00000
10	7	3	5	1	2	2	4	5	5	1	3						
11	8	4	1	5	2	5	6	3	1	6	2	sum	1000	1.0000		1.00000	
12	9	2	1	5	1	4	6	3	3	4	3						
13	10	6	2	5	3	5	4	1	6	3	2	B4:	=ROUND(RAND()*(6.5-0.5)+0.5,0)				
14	11	6	2	5	6	5	2	5	6	2	2	M4:	=FREQUENCY(sample, bin)				
...	...	...															
103	100	5	1	1	3	5	5	1	6	2	1						

Generating $100 \times 10 = 1000$ *random numbers*

Enter formula (6.14) in cell B4. At B4, use fill-handle to copy B4 to K4. Then, use fill-handle again to copy B4:K4 to B103:K103. This will give you 1000 random integers from 1 to 6.

The column width is too large and the data spread out to the right-hand side of the window. **Optimize the column width** by selecting columns A–K and clicking any column separator twice (see Chapter 1).

This is equivalent to tossing a die 1000 times, but it occurs instantly! Note how easy it is, how wonderful a computer is. This set of data is a sample of size 1000 from tossing a fair die. In this case, the "population" contains infinite numbers of chips numbered from 1 to 6, and tossing a die and recording the number is equivalent to drawing a chip from the population. Enclose the data with a heavy border and name the data range, B4:K103, "sample".

6.5.2 *Counting the frequencies — The range copy command (RCC)*

We now find the frequency of the appearance of each number from 1 to 6:

(1) In L1:Q3 of Table 6.3, enter column labels: bin, f, rf, crf, and P for frequency, relative frequency, cumulative relative frequency, and probability. These terms will be explained below.
(2) In L4–L9, enter the bin numbers from 1 to 6. Name range L4:L9 **bin**.
(3) To enter the *frequency* of each number in the bin, we have to perform a special **range copy command (RCC)** to enter the array formula[5] as follows:

Method I. The range–formula method

Define the range first and then enter the formula as in Table 6.4.

(a) Select range M4:M9, which has the same size as the bin range. The range M4:M9 will be shaded, like Table 6.4.
(b) Enter the following frequency formula in M4:

=frequency(sample,bin). (6.16)

As before, no space is allowed after the comma. See the formula in cell M4 in Table 6.4.
(c) Enter **RCC** as follows:

Ctrl + Shift + <Enter>	(6.17)

That is, holding the control key and shift key at the same time, press the return key. The frequency of each number in the bin appears in M4:M9, as in Table 6.3.

[5]When a formula is entered in an array, it is a **array formula**. In contrast, the single formula we have been using is called the **standard formula**. To implement (6.17), use the left forefinger to press the ctrl key and left middle finger to press the shift key, at the same time, use the right forefinger to press the return key simultaneously.

Table 6.4 The range copy command — Method I

	L	M	N	O
1	The frequency table			
2	Bin	Freq		
3		f		
4	1	=FREQUENCY(sample,bin)		
5	2			
6	3			
7	4			
8	5			
9	6			

Table 6.5 The range copy command — Method II

	L	M			M	N
1	The frequency table					
2	Bin	Freq			Freq	
3		f			f	
4	1	=FREQUENCY(sample, bin)			=FREQUENCY(sample,bin)	
5	2					
6	3		→			
7	4					
8	5					
9	6					

Method II. The formula–range method

Enter the formula first and then define the range. If you forget to select a range, you still can use the RCC as follows:

(a) Enter the frequency formula in the first cell, say, M4, as in Table 6.5.
(b) Select range M4:M9. The range will be shaded. See the RHS of Table 6.5.
(c) Click anywhere in the formula bar to let Excel know that you are dealing with the formula. The window changes from the ready mode to the edit mode (you can see "Edit" at the bottom left corner of the screen), and the X and check marks appear on the left side of the formula box.
(d) Use the RCC as explained in Method I above.

In both methods, after using the RCC, the frequency distribution of the sample of 1000 appears in the range, and each cell is enclosed with $\{\cdots\}$, as in:

$$\{=\text{FREQUENCY(sample, bin)}\}. \tag{6.18}$$

If $\{\cdots\}$ is missing, it means RCC was not performed.[6]

The frequency column in Table 6.3 shows that there are 172 aces in the sample, 179 2's... and 165 6's, a total of 1000 numbers as shown in cell M11. In the old days, you have to tally the 1000 numbers using a notation like

~~////~~ ~~////~~ ///

and counting the tallies. It would take many hours to count in the old way.

Counting the sample size

To make sure that the frequencies add to 1000 in Table 6.3, use AutoSum Σ from the menu and enter the sum two cells below the frequencies, M11. Note that sometimes it may give a number close to 1000, say, 998. See Appendix 6D for the explanation.

Example 6.8 The frequency command and the manual tallying method. To see the difference between the two methods, copy A3:K5 of Table 6.3 to a new sheet. Call the truncated table Table 6.6. Call the sample set B4:K4 of size 10 "s10", and call the sample set B4:K5 of size 20 "s20". Enter bin 1–6 in L4:L9, the frequency of s10 in M4:M9, and that of s20 in N4:N9. Use the RCC at

M4: =frequency(B4:K4,L4:L9),

N4: =frequency(B4:K5,L4:L9).

Table 6.6 The tallying method of counting frequency

	A	B	C	D	E	F	G	H	I	J	K	L	M	N	O	P
1												Frequency			Manual count	
2	Sample of sizes 10 and 20											Bin	s10	s20	s10	s20
3	No.	1	2	3	4	5	6	7	8	9	10		f	f	f	f
4	s10	2	4	2	5	4	4	5	5	5	2	1	0	1	0	1
5	s20	5	4	5	4	6	2	1	5	6	6	2	3	4	3	4
6												3	0	0	0	0
7												4	3	5	3	5
8												5	4	7	4	7
9												6	0	3	0	3
10												sum	10	20	10	20

[6] After performing the RCC, the formula in the range cannot be changed. To edit the formula in whole range, select the range, then, go to the first element in the range, enter <F2(edit)>, change the formula, and then use RCC. To change part of the formula or add or delete some rows or columns in the range, you have to reenter the data and redo the RCC.

Frequencies are entered in M4:N9. Make sure column M sums to 10, and column N sums to 20. Then, manually count the numbers in s10 and s20 and enter the frequencies in columns O and P, respectively, using the manual tallying method. The frequency formula results should match the results from the tallying method. Enter <F9> (the recalculation key) several times and make sure that the frequencies obtained from the RCC match those obtained from the manual counting. □

6.5.3 *Relative frequency, relative and absolute references*

As defined in Sec. 6.2.2, the **relative frequency** is f_i/n, where n is the sample size (total observations), and f_i is the number of observations at each point or value of the random variable.

Relative references

Since it may happen that the total number of tosses may be less than 1000, we need to use the cell address of the sum of frequencies in M11 of Table 6.3 to find the relative frequency of occurrence of each number. In N4, enter M4/M11 and copy cell N4 down to cell N9. What we have will be the result shown in N4:N9 of Table 6.7. The formulas in range N4:N9 are shown in NN4:NN9. What the **copy command** did was to duplicate the relative cell positions of the formula in cell N4: one cell (M4) to the left of the current cell N4, divided by the seventh cell below and one cell to the left (M11). These relative positions are maintained throughout when cell N4 is copied down (see column NN of Table 6.7). This is called **the relative references** of the formula. Since M12, M15, and M16 are empty, the results indicate that the formula repeats the error of dividing by zero (#DIV/0!) in N5, N8, and N9. Since M13 and M14 have the text entries, the numbers in N6 and N7 are divided by text, resulting in an error of #VALUE!. To avoid these errors, we have to "fasten" or

Table 6.7 Relative reference

	L	M	N	NN
1	The frequency table			
2	Bin	Freq	Relative freq.	
3		f	rf	Formula in rf
4	1	172	0.1720	=M4/M11
5	2	179	#DIV/0!	=M5/M12
6	3	171	#VALUE!	=M6/M13
7	4	161	#VALUE!	=M7/M14
8	5	152	#DIV/0!	=M8/M15
9	6	165	#DIV/0!	=M9/M16
10				
11	sum	1000		
12				
13	B4:	=ROUND(RAND()*(6.5-0.5)+0.5		
14	M4:	=FREQUENCY(sample,bin))		

"lock" the position of cell M11, so that the denominator does not move when the formula is copied down.

Absolute and mixed references

There are three methods of **fastening** or **locking the cell**. One is locking the row but not the column, one is locking the column but not the row, and one is locking both row and column. A row or column is locked by adding $ before the row or column address, as shown below.

Click N4 of Table 6.3, enter =M4/$M$11 (prefix $ to both M and 11) in N4, and copy N4 down to N9. We have the correct entries in N4:N9 in Table 6.3. The formulas in range N4:N9 are shown in NN4:NN9 of Table 6.8. In this example, the numerator remembers its relative position with M4, that is, one cell to the left of cell N4, but the denominator is locked at cell M11 with respect to both row (row 11) and column (column M). Hence, the cell address M11 is locked when the cell is copied. In this case, we say that cell M11 is **copied with absolute reference**.

On the other hand, we may also enter =M4/M$11 (prefix $ to 11 only) in N4, as demonstrated in cell NNN4 of Table 6.8. In this case, M11 is **copied with relative reference** when N4 is copied down. M may change if N4 is copied to the right or left. In this example, however, since N4 is copied down instead of to the right or left, the effect of M$11 is the same as $M$11. Thus, in this example, what we need is to lock the row. The formulas in columns NN and NNN of Table 6.8 will show the same results.

You may also lock the column only, for example, $M11 will lock the column M, but not row 11. In general, if only the row or the column is locked, then we have a **mixed reference**.

The F4 (abs) key

Instead of adding $ each time by hand, Excel has a convenient **<F4> key**, which is called the "**abs**" **key**. It toggles between the four references as in Table 6.9. As an example, enter

Table 6.8 Formulas for the relative frequency and the cumulative relative frequency

	M	N	NN	NNN	N	O	OO
			Formula in N4:N9				Formula
3		rf	rf	rf	rf	crf	
4	172	0.1720	=M4/$M$11	=M4/M$11	0.1720	0.1720	=+N4
5	179	0.1790	=M5/$M$11	=M5/M$11	0.1790	0.3510	=O4+N5
6	171	0.1710	=M6/$M$11	=M6/M$11	0.1710	0.5220	=O5+N6
7	161	0.1610	=M7/$M$11	=M7/M$11	0.1610	0.6830	=O6+N7
8	152	0.1520	=M8/$M$11	=M8/M$11	0.1520	0.8350	=O7+N8
9	165	0.1650	=M9/$M$11	=M9/M$11	0.1650	1.0000	=O8+N9
10		value	formula for rf	formula for rf		value	formula for crf
11	1000						

Table 6.9 The absolute key <F4>

A1	A2	A2	A2	A2
Press <F4> to fix		once	twice	thrice
	none	both	row	col
20	=A1	=$A$1	=A$1	=$A1

any number, say, 20 in A1, and in A2 enter a formula, say =A1 to copy 20 into A2. Click A2, and press <Edit(F2)> to show the formula. The vertical line in cell A2 flashes. When the vertical line is near the letters =A2 or on =A2, press <abs (F4)> key once, and the formula will change to $A$1 (both row 1 and column A are locked). Press <F4> a second time, and the formula will change to A$1 (row 1 is locked). Press <F4> a third time, and we will have $A1 (column A is locked). Press it a fourth time, and we will go back to the original formula, A1 (none is locked). It is something like the four-step waltz: both, row, column, and return. The process is shown in Table 6.9.

6.5.4 *The cumulative relative frequency*

The **cumulative relative frequency** is the proportion of observations up to and including a particular value of a random variable and ALL previous values:

$$\sum_{i \leq j} f_j/n = f_1/n + f_2/n + \cdots + f_j/n, \tag{6.19}$$

for $j = 1, 2, \ldots, n$. For tossing a die 1000 times, the relative frequencies (rf) are found in N4:N9 of Table 6.3. We now apply the formula (6.19) as follows:

For $j = 1, i = 1$ $\quad f_1/n = 0.172;$

For $j = 2, i = 1, 2$ $\quad (f_1/n) + f_2/n = 0.172 + 0.179 = 0.351;$

For $j = 3, i = 1, 2, 3$ $\quad (f_1/n + f_2/n) + f_3/n = 0.351 + 0.171 = 0.522;$

For $j = 4, i = 1, 2, 3, 4$ $\quad (f_1/n + f_2/n + f_3/n) + f_4/n = 0.522 + 0.161 = 0.683,$

etc. It is clear that (6.19) is the total sum of the relative frequencies for $j = 6$. The sum should be equal to 1, as shown in cell O9 of Table 6.3.

To implement (6.19) on the spreadsheet, the formulas are shown in column OO of Table 6.8. In Table 6.3 or Table 6.8, enter in

O4: =N4, Repeat cell N4,

O5: =O4+N5, Upper cell plus left cell.

Copy O5 to O9. When O5 is copied down to O9, Excel remembers the relative position of the formula operation in O5, that is, the cell in the immediate row above plus the cell at the immediate left. Thus, we have **copied O5 with relative reference**. The relation is shown in Table 6.8 between columns N and O, and the formulas are shown in column OO. Note that cell O9 must show 1.00 by the property of the relative frequency distribution.

6.5.5 *Showing the cell formula embedded in the cell or range of cells*

On many occasions, especially during examinations, we may like to show how the 1000 numbers were generated and what the underlying cell formulas are. In Table 6.3, the cell formula of random number in B4 is copied to M13 and the cell formula for finding the frequency in M4 is copied to M14. You may enter the formula manually, but it is better to use the copy command to avoid errors. The copy command can be applied in the following two ways.

A formula in a regular cell

For a regular cell, you can edit either in the formula box or in the cell directly.

To edit in the formula box, select B4 in Table 6.3 and the formula in B4 appears in the formula box. In the formula box, insert a space before the = sign to convert the entry to a text. Select (shade) the formula, including the space, and enter <^C>. Move the pointer to M13, and enter <^V>. The formula in B4 is copied to M13. Move the pointer back to B4 and select B4. In the formula box, delete the space before the = sign and press <Enter> to restore the original formula entry.

To edit in the cell directly, select B4 and press

<F2(edit)><home><space><return>. (6.20)

This will create a space before the = sign, and change the value into the corresponding formula text. Since it is a text, copy B4 to M13. Go back to B4 and delete the space before the = sign. This will show the previous value in cell B4.

Formula in an array

For the formula of M4, the above methods do not work, as M4 in Table 6.3 is a cell in an array and "you cannot change part of array" by inserting a space before the cell formula. In this case, click M4, and simply enter <^C>. Move the pointer to M14 and enter <^V>. The original number appears in M14 with the underlying formula. We then change the formula to value by following the procedure in (6.20).

Note that this method can also be applied to a formula in a regular cell. The drawback is that if the formula is written with relative references, the cell addresses will change and the original cell addresses will be lost.

To show the formulas in the whole worksheet

To **show the formulas** in the whole worksheet, enter <^ ~>, that is, hold the control key and press the "similar" key "~", which is the left-most key above the Tab key on the keyboard. The whole worksheet will show the formula mode, as Table 6.10.

Copy B4 to M13 and M4 to M14. Select M13 and the formula will be shown in the formula box. Enter a space before the = sign, and press the <Enter> key. The formula is shown. Exit the formula mode by entering <^ ~ > again. Another method of showing a formula embedded in many cells is to use the "find and replace" command. This is explained in Appendix 6E.

Table 6.10 The formula mode of the worksheet of Table 6.3

	A	B	C	D	E	F	G	H	I	J	K	L	M	N	O	P
1	Sa											The f				
2	To											Bin	Freq	Relative freq.	Cum.rel.freq.	Probability
3		1	2	3	4	5	6	7	8	9	10		f	rf	crf	P
4	1	=ROI	=ROI	=ROI	=ROI	=ROI	=ROI	=ROI	=ROI	=ROI	=ROU	1	=FREQUENC'	=M4/$M$11	=N4	0.1666666
5	2	=ROI	=ROI	=ROI	=ROI	=ROI	=ROI	=ROI	=ROI	=ROI	=ROU	2	=FREQUENC'	=M5/$M$11	=O4+N5	0.1666666
6	3	=ROI	=ROI	=ROI	=ROI	=ROI	=ROI	=ROI	=ROI	=ROI	=ROU	3	=FREQUENC'	=M6/$M$11	=O5+N6	0.1666666
7	4	=ROI	=ROI	=ROI	=ROI	=ROI	=ROI	=ROI	=ROI	=ROI	=ROU	4	=FREQUENC'	=M7/$M$11	=O6+N7	0.1666666
8	5	=ROI	=ROI	=ROI	=ROI	=ROI	=ROI	=ROI	=ROI	=ROI	=ROU	5	=FREQUENC'	=M8/$M$11	=O7+N8	0.1666666
9	6	=ROI	=ROI	=ROI	=ROI	=ROI	=ROI	=ROI	=ROI	=ROI	=ROU	6	=FREQUENC'	=M9/$M$11	=O8+N9	0.1666666

6.5.6 *The probability distribution*

The frequency function (6.2) of tossing a die is shown in column P of Table 6.3. Since 1/6 is a fraction, we enter

$$\text{P4: } =1/6, \tag{6.21}$$

with an equal sign in front of the fraction. 0.16666667 will show in P4. Equation (6.21) is a probability mass function. Copy P4 to P5:P9, as shown in Table 6.3.

Note that, if 1/6 is entered directly, by Excel's default setting, Excel will take 1/6 as a date and returns 6-Jan, and the formula bar will show 1/6/2008. If you want to enter the number as a fraction, enter 0 1/6 in P4, that is, enter 0 first and a space and then 1/6. P4 will show 1/6, and the formula bar will show 0.16666667. To disable Excel's default date format, see Appendix 6C.

Example 6.9 Entering fractions in an Excel worksheet. In column B of Table 6.1, to enter 1/10 directly in B3:B12, we enter 0 1/10 in B3, and copy it down to B12. We have another discrete uniform probability distribution. □

As explained in Sec. 6.2.2, according to the frequency theory of probability, p = 1/6 in (6.21) is the limiting value of the relative frequency of the appearance of, say, the ace, when a die is tossed infinite times. The ordered list of $p(x) = 1/6$ for all $x = 1, 2, \dots, 6$, is **the (theoretical) discrete uniform probability distribution** of the numbers obtained tossing a die, in contrast to the empirical frequency distribution when a die is tossed a finite number of times.

6.5.7 *Diagrammatic illustration of distributions*

The relative frequency distribution of column N and the probability distribution of column P in Table 6.3 are illustrated in Fig. 6.4. To draw the chart, first select L3:L9 (including the cell L3 and making sure that the X-axis has correct labels), and then holding the control key, use the mouse to select N3:N9 and P3:P9. Then, invoking the chart wizard, the data range will be automatically selected. Select the line chart. We have **two lines** in the chart.

Changing a line to columns

Click the horizontal line representing P in Fig. 6.4, and click <RM>. Select

<Change Series Chart Type···> (select the first column) <OK>.

The graph of the probability distribution will change to columns as shown in Fig. 6.4. Since this chart has both a line and columns, it is called a **combination chart** (see Appendix 6B for details). Format the chart as shown in Fig. 6.4. Enter the <F9(Recal)> key several times and observe how the line dances at the top of the columns.

Similarly, we can draw a cumulative relative frequency distribution and a cumulative probability distribution as in Fig. 6.5. It is interesting to note that while the relative frequency distribution shows a large fluctuation as compared with its probability distribution, the cumulative relative frequency distribution appears to be almost the same as the

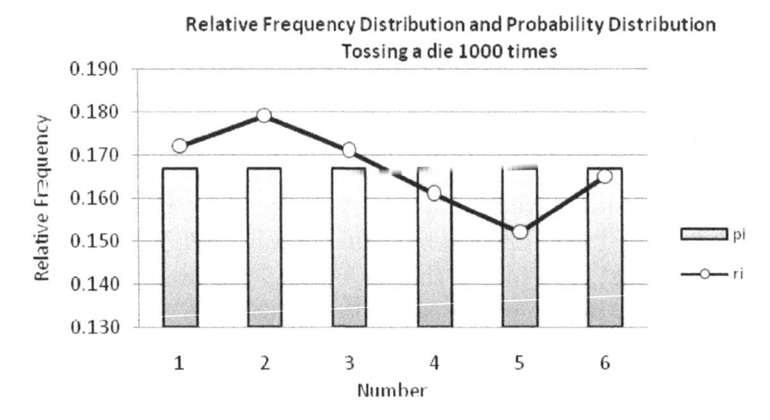

Fig. 6.4 Relative frequency distribution and probability distribution

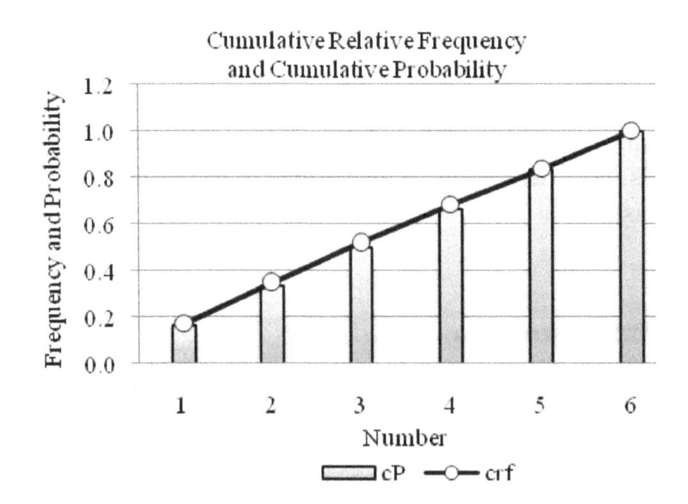

Fig. 6.5 Cumulative relative frequency distribution and probability distribution

cumulative probability distribution. The chart needs to be enlarged to see the differences between the two cumulative distributions.

6.6 The Mean and Variance of the Sample and Population

We first show that for the discrete random variable x (in this section, unlike Chapter 5, we use small letter x_i, instead of capital letter X_i, to denote the random variable), the mean and variance of grouped data are the value of the random variable weighted by the relative frequencies, and are exactly the same as the ungrouped data. Since the probability is an idealized or conceptual form of relative frequency, the mean and variance of population can be derived in a similar way to the mean and variance of sample.

6.6.1 *The mean and variance of grouped data*

To calculate the mean and variance, we first copy Table 6.3 to another sheet, say sheet2, and reconstruct a calculation table like that shown in Table 6.11. Leave the columns of bin, frequency, and relative frequency the same as in Table 6.3. Enter <F9> several times to make sure that the range names of sample and bin carry over to the new sheet (the numbers in the sample range will dance when you press <F9>). Since B4:K103 are still defined as Sample, enter =average(sample) in M13 and =var(sample) in M14. We instantly obtain the mean and variance of the sample of 1000 tosses of a die.

The mean of grouped data

Since we already have the summary table of the sample in Table 6.3, we may also calculate the mean and variance directly from the table. It shows that ace occurred 172 times, 2 occurred 179 times, ..., and 6 occurred 165 times. The sum of all the dots in the 1000

Table 6.11 Mean and variance of tossing a die 1000 times

	L	M	N	O	P	Q	R	S	T	U
1	The frequency table									
2	Bin	Freq	Relative Freq. (empirical)				Probaility (theoretical)			
3		fi	ri	xiri	(xi-m)ri	(xi-m)^2ri	pi	xipi	(xi-mu)pi	(xi-mu)^2pi
4	1	172	0.172	0.172	-0.41916	1.02	0.16667	0.16667	-0.41667	1.04167
5	2	179	0.179	0.358	-0.25722	0.37	0.16667	0.33333	-0.25000	0.37500
6	3	171	0.171	0.513	-0.07473	0.03	0.16667	0.50000	-0.08333	0.04167
7	4	161	0.161	0.644	0.09064	0.05	0.16667	0.66667	0.08333	0.04167
8	5	152	0.152	0.760	0.23758	0.37	0.16667	0.83333	0.25000	0.37500
9	6	165	0.165	0.990	0.42290	1.08	0.16667	1.00000	0.41667	1.04167
10										
11	sum	1000	1.0000	3.43700	0.00000	2.93003	1.00000	3.50000	0.00000	2.91667
12				m		varp=sn^2		mu		sig^2
13	mean	3.43700	=AVERAGE(sample) = m							↙
14	var	2.93018	=VAR(sample) = s^2						(6.7b)	=(36-1)/12

sample should be

$$(1)(172) + (2)(179) + \cdots + (6)(165) = 3437. \tag{6.22}$$

Dividing both sides of (6.22) by the total sample points, 1000, we have the average of 3.437, which is exactly the same as the average in M13. Thus, there is no difference in using the raw data ("sample") or the grouped data (frequency).

For the notation, let us denote the ace x_1, the 2 x_2, etc., and the corresponding frequencies f_1, f_2, etc. Then (6.22) is

$$(x_1)(f_1) + (x_2)(f_2) + \cdots + (x_6)(f_6) = \sum_i x_i f_i.$$

Thus, dividing both sides by n, and denoting the relative frequency r_i, we have the formula for the mean of the grouped data as:

$$\boxed{(x_1)(r_1) + (x_2)(r_2) + \cdots + (x_6)(r_6) = \sum_i x_i r_i = \overline{X} = m,} \tag{6.23}$$

where we also use $\overline{X} = m$ (sometimes we write mX or m(X) to indicate which variable we are referring to) to denote the mean of grouped data.

The actual calculation is shown as $x_i r_i$ in O4:O9. Its sum in cell O11 gives the grouped mean which is the same as the ungrouped mean in M13. Equation (6.23) can be extended to any finite number of the values of a discrete random variable.

For a grouped continuous random variable, x_i should be taken as the mid-point value of the ith interval to represent the values in that interval. Thus, in this case, the mean (6.23) will only be an approximation.

The sum of mean deviations of grouped data

For the example in tossing a die in Table 6.11, the mean deviations for each x_i are $x_i - m$, $i = 1, 2, \ldots, 6$. The mean deviations of the ace appeared 172 times in the sample, hence, it must be multiplied by 172. Similarly, 2 appeared 179 times, so its mean deviations have to be multiplied by 179, and so on. The result is

$$(1 - 3.437)(175) + (2 - 3.437)(179) + \cdots + (6 - 3.437)(165) = 0.$$

The sum must be zero, as shown in P11, and is expected from Chapter 5. For the notation, dividing both sides by the sample size n, we may write

$$\boxed{\begin{aligned}&(x_1 - m)(r_1) + (x_2 - m)(r_2) + \cdots + (x_6 - m)(r_6)\\ &\quad = \sum_i (x_i - m)r_i = 0,\end{aligned}} \tag{6.24}$$

where m denotes the mean in O11. In P4 of Table 6.11, enter =(O4–$O$11)*N4, which will give the mean deviation of ace. Equation (6.24) is shown in P4:P9 of Table 6.11. The sum of the range is zero, as shown in cell P11. Note that if we move the summation sign in (6.24) inside the parentheses and equate to zero, (6.24) reduces to the definition of mean in (6.23) (since $\sum r_i = 1$).

The variance of grouped data

The numerator of the variance is the sum of squared mean deviations of a random variable. Each squared mean deviation must be multiplied by its corresponding frequency, and hence we have

$$(1 - 3.437)^2(172) + (2 - 3.437)^2(179) + \cdots + (6 - 3.437)^2(165) = 2930.03.$$

Taking the average by dividing both sides by the sample size, 1000, we have the variance 2.93003, as shown in Q11 of Table 6.11. For the notation, writing the equation in terms of relative frequencies, we have the formula of the variance for grouped data:

$$\begin{aligned}(x_1 - m)^2(r_1) + (x_2 - m)^2(r_2) + \cdots + (x_6 - m)^2(r_6) \\ = \sum_i (x_i - m)^2 r_i \equiv s_n^2\end{aligned} \tag{6.25}$$

Note that, unlike (6.23), the division by the sample size is already included in the relative frequencies.

As in the case of the mean, for the grouped continuous data, x_i should take the midpoint of the ith interval to represent the whole interval. Thus, the result in (6.25) will only be an approximation for grouped continuous data.

6.6.2 *The mean and variance of the population*

Based on the frequency theory of probability in Sec. 6.2.2, the mean and variance for grouped data defined in the previous section can also be applied to finite population. Substituting p_i for r_i in (6.23), where p_i is the probability of the occurrence of the ith event x_i, define

$$E(x) = \sum_i x_i p_i \equiv \mu, \tag{6.26}$$

which is the **population mean**. For continuous grouped data, x_i should be taken as the midpoint value of the ith interval. Equation (6.26) is also called the **expected value** of random variate x. In Table 6.11, the probability for each x_i is entered in R4:R9, and $x_i p_i$ is entered in S4:S9. Their sum, 3.5, is the population mean μ in S11.

Example 6.10 Lottery drawing. In the lottery drawing of Table 6.1, the population mean (expected value) is the sum of the entries in C3:C12, and is 4.5 in cell C14. □

Similarly, for the population data, we write (6.24) as

$$\boxed{E(x - \mu) = \sum_{i} (x_i - \mu)p_i = 0,} \tag{6.27}$$

which is the sum of the mean deviations of all x_i, $i = 1, 2, \ldots, 6$, of the random variable, and is entered at T11 in Table 6.11. For continuous grouped data, x_i should be taken as the midpoint of the ith interval.

The **population variance** is written as

$$\boxed{V(x) = E(x - \mu)^2 = \sum_{i} (x_i - \mu)^2 p_i \equiv \sigma^2,} \tag{6.28}$$

which is the sum of weighted squared deviations from the population mean, the weight being the probability. It is calculated at U11 in Table 6.11 as 2.91667, which is the sum of the values in U4:U9. Note that we use V(.) to distinguish it from Excel's sample variances var(.) or varp(.).

The positive square root of σ^2 is the **population standard deviation** and denoted by $\sigma \equiv \text{sqrt}(V(x))$.

Example 6.11 Lottery drawing. In Table 6.1, the squared mean deviations are calculated in D3:D12, and its sum is the **population variance**, $\sigma^2 = 8.25$ in D15. Its positive square root is the **population standard deviation** $\sigma = 2.87$. □

6.6.3 *Comparison of sample and population characteristics*

For a discrete random variable, the mean and variance calculated directly from a sample (for example, a sample obtained by tossing a die 1000 times) or indirectly from grouped data should give the same results, as shown in M13 and O11 in Table 6.11. Both cases give the mean 3.437. However, there is a slight difference in variance, 2.93018 in M14, and 2.93003 in Q11. The difference is due to the difference in var and varp, and may also be due to decimal rounding. We consider that the direct calculation is more reliable, although it may be costly, and in many other experiments, impossible, due to a large sample size.

Both sample mean and sample variance can be compared with the population mean, $\mu = 3.500$, and population variance, $\sigma^2 = 2.91667$. If sample characteristics are used to approximate the population characteristics, then, for this set of samples, we can conclude that the sample mean ($m = 3.437$) underestimates the population mean ($\mu = 3.5$), and the sample variance ($s^2 = 2.93003$) overestimates the population variance ($\sigma^2 = 2.91667$). Extending the frequency theory of probability, if the sample size goes to infinity, the sample mean and sample variance will eventually approach the population mean and population variance, respectively.

6.7 The Normal Distribution

In Secs. 6.4 and 6.5, we have worked on a discrete random variable, using the occurrence of numbers in tossing a die as an example. This section introduces the normal distribution, the most important distribution of continuous random variables, like the distributions of SAT, GPA, and other test scores, IQ, factory defective parts, rainfall, temperature, etc.

6.7.1 *The normal probability density function (pdf)*

As shown in Fig. 6.6, a standard normal distribution is a bell-shaped, unimodal, and symmetric frequency distribution of a random variable z, which ranges from $-\infty$ to $+\infty$, with the maximum point at its mean 0, and its standard deviation $\sigma = 1$. It is customary to denote the random variable with standard normal distribution as z, and z is simply called the **standard normal variate**. The **probability density function** (pdf) **for the standard normal distribution** is given by

$$\mathrm{p(z)} = \frac{1}{\sqrt{2\pi}} \mathrm{e}^{-\frac{\mathrm{z}^2}{2}} \equiv \mathrm{N}(0,1) \quad \text{for} \ -\infty < \mathrm{z} < \infty, \tag{6.29}$$

where $\pi = 3.14159\ldots$ and $\mathrm{e} = 2.718281\ldots$. They are mathematical constants obtainable in Excel as =pi() (that is, pi and parentheses) and =exp(1), accurate to 15 digits. The normal pdf has the following mean and variance:

$$\mathrm{E(x)} = \mu = 0, \quad \mathrm{V(x)} = \mathrm{E}((\mathrm{x} - \mu)^2) = \sigma^2 = 1. \tag{6.30}$$

6.7.2 *The standard normal distribution in Excel*

Excel has two formulas for the normal distribution. One is the **standard normal cumulative distribution function** (**cdf**)), and the other is the **general normal cdf**, which can be converted to a **general normal pdf**.

The standard normal **cdf** is illustrated as the dotted cdf curve in Fig. 6.6. It corresponds to the empirical cumulative relative frequency distribution we have seen in Sec. 6.5.4. The Excel formula is

$$=\mathrm{normsdist(z)}, \tag{6.31}$$

where s in the middle stands for "standard". Another formula is the **general normal distribution**, which will be explained in the next subsection. Formula (6.31) is implemented in column D of Table 6.12, and is explained below.

To construct Table 6.12, the standard normal variate z is taken as $\mathrm{z} = -4, -3.9, \ldots, 4$, with an interval of 0.1 in column B (use the fill-handle box to fill the column). Number the cells from 1 to 81 in column A. Since the variate is continuous, each interval of length 0.1 contains infinite numbers of values. Thus, we take the **midpoint** of the interval as the

Table 6.12 The normal distribution and the cumulative normal distribution

	A	B	C	D	E	F	G
1					mu	sig	
2					0	1	
3	Excel formulas			(6.29)	(6.30)normdist(x,mu,sig,cumulative)		
4		x	mid	normsdist(z)	false	true	area
5				cumulative	density	cumulative	density*0.1
6	1	-4.0					
7	2	-3.9	-3.95	0.000039	0.000163	0.000039	0.000016
8	3	-3.8	-3.85	0.000059	0.000241	0.000059	0.000024
9	4	-3.7	-3.75	0.000088	0.000353	0.000088	0.000035
10	5	-3.6	-3.65	0.000131	0.000510	0.000131	0.000051
11	6	-3.5	-3.55	0.000193	0.000732	0.000193	0.000073
86	81	4.0	3.95	0.999961	0.000163	0.999961	0.000016
87	sum			40.000000	9.999371	40.000000	0.999937

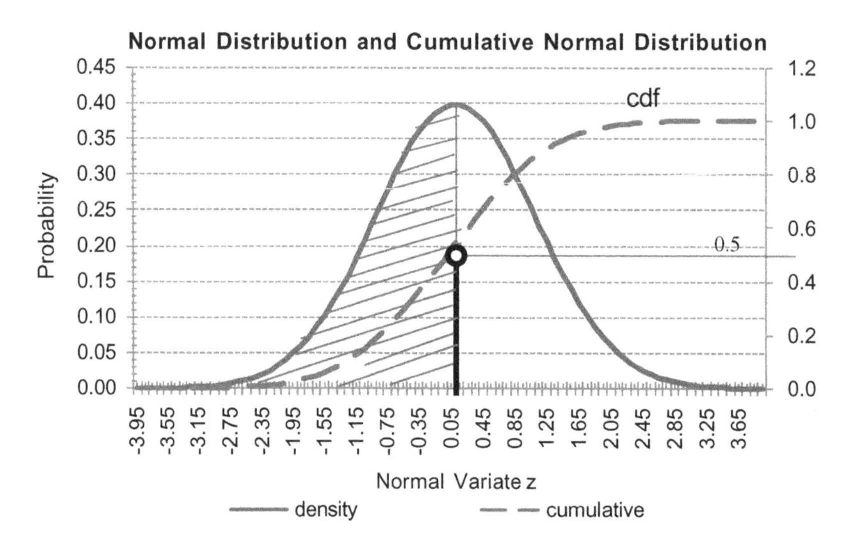

Fig. 6.6 The standard normal distribution and standard normal cumulative distribution

representative of the interval. This is done in columns C and D as follows: we enter

C7: =(B7+B6)/2, The value at midpoint,
D7: =normsdist(C7), Cumulative distribution (cdf),

and copy C7:D7 down to C86:D86. For this long-range copying, we use the method in Box 6.1.

> Box 6.1: **To copy C7:D7 to C86:D86**, use the horizontal split line to split the window as shown by the arrow in Table 6.12. Select C7:D7, enter <^C>. Clicking C8 and holding the shift key, click D86 to select and shade C8:D86, and enter <^V>. C7:D7 will be copied to range C8:D86. This method of copying can be applied to copying all other long columns.

6.7.3 *The general normal distribution*

For the general normal distribution with mean μ and variance σ^2, the **general normal probability density function** (pdf) is

$$\mathrm{f(x)} = \frac{1}{\sqrt{2\pi}\sigma}\mathrm{e}^{-\frac{1}{2}\left(\frac{\mathrm{x}-\mu}{\sigma}\right)^2} \equiv \mathrm{N}(\mu, \sigma^2), \quad \text{for } -\infty < \mathrm{z} < \infty, \tag{6.32}$$

where σ is the standard deviation. The general normal distribution has the following mean and variance,

$$\mathrm{E(x)} = \mu, \quad \mathrm{V(x)} = \mathrm{E((x-\mu)^2)} = \sigma^2. \tag{6.33}$$

The Excel formula for this distribution is

=normdist(x,mean,standard_dev,cumulative),

False → pdf(density function) (6.34)
True → cdf(cumulative function)

where "cumulative" is a logical value. If "false" is entered for "cumulative", Excel returns the **normal pdf**. If "true" is entered, Excel returns the **normal cdf**. In Excel spreadsheets, it is difficult to enter Greek letters. We write the mean as **mu** and the standard deviation as **sig**, as shown in E1:F2 in Table 6.12. To be flexible in choosing various means and standard deviations, we name E2 "mu" and F2 "sig". To compare this with the standard normal distribution, we enter mu = 0 and sig = 1 in cells E2 and F2, respectively.

In Table 6.12, after naming mu and sig, we enter the formulas as follows:

E7: =normdist(C7, mu,sig,false)), Density function (pdf);
F7: =normdist(C7, mu,sig,true)), Cumulative function (cdf),
G7: =E7*0.1, Probability element in the interval.

The range E7:G7 is then copied to E8:G86 by the method suggested in Box 6.1. As expected, column D, which presents the results obtained from Excel formulas of **standard** normal distribution, and column F, which presents the results obtained from formulas of **general** normal distribution, are identical. In both cases, the cumulative distribution is close to 1 in cells D86 and F86.

6.7.4 *Illustrations*

The density function and the cumulative function

The density function (pdf) and the cumulative function (cdf) in Table 6.12 are illustrated in Fig. 6.6 for the midpoints of $z = -3.95, -3.85, \ldots, 3.95$. The cumulative function (cdf) uses the secondary Y-axis (on the right-hand side of the chart) and shown as a dashed line. It ranges from 0 to 1 in the S-shaped curve as z changes from $-\infty$ to $+\infty$. The entry in column G, say 0.000016 in G7, shows the probability element (area) for the standard normal distribution at midpoint -3.95 (in C7) of the corresponding interval between -4.0 and -3.9 (in cells B6 and B7). The sum of the probability elements is the approximation of the area under the density function, and is very close to 1 as indicated in cell G87, which is the sum of values in G7:G86.

It should be noted that the height of the cumulative curve at any point of z is the same as the area under the normal curve to the left of that point. For example, at $z = 0.0$ (the labels of the X-axis in Fig. 6.6 are too crowded to see the tick mark of 0 clearly), the height of the cumulative curve is 0.5, as measured from the secondary Y-axis. The height is equivalent to the shaded area under the normal curve left of point 0. In Table 6.12, this can be seen from the cdf at $z = 0$ in column D or F, and also from the sum of area between $z = -\infty$ and 0 in column G (not shown).

6.8 The Normally Distributed Random Variable

The reverse procedure of Sec. 6.7 is to find the normally distributed random variable.

6.8.1 *The inverse function method*

Figure 6.7 illustrates the cumulative normal distribution. At $z = 0.35$, for example, we find the cumulative probability to be 0.636831 (from G50 in Table 6.12, not shown). Thus, the functional relationship is $p = F(z)$ in Fig. 6.7. The normal variate z is the independent variable, and p is the dependent variable. Conversely, we may ask, given a probability, say $p = 0.911$, what is the corresponding z if z generates a normal distribution? From Table 6.12, the value of the normally distributed random variate is approximately $z = 1.35$ (from F60 of Table 6.12, not shown). This method of finding z is called the **inverse function method, or the inverse method of finding normally distributed random variable**. In Fig. 6.7, the functional relationship is reversed: $z = F^{-1}(p)$, where p is now the independent variable and z is the dependent variable.

The Excel formula for the inverse function

Corresponding to (6.31), to generate a random variate which has the standard normal distribution, Excel uses the following formula:

$$=\text{normsinv(probability)}, \tag{6.35}$$

where s denotes "standard". Corresponding to (6.34), to generate a random variate which has the general normal distribution with mean μ and standard deviation σ, Excel uses the

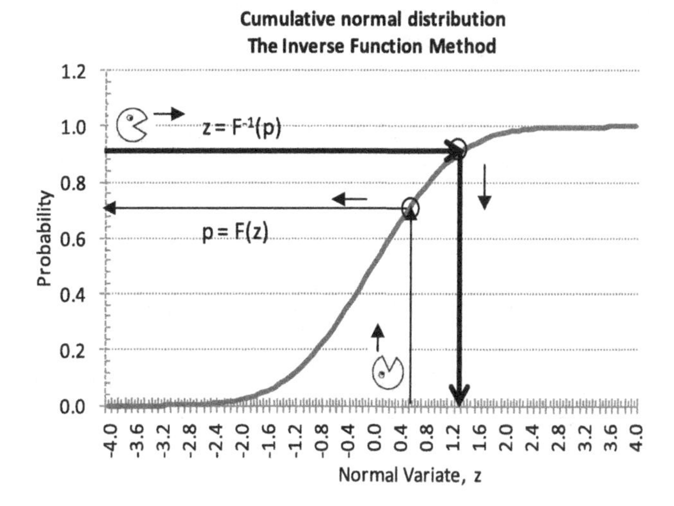

Fig. 6.7 Derivation of a normally distributed random variate

following formula:

$$=\text{norminv}(\text{probability},\text{mean},\text{standard\_dev}), \tag{6.36}$$

where "probability" is a random number from 0 to 1. The method is similar to the appearance of numbers in tossing a die, which has the discrete uniform distribution from 1 to 6 and was generated in Table 6.3. The difference is that in tossing a die, there are six chips in the discrete population of a die, while in this case, we have infinite numbers of imaginary chips in the continuous normal population.

The data table

Table 6.13 draws 1000 imaginary "normal chips" (see HW6-10) from an infinite population with the standard normal distribution. The construction of the table is similar to Table 6.3, and can be reproduced from the "sample" in Table 6.3. To do so, copy Sheet1, in which you have Table 6.3, to Sheet2. In Sheet2, rename the table Table 6.13, and delete row 1 to obtain the entries in Table 6.13. Then, in cell B3, enter =normsinv(rand()) and copy B3 to B3:K102 using the copying method expounded in Box 6.1. Name B3:K102 as "sampleN".[7]

6.8.2 *Derivation of frequency, relative and cumulative relative frequencies*

Table 6.14 shows the same table as Table 6.13, starting from column L. To find the frequencies of the 1000 sample points generated in Table 6.13, let the bin range be $-4.00, -3.50, \ldots, 4.00$, as shown in L3:L19 in Table 6.14. Name it "binN". We find the

[7]If you use the old file, the range name "sample" has been used in Sheet1 for the sample of 1000 chips from the uniform distribution. Hence, you have to use different range name in Sheet2.

Table 6.13 Drawing 1000 chips from a normal population

	A	B	C	D	E	F	G	H	I	J	K
1											
2		1	2	3	4	5	6	7	8	9	10
3	1	-0.199	0.2367	-1.092	-0.035	-0.694	0.4289	-0.652	0.4949	2.0882	1.054
102	100	0.7847	-0.808	-1.694	-0.008	-1.186	0.4305	-2.226	0.6014	-0.392	-1.129

Table 6.14 Data mining from the normal sample

	A	L	M	N	O	P	R	S	T	U	V	W
1		Bin	mid	Freq distr		Prob distr	Sample mean and variance			Population mean and variance		
2				f	r	p=den*0.5	mid*r	(mid-m)r	(mid-m)^2r	mid*p	(mid-mu)p	(mid-mu)^2p
3	1	-4		0			-	-	-		-	-
4	2	-3.5	-3.75	0		0.0002	-	-	-	(0.0007)	(0.0007)	0.0025
5	3	-3	-3.25	1	0.0010	0.0010	(0.0033)	(0.0033)	0.0106	(0.0033)	(0.0033)	0.0107
17	15	3	2.75	5	0.0050	0.0045	0.0138	0.0137	0.0377	0.0125	0.0125	0.0344
18	16	3.5	3.25	0	-	0.0010	-	-	-	0.0033	0.0033	0.0107
19	17	4	3.75	0	-	0.0002	-	-	-	0.0007	0.0007	0.0025
20	18						m	meandev	sn^2	mu	meandev	sig2
21	19	sum		1000	1.0000	0.9998	0.0050	0.0000	0.9910	0.0000	0.0000	0.9990
22	20	stdevp (calculated from grouped data)						sn=	0.9955		sig=	0.9995
23	21	Direct calculation from SampleN					0.0063	0.98619	0.9857			
24	22						=m	stdev=s	stdevp=sn			

midpoints of the intervals, $(-4.0, -3.5], (-3.5, -2.5], \ldots, (3.5, 4.0]$ in M4:M19, similar to those in Table 6.12. Using the RCC to enter the frequencies (f) for each interval in N3:N19, make sure that the absolute frequencies f sum to 1000, or very close to 1000, in cell N21. Enter

O3 : N3/$N$21, Relative frequency (empirical distribution),
P4 : =normdist(M4,0,1,FALSE)*0.5 Probability element (theoretical distribution).

Copy O3 and P4 down to O19 and P19. The column sum in O21 must be 1 and that in P21 should be very close to 1 (why is it not 1?).

The relative frequency Distribution (r in column O) and the normal distribution (column P) of Table 6.14 are drawn in Fig. 6.8 as columns (probability) and the line (frequency). We close the gap width between the columns by selecting the columns, then, clicking <RM> and

<Format Data Series><Series Options><Gap Width, No Gap>.

Move the down arrow to 0%. When the gaps between the columns are closed and the total area of columns is 1, statisticians call the diagram a **histogram**.

Enter <F9> several times and enjoy the frequency curve dancing on the top of the probability columns, showing that the approximation of the frequency curve is indeed very close to the probability columns.

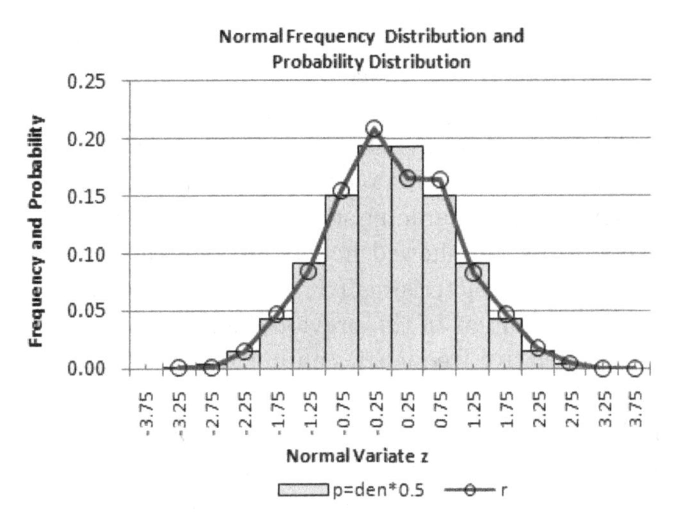

Fig. 6.8 The normal frequency distribution and the probability distribution

6.8.3 *Mean and standard deviation of the normal distribution*

In Table 6.14, range M3:M19 is named "mid" (for midpoint). In R3:R19, enter mid*r and calculate the weighted sum (6.23) in R21, which is the sample mean = 0.0050. In S3:S19, enter weighted mean deviations (mid − m)r and make sure that its sum, which is the same as (6.24), equals zero in S21. In T3:T19, enter the weighted squared mean deviations and calculate the sum, which is the same as (6.25), in T21. We have the "population" variance varp = $s_n^2 = 0.9910$ in T21, and the standard deviation $s_n = 0.9955$, in T22.

In columns U–W, we calculate the population mean mu and population variance sig^2. The weighted sum of the midpoints is the mean of population (6.26) and is calculated in U21, which is zero. The sum of weighted mean deviations of population (6.27) is 0 in V21, as expected. The sum of weighted squared mean deviations of population (6.28) is entered in W21, which is the population variance, $\sigma^2 = 0.9990$, and the population standard deviation is $\sigma = 0.9995$ in W22, which is almost equal to 1. The fact that the sample mean is close to 0 and sample variance is close to one, and that the population mean equals 0 and the population variance is close to 1, is not surprising, since we have generated the sample from a population which is normally distributed as N(0, 1).

To calculate the sample mean and variance directly from the 1000 sample, enter =average(sampleN) in R23 and =stdev(sampleN) in S23, and =stdevp(=sampleN) in T23. Whether it is grouped data or ungrouped data, both results appear to be consistent with the population characteristics, and we can say that the sample characteristics are a good approximation of the population characteristics.

6.9 Summary

In this and previous chapters, we have introduced some basic concepts of statistics. Their relations are illustrated in Fig. 6.9. In Chapter 5, we introduced the definitions and interpretations of mean, variance, standard deviation, coefficient of variation, covariance, and correlation coefficient, among other concepts. These definitions and concepts are mainly related to organizing and manipulating sample data[8]; they are in the realm of **Descriptive Statistics**, on the right side of Fig. 6.9. In this chapter, we have introduced population distribution and its characteristics (that is, mean and variance) to contrast them with sample frequency distribution and sample characteristics. Taking advantage of the spreadsheets, we have generated a large sample and showed how to make comparisons. For a large sample, we may calculate the sample characteristics (the arrow on the right-hand side of Fig. 6.9) either directly by using the formulas of the previous chapter, or indirectly by grouping and using relative frequencies. Relative frequencies and probabilities are linked by the frequency theory of probability.

In many empirical studies, however, the nature of population, including population distribution and population characteristics, are unknown, and we have to use the theory of **Statistical Inference** to estimate and test hypotheses about the population characteristics, based on the sample distribution and sample characteristics. The linkage between population and sample is random sampling, namely, the theories of random variable and probability. These relations are shown on the left-hand side of Fig. 6.9. Due to the limited space and

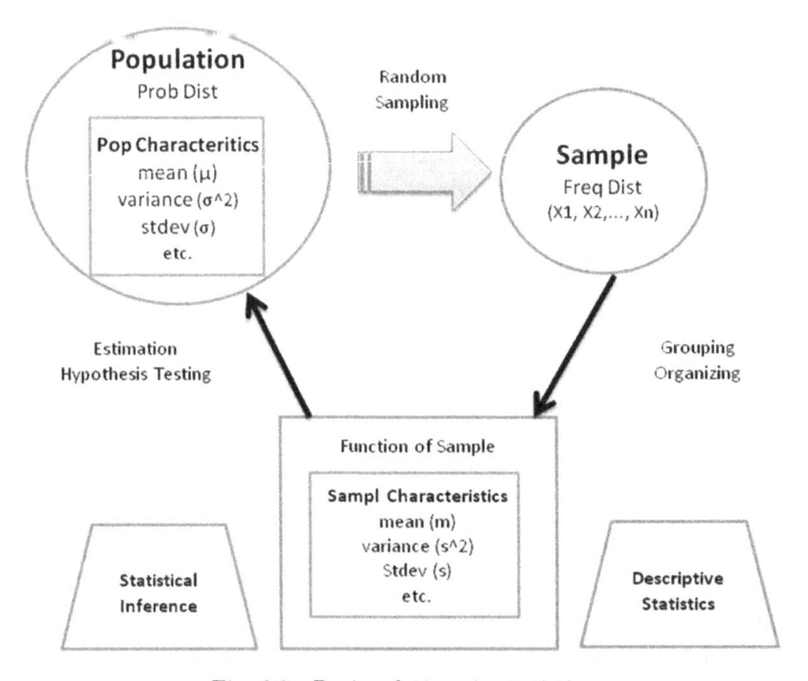

Fig. 6.9 Basic relations in statistics

[8]Although varp and stdevp are called the "population variance" and the "population standard deviation", they are merely used to denote that the sum of squared deviations are divided by n rather than by the degree of freedom $n - 1$. Varp and stdevp are still used in the calculation of sample mean, since variance has its appeal as an average.

limited purposes of this book, we are unable to cover the left-hand side of the diagram, the part on statistical inference.

Since the random numbers are continuous numbers, we have first showed how to convert continuous random numbers into a discrete random variable through finite sample corrections. We have introduced the example of tossing a die 1000 times as an example of sampling from the population of the uniform distribution. We then introduced the example of drawing a chip 1000 times as a sampling from the population of the normal distribution. For each distribution, we have calculated and compared the sample distribution and its characteristics with the population probability distribution and its characteristics (the arrow on the left-hand side of Fig. 6.9).

Table 6.15 lists a summary of basic sample and population distributions and their means and variances covered in this chapter. We have listed the corresponding Excel formulas and the equation numbers in the book for reference. The standard distributions are special cases of the general distributions and are so noted. We have not discussed the continuous case of the general theoretical distribution. It is listed here to show the similarity among pmf, pdf, and the relative frequency of empirical grouped data.

Several new Excel graphic methods and procedures are introduced in this chapter. To manage a long vertical range of data, we have used the horizontal split bar; to find the frequency of observations, we have used the method of the RCC; to find relative frequencies and cumulative frequencies, we have explained in detail the relative and absolute references (that is, using the $ sign to lock a row or column or both); and finally, to illustrate and compare the sample frequency distribution and population probability distribution, we have used the composite charts. These are some of the techniques that will be used in the later chapters.

Appendix 6A. Probability Density Function for a Continuous Random Variable

A pdf for a continuous random variable x is a function f(x) such that

$$f(x) \geq 0, \quad \int_{-\infty}^{\infty} f(x)dx = 1, \quad \int_{a}^{b} f(x)dx = P(a < x < b),$$

where $a < b$ and are any two values of x, and $P(a < x < b)$ is a notation indicating the probability of the occurrence of x in the interval (a, b). $\int$ sign is a continuous counterpart of the discrete (and scripted version) of the sum $\sum$ sign.

Appendix 6B. Combination Charts and Trendlines

When two chart types are mixed in one graph we have a combination chart. The most common combination chart probably is the line-column chart.

Let the quarterly sales and yearly sales of the GiGo Company be shown as in Table 6B.1, in which the range of the sales is assumed in row 2, that is, Q1 ranges from 0 to 10, Q2

Table 6.15 Summary table of basic sample and population distributions covered in Chapter 6

Empirical		Name	Distribution	Variable range	Mean	Variance
ungrouped	$\{X_i\}$	rf	$r_i = 1/n$	$i = 1, 2, \ldots, n$	$m = \Sigma X_i/n$	$sn^2 = \Sigma (X_i - m)^2/n$
data {Xi}	(relative freq)				=average(x)	=varp(X) = s_n^2
			(Table 5.2)		*(5.9)*	*(5.12)*
grouped data	$\{x_i\}$ midpt	rf	$r_i = f_i/n$	$i = 1, 2, \ldots, k$	$m = \Sigma r_i x_i$	$sn^2 = \Sigma r_i (x_i - m')^2$
(k intervals)			*(Table 6.11)*		*(6.23)*	*(6.25)*
Theoretical	**General**					
Discrete		pmf	$p(x_i) = p_i$	$I = 1, 2, \ldots, n$	$\mu = \Sigma p_i x_i$	$\sigma^2 = \Sigma p_i (x_i - \mu)^2$
			(6.4)		*(6.26)*	*(6.28)*
Continuous		pdf	$f(x)$	$-\infty < x < \infty$	$\int x f(x) dx$	$\int (x - \mu)^2 f(x) dx$
			(Appendix 6A)		*(not discussed)*	
	Uniform Distribution					
Discrete	General	pmf	$p(x) = 1/(b-a+1)$	a, b integers	$(a+b)/2$	$(n^2-1)/12$
			(6.7a)	$a < b$	*(6.7b)*	$n = b-a+1$ *(6.7b)*
Continuous	General	pdf	$f(x) = 1/(b-a)$	$a \le x \le b$	$(a+b)/2$	$(b-a)^2/12$
			U(a,b) *(6.10a)*		*(6.10b)*	*(6.10b)*
	standard	pdf	U(0,1) =1		1/2	1/12
			(6.8a)		*(6.8b)*	*(6.8b)*
	Normal Distribution					
Continuous	general	pdf	$f(x) = N(\mu, \sigma^2)$	$-\infty < x < \infty$	μ =mu	σ^2 =sig^2
			=normdist(x;μ,σ,false). *(6.32), (6.33), (6.34)*			
	standard	pdf	$f(z) = N(0,1)$	$-\infty < z < \infty$	$\mu = 0$	$\sigma^2 = 1$
			=normsdist(z), $z = (x-\mu)/\sigma$. *(6.29), (6.30), (6.31).*			

from 0 to 30, etc. When you draw the chart of this table for column range A3:F9, the total column is exceptionally tall, and does not balance well with the other columns. Using Table 6A.1, reproduce Fig. 6B.1 next to the table as follows.

We first select A3:F9, and draw the column chart for the quarterly sales and yearly total. Click the year total column in the chart and <RM>, and

<Change Series Chart Type···><Line>
<Line with Markers><OK>.

The "total" column changes to a line with markers, similar to Figs. 6.4 and 6.5. This will distinguish the quarterly data from the yearly total. Thus, we have **a combination chart**.

Table 6B.1 A combination chart and trend line

	A	B	C	D	E	F
1	Yearly and Quarterly Sales					
2	Range	10	30	50	70	year
3		Q1	Q2	Q3	Q4	total
4	2001	2	22	17	22	64
5	2002	1	14	8	4	27
6	2003	2	23	18	46	90
7	2004	5	3	4	31	43
8	2005	5	20	26	40	92
9	2006	8	28	29	32	98
10	Q total	24	111	102	177	414

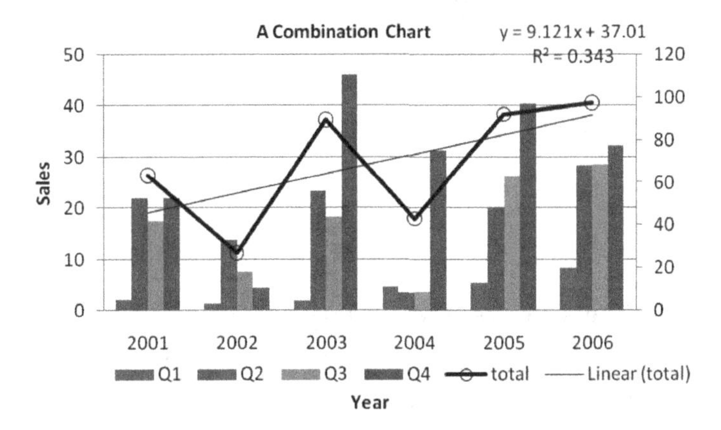

Fig. 6B.1 A combination chart with trend line

a. To make the line of the "total" numbers closer to the quarterly numbers, we may move the axis of the total numbers from **the primary axis** to **the secondary axis**. Click the line, press <RM>, and

< Format Data Series >< Series Options,
Plot Series On @Secondary Axis >< Close > .

Edit the secondary axis if necessary.

b. To **add a trendline** to the "total" line, click the line, enter <RM>, and <Add trendline··· ><linear><OK>. You may add the equation for the regression line by clicking the trendline, <Format Trendline> and click < xDisplay Equation on chart>, < xDisplay R-squared value on chart>, and <Close>.
c. Enter <F9> several times and enjoy the columns and the line dancing together.
d. Do the same for rows by charting the range A3:E10.
e. Using Excel help, find the formula/equation for the **six types of trendlines** given in <Add trendline··· ><Types> window.

Table 6C.1 Disabling the date format in Excel

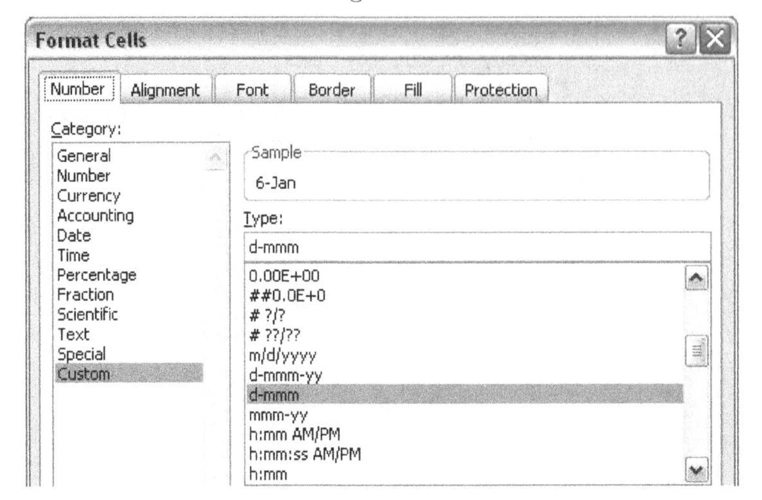

f. Experiment with other trendlines. Fit the “year total” line with a polynomial of degree 3, for example.

Appendix 6C. Disabling Excel’s Time/Date Default

Another method of entering a fraction is to disable the date format of Excel by converting 6-Jan from date to number. Selecting cell P4 of Table 6.3, or any blank cell on the spreadsheet which shows 6-Jan, enter

<“Home”> <Cells, Format!> <Format Cells>.

The “Format Cells” dialog box appears (Table 6C.1). Under the “Category” box, “Custom” is highlighted and in the “Type:” box “d-mmm” is highlighted. Select “Number” in the category box and the number box open to the right. Change Decimal places to 4, and select <OK>. 6 Jan will change to a number (like 40184.0000). Enter 1/6 again and the cell changes to 0.1667 or 0.166667, depending on the decimal points assigned to the cell.

Appendix 6D. The Class Boundaries of a Continuous Variable

In the case of the normal distribution, we have taken the numbers in the bin from $-4.0, -3.5, \ldots, 0, \ldots, 3.5, 4.0$, as shown in column K of Table 6D.1. In column L, manually enter the frequency of the 10 observations in A3:J3. We use a down arrow to show the classification of the observations in the bin, which is classified into 17 class intervals as shown in column O. In Excel, any two consecutive numbers in Bin are a class interval, for instance, $(-4.0, -3.5)$ is a class interval. The left number is called the **lower boundary** and the right number the **upper boundary**. We denote $(-4.0, -3.5]$ to mean the lower boundary is EXCLUDED (“(”) but the upper boundary is INCLUDED (“]”) in the interval.

Table 6D.1 On the effects of class boundaries

	A	B	C	D	E	F	G	H	I	J	K	L	M	N	O
1	Sample #												Boundary		Interval
2	1	2	3	4	5	6	7	8	9	10	Bin	f	lower	upper	Class#
3	-5	-3	blk	4	3.5	5	6	-5	2.5	3	-4.0	2	&below	-4]	#1
4	↓	↓	↓	↓	↓	↓	↓	↓	↓	↓	-3.5	0	(-4	-3.5]	#2
5											-3.0	1	(-3.5	-3.0]	#3
6	#1	#3	?	#17	#16	?	?	#1	#14	#15	-2.5	0	(-3	-2.5]	#4
16											2.5	1	(2	2.5]	#14
17											3.0	1	(2.5	3.0]	#15
18											3.5	1	(3	3.5]	#16
19											4.0	1	(3.5	4]	#17
20										sum		7			

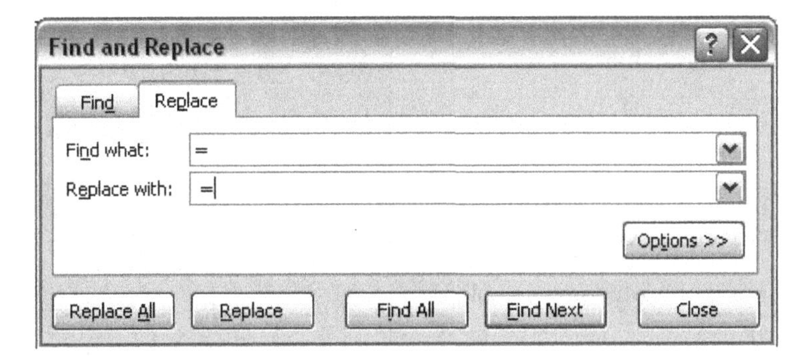

Fig. 6E.1 The "Find and Replace" window

The number in Bin shows only the upper boundary included in the interval. Thus, the numbers in Bin should read as in columns M and N.

The problem is that if we enter 4.0 in K19, the last number in the bin shows the upper boundary of the last class is 4 (class #17, see cell N19), and so any observation larger than 4.0 will not be recorded. This is the case of observations (sample points) #6 (in F3) and #7 (in G3). They are not counted in the sum. Thus, to include these two observations, we should simply leave K19 blank.

Note that the frequency formula ignores cells that are blank or that have text. Thus, since the entry of C3 is a text, the sum of the frequencies in cell L20 is 7: one blank due to the text entry, and two blanks due to two observations exceeding the upper boundary of the last class interval.

Appendix 6E. Showing a Group of Cell Formulas Embedded in the Table

We use Table 6.3 as an example. After entering the formulas in M4:O9, we want to show the formulas in this range. Select M4:O9 in Table 6.3. Enter <^F> to invoke the "Find and Replace" window. Click <Replace>, see Fig. 6E.1.

In the "Find what:" box, enter the equality sign, =, and in the "Replace with:" box, enter <space> and the = sign. Click the <Replace All> button. The selected range M4:O9 in Table 6.3 changes to formula entries as shown in M4:O9 of Table 6.10, with a space before all the = signs. Copy the formulas in M4:O9 to a blank space to preserve them.

To restore the original format of Table 6.3, enter <^F> again. Then, in the "Find what:" box, enter <Space> and the = sign, and in the "Replace with:" box, enter the = sign without a space. Click the <Replace All> button. The underlying formulas disappear and range M4:O9 of Table 6.10 returns to range M4:O9 of Table 6.3.

Review of other basic equations and formulas

Frequency theory of probability

(6.3) $\lim_{n\to\infty} \frac{f_1}{n} = p_1$

Axiomatic theory of probability

(6.4) $p(x_i) \geq 0$,

(6.5) $\sum_{i=-\infty}^{i=\infty} p(x_i) = 1$,

(6.6) $\sum_{i=a}^{i=b} p(x_i) = \text{Prob}(a \leq x_i \leq b)$,

Liner transformation of a random variable x

(6.9) $y = (b - a)x + a$

Cumulative relative frequency

(6.19) $\sum_{i\leq j} f_j/n = f_1/n + f_2/n + \cdots + f_j/n$

(6.24) $\sum_i (x_i - m)r_i = 0$

(6.27) $E(x - \mu) = \sum_i (x_i - \mu)p_i = 0$

(6.29) Standard Normal distribution

$$p(z) = \frac{1}{\sqrt{2\pi}} e^{-\frac{z^2}{2}} \equiv N(0,1)$$

$$\text{for } -\infty < z < \infty$$

(6.32) $f(x) = \frac{1}{\sqrt{2\pi}\,\sigma} e^{-\frac{1}{2}\left(\frac{x-\mu}{\sigma}\right)^2} \equiv N(\mu, \sigma^2)$,

for $-\infty < x < \infty$

Key terms: Economics and Business

Key terms: Excel

Homework Chapter 6

Tossing a fair die

6-1 Tossing a die Construct Table 6.3 for a 100×20 range and draw the relative frequency and cumulative relative frequency distributions. This distribution is called the **sampling distribution of a sample** for tossing a die 2000 times.

(a) In this homework, the sample size is twice larger than that in Table 6.3. Draw the relative frequency curves of the samples of 2000 and 1000, along with the probability distribution of the die tosses (you should draw two lines in Fig. 6.4 in addition to the columns). Is the frequency distribution of the sample of 2000 a better approximation to the population probability distribution than that of the sample of 1000? How do you know?
(b) Calculate the cumulative relative frequency distribution of this sample of 2000 and compare it with the population cumulative probability distribution, as shown in Fig. 6.5.
(c) Calculate the mean, the variance, and the standard deviation of the grouped and ungrouped data, and compare them with the population mean and variance as in Table 6.11.
(d) Show the cell formula of the first entry of random sample and relative frequency distribution, and the first two entries of cumulative frequency distribution below the table, as in Table 6.3.

6-2 Lottery drawing Construct Table 6.3 for the example of lottery drawing in Table 6.1 for a sample based on drawing 1000 times. Answer the same questions as a (in this case, you have only one line in Fig. 6.4), b, c, and d, as posed in the above question HW6-1.

6-3. The total sum of frequencies We reproduce the result in column L of Table 6D.1 in Appendix 6D.

(a) Using the frequency formula (6.18), verify the entries in column L. The data set is A3:J3, and the bin is K3:K19.
(b) Reproduce columns K and L in a blank space, but this time delete the entry 4 in K19. What is the frequency distribution of the data set? What is the column sum of the frequency distribution?
(c) In cell C3 of the data set, instead blk (blank), enter 2.75. Which interval class (or bin number) this entry should go? When you enter 2.75 in C3, the frequency table is "alive" in the sense that

the frequency distribution of (a) and (b) will change automatically. What is the column sum of the frequency distribution in the table of (a) and (b) above?

(d) In tossing a die 2,000 times, why may the total sum of frequencies be smaller than 2,000? Can it be higher? What is the method of correcting this situation?

6-4 The uniform distribution Given U(0, 1), U(2, 4), and U(1.4, 3.0), draw the three uniform distributions in one diagram with the random variable calibrated at $0, 0.2, \ldots, 5$. Make sure that

(a) U(1.4, 3.0) is drawn behind the other two and is painted in yellow,
(b) the other two distributions are painted in blue gradient color,
(c) all three distributions show black borders for each column (see Fig. 6.8),
(d) the size of the chart is adjusted so that the standard uniform distribution U(0, 1) really looks like a square,
(e) there are no gaps between the columns (see Fig. 6.8), and series can be overlapped 100%.

(Hint: Open <Format Data Series> and <Select Data> in its dialog box).

Applications of mixed references — Probability theory

6-5 Mixed references In B2 of Table HW6-5, enter $A2*B$1, which means that when B2 is copied to B2:D4, column A and row 1 are locked.[9]

Let the row labels be from 1 to 10 and column labels be from 1 to 5. Reproduce the format of the table. Please make sure that the product in the last cell is $10 \times 5 = 50$. Show the cell formulas of the whole table below the original table.

6-6 The application of mixed references Table HW6-6 below shows the relationship between the market survey of expected stock price change (called **event** A) and the economic conditions (called **event** B). Event A consists of Decrease (A1), Increase (A2), and No change (A3) of stock prices, and event B consists of Bad (B1), Good (B2), and Same (B3) economic conditions. The **joint occurrence** of the two events is denoted (A_i, B_j). Any set of nonnegative fractions which add to unity is called a **probability distribution**. The fractions in range D5:F7 add to 1, hence the numbers are called the **joint probability distribution** of events A and B, and denoted $P(A_i, B_j)$. For example, 0.075 in D5 is the joint probability of the simultaneous occurrence of bad economic conditions (A1) and the expectation of a price decrease (B1). The joint probability of the occurrence of these two

Table HW6-5 The mixed references

	A	B	C	D
1		3	2	1
2	3	9=3×3	6=3×2	3=3×1
3	2	6=2×3	4=2×2	2=2×1
4	1	3=1×3	2=1×2	1=1×1

[9] In linear algebra, Table HW6-5 is called the **scalar** (or **inner**) **product** of column A2:A4 to row B1:D1, which will give a 3 × 3 matrix. See Chapter 10.

Table HW6-6 The stock market and economic conditions

	A	B	C	D	E	F	G
1				**Condition of the Economy**			
2				Bad	Good	Same	
3	**Stock**			B1	B2	B3	sum
4	**Price**	Event		0.30	0.20	0.50	1.00
5	Decrease	A1	0.25	0.075	0.050	0.125	
6	Increase	A2	0.40	0.120	0.080	0.200	
7	No change	A3	0.35	0.105	0.070	0.175	
8		Sum	1.00				1.000

events is 7.5%. The numbers in column C5:C7 are the row sum of the joint probability distribution. They also add to 1, as in C8. This set of numbers also represents a probability distribution, called the **marginal probability distribution of event A**, denoted $P(A_i)$. For example, $P(A_1) = 0.25$ (in C5) means that 25% of people surveyed expects a price decrease, regardless of economic conditions. The numbers in row D4:F4 are the column sum of the same distribution. They also add to 1, as in G4, and are called the **marginal probability distribution of event B**, denoted $P(B_j)$. For example, $P(B_1) = 0.3$ (in D4) means that the probability of occurrence of bad economic conditions is 30% regardless of people's expectations regarding stock prices.

When the joint probability distribution of events A_i and B_j can be decomposed into the product of its corresponding marginal probability distributions, that is, if

$$P(A_i, B_j) = P(A_i)P(B_j),$$

then we say the two events are **statistically independent**.

(a) Reproduce the table. Using "autosum", verify that the numbers in the tables are indeed a probability distribution.
(b) Using the method of mixed references, show that the joint occurrence of events A and B as presented in the table are indeed statistically independent.
(c) Duplicate the table below the original one and call the duplicate Table HW6-6(b). Using the "find and replace" command, show all the cell formulas of the table (see Appendix 6E for the command). Place the formula table on the right of the original table.
(d) What is the economic meaning of the numbers 0.50, 0.40, and 0.070?

6-7 Conditional Probability In the above exercise, A and B are two events in a sample space, as we have in Table HW6-6. Now we may ask, if or when one of the events occurs, what the probability is that the other event occurs. For example, supposing we already know that economic conditions are bad, what is the probability that stock prices will decrease? Since the probability of economic conditions being bad is 0.30, and the probability of economic conditions being bad AND stock prices decreasing is 0.075, the **conditional probability** that stock prices will decrease, given that economic conditions are bad is $0.075/0.3 = 0.25$.

More generally, the **conditional probability** of the occurrence of event A (or B) given the occurrence of event B (or A) is given by

$$P(A|B) = P(A, B)/P(B), \quad P(B|A) = P(A, B)/P(A).$$

Table HW6-7(a) Conditional probability table for a given condition of the economy

	A	B	C	D	E	F
12				Condition of the Economy		
13				Bad	Good	Same
14	Stock			B1	B2	B3
15	Price			P(Ai |B1)	P(Ai |B2)	P(Ai |B3)
16	Decrease	A1	0.25	0.25	0.25	0.25
17	Increase	A2	0.40	0.40	0.40	0.40
18	No change	A3	0.35	0.35	0.35	0.35
19		Sum	1.00	1.00	1.00	1.00

Table HW6-7(b) Conditional probability table for given stock price expectations

	A	B	C	D	E	F	G
23				Condition of the Economy			
24				Bad	Good	Same	
25				B1	B2	B3	sum
26	Stock Price			0.30	0.20	0.50	1.00
27	Decrease	A1	P(Bj |A1)	0.30	0.20	0.50	1.00
28	Increase	A2	P(Bj |A2)	0.30	0.20	0.50	1.00
29	No change	A3	P(Bj |A3)	0.30	0.20	0.50	1.00

In Table HW6-6, event B consists of three conditions of the economy. We may construct three conditional probabilities for each economic condition, as in Table HW6-7(a), where P(Ai|Bj) denotes the **conditional probability of Event Ai (change of stock prices) given the occurrence of Event Bj (economic condition)**, $j = 1, 2, 3$. Thus, row D16:F16 shows that the probability of price decrease given the economic condition is bad, good, or the same.

Similarly, Event A consists of three stock price changes. We may also construct three conditional probabilities for each stock price, as in Table HW6-7b. P(Bj|Ai) denotes the **conditional probability of Event Bj (economic condition) given (the occurrence of) Event Ai**, $i = 1, 2, 3$ (stock prices). Thus, column D27:D29 shows the probability of bad economic condition if stock prices decrease, increase, or remain the same.

(a) Reproduce the above two tables of conditional probabilities below the original tables, that is, Tables HW6-7(a) and HW6-7(b).
(b) Using the "Find and Replace" command, show the formulas of the tables by reproducing the tables with cell formulas next to each table.
(c) Why are all the conditional probabilities in the first and the second tables the same as the marginal probabilities? Show your proof analytically.
(d) In question (c) above, does it make economic sense? Why?

6-8 The frequency theory of probability In a survey of the expenditures of 10,000 people on seven different commodities, we obtained the following data, shown in Table HW6-8. Row 3 shows the number of people who bought the seven commodities in year 2009 by age group (event A),

Table HW6-8 Probability of independent events

	A	B	C	D	E	F	G	H	I
1				Age group					
2			A	Under 20	20 to 30	30 to 40	40 to 50	over 50	sum
3		B		1149	2854	1400	2518	2079	c
4	Com	Sold$	mp						d
5	1	41							
6	2	129							
7	3	130							
8	4	113							
9	5	205							
10	6	284							
11	7	365							
12	sum	a	b						e

and column B shows the value (in $1000) of the commodity sold to these people in the same year (event B). In this example, the sample size is so large that we may consider the frequencies as the probabilities (by the frequency theory of probabilities).

(a) Find the marginal probability distribution of the number of people who bought the seven commodities in 2009 and show them in row 4. Find the marginal probability distribution of the value of the commodities sold to these people in 2009 and show them in column C.
(b) If the two events are independent, find the joint distribution of the two events. Make sure that the joint probability distribution sums to one. (Fill in the sum in cells a to e).
(c) Give an economic interpretation of the numerical value of the marginal probability in cells E4 and C6, and the numerical value of the joint probability in E6 and H5, using the number calculated in this example.
(d) Using the "Find and Replace" command, show the cell formulas in the range C4:I12 of the table. Place the table below the original table.

6-9 Tossing a Crooked Die Show that, in Table 6.3, if the formula is given as

=round((6−1)*rand()+1,0),

then the die is crooked. Draw a chart similar to Fig. 6.4 and explain in what sense the die is crooked.

Normal Distribution

6-10 The general normal distribution In Table 6.12, the normal distribution $N(\mu, \sigma^2)$ is constructed for $\mu = 0$, and $\sigma = 1$, where μ and σ are named in E2:F2 as mu and sig.

(a) Find the normal curves for $N(0, 1)$, $N(1, 1)$, and $N(2, 1)$. Draw the three curves in one chart. Do the shapes of the curves differ? What conclusions do you draw?
(b) Draw the normal curves for $N(0, 1^2), N(0, 2^2)$, and $N(0, (2.5)^2)$. Draw the three curves in one chart. Do the shapes of the curves differ? What conclusions do you draw?

6-11 The probability between two points (a) Reproduce Table 6.12. Using the standard normal cumulative probability function (cpf), find the probability between the following intervals of z, $\text{Pro}(a \leq z \leq b)$: $\text{Pro}(-1 \leq z \leq 1)$, $\text{Pro}(-2 \leq z \leq 2)$, $\text{Pro}(0.6 \leq z \leq 1.4)$, $\text{Pro}(1.4 \leq z)$, and $\text{Pro}(z \leq -1.5)$. (b) Instead of using the cumulative probability distribution, use the area (density*0.1) calculated in Table 6.12 to calculate. (c) Will the results be the same if you use the area? Why?

6-12 Find an interval Reproduce Table 6.12. (a) Using the cumulative distribution function, find the value of a such that $\text{Pro}(-a \leq z \leq a) = 0.90, 0.95$, and 0.99. (b) Using the area, answer question (a).

6-13 Distribution characteristics In Table 6.14, we have calculated the population mean and variance of $N(0, 1)$ and verified that the standard normal distribution indeed has a mean of 0 (cell U21) and a standard deviation very close to 1 (cell W22). Using a similar method, verify that

(a) $N(2, 1^2)$ has a mean of 2 and a standard deviation of 1.
(b) $N(0, 2^2)$ has a mean of 0 and a standard deviation of 2.
(c) $N(0, 3^2)$ has a mean of 0 and a standard deviation of 3.

6-14 Normal Chips of 1000 In Fig. 6.8, change the axis label at each minor tick mark consecutively from 1 to 16, that is, −3.75 changes to 1, −3.25 to 2, etc. Suppose you have 1000 chips, and you want to mark the chips with $1, 2, \ldots$, etc. such that the chips follow the shape of the normal pdf (the rectangles in Fig. 6.8). This collection of chips is called the **normal chips of 1000**.

(a) What is the possible range of the numbers (may not be 16. Why?)?
(b) How many chips should be associated with 1, with 2, etc.?
(c) What is the probability of drawing a chip from this collection of chips?

(This was the way statisticians experimented on drawing a chip from the normal population in the old days.)
(Hint: Multiply 1000 to P3:P19 in Table 6.14. Since $P3 * 1000 = 0$, $P4 * 1000 = 0$ (that is, zero chips), the range of integers should start from P5. Thus, from P5, there should be one chip marked 1 ($0.001 \times 1000 = 1$), 5 chips marked 2 (in P6 (not shown), we have 0.005), etc. The marked chips have the normal probability distribution shown in Fig. 6.8).

6-15 Normal chips of 100 and 10,000

(a) Using Table 6.14, construct normal chips of 100 and 10,000. Explain your result by answering the same questions as in HW6-14.
(b) Supposing, in Table 6.14, that the width of the interval is taken as 0.2 instead of 0.5, what is the distribution of 10,000 chips?

The student t-distribution

6-16 The sampling distribution of a sample from the Student t-distribution In this exercise, we draw a random sample of 1000 from a population that has the **Student t-distribution**.

This distribution is similar to the standard normal distribution in Fig. 6.8. It has the bell-shaped symmetric frequency distribution of a random number t, which ranges from $-\infty$ to $+\infty$, with the maximum point at $t = 0$. The difference is that the t-distribution has fatter tails in both sides, and its distribution depends on the degrees of freedom (df).

In Excel, the formula for t is given as

=tinv(rand(),df),

where the range is a random number from 0 to 1 and df is the "degrees of freedom". Unlike =normsinv(rand()), this formula only gives the positive part of the random t-variate. Thus, in order to generate the negative part of the t-variate, we should generate a negative version of the above formula by attaching a minus sign (−tinv(rand(), df)).

Let df = 10. Draw the relative and cumulative frequency distribution of the sample of 1000 from the t-distribution by constructing a table similar to Table 6.13 (or Table 6.3) based on drawing 1000 chips from the positive and negative sides, respectively.
(Hints: Add columns #11 to #20 on the right of the original table (Table 6.13) and enter the negative variate. Thus, you have 2000 values of the t-variate. Give the range name "samplet" and "bint". Letting the bin range from −3, −2.5, to 3, draw a line chart).

6-17 The Theoretical t-Distribution The question above derives the empirical t-distribution. We may also derive the theoretical t-distribution using a table similar to column P of Table 6.14. Suppose column P (or any other column) is the place to enter the probability density function (pdf) for the t-distribution. Then in P4, enter the following equation:

=ABS((TDIST(ABS(mid+0.001),10,1)−TDIST(ABS(mid),10, 1)))/0.001,

where "mid" is the range name of M4:M19 in Table 6.14, and Abs() represents the absolute value of the return of the value of the function in parentheses.[10] The tdist($\cdots$) inside the formula returns the **student t cumulative distribution function (cdf)**. See Excel Help. The syntax is

=tdist(t,degrees of freedom,tails),

where t is the numeric value at which distribution is evaluated. In this example, we take abs(mid + 0.001) as t in the first term, "mid" as the midpoint of the interval, and 0.001 as the increment in t. The idea here is to take the difference quotient $\Delta F(t)/\Delta t$ of the cumulative function F(t). As in HW6-16, df = 10. "Tails" specifies the number of distribution tails to return. Tails may equal 1 or 2, but here tails = 1. Copy P4 to P5:P19. This is the **student t probability density function (pdf)**. Multiply it by the interval width 0.5, either in P4:P19 or in a separate column, column Q. This will give the t probability elements. Note that the frequencies do not add to 2000, and due to the rounding errors and truncation at −4 and 4, the area also does not add to 1, but is very close.

Question: Naming the "mid" column as "mid", construct a table similar to Table 6.14. Add the theoretical t-distribution to the empirical t-distribution derived in HW6-16.

---

[10]The t-distribution in Excel is given in cumulative form. Hence, this formula is the derivative of the cumulative t-distribution function (also see Sec. 7.5).

Table HW6-18 The normal distribution and the t-distribution

	A	B	C	D	E	F
1	Comparison of the t-distribution and the Normal Distribution					
2		The t-distribution				Normal dist
3	t	df				
4		2	10	30	60	z
5	-4	0.01314	0.00205	0.00053	0.00030	0.00013
6	-3.5	0.01866	0.00483	0.00199	0.00140	0.00087
20	3.5	0.01852	0.00474	0.00194	0.00135	0.00087
21	4	0.01305	0.00201	0.00052	0.00029	0.00013

6-18 The relationship between the t-distribution and the normal distribution In Table HW6-18, enter a bin range from $-4, -3.5, \ldots, 4$, and name the range t. Enter the degree of freedom for the t-distribution in range B4–E4. In B5 enter the formula

=ABS((TDIST(ABS(t+0.01),B$4, 1)−TDIST(ABS(t),B$4,1)))*0.5/0.01,

where B4 is the degree of freedom. Copy B5 down to E21. This gives the t-distribution with the various degrees of freedom indicated in row 4.

In F5, enter the standard normal distribution,

=NORMDIST(t,0,1,FALSE)*0.5

and copy it down to F21. Draw the five curves, A4:F21. Note the tails of the distributions, the heights of the distributions in the middle, and the increasing degrees of freedom, and give at least three features of the relationship between the two distributions.

6-19 Comparisons of the Cumulative Distribution From Table HW6-18, construct a cumulative t-distribution and a cumulative normal distribution. Draw the charts for both cumulative distributions in one chart. Do the three features in HW6-18 still hold?

Advertisement

6-20 Response Functions (Mizrahi and Sullivan) A simple rule of the effect (x) of advertisement after t days is given by

$$P(x < t) = 1 - \exp(-at),$$

where x is the consumer responses to advertisement of a new product, t is the time passed, say certain years, and a is a parameter showing the degree of response. $\exp(-at)$ is an exponential function with exponent $-at$, which in Excel is entered as = exp(−at). The above probability implies the fraction of consumer response to the advertisement (that is, consumers buying the product) after the time elapsed in t periods. If the product is popular, a will be large, and vice versa. For example, if $a = 0.2$, then the response of consumers to the new product after 10 years is

$$P(x < t) = 1 - \exp(-0.2^*10) = 0.865,$$

that is, 86.5%. If the market consists of 1 million potential consumers, then 865,000(= 1,000,000*0.865) people are expected to buy the product.

(a) Draw the consumer response curves for $a = 0.2, 0.4, 0.6$, and $t = 0, 1, \ldots, 20$. Give chart and axis titles.
(b) Will probability converge to 1? If so, what is the economic interpretation?
(c) If the unit revenue (without cost) of the product is \$0.5, what is the total revenue for the firm after 5 years if $a = 0.2$ and the market consists of 1 million potential customers.
(d) Suppose the firm's fixed cost is FC = \$20,000 and the variable cost is VC = \$12,000 per year. Draw the TR, the TC, and the profit curves of the firm over the years.
(e) Find the maximum profit.
(f) What is the profit at $t = 10$? $t = 20$? What is the maximum profit?

6-21 The Logistic distribution (Mizrahi and Sullivan) Continuing from HW6-20, in most cases, the response to advertising accelerates at the beginning, and then, after a certain time, tapers off. This is expressed by the following formula:

$$P(x < t) = \frac{1}{1 + be^{-at}}.$$

Assuming $b = 1$, answer the same questions as in HW6-20.

Chapter 7

Regression Analysis — Excel Commands

Chapter Outline

Objectives of this Chapter

In Chaps. 5 and 6, we have introduced the basic concepts of some statistical terms, methods of calculation, sampling, random variables, and empirical and theoretical distributions. In this chapter, we continue from the previous chapters and introduce the basic method of the least squares regression using spreadsheets. We explain the relations between population and sample regression functions by using Excel simulation; we give definitions, explanations, and meaning to terms used in Excel in the context of regression theory; and we show some test statistics that can be derived readily from the statistics provided in the Excel regression function. For more detail and advanced topics in regression analysis, the readers should consult with some statistics or econometrics textbooks.

7.1 Introduction to the Least Squares Regression

In Chap. 5, from the trendline equation of the scatter chart (Fig. 5.11), we obtained the following equation:

$$y = 0.713x - 254.950. \tag{7.1}$$

In terms of relation between consumption (C) and income (X), omitting the subscript i for simplicity, the regression line can be rewritten as

$$\hat{C} = -254.950 + 0.713X, \quad (7.2)$$

which is obtained from regressing consumption on income based on the observed data in the United States. As in Chapter 5, we use a **caret mark** over C to indicate that consumption is an estimated value. Below, we answer some questions, such as what is the theory behind the derivation of the least squares estimate, a and b? Why is it called the **least squares estimate**?

In parametric form, the estimated regression line (7.2) can be written as

$$\boxed{\hat{C}_i = a + bX_i,} \quad (7.3)$$

where a and b are unknown values, which are to be estimated like (7.2) using the observed data. This is different from numerical models in Chapters 3 and 4, where the parameters are given arbitrarily. Following general convention, in this chapter, unless specifically noted, we use capital letters like X and Y for the original levels of variables and reserve lower case letters like x and y for the mean deviations. Equation (7.3) is in parametric form, like those parametric forms in microeconomics or macroeconomics we have seen in Chapters 3 and 4. The most common method of estimating a and b, using the observed data (C_i, X_i), $i = 1, 2, \ldots, n$, where n is the number of observations in the sample, is by using some form of **regression analysis**. In economics, the least squares estimation method is the most popular method of regression analysis. We explain the method briefly in this chapter.

7.2 The Method of Least Squares

Population regression line

Implicitly in the regression equation (7.2) or (7.3) is that the observed data (C_i, X_i) are taken from a population of consumption and income. This population is assumed to have a probability distribution (usually normal) of (C_i, X_i). For a given Xi, we can find a distribution of C_i as shown in Fig. 7.1 for each income group, say, $X = 40, 50, \ldots, 100$, there is a distribution of consumption associated with that group. The distribution is called the **conditional distribution**[1] of C given X. The basic assumption of the regression analysis is as follows:

(a) that the mean of the consumption C given X_i lies on a straight line, denoted as,

$$\boxed{E(C|X_i) = \alpha + \beta X_i,} \quad (7.4)$$

which is called the **conditional mean** of consumption C given income X. Equation (7.4) is called the **population regression line.** It is further assumed that

[1]We have encountered conditional probability in HW6-7.

(b) The conditional distributions have the **same conditional variance** σ^2 (and also the same standard deviation):

$$V(C|X) = \sigma^2$$

for all X. This means that the shape of the distribution of consumption is the same for all income levels, as shown in Fig. 7.1.

(c) The values of random variable C_i are **statistically independent**, that is, a large value of consumption C_i does not make the next value of C_j large, and vice versa.

(d) Later, when we discuss hypothesis testing, the assumption of normality of the conditional distribution will be added.

Drawing of conditional distribution

Figure 7.1 is based on Table 7.1. We assume that the family income is grouped in seven categories, as shown in row 2, and for each category, consumption is distributed normally[2] with mean 0 and standard deviation 1. The conditional mean, the population regression line, is then drawn manually by connecting the highest points (the cells with 20), as shown in Fig. 7.1.

The population disturbance term

Similar to the mean deviation in the previous chapter, the deviation of individual consumption C_i from its conditional mean is denoted as

$$C_i - E(C|X_i) = \varepsilon_i. \tag{7.5}$$

Table 7.1 Conditional distribution of consumption given income

	A	B	C	D	E	F	G	H
1		Family income group (unit $1,000)						
2		40	50	60	70	80	90	100
3	20	1						
4	30	3						
5	40	6		1				
6	50	12		3				
7	60	18		6		1		
8	70	20		12		3		
9	80	18		18		6		
10	90	12		20		12		
11	100	6		18		18		
12	110	3		12		20		
13	120	1		6		18		
14	130			3		12		
15	140			1		6		
16	150					3		
17	160					1		
18	170							
19	180							

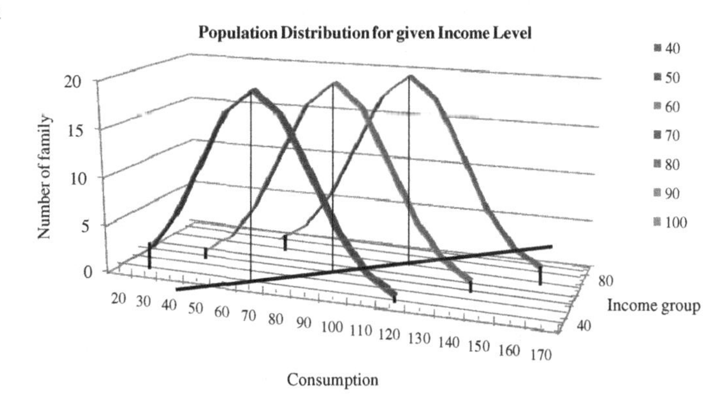

Fig. 7.1 Conditional distribution of consumption given income

[2]The normal distribution of columns B, E, and F are obtained from HW6-15.

ε_i is called the **stochastic disturbance term** or **stochastic error term**. Substituting (7.4) into (7.5), we have

$$C_i = \alpha + \beta X_i + \varepsilon_i, \tag{7.6}$$

where the error term ε_i is due to error in measurement of consumption, such as reporting error, human ignorance, negligence, omitted variables, wrong functional form, etc. α is the **intercept** or **constant** term of the population regression line, and β is the **slope** of the population regression line.

Sample regression line

We want to show now how to use a sample to estimate the population regression line. Since a straight line is determined when the intercept and the slope are known, this is equivalent to estimating α and β in (7.6). For a given sample, if the data are drawn in a scatter diagram like Fig. 5.9 in Chap. 5, we might as well draw a straight line manually so that the line can best represent the data. However, this is not scientific, as different people may have different "best fittings" of the straight line. The great German mathematician Karl Gauss discovered the method of **least squares** around 1794 when he was a student.

In this section, we will continue the discussion of the relation between consumption expenditure and national income in Chapter 5. The consumption and national income data of Table 5.7 are reproduced in Table 7.2. The scatter diagram and the regression line are also reproduced in Fig. 7.2.

The interpretation of this set of data is the same as Sec. 7.1. The income level X, say \$2784.2 (in US\$ billion) in C5 of Table 7.2, may be taken as given, and the consumption expenditure, \$1760.4 (in US\$ billion) corresponding to that income level is a random sample of size one from an imaginary infinite population of consumption expenditure at that income level. The conditional mean of consumption at income level \$2784.2, $E(C|X = 2784.2)$, can

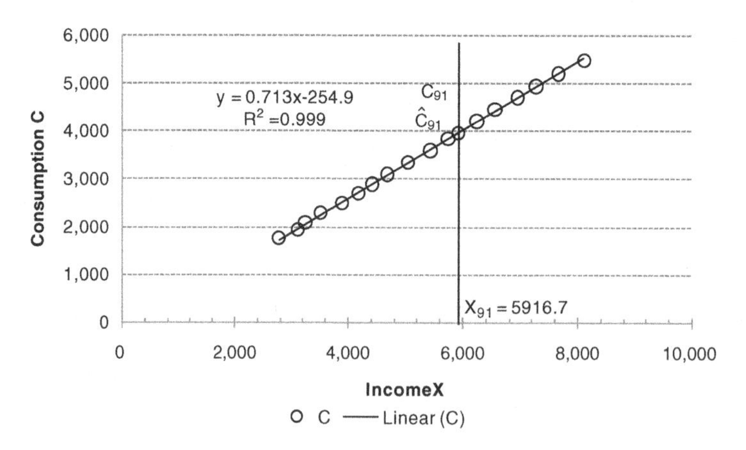

Fig. 7.2 The regression line and least squares equation

be estimated, and similarly at all other income levels, under the three basic assumptions about the consumption expenditure probability distribution[3] in Sec. 7.2.

7.2.1 *The error term*

Assume that we have an estimated equation like (7.2). The **estimated error term** or **residual** e_i is, from (7.5),

$$e_i = C_i - \hat{C}_i. \tag{7.7}$$

For example, from (7.2), at i = 1991, $X_{91} = 5916.7$ (C16 in Table 7.2), and the actual consumption is $C_{91} = 3975.1$ (for C at 1991, see B16 of Table 7.2). Substituting into (7.2), the estimated consumption is

$$\hat{C}_{91} = -254.95 + 0.71368(5916.7) = 3967.71. \tag{7.8}$$

Thus, the **estimated error**[4] is $e_{91} = C_{91} - \hat{C}_{91} = 7.4$. We can obtain this estimated error since we have already estimated a and b through Excel graphics and so the sample regression line $\hat{C}$ is known.

7.2.2 *The normal equations*

In general, the population regression line is unknown. Hence, we draw a sample (C_i, X_i) of size n from the population to estimate the population regression line. The most commonly used method is known as the **method of least squares**. It minimizes the sum of the n squared deviations of observed value C_i from its conditional expectation (7.5). Mathematically, given the sample size n, we want to minimize the function[5]

$$\begin{aligned} g(\alpha, \beta) &= \varepsilon_1^2 + \varepsilon_2^2 + \cdots + \varepsilon_2^2 \\ &= \sum_{i=1}^{n} (C_i - \alpha - \beta X_i)^2 \end{aligned} \tag{7.9}$$

with respect to α and β, namely,

$$\boxed{\min_{\alpha,\beta} \sum_{i=1}^{n} \varepsilon_i^2 = \min_{\alpha,\beta} \sum_{i=1}^{n} (C_i - \alpha - \beta X_i)^2.} \tag{7.10}$$

[3]If both C and X are random variables, we may assume bivariate (normal) distribution and the above argument still holds. See Hoel (1956).

[4]The estimated errors are calculated later in column E of Table 7.4.

[5]In descriptive statistics and many textbooks, it is stated that given a sample (C_i, X_i), the least squares method is to minimize $g(a, b) = \sum e_i^2 = \sum (C_i - a - bX_i)^2$, which is the sum of the squared deviations from the "fitted" regression line. In this case, we lose the connection between population regression and sample regression.

The general practice is to take the first-order partial derivatives of (7.9) with respect to α and β, and equate the two partial derivatives to zero. We then have

$$\sum C_i = \alpha n + \beta \sum X_i \tag{7.11}$$

$$\sum C_i X_i = \alpha \sum X_i + \beta \sum X_i^2 \tag{7.12}$$

which is a system of two equations with two unknown coefficients, α and β, and C_i and X_i are constants because their values come from the observed data. This system of equations is called the **normal equations** of the least squares method. Solving for α from the first equation and then substituting it into the second equation, and denoting the solution as a and b, we have,

$$a = \bar{C} - b\overline{X} \tag{7.13}$$

$$b = \frac{n\sum C_i X_i - \sum C_i \sum X_i}{n\sum X_i^2 - (\sum X_i)^2} = \frac{\sum (X_i - \overline{X})(C_i - \bar{C})}{\sum (X_i - \overline{X})^2} = \frac{SSxy}{SSx} \tag{7.14}$$

where $\bar{C} = \sum C_i/n$, the mean of consumption C, and $\overline{X} = \sum X_i/n$, the mean of income X. SSx and SSxy are defined[6] in Chapter 5 (see (5.12) and (5.10)). Thus, it turns out that the intercept a is the mean of observed consumption (dependent variable) minus the MPC (marginal propensity to consume) multiplied by the mean of observed income (independent variable). In (7.14), dividing the numerator and denominator by n, we see that b is the covariance between consumption and income divided by the variance (varp) of income.

The function a is called the **estimator** of population intercept α and the function b is called the **estimator** of population slope β of the population regression line (7.6). Both a and b are themselves functions of the sample points (C_i, X_i). When a particular sample is taken and a and b are calculated from (7.13) and (7.14), the resulting particular numerical values for a and b are called the **estimates** of α and β, respectively.

Substituting a and b derived above into (7.3), we have $\hat{C}_i$, which is the **least squares estimator of $\mathbf{E(C|X_i)}$** in (7.4) or C_i in (7.6). Given the value of $\hat{C}_i$ like (7.8), the sample residual $e_i = C_i - \hat{C}_i$ is the **least squares estimator of the population error term ϵ_i** in (7.5).

7.2.3 *The calculation table*

Table 7.2 shows the calculation of a and b using the data given in Table 5.7. In fact, the necessary calculation and the numbers for the formulas (7.13) and (7.14) are already given in Table 5.7. In Table 7.2, columns D and E repeat column G and H in Table 5.7. We enter $x^2 = (X - \overline{X})^2$ directly in D5:D22, and $c * x$ in E5:E22. The sum of $c * x$ in E24 gives the numerator SSxy of (7.14), and the sum of x^2 in D24 gives the denominator SSx of (7.14).

[6]An expression like $b = SSxy/SSx$ is simpler and more useful in the discussion in Sec. 7.4.1. In some textbooks, a and b are denoted $\hat{\alpha}$ and $\hat{\beta}$, respectively.

Table 7.2 Estimation of the intercept and the slope.

	A	B	C	D	E
1	**Estimation of the intercept and the slope**				
2		US$ bil	US$ bil		
3		dep var	ind var	(5.12)	(5.19)
4	Year	C	X	x^2	c*x
5	1980	1,760.4	2,784.2	6161179	4326867
6	1981	1,941.3	3,115.9	4624531	3359635
7	1982	2,076.8	3,242.1	4097678	2888187
8	1983	2,283.4	3,514.5	3069056	2137596
9	1984	2,492.3	3,902.4	1860420	1379355
10	1985	2,704.8	4,180.7	1178684	867211
11	1986	2,892.7	4,422.2	712627	515686
12	1987	3,094.5	4,692.3	329559	234840
13	1988	3,349.7	5,049.6	46990	33356
14	1989	3,594.8	5,438.7	29697	15720
15	1990	3,839.3	5,743.8	227937	160283
16	1991	3,975.1	5,916.7	422926	306644
17	1992	4,219.8	6,244.4	956538	700485
18	1993	4,459.2	6,558.1	1668561	1234404
19	1994	4,717.0	6,947.0	2824510	2039311
20	1995	4,953.9	7,269.6	4012922	2905326
21	1996	5,215.7	7,661.6	5737116	4100923
22	1997	5,493.7	8,110.9	8091338	5660958
23	Avg	3,503.6	5,266.4		
24	Sum			2,558,459	1,825,933
25		m(C)	m(X)	SSx	SSxy
26	Equ #		**Manual calculation**		
27	(7.22) b = covar(C,X)/varp(X) = SSxy/SSx				0.71368
28	(7.21) a = m(C)-b		-254.95		
29					
30		**Calculation using Excel equations**			
31	b = slope(C,X)				0.71368
32	a = intercept(C,X)		-254.95		

Dividing E24 by D24, we have the estimation of b in E27. After b is estimated, substituted into (7.13), and multiplied by $\overline{X}$ in C23, subtract the product $b\overline{X}$ from $\overline{C}$ in B23 and we have the estimate of a in C28.

For a simple regression analysis, with only one independent variable, Excel provides functions for a and b. The command for estimating a is **=intercept(C,X)**, and the command for estimating b is **=slope(C,X)**, where C is the range of data C, B5:B22, and X is the range of data X, C5:C22. The estimates for a and b are calculated directly in C32 and E31, respectively, in Table 7.2. The manual calculation and calculation using Excel equations are exactly the same, as expected.

7.3 Simple Least Squares Regression Functions in Excel

In Excel, following the mathematical convention, the dependent variable is denoted Y and the independent variable is denoted X, and the regression equation is written

$$\boxed{Y = a + bX.} \tag{7.15}$$

The least squares regression coefficients a and b may be estimated simultaneously in Excel, and can be extended to the case of more than one independent variable.

7.3.1 *Excel functions*

The Excel function for estimating α and β simultaneously is the **linear estimation = LINEST**, which has the following arguments:

$$=\text{LINEST}(\underbrace{\text{known\_y's}}_{\text{Dep var Y}}, \underbrace{\text{known\_x's}}_{\text{Indep var X,}}[, \underset{0/1}{\text{const}}]\ [, \underset{0/1}{\text{stats}}])$$

where [·] denotes the optional entry, either 0 or 1, as explained below.

Known_y's : a known set of the values of the dependent variable, like C; We simply use Y to show the range of dependent variable.

Known_x's : a known set of the values of the independent variable(s), like X; We simply use X to show the range of independent variable(s).
X may consist of several variables, and can be taken as a matrix (see Chapter 10 for matrices): like time, income, assets, and interest rate.

Const : a constant term; If const = 1 (default), then a $\neq$ 0; if const = 0, then, a = 0, that is, a is forced to take the value of 0 (the regression line passes through the origin)

Stats : some statistical characteristics of regression (see below); If stats = 1, show the stats; if stats = 0 (default), ignore the stats.

Since const = 1 and stats = 0 are defaults, if we accept the defaults, then we can write the function as

$$\text{LINEST}(\text{Y}, \text{X}, 1, 0) \equiv \text{LINEST}(\text{Y}, \text{X}), \tag{7.16}$$

where we simply denote the argument (Y, X). The stats consist of the following information.

a, b: Regression coefficients
s_a, s_b: Standard errors of coefficient a and b
R^2: Coefficient of determination
s_e: Standard error of regression
F: F-statistic

These terms will be explained in Secs. 7.3.3 and 7.4.1

7.3.2 *An application*

We now use the (C, X) data in Table 7.2 to explain the use of Excel functions. Note that we denote the dependent variable C instead of Y, to be consistent with our previous exposition.

Table 7.3 Excel least squares method.

	F	G	H	I
1		Y = a + bX		
2	Case			
3	1	=LINEST(B5:B22,C5:C22)		
4			0.71368441	-254.949977
5			b	a
6				
7	2	=LINEST(B5:B22,C5:C22,1,1)		
8			b	a
9		coeff	0.71368441	-254.949977
10		s_b s_a	0.00361864	19.91671948
11		R^2, se	0.99958883	24.55676807
12		F, df	38897.4417	16
13		SSR, SSE	23456513.3	9648.557726
14				
15		t-statistic	197.224344	-12.8008017
16		MSR, MSE	23456513.3	603.0348578
17		F-statistic	38897.4417	
18				
19	3	=LINEST(B5:B22)		
20			219.765325	

There are three cases for the LINEST(·) function:

1. If **no stats** are needed, we only derive the values of the regression coefficients a and b. In this case, we need two cells in which to put the values of a and b. Hence, in Table 7.3, which is continued from Table 7.2, we select a blank cell range, say H4:I4, and in H4, enter the function as shown in G3. Selecting H4:I4, use the **range-copy command** (see (6.17)), and you will see the results in H4:I4. When any one of the two cells is clicked, the function box will show the function in braces, { }:

   $$\{=\text{LINEST(B5:B22,C5:C22)}\}.$$

   Note that the arrangement of the coefficients is reversed in Excel. That is, the estimate of a is in I4 and that of b is in H4. To avoid confusion, it is suggested that readers enter the coefficient name below the number as shown in H5:I5 in Table 7.3.

   From the table, we can write the estimated regression equation as in (7.2).
2. **If "stats" are needed**, we select a blank cell range of 5 × 2, say H9:I13 in Table 7.3, and at the northwest corner of the range, that is, at H9 (or anywhere in the range), we enter the function shown in G7. Using the range-copy command, we obtain 10 numbers in H9:I13 instantly. Excel does not provide the row and column labels. Thus, it is not clear which number shows which statistic. The ten regression statistics come in pairs and

have a predetermined arrangement. They should be entered by the readers as shown in G9:G13 and H8:I8, either before or after performing the range-copy command when we use "stats".

In particular, H10 shows the standard error of b, denoted s_b, I10 the standard error of a, denoted s_a, and H11 the coefficient of determination R^2, etc. These terms will be explained in Sec. 7.4.

3. If only the **trendline** of C is needed, the independent variable X can be omitted. We enter only the dependent variable C, as shown in H20, using the function in G19. In this case, we obtain only the coefficient of the time trend b in

$$\hat{C} = a + bT, \tag{7.17}$$

and a is ignored.

Note that the functions in range H9:I13 are "alive" in the sense that when the original data in columns A and B in Table 7.2 change, the estimated results will change accordingly. There is no need to retype the Excel function. Thus, Table 7.3 can be reserved and used as a template for simple regression analysis in the future.

Example 7.1 The trendline. In (7.17), let T be $T = 0, 1, \ldots, 17$, and let C be the same as those in Table 7.2. Enter T in, say, column F in Table 7.2. We estimate $C_i = \alpha + \beta T_i + \epsilon_i$.

(a) Copying Table 7.2 to another sheet and using it as a template, draw the scatter diagram and estimate a, b, and R^2 in (7.17). Make sure that T is on the X-axis.
(b) Verify your result by using manual calculation as in Table 7.2
(c) Add the linear trendline to your scatter diagram. Make sure that the chart shows the equation and the R^2 correctly.
(d) Write the regression equation with R^2.

Answer: (a) and (c) The average of T, $m(T) = 8.5$. No change in the estimate b. Hence the trendline is $C = 1635.573 + 219.765T$, $R^2 = 0.99717$. □

Example 7.2 The intercept and slope of a trendline. In the above example, let variable U in (7.17) take the years from 1980 to 1997, as shown in column A of Table 7.2. Then, $U = T + 1980 = T + K$. We now estimate $C_i = \alpha + \beta U_i + \epsilon_i$.

Answer questions (a) to (d) in Example 7.1 above, and the following questions.

(e) Does the estimate of intercept a change? Why? Let the two intercepts be denoted as a(C, T) and a(C, U). How are the two intercepts of regression equations related? Can you derive one equation from the other? Show how.
(f) Does b change? How about R^2? Why?

Answer:

(a) and (c) The average of U is, $m(U) = 1988.5$. No change in the estimate of slope b and R^2. The trend line is $C = 219.765U - 433499.771$, $R^2 = 0.99717$.

(e) Yes. a will change. Using the notation in (e) above, (7.13) is, $a(C, U) = m(C) - bm(U)$. But, $m(U) = m(T) + k$. Hence,

$$a(C, U) = m(C) - bm(T) - bk = a(C, T) - bk. \quad (7.18)$$

Thus, the difference between a(C,U) and a(C,T) is a constant bk. In this example, $bk = 219.765 * 1980 = 435135.3437$. Therefore, $a(C, U) = 1635.573 - 435135.344 = -433499.771$.

(f) No. b and R^2 do not change. Since $T = 0, 1, \ldots, 17$ and $U = 1980, 1981, \ldots, 1997, U = T + 1980$. This means that a constant k is added to T. The formula for b in (7.14) is,

$$b(C, U) = SScu/SSu = \sum(C - m(C))(U - m(U))/\sum(U - m(U))^2, \quad (7.19)$$

where m(U) is the mean of U. Let $U = T+k$, where $k = 1980$. Then, $m(U) = m(T+k) = m(T) + m(k) = m(T) + k$, since k is a constant. Substituting into b(C, U), the constant term k cancels out, hence $b(C, U) = SS_{cu}/SS_u = SS_{ct}/SS_t = b(C, T)$. The estimate b should not change. R^2 does not change since it is independent of the independent variable T, see its definition in (7.23) in Section 7.3.3. □

7.3.3 *The coefficient of determination — A measure of goodness of fit*

The coefficient of determination, R^2, is probably the most often used measure of the goodness of fit in the least squares regression analysis.

Decomposition of the total deviation

In Fig. 7.2, as can also be read from Table 7.2, when the independent variable, income X, is at \$5,916.7 billion in 1991, the observed consumption is $C_{91} = \$3,975.1$ billion, and the estimated consumption is calculated at $\hat{C}_{91} = \$3,967.7$ billion (see (7.8)). This means that we can use $\hat{C}_{91}$(See (7.8)), to "predict" consumption at the income level $X = \$5,916.7$ billion (that is, to predict the conditional mean at X). Otherwise, without the least squares estimated consumption $\hat{C}_{91}$, the best way we can do the prediction is to use the average consumption $\overline{C} = 3,503.6$ (which is the sample unconditional mean). In this sense, the deviation $\hat{C}_{91} - \bar{C} = 464.1$ is "explained" by the regression line, and therefore is called the **explained deviation**. The rest of the deviation from the mean, that is, the difference between the observed consumption and the estimated consumption, $C_{91} - \hat{C}_{91} = 7.4 = e_{91}$ is unpredictable, since it depends on the sample point and cannot be explained by the regression line. This deviation is called the **unexplained deviation**, which in (7.7) has been called the **estimated error**.

Decomposition of the total variation

Since the three deviations in Fig. 7.2 are so close to each other that it is hard to distinguish the three points, we have reproduced the three deviations in Fig. 7.3 for an X, say X_i. Clearly, from Fig. 7.3, we can see the explained deviation plus the unexplained deviation

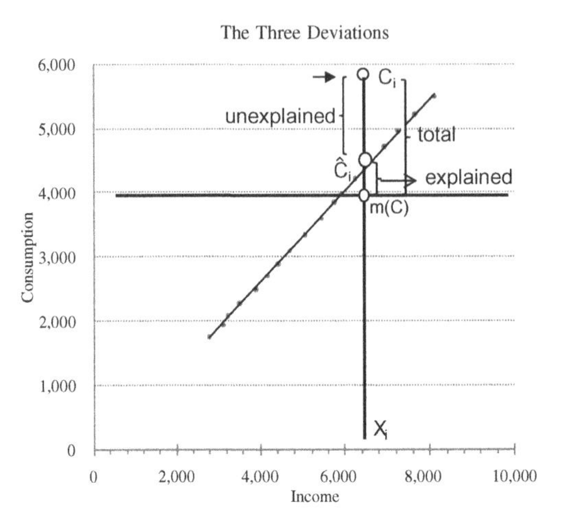

Fig. 7.3 The three deviations in regression analysis

equals the deviation of the observed consumption from the mean consumption, $C_i - \overline{C}$, which is called the **total deviation**. Thus, we have the following identity for any ith observation:

$$\underbrace{C_i - \overline{C}}_{\text{Total deviation}} = \underbrace{(C_i - \hat{C}_i)}_{\text{Unexplained deviation}} + \underbrace{(\hat{C}_i - \overline{C})}_{\text{Explained deviation}}, \tag{7.20}$$

It is one of the basic results in regression analysis that this identity still holds when each term is squared and summed, as shown below.

$$\boxed{\underbrace{\sum_i (C_i - \overline{C})^2}_{\text{Total Variation}} = \underbrace{\sum_i (C_i - \hat{C}_i)^2}_{\text{Unexplained Variation}} + \underbrace{\sum_i (\hat{C}_i - \overline{C})^2}_{\text{Unexplained Variation}}} \tag{7.21}$$

The squared terms are called **total variation**, **unexplained variation**, and **explained variation**, as shown in (7.21). For simplicity, they are also written as SST, SSE, and SSR, and called the **sum of squares total** or the **total sum of squares** (SST) of the dependent variable, the **sum of squares error** or **the error sum of squares** (SSE), and lastly, the **sum of squares due to regression** or the **regression sum of squares** (SSR).[7] Note that SST is closely related to varp; in fact, SST $= (n) * (\text{varp}(C))$. The relation (7.21) is also known as the **decomposition of the total variation** of the dependent variable C.

[7]Some textbooks use SSE to denote the "sum of squares explained" and SSR to denote the "sum of squares residual", just opposite to our usage. Our notations, mostly used in business statistics, are adopted for mnemonic reasons, to correspond SSR to R^2 in (7.23).

The coefficient of determination

Moving the unexplained variation to the left-hand side of the equation and then dividing both sides by total variation, we have

$$\frac{\sum_i (\hat{C}_i - \overline{C})^2}{\sum_i (C_i - \overline{C})^2} = 1 - \frac{\sum_i (C_i - \hat{C}_i)^2}{\sum_i (C_i - \overline{C})^2}, \tag{7.22}$$

$$\boxed{R^2 \equiv \frac{SSR}{SST} = 1 - \frac{SSE}{SST},} \tag{7.23}$$

where the ratio SSR/SST is called the **coefficient of determination**, and is denoted R^2. It is a fraction (the percentage or proportion) of the total variation of the dependent variable explained (or determined) by the least squares regression line, and is a measure of the **goodness of fit** of the regression line to the sample data.

In general, it can be shown that $0 \leq R^2 \leq 1$. The numerators of (7.23) give two extreme cases.

a. SSE = 0. If the regression line fits the data perfectly, that is, if all the data points (the observed cases of consumption) lie on the regression line, $C_i = \hat{C}_i$, then SSE = 0 and $R^2 = 1$. In this special case, the regression line is a perfect fit to the data and has the perfect power of explaining the change of the dependent variable.
b. SSR = 0. When the estimated regression line coincides with the mean line, $\hat{C}_i = \bar{C}$ for all i, then we have $R^2 = 0$ and SSE/SST = 1.

In Table 7.3, cell H11 shows $R^2 = 0.9996$; that is, 99.96% of the total variation in consumption C has been explained by the least squares regression line. This regression line is a near perfect fit to the observed consumption data.

7.3.4 *Calculation of the coefficient of determination*

The statistics in Table 7.3 are obtained from the Excel function. As in Chapter 5, we would like to verify the calculation using definitions of the functions so that we can gain insight into the meaning behind the numbers. This is done in Table 7.4. Rows 24–27 of Table 7.4 show the calculation of the three variations in (7.21) and the coefficient of determination (7.23), using the following definitions. For each i of the independent variable X_i in column C, we calculate the estimated value of consumption $\hat{C}_i$ in column D (see (7.2) or (7.3)), the estimated error (7.7) in column E, the squared unexplained deviation for each i (SSEi) in column F, the squared explained deviation (SSRi) for each i in column G, and the squared total deviation for each i (SSTi) in column H. Since it is not easy to enter a bar or a cap above the letter in Excel, we use C^ to denote $\hat{C}_i$ and mC to denote $\overline{C}$. The method of entering the functions is the same as in Table 5.7, and will not be repeated here.

Table 7.4 Various sums of squares

	A	B	C	D	E	F	G	H	I	J
1	**Regression statistics**									
2		US$ bil	US$ bil	estimate						
3		dep var	indep var	indep v	error	Calculation of four SS				
4	coeff	b	a	a+bX	C-C^	SSE	SSR	SST	SSx	SSX
5		C	X	C^	e	e^2	(C^-mC)^2	(C-mC)^2	x^2	X^2
	Year	(Table 5.7)		(7.3)	(7.7)	(7.21)	(7.21)	(7.21)	(7.3)	
6	1980	1,760.4	2,784.2	1,732.1	28.3	801	3,138,168	3,038,668.76	6,161,179	7,751,770
7	1981	1,941.3	3,115.9	1,968.8	(27.5)	757	2,355,484	2,440,711.85	4,624,531	9,708,833
21	1995	4,953.9	7,269.6	4,933.3	20.6	426	2,043,963	2,103,434.55	4,012,922	52,847,084
22	1996	5,215.7	7,661.6	5,213.0	2.7	7	2,922,174	2,931,362.50	5,737,116	58,700,115
23	1997	5,493.7	8,110.9	5,533.7	(40.0)	1,598	4,121,286	3,960,586.46	8,091,338	65,786,699
24	sum of squares (SS)				0.0	9,649	23,456,513	23,466,162	46,052,269	545,276,444
25	(7.21) identity					SSE	SSR	SST	SSx	SSX
26	(5.9) A 3,503.6		5,266	3,503.6		Identity		23,466,162	(7.32)	(7.28)
27	(7.29) R2 = SSR /SST						0.99958883			
28	n = 18					n-2	1	n-1	nSSx	
29	df					16	1	17	828940847	
30						MSE	MSR	MST		
31	(7.31) mean square (MS)					603.0	23456513	1380362.5		
32	(7.27) se = sqrt(MSE) →					24.5568	(7.37)	(7.37)		
33	Derivation of sa and sb									
34	(A) =sqrt(SSX/(nSSx))								(A)	
35	(7.28) sa =se*(A) →					19.9167			0.8110481	
36	(7.29) sb =se/sqrt(SSx) →					0.00362			0.0001474	
37	(7.45) F =MSR /MSE →						38897		sqrt(1/SSx)	

The sum of estimated errors

Note that the sum of estimated errors e_i equals zero (or near zero due to rounding errors), as in cell E24 of Table 7.4:

$$\boxed{\sum (C_i - \hat{C}_i) = \sum e_i = 0,} \tag{7.24}$$

which is similar to Eq. (5.10) in the case of mean deviations. The proof of (7.24) follows algebraically from the definition of $\hat{C}_i$ in (7.3) and the first normal equation in (7.11) when α and β are substituted by their estimates a and b (exercise). The SSE (= 9649) in F24 is the minimum value of the function posted in (7.10) (since $\hat{C}_i$ is derived from the normal equations). From (7.24), we have

$$\sum \hat{C}_i = \sum C_i. \tag{7.25}$$

Equation (7.25) means that, although the estimated dependent variable $\hat{C}_i$ and the observed dependent variable C_i differ individually in each i, the sum of each variable turns out to be the same.

There is another way to look at (7.25). When both sides of (7.25) are divided by sample size n, the average value of the observed dependent variables is equal to the average value of the estimated dependent variables. We may also state that the mean of the actual consumption equals the mean of the estimated consumption.

Decomposition of the total variation

We have calculated SSE = 9,649 in F24 and SSR = 23,456,513 in G24. They are exactly the same as I13 and H13 in Table 7.3 obtained directly from the Excel LINEST function. The **identity of the decomposition of the variations**, (7.21), is shown in rows 25 and 26: in H26, we calculate SSE + SSR = 23,466,162, which is exactly the same as SST = 23,466,162 in H24, which is calculated as the sum of squares of column H6:H23. Hence,

$$\begin{array}{ccc} \text{SST} = & \text{SSE} + & \text{SSR} \\ \text{H24} & \text{F24} & \text{G24;} \end{array}$$

The coefficient of determination $R^2 = 0.99958883$ in G27 is obtained by dividing SSR = 23,456,513 in G24 by SST = 23,466,162 in H24, as linked by an arrow. This R^2 is the same as the R^2 derived from the Excel function in cell H11 in Table 7.3.

7.4 Standard Errors of Least Squares Coefficients

7.4.1 *Definitions and formulas of standard errors*

Each element of a sample of size n may be considered as a random variable, and the characteristics of the sample distribution, such as mean and variance, are functions of random variables. This is also the case of the regression coefficients a and b, both are, as shown in formulas (7.13) and (7.14), function of random variables, and therefore, are themselves random, and the estimated value of a or b may be considered as a single sample from the (population) probability distribution of a or b. The probability distribution of a or b is called the (theoretical) **sampling distribution of the least squares estimator** (or, more generally, **sample statistics**) (see the second part of Fig. 7.7 at the end of this Chapter). For the regression analysis specified by conditional mean (7.4), it can be proved[8] that, using the notations of E and V defined in Sec. 6.6.2 of Chap. 6,

$$\begin{aligned} E(a) = \alpha, \quad V(a) &= \sigma^2 \frac{\sum X_i^2}{n \sum x_i^2}, \\ E(b) = \beta, \quad V(b) &= \frac{\sigma^2}{\sum x_i^2}, \end{aligned} \tag{7.26}$$

where σ^2 is conditional variance of the conditional distribution of C given X in (b) of Sec. 7.2.

[8]See any textbook on Econometrics, for example, Gujarati (2003).

Since σ^2 is unknown, it can be estimated by the variance of the sample errors e_i defined as s_e^2, where[9]

$$s_e^2 = \frac{\sum (Y_i - \hat{Y}_i)^2}{n-2} = \frac{SSE}{n-2} = MSE. \tag{7.27}$$

The square root of s_e^2, s_e, is called the **standard error of estimate** or the **standard error of the regression**. When s_e^2 is substituted into V(a) and V(b), and the positive square root is taken, using more general notation of the dependent variable Y instead of C, we have (7.28) and (7.29):

$$s_a = s_e \sqrt{\frac{\sum X_i^2}{n \sum x_i^2}} = s_e \sqrt{\frac{SSX}{nSSx}} = s_e \sqrt{\frac{1}{n} + \frac{\overline{X}^2}{SSx}}, \tag{7.28}$$

$$s_b = s_e \frac{1}{\sqrt{\sum x_i^2}} = s_e \sqrt{\frac{1}{SSx}}. \tag{7.29}$$

s_a and s_b depend only on the sample (X_i, Y_i), and their squares can be calculated from the sample to estimate $\hat{V}(a)$ and $\hat{V}(b)$, respectively. s_a (or s_b) is called the **standard errors of regression coefficient** a (or b), respectively. It is the standard deviation of the (empirical) sampling distribution of regression coefficient, namely, it is a sample characteristic of the frequency distribution of a sample statistic, not the sample itself. Hence, s_a and s_b are called the standard errors instead of standard deviations. In the following Section, we explain the method of calculation.

Example 7.3 Samples of sample statistics. Copy Table 7.2 to another sheet, say Sheet2. Rename the copied table Table 7.2(a), for simulation. In Table 7.2(a), enter the simulated consumption and income of HW5-5. Thus, in B5 enter 1500, and in C5, enter 2000, and enter in

B6: =400+B5+2000*rand(),

C6: =500+C5+3000*rand(),

where the initial values are assumed as $C(0) = 1500, X(0) = 2000$. Copy B6:C6 to B7:C22. Each time when <F9> is pressed, you obtain a different set of sample. Press <F9> ten times, each time <copy> and <paste in value> the regression result in G7:I7 of Table 7.3 to a blank space (Tables 7.2 and 7.3 are on the same spreadsheet). Thus, you have collected 10 sample points of sample statistics a, b.

[9]Since σ^2 is the variance of the stochastic disturbance terms ε_i, it is plausible to use the variance of the error terms e_i to estimate σ^2. By definition, $var(e) = \sum (e_i - \bar{e})^2/n = \sum e_i^2/n$, from (7.24). Like the case of sample variance, the unbiased estimator of σ^2 is $s_e^2 = \sum e_i^2/(n-2) = SSE/(n-2)$ in (7.27).

For a sufficiently large sample, the empirical frequency distribution of a or b will approach the theoretical probability distribution of a or b, just the same as tossing a die 1000 times and the empirical frequency distribution of the numbers of the face will approach the theoretical distribution we have discussed in Chapter 6. The theory says that, the standard errors of a in (7.28) (or b in (7.29)) are the estimate of the standard deviation of the (theoretical) probability distribution of a (or b) when n is sufficiently large. □

7.4.2 *Calculation of standard errors*

Degrees of freedom and mean square error

In order to calculate s_a and s_b, we have to calculate s_e first. In (7.27), SSE is divided by n − 2, which is called the number of **degrees of freedom (df)** in calculating SSE. The intuitive explanation is as follows. To calculate SSE, we have to calculate $\hat{Y}$ (which is $\hat{C}$ in our example, see (7.3)), but $\hat{Y}$ can be calculated only if a and b are estimated first. This means that the number of independent observations in the sample loses two degrees of freedom. Hence, the degrees of freedom of the sample in calculating SSE equals n − 2. In general, when the SSE is divided by the degrees of freedom, we call it the **mean square error** (MSE), which is

$$\mathrm{MSE} = \frac{\mathrm{SSE}}{n-2}. \tag{7.30}$$

Standard error of regression

In (7.27), the standard error of regression s_e is the positive square root of the MSE.[10] In Table 7.4, SSE = 9649 is in cell F24, and its degrees of freedom, df = n − 2 = 18 − 2 = 16, is in F29. Dividing SSE by its df, we have the MSE = 603.0, in F31. Taking the positive square root of MSE, we have s_e = 24.5568 in F32.

In Table 7.3, MSE in I16 is derived from dividing SSE in I13 by its df in I12.

As the formula (7.30) shows, MSE is a measure of the scattering of the observed values of dependent variable around the regression line. If MSE is small, the fit of the regression line to the dependent variable is good. On the other hand, if it is large, the fit is not good. It is similar to the coefficient of determination, R^2, but R^2 is a better measure of goodness of fit than MSE, as it is a ratio independent of the unit of measurement.

Mean squares regression (MSR) and mean squares total (MST)

Other sum of squares formulas, SSR and SST, are also associated with the number of **degrees of freedom**. As we have seen in (7.21), the SST on the left side of identity is the numerator of the variance of dependent variable Y (note that we use Y here instead of C), the variance of Y has n − 1 degrees of freedom, and the SSE on the right side of identity has n − 2 degrees of freedom. Thus, to make the degrees of freedom of the same variable

---

[10]See Harnett and Murphy (1985), p. 610.

on both sides of identity consistent, the SSR must have one degree of freedom.[11] Therefore, the **mean squares regression** (MSR) and the **mean squares total** (MST) are

$$\mathrm{MSR} = \frac{\mathrm{SSR}}{1} = \mathrm{SSR} \quad \text{and} \quad \mathrm{MST} = \frac{\mathrm{SST}}{\mathrm{n}-1}. \tag{7.31}$$

For the example in Table 7.4, the degrees of freedom (df) for SSR and SST are shown in row 29. If the SSR and the SST are divided by their respective degrees of freedom, we obtain MSR and MST. They are shown in G31 and H31. In Table 7.3, the MSR is the same as SSR in H13, which is copied to H16.

Standard error of regression coefficient a

To calculate the standard error of coefficient a, s_a, as defined in the last expression in (7.28), we should calculate the **sum of squares of mean deviations** of X, namely,

$$\mathrm{SSx} \equiv \sum (X_i - \overline{X})^2. \tag{7.32}$$

which is given in I24 of Table 7.4, which is multiplied by n = 18 in I29. On the other hand, the denominator of s_a, the sum of squared independent variable X, is calculated in J24.

Thus, in I35, we define

$$\text{(A)=sqrt(SSX/(nSSx))}. \tag{7.33}$$

Finally, multiplying (A) by the standard error of regression s_e in F32, we have the **standard error of regression coefficient a** in F35,

$$s_a = s_e \times (A) = 19.91767, \tag{7.34}$$

which is the same as that we obtained using the Excel function in I10 of Table 7.3.

Standard error of regression coefficient b

From (7.29), the standard error of regression coefficient b, s_b, can be derived from the standard error of regression, $s_e = 24.5568$ in F32, divided by the square root of SSx, giving us the **standard error of regression coefficient b**, $s_b = 0.00362$, in F36. This result conforms to the entry of Excel function in H10 of Table 7.3.

7.5 The t-Distribution and Hypothesis Testing of Regression Coefficients

This section introduces some basic concepts of hypothesis testing. As we have mentioned in Sec. 7.2, the fourth condition of the regression analysis must be satisfied, namely, the data (either Y or Y and X) must be a random sample and the conditional probability

[11] Note that, like the decomposition of the total variation, the degrees of freedom for SST must be equal to the sum of the degrees of freedom for SSE and the degrees of freedom for SSR. This is indeed the case for our example.

distribution function of the dependent variable (in the previous example, C_i), given the independent variable (in the previous example, X_i), must be normally distributed with a given conditional mean and conditional variance, $N(\mu, \sigma^2)$.

We have seen the Student t-distribution in the homework of Chapter 6. The **t-statistic**, along with the **F-statistic**, is used in statistical inference under the assumption that either the conditional probability distribution of the dependent variable or the error terms follow a normal distribution. The t-statistic is useful in statistical inference testing the null hypothesis that the regression coefficient a or b is zero. The F-statistic is used in testing the null hypothesis about goodness of fit of the regression line. This section explains the method of testing hypotheses. These test statistics can be calculated easily from the information provided by the Excel Regression Table in Table 7.3.

7.5.1 *The t-statistic*

Given the least squares estimates of coefficients a and b for the population intercept α and the population slope β, the t-statistics can be written as

$$\boxed{t_a = \frac{a - \alpha}{s_a}, \quad t_b = \frac{b - \beta}{s_b},} \tag{7.35}$$

and t_a or t_b has the t-distribution with $n - 2$ degrees of freedom, where n is the sample size. It turns out that t_a and t_b are similar to the standardized variable (see Sec. 5.3.2) of a and b, respectively.

The t-distribution

Excel gives a function for the t-distribution as

$$=\text{tdist}(|t|,\text{df},\text{tails}), \quad -\infty < t < \infty, \tag{7.36}$$

which gives a cumulative distribution function of the t-distributed random variable t. **Tails** refer to the extreme portion of a small area under the (usually bell-shaped) probability density function. For example, Fig. 7.4 shows **two tails** of the t-distribution. The **right tail** is on the right of $t_0 = 2.12$, such that $P(t > t_0) = 0.025$, and the **left tail** is on the left of $t_0 = -2.12$, such that $P(t < t_0) = 0.025$. We call either of these two cases **one-sided tail**, and we enter 1 in **tails** portion in (7.36). If both cases are considered, then we have **two-sided tails** $P(t > t_0 \text{ and } t < -t_0) = 0.05$, and enter 2 in **tails** portion in (7.36).

The cumulative probability distribution of t

In general, in terms of the probability statement, we can write

$$\text{tdist}(t, \text{df}, 1) = P(t > t_0) \quad \text{and} \tag{7.37}$$

$$\text{tdist}(t, \text{df}, 2) = P(t > t_0, t < -t_0). \tag{7.38}$$

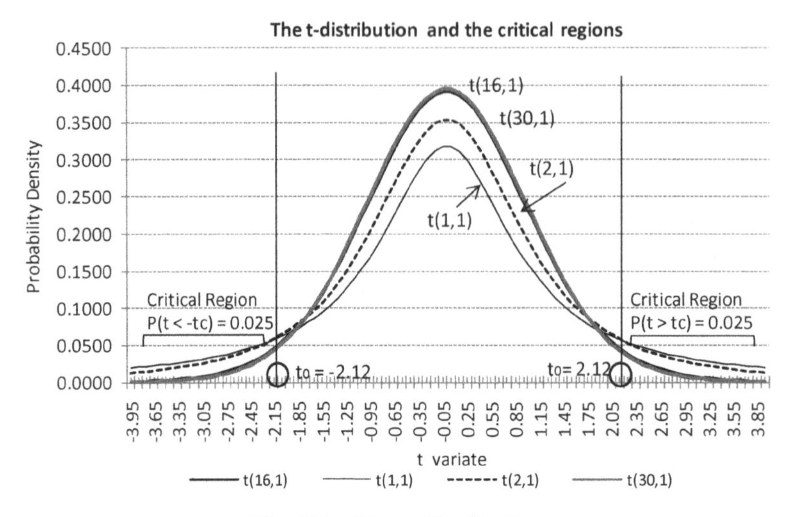

Fig. 7.4 The t-distribution

As with the normal cumulative probability distribution (cdf) in column D of Table 6.12, enter in D7 of Table 6.12,

$$=\text{tdist}(\text{abs}(t),16,1), \qquad \text{for } t = \text{midpoint from } -3.95 \text{ to } -0.05, \tag{7.39}$$

$$=1-\text{tdist}(\text{abs}(t),16,1), \quad \text{for } t = \text{midpoint from } 0.05 \text{ to } 3.95, \tag{7.40}$$

where, as in Table 7.3, we take df = 16, and tail = 1, and "abst(t)" is the absolute value of t, as t must be positive. (7.39) and (7.40) give a cdf of t, the shape of which is very similar to the normal cdf as shown by the dashed curve in Fig. 6.6.

7.5.2 *The t-distribution*

Since (7.37) gives a cdf of t, to draw the density function, we have to find the effect of a small change (say, 0.001) in t on the cumulative function (7.37). As shown in the homework question HW6-17, the **probability density function of the t-distribution** is shown in Fig. 7.4. Note that the curve for t(16, 1) and t(30, 1) almost overlap each other; there is not much difference between the two curves. The t-distribution is similar to the normal distribution, having the mean at 0 and being symmetric around the mean, but unlike the normal distribution, it has fatter (or thicker) tails. The t-distribution approaches the standard normal distribution when n goes to infinity.

7.5.3 *Hypothesis testing of population regression coefficients*

The t-statistics in (7.35) will be used to test the significance of a single coefficient being different from zero. For this purpose, statisticians set up a **null hypothesis** H_0 about the population parameter(s) such that H_0: $\alpha = 0$ (i.e. there is no intercept), against the alternative hypothesis that H_1: $\alpha \neq 0$. They then make a criterion for accepting or rejecting the null hypothesis. We first calculate the estimate a of α from the sample, and standardize

Table 7.5 Two methods of hypothesis testing

	G	H	I	J	K	L
7	=LINEST(B5:B22,C5:C22,1,1)					
8		b	a			
9	coeff	0.71368441	-254.949977			
10	sb sa	0.00361864	19.91671948			
11	R2, se	0.99958883	24.55676807			
12	F, df	38897.4417	16			
13	SSR, SSE	23456513.3	9648.557726			
14						
15	t-stat	197.938028	-12.8008017			
16	Critical value approach (tinv)			p-value approach (tdist)		
17	coeff tinv(p,df)			coeff	tdist(t,df,tails)	
18	0.05/2	2.11990529		0.05/2	1.5144E-28	8.02554E-10
19		s	s		s	s
20	0.05/1	1.74588367		0.05/1	7.5722E-29	4.01277E-10
21		s	s		s	s
22	goodness-to-fit Finv(p,df1,df2)			goodness-to-fit Fdist(F,df1,df2)		
23	0.05/1	4.49399842		0.05	1.6045E-28	
24		s			s	

a to obtain the **test statistic**, t_a as in (7.35), which is known to have a t-distribution with df = 18 – 2 (the size of sample minus the number of coefficients including the constant). Under the null hypothesis of $\alpha = 0$, dividing the estimate a by its standard error s_a, we have

$$t_a = a/s_a = -254.94998/19.91672 = -12.80080, \tag{7.41}$$

which is obtained in I15 of Table 7.3, and is reproduced in I15 of Table 7.5. If t_a falls into one or two tails, say, $t > t_0$ or $t < -\ t_0$, called the **critical region**, then we **reject** the null hypothesis, and t_0 is called the **critical value** (see 7.42 below) of the t variable. For example, in Fig. 7.4, if the two-sided tails are set at the value of $t_0 = 2.12$, such that P(t > 2.12 and t < –2.12) = 0.05, then if t_a falls inside the critical region (either in the left or right tail, as shown in Fig. 7.4), there is only a 5% probability that this can happen. We consider that this chance is very small and reject the null hypothesis, saying that the value of statistic t_a is **significant** (namely, significantly different from zero). The probability P(t > t_0) used to reject the null hypothesis is called the **level of significance**. The level of significance is usually set at 1%, 5% or 10%. In (7.41), t_a falls far to the left of the critical value $t_0 = -2.1199053$, hence we can safely reject the null hypothesis and say that the estimate a is **significant at the 5% level of significance**. In I19 of Table 7.5, we enter s to indicate that a is significant for the two-tailed test.

One-tailed test

The above testing has two-sided tails and is called a **two-tailed test**. This is usually the case if parameter α can be positive or negative. If there is a theoretical or empirical reasoning or evidence to think that α should be positive (or negative), then we can set up a critical region on the right (or left) tail, and perform a **one-sided test**. It can be shown from (7.42) below that the critical value of the 5% level of significance for one-sided test is 1.745884, which is located at the left of $t_0 = 2.12$. The interpretation of the result of testing is the same as the case for two-tailed test.

Finding the critical values of the t-distribution

The problem of finding the critical value of the t-distribution is the inverse of the operation in (7.36), in which we are given a value of t_0 and find the probability that $P(t > t_0)$. Its inverse is that given a probability, say 0.05, find the value of t_0 such that $P(t > t_0) = 0.05$. We have encountered this problem for the normal distribution in Fig. 6.7 in Sec. 6.8.1. In Excel, the **inverse function of the t-distribution** is

$$=\text{tinv}(\text{probability},\text{df}), \tag{7.42}$$

which gives the two-tailed t value, t_0, of (7.38). Thus, at the 5% level of significance in a two-tailed test, the critical value is $t_0 = \text{tinv}(0.05,\ 16) = 2.119905$, as shown by the two critical values in Fig. 7.4.

If the problem is to find the critical value at the 5% level of significance for a one-tailed test, then the function is tinv(2*0.05, 16) = 1.745884, and the probability (area) of the right-side critical region $P(t > 1.75)$ or the left-side critical region, $P(t < -1.75)$ in Fig. 7.4 should be doubled.

Hypothesis testing of slope coefficient b

Similarly, we have estimated b as 0.7136844 in H9 of Table 7.3 or Table 7.5. The standardized test statistic for b for testing the null hypothesis that $\beta = 0$ is, from (7.35),

$$t_b = b/s_b = 0.71368/0.00362 = 197.22434. \tag{7.43}$$

Compared with $t_0 = 2.1199053$, t_b falls far to the right of the critical value in Fig. 7.4. Hence, like t_a, we can reject the null hypothesis H_0: $\beta = 0$ at the 5% level of significance for the two-tailed test, and accept the alternative hypothesis that H_1: $\beta \neq 0$. The test result is significant and we also enter s in H19 in Table 7.5.

7.5.4 *The general format for writing a regression equation*

The usual practice in econometrics is to list the t-statistic under the null hypothesis (7.41) and (7.43) right below the **estimated coefficients**, along with the coefficient of determination, as follows:

$$\boxed{\begin{aligned} \hat{C}_i &= -254.950 + 0.714X_i, \quad R^2 = 0.99959, \\ &\quad (-12.801)\ (197.938) \quad (t), \end{aligned}} \tag{7.44}$$

where $\hat{C}_i$ is the estimated value of C_i at a given observed income X_i. Note that, in earlier economics and statistics literature, it was customary to present the standard error of coefficient, instead of the t-statistic, below the regression coefficient. Thus, to avoid confusion, the writer should explain what is inside the parentheses, like "the numbers in the parentheses are the t-statistics,[12]" as well as indicating that with (t).

7.6 The F-Distribution and Hypothesis Testing of Goodness of Fit

The F-statistic for testing the overall significance of a model is especially useful in multiple regression analysis when there are two or more independent variables in the regression model (see Sec. 7.8). The F-test is a joint test of all slope coefficients. We will briefly explain the F-distribution and F-test procedure in the context of a simple regression analysis, using our consumption and income example.

7.6.1 *The F-statistic*

The F-statistic in regression analysis of a simple regression is given as

$$\boxed{\mathrm{F}(1, \mathrm{n}-2) = \frac{\mathrm{MSR}}{\mathrm{MSE}} = (\mathrm{n}-2)\frac{\mathrm{SSR}}{\mathrm{SSE}} = (\mathrm{n}-2)\frac{\mathrm{R}^2}{1-\mathrm{R}^2},} \tag{7.45}$$

where the first argument in $\mathrm{F}(\cdot)$ is the degrees of freedom of the numerator (SSR) and the second argument is the degrees of freedom of the denominator (SSE). The last expression is derived by dividing the SSR and SSE by SST and using the definition of R^2 in (7.23).

Clearly, from (7.45), a large value of F shows that SSR is larger than SSE, or a large value of R^2. This then implies that the regression line has a good fit to the observed data.

7.6.2 *The F-distribution*

In general, the Excel function for the F-distribution is

$$=\mathrm{Fdist}(\mathrm{F}, \mathrm{df1}, \mathrm{df2}) = \mathrm{P}(\mathrm{F} > \mathrm{F}_0), \quad \mathrm{F} > 0. \tag{7.46}$$

Like the t-distribution, (7.46) gives a cumulative distribution function of the F-distributed random variable, F being greater than a certain value F_0 of F. In (7.46), we have $\mathrm{df1} = 1$, as SSR has 1 degree of freedom (see (7.31)), and $\mathrm{df2} = \mathrm{n} - 2$, as SSE has $\mathrm{n} - 2$ degree of freedom (see (7.30)). In the case of the consumption–income regression analysis, we have $\mathrm{n} - 2 = 16$, as shown in I12 in Tables 7.3 and 7.5.

Four probability density function of the F-distribution are drawn in Fig. 7.5. It appears that the shape of the curves changes considerably when df1 and df2 are small, but it approaches a normal distribution as both degrees of freedom increase to infinity. This can be proved mathematically.

[12]If the numbers in the parentheses show the standard errors of coefficients, then (s.e.) instead of (t) may be used.

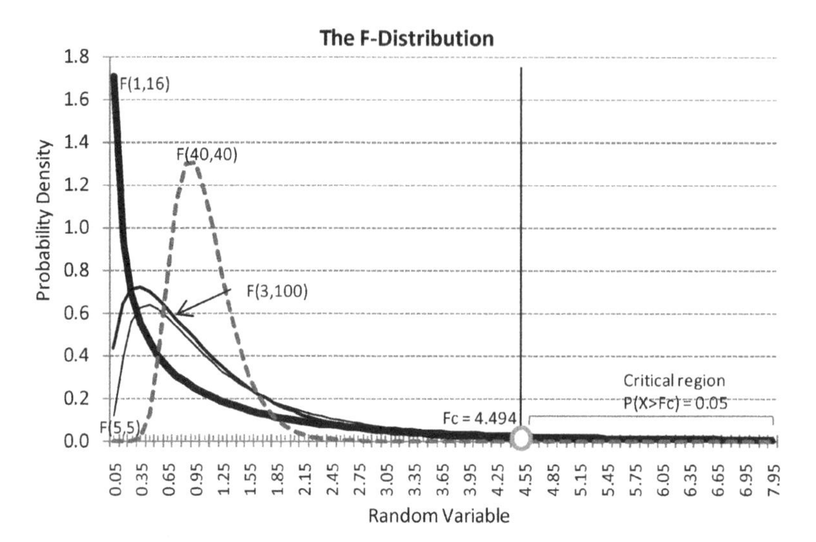

Fig. 7.5 The F-distribution F(m,n)

Application to the consumption–income example

In Table 7.3 or Table 7.5, the F(1,16) statistic is given directly in cell H12. It can also be derived or verified directly from Table 7.3. In Table 7.3, the MSR in H16 is obtained from H13 (since df = 1), and MSE in I16 is obtained from I13 by dividing its df = 16 in I12. The F-statistic is MSR/MSE and is entered in H17, which is the same as H12 in the regression table. Similarly, in Table 7.4, we take MSR = SSR = 23,456,513 in G31 and divide it by MSE = 603 in F31, and we have in G37

$$F(1, n-2) = F(1, 16) = 23{,}456{,}513/603 = 38897. \tag{7.47}$$

7.6.3 *Hypothesis testing of the goodness of fit*

The F-statistic in (7.45) can be used to test the **overall significance** of an estimated regression line to test whether the dependent variable Y is linearly related to the independent variable(s). In the current example of one independent variable X, we test the null hypothesis that H_0: $\beta = 0$, i.e. the slope coefficient equals zero, against the alternative hypothesis that H_1: $\beta \neq 0$, i.e. the slope coefficient is not equal to zero (in the case of a single independent variable, this is the same as the hypothesis test using t-statistic). From formula (7.45), we know the F-statistic is always positive and the F test is a right-tailed test in the F-distribution.

The calculated F-statistic from (7.45) is then compared with a **critical value** of the F-distribution at, say, a 5% **level of significance**. In our example, if we select a level of significance of 5% = 0.05, as in the case of the t-distribution in (7.42), we can use the **inverse function of the F-distribution**,

$$=\text{Finv(probability,df1,df2)}. \tag{7.48}$$

To find the critical value, enter probability = 0.05 in (7.48), which gives a critical value F_0 =Finv(0.05,1,16) = 4.49. This is shown in H23 of Table 7.5 and indicated by a circle in Fig. 7.5. Since the calculated value of F(1, 16) = 38897, which is greater than $F_0 = 4.49$, it falls in the critical region in Fig. 7.5, and we reject the null hypothesis that H_0: $\beta = 0$ in favor of the alternative hypothesis that H_1: $\beta \neq 0$. This result is the same as the one using the t-statistic t_b in (7.43): It implies that we have additional evidence that the least squares estimate b = 0.714 of β is a good estimate, and the linear regression model (7.2) offers a good estimate of the population regression model.[13] For convenience, we enter s (significant) in H24 of Table 7.5.

On the other hand, if we have calculated $F < F_0$, the result implies that we have misspecification in the regression model. In this case, we need to improve the regression model by including more important independent variable(s), deleting wrong independent variable(s), using the nonlinear regression model, etc. Some of the remedies are explained in the econometrics textbooks.

7.7 The P-Value Approach

In the above method of testing a hypothesis, we set up a critical value (t_0 for a one-tailed test) or critical values (t_0 and $-t_0$ for a two-tailed test) and compare the critical value(s) with the actual (or observed) value t_a (or t_b) of the test statistic, and decide whether to reject or accept the null hypothesis. This is the classical **critical value approach** of hypothesis testing we have done in G16:I24 in Table 7.5. In this approach, if you change the level of significance, you have to recalculate the critical value each time.

Note that there is a one-to-one correspondence between the size (or probability) of the tail area and the value of the test statistic. For example, $P(t > t_a) < P(t > t_0)$ (or $P(t < -t_a) < P(t < -t_0)$) if, and only if, $t_a > t_0$ (or $-t_a < -t_0$). This probability, $P(t > t_a)$ (or $P(t < -t_a)$), is called the **p-value**. We then compare this p-value with the chosen level of significance (1%, 5%, or 10%), and make a decision about the null hypothesis. This approach will spare us from recalculating the critical value.

The p-value approach in table format

In Table 7.5, we set up a calculation table in range J16:L24. The actual values of t_a and t_b are H15 and I15 in row 15. For a two-tailed test, the p-values of $t_a$and for t_b are

$$\text{K18: =tdist(H15,16,2)} = P(t > 197.94 \text{ or } t < -197.94) = 1.51\text{E-28}, \quad (7.49)$$

$$\text{L18: =tdist(I15,16,2)} = P(t > 12.80 \text{ or } t < -12.80) = 8.0255\text{E-10}, \quad (7.50)$$

which means that the probability of observing a t-statistic as large as abs(197.94) or as small as abs(−12.80) is almost zero. Hence we can reject the null hypothesis H_0: $\alpha = 0$ and

[13] In fact, it can be shown that for the simple regression model, the t test of the individual significance of the slope coefficient is equivalent to the F test of the overall significance of the regression line. See Dougherty (2007), p.117. For the difference between individual and overall significances, see Gujarati (1995), Chapter 8.

Table 7.6 Multiple regression

	A	B	C	D	E	F	G	H	I
1		US$ bil	US$ bil			Y = a + bX + ct			
2		dep var	ind var1	var2					
3	coeff	a	b	c	Case				
4	Year	C	X	t	1	=LINEST(B5:B22,C5:D22)			
5	1980	1,760.4	2,784.2	0			34.82051	0.60094	42.81945
6	1981	1,941.3	3,115.9	1			c	b	a
15	1990	3,839.3	5,743.8	10	2	=LINEST(B5:B22,C5:D22,1,1)			
16	1991	3,975.1	5,916.7	11			c	b	a
17	1992	4,219.8	6,244.4	12		coeff	34.82051	0.60094	42.81945
18	1993	4,459.2	6,558.1	13		s_b s_a	17.16771	0.05568	147.93716
19	1994	4,717.0	6,947.0	14		R^2, s_e	0.99968	22.46764	#N/A
20	1995	4,953.9	7,269.6	15		F, df	23236	15	#N/A
21	1996	5,215.7	7,661.6	16		SSR,SSE	23458590	7572	#N/A
22	1997	5,493.7	8,110.9	17					
23	Avg	3,503.6	5,266.4	8.5		t-statistic	2.02826	10.79192	0.28944
24			tdist(t,n-k,2)			p-value t	0.060678	1.8E-08	0.7762043
25						0.05/2tl	ns	s	ns
26			Fdist(df1,df2)			p-value F	6.54E-27		
27			df1 =k-1			0.05	s		
28			df2 =n-k						

Statistical inference

H_0: $\beta = 0$. For a one-tailed test, we change "tails" to 1 in the above function in K20 and L20.

For the test of goodness to fit, enter

$$\text{K23: =Fdist(H12,1,16)} = P(F > 38897.442) = \text{1.6E-28}, \tag{7.51}$$

which means that the probability of obtaining an F-statistic as large as 38897 is almost zero. Hence, from this test, we can also reject H_0: $\beta = 0$.

If only the p-value approach is used, range J16:L24 can replace G16:I24. This is done in Table 7.6.

7.8 Multiple Regression Analysis

If a regression model contains two or more independent variables, like Z_i, that influence the dependent variable, Y_i, then we have a **multiple regression model**. We may write the general multiple regression model as

$$Y_i = \alpha + \beta X_i + \gamma Z_i + \varepsilon_i. \tag{7.52}$$

7.8.1 *An example*

In our consumption and income relationship, consumption may also depend on the time factor (t), which represents unknown change in taste and fashion. In this case, the least squares consumption regression line can be written as

$$\hat{C}_i = a + bX_i + ct_i, \tag{7.53}$$

where a is the intercept, and b and c are the slope estimates of the population parameters β, and γ. The time factor (t) takes the values 0, 1, 2, ...,17 in column D of Table 7.6.

(a) Since we have three coefficients to estimate, select three cells, G5:I5. In cell G5, entering the function shown in F4, and using the range-copy command, we have the estimated intercept and slope coefficients. Note again that the sequence of the estimated coefficients is in a reversed order: c, b, and then a. This can be reminded (and strongly recommended) by writing the regression equation (7.53) in cell F1 of Table 7.6.
(b) If more information about the regression estimates are needed, we may use Case 2. Select any blank 5×3 range, such as G17:I21 in Table 7.6. At G17, entering the function shown in F15 (also, see Sec. 7.3.1), and using the range-copy command, we have the information as listed inside the range. Note that the estimated coefficients are identical in Cases 1 and 2. Note also that the range of the independent variables in the LINEST function contains two columns (columns C and D).
(c) The #N/A appears in the blank space I19:I21 of the 5 × 3 matrix, since additional information is provided only in G18:H21. Other parts are blank. This is also the case if the matrix is defined as larger than 5 × 3.
(d) Calculate the t-statistics for the estimated coefficients in row 23.
(e) The estimated equation for (7.53) is

$$\boxed{\begin{array}{llllll} \hat{C}_i = & 42.81945 & +\,0.60094X_i & +\,34.82051t, & R^2 = 0.99968, \\ & (0.28944) & (10.79192) & (2.02826) & (t) \end{array}} \tag{7.54}$$

where the numbers in parentheses are the t-statistics.
(f) Statistical inference will be explained in the next section.

7.8.2 *The t-statistic and the F-statistic using the p-value approach*

For multiple regression analysis with k coefficients, including the intercept, and k − 1 slope coefficients, each of the k t-statistics in (7.35) has a t-distribution with n − k degrees of freedom. Hence, we have k = 3, n = 18, and df = n − k = 15 for regression (7.54). For the two-tailed test, we find the p-values in row 24 in Table 7.6. In G24, enter tdist(G23, 15, 2), copy G24 to H24:I24, and we have p-values for each corresponding t-statistic. Comparing the p-values with the 5% level of significance, denoted as "0.05/2" at F25, we find that α and γ are not significant (denoted as ns in G25 and I25), and β is significant (denoted as s in H25). The results are indicated in row 25.

The F-statistic

For multiple regression with k coefficients, SSR has df1 = k − 1 degrees of freedom and SSE has df2 = n − k degrees of freedom, where k is the number of coefficients (including the constant) in the regression equation. Hence, MSR = SSR/(k − 1), and MSE = SSE/(n−k). The F(k−1, n−k) statistic takes the form of

$$F(k-1, n-k) = \frac{MSR}{MSE} = \frac{SSR/(k-1)}{SSE/(n-k)} = \frac{SSR}{SSE}\frac{n-k}{k-1} = \frac{R^2}{(1-R^2)}\frac{n-k}{k-1}. \tag{7.55}$$

In our three-parameter consumption equation (7.54), k = 3, k−1 = 2, n = 18, and n−k = 15, hence,

$$MSR = SSR/2 = 11,729,295, \quad MSE = SSE/15 = 504.79.$$

The F-statistic is calculated as F(2, 15) = MSR/MST = 23235, which is the same as the Excel result in G20 of Table 7.6. Thus, no calculation like (7.55) is needed if we use the Excel Regression Table.

Hypothesis testing of goodness of fit

In multiple regression, the F-statistic is used to test the **overall significance** of the estimated regression. Thus, we test the null hypothesis that

$$H_0: \beta = \gamma = 0 \quad \text{and} \quad H_1: \beta \neq 0 \text{ or } \gamma \neq 0, \tag{7.56}$$

that is, that all slope coefficients are equal to zero, against the alternative hypothesis that at least one of the slopes, β or γ, is nonzero. In our example, if we select the level of significance at 5%, for the one-tailed test, we find the p-value in Table 7.6,

G26: =Fdist(G20,2,15) = 6.54E-27,

which is much smaller than 0.05. Thus, we reject the null hypothesis that H_0: $\beta = \gamma = 0$ in favor of the alternative hypothesis that H_1: not all slope coefficients are equal to zero. This result implies that our multiple regression equation fits the data very well in general, and the estimated equation (7.54) is a valid equation.

More general cases

In a more general case, say k = 4, we have four coefficients a, b, c, and d, to be calculated. The least squares estimation and the hypothesis testing procedures can be extended in a straightforward manner.

7.9 Summary

In this chapter, we have introduced the basic concepts and methods of using the least squares regression analysis. To put all the topics in perspective, we present the structure of the regression analysis in the relational chart of Fig. 7.7. Since statistical analysis deals

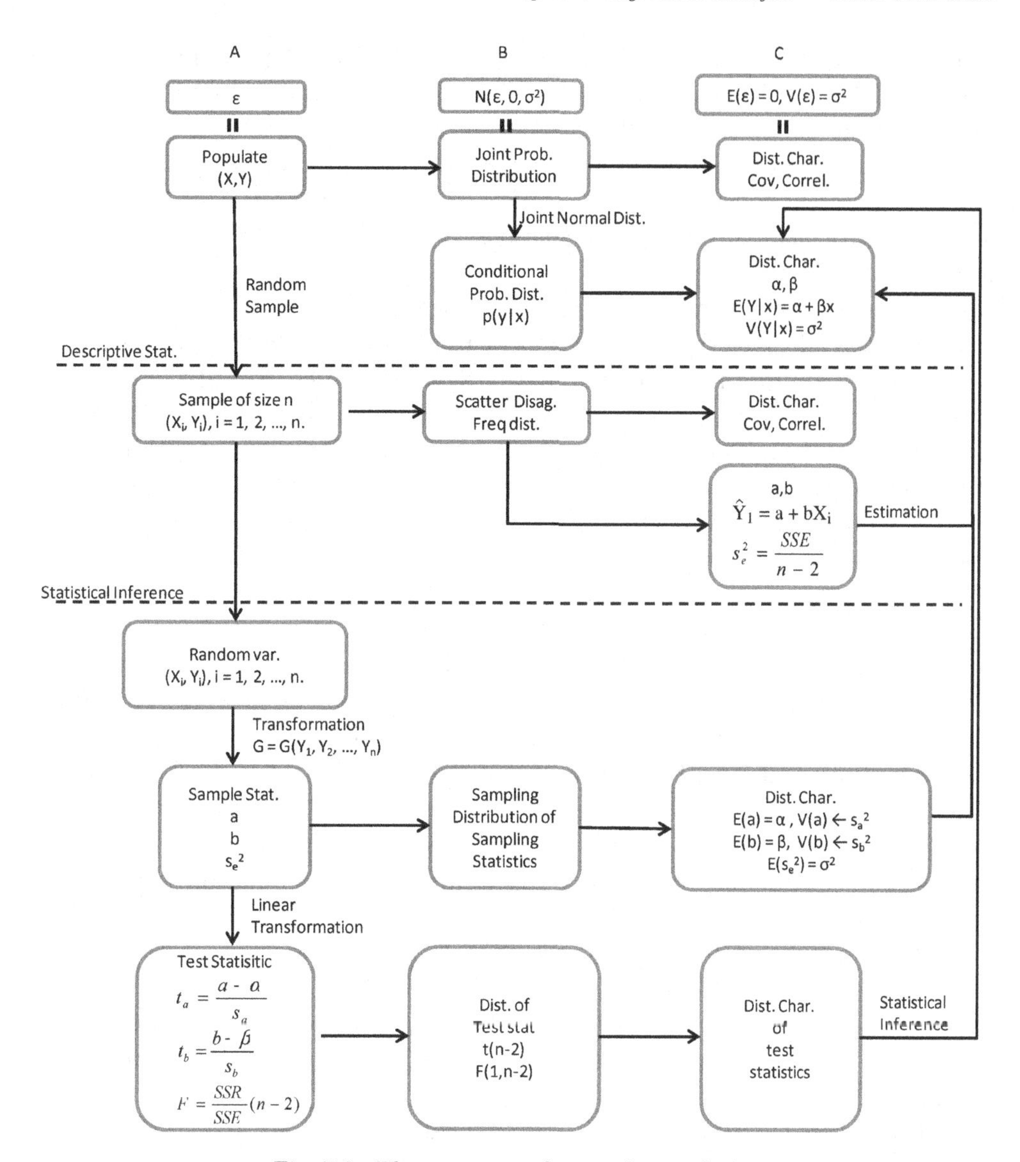

Fig. 7.6 The structure of regression analysis

with random variables (the first column block A), they are associated with probability distributions (the mid-column block B), and for every distribution, we can find distribution characteristics (the last column block C). This chart is a detailed illustration of the structural chart of Fig. 6.9 in the Chap. 6 applied to regression analysis.

For least squares regression analysis, we have used two variables, which may be considered to have a bivariate joint probability distribution (either a normal or a nonnormal distribution), with distribution characteristics such as the covariance and the correlation coefficient. The regression analysis is based on the conditional probability distribution[14]: the distribution of a variable (Y), given the level of another variable (X), as shown in Fig. 7.1. The basic conditions on the regression analysis are imposed on its distribution characteristics: the conditional mean is linear in X (with intercept α and slope β) and the conditional variance is constant for all X. For statistical inference, we need assumptions such that Y_i is stochastically independent, and the conditional distribution of Y at each level of X is normal.

The upper part of the chart shows that the above specification of conditional probability distribution is equivalent to the specification using the stochastic error terms ε.

Based on the assumptions about population characteristics, we take a sample of size n in two variables, draw a scatter diagram, and compile a frequency distribution. We calculate the sample covariance and correlation coefficient to estimate the population covariance and correlation coefficient. We also set up a linear regression model (7.6) (with intercept α and slope β), and use the least squares method (7.10),[15] to estimate α by a, β by b, the conditional mean E(Y|X) by $\hat{Y}$, and conditional variance σ^2 by s_e^2.

How do we know that these estimations are "good estimates"? That is the field of statistical inference. To have a good estimate, the sample must be random, and the variables must be random variables. The sample statistics, like a, b, and s_e^2, are considered (nonlinear) transformations of random variables consisting of samples, and as such, they are also random variables. As random variables, they also have frequency or probability distributions, called the **sampling distribution of sample statistics** (see Sec. 7.4.1), and their distribution characteristics. It can be proved that the expected value of a is α, the expected value of b is β, and the expected value of s_e^2 is σ^2. We can also find variances of a, V(a), and b, V(b). The process is quite complicated. They are also functions of population variance σ^2. When σ^2 is replaced by s_e^2 in V(a) and V(b), we have a sample variance of a, s_a^2, and a sample variance of b, s_b^2, as shown by the left arrows in Fig. 7.7. When the square root is taken, we can estimate the population standard deviation of a, by the **standard error of a**, s_a in (7.28), and also estimate the population standard deviation of b, by the **standard error of b**, s_b in (7.29). They can then be used to construct the test statistics t_a and t_b in (7.35), as shown in the last row of Fig. 7.7.

The test statistics t_a and t_b in (7.35) have a t-distribution with $n-k$ degrees of freedom (k being the number of parameters in the regression equation). We then find the critical value or the p-value to test the null hypothesis, whether the intercept α or the slope β is zero, against the alternate hypothesis that they are not equal to zero, by observing whether the observed test statistic falls into the critical region. To test the goodness of fit of the regression model to the data, we construct the F-statistic in (7.45), which has the

[14]Another method of specifying population regression function is through the equivalent specification on the error terms, as in (7.6).

[15]Other methods include the generalized least squares method, orthogonal regression, maximum likelihood, and other methods.

F-distribution with (k − 1, n − k) degrees of freedom. These relations are shown by the direction of the heavy arrow on the right-hand side of Fig. 7.7.

The statistical inference mentioned above is summarized in the lower part of Table 7.6. After we understand the theory behind the statistical inference, it is a simple matter to add t and F-statistics for the estimated coefficients, find the p-values for the t-statistics and the F-statistic, and test the significance of the regression coefficients and the goodness of fit of the regression line by choosing the percentage level of significance.

In the process of explaining the structure, we have effectively used Excel functions, tables, and charts to explain the definitions and relations in calculation. This completes the descriptions of the essential topics in economic and business statistics.

Appendix 7A. The Add-In Regression Package

Installing the Analysis ToolPak

The Excel Program also comes with an add-in regression package. To use this package, we first retrieve the ToolPak using the following steps:

<“Office”><Excel Options><Add-Ins><Analysis ToolPak>

<Manage: [Excel Add-ins]><Go···>.

The add-ins dialog box appears. Enter <Add-Ins available: xAnalysis ToolPak><OK>. (If Analysis ToolPak is not installed in your Excel program, you should install it from the original Excel CD).

Data Analysis

Enter <“Data”><Analysis, Data Analysis>. The following window appears (Fig 7A.1):

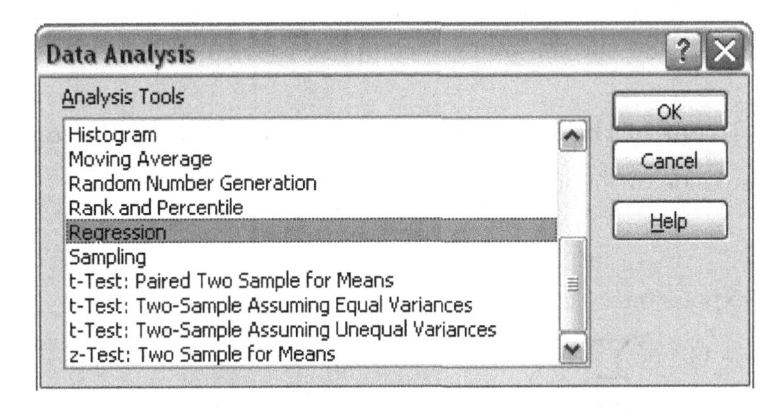

Fig. 7A.1 Data analysis dialog box

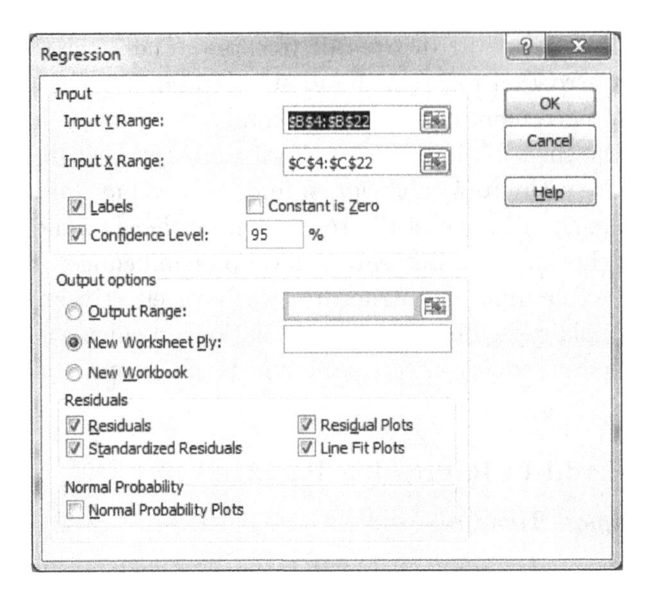

Fig. 7A.2 Regression dialog box

Select <Regression>. The regression dialogue box appears (Fig 7A.2).

The regression dialog window

We use the data in Table 7.2. In the "Regression" dialog box, in the "Input Y-range" enter B5:B22 (the dependent variable), in the "Input X-range", enter C5:C22 (the independent variable(s)). Note that, unlike in the Excel regression function, *both ranges should include the column titles.* In other parts of the dialog box, you may fill in all the options except "Constant is Zero", click <OK>, and the results will appear in the new worksheet (which is recommended, especially if more items are checked in the regression window), as in Table 7A.1 (this table has been edited to reduce the decimals and to be compact). It shows the "X Residual Plot", which is the scatter diagram of Predicted Cs and the Residuals. Note that, compared with Table 7.4, the Data Analysis only shows the standard error of regression s_e in B6 of Table 7A.1; the "predicted" column in B18: B35 contains the estimated values $\hat{C}$ in Table 7.4; and the "Residuals" comprise the errors $e = C - \hat{C}$ in Table 7.4.

Appendix 7B. Proof of the Decomposition of the Total Variation

From the left-hand side of the identity (7.21), we have,

$$\sum(C_i - \bar{C})^2$$

in the following expression

$$\begin{aligned}\sum(C_i - \bar{C})^2 &= \sum[(C_i - \hat{C}_i) + (\hat{C}_i - \bar{C})]^2 \\ &= \sum(C_i - \hat{C}_i)^2 + \sum(\hat{C}_i - \bar{C})^2 + 2\sum(C_i - \hat{C}_i)(\hat{C}_i - \bar{C}).\end{aligned}$$

Table 7A.1 Summary output from data analysis

	A	B	C	D	E	F	G	H	I
1	SUMMARY OUTPUT								
2	*Regression Statistics*								
3	Multiple R	0.9998							
4	R Square	0.9996							
5	Adjusted R Square	0.9996							
6	Standard Error	24.557							
7	Observations	18							
8	ANOVA								
9		*df*	*SS*	*MS*	*F*	*ignificance F*			
10	Regression	1	23456513	23456513.3	38897.4	0.00			
11	Residual	16	9649	603.0					
12	Total	17	23466162						
13		*Coefficient*	*andard Err*	*t Stat*	*P-value*	*Lower 95%*	*Upper 95%*	*Lower 95.0%*	*Upper 95.0%*
14	Intercept	-254.95	19.92	-12.80	0.00	-297.17	-212.73	-297.17	-212.73
15	X	0.71	0.00	197.22	0.00	0.71	0.72	0.71	0.72
16	RESIDUAL OUTPUT								
17	*Observation*	*redicted (*	*Residuals*	*ndard Residuals*					
18		1	1732.1	28.3	1.2				
19		2	1968.8	-27.5	-1.2				
20		3	2058.9	17.9	0.8				
21		4	2253.3	30.1	1.3				
22		5	2530.1	-37.8	-1.6				
23		6	2728.8	-24.0	-1.0				
24		7	2901.1	-8.4	-0.4				
25		8	3093.9	0.6	0.0				
26		9	3348.9	0.8	0.0				
27		10	3626.6	-31.8	-1.3				
28		11	3844.3	-5.0	-0.2				
29		12	3967.7	7.4	0.3				
30		13	4201.6	18.2	0.8				
31		14	4425.5	33.7	1.4				
32		15	4703.0	14.0	0.6				
33		16	4933.3	20.6	0.9				
34		17	5213.0	2.7	0.1				
35		18	5533.7	-40.0	-1.7				

X Residual Plot

Residuals

X

Thus, we only need to show that the last term vanishes. Denoting the last term in the summation sign as L, and multiplying out, the equation reduces to

$$L = \sum (C_i - \hat{C}_i)\hat{C}_i - \sum (C_i - \hat{C}_i)\overline{C}.$$

Since $\overline{C}$ is constant, the second term is zero, due to (7.24). Hence, using (7.11),

$$\begin{aligned} L &= \sum (C_i - \hat{C}_i)\hat{C}_i = \sum (C_i - \hat{C}_i)(a + bX_i) \\ &= a\sum (C_i - \hat{C}_i) + b\sum X_i(C_i - a - bX_i) = 0. \end{aligned}$$

The first term is zero due to (7.24), and the second term is zero due to the second normal equation (7.12). This proves the identity (7.21). The proof is very useful in understanding the nature of the least squares method.

Review of Equations and Functions

(7.4) $E(C|X_i) = \alpha + \beta X_i,$

(7.5) $C_i - E(C|X_i) = \varepsilon_i$

(7.6) $C_i = \alpha + \beta X_i + \varepsilon_i$

(7.3) $\hat{C}_i = a + bX_i$

(7.7) $C_i = a + bX_i + e_i.$

(7.7) $e_i = C_i - \hat{C}_i.$

(7.9) $$g(\alpha, \beta) = \varepsilon_1^2 + \varepsilon_2^2 + \cdots + \varepsilon_2^2 = \sum_{i=1}^{n}(C_i - \alpha - \beta X_i)^2$$

(7.11) $\sum C_i = n\alpha + \beta \sum X_i$

(7.20) $\sum C_i X_i = \alpha \sum X_i + \beta \sum X_i^2$

(7.13) $a = \bar{C} - b\overline{X}$

(7.14) $$b = \frac{n\sum C_i X_i - \sum C_i \sum X_i}{n\sum X_i^2 - (\sum X_i)^2} = \frac{\sum (X_i - \overline{X})(C_i - \bar{C})}{\sum (X_i - \overline{X})^2} = \frac{SSxy}{SSx}$$

(7.20) $C_i - \overline{C} = (C_i - \hat{C}_i) + (\hat{C}_i - \overline{C})$

(7.21) $$\sum_i (C_i - \overline{C})^2 = \sum_i (C_i - \hat{C}_i)^2 + \sum_i (\hat{C}_i - \overline{C})^2$$

(7.23) $$R^2 \equiv \frac{SSR}{SST} = 1 - \frac{SSE}{SST}$$

(7.24) $\sum(C_i - \hat{C}_i) = \sum e_i = 0$

(7.25) $\sum \hat{C}_i = \sum C_i$

(7.27) $$s_e = \sqrt{\frac{\sum (Y_i - \hat{Y}_i)^2}{n-2}} = \sqrt{\frac{SSE}{n-2}} = \sqrt{MSE}$$

(7.28) $$s_a = s_e\sqrt{\frac{\sum X_i^2}{n\sum x_i^2}} = s_e\sqrt{\frac{SSX}{n\,SSx}} = s_e\sqrt{\frac{1}{n} + \frac{\overline{X}^2}{SSx}}$$

(7.29) $$s_b = s_e\frac{1}{\sqrt{\sum x_i^2}} = s_e\sqrt{\frac{1}{SSx}}$$

(7.31) $$MSR = \frac{SSR}{1} = SSR, \quad MST = \frac{SST}{n-1}$$

(7.35) $$t_a = \frac{a - \alpha}{s_a} \quad t_b = \frac{b - \beta}{s_b}$$

(7.45) $$F(1, n-2) = \frac{MSR}{MSE} = (n-2)\frac{SSR}{SSE} = (n-2)\frac{R^2}{1 - R^2}$$

(7.53) $\hat{C}_i = a + bX_i + ct_i$

(7.55) $$F(k-1, n-k) = \frac{MSR}{MST} = \frac{SSR}{SSE}\frac{n-k}{k-1} = \frac{R^2}{(1-R^2)}\frac{n-k}{k-1}$$

Key terms: Economics, Business, and Statistics

Key terms: Excel

Homework Chapter 7

7-1 Exercise on summations Derive algebraically the least squares estimates a and b in (7.13) and (7.14) using the normal equations (7.11) and (7.12).

7-2 Review of general concepts To evaluate the firm's marketing efforts of selling Click brand ball point pens, an economist collected the following data set, where sales (S) are measured in thousand of the pen sold, advertising (A) is the number of spot TV advertising (Table HW7-2).

The economist is interested in fitting the data to a linear regression line.

Table HW7-2 Sales of Click ball point pen

	S	A
2003	260	5
2004	286	7
2005	279	6
2006	500	9
2007	438	12
2008	315	8
Avg		
sum		

Table HW7-9 Demand for Asian pears

Unit Name Variable	Pears D (1000 dozens)	Price P (per dozen)	Weekly Income X
2000	80	5	1400
2001	100	3	1500
2002	90	4	1600
2003	95	3	1700
2004	85	4	1800

Source: Dominick Salvatore (1982), 156–159.

(a) Draw the scatter diagram with regression equations and the coefficient of determination. Make sure that the advertisement is on the horizontal axis. The labels of the horizontal axis should be 0, 1, ..., 13 (with the interval of 1), the Major tick mark type is "Cross", and the Minor tick mark type is "Inside".

(b) From the scatter diagram, what is the relationship between the two variables S and A? Does it make sense? Why?

(c) The scatter diagram gives the variables in X and Y. Rewrite the linear equations in the scatter diagram in terms of Sales and Advertisement. Which variable is independent, which is dependent?

(d) The OLS formulas for a and b are given as

$$a = \overline{S} - b\overline{A}$$

$$b = \frac{\sum_{i=1}^{6} (A_i - \overline{A})(S_i - \overline{S})}{\sum_{i=1}^{6} (A_i - \overline{A})^2}.$$

Using the above formulas, calculate a and b step by step. Place your calculation on the right-hand side of Table HW7-2.

(e) Choose the point at $A = 9$, $S = 500$ on the scatter diagram, and write below the identity relation among the explained deviation, unexplained deviation, and total deviation (draw a line of the Sale average to make your point).

(f) Why is it called "explained" deviation? "Unexplained" deviation?

(g) Which deviation is related to the method of least squares? What is the method of least squares? "Least" of what? "Squares" of what? Can you do without "squares"? Why?

(h) The decomposition of the total variation is given by

$$\underset{\text{SST}}{\sum (S_i - \overline{S})^2} = \underset{\text{SSE}}{\sum (S_i - \hat{S}_i)^2} + \underset{\text{SSR}}{\sum (\hat{S}_i - \overline{S})^2},$$

which is an identity. Using Table HW7-2, verify the identity.

(i) SST is the abbreviation of ______________________________
SSE is the abbreviation of ______________________________
SSR is the abbreviation of ______________________________

(j) The coefficient of determination is defined as $R^2 = SSR/SST$ Manually calculate R^2.

(k) Is your result the same as the one on the scatter diagram? If not, why? What is the range of the value of R^2? How do you judge the value of your R^2? Is it relatively large or small? Does it make economic sense? Why?

(l) What is the use of R^2? What is the meaning of "determination"?

7-3 A numerical example From a sample of 100 (X, Y) pairs of observations, the following sums are calculated:

$$\sum X = 21.23, \quad \sum Y = 30.51, \quad \sum X^2 = 24.95, \quad \sum Y^2 = 97.43, \quad \text{and} \quad \sum XY = 56.69.$$

Find the least squares equations $Y = a + bX$ and $X = c + dY$.

7-4 The sum of squares regression

(a) Show that for the simple linear regression model $Y = a + bX$,

$$SSR = \sum (\hat{Y}_i - \bar{Y})^2 = b^2 SSx = r^2 SSy,$$

where SSx is defined in (7.32) and SSxy is defined in (5.19). r is the correlation coefficient between X and Y, see (5.21).

(b) Using the information obtained in the above question (HW7-3), find the value of SSR.

(Hint: (a) Using the definition of $\hat{Y}$ in (7.3) and $\bar{Y}$ in (7.13) (note that $Y = C$ in these equations), and b in (7.14), you will have (a). (b) Use the first equality expression of b in (7.14)).

7-5 Test statistic Given information on sums HW7-3, and the results from HW7-4, find the following:

(a) As with SSx, show that $SSy = SST = n\sum Y_i^2 - (\sum Y_i)^2$.
(b) Find SSE and R^2.
(c) Find the standard error of regression s_e, of coefficients, s_a and s_b.
(d) Find the test statistics, t_a, t_b and F.

7-6 A simulation model of HW5-5

(a) Reconstruct Table 7.2 (Estimation of the intercept and slope) for HW5-5.
(b) Construct Table 7.3 (Excel least squares method) using Excel commands.
(c) Construct Table 7.4 (Regression statistics) to calculate various sum of squares.
(d) Write the regression equation in the standard form as shown in (7.44).
Make sure that the t-statistics are included.
(e) What are the values of the test statistics, t_a, t_b and F?
(f) What is the use of these test statistics?
(g) Test the null hypotheses that population intercept and slope coefficients being zero, and also the goodness of fit. Do the test results make economic sense?

7-7 Demand analysis of HW5-6 Copy HW5-6 to Sheet2 and change the data into values. Answer the same questions in (a) to (g) in HW7-6.

7-8 The estimation of supply function of HW5-7 Copy Table HW5-7 to Sheet2 and change the data into values. Answer the same questions (a) to (g) in HW7-6.

7-9 Demand for Asian pears Demand for Asian pears in the Denver area from 2000 to 2004 are given in Table HW7-9. D is the quantity of pears measured in 1000 dozens, P is the price per dozen, and X is the average weekly income of the buyers. Let the OLS regression equation be

$$D = a + bP \quad \text{and} \quad D = a + bP + cX.$$

In this question, we use the Excel Regression Table.

(a) Draw scatter diagrams separately for (D, P) and (D, X), where D is the dependent variable.
(b) Fit the LS regression to the above two linear equations.
(c) What are the standard errors of coefficient for b and c? What are the standard errors of regression? Calculate t-statistics for a, b, and c. Explain the results.
(d) What is R^2? What does it indicate about the regression results? How do we test whether R^2 is significant?
(e) Find the estimated (or predicted) regression equation. Find the estimated value of D and the error of estimates for each year. Do the errors of estimates sum to zero? Why?

7-10 Jamaican coffee supply Table HW7-10(a) shows the data on Jamaican coffee supply from 1962 to 1968. Output of coffee is measured in pounds. The prices of coffee (Pcoffee), of cocoa (Pcoco), of bananas (Pbana), and the average weekly wages in agriculture (Wwk) are in US$.
Let the OLS regression equation be

(i) Coffee = a + bPcoff
(ii) Coffee = a + bPcoff + cPcoco.
(iii) Coffee = a + bPcoff + cPcoco + dPbana,
(iv) Coffee = a + bPcoff + cPcoco + dPbana + eWwk.

(a) Draw four scatter diagrams of coffee and four prices, each price being on the X-axis.

Table HW7-10 (a) Data on Jamaican Coffee supply (b) OLS Results

(a)

Unit Var	Pound Coffee	US$ Pcoff	Pcoco	Pbana	Wwk
1962	34486.32	1.03	0.83	1.02	1.07
1963	27093.00	1.08	0.67	0.90	1.09
1964	28492.20	1.11	0.69	0.90	1.51
1965	28390.56	1.16	0.77	0.96	1.57
1966	32068.40	1.22	0.62	0.87	1.75
1967	24360.61	1.35	0.62	0.91	1.79
1968	31334.00	1.39	0.67	0.98	1.86

Source: Adopted from Lotty and Ray (1992), 183–187.

(b)

equ #	(i)	(ii)	(iii)	(vi)
Const				
t-value				
Pcoff				
t-value				
Pcoco				
t-value				
Pbana				
t-value				
Wwk				
t-value				
R^2				

(b) Estimate the four equations and fill in the value of the coefficients in Table HW7-10(b). The t-value of each coefficient is placed in parentheses just below the coefficient and should be in a smaller font. Specify the degree of freedom for each case.
(c) From Table HW7-10(b), what conclusion can you make from the four equations? Which good is complementary and which is a substitute for coffee? What is the effect of the weekly wage?
(d) What is the F-statistic for the above four cases? Specify the degrees of freedom in each case.

7-11 SAT scores of HW5-8 Using the table and results from HW5-8, answer the same questions (a) to (g) in HW7-6.

7-12 US patents and R&D expenditure Using the table and results from HW5-10, answer the same questions in (a) to (g) in HW7-6.

7-13 Relations among the sum of squares Show that, for the simple linear regression model $Y = a + bX$,

(a) $SSE = (1 - r^2)SSy$ and $SSy = r^2SSy + (1-r^2)SSy$.
(b) $r^2 = b^2(varp(X)/varp(Y))$.
(c) The F-statistic equals the square of the t-statistic t_b when $\beta = 0$, that is, $F = (t_b)^2$.

(Hints: (a) Note that $SST = SSy$ and $SSR = r^2$ SSy. Substituting into the identity (7.21) by changing C_i to Y_i and using HW7-4 above, we have (a). (b) From HW7-4. (c) Use the definition of F in (7.45) and the definitions of s_e^2 and s_b^2 in (7.27) and (7.29)).

Multiple regression

7-14 Multiple regression Continuing from HW7-2, suppose the economist has additional data on the number (N) of salespersons, as listed below. He also added the time trend on the last column. (copy the data part of Table HW7-2, and add the data on salesperson and the time trend as shown in Table 7-14(a)). From Table HW7-14(a), let the OLS regression equations be

$$\text{(i) } S = a + bA$$

$$\text{(ii) } S = a + bA + cN + dt$$

(a) Draw the chart of the four curves. To show all curves, draw the sale variable using the secondary axis. Make sure that you add the chart titles and the axis titles.
(b) Find the regression coefficients of the two equations and fill in the value of the coefficients in Table HW7-14(b). The t-value of each coefficient should be placed in parentheses below the coefficient and should be in a smaller font.
Note that the extended regression command in Excel is =LINEST(Y,X,1,1), which gives the Excel Regression Table for information on the values of estimated coefficients, s_b, s_a, R^2, df, SSR, and SSE.
(c) Following the convention, rewrite the two equations below with the estimated coefficients. Write all the t-statistics and R^2 in parentheses just below each coefficient.

Table HW7-14 Extended data of Table HW7-2

(a)

	S	A	N	t
2003	260	5	3	0
2004	286	7	5	1
2005	279	6	3	2
2006	500	9	4	3
2007	438	12	6	4
2008	315	8	3	5

(b)

		(i)	(ii)
a	const		
	t-value		
b	Advertising		
	t-value		
c	Number		
	t-value		
d	Time trend		
	t-value		

(d) What conclusion can you make from the t-statistics of the second equation (ii)? Which coefficients are "significant," which are not?
(e) What is the value of R^2 for equation (ii)? Is it high or low? What is the meaning of R^2 in (ii)
(f) Comparing the R^2 for equations (i) and (ii), which one is larger? Does it make sense? Why? Give a justification.
(g) In the second equation (ii), what is the economic meaning of a? Does it make sense? Why?
(h) In the second equation (ii), what is the economic meaning of b? Does it make sense? Why?
(i) In the second equation (ii), what is the economic meaning of c? Does it make sense? Why?

7-15 Determination of the birth rates In the field of economic development, economists are interested in the relations among birth rates (B, birth per 1000 population), life expectancy (E,

Table HW7-15 Determination of the birth rates

(a)

Ctry	B	E	L	B^	SSE	SSR	SST
1	14	76	94				
2	39	51	68				
3	16	69	94				
4	17	71	85				
5	11	78	99				
Avg							
Sum							

(b)

	c	b	a
coeff			
sc, sb, sa			
R2, se			
F, df			
SSR, SSE			
t statistic			
tdist(t,df,2)			
Fdist(F,df1,df2)			
df1=k−1=2 df2=n−k=5−3=2			

years at birth), and literacy (L, percent of population who finish primary schools). Five countries are randomly chosen and listed in Table HW7-15(a) below. In answering questions, all the tables, charts, titles must be properly labeled.

(a) Draw two scatter diagrams, one between birth rates and life expectancy, one between birth rates and literacy. Make sure the birth rate is on the horizontal axis. Give appropriate titles or labels.
(b) Fit a linear trend line to the scatter diagrams and show the trend equation and R^2 for both scatter diagrams.
(c) Let the least squares regression equation be

$$B = a + bE + cL$$

Fit the LS regression line to the above linear equation (Note: use the extended formula: LINEST(Y,X,1,1)). The format of the regression table should be the same as Table HW7-15(b).
(d) The decomposition of the total variation is given by

$$\underset{\text{SST}}{\sum(B_i - \overline{B})^2} = \underset{\text{SSE}}{\sum(B_i - \hat{B}_i)^2} + \underset{\text{SSR}}{\sum(\hat{B}_i - \overline{B})^2}$$

which is an identity. Setting up a calculation table as shown by Table HW7-15(a), verify the identity empirically.
(e) Write equation in (c) in the text box below with the estimated coefficients obtained from Table HW7-15(b). Write the t-statistics under the null of coefficient being zero in parentheses just BELOW each coefficient.
(f) Calculate tdist(t,df,2) for t distribution of each coefficient and Fdist(F,dfn,dfd) for the regression and write them as in Table HW7-15(b).
(g) What conclusion can you make from the t statistics? Which coefficients are "significant," which are not, at the 5% level of significance?
(h) What conclusion can you make from the F statistics? Is it significant at 5% level?
(i) What is the meaning of R^2? What is the range of R^2 ? Should the R^2 in the scatter diagrams in (a) and the R^2 calculated in Table 7-15(b) match? Why?
(j) What is the economic meaning of a? Does *your* result make sense? Why?
(k) What is the economic meaning of b? Does *your* result make sense? Why?
(l) What is the economic meaning of c? Does *your* result make sense? Why?

Part III

Private and Public Decision Making

Chapter 8

Future Value Problems — Exponential and Logarithmic Functions

Chapter Outline

Objectives of this Chapter

Economics and business deal with money and assets, in contexts of national or regional cooperation or of personal finance. On many occasions, the future value of a given amount of money or an asset over a period of time is the most important factor in making an economic or business decision. A dollar we hold today is worth more than a dollar we will receive in the future, because, in general, there is a price inflation or time preference in our economy. Therefore, today's dollar grows with interest in the bank.

On the other hand, a dollar received in the future is worth less than a dollar received today since less than a dollar ("discounted" dollar) can be deposited in the bank to grow to a dollar in the future. Economists have often referred to this as the **objective time preference**. Business economists referred to this as the **time value of money**. People also like to enjoy the good life that money can buy today rather than in the future. This is referred to as the **subjective time preference**.

In this chapter, we examine the future value of a given amount of money or an asset. We calculate the future value problems without annuity in this chapter and with annuity and the present value problems in the next chapter. These problems occur when a person invests, lends, or borrows money. They also occur in studies of economic growth, inflation rate, and finance related to future planning, when the value of money, utility, welfare, etc., will change over time. Here we show the basic concepts and terminologies related to the personal finance of saving, investment, borrowing, and the future values or future income stream from a current saving program.

The concept of continuous compounding in a given period (year) is related to the mathematical concept of the base (e) of natural exponential and logarithmic functions. Thus, we discuss the rules of exponentials and logarithms. As in previous chapters, we introduce the concepts using mathematical formulas and definitions and then show the equivalent Excel formulas, so that we will understand the concepts behind the formulas. Because of the nature of the topics, we give ample examples to demonstrate the calculation procedures. In addition to personal financial problems, the applications of the topics to the national economy, including the growth of national products, price inflation, investment, etc. are included in the homework.

8.1 Basic Definitions and Examples

Interest is the amount of money paid for the use of borrowed money (or capital) in a period of time (t). We may say that interest is money paid for the use of money. The amount of borrowed money (or capital) is called the **principal** (P) or the **present (initial) value (PV)**, and the period of time is called the **term of borrowing**. In this chapter, we use period and year interchangeably. The **final value** or the future value (FV), is the sum (S) of the principal and the amount of interest that grows during t periods, $S = P + I$, where I is the total interest paid at the end of t periods.

Simple interest (I) is the interest calculated only on the original principal (P) for the entire period (t), not including the interest on the accrued interest. $I = Pit$, where i is the interest rate, and the interest on P, iP, is simply added to the next period; the interest accrual on iP is not calculated in subsequent periods. The future value F after t periods (or years) is

$$F(t) = S = P + I = P(1 + it). \tag{8.1}$$

For example, if we have $P = \$100$ to deposit in a saving account of a bank with an interest rate, say, of $i = 5\% = 0.05$ per period, then the interest (P)(i)(t) will be paid by the bank at the end of each period (t) by multiplying the interest by the number of period t. The future value of the deposit,

$$F(t) = 100(1 + 0.05t),$$

at the end of $t = 1, 2, 3, 4$, is illustrated in Fig. 8.1.

In real-world practice, the simple interest method may be used for short-term transactions. For long-term transactions, the interest amount earned in the previous period(s)

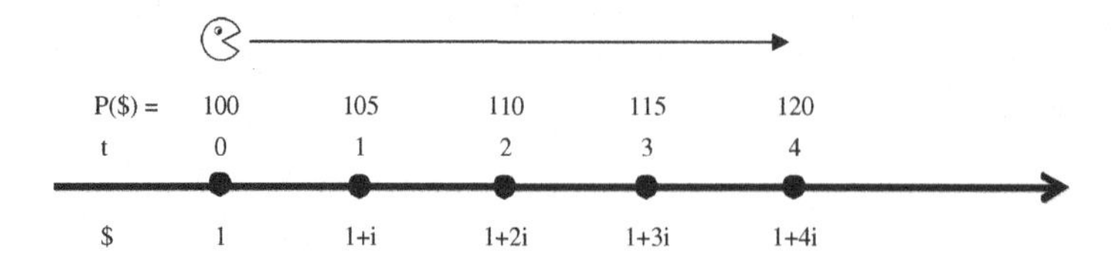

Fig. 8.1 Simple interest method

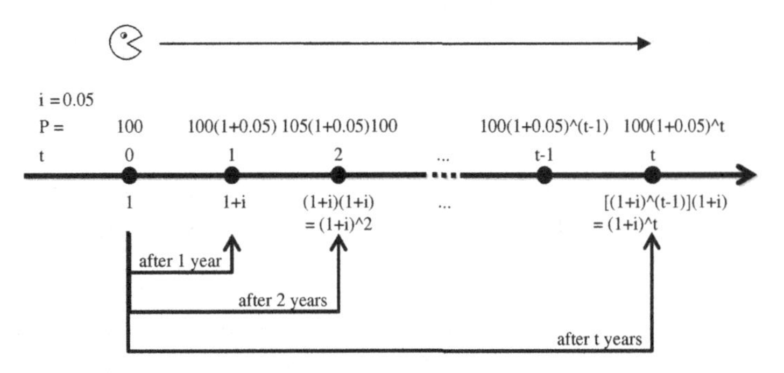

Fig. 8.2 Compound interest method (i = 0.05)

can be added to the initial principal to grow additional interest in the next periods. This is called the **compound interest** method.[1] The interest in the next period is calculated on the initial principal combined with the interest from the previous period(s). This is illustrated in Fig. 8.2 (see Appendix 8A for drawing instruction). The time that the interest is converted (added) into principal is called the **conversion period**. It can be daily, monthly, quarterly, semi-annually, or annually (end of a year). The future (final) value, F(t), at the end of the tth period is

$$\boxed{F(t) = P(1+i)^t} \tag{8.2}$$

or F for simplicity, and is called **the compound amount**, or the **future value** (FV) at the end of tth period. The difference F − P is called the **compound interest** for t periods. In (8.2), i is the **compound interest rate**[2] at each period. We say that P has **accumulated to the future value** F with $(1+i)^t$ as the **future value interest factor (FVIF)** (or **accumulation factor**).

Example 8.1 Simple and compound interest methods. Upon graduation, you obtain a job and your annual salary is $45,000. You are told that your annual increase will be the year's rate of inflation plus 3% per year. Assuming that the inflation rate is 2% per year, what will be your salary at the end of the fourth year if you calculate it by

(a) the simple interest method? What is the future value interest factor?
(b) the compound interest method? What is the accumulation factor?
(c) Set up a table like Table 8.1 without the parameter table, and calculate your salary increase in each year for the next four years after graduation.
(d) Set up another table like Table 8.1, with the parameter table, and name the cell of interest i, the cell of principal P, and the column of the term t (in A8:A12).

[1] "Compound" means "combine", "to put together into a whole". The initial principal and interest of previous periods are put together as a whole to calculate the interest in the next period.
[2] In this chapter, the interest rate is assumed to be constant in each period. If it varies with time t, then it is called a **variable interest rate**. Note that if t = 1 in (8.2), we can write i = (F/P) − 1 = (F − P)/P, which is the percentage change or the growth rate (in this case, of money) that we have studied in Sec. 5.4.1.

(e) Using the current example, illustrate your answers with a time line diagram similar to Fig. 8.2.

Answer:

(a) Your salary's total percentage increase is 2% + 3% = 5% per year, which is similar to the interest rate i = 5% = 0.05 per year. Thus, the salary at the end of fourth year (i.e. t = 4) will be, from (8.1),

$$F = P + I = P(1 + it) = 45,000(1 + 0.05^*4) = 54{,}000,$$

where $(1 + 0.05^*4) = 1.20$, which is FVIF. Thus, your salary at the end of fourth year will be $54,000.

(b) You are earning at the compound rate of i = 5% per year. At the end of the fourth year, your salary will be, from (8.2),

$$F(t) = P(1 + i)^t = 45,000(1 + 0.05)^4 = 54,697.8,$$

where the FVIF is $(1 + 0.05)^4 = 1.215506$.

(c) For the simple interest method, enter in Table 8.1

B8: 45000 B9: =B$8*(1+0.05*$A9).

For the compound interest method, enter

C8: 45000 C9: =C$8*(1+0.05)^$A9.

And then copy B9:C9 to B10:C12. Note the position of the $ sign of the mixed references in the above formulas.

(d) In Table 8.1, cells C2, C3 and column A8:A12 should be named i, P, and t, respectively. In B8, enter=P*(1+i*t), in C8, enter=P*(1+i)^t, and copy B8:C8 to B12:C12. Note the difference in the methods of (c) and (d).

Table 8.1 Simple and compound interest

	A	B	C
1		Parameter Table	
2		i	0.05
3		P	45000
4			
5	Future values		
6	Year	Interest calculation	
7		simple	Compound
8	0	45000.00	45000.00
9	1	47250.00	47250.00
10	2	49500.00	49612.50
11	3	51750.00	52093.13
12	4	54000.00	54697.78

Table 8.2 The Table of future value interest factors (FVIF)

	A	B	C	D	E	F	G	H
1	Calculation of the Future values							
2		simple	Compound Interest					
3	t/i	1+it	(1+i)^t					
4		0.05	0.01	0.05	0.1	0.15	1	1.5
5	0	1.00	1.00	1.00	1.00	1.00	1	1
6	1	1.05	1.01	1.05	1.10	1.15	2	2.5
7	2	1.10	1.02	1.10	1.21	1.32	4	6.25
8	3	1.15	1.03	1.16	1.33	1.52	8	15.625
9	4	1.20	1.04	1.22	1.46	1.75	16	39.0625
10	5	1.25	1.05	1.28	1.61	2.01	32	97.65625
11	6	1.30	1.06	1.34	1.77	2.31	64	244.14063
12	7	1.35	1.07	1.41	1.95	2.66	128	610.35156
13	8	1.40	1.08	1.48	2.14	3.06	256	1525.8789
14	9	1.45	1.09	1.55	2.36	3.52	512	3814.6973
15	10	1.50	1.10	1.63	2.59	4.05	1024	9536.7432
16	11	1.55	1.12	1.71	2.85	4.65	2048	23841.858
17	12	1.60	1.13	1.80	3.14	5.35	4096	59604.645
18	13	1.65	1.14	1.89	3.45	6.15	8192	149011.61
19	14	1.70	1.15	1.98	3.80	7.08	16384	372529.03
20	15	1.75	1.16	2.08	4.18	8.14	32768	931322.57

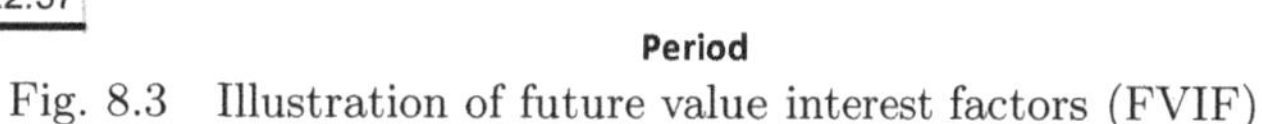

Fig. 8.3 Illustration of future value interest factors (FVIF)

(e) The illustration will be similar to Fig. 8.1 for (a) and to Fig. 8.2 (see Appendix 8A) for (b) with t = 4. □

Table of future value interest factors

In Table 8.2, column B shows the calculation of future values F(t) for P = \$1 and i = 0.05, using the simple interest method (8.1). Using the compound interest method (8.2), for $t = 0, 1, \ldots, 15$, columns C–H show the calculation of future values F(t) at P = \$1 and i = 1%, 5%, 10%, 15%, 100%, and 150%.

The seven curves of future values are graphed in Fig. 8.3. The last two curves with the hollow circle (i = 100%) and the filled black circle (i = 150%) are measured by using the secondary Y-axis so that all curves can be in the same chart. Note that the curves from compound interest methods show an accelerating upward trend as the interest rate rises to a higher level. If $i \geq 10\%$, as shown in the graph, the values of F(t) virtually explode after t = 10. For comparison, as shown in Fig. 8.3, the straight line drawn from column B using the simple interest method is a linear function with slope i = 0.05.

A table such as Table 8.2, which shows different values of the dependent variable (the **FVIF**) for each different value of the independent variables (i, t), is called the **sensitivity analysis table**, or simply the **sensitivity table**. Since one of the independent variables is time, we may call this method of presentation the **time profile** of changes. By contrast, we may calculate compound interest, say at period 10, by simply entering the formula at t = 10, that is, =(1 + 0.05)^10, without showing the time profile from t = 0, 1, to 15.

Example 8.2 Sensitivity table. Using Table 8.2 as a **template**, calculate future values at P = \$1 by simple and compound interest methods, using part (a) and part (b) of Example 8.1 (without directly using formulas (8.1) and (8.2)). What are the differences in answers from using formulas? Why?

Answers:

(a) Table 8.2 shows that at P = \$1.00, i = 0.05, and t = 4, the future value by the simple interest method is \$1.20. Multiply 1.20 by the present value of P = \$45,000, and we find F = \$54,000, which is the same answer we get using formula (8.1).

(b) Table 8.2 shows that at P = \$1.00, i = 0.05, and t = 4, the future value by the compound interest method is \$1.22. Multiplying 1.22 by the present value of P = \$45,000, we find F = \$54,900. This amount is greater than \$54,697.8 using formula (8.2) directly, because the formula gives the more accurate value of $\$1(1+0.05)^4 = \$1.215506$, which is less than \$1.22 in the table due to rounding errors. □

Example 8.3 Construction and Charting of Table 8.2. To construct the table, name A5:A20 as "term", and B4:H4 as "rate". Then, enter in

B5: =1+rate*term,

C5: =(1+rate)^term,

and copy B5 to B6:B20 and C5 to C6:H20. Note that Excel detects the appropriate row and column automatically.

To draw the chart, select the range A4:H20 and use <"insert"><Chart, Line><Line with Markers>, and we have the chart with lines. Click the chart and identify the line for i = 1.5, click the line to open the Format Data Series window, and click <Format Data Series, Series Options> <Plot Series on @Secondary Axis>. The numbers on the secondary Y-axis will appear. Similarly, click the line for column i = 1, do the same to move the axis to the secondary axis, then click <OK>. Clicking the chart and using the text box to enter the line labels, editing the other chart elements, we have Fig. 8.3. □

8.2 Multiple Conversions per Period

In the compound interest method explained in Sec. 8.1, the future value is calculated by assuming the interest earning is converted to additional principal only once at the end of a year. In real-world practice, it is often the case that the earned interest will be converted into additional principal many times, say, m times a year. Then, m is called the **frequency of conversion** within a year. In this case, the **term** n of investment, loan, or contract, is the number of conversion m multiplied by the number of total years t, that is, n = mt, when future values are calculated. Thus, if a person invests a certain amount of money for t = 3 years, then the interest earned is converted monthly, or quarterly, or semi-annually to additional principal, and the term n (of investment, loan, or borrowing) represents **mt** = $12 \times 3 = 36$, or $n = 4 \times 3 = 12$, or $n = 2 \times 3 = 6$ periods, respectively.

Table 8.3 Interest per annum converting m times a year

	A	B	C	D	E	F	G	H	I
1	**Interest Converting m times a year**								
2	Interest rate per annum i =		0.05	Accumulation factor (t=1) y = (1+ i/m)^h, h<=m				Convergence	
3	Illustration			Fig.8.4	Fig. 8.5		Fig.8.6	(8.6)	(8.7)
4	Name of convertion		year	semi-ann	quarter	day	minute	=(i+i/m)^m	=(i+1/m)^m
5	Naming label		m1	m2	m3	m4	m5	i=	i=
6	Freq of conv	h	1	2	4	365	525600	0.05	1
7	Principal P	0	1.00	1.00	1.00	1.00	1.00	1.000000	1.000000
8	**Annual (m=1)**	1	1.0500	1.025000	1.0125	1.0001	1.0000001	1.050000	2.000000
9	**Semi-annually**	2		1.050625	1.0252	1.0003	1.0000002	1.050625	2.250000
10	(m=2)	3			1.0380	1.0004	1.0000003	1.050838	2.370370
11	**Quarterly**	4			1.050945	1.0005	1.0000004	1.050945	2.441406
12	(m=4)	5				1.0007	1.0000005	1.051010	2.488320
13		6				1.0008	1.0000006	1.051053	2.521626
14		7				1.0010	1.0000007	1.051084	2.546500
15		8				1.0011	1.0000008	1.051108	2.565785
16		9				1.0012	1.0000009	1.051126	2.581175
17		10				1.0014	1.0000010	1.051140	2.593742
18		11				1.0015	1.0000010	1.051152	2.604199
19	**monthly**	12				1.0016	1.0000011	1.051162	2.613035
20	(m=12)	...							
21	**Daily**	365				1.051267	1.0000347	1.051267	2.714567
22	(m=365)	...							
23	**minute**	525,600					1.0512711	1.0512711	2.718279
24	(m=525600)	...							
25	**million times**	1,000,000						1.051271	2.718280
26									
27		...							
28	**Instant**	infinity	(continuous)					1.051271	2.718282
29	End formula		1+0.05	(1+0.05/2)^2	(1+0.05/4)^4	(1+0.05/m)^m	(1+0.05/m)^m	=EXP(0.05)	=exp(1)
30						m=365	m=525600		
31	P = 10,000		10,500.00	10,506.25	10,509.45	10,512.67	10,512.71	10,512.71	27,182.82

Table 8.3 shows the calculation of future value, F, under the assumption that the interest rate i is converted m times a year, the frequency of conversion being $m = 1, 2, 4, 365, 525600$. i/m is the **interest rate per conversion period** or the **periodic interest rate**. Row 6 shows the frequency of conversion, and column B shows conversion at the end of the interval up to 12, $h \leq m$. Throughout the table, the interest rate is assumed to be $i = 5\%$ per year (see C2), and the principal is assumed to be $P = \$1$ (see row 7). We explain the structure of Table 8.3 column by column in Secs. 8.2.1–8.2.4. Note that, in column C, \$1.00 in C7 is compounded once at the end of the year, at which the future value is \$1.05 in C8.

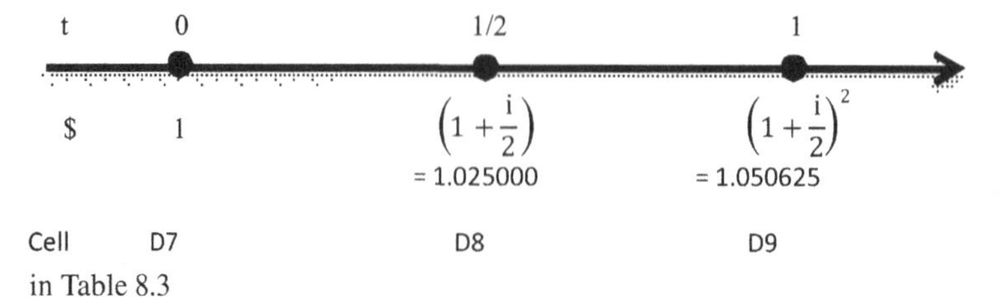

Fig. 8.4 5% interest rate per annum convertible semi-annually.

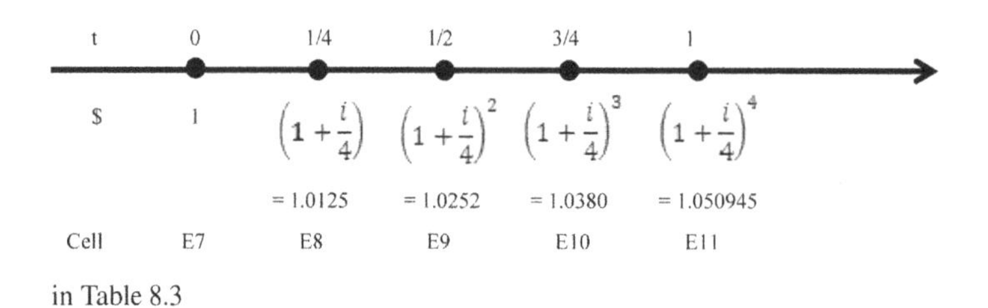

Fig. 8.5 5% interest rate per annum convertible quarterly

8.2.1 *Interest is convertible semi-annually*

This is shown in column D of Table 8.3 and Fig. 8.4. When interest rate per year is $i = 0.05$ and the interest is convertible semi-annually ($m = 2$), the interest rate for each half year is $i/2 = 0.05/2$. Thus, the future value for $P = \$1$ and the term $n = mt = 2 \times 1 = 2$ will be $F = \$1 * (1 + 0.05/2)\hat{}1 = 1.025$ at the end of the first half year ($h = 1$, see D8), and $F = \$1 * (1 + 0.05/2)\hat{}2 = \$1.050625$ at the end of the second year at $h = m = 2$. They are shown in D9 of Table 8.3 and also illustrated in Fig. 8.4.

8.2.2 *Interest is convertible quarterly*

This is shown in column E of Table 8.3 and Fig. 8.5. In this case, the interest is converted four times in a year and the interest rate for each quarter is $i/4 = 0.05/4$. Thus, the future value for $P = \$1$ and the term $n = mt = 4 \times 1 = 4$ will be $F = \$1(1 + 0.05/4)\hat{}4 = 1.050945$ at the end of one year. This is shown in E11 of Table 8.3, and the accumulated value of each quarter is shown in E8:E11, and also illustrated in Fig. 8.5.

8.2.3 *Interest is convertible daily*

This is shown in column F of Table 8.3. Often, our saving accounts or certificate of deposit (CDs) in commercial banks or saving institutions earn interest that is compounded daily and credited to the account monthly. In this case, the calibration of the line will be $m = 365$ intervals within a year ($t = 1$) and $n = mt = 365 \times 1 = 365$. The future value at the end of the first interval, $h = 1$ of 365, will be $(1 + i/365)^1 = 1.0001$ in F8 of Table 8.3; at the end of the second interval, $h = 2$ of 365, the future value will be $(1 + i/365)^2 = 1.0003$ in F9, etc. Thus, up to the end of the $h = m = 365$th interval (i.e. end of the year), the future

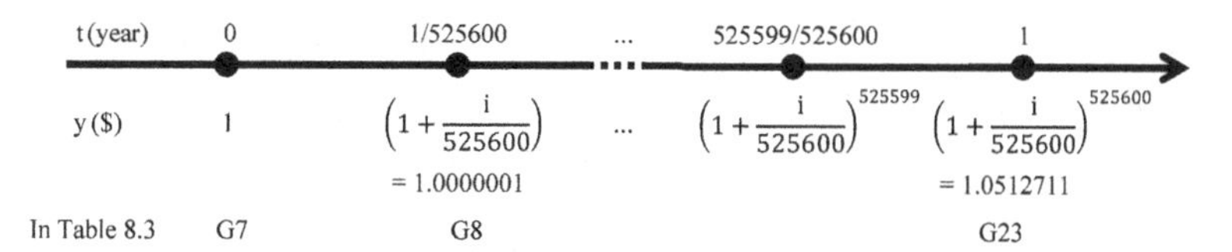

Fig. 8.6 5% interest rate per annum convertible at each minute

value for P = \$1.00, i = 0.05, and mt = 365 × 1 = 365 will be

$$F = \$1(1 + 0.05/365)\hat{}365 = 1.051267.$$

This is shown in F21 of Table 8.3.

8.2.4 *Interest is convertible every minute*

This is shown in column G of Table 8.3 and Fig. 8.6. Just to show the fun of using mathematics, in this case, we have n = mt = (60 × 24 × 365)(1) = 525,600 min in a year. Then, the future value for P = \$1 and i = 0.05 per year will be F = \$1(1 + 0.05/525600)^525600 = 1.051271 at the end of the year, as shown in G23 of Table 8.3 and also illustrated in Fig. 8.6. The accumulated interests for the first 12 min (h = 1, 2, ..., 12) are shown in G8:G19. Note how small the increment of interest is in each interval.

8.2.5 *A summary table of interest converting* m *times within a year*

In general, if the interest is **convertible** (or **compounded**, or **payable**) m equal times per year (t = 1), the accumulated future value of P = \$1.00 at the end of one year is

$$\boxed{F = \left(1 + \frac{i}{m}\right)^m.} \quad (8.3)$$

Construction of Table 8.3

To construct Table 8.3 using Excel, we suggest the following steps:

(a) Enter the labels in rows 1–6, name cell C2 as i, and C6:G6 m1_, ... m5_, as listed in C5:G5.
(b) In row 7 enter principal \$1.00, and in C8, enter

C8: =(1+i/m1_)^\$B8,

holding column B constant.
(c) Copy C8 to D8:G8.
(d) At first, the formula in D8:G8 is the same as that in C8. Change m1_ to m2_ in the D8 formula. Similarly, edit the formulas in cells E8, F8, and G8 to the corresponding numbers of m (compounding frequency) in row 6.

(f) Copy D8 to D9, ..., G8 to G23, column by column up to the enclosed square of each column, as shown in the table.
(g) Erasing #VALUE! or #DIV/0! between the numbers, we have Table 8.3.

Note that the interpretation of the conversion frequency (m) in column B is the sequence of the conversion intervals up to conversion frequency m, $h \leq m$. For example, in E5, the m3 column is entered quarterly. For $h = 1$, at the end of the first quarter, the future value is \$1.0125 in E8; at the end of the 2nd quarter, $h = 2$, and the future value is \$1.0252 in E9; at the end of the third quarter, $h = 3$, and the future value is \$1.0380 in E10; and at the end of the 4th quarter, $h = 4$, and the future value is \$1.050945 in E11. The column m5 is entered in minutes. The future value at the end of $h = 365$th minute is \$1.0000347 in G21, and after $h = 525{,}600$ min, at the end of the year, the future value is \$1.051271094 in G23. All the end-of-the-year future values are enclosed with borders and shaded, and the applications of Eq. (8.3) at the end of the year is shown in row 29 for $h = m$. The last two columns of Table 8.3 will be explained in the next section.

The table shows that as the conversion frequency (m) in a year increases from m1 to m5 in row 5, the future value of one dollar at the end of the year will also increase, but it will increase slowly, as shown by the bordered yellow cell at the end of each column. Figure 8.7 illustrates the range of B6:E11 when P = investment of \$10,000 at $i = 5\%$. For $m = 1$, 2, 4, the interest compound increases from \$500 to \$506.25 to \$509.45, respectively. This indicates that although the conversion frequency m doubles each time, the interest earning increases only gradually by a much smaller amount. For the method of drawing Fig. 8.7, see Appendix 8A.

The last row of Table 8.3 shows how much we will receive at the end of one year when the interest is compounded once a year, semi-annually, or quarterly, if we have deposited \$10,000 in a bank. Note that the differences in the future values are small. For example, at $m = 365$, $F = \$10{,}512.67$ (see F31); and at $m = 525600$, $F = \$10{,}512.71$ (see G31). It is rather surprising that the difference at the end of the year between future values resulting from compounding daily and compounding each minute is only four cents! This probably is the reason that our bank can afford to use daily compounding to attract customers. We can see that as m continue to increase, F trickles down and **converges** slowly to a number. We will discuss the property of convergence below.

8.2.6 *Compounding for t years*

In formula (8.3), the interest is converted m times in one year only. If the interest is converted m times a year and compounded for t years, then we can find the future value of principal \$P at the end of t years by extending formula (8.3) into a general formula (8.4) below:

$$\boxed{F_t = P\left(1 + \frac{i}{m}\right)^{mt}.} \tag{8.4}$$

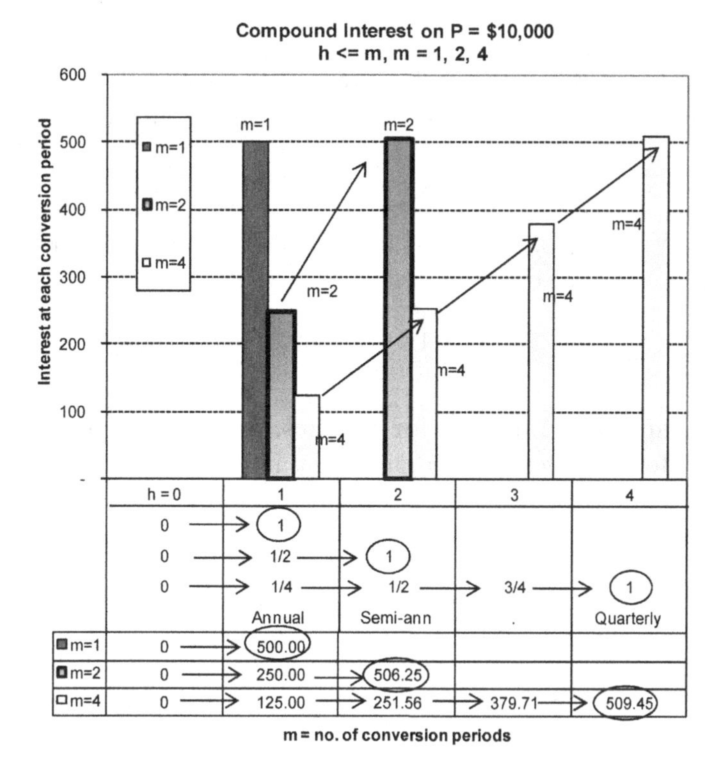

Fig. 8.7 Conversion of interest annually, semi-annually, and quarterly within a year

Note that this formula can be combined with (8.2) and rewrite as

$$\boxed{F_n = P(1+r)^n,} \tag{8.5}$$

where

$$r = i/m \quad \text{and} \quad n = mt.$$

Formula (8.5) is called the **discrete growth equation**. In the following sections, we will use r and n, with the understanding that r is the interest rate at each conversion, that is, r is the compound interest rate which takes into account the conversion frequency, and n is the term of the transactions, the total number of conversion (or compounding) for t years. When m = 1, formulas (8.4) and (8.5) reduce to (8.2). This is the way the future value table is constructed. We call (8.2) or (8.5) the **business form** of writing the future value equation. Other forms are the **difference equation form** and the **solution form.** See Appendix 8B.

Example 8.4 In Example 8.1, what will your salary be after your first year of employment if your rate of salary increase is calculated by (a) converting quarterly? (b) converting weekly with 52 weeks per year?

Answer:

(a) For $P = \$45{,}000$, $m = 4$, $t = 1$, $i = 5\%$, and $r = 0.05/4$, your salary at the end of the first year (or at the beginning of the 2nd year), using formula (8.4), will be

$$F = \$45,000(1 + 0.05/4)^4 = \$47,292.53.$$

Note that the FVIF $(1 + 0.05/4)^4 = 1.050945$ can be found from E11 in Table 8.3.

(b) If the salary is compounded weekly for a year (52 weeks), your salary will be

$$F = \$45,000(1 + 0.05/52)^{52} = \$47,306.06.$$

□

Example 8.5 Compound interest for t years. In (b) of Example 8.1, what will your rate of salary be after four years of employment if the salary increase is calculated by (a) converting quarterly? (b) converting weekly?

Answer:

(a) With an interest rate of 5% per annum, convertible quarterly, $m = 4$, after four years ($t = 4$) of employment, the salary will be

$$F = \$45{,}000(1 + 0.05/4)^{4^*4} = \$54{,}895.03.$$

(b) With the interest rate convertible weekly, $m = 52$, after four years of employment ($t = 4$), the salary will be

$$F = \$45{,}000(1 + 0.05/52)^{52^*4} = \$54{,}957.84.$$

□

8.3 Continuous Compounding within a Period

Extending the idea of converting the interest every minute in Sec. 8.2.4, we may imagine converting the interest every second, or every instant of time. In mathematics, this is the case when converting frequency m goes to infinity. We have the following basic mathematical proposition.

8.3.1 *A basic proposition on e*

Proposition 8.1 *The value of an investment* $P = \$1$, $t = 1$, $i =$ *annual interest rate, with the interest compounded continuously, that is,* $m \rightarrow$ *infinity, the future value at the end of one year will be*

$$\boxed{\lim_{m\to\infty}\left(1+\frac{i}{m}\right)^m = e^i.} \tag{8.6}$$

Proof: Let $x = \frac{m}{i}$ in (8.6). Then, $m \to \infty$ implies $x \to \infty$. Hence

$$\lim_{m\to\infty}\left(1+\frac{i}{m}\right)^m = \lim_{x\to\infty}\left(1+\frac{1}{x}\right)^{ix} = \left[\lim_{x\to\infty}\left(1+\frac{1}{x}\right)^x\right]^i = e^i.$$

□

e is a number, which is called the **base of natural logarithm**, and has any desired number of decimal places (hence, e is called an **irrational number**). The value of e to the 15 decimal places given by Excel is

$$e = 2.7\ 1828\ 1828\ 45\ 90\ 45\ \ldots$$

In the Excel spreadsheet, the formula for e is

$$=\exp(1), \tag{8.7}$$

and the formula for e^x is $= \exp(x)$, where x is any number.

Illustration

Column H of Table 8.3 implements Eq. (8.6) and calculates e^i at $i = 0.05$, as the interest is compounded continuously over a year. In cell H8, enter the formula

H8: =(1+i/B8)^B8.

Copy H8 down to H9:H25 for $h = 1, 2, \ldots$ 1,000,000, and then erase #DIV/0! between the numbers because we cannot calculate the cells with $h = 0$ or a missing value of h. When h is very large, i.e. when the division of the conversion interval within a year becomes infinitesimally small, the value of (8.6) converges to $e^i = e^{0.05} = \exp(0.05) = 1.051271096$ as its limit. This is shown in H28 of Table 8.3. This value is very close to the future value at h = 1,000,000 in H25.

F = (1 + 0.05/1,000,000)^1,000,000 = 1.051271095.

Thus, in this case, the discrete formula can be approximated by the continuous formula, and vice versa. In fact, we may also notice that in terms of depositing $10,000 in a bank, as shown in row 31 of Table 8.3, whether the interest is compounded daily, or every minute, or even continuously at each moment of time does not make much of the difference in the future values.

8.3.2 *Illustration of e*

We now show algebraically and graphically the expression

$$\lim_{m\to\infty}\left(1+\frac{1}{m}\right)^m = e, \tag{8.8}$$

which is the special case of (8.6) when $P = 1$, $i = 1$, that is, when the interest rate is 100% per year and converted continuously within the year. This can be proved mathematically using L'Hopital's rule (see any calculus textbook).

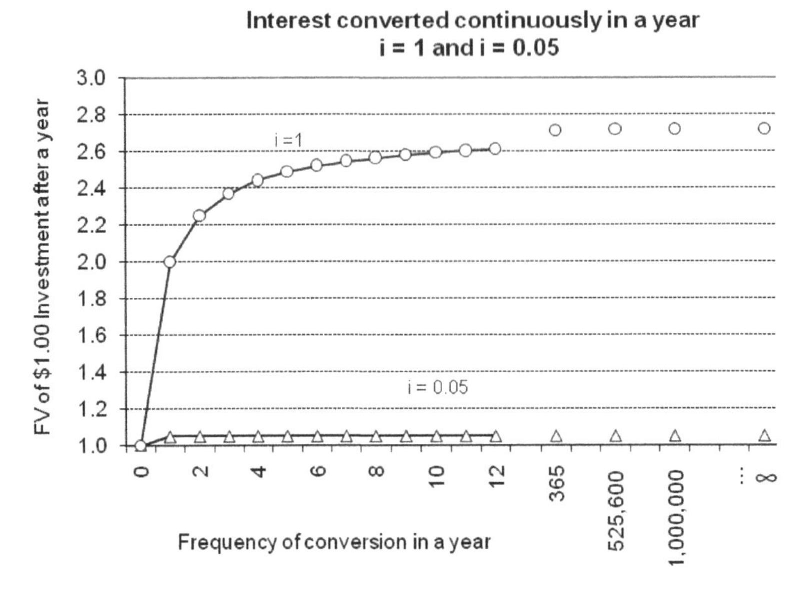

Fig. 8.8 5% and 100% interest converted continuously for a year

Column I of Table 8.3 shows the calculation of the future value $F = (1 + 1/m)^m$ for $m = 1, 2, 4, \ldots$ 1,000,000, and is graphed in Fig. 8.8. When m goes to infinity, we have $F = e = 2.718282$ in cell I28. The calculations show that F **converges** to $e = 2.71828\ldots$, rather rapidly from $m = 1$ to 12, slowly from $m = 12$ to 365, very slowly from $m = 365$ to 525,600, and extremely slowly from 525,600 to 1,000,000. It trickles down to e when m goes to infinity. The irregular shape in Fig. 8.8 is due to jumps in the value of m, as it jumps conspicuously from $m = 12$ to $m = 365$, then from 365 to 525,600, etc.

8.4 The Exponential Growth Equation and its Applications

If the principal is \$P, the annual interest rate is i, and the interest is compounded continuously for t years, then its accumulated future value is called the **exponential growth equation**. The formula is

$$\boxed{F(t) = Pe^{it},} \tag{8.9}$$

which is the continuous counterpart of (8.4) of discrete compounding. In (8.9), we say that the interest i is **compounded continuously (or instantaneously)** for t years.

Illustration

The interpretation and derivation of (8.9) are illustrated in Fig. 8.9. In continuous compounding, for $P = \$1$ with the interest rate being i per year, the future value is $\$e^i$ at the end of the first year. Using this $\$e^i$ as the principal at the beginning of the second year, it will

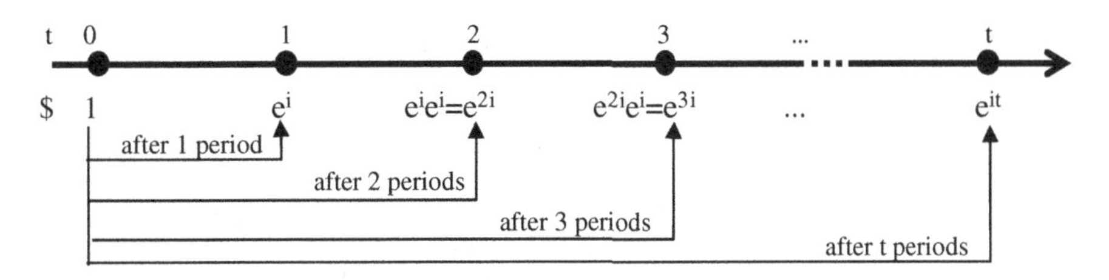

Fig. 8.9 Interest is compounded continuously

be compounded to $\$e^i(e^i)$ at the end of the second year, and so on.[3] After t years, the total amount of principal and interest will be e^{it}.

Example 8.6 Continuously compounding interest. In Example 8.1, what will your salary be if the salary increase is calculated by continuous compounding? (a) at the end of one year of employment; (b) at the end of four years.

Answer:

(a) In this case, after one year (t = 1), your salary will be $F = \$45{,}000e^{0.05} = \$47{,}307.20$. (Note: In Excel, $e^{0.05}$ is entered as =exp(0.05)).

(b) After four years (t = 4), your salary will be $F = \$45{,}000e^{0.05^*4} = \$54,963.12$. (Note: In Excel, $e^{0.05^*4}$ is entered as =exp(0.05*4))

□

Example 8.7 Columbus. If Columbus had an opportunity to deposit \$1.00 in a bank when he "discovered" the new world in 1492, with the interest rate staying at i = 5% per year, calculate the future values at the end of 500 years:

(a) assuming the interest is compounded continuously;

(b) assuming the interest is compounded m times a year, where m = 1, 2, 4, 12, 365, 525600, and 1,000,000. Construct a table to show your calculation and illustrate it in a bar diagram.

Answer:

(a) From (8.9), $F = e^{0.05^*500} = e^{25} = 72{,}005{,}000{,}000$. It will be more than 72 billion dollars!! His offspring would be the richest person in the world today (if the bank still existed today).

(b) Using formula (8.4) for P = \$1, i = 0.05, t = 500, and m = 1, or 2, or 4, ..., we may construct Table 8.4 below. It shows the importance of the frequency of compounding in the long run. In the last entry, instant compounding is obtained as = exp(0.05*500). □

Example 8.8 Inflation rate. Show that an inflation rate of 6% per year and continuously for 70 years will cause the prices to increase by a factor of about 67. Hence, a \$0.40 loaf of bread fed to a toddler today will cost about \$27.00 when the toddler retires at age 70. Show the calculation.

[3]Note that, from the law of exponentials in Sec. 8.5.2, exp(r)* exp(r) = exp(r + r) = exp(2r).

Table 8.4 Columbus deposit of $1.00

m	=(1+0.05/m)^(m*500) FV (after 500 yr)
1	$39,323,261,827.22
2	$52,949,930,178.99
4	$61,668,555,160.64
12	$68,360,491,974.23
365	$71,881,720,016.48
525,600	$72,004,813,650.85
1,000,000	$72,004,850,248.29
instant	$72,004,899,337.39 =EXP(0.05*500)

Answer: From (8.9),

$$F = 0.4e^{0.06*70} = 0.4e^{4.2} = (0.4)(66.69) = 26.8.$$

Thus, the factor is $66.69 \cong 67$. □

Nominal, effective, and force of interest

The following definitions are informative in understanding the nature of interest rates.

$$\begin{cases} \mathrm{i} & \text{is called the } \textbf{nominal rate of interest (for one year)}. \\ \mathrm{j} = \left(1 + \frac{\mathrm{i}}{\mathrm{m}}\right)^{\mathrm{m}} - 1 & \text{is called the } \textbf{effective rate of interest (for one year)}. \\ \mathrm{g} = \mathrm{e}^{\mathrm{i}} - 1 & \text{is called the } \textbf{force of interest (or instantaneous rate of interest) (for one year)}. \end{cases}$$

In the above definitions, force of interest is seldom used in business practice, except in large business transactions. As we will see in the exercises, it is very important in economic theory. Note that the original principal $P = \$1$ is deducted from the future value in the expression calculating the effective and instantaneous rates of interest. In practice, the interest rate is generally quoted in nominal rate for one year. The effective rate is also quoted for one year.

Example 8.9 Nominal rate and effective rate of interest. In Example 8.1, if your salary increases for one year, find (a) the nominal rate of increase in the salary, (b) the effective rate of salary increase if the interest is compounded quarterly, and (c) the instantaneous rate of interest.

Answer:

(a) The nominal rate of interest is $i = 5\%$.

(b) The effective rate of interest for $m = 4$ is $j = (1 + 0.05/4)^4 - 1 = 0.050945$, or about 5.10%.
(c) The instantaneous rate of interest is $g = e^{0.05} - 1 = 0.051271$ or about 5.13%. □

Example 8.10 Nominal rate and effective rate of interest. If $P = \$1.00$ is invested at a nominal rate of interest $i = 8\%$ per year, what is the effective rate of interest for 5 years if it is compounded (a) semi-annually? (b) quarterly? (c) What are the instantaneous rates of interest in the above two cases?

Answer:

(a) Given $i = 0.08$, $m = 2$, and $P = \$1.00$, we have

$$j_2 = \left(1 + \frac{0.08}{2}\right)^2 - 1 = (1.04)^2 - 1 = 0.0816.$$

So the effective rate of interest is 8.16% per year when $m = 2$.

(b) Given $i = 0.08$, $m = 4$, and $P = \$1.00$, we have

$$j_4 = \left(1 + \frac{0.08}{4}\right)^4 - 1 = (1.02)^4 - 1 = 0.0824.$$

So the effective rate of interest is 8.24% per year when $m = 4$.

(c) $g = e^{0.08} - 1 = 0.0833$. The force of interest or the instantaneous rate of interest is 8.33% per year in continuous compounding for a year. □

Discrete and continuous time analysis

Note again the difference in interest is relatively small while the difference in the perception of time is huge. This is part of the reason why, in practice, banks may advertise that their interest is compounded daily (**discrete time analysis**), instead of yearly or monthly. Even if the bank compounds the interest every minute or every second, the resulting calculation of interest earned is not too different from **continuous time analysis**. This is the main reason that economists prefer to use "continuous time" analysis in the theoretical framework of economics and finance, to take advantage of the elegant and systematic methods developed in calculus. The following sections introduce continuous time analysis.

8.5 Natural Exponential and Logarithmic Functions

Since we have introduced $e = 2.71828\ldots$ as the base of the natural logarithm in the previous section, in this section, we will discuss some important properties of e and logarithms.

8.5.1 *Some definitions and relations*

A **power function** is written as $y = x^a$. The base x is a variable, and the exponent a is a constant. If the positions are interchanged, and variable x is the exponent, then we have an

exponential function of x with base a:

$$y = a^x = f(x), \quad a > 0.$$

The dependent variable y is also called the **x^{th} power of a**. If e is substituted for a, we have the growth equation (8.9) for P = 1 and it = x:

$$\boxed{y = e^x = f(x).} \tag{8.10}$$

This is called the **natural exponential function**, or simply, the **exponential function**. In this case, x is the independent variable and y is the dependent variable. Thus, (8.10) is a special case of exponential function when a = e.

The inverse function of the natural exponential function

Its inverse function[4] changes the role of the variables, and is written:

$$\boxed{x = \ln y = f^{-1}(y).} \tag{8.11}$$

Equation (8.11) is called the **natural logarithmic function, or simply, log function**. Using the conventional notations, y denotes the dependent variable and x denotes the independent variable, and thus (8.11) above can be rewritten as

$$\boxed{y = \ln x = f^{-1}(x).} \tag{8.12}$$

Column A of Table 8.5 enters x from 0 to 2.0 for y = exp(x), and for y = ln x.

A property of the inverse function

In Table 8.5, given the independent variable x in column A, the exponential function $y = e^x = \exp(x)$ is shown in column B for x = 0.0 to 2.0, and x = 2.7. It is graphed in Fig. 8.10 showing the inverse function relation using the exponential function. The relation is similar to Fig. 6.7 in Sec. 6.8.1, the inverse function method. Note that there is a one-to-one correspondence between the X-axis and the Y-axis along the exponential curve. A property of the inverse relationship is that if (a, b) is a point (coordinate) of the function in the (X, Y) space, then (b, a) must be a point (coordinate) of its inverse function. Thus, in Fig. 8.10, viewed from the X-axis, the point (1, 2.7) is on exp(1) = 2.71828, but the point is also on the inverse function ln(2.71828) = 1, measured from the Y-axis.

This property of the inverse function means that the logarithmic function is the reflection of the exponential function along the 45-degree line through the origin. Figure 8.11 shows the exponential function and the logarithmic function separately. It is drawn from the range

[4]We have encountered the inverse function in Chapters 3, 6, and 7.

A2:C24 of Table 8.5. There is a jump from x = 2.0 to x = 2.7 (not shown in Fig. 8.11). The upper curve shows the exponential function (8.10). The curve is convex and extends to infinity very quickly. Its inverse function, the logarithmic function y = ln(x), is given in column C and graphed as the lower curve of Fig. 8.11. It is concave, and tends to infinity very slowly as x goes to infinity.

Identity

Since the exponential function and the logarithmic function are inverse to each other, we have

$$\boxed{e^{\ln x} = x, \quad \text{and} \quad \ln e^{x} = x.} \tag{8.13}$$

That is, the inverse of the inverse of x is x itself. In words, it says that any variable x is identically equal to its logarithmic power with base e and any variable x is also the logarithm of its exponential function. Thus, the two functions are symmetric along the 45-degree line through the origin. In Fig. 8.11, we have manually drawn the 45-degree line through the origin. For example, in Fig. 8.10, point (1, 2.7) on the exponential function is the reflection point of (2.7, 1) on the logarithmic function. Similarly, in Fig. 8.11, point $a' = (1, 0)$ for $\ln(1) = 0$ corresponds to point $a = (0, 1)$ for $\exp(0) = e^0 = 1$; and point $b' = (2, 0.7)$ for $\ln(2) = 0.69 \approx 0.7$ corresponds to point $b = (0.7, 2)$ for $e^{0.7} \approx 2$.

In the Excel spreadsheet, Column D of Table 8.5 shows

$$\exp(\ln(x)) = x, \tag{8.14}$$

for all x, and column E shows that

$$\ln(\exp(x)) = x, \tag{8.15}$$

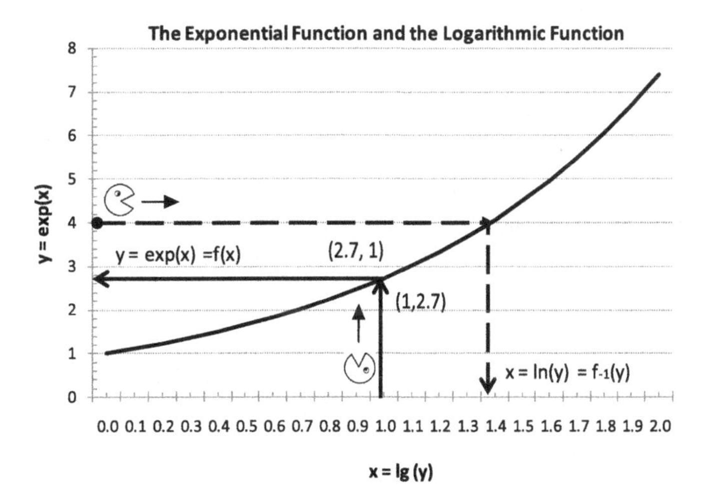

Fig. 8.10 Inverse relation of the exponential and logarithmic functions

Table 8.5. Exponential and log functions

	A	B	C	D	E
1	x	Exp	Log	Identity	
2		y=exp(x)	y=lnx	exp(ln(x))	ln(exp(x))
3	0.0	1.0000		#NUM!	0.0
4	0.1	1.1052	-2.3026	0.1	0.1
5	0.2	1.2214	-1.6094	0.2	0.2
6	0.3	1.3499	-1.2040	0.3	0.3
7	0.4	1.4918	-0.9163	0.4	0.4
8	0.5	1.6487	-0.6931	0.5	0.5
9	0.6	1.8221	-0.5108	0.6	0.6
10	0.7	2.0138	-0.3567	0.7	0.7
11	0.8	2.2255	-0.2231	0.8	0.8
12	0.9	2.4596	-0.1054	0.9	0.9
13	1.0	2.7183	0.0000	1.0	1.0
14	1.1	3.0042	0.0953	1.1	1.1
15	1.2	3.3201	0.1823	1.2	1.2
16	1.3	3.6693	0.2624	1.3	1.3
17	1.4	4.0552	0.3365	1.4	1.4
18	1.5	4.4817	0.4055	1.5	1.5
19	1.6	4.9530	0.4700	1.6	1.6
20	1.7	5.4739	0.5306	1.7	1.7
21	1.8	[illegible]	[illegible]	1.8	1.8
22	1.9	6.6859	0.6419	1.9	1.9
23	2.0	7.3891	0.6931	2.0	2.0
24	2.7	14.8797	0.9933	2.7	2.7

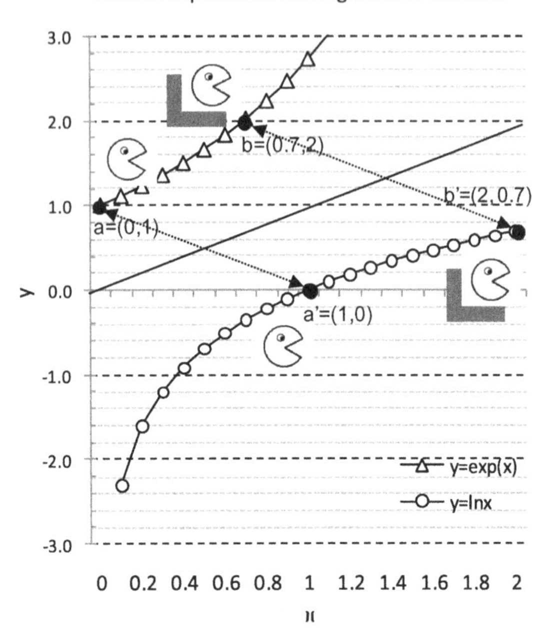

Fig. 8.11 Exponential and logarithmic functions

for all x. Thus, the mathematical operations of function "exp" and function "ln" will "cancel" each other, and yield the same result x.

In Fig. 8.11 we see that $\ln(0) = -\infty$, and in both Table 8.5 and Fig. 8.11, we see that $\exp(0) = 1$, $\ln(1) = 0$, and $\exp(1) = 2.7183 = e$. We will apply some basic rules of exponential and logarithmic functions in Example 8.11.

Example 8.11 Rule of 70 in doubling time. An important and practical application of the inverse relationships between the exponential function and logarithmic function is the Rule of 70. In the exponential growth equation, if the present value P = \$1 and we would like to know how long it will take to double the value of P, i.e. F = \$2, at a given interest (or growth) rate r100% per year. Then, we use the exponential growth equation (8.9):

$$F = e^{(rt)} = \exp(rt) = 2,$$

and solve for the unknown time period t. Taking the logarithm of both sides, we have $\ln(\exp(rt)) = \ln 2$, or $rt = \ln 2$, so we have $t = \ln 2/r$. If we know r, we can use it to solve t, or if we know t, we can use it to solve r. From Excel, $\ln 2 = 0.6931$, and if we assume $r = 0.05$, we can rewrite the formula as $t = 100^*(0.6931)/(r^*100)$, where $100^*(0.6931) \cong 70$

and $r^*100 \equiv R = 5$, we have an approximation formula for doubling time t given R as

$$t = \frac{70}{R}. \tag{8.16}$$

For $R = 0.05^*100 = 5$, the doubling time is $t \cong 14$ years. For $r = 1\% = 0.01$ and $R = 0.01^*100 = 1$, we have $t = 70$ (years), i.e. if interest (growth) rate is 1% per year, it will take about 70 years to have the future value twice larger than the original principal, which is why formula (8.16) is called the **rule of 70** in doubling time. □

8.5.2 *Rules of exponents*

We find the rules of exponential function using base a, instead of e:

$$(1)\ a^x a^y = a^{x+y}, \quad (2)\ (a^x)^y = a^{xy},$$
$$(3)\ a^x/a^y = a^{x-y}, \quad (4)\ a^0 = 1,$$

where base 'a' can be any real number.

Proof of the rules of exponents

Intuitively, the proof of each rule is simple. Without loss of generality, let $x = 2$ and $y = 3$. Then, by algebra, we have

(1) $a^2\, a^3 = (aa)(aaa) = aaaaa = a^5$;
(2) $(a^2)^3 = (a^2)\,(a^2)\,(a^2) = (aa)(aa)(aa) = a^6 = a^{2\times3}$;
(3) $a^2/a^3 = aa/aaa = 1/a = a^{2-3} = a^{-1}$;
(4) $a^0 = a^{2-2} = a^2/a^2 = 1$. □

To help readers get a feeling for the above rules, Table 8.6 shows the first three rules in Excel. We name B4:B8 "x" and C4:C8 "y", and name cell C2 "a". x and y are random variables from 0 to 1, and a is taken to be 10*rand(). The formulas in D4:I8 are the same

Table 8.6 Rules of exponents

	A	B	C	D	E	F	G	H	I
1	Rules of Exponents								
2		a =	8.748	(1) a^x*a^y		(2) (a^x)^y		(3) a^x/a^y	
3		x	y		=a^(x+y)		=a^(xy)		=a^(x-y)
4	1	0.0	0.1	1.4	1.4	1.0	1.0	0.8	0.8
5	2	0.1	0.8	6.4	6.4	1.1	1.1	0.2	0.2
6	3	0.7	0.8	25.2	25.2	3.3	3.3	0.8	0.8
7	4	0.1	0.0	1.4	1.4	1.0	1.0	1.2	1.2
8	5	0.2	0.3	2.7	2.7	1.1	1.1	0.7	0.7
9		a may be taken as exp(1)							

as the formulas in rows 2 and 3. For example,

D4: (=a^x*a^y) is equal to E4: (=a^(x+y)).

We apply the formula, and find they are identically equal. Similarly, F4 is equal to G4, and H4 is equal to I4. Copying D4:I4 to D5:I8, we have the table. In a special case, when the base "a" = exp(1) = e, we have the **laws of natural exponential functions**. Enter <F9> several times to make sure that the equality for each law holds for any numbers.

8.5.3 *Rules of logarithms*

Based on the above properties, we may derive the properties of the logarithmic functions.

(1) $\ln(uv) = \ln u + \ln v$	(product rule).
(2) $\ln(u/v) = \ln u - \ln v$	(quotient rule).
(3) $\ln u^a = a \ln u$	(power rule).

Proof of rules of logarithms

Derivation of these properties is based on the definition of the logarithmic function. In the course of explaining this, we will also review our understanding of exponential functions. If $\ln u = x$ and $\ln v = y$, then by definition, $u = e^x$ and $v = e^y$. From the properties of exponential function, if the base $a = e$, we have

(1) $uv = e^x e^y = e^{x+y}$. Hence, taking the log on both sides we have $\ln uv = \ln(e^{x+y}) = x+y = \ln u + \ln v$.
(2) $u/v = e^x/e^y = e^{x-y}$. Hence, taking the log on both sides, we have $\ln(u/v) = \ln(e^{x-y}) = x - y = \ln u - \ln v$.
(3) Since $u = e^x$, we have $u^a = e^{xa}$. Taking the log on both sides, we have $\ln u^a = \ln e^{ax} = ax = a \ln u$.

The rules of logarithmic functions are demonstrated in Table 8.7.

In Table 8.7, we name B4:B8 "u" and C4:C8 "v". u and v are taken to be random variables from 0 to 1. The formulas in D4:I8 are the same as the formulas in rows 2 and

Table 8.7 Rules of logarithms

	A	B	C	D	E	F	G	H	I
1	Rules of Logarithm								
2				(1) ln uv		(2) ln(u/v)		(3) ln (u^v)	
3		u	v	=ln u + ln v		=ln u - ln v			=v ln u
4	1	0.2	0.0	-5.7	-5.7	2.5	2.5	0.0	0.0
5	2	0.7	0.3	-1.7	-1.7	1.0	1.0	-0.1	-0.1
6	3	0.0	0.6	-5.3	-5.3	-4.4	-4.4	-3.1	-3.1
7	4	0.2	0.7	-2.1	-2.1	-1.3	-1.3	-1.1	-1.1
8	5	0.9	0.8	-0.4	-0.4	0.1	0.1	-0.1	-0.1

3. Columns D, F, and H are the values of the left-hand side of the equation of rules (1), (2), and (3), and those in E, G, and I are the values of the right-hand side of the equation of rules for the corresponding values of u and v in columns B and C. Note that the values in column D are identical to the values in column E, showing the validity of the first rule. Similarly, column F = column G, and column H = column I. In all three cases, the equalities hold for all three rules. Note, in Rule (3) of Table 8.7, that the exponent of the base u is v, instead of a. Since v can be taken as a constant, this does not affect the validity of Rule (3).

Example 8.12 Application of logarithms. Find the logarithmic expression of the following production functions:

$$\text{(a) } Q = AN^aK^{1-a} \quad \text{(b) } Q = A(aN^{-s} + bK^{-s})^{-1/s} \quad \text{(c) } Q = A_0e^{rt}N^aK^b$$

Answer: Take the logarithms on both sides of each equation and then apply the rules:

(a) $\ln Q = \ln(AN^aK^{1-a}) = \ln A + \ln N^a + \ln K^{1-a} = \ln A + a \ln N + (1-a)\ln K$.
This is called the **Cobb–Douglas production function**. We will discuss the property of this function extensively in Chapter 11.

(b) $\ln Q = \ln A + (-1/s)\ln(aN^{-s} + bK^{-s})$. This type of function is called the **constant elasticity of substitution (CES) production function** or **utility function**. This function will also be discussed in Chapter 11. Note that the rule of logarithm does not imply $\ln(x+y) = \ln x + \ln y$. Thus, we have to leave $\ln(aN^{-s} + bN^{-s})$ as it is. No more simplification or operation is possible.

(c) $\ln Q = \ln A_0 + rt + a \ln N + b \ln K$. This is an extended version of the Cobb–Douglas production function in (a). The constant r shows that output increases at the rate of r over time even if labor and capital are held constant. Economists call it the **rate of unembodied technological progress** (not embodied in labor or capital). □

Example 8.13 Continuous growth rate. Let Y_t be Gross Domestic Product (GDP) at time t. Let it grow at r∗100% per year. From the growth equation $F(t) = Pe^{rt}$ in (8.9), let $P = Y_{t-1}$ and $F(t) = Y_t$, and let the growth equation be given for one period only ($t = 1$ in e^{rt}). Then,

$$Y_t = \exp(r)Y_{t-1}.$$

is the GDP at time t, when GDP is compounded continuously throughout the period from $t-1$ to t at the constant growth rate r. Exp(r) is called the **growth factor**. If the growth rate differs in each period, then we date r by adding time:

$$Y_t = \exp(r_t)Y_{t-1}. \tag{8.17}$$

Taking a natural logarithm on both sides, we have

$$\boxed{r_t = \ln Y_t - \ln Y_{t-1}.} \tag{8.18}$$

Thus, the logarithmic difference between consecutive years of GDP is the continuous growth rate of GDP. Equation (8.18) is often used in econometric analysis of the (continuous) growth of GDP, exports, employment, prices (inflation), foreign exchange rates, etc. □

Example 8.14 The relation between discrete and continuous growth rates. In Sec. 5.4.1, we defined the discrete growth rate. Let the discrete growth rate for one period be g_t, such that $g_t = (Y_t - Y_{t-1})/Y_{t-1}$, or, rearranging,

$$Y_t = (1 + g_t)Y_{t-1},$$

which is the same as (8.5) when $n = 1$, $F_n = Y_t$, and $P = Y_{t-1}$. Taking the natural logarithm on both sides, we have

$$\ln Y_t - \ln Y_{t-1} = \ln(1 + g_t).$$

Comparing this with the continuous growth rate (8.18), we have

$$\boxed{r_t = \ln(1 + g_t).} \tag{8.19}$$

Thus, the continuous growth rate r_t is the logarithm of one plus the discrete rate g_t. When g_t is very small (close to 0), we have

$$\ln(1 + g_t) \cong g_t.$$

Thus, we may also say that g_t is an approximation of r,[5] and that

$$\boxed{r_t < g_t.} \tag{8.20}$$

Hence, if both growth rates are positive in a period, and have the SAME accumulation from Y_t to Y_{t+1}, the continuous growth rate is smaller than the discrete growth rate. (However, for the same principal, the future value is larger if the same interest rate is compounded continuously rather than discretely.) □

Example 8.15 The rule of doubling time. Instead of using the continuous time model in Example 8.11, derive the rule of doubling time by the discrete time model.

Answer: From formula (8.5), we use $F = 2 = (1 + r)^n$. Taking the log on both sides, we have $\ln 2 = n * \ln(1 + r)$ and $n = (\ln 2)/\ln(1 + r)$. Since $\ln(1 + r) \cong r$ (see Footnote 5), $n = 100^*0.6931/100^*r = 70/R$. This is the same as formula (8.16). □

[5]Expanding by Taylor's series, $r = \ln(1 + g) = g - g^2/2 + g^3/3 - g^4/4 + \ldots$ where $|g| < 1$. When g is very small, the first and the second terms become the dominant term. Since $g > 0$, we have $r - g < 0$.

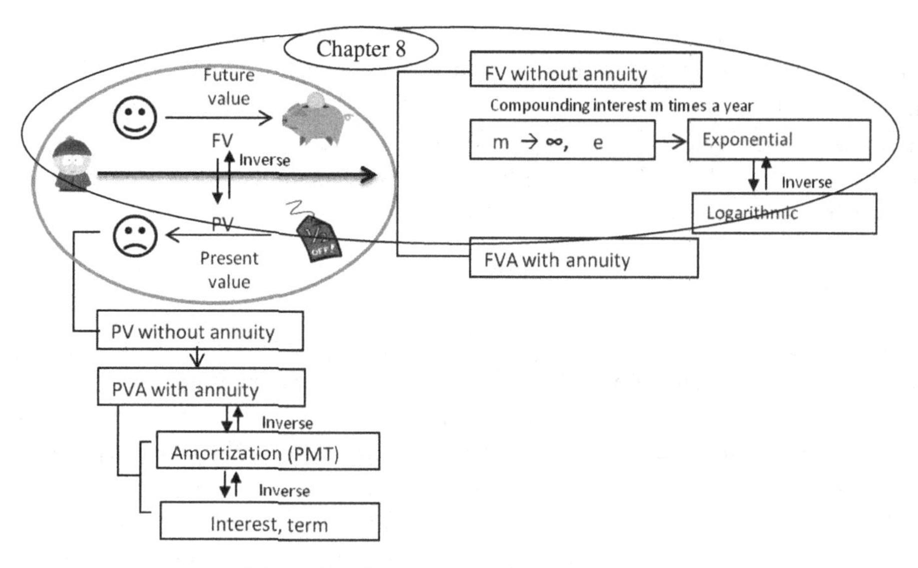

Fig. 8.12 Summary of Chapters 8 and 9

8.6 Summary

Figure 8.12 illustrates relations among the topics covered in this chapter and the next chapter. We started with the definitions of simple and compound interest (Sec. 8.1). An interesting case was the conversion of interest multiple times within a period. We explained the process diagrammatically by starting from the case in which interest is compounded semi-annually, and then quarterly, daily, every minute, and even at every instant (Sec. 8.2). This led to the abstract concept of the base of the natural logarithm, e, which was derived and illustrated using the powerful Excel spreadsheet program (Sec. 8.3). Since the difference between the interest in discrete time and in continuous time is very small, and the continuous time model is easy to handle mathematically, the derivation of e also highlighted the justification for the use of continuous time analysis in economics.

One of the most important applications of e is in the field of economic growth and development (Sec. 8.4). The introduction of e also led to the study of the properties of e and its inverse function, the natural logarithm. Using spreadsheets, we then showed the validity of the laws of exponentials and logarithmic functions (Sec. 8.5).

Chapter 9 will return to the discussions of time preference (in Economics) or time value of money (in Business).

Appendix 8A Drawing Charts: Fig. 8.4 and Fig. 8.7

Figure 8.4 5% interest rate per annum convertible semi-annually

Drawing the line

To draw Fig. 8.4, use <"insert"> <Illustrations, Shapes> <Arrow>, then click anywhere in the worksheet and draw an arrow. To get the horizontal arrow, click the arrow and

<"Drawing Tools", "Format"> <Size>, adjust the vertical unit to 0 and the horizontal unit to the length you desire. For Fig. 8.4, use 4 for the horizontal unit. Now right click on the arrow and choose <Format Shape ...> at the bottom of the dialogue box. Choose <Format Shape, Line Style> <Line Style, Width!> [2pt], click <Line Color>, and change the color to black. To add a shadow under the line, select <Shadow> and choose <offset bottom> (second shadow) under <Presets!><Close>.

The circles

To add a circle, use <"insert"> <Illustrations, Shapes> <Oval> and click anywhere in the worksheet. A circle will come up. To adjust the circle radius, use vertical unit and horizontal unit as above under <"Drawing Tools", "Format"> <Size>. Choose 0.1 for both cases (Note: to get a circle, enter vertical unit = horizontal unit; otherwise, you will get an oval). Select the circle, <RM> to get the dialog window. Select <Format Shape>. Under <Format Shape> dialog box, click <Format Shape, Fill> <@Solid fill> <Color, Black><Close>. Copy that circle and paste it to a total of three circles and put them on the arrow evenly as in the figure.

Adding labels above the line

To label the figure, use the text box so that all the text will be on the same horizon. To add the label on top of the arrow, click <"insert"> <Text, Text Box>, click the text box once, move it to a blank space, and enter <RM> twice rapidly. A vertical line appears and you can continue entering the numbers. The text box is transparent and not bordered. (If, instead of simply selecting <RM>, you drag the mouse while holding the RM, then the text box will be opaque and bordered). Adjust the length and size of the text box and type the numbers and the text. Moving the text box above the line, we have the entries above the line.

Entering equations

We first create a transparent unbordered text box as noted above. To add the equations below the line, click <"insert"><Text, Object>, select <Microsoft Equation 3.0> in the pop up window, and select <OK>. (Microsoft Equation 3.0 is available in Excel 2003 and 2007). An equation box and the menu will appear on the spreadsheet. Select the parentheses, type in the equation in the parentheses, and click anywhere on the spreadsheet. The equation menu disappears, leaving the equation inside a bordered box. Place the equation below the line as shown, making some adjustment of the equations manually to inline with the other text.

To remove the equation borders, click a side of the border to select the box, and select <RM>. A dialogue box appears. Select <Format Object>. The <Format Object> dialogue box appears. Select <"Colors and Lines><Line, Color!><No Line><OK>. The borders disappear. You may make the box transparent by choosing <"Colors and Lines><Fill, Transparency:, [100%]>.

Table 8A.1 Drawing compound interest

	A	B	C	D	E	F	G	H
1	Compound Interest in a year m = 1, 2, 4							
2						Interest rate i = 0.05		
3		Conversion frequency m cov				F = P*(1+ i/m)^h, h <= m		
4		4	2	1	interval	i	i/2	i/4
5						m=1	m=2	m=4
6		0	0	0	h = 0	0	0	0
7	Annual	1/4	1/2	1	1	500.00	250.00	125.00
8	Semi-ann	1/2	1		2		506.25	251.56
9	.	3/4			3			379.71
10	Quarterly	1			4			509.45

Microsoft Excel 2007

If you have Microsoft Word 2007 you can use <“insert”><Symbols, Equation> (if the Equation Add-In is installed) and type in the equation. Then copy the equation and paste it in the Excel worksheet. <“Insert”><Symbols, Equation> is not available in Excel 2003.

Figure 8.7 Interest converted annually, semi-annually, and quarterly within a year

Multiple independent variable columns

The table for Fig. 8.7 is given in Table 8A.1. Select range A5:H10, and then choose <“Insert”><Charts, Column><2-D Column, Clustered Column>. The “independent variable” columns B–E will show at the bottom of the chart in reversed order.

To add the data below the chart, click the chart and choose <“Chart Tools”, “Design”> <Chart Layouts, Layout 5>. The data part for m = 1, m = 2, and m = 4 will show at the bottom of the chart. Use <“Insert”, Shapes> to edit the chart. This figure is different from previous charts since we have several layers of the X-axis.

Appendix 8B. Three Forms of the Future Value Problem

In expression (8.2) or (8.5),

$$F_t = P(1 + r)^t,$$

is called the **business form**, since it is mostly used in business. For simplicity, we write $i = r$ in (8.2). In most mathematical economics textbooks, P is taken as the amount at t, denoted X_t, in this case, F is the accumulated amount at $t + n$, denoted X_{t+n}, where n is the time elapsed since t. In this notation, (8.5) becomes $X_{t+n} = (1 + r)^n X_t$. When $n = 1$, we have

$$X_{t+1} = (1 + r)X_t, \tag{8A.1}$$

which is in the **difference equation form**. If current time is taken as $t = 0$, then, the above equation becomes[6]

$$X_t = (1 + r)^t X_0, \tag{8A.2}$$

which is the **solution form** for the discrete future problem. It is a "(particular) solution" of the difference equation form, as it can be derived from (8A.1) by substitution of $t = 0, \ldots$. (see Chap. 15): Let $t = 0$; then $X_1 = (1 + r)X_0$, $X_2 = (1 + r)X_1 = (1 + r)^2 X_0$, etc. For X_t, we have (8A.2). All three forms are equivalent.

Other methods of writing the continuous growth equation

As in (8.4), if we write F and P in difference equation form, and we starts from year or period t up to $t + n$ period converting m times a year, we have

$$X_{t+n} = \left(1 + \frac{r}{m}\right)^{mn} X_t. \tag{8A.3}$$

For $m \to \infty$, we have the continuous form

$$X_{t+n} = e^{rn} X_t. \tag{8A.4}$$

If we start from $t = 0$, we have another expression called the solution form,

$$X_n = e^{rn} X_0.$$

The Three Faces of the FV Problem: for discrete and continuous variables

Figure 8B.1 illustrates the above three faces of the further value (FV) problem for the discrete and continuous models. All three are equivalent expressions.

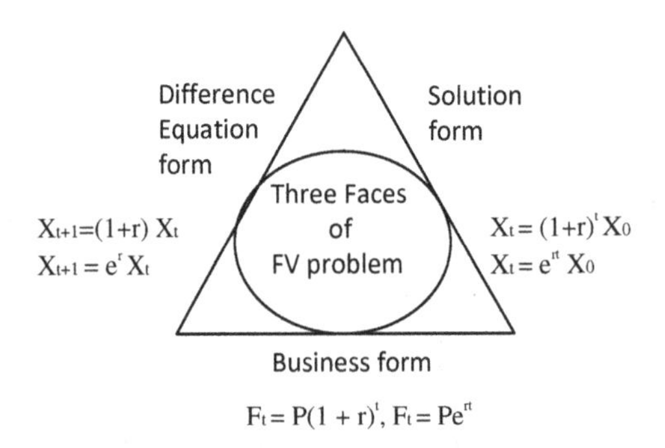

Fig. 8B.1 Three faces of the FV problem

[6] In Appendix 8B, we have shown that (8A.1) implies (8A.2). Conversely, we show (8A.2) implies (8A.1), and thus, (8A.2) is a solution to the difference equation (8A.1). To show this, we first note that the LHS of (8A.2) can be written as $X_{t+1} = (1 + r)^{t+1} X_0$. Substituting this and (8A.2) into both sides of (8A.1), we have identity in both sides of (8A.1). Hence, (8A.2) is a solution to (8A.1). (see Chapter 15).

Review of Equations and Formulas

Mathematical formula	Equivalent Excel formulas
The Future Value Problem	r = i/m, n = mt
The Discrete Case:	(9.9) =FV(r,n,pmt,[pv],[type])
(8.2) $F(t) = P(1+i)^t$	(9.10) =FV(r,n,0,1), m=1 (Section 9.4.1)
(8.4) $F_t = \left(1 + \frac{i}{m}\right)^{mt}$	
More generally,	
(8.5) $F_n = P(1+r)^n$, $r = i/m$, $n = mt$	
The continuous case:	
(8.9) $F_n = Pe^{nr}$	= exp(r*n)
Rules of exponents (Section 8.5.2)	=exp(x)
Rules of logarithms (Section 8.5.3)	=ln(x)

Key terms: Economics and Business

Key terms: Excel

Homework Chapter 8

Sensitivity tables

8-1 Reproduction of Fig. 8.2 Using shapes and text box, reproduce the time profile of Fig. 8.2 on compound interest for t periods.

8-2 Sensitivity table of future values The current checking account interest rates are 0.1% and 0.4%, depending on the average minimum cash in the account in a year. The savings account interest rates are 3% and 4% a year, also depending on the minimum amount in the account. Construct a sensitivity table, like Table 8.2, of the future values of one dollar for these four rates for 15 years, compounded monthly. From the sensitivity table, find the future interest accrued in five years. Draw a composite line chart for the table. Enter the chart and axis titles, and use text boxes to label the lines.

8-3 The base of natural logarithm Construct the time profile of e as shown in column B and Column I in Table 8.3, and draw the chart (see Fig. 8.8). Label the chart and axis clearly. Can you show, using the percentage changes, how the value of $(1 + 1/m)\hat{}m$ in (8.8) increases as m increases?

Growth equation

8-4 Finding the length of time Using both the equation method and the time profile method $(t = 0, 1, \ldots)$ find the following solutions.

(a) The evening news has announced that the consumer price index has increased 9/10 of 1% this month. What is the yearly rate if index is compounded continuously, discretely? After two years? (Hints: Find $r = 0.9 \times 0.01$ per month and compound for 12 months)
(b) The world population today is 4 billion and is growing at 1.9% per year. Suppose the population is increasing continuously.

   (i) How long will it take to increase by 1 billion?
   (ii) How long will it take to double?

(c) The conditions of a will dictate that a sum of \$200,000 is to be held in trust by the benefactor's guardian until it amounts to \$350,000. How long will this take if the fund is invested at 6% convertible quarterly?

(Hints: (b) Set up an exponential function and take the logarithm on both sides. (c) Set up a t-year convertible interest model and take the logarithm both sides)

8-5 More general results of doubling time

(a) The time required to triple a variable is called the Rule of 110 and to quadruple a variable is called the rule of 140. Derive these two rules.
(b) The general rule of the time required for a variable to increase a factor of F is the Rule of log F. Derive the general rule.
(c) Derive the rule of halving time, that is, if the present value is only one-half of the future value, how long will it take to reach the full value? (Hints: $1/2 = \exp(-rt)$, or $1/2 = (1 + r)^{-t}$, where r is given).

8-6 Doubling time The Economic Planning Bureau of Slovia has announced that it plans to double manufacturing output from current value of \$2 billion. It is known from recent experience that the growth rate of the manufacturing output is 10% converted quarterly.

(a) Using the time profile method (enter $t = 0, 1, \ldots$), find the amount of manufacturing output produced at the end of each year until the output is doubled.
(b) How long will it take to achieve the goal of doubling the output?
(c) If the planner sets the goal at \$5 billion, how long will take to achieve the goal?
(d) The planner is impatient and would like to achieve the \$5 billion goal in seven years. This means that the economy has to increase its yearly growth rate of output. If the growth rate of manufacturing output is still converting quarterly, how large should the yearly rate of growth

be? Use the time profile method (Note: In this question, time is given as 7, and the growth rate i is now the unknown variable. Let i range from 0.00, 0.01, to 0.20).

(e) Using the equation method, solve the problem (d) algebraically.

8-7 Continuous and discrete compounding The CEO of a small company announced that the company sales in the previous month had increased 1.1%. Using the time profile method $(0, 1, \ldots 12, 13, \ldots 24)$ and the equation method, answer the following questions.

(a) What is the yearly rate (compounded continuously)?
(b) If this trend lasts until the second year, what is the combined growth rate of sales by the end of two years?
(c) Instead of compounding continuously, assume the sales will grow only at the end of each month. What is the growth rate of the sales at the end of the second year?
(d) Comparing the results of (b) and (c), does it matter whether you use continuous compounding or discrete compounding? Why?

8-8 Finding the time to double the population The US population just reached 300 million in October, 2006, and is growing at 0.91% per year in 2006 (according to the CIA FactBook). Assuming that the population is increasing continuously, answer the following questions.

(a) Setting up the time profile of population change each year for 15 years, find the year in which the US population will increase by a quarter million. Write the first cell formula you use next to the table.
(b) Using the property of exponential and logarithmic functions (not the time profile), compute the year the US population doubles. Write the cell formula you use in a text box.
(c) Can you also use the time profile method to answer the equation method in (b)? Justify your answer. In general, what are the pros and cons of these two methods?
(d) Using a chart to illustrate the time profile of population change in (a), and use a dotted line to show the point at which the US population increases by a quarter million.

Future Value Problems

8-9 Ben Franklin's Plan This actually happened! Knowing the power of interest compounding, in January 1794, Franklin planned to invest 1000 pounds for 100 years at 5% (discrete) interest rate to benefit education and social welfare.

(a) What was the amount at the end of 1894?
(b) From the amount accrued at the end of 1894, in addition to the original investment of 1000 pounds, 31,000 pounds were reinvested for the next 100 years in January 1895. What was the amount of total investment from these two projects at the end of the next 100 years?
(c) What is the amount in questions (a) and (b) if the interest was calculated quarterly?
(d) What is the amount in questions (a) and (b) if the interest was calculated continuously?
(e) At the end of the 200 years, how much better off would Franklin's offspring be if the interest was compounded quarterly? Continuously?

8-10 Country growth rates (Klein, 1997, p. 49) Indonesia is one of the most highly populated countries in the world, and it also has one of the fastest-growing populations. Its population in 1994 was 192 million. Indonesia's population is expected to grow at a 1.75% average annual rate through 2010. Although China has the world's largest population (1201 million in 1994), its population is expected to grow at an average annual rate of only 1.04% through 2010.

(a) Draw their growth curves (time profile) and determine what each country's population will be in 2010 starting from 1994. Compare the ratio of the two countries' population *levels* from 1994 to 2010 to examine the impact (in terms of percentage change) of different population *growth rates*. Show diagrammatically that the ratios decrease steadily.
(b) With this pace of economic growth, other things being equal, how long will it take for the Indonesian population to catch up with the Chinese population? Calculate by using the discrete growth equation.
(c) Using the data for China and Indonesia and assuming continuous compounding, calculate the level of each country's population in 2010. What is the ratio of the population of the two countries in 2010?

8-11 Sensitivity table for future values Assuming that P = 80, what is the future value of the following combinations of n and r using the discrete compounding formula and a 5×5 sensitivity table? Please present the answer using

(a) the equation method,
(b) the Excel formula method, and
(i) n = 4, r = 8% (ii) n = 0.5, r = 4% (iii) n = 3, r = 15%
(iv) n = 0.25, r = 8% (v) n = 0.75, r = 3%

(Hints: Do not calculate one by one. Setting up a 5×5 sensitivity table, the answer should be shown along the diagonal.)

8-12 Health care (Klein, 1997, p. 69) The United States' health care costs have grown at an average rate of 6% since 1998. They have accounted for 15% of the United States' Gross Domestic Product (GDP). If health care costs continue to grow at this rate while GDP expands at a 2.5% average annual rate, what will the health care share of GDP be in 2028 (t = 30)? Is this ratio sustainable over time? Why? What economic policy should the government follow? Answer the questions in two ways:

(a) Using the equation method and
(b) Using the time profile method (t = 1998, 1999, ... 2028).
(Hints: Set up the growth equations for GDP and health care separately and find the ratio)

Advertisement

8-13 Response Functions (Mizrahi and Sullivan, p. 631) A simple rule of the effect (x) of advertisement after t days is given by

$$\text{Prob}(x < t) = 1 - \exp(-at),$$

where x is the consumer responses to advertisement of a new product, t is the time passed, say certain years, and a is a parameter showing the degree of response. This formula shows the fraction of consumer response to the advertisement (that is, whether consumers buy the product) after the time elapsed by t periods. If the product is popular, a will be large, and vice versa. For example, if $a = 0.2$, then the response of consumers to the new product after ten years will be

$$\text{Prob}(x < t) = 1 - \exp(-0.2^*10) = 0.865,$$

that is, 86.5%. If the market consists of 1 million potential consumers, then 865,000 (=1,000,000*0.865) people are expected to buy the product.

(a) Draw the consumer response curves for $a = 0.2$, 0.4, and 0.6 for $t = 0, 1, \ldots 20$. Give chart and axis titles.
(b) Will Prob converge to 1? If so, what is the economic interpretation?
(c) If the unit revenue (without cost) of the product is $0.5, what is the total revenue for the firm after 5 years if $a = 0.2$ and the market has 1 million potential customers.
(d) Suppose further that the firm's fixed cost is FC = $20,000 and the variable cost is VC = $12,000t per year. Draw the TR, the TC, and the profit curves of the firm over the years.
(e) Find the maximum profit.
(f) What is the profit at $t = 5$?, $t = 15$? Indicate the two profit levels by vertical lines.

8-14 The Logistic distribution (Mizrahi and Sullivan, p. 641) In most cases, the response to advertisement accelerates at the beginning, and after a certain time, the response will taper off. This case is shown by the following formula:

$$P(x < t) = \frac{1}{1 + be^{-at}}.$$

Assuming $b = 1$, answer the same questions as (a) to (f) in HW8-13.

Chapter 9

Present Value Problems — Making Financial Decisions

Chapter Outline

Objectives of this Chapter

In the previous chapter, we introduced the concept and calculation of the simple future value of money. This leads to the natural exponential function and its inverse, the natural logarithmic function. We now continue from future value problems and introduce their inverse function, the present value problems (Sec. 9.2). In reality, however, many of the financial transactions occur in installment payments, as when a person rents a house and pays the rent every month under the lease, or when a person borrows a certain amount of money from a bank to buy a car or a house and returns the money in monthly installments, etc. This is the future value problem with annuity or the present value problem with annuity (to be defined in Secs. 9.2 and 9.3). As in the previous chapters, we first introduce the theoretical background of problems and illustrate the problems with tables and charts (Secs. 9.1–9.3). We then show how Excel functions can be substituted for the mathematical formulas (Sec. 9.4). Finally, we present the loan amortization table and show how the table is related to various Excel functions (Sec. 9.5).

9.1 Present Value Problems

9.1.1 *The discrete case*

In the previous chapter, we have shown that if interest rate per year (period) is i, converted m times a year, then the future value F of principle \$P at the end of t years will be

$$F_t = P\left(1+\frac{i}{m}\right)^{mt} \quad \text{or} \quad FV = PV(1+r)^n, \tag{9.1}$$

where r = i/m, interest rate per conversion period, and n = mt is the total periods (see (8.5)). The second part with parentheses is the **future value interest factor** (FVIF) for n periods. It is a multiplier of P. To find its inverse function, we solve P for F, and we have (9.2).

Proposition 9.1 The present value. *The present value P of future value F with interest rate r per conversion period for n period is given by the formula*

$$\boxed{\begin{aligned} P_t &= \frac{F}{(1+i/m)^{mt}} = F\left(1+\frac{i}{m}\right)^{-mt}, \\ \textit{or} \quad P_n &= F(1+r)^{-n} = FV * PVIF_n \end{aligned}} \tag{9.2}$$

□

The interpretation of (9.2) is that if we want to save \$F to buy, say, a car at the end of n periods, what is the original principal \$P that we should save, if the interest rate is r100% per period and r is a small fraction. Written as (9.2), the P in (9.2) is called the **present value** (or **discounted value**) of the future value \$F available at the end of n periods. $1/(1+r)^n = (1+r)^{-n}$ is the **present value interest factor** ($PVIF_n$) (or **discount factor**) **of n periods**. The interest rate r used in calculating (9.2) is also called the **discount rate**, because the amount of principal (present value) P should be less than the amount of its future value F if r > 0. P is also called **discounted value**.

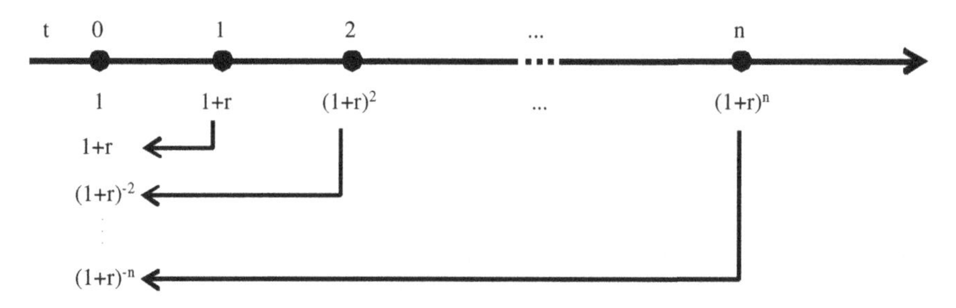

Fig. 9.1 Simple present value problem

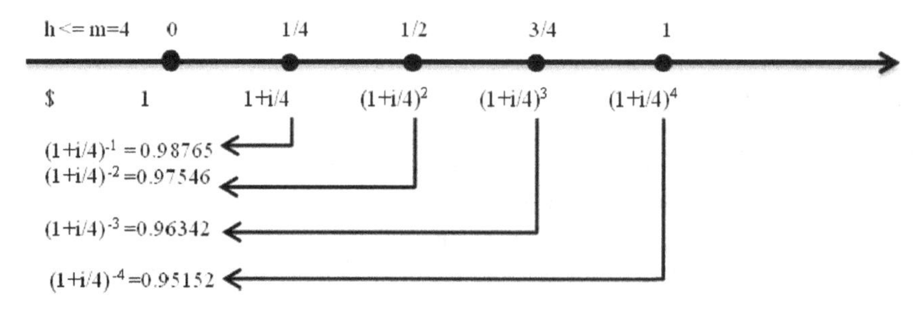

Fig. 9.2 The present value of \$1.00 when interest is converted quarterly for a year (PVIF)

Equation (9.2) is illustrated in Fig. 9.1, which is comparable to Fig. 8.2, except that we use the more general expressions r and n based on (8.5). Figure 9.2 illustrates the present value of \$1.00 when interest is compounded quarterly ($m = 4$) per year. Example 9.1 gives cases in which interest is compounded quarterly for more than a year.

9.1.2 *The continuous case*

If the interest rate is i*100% per year and the interest is compounded continuously for t years, then the present value P for a given amount of future value F can be calculated from formula (8.9) as,

$$\boxed{P(t) = Fe^{-it},} \tag{9.3}$$

where $\exp(-it)$ is the continuous PVIF.

Example 9.1 The present value of a future expense. Suppose that you plan to buy a luxury car four years from now and that car will cost \$55,000 at that time. If your bank's savings deposit interest rate is 5% per year, what is the amount that you have to deposit now (present value) in your savings account in order to grow enough interest and with the principal to buy your dream car in the future? Assume the interest is compounded: (a) annually; (b) quarterly; (c) weekly; or (d) continuously, for a year. (e) Show your results in a table of PV calculation with separate FV, PVIF, and PV columns.

Answer: We use formula (9.2) for questions (a) to (d):

(a) annually $\quad P = \$55,000(1 + 0.05)^{-4} = \$45,248.64;$
(b) quarterly $\quad P = \$55,000(1 + 0.05/4)^{-4*4} = \$45,086.05;$
(c) weekly $\quad P = \$55,000(1 + 0.05/52)^{-52*4} = \$45,034.52;$
(d) continuously $\quad P = \$55,000e^{\ 0.05*4} = \$45,030.19;$
(e) In table format, naming m and FV, we have Table 9.1. The equation is shown in the first row of the table.

Table 9.1 Given the FV to find the PV

i = 0.05 m	FV	=((1+i/m)^(−m*4)) PVIF	= FV*PVIF PV
1	55,000	0.82270247	45248.64
4	55,000	0.81974635	45086.05
52	55,000	0.81880943	45034.52
Continuous (instant)	55,000	0.81873075	45030.19

In general, the present value is the amount of money needed to deposit in a bank to grow interest under different compounding arrangements, such that the future value will be $55,000 after four years. From Table 9.1, we can see that, as the frequency of conversion m within a year increases, the present value decreases. □

9.1.3 *Present value interest factor (PVIF)*

For clarity, since n = m*t, let m = 4 and t = 4 in (b) in Example 9.1, we have

$$\mathrm{PVIF}_{(m*t)} = \mathrm{PVIF}_{(4*4)} = (\mathrm{PVIF}_{(4*1)})^4 = 0.819746,$$

where $\mathrm{PVIF}_{(4*1)} = 0.95152$, as shown in the last entry of Fig. 9.2.

More generally, we can write $\mathrm{PVIF}_{(m*t)} = (\mathrm{PVIF}_{(m*1)})^t = ((1+i/m)^{-m})^t$. This is the same as (9.2), but in a different format. Thus, if we know the frequency of conversion m within a year, PVIF of other years can be obtained by raising it to the power of t.

Example 9.2 Columbus. This example shows the nature of inverse relationship between the FV and PV problems. Let us pretend that Columbus knew that after 500 years the future accumulated value FV of his deposit would be the amount shown in Table 9.2, and asked how much he would have to deposit in the bank if the interest was compounded m times a year for 500 years, i = 0.05, m = 1, 2, 4, 12, 365, 525,600, 1,000,000, and continuously.

Answer: The problem now is exactly reversed from Example 8.7. For a given future value, using formula (9.2), the present values for all the cases show $1.00, as expected. The last PV column calculates the PV directly as PV = FV*PVIF_n. □

9.1.4 *A Sensitivity table for future and present values*

Table 9.3 compares accumulation factor FVIF_n and discount factors PVIF_n for ten periods for three levels of interest rate: r = 0.05, 0.07, and 0.10. The formulas for B4 and F4 are

Table 9.2 If Columbus knew the FV and to find the PV.

i = 0.05 m	FV(now)	=FV/((1+i/m)^(m*500)) PV (500 ago)
1	39323261827	1.00000000
2	52949930179	1.00000000
4	61668555161	1.00000000
12	68360491974	1.00000000
365	71881720016	1.00000000
525,600	72004813651	1.00000000
1,000,000	72004850248	1.00000000
Cont.(instant)	72004899337	1.00000000 = FV*EXP(−0.05*500)

Table 9.3 The sensitivity table — the FVIF and the PVIF

	A	B	C	D	E	F	G	H
1		FVIF, P=\$1.00				PVIF, F=\$1.00		
2	n	Rates	B4=(1+B\$3)^\$A4			Rates	F4=(1+F\$3)^(-\$A4)	
3		5%	7%	10%		5%	7%	10%
4	0	1.000	1.000	1.000		1.000	1.000	1.000
5	1	1.050	1.070	1.100		0.952	0.935	0.909
6	2	1.103	1.145	1.210		0.907	0.873	0.826
7	3	1.158	1.225	1.331		0.864	0.816	0.751
8	4	1.216	1.311	1.464		0.823	0.763	0.683
9	5	1.276	1.403	1.611		0.784	0.713	0.621
10	6	1.340	1.501	1.772		0.746	0.666	0.564
11	7	1.407	1.606	1.949		0.711	0.623	0.513
12	8	1.477	1.718	2.144		0.677	0.582	0.467
13	9	1.551	1.838	2.358		0.645	0.544	0.424
14	10	1.629	1.967	2.594		0.614	0.508	0.386

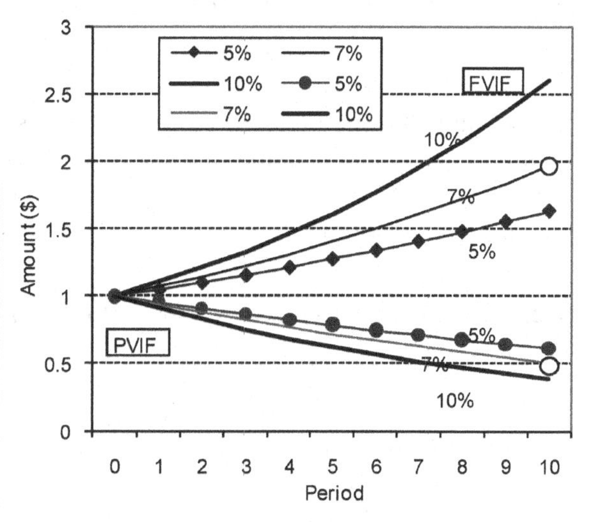

Fig. 9.3 FVIF and PVIF

indicated in row 2 (note the relative references of the cell). To construct the table, name column A4:A14 "term" and name row B3:H3 "rate". Enter in cells

B4: =(1+rate)^term and F4: =(1+rate)^(−term).

B4 is FVIF and F4 is PVIF. Once B4 is entered, copy the formula to B4:D14. Similarly, copy F4 to F4:H14. Note that the names eliminate the need for using the \$ sign. Figure 9.3 illustrates Table 9.3. The upper three curves show FVIF, the future values of \$1.00 from period 1 to period 10 at the interest rates of 5, 7, and 10%. We can see that the higher the interest rate, the higher the future value. Thus, with the interest rate at 10%, the present value of \$1.00 will grow to \$2.59 (cell D14) after 10 periods.

On the other hand, the lower three curves show PVIF, the present values of \$1.00 from period 1 to period 10 at interest rates of 5, 7, and 10%. We can see that the higher the interest rate, the lower the present value. For example, with the interest rate at 10%, the \$1.00 at the 10th period will only be worth 39 cents (cell H14) at the present. Figure 9.3 shows that both the future value curves and the present value curves are convex with respect to time periods. However, the former goes to infinity, the latter to zero.

In old days, when calculators and computers were not available, most textbooks on business mathematics provided a table of FVIF $(1+r)^n$ for various interest rates and the terms of loans, as well as a table of PVIF $(1+r)^{-n}$ for various interest rates to facilitate calculations. The tables are similar to our Table 9.3, which only shows the interest rates for r = 5, 7, and 10%, with 10 periods, n = 1, . . . 10.

9.2 Future Value Problems with Annuity

So far, we have discussed the problem of the future and present values of one-time investments. In many applications, we receive or pay a certain amount of money (an installment) periodically; for example, we receive social security benefits each month, make mortgage payment monthly for home loans or car loans, etc. These are all related to the concept of annuity, as compared to the simple future and present value problems that we discussed in the previous section.

9.2.1 *Some definitions and an illustration*

Some definitions

1. An **annuity** is a set of periodic payments (pmt) that are payable or receivable at equal **intervals of time**, usually in equal amounts. For example, if you receive $1000 per month for three years, $n = 3 \times 12 = 36$. This is different from the simple future value problem, in which only a single cash flow is allowed at the beginning or end of the period.
2. If the payments begin and end at fixed dates for n periods, the arrangement is called an **annuity certain**.
3. The time range n between the beginning and end of the payments is called the **term of annuity**.
4. If the payments are made at the ends of intervals, the arrangement is called an **ordinary annuity**.
5. The **amount** (or **final value**) of an ordinary annuity is the sum of the annuity at the end of the term.

These definitions are illustrated in Fig. 9.4. In this chapter, we are interested in ordinary annuity and annuity certain.

Example 9.3 A future value problem with annuity. John has graduated from a college and started his job. He plans to set aside $4000 a year to make a down payment on a house after 6 years. If he deposits the money at the end of each year at an interest rate of $r = 5\%$ per year, as illustrated in Fig. 9.4, what is the total future value of the six payments?

Answer:

(a) At the end of the first period, pmt = $4000 is deposited. At the end of the sixth period, this $4000 will be compounded to

$$FV_5 = \$4000^*(1 + 0.05)^5 = \$4000^*1.27628 = \$5105.13.$$

(b) $4000 paid at the end of the second period will have compounded to

$$FV_4 = \$4000^*(1 + 0.05)^4 = \$4000^*1.21551 = \$4862.03.$$

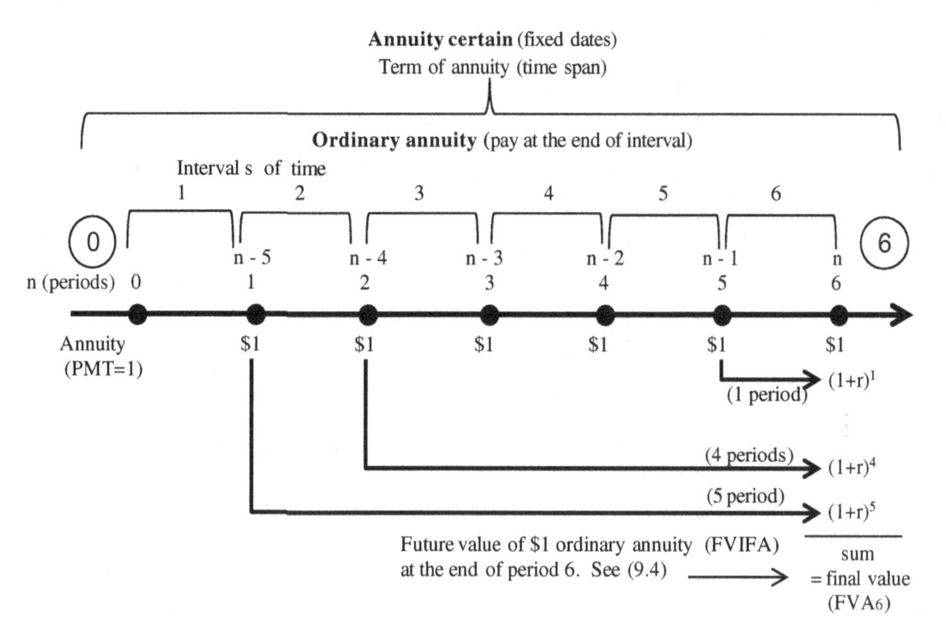

Fig. 9.4 Illustration of a future value of annuity problem with annuity and the derivation of FVIFA

(c) \$4000 paid at the end of the third period will have compounded to

$$FV_3 = \$4000^*(1 + 0.05)^3 = \$4000^*1.15763 = \$4630.50.$$

(d) \$4000 paid at the end of the fourth period will have compounded to

$$FV_2 = \$4000^*(1 + 0.05)^2 = \$4000^*1.10250 = \$4410.00$$

(e) \$4000 paid at the end of the fifth period will have compounded to

$$FV_1 = \$4000^*(1 + 0.05) = \$4000^*1.0500 = \$4200.00.$$

To find the future value with annuity, FVA_6, we decompose FVA into two parts: the payment (pmt) and the FVIF at each period. After six payments, the sum of six future values will be

$$\begin{aligned} FVA_6 &= pmt[1 + (1 + r) + (1 + r)^2 + (1 + r)^3 + (1 + r)^4 + (1 + r)^5] \\ &= pmt \sum_{k=0}^{k=5} (1 + r)^k \\ &= 4000(1 + 1.10500 + 1.10250 + 1.15763 + 1.21551 + 1.27628) \\ &= 4000(6.80191) = 27207.65, \end{aligned} \quad (9.4)$$

which is the sum of a geometric progression. Note that the dummy variable k in the sum starts from $k = 0$ and ends at $k = n - 1 = 5$, and the sum must be larger than 6, since you have contributed \$1.00 for each of the six periods. □

9.2.2 *The future value interest factor with annuity (FVIFA)*

For the general case, if the number of payment periods is n, then the sum in (9.4) reduces to a compact formula as in (9.5) below (see Appendix 9A for the derivation):

Proposition 9.2 Amount of an annuity of $1.00 per period (FVIFA). *If an ordinary annuity certain is denoted as payment, PMT, and the interest rate is r per period, then the future value FV of the annuity at the end of* n *periods, denoted as* FVA_n, *is*

$$\begin{aligned} FVA_n &= PMT\left(\frac{(1+r)^n - 1}{r}\right) \\ &= PMT * FVIFA_n \\ &= \textit{FV of annuity of \$PMT for n periods.} \end{aligned} \tag{9.5}$$

□

Equation (9.5) is called the **amount of annuity of $PMT per period**, the term in the parenthesis is called the **future value interest factor with annuity (FVIFA)**. In Example 9.3, substituting PMT = 4000, r = 0.05, n = 6, we have FV = 4000(6.801.91) = 27207.65, which is the same as the straightforward calculation in (9.4).

Example 9.4 John deposits $100 per month (PMT) in a savings plan for a down payment to buy a car in three years. The bank's interest rate is i = 6% per year. What is the total future cash value of his savings after three years?

Answer: Since PMT= 100, the monthly interest rate is $r = i/m = 0.06/12 = 0.005$, and $n = mt = 12 * 3 = 36$. Substituting these values into formula (9.5), we have

$$FVA_{36} = 100\frac{(1+0.005)^{36} - 1}{0.005} = 3933.61.$$

The cash value of John's savings in the bank grows to almost $4000. □

9.3 Present Value with Annuity Problems

A different condition in the future value with annuity problem might be that John expects that when he goes to graduate school next month, room and board will cost him $1000 per month for the next three years. Suppose the bank's saving account interest rate is i = 6% per year. John would like to find the present value of his future room and board expenditure for the next three years so that he can make deposits in the savings account now to grow interest, and withdraw $1000 per month for his future room and board expenditure during the next three years. This is illustrated in Fig. 9.5.

9.3.1 *Some definitions and an illustration*

Let us first derive the equation for solving the above problems. From Fig. 9.5, we can see that the present value of the first month's cost is $\$1000(1+0.06)^{-1}$, that of the second month's cost is $\$1000(1+0.06)^{-2}$, etc., until that of the 36th month is $\$1000(1+0.06)^{-36}$.

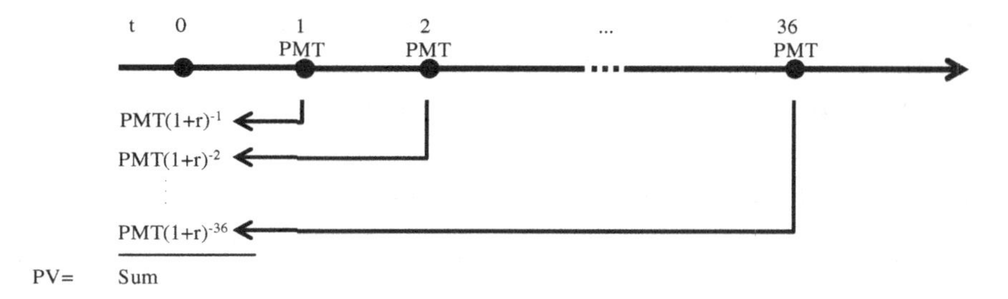

Fig. 9.5 Present value problems with annuity (PVA) and interest factor (PVIFA).

The sum of the present values of the 36 monthly payments (n = 36) of \$1000 is

$$\begin{aligned}\mathrm{PVA}_{36} &= 1000[(1+\mathrm{r})^{-1} + (1+\mathrm{r})^{-2} + \cdots + (1+\mathrm{r})^{-36}] \\ &= 1000(0.9950 + 0.9901 + 0.9851 + \cdots + 0.8356) \\ &= 1000(32.87102) = 32871.02, \end{aligned} \tag{9.6}$$

where $r = i/12$ is the interest rate per month. We add A after PV to show the PV with annuity.

9.3.2 *Present value interest factor with annuity (PVIFA)*

More generally, we may write

$$\mathrm{PVA}_{\mathrm{n}} = \mathrm{PMT}[(1+\mathrm{r})^{-1} + (1+\mathrm{r})^{-2} + \cdots + (1+\mathrm{r})^{-\mathrm{n}}]. \tag{9.7}$$

The sum in the brackets can be reduced to the right-hand side fraction of (9.8) below. Thus, we have the following proposition.

Proposition 9.3 The present value of an annuity of \$1.00 per period. *The present value of an installment payment (PMT) at the interest rate r per installment period for n periods is given by*

$$\begin{aligned}\mathrm{PVA}_{\mathrm{n}} &= \mathrm{PMT}\frac{1-(1+\mathrm{r})^{-\mathrm{n}}}{\mathrm{r}} \\ &= \mathrm{PMT} * \mathrm{PVIFA}_{\mathrm{n}} \\ &= \textit{PV of annuity of \$PMT for n periods.}\end{aligned} \tag{9.8}$$

□

The ratio in (9.8) is called the **present value interest factor with annuity (PVIFA).**

Example 9.5 Present value with annuity. Now we can apply formula (9.8) to solve the present value of John's total room and board expenditure for the next three years that we stated at the beginning of this section.

With pmt = \$1000, r = 0.06/12, and n = 36, we have

$$\mathrm{PVA}_{36} = 1000(1-(1+0.06/12)^{-36})/(0.06/12) = 32871.02,$$

which is the same as the straightforward calculation in (9.6). Thus, John has to deposit \$32,871.02 in the bank now to grow the interest, so that he can withdrawal \$1000 per month to pay for his room and board expenditure in the next three years. □

9.3.3 *The ordinary annuity with unequal annuity*

In the above formulation, the amount of annuity is assumed to be equal in each period. If the amount is not the same, then we cannot use Propositions 9.2 or 9.3. A direct manual calculation is required, as shown in the following example.

Example 9.6 Unequal annuity. The dividend payments of a company share for the next three years are expected to be as follows. After the fourth year, the shares will be sold at \$120 without any dividend. What is the present value of the shares if interest rate is 5% compounded once a year?

t	1	2	3	4	time period
\$	2	3	5	120	expected dividends at the end of period

Answer: The present value, from (9.7), is

$$\text{PVA} = \frac{2}{1.05} + \frac{3}{1.05^2} + \frac{5}{1.05^3} + \frac{120}{1.05^4} = 107.67$$

Thus, at the present time, the shares are worth \$107.67. □

9.3.4 *Comparison of present and future values with annuity*

Table 9.4 compares "the future values of an annuity of \$1.00" (FVIFA) with "the present values of an annuity of \$1.00" (PVIFA) for periods ranging from 0 to 10 for three selected interest rates, r = i = 5, 7, and 10%. To construct the table, we first name A4:A14 as termwa (term with annuity), and B3:H3 as ratewa (interest rate with annuity). Then, in cell B4, we enter formula (9.5), and in F4, we enter formula (9.7), as follows:

```
B4: =((1+ratewa)^termwa−1)/ratewa
F4: =((1−(1+ratewa)^(−termwa))/ratewa.
```

Copy B4 to C4:D4, then copy B4:D4 to B5:D14. Similarly, copy F4 to G4:H4, then copy F4:H4 to F5:H14. Drawing the chart, we have Fig. 9.6.

Thus, at the end of the 10th period, the future value of the annuity of \$1.00 with an interest rate at 5% will grow to \$12.58 (in B14), and the present value of the annuity of \$1.00 with an interest rate at 5% is \$7.72 (in F14). In Fig. 9.6, the three FVIFA curves are convex to the period axis; they increase with an increasing rate as n increases, and eventually become infinity. The three PVIF curves are concave to the X-axis: they increase at a decreasing rate as n increases, and converge to 1/r.

Table 9.4. Future and present values of a \$1.00 ordinary annuity

	A	B	C	D	E	F	G	H
1		FVIFA				PVIFA		
2	n	Rates	B4=see (9.5)			F4=see (9.7)		
3		5%	7%	10%		5%	7%	10%
4	0	0.000	0.000	0.000		0.000	0.000	0.000
5	1	1.000	1.000	1.000		0.952	0.935	0.909
6	2	2.050	2.070	2.100		1.859	1.808	1.736
7	3	3.153	3.215	3.310		2.723	2.624	2.487
8	4	4.310	4.440	4.641		3.546	3.387	3.170
9	5	5.526	5.751	6.105		4.329	4.100	3.791
10	6	6.802	7.153	7.716		5.076	4.767	4.355
11	7	8.142	8.654	9.487		5.786	5.389	4.868
12	8	9.549	10.260	11.436		6.463	5.971	5.335
13	9	11.027	11.978	13.579		7.108	6.515	5.759
14	10	12.578	13.816	15.937		7.722	7.024	6.145

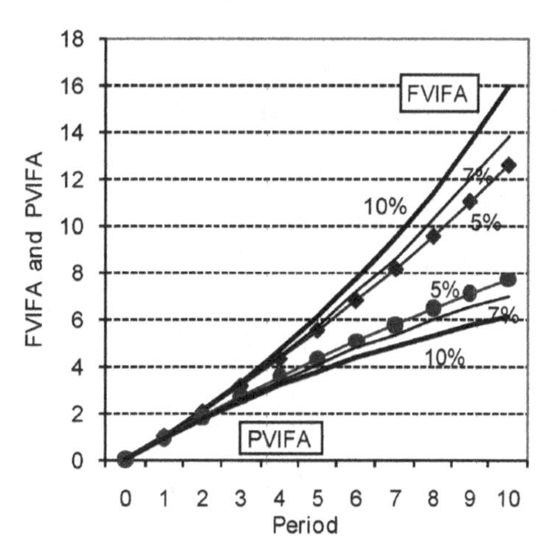

Fig. 9.6 Future and present values of annuity of \$1.00

9.4 Excel Function for Financial Analysis

The above equations and formulas appear to be complicated. Excel provides simpler financial functions. In this chapter, we introduce some of them that are related to the time value of money.

9.4.1 *Future value problems using Excel*

Future value problems using Excel

The Excel program combines the simple future value problem (FV) and the future value problem with annuity (FVA) in one function . The function is given as

$$\boxed{\begin{array}{l} \mathrm{FV} = \mathrm{FV}(\mathrm{r}, \mathrm{n}, \mathrm{pmt}[, \mathrm{pv}][, \mathrm{type}]) \\ \qquad\qquad\qquad\quad \downarrow \qquad\quad \downarrow \\ \qquad\qquad\qquad\quad \pm \qquad\ 0/1 \\ \qquad\qquad\qquad\quad \downarrow \\ \qquad\qquad\qquad\quad \mathrm{pmt} = 0 \quad \mathrm{FV} \\ \qquad\qquad\qquad\quad \mathrm{pmt} \neq 0 \quad \mathrm{FVA}, \end{array}} \tag{9.9}$$

where space is not allowed between the variables. r = i/m is the interest rate at each conversion interval with nominal interest rate i and the conversion frequency m in a year, as we have discussed in Chapter 8. n = mt is the term or total number of periods in t years, and the **pmt** is the constant payment amount made at the end of each period. From an agent's point of view, if a payment is paid out, then it is negative, and if a payment is received, it is positive.

The variable in the brackets is optional items. **pv** is the present value, and **type** takes the number 0 or 1, 0 meaning the payment is due at the end of the period (default), and 1 meaning the payment is due at the beginning of the period. If the pv and type are not specified, then they will be assumed to be 0 as default. If they are specified, they should be entered with a comma as shown in (9.9).

Simple future value problems

In the special case of a simple **future value problem** (9.1), the Excel command is

$$\boxed{\mathrm{FVA_n} = \mathrm{FV(r, n, 0, pv)} \equiv \mathrm{PV{*}FV(r, n, 0, 1)},} \tag{9.10}$$

where subscript n denotes the number of periods. Note that $\mathrm{FV(r, n, 0, 1)} = \mathrm{FVIF_n}$.

Example 9.7 A simple FV problem using Excel. John borrows $1000 on May 1, 2000 and agrees to pay interest at 7% per year.

(a) If the interest is compounded at the end of the year, how much does he have to pay back on the loan with interest on April 30, 2005?
(b) If the interest is converted quarterly in a year, how much will he need to pay back on the loan with interest on April 30, 2005?

Answer:

(a) Using Excel function (9.10) for $\mathrm{r} = \mathrm{i} = 0.07, \mathrm{n} = 5$, pmt= 0, pv= 1000, we have

$$\mathrm{FV}(0.07, 5, 0, 1000) = -\$1{,}402.55.$$

He has to pay back $1,402.55. We may also verify this from the sensitivity table, Table 9.3. From cell C9, we have $\mathrm{FVIF_5} = 1.403$, hence, he has to pay back $\mathrm{pmt{*}FVIF_5} = 1{,}000{*}1.403 = \$1403$.
(b) In this case, we use the formula $\mathrm{r} = 0.07/4, \mathrm{n} = 4{*}5$, pmt = 0, and pv = 1000,

$$\mathrm{FV}(0.07/4, 4{*}5, 0, 1000) = -\$1414.78,$$

which is the same as in (9.1): \$1000*(1+0.07/4)^(4*5). He has to pay back \$12.23 more. □

The future value problem with annuity using Excel

For future value problem with annuity payment of the amount $pmt with interest rate r per period and the term n, the Excel function is

$$\boxed{\mathrm{FVA_n} = \mathrm{FV(r, n, pmt)} \equiv \mathrm{PMT{*}FV(r, n, -1)},} \tag{9.11}$$

Note that, since it is a payout, pmt is negative, and pv = 0. Note that $\mathrm{FV(r, n, -1)} = \mathrm{FVIFA_n}$

Example 9.8 A FV problem with annuity using Excel. A man 40 years old decides to set aside $1000 per year to be invested in a savings program with an annual interest rate of 7%. The interest will be left in the account to accumulate.

(a) What amount will he receive at age 50? Use Eq. (9.5) and the Excel function (9.11) to solve the problem. Can you use sensitivity table, Table 9.4, to solve this problem?
(b) How much will he accumulate if he sets aside $100 per month instead?

Answers:

(a) From Eq. (9.5), we have

$$\mathrm{FVA}_{10} = \$1000^{*}((1 + 0.07)\hat{\ }10 - 1)/0.07 = \$13{,}816.45.$$

From Excel function (9.11), we have

$$\mathrm{FVA}_{10} = \mathrm{FV}(0.07, 10, -1000) = \$13{,}816.45.$$

From C14 of Sensitivity Table 9.4, $\mathrm{FVIFA}_{10} = 13.816$. The FV is approximately equal to $\$1000^{*}13.816 = \$13,816$.

(b) If he sets aside $100 per month, then the term is $n = 12 \times 10 = 120$ months (periods) and $r = 0.07/12$. From Eq. (9.5), we have

$$\mathrm{FVA}_{120} = \$100^{*}((1 + (0.07/12))\hat{\ }(12^{*}10) - 1)/(0.07/12) = \$17,308.48.$$

From Excel function (9.11), we have

$$\begin{aligned}\mathrm{FVA}_{120} &= \mathrm{FV}(0.07/12{,}120, -100) = \$17{,}308.48\\ &= \mathrm{FV}(0.07/12{,}120, -1)^{*}100.\end{aligned}$$

To solve the problem by the table-based method, we need to construct a sensitivity table which includes $n = 120$ and $r = 0.07/12 = 0.005833 = 0.5833\%$. This is not the case for Table 9.4. The readers may try to construct such a table.

□

Example 9.9 Construction of the sensitivity tables using Excel functions

(a) Reproduce the future values of $1.00 (FVIF) in Table 9.3 (Columns A–D) using the Excel formula (no annuity).
(b) Construct future values of $1.00 with annuity (FVIFA) like Table 9.4 (Columns A–D) using the Excel function.

Answer:

(a) Use (9.10) for pmt = 0. Use pv = −1 to get positive numbers. In B4 of Table 9.3, enter FV(B$3, $A4, 0, −1), and copy B4 to B4:D14.
(b) Use Excel function (9.11) for pmt = 1, pv = 0. In B4 of Table 9.4, enter FV(B$3,$A4,−1,0), and copy B4 to B4:D14.

□

Three methods of solution

The above examples show that to solve the future value problems, we have three methods of solution. They are the mathematical equation method, the Excel function method, and the table-based method. All three should give the same answers. The mathematical method may be too complicated, the table-based method may be too limited in applicable r and n. If a spreadsheet is readily accessible, the Excel function method may be the easiest.

9.4.2 *Present value problems using Excel*

Like the future value problems given in Excel function (9.9) , Excel also combines the simple present value problems and the present value problems with annuity in one function. The formula is given by

$$\boxed{\begin{array}{lll} \mathrm{PV} = \mathrm{PV}(\mathrm{r}, \mathrm{n}, \mathrm{pmt}[, \mathrm{fv}][, \mathrm{type}]) & & \\ & \downarrow & \downarrow \\ & \pm & 0/1 \\ & \downarrow & \\ & \mathrm{pmt} = 0 & \mathrm{PV} \\ & \mathrm{pmt} \neq 0 & \mathrm{PVA}, \end{array}} \tag{9.12}$$

where, except for the exchange of pv (present value) and fv (future value) positions, the parameters are the same as the FV in (9.9). As before, the difference between the two problems is the payment (pmt). For a simple present value problem, pmt $=$ 0. For the present value problem with annuity of \$1.00, pmt $= 1$ for each payment period.

Simple present value problems

In the special case of a **simple present value problem** (9.2), the Excel command is,

$$\boxed{\mathrm{PV_n} = \mathrm{PV}(\mathrm{r}, \mathrm{n}, 0, \mathrm{fv}) \equiv \mathrm{FV} * \mathrm{PV}(\mathrm{r}, \mathrm{n}, 0, 1),} \tag{9.13}$$

where subscript n denotes the number of periods. $\mathrm{PV}(\mathrm{r}, \mathrm{n}, 0, 1) = \mathrm{PVIF_n}$.

Example 9.10 Simple PV problem using Excel function. Parents want to invest at the birth of a daughter in a fund that will accumulate to \$10,000 by the time the daughter is 10 years old. Answer the following questions by using the mathematical equation method, the Excel function method, and the table-based method, with the following assumptions.

(a) If the nominal interest rate is 7% per year and compounded annually, how much money must the parents invest at the time of the child's birth?
(b) If interest is compounded quarterly at the same 7% per year, how much money must be invested?

Answer:

(a) From Eq. (9.2), we have 10,000/(1 + 0.07)^10 = 5083.49.
From Excel function (9.13), for r = i = 0.07, n = 10, pmt = 0, fv = 10,000, we have

$$\mathrm{PV}_{10} = \mathrm{PV}(0.07, 10, 0, 10000) = -5083.49,$$

which says that the parents have to deposit (pay out) \$5083.49 in the bank to receive \$10,000 after 10 years.
Note that this result may also be approximated from a sensitivity table, Table 9.3. Since for FV = \$1.00 at r = i = 7%, the present value for n = 10 is $\mathrm{PVIF}_{10} = 0.508$ (cell G14 in Table 9.3), the amount needed to invest now is only \$10,000*0.5083 = \$5083 (Table 9.3 should have five decimal places to be accurate).
In addition, we also can see that the Rule of 70 holds in this case. At an interest rate of 7% per year, an initial value of \$5000 will be doubled to about \$10,000 in ten years.

(b) From Eq. (9.2), we have PV = \$10,000/(1 + 0.07/4)^(4*10) = \$4996.01. From Excel function (9.13), for r = 0.07/4, n = 4*10 = 40, pmt = 0, fv = 10,000, we have

$$\mathrm{PV}_{40} = \mathrm{PV}(0.07/4, 40, 0, 10000) = -\$4996.01.$$

If the interest is compounded quarterly instead of yearly, the amount of present investment can be reduced slightly.
Unless we construct a more detailed sensitivity table, including r = 0.07/4 = 0.0175 and n = 40, we cannot have a table-based solution. □

Present value problems with annuity

For a present value problem with an annuity payment of the amount \$pmt, with interest rate r per period, and with term n, the Excel function is,

$$\boxed{\mathrm{PVA_n} = \mathrm{PV(r, n, pmt)} \equiv \mathrm{PMT} * \mathrm{PV(r, n, 1)}} \tag{9.14}$$

where $\mathrm{PV(r, n, 1)} = \mathrm{PVIFA_n}$.

Example 9.11 Present value with annuity. John retired recently at age 65 and expects to receive a social security benefit payment of \$1500 per month. Assuming that money in a bank will grow interest at 7% per annum, what is the present value of the annuity that John will receive (a) after 10 months of his retirement? (b) after 10 years of his retirement? (c) What is the future value after 10 years? Answer in two ways: by the mathematical equation method (9.8) and by the Excel function method (9.12).

Answer:

(a) In this case, using Eq. (9.8), for n = 10, r = 0.07/12, and pmt = \$1500, we have

$$\mathrm{PVA}_{10} = 1500(1 - (1 + 0.07/12)\hat{}(-10))/(0.07/12) = 14529.77.$$

From Excel function (9.12) or (9.14), for r = 0.07/12, n = 10, pmt = −1500 (negative, since John has to deposit pmt to the bank to receive interests) and fv = 0,

$$\text{PVA}_{10} = \text{PV}(0.07/12, 10, -1500) = \$14{,}529.77$$
$$= \text{PV}(0.07/12, 10, -1)^{*}1500.$$

The present value of the annuity after 10 months of retirement is worth $14,529.77.

(b) After 10 years of John's retirement, the present value of the total annuity can be calculated from Eq. (9.8) for r = 0.07/12, n = 10 × 12, and pmt = $1500. We have

$$\text{PVA}_{120} = 1500(1 - (1 + 0.07/12)\hat{}(-12^{*}10))/(0.07/12) = 129189.53.$$

Or, from Excel Eq. (9.14),

$$\text{PVA}_{120} = \text{PV}(0.07/12, 10^{*}12, -1500, 0) = \$129{,}189.53.$$

(c) The future value of the annuity after 10 years of retirement can be calculated from Eq. (9.5):

$$\text{FVA}_{120} = \$1500 * ((1 + 0.07/12)\hat{}(12^{*}10) - 1)/(0.07/12) = \$259{,}627.21.$$

Or, using the Excel function (9.11), we have

$$\text{FVA}_{120} = \text{FV}(0.07/12, 10^{*}12, -1500) = \$259{,}627.21. \qquad \square$$

Example 9.12 A Decision making problem. John plans to buy a car and needs an extra $10,000 to pay for it. The car dealer offers him two options of payment: option 1, pay $10,000 in cash now; option 2: pay a $1000 down payment now and then pay $350 per month in installments at the end of each month for the next three years. Suppose the current bank interest rate is 10% per year and compounded monthly. (a) Which option of payment is better off for John? (b) Other things being equal, under what interest rate would John's decision be reversed?

Answer:

(a) In option 2, using Excel function (9.12) or (9.14), the present value of the installment payments at $350 per month for the next three years is

$$\text{PVA}_{36} = \text{PV}(0.1/12, 12^{*}3, -350) = \$10{,}846.93.$$

Hence, the total present value of the car payment is $\$1000 + \$10,846.93 = \$11,846.93$, which is greater than buying the car with option 1 by paying $10,000 in cash now, if he has that amount of money.

(b) We need to find the interest rate that will make the present value of installation payment of $350 per month for 3 years plus the down payment of $1000 become less than $10,000. We construct a sensitivity analysis table, such as Table 9.5. Let the interest rate ranges be from 0.10 to 0.25, as in column A. In B3, enter PV(A3/12,12*3,−350), and in C3 enter B3+1000, and copy B3:C3 down to B4:C18. Then, the table shows that John should choose option 2 to buy the car if the interest rate is more than 24%. □

Table 9.5 Sensitivity of the interest rate r

	A	B	C
1	Sensitivity of the interest rate r		
2	r	Present value	+dwnpmt
3	0.1	$10,846.93	$11,846.93
4	0.11	$10,690.71	$11,690.71
	...		
16	0.23	$9,041.66	$10,041.66
17	0.24	$8,921.09	$9,921.09
18	0.25	$8,802.86	$9,802.86

9.5 The Amortization Table

In Example 9.12, suppose that John has a different strategy. He has decided to borrow $10,000 (=PV) from a bank, not from the car dealer, and pay for the car in cash now. He will pay off the bank loan in equal monthly installment (pmt) during the next three years, at the bank's lending interest rate of 6% per year. What is his monthly payment to the bank? What is his total financial cost at the end of three years?

9.5.1 *Some definitions and illustrations*

The above problem is a special case of the present value problems with annuity. In this case, we know the present value is $10,000, $\mathrm{r} = \mathrm{i}/12 = 0.06/12$, and we are looking for the fixed amount of the monthly installments (pmt) for $\mathrm{n} = 36$ months. From (9.7), the present value of each monthly payment for 36 months must sum to the amount of the loan, PVA=$10,000. Hence,

$$\mathrm{PMT}[(1+\mathrm{r})^{-1} + (1+\mathrm{r})^{-2} + \cdots + (1+\mathrm{r})^{-36}] = 10{,}000, \tag{9.15}$$

which is the same as (9.8), except that PMT is the unknown variable. Thus, solving for PMT in (9.15), or equivalently, (9.8), as a function of PVA, we have the following proposition.

Proposition 9.4 *The installment payment (PMT) at interest rate r per conversion period for n periods to pay off the present value (current amount) PVA is*

$$\boxed{\mathrm{PMT} = \mathrm{PVA}\frac{\mathrm{r}}{1-(1+\mathrm{r})^{-\mathrm{n}}}.} \tag{9.16}$$

□

Equation (9.16) is the inverse function of (9.8). The set of payments based on (9.16) is said to be an **amortization of the debt**.

Example 9.13 Installment payments. If John borrows $10,000 (=PV) from a bank for three years at the interest rate of 6% compounded monthly, what is his monthly payment

to the bank to repay the loan?
Answer:
The pmt can be calculated using formula (9.16):

$$\text{PMT} = 10,000\frac{0.005}{1-(1+0.005)^{-36}} = 304.22. \tag{9.17}$$

John's installment payment will be \$304.22 per month, which is smaller than the car dealer's offer of \$350 per month, and he need not pay the \$1,000 down payment to the dealer. He will pay out a total of \$304.22*36 = \$10,951.92 at the end of 36 months, and his total financial cost for the loan will be \$10,951.92 – \$10,000 = \$951.92. □

9.5.2 *The amortization table using Excel equations*

Excel function for PMT

In the present value problem with annuity, we have a given amount of monthly payment (pmt) over a period of time and we calculate the sum of present value of the payment stream. Conversely, in the amortization problem,[1] the present value is given and we calculate the amount of monthly payment. Formula (9.16) of Proposition 9.4 applies. In Example 9.13, using (9.16), we calculated John's amortization or installment payment to be \$304.22 in (9.17).

Excel has an equivalent function which is called PMT,

$$\boxed{\text{PMT} = \text{PMT}(\text{r}, \text{n}, \text{pv}[, \text{fv}][, \text{type}]).} \tag{9.18}$$

This is an **inverse function** of the PV function (9.12). In (9.18), PMT is solved as a function of pv. The arguments in function (9.18) are almost the same as in the PV function in (9.12), except that PV now is the independent variable, and PMT is a dependent variable, the value of which we want to solve.

In (9.18), ignoring the brackets, $\text{PMT} = \text{PMT}(\text{r}, \text{n}, \text{pv})$, where n is the number of payment periods and is usually expressed in terms of months. The pv is the present value of the loan. If we enter pv as a positive value (receiving \$10,000), then, $\text{PMT} = \text{PMT}(0.06/12, 12*3, 10000) = -304.22$, which yields a negative value. It means the amount of monthly payment by the borrower. If we enter a negative pv, i.e., -10000, in the function, we will obtain a positive PMT value. The distinction is sometimes confusing.

Example 9.14 Construction of an Amortization Table. In Example 9.13, John borrows \$10,000 (=PV) from the bank for 3 years at the interest rate of 6% per year and the interest is compounded monthly. We found that his equal monthly payment (pmt) is

[1] An amortization table is a schedule of amortizing a debt by listing the amount of payment, interest, principle payment, and the balance of the debt outstanding. The word means to (=a) deaden or extinguish (=mortise) the debt by installment payments.

\$304.22 from (9.17). Equivalently, we can apply Excel function (9.18) as

$$\text{PMT} = \text{PMT}(0.06/12, 12^*3, 10000) = -304.22. \tag{9.19}$$

We have the same results as in Example 9.13 using Eq. (9.18). □

Amortization Table

Using the above example, we construct an **amortization table** showing the monthly payment, interest paid, principle paid, and the balance of the loan outstanding at every period. We also make sure that John's total payment and financial cost at the end of 36 months are the same as those calculated above. We then draw a chart showing the above four variables. This is a good exercise to understand financial calculations.

Variables

Rows 2–6 of Table 9.6 show the information about the loan. Cells E2:E5 are named as the corresponding symbols in column D. In column A, A11:A47 shows the payment period t.[2] Row 7 shows the four variables, rows 8 and 9 show formulas, and row 10 has the column labels. The formula in E8:E9 is $= \text{B}[\text{t}-1] - \text{PMT}[\text{t}]$, which is the balance of the outstanding loan in the previous period $(\text{t}-1)$ minus the current payment. In row 10, B10 = PMT[t] is the installment payment at time t; C10 = I[t] is the interest payment at t; D10 = Ppaid[t] is the principal paid at t; and E10 = B[t] is the amount of principle outstanding at period t. We use brackets [t] to denote the period domain of the variable.[3]

Enter the following formulas and functions:

B12:	= PMT(r, n, −P)	r = i/12
C12:	= r*E11	interest paid, same as iPMT(.)
D12:	= B12-C12	principal paid, same as pPMT(.)
E12:	= E11-D12	balance of P outstanding

Column G gives the related Excel functions.

Calculation

The loan starts from \$10,000 in E11 in period t = 0, shown in A11. In the next period, t = 1, John pays

$$\text{PMT}(0.06/12, 36, -10000) = 304.22,$$

which is shown in B12. The payment includes

$$\text{I}[1] = \text{r}^*\text{B}[0] = (0.06/12)^*10{,}000 = 50.$$

[2]If actual dates of payment are desired, simply enter the date of borrowing, like 3/15/2008, in A11, and the first day of payment, 4/15/2008 in A12. Select both dates and use fill-handle to drag them down to A47.
[3]Thus, PMT(.) is different from PMT[t] in its implication.

Table 9.6 Loan amortization schedule

	A	B	C	D	E	F	G
1	**Loan Amortization Schedule**						Explanation
2		Loan Amount		P	10000	10000	=PV(r,n,-pmt), see (9.12), pmt = 304.22
3		Interest rate		i	0.06	0.06	=RATE(n,-pmt,P)*12, see (9.22)
4		Monthly Interest rate		r	0.005	0.5%	=RATE(n,-pmt,P), see (9.22)
5		Months		n	36	36	=NPER(r,-pmt,P), see (9.23)
6		Beginning date					r = i/12. P = pv
7		Payment	Interest	Principal	Bal of P		P=pv
8	Formula	PMT(.)	iB[t-1]/12	PMT[t]	B[t-1]		
9	t			-I[t]	-PMT[t]		
10		PMT[t]	I[t]	Ppaid[t]	B[t]		
11	0				10000		Formulas and equations
12	1	304.22	50.00	254.22	9745.78	B12	=PMT(r,n,-P)=PMT[1]
13	2	304.22	48.73	255.49	9490.29	C12	=r*E11 =iPMT(1)
14	3	304.22	47.45	256.77	9233.52	D12	=B12-C12 =pPMT(1)
15	4	304.22	46.17	258.05	8975.47	E12	=E11-D12
16	5	304.22	44.88	259.34	8716.13		
...	...	...	...	...	...		
32	21	304.22	23.33	280.89	4385.82		
33	22	304.22	21.93	282.29	4103.53		Example 9.14
34	23	304.22	20.52	283.70	3819.82	(C36) =iPMT(0.06/12,25,36,-10000)=iPMT[25]	
35	24	304.22	19.10	285.12	3534.70	(D36) =pPMT(0.06/12,25,36,-10000)=pPMT[25]	
36	25	304.22	17.67	286.55	3248.16	$17.67	$286.55
37	26	304.22	16.24	287.98	2960.18	C39 =iPMT(r,t,n,-P), see (9.20), t=28=A39	
38	27	304.22	14.80	289.42	2670.76	D39 =pPMT(r,t,n,-P) see (9.21), t=28 =A39	
39	28	304.22	13.35	290.87	2379.90	$13.35	$290.87
40	29	304.22	11.90	292.32	2087.58	$11.90	$292.32
41	30	304.22	10.44	293.78	1793.79	$10.44	$293.78 calculated from Excel formulas
42	31	304.22	8.97	295.25	1498.54	$8.97	$295.25 PMT[t] = iPMT[t] + pPMT[t]
43	32	304.22	7.49	296.73	1201.82	$7.49	$296.73 (9.18) (9.20) (9.21)
44	33	304.22	6.01	298.21	903.61		
45	34	304.22	4.52	299.70	603.91		
46	35	304.22	3.02	301.20	302.71		
47	36	304.22	1.51	302.71	0.00		
48	Sum	10951.90	951.90	10000			
49		PMT[t]	I[t]	Ppaid[t]	B[t]		

This amount is shown in C12. It is the interest that John paid on the beginning balance B[0] = 10,000. Here, B[t − 1] = B[0] denotes the balance in the previous period. The difference between PMT and the interest paid is

$$\text{PMT}[1] - \text{I}[1] = 304.22 - 50 = 254.22.$$

This amount is shown in D12. It is the part of the principal of the loan that is paid back to the bank in t = 1. Hence, the outstanding balance of the loan in period t = 2 is:

$$\text{B}[0] - \text{P}[1] = \$10{,}000 - \$254.22 = \$9745.78.$$

It is shown in E12 at the end of the first period. Other rows for t = $2, 3, \ldots 36$, have the same interpretations.

Completing the amortization table: The total payment and financial cost

After entering row 12, copy B12:E12 to B13:E47. Note that we have hidden the rows from 17 to 31. The outstanding balance of the loan at t = 36 in E47 must be 0, showing that John has paid back the total amount of the loan. The sum $10,951.90 in B48 is the total amount of the 36 payments. It includes the **total financial (or interest) cost** of $951.90 in C48, and **the total principal paid,** $10,000 (the loan that John paid back to the bank) in D48. For convenience of reading, the column titles in row 10 are reproduced in row 49.

Chart

Figure 9.7 illustrates the calculation results in Table 9.6: PMT (column B), interest paid (column C), principal paid (column D), and the outstanding balance in italic (column E), which is measured from the secondary Y-axis, the axis labels of which are also shown in italic fonts. Note that the interest paid decreases steadily over time from $50.00 to $1.51. The principal paid increases steadily over time from $254.22 to $302.73. The outstanding balance decreases steadily from the loan of $10,000 to zero at the last period . The monthly payment (line PMT) remains constant at $304.22 over the 36 periods. Thus, the shaded part at each period has the same length (PMT(t)−Principal Paid(t) = I(t)) .

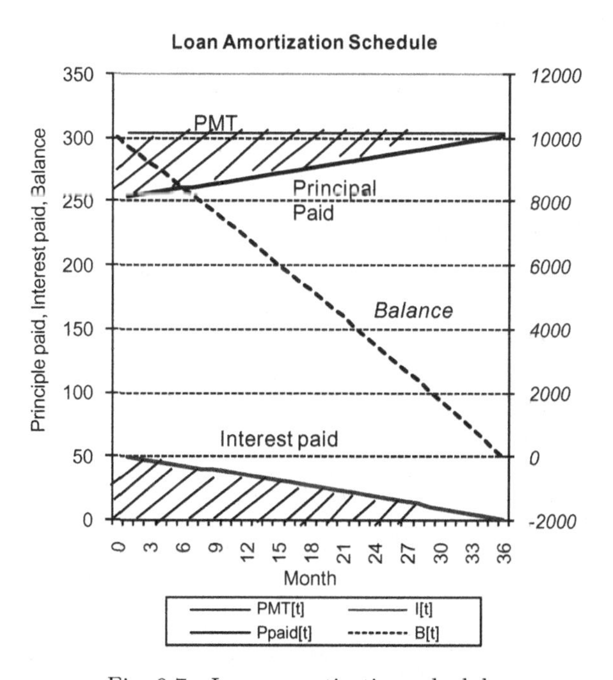

Fig. 9.7 Loan amortization schedule

9.5.3 *Other related financial functions in Excel*

In addition to the PMT(.) function (9.18) , Excel provides the following inverse functions of PMT(.). If function (9.18) is given, the independent variables, like r, n, pv, in the parentheses on the right-hand side can be "moved out" to become the dependent variables, and the dependent variable, shown as a function name (like FV, PMT, etc.), moves inside the parentheses. This is shown in (9.22) and (9.23) below.

iPMT, pPMT and installment payment decomposition

As shown in columns C and D in Table 9.6, the PMT function (9.18) can be decomposed into interest paid (I) and principal paid (P) for each period. Hence, (9.18) is the sum of (9.20) and (9.21):

$$\begin{array}{llll} & \text{PMT} = & \text{iPMT} + & \text{pPMT} \\ & 304.22 = & 13.35 + & 290.87, \\ \text{In Table 9.6} & \longrightarrow & \text{C39} & \text{D39} \end{array}$$

where iPMT is Excel's interest paid function,

$$\text{iPMT} = \text{iPMT}(\text{r}, \text{t}, \text{n}, \text{pv}), \tag{9.20}$$

and pPMT is the Excel's principal paid function,

$$\text{pPMT} = \text{pPMT}(\text{r}, \text{t}, \text{n}, \text{pv}). \tag{9.21}$$

They are functions of the same independent variables. These functions are useful, since whenever period t, along with r, n and pv, are known, then we can find out the interest paid and principal paid at any intermediate period t, without constructing a complete amortization table like Table 9.6. To show the use of the identity, we enter the formulas in F37 and F38 into F39 and G39 of Table 9.6, and copy down to F43:G43. Note that to obtain positive numbers, we enter pv = −P as a negative number. The entries in Range F39:G43 demonstrate that the calculations from the table and the calculations from functions (9.20) and (9.21) yield exactly the same results.

Excel functions for the interest rate and the term of loan

From the PMT function (9.18), PMT=PMT(r, n,pv), solving for the interest rate r, as a function of n, PMT, and pv, we have:

$$\boxed{\text{r} = \text{RATE}(\text{n}, \text{PMT}, \text{pv}),} \tag{9.22}$$

which is given in cells G3 and G4 in Table 9.6. For example, F4 can be derived from

$$\text{RATE}(36, -304.22, 10000) = 0.005 = 0.5\%,$$

which gives r, the monthly interest rate. Multiplying r by 12, we have the nominal interest rate of i = 6% per year in cell F3. Note that pmt must be negative, since it is the amount the borrower must pay out. If pmt is entered as positive, Excel will show an error #NUM!.

In addition, we can solve for n, the term of the loan, as a function of other variables:

$$\boxed{\text{n} = \text{NPER}(\text{r}, \text{PMT}, \text{pv}),} \tag{9.23}$$

which is given in cell G5 of Table 9.6. For example, F5 can be derived as

$$\text{NPER}(0.005, -304.22, 10000) = 36,$$

which is the term of loan n = 36 periods. Note that pmt must be entered as a negative number.

Example 9.15 iPMT and pPMT. In Example 9.13, as shown in Table 9.6, John would like to know, for tax purposes, after paying $304.22 for the 25th month of installment payments, what the amount of interest and the amount of principal that he paid in that month were. How should he calculate without constructing the whole loan amortization schedule?

Answer: We can compute by using the iPMT and pPMT functions, and the results should be the same as those in Amortization Table 9.6. From function (9.20), we have

$$\text{iPMT}(0.06/12, 25, 36, 10000) = -\$17.67,$$

and from function (9.21), we have

$$\text{pPMT}(0.06/12, 25, 36, 10000) = -\$286.55.$$

Since John received his 10,000 loan, pv = P has been positive, and results have been negative, showing that he paid $17.67 as interest and $286.55 as principal. They sum up to −$304.22. In F36 and G36 of Table 9.6, however, we enter pv = −P as negative numbers to obtain positive outputs. The results are the same as those in cells C36 and D36 in Table 9.6. □

Example 9.16 NPER. In order to encourage customers to pay off their debts in credit card accounts, a credit card company reduces the interest rate from 3% per month to 1% per month. If a family has a debt of $8000 and plans to pay back $500 per month to the credit card company, calculate the number of months that is needed for the family to pay off the debt at interest rates of 3% and of 1% per month, and then compare the results.

Answer: From function (9.23), at r = 0.03, we have,

$$\mathrm{NPER}(0.03, -500, 8000) = 22.$$

That is, the family will need 22 months to pay off the debt at r = 0.03. On the other hand, at r = 0.01, we have

$$\mathrm{NPER}(0.01, -500, 8000) = 17.5.$$

In the case of r = 1%, the family will need only 18 months to pay off the debt. They will be free of debt four and half months earlier. □

9.6 Summary

Figure 9.8 illustrates the relations among the topics covered in this chapter. This chapter returns to the discussions of time preference (in Economics) or time value of money (in Business). See the left-hand side of Fig. 9.8. In contrast to the simple future value problems, in which money can accumulate in the future, we also have studied present value problems, to measure the degree to which the future money is not as valuable as the current money (Sec. 9.1). Instead of simple future growth or present values (FV or PV), more interesting and useful cases are future or present value problems with annuity (FVA or PVA). It is very common in practice that people buy goods and services and pay by installments, or receive money or income by monthly payments, or invest in a project by issuing bonds with a fixed term of maturity to be paid off in the future. While the mathematical formulas for the future value and present value problems with annuity are quite complicated and hard to interpret intuitively (Secs. 9.2 and 9.3), the Excel spreadsheet program provides

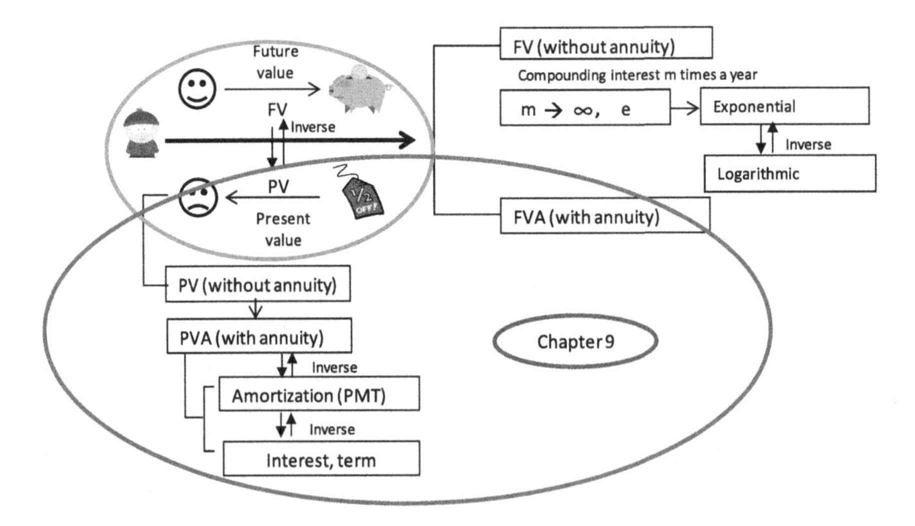

Fig. 9.8 Summary of Chapter 9

one convenient formula to cover future value problems, either with or without switching the "pmt", namely, the installment payment, on (entering pmt) or off (letting pmt equal zero). Similarly, Excel also has one formula for present value problems with or without annuity (Sec. 9.4).

We then introduced the Excel functions of the interest payment and the principal payment at each time period. Finally, using the concept of inverse function, these Excel formulas can be reversed to find the amount, the interest rate, and the term of payment, given the installment payment. The applications of these commands can be found in the loan amortization schedule Table 9.6, which completes our discussions in this chapter.

Appendix 9A. Derivation of FVIFA and PVIFA

Derivation of the future value with annuity interest factor (FVIFA) (Eq. (9.5))

If an annuity of \$1.00 per period is paid and accrues interest at a rate of r per period for n periods, the total sum is, from the terms inside the brackets of (9.4),

$$y_n = 1 + (1 + r) + (1 + r)^2 + \cdots + (1 + r)^{n-1}. \tag{9A.1}$$

Let $x = 1 + r$. Then, since

$$y_n = 1 + x + x^2 + \cdots + x^{n-1},$$

we multiply both sides by x,

$$xy_n = x + x^2 + \cdots + x^n,$$

subtract xy_n from y_n, and we have

$$y_n = (x^n - 1)/(x - 1). \tag{9A.2}$$

Substituting $x = 1 + r$ back into (9.A2), we have (9.5). □

Derivation of the present value problem with annuity interest factor (PVIFA) (Eq. (9.8))

In (9.7), let $x = (1 + r)^{-1}$. Then the terms inside the braces in (9.7) can be written, as in (9A.2), as

$$[.] = x\left(1 + x + x^2 + \cdots + x^{n-1}\right) = x\frac{x^n - 1}{x - 1}.$$

Substituting x back into the above expression, we have

$$[.] = (1 + r)^{-1}\frac{(1 + r)^{-n} - 1}{(1 + r)^{-1} - 1} = \frac{1 - (1 + r)^{-n}}{r},$$

which is (9.8). □

Review of Basic Equations and Formulas

Mathematical equations	Equivalent Excel functions
Future value problem	
(9.1) $F_t = P\left(1+\frac{i}{m}\right)^{mt}$	(9.9) FV = FV(r,n,pmt[,pv][,type])
or $F_n = P(1+r)^n = PV * FVIF_n$	(9.10) $FV_n = FV(r, n, 0, pv)$
$r = i/m$, $n = mt$.	$= PV * FV(r, n, 0, -1)$
	$\equiv PV * FVIF_n$
Present value problem	
(9.2) $P_t = \frac{F}{(1+i/m)^{mt}} = F(1+i/m)^{-mt}$	(9.12) PV = PV(r, n, pmt[, fv.][, type])
or	(9.13) $PV_n = PV(r, n, 0, fv)$
$P_n = \frac{F}{(1+r)^n} = F(1+r)^{-n} = FV * PVIF_n$	$= FV * PV(r, n, 0, 1)$
	$\equiv PVIF_n$
(9.3) $P = Fe^{-rn}$	(9.3) $P = \exp(-r * n)$
Future value problem with annuity	
(9.5) $FVA_n = PMT\frac{(1+r)^n - 1}{r}$	(9.11) $FVA_n = FV(r, n, pmt)$
	$= PMT * FV(r, n, -1)$
$FVA = PMT * FVIFA_n$	$\equiv PMT * FVIFA_n$
Present value problem with annuity	(9.14) $PVA_n = PV(r, n, pmt)$
(9.8) $PVA_n = PMT\frac{1-(1+r)^{-n}}{r}$	$= PMT * PV(r, n, 1)$
$= PMT * PVIFA_n$	$\equiv PVIFA_n$
Amortization of the debt	(9.18) PMT = PMT(r, n, pv[, fv][, type])
(9.16) $PMT = PVA\ \frac{r}{1-(1+r)^{-n}}$	= PMT(r, n, pv)
	Decomposition of PMT
	PMT(r,n,pv)
	= iPMT(r, t, n, pv) + pPMT(r, t, n, pv)
	(9.22) r = RATE(n, PMT, pv)
	(9.23) n = NPER(r, PMT, pv)

Key terms: Economics and Business

Key terms: Excel

The Excel functions are listed in Review of Basic Equations and Formulas

Homework Chapter 9

Present value problems

9-1 Present value In your freshmen year in high school, your parent decided to buy a CD (certificate of deposit) for your college expenses. They purchased a CD that would pay $40,000 in 4 years, with the bank charging an interest rate on the loan at 7% compounded continuously. (a) What is the present value of this CD? (b) If the interest rate is 5%, what is the present value of the CD? (c) What is the present value if the interest rate has risen to 9.5% ? Construct a table to show the three cases. (d) What is the meaning of "present value" in these cases?

9-2 A Simple Loan Analysis John plans to buy a Saturn for $14,000 with a $4000 down payment. He wants to borrow $10,000 from the Arapahoe State Bank in town, with monthly installments of about $300 for 3 years. The current lending rate at the Arapahoe is 6% per year. Answer in the blank on the right-hand side of Table HW9-2 by changing "current" column and copying it to the column of (a), (b), and (c), respectively.

(a) If he makes an extra effort to shop at another bank for a lower interest rate, say, 5%, how will his monthly payment be reduced?
(b) He also wonders, if at the same time the term of payment increases to 4 years, how much more would his monthly payment be reduced.
(c) With reduced monthly payments, he now figures that he can afford to decrease his down payment from $4000 to $3000, and borrow $11,000 from the bank. What is his monthly payment now?

(Hints: Set up Table HW9-2, change the original calculation to values, and obtain the current calculation, copy it to the RHS of the table to answer the questions systematically by denoting (a), (b), (c) for comparison. We use rr to avoid the use of r_. nn is used to correspond to the form of rr)

Table HW9-2 Mortgage loan analysis

Title	Mortgage Loan Analysis - Sensitivity Analysis									
Date	8/11/2010		=TODAY()	i = 0.06	i=5%		i=5%, t=4		i=5%, t=4, pv=11000	
Name	Input	Current	Original	diff	(a)	diff	(b)	diff	(b)	diff
Interest Rate	r	0.0041667	0.005	-0.08%						
# of Months	n	36	36	0.00						
Loan Amount	Pv	$10,000.00	$10,000.00	0.00						
Loan Payment	pmt	($299.71)	($304.22)	4.51						
	Formula	=PMT(r,n,pv)	=PMT(i/12,3*12,pv)							
Current = current calculation of each problem, diff = current - original										

9-3 The formula method for the sensitivity table Table HW9-2 is rather fragmental. To make a more systematic comparison of the payments, we may construct the following sensitivity table. Enter interest rates from 5, 5.25, ... to 8% in the leftmost column, and term of payment $n = 30, 32, \ldots 48$ in the upper row.

(a) Enter the PMT(.) formula and use the method of relative references to construct the table.
(b) Use the naming method (see Table 9.3) to construct the table.
(c) How do you find the answer to question (c) in HW9-2?

Loan amortization problems

9-4 A car loan John has decided to borrow $20,000 from the Boulder Saturn dealer to buy a Saturn. He then pays off the loan in equal monthly installments for 4 years at 3.9% per year.

(a) Construct a loan amortization table showing his monthly payment, interest paid, the principal paid, and the remaining principal, for each month for 4 years.
(b) What is his monthly payment to the dealer?
(c) What is his total payment over 4 years?
(d) What is the financial cost of his borrowing?
(e) Draw the combination chart showing interest paid and principal paid on the primary axis, and the outstanding balance on the secondary axis. Make sure all the axes and titles are properly labeled.
(f) Boulder Saturn stipulates that after 20 months, the borrower may return the remaining loan with a penalty of 2% of the remaining loan. Let us say that when he graduates, after two years, John gets a well-paying job, and decides to change the car after he has paid installments for 24 months. How much should he pay back to the dealer? Please show your calculation.
(g) Is it worthwhile to pay back the loan after 2 years with the penalty? Please show your calculation.
(h) Given the penalty of 2% of the remaining loan, after which month will it become worthwhile for John to keep the loan?

9-5 Buying a house Dan has just moved to a college town as a freshman. His parents have decided to buy a two-bedroom condominium, listed as $185,000, east of campus. They borrow $120,000 from a bank to buy the property. Assuming a 30-year mortgage at 6% fixed interest rate, answer the following questions.

(a) Construct a loan amortization table showing his monthly payment, the interest paid, the principal paid, and the remaining balance of the principal for each month of the first four (4) years.
(b) What is his monthly payment to the bank?
(c) What is his total payment over the first 4 years?
(d) What is his interest cost of borrowing over the first 4 years?
(e) How much is the balance of the principal to be paid back after 4 years?
(f) Draw a chart showing the monthly payment, interest paid, and principal paid on the primary Y-axis, and the outstanding balance on the secondary Y-axis. Make sure all three axes and four lines are properly labeled.
(g) If the house price is expected to increase at 2% per year for 4 years, and Dan will sell the house after 4 years, how much can he sell the house for? Calculate by both discrete and continuous compounding methods.
(h) Using the discrete method, what is his capital gain (the sale price minus the purchase price) from owning the house, assuming no other costs?

9-6 Installment payment decomposition Continuing from HW9-5, in April of the third year, Dan would like to find the total interest payment during the second year to make an itemized deduction of his housing cost for his tax returns. (a) Find the total interest paid during the second year WITHOUT constructing the loan mortgage schedule. (b) How much of the loan does he return to the bank during the second year? (c) Are the interest cost and principal returned to the bank the same as those derived from the mortgage schedule in HW9-5?

9-7 Renting out the house Continuing from question HW9-5, Dan has made a down payment of $65,000. If he had put the down payment in the bank, he could have earned 3% interest rate per year on it.

(a) What is his opportunity cost of four years for not putting $65,000 in the bank?
(b) What is his total financial cost (interest paid plus interest lost from the down payment)?
(c) Is it worthwhile for Dan to own the house?
(d) If Dan rents out one bed room for $500 per month from the first month when he bought the house, and the rent does not change over the 4 years, what is his net financial cost of owning the house?

9-8 To rent or to buy Continuing from HW9-5, if Dan does not buy a house, he has to rent an apartment for $600 a month for the first 2 years and for $800 a month for the next 2 years. If the house price does not change, and the bank lending rate is still 6%, and he does not rent part of the house out, should he rent or buy?

9-9 A short amortization table (a) To avoid a long table, reconstruct Table 9.6 for t = $1, 2, \ldots$ 12, 24, 36, omitting the interim time periods. (b) Without using the amortization table, can you calculate the total amount of payments and financial cost? (Hints: (b) use the PMT function).

9-10 An annuity plan An investment company offers retirement annuity plans as follows. For $100,000, it pays $545 per month until age 70, and after age 70, it pays $605 per month. The term of payment is guaranteed for 20 years, that is, if the person dies before the end of 20 years, the annuity goes to the beneficiaries until the end of 20 years, at which time the principal will be exhausted. If the person retires at age 65 and survives for 20 years, the annuity continues until the person dies. This is the same problem as amortization, in which the company borrows $100,000 and pays it back to the retiree by installments.

(a) What are the interest rates the company pays before and after age 70?
(b) Set up the company's amortization table for the retiree before age 70. How much principal should the company return to the retiree at age 70?
(c) Set up the company's amortization table for the retiree after age 70. Would the company return all the principal to the retiree at age 85?
(d) How much can a retiree gain from lending money to the company?
(e) From your understanding of the mechanism of annuity calculation, what are the pros and cons of buying an annuity plan?
(f) How can the investment company earn profits from such arrangement?

(Hints: (a) Apply the Excel RATE(.) function. 2.8% per year from 60 to 70, 4.33% for 70 to 85. (b) $79,988.16. This amount will be reinvested and continue to age 85.)

9-11 Investment Instead of the annuity plan in Question HW9-10, suppose the person deposits the principal of $100,000 in a bank, which offers a 5% interest rate for 20 years converting monthly. Will he still be better off putting money in the investment company, and withdrawing $600 per month for living expenses? How much will he gain or lose? Show your calculation.

(Hints: For n = 12 × 20, from (9.2) or (9.10), FV = 271,264.)

Chapter 10

Economic Policy Analysis — Vectors and Matrices

Chapter Outline

Objectives of this Chapter

Up to now, we have been dealing with economic and business models with only a small number of variables. When the number of variables becomes large, it is almost impossible to write each equation and each variable. In this case, vectors and matrices are very useful. They are a compact way of writing a large system of simultaneous linear equations. Since vectors and matrices are new mathematical entities with their special operations, this chapter starts from an introduction of some basic operations of vectors and matrices. Analogy with the real number system should be noted. For those readers who are already familiar with them, they may review the basic operations quickly, and familiarize themselves with powerful Excel functions, such as sum and product of matrices, matrix multiplication, and matrix inversion. For those who are not familiar with matrix algebra, this chapter will give an excellent learning experience, especially from the Excel point of view. Readers may like to consult with other textbooks on matrix algebra for more advanced topics. After a brief introduction of the vector and matrix operations, we show how to use Excel functions to find the solution to a system of linear equations, represented by simple market models and macroeconomic models. These come very easily and instantly with Excel functions. In fact, not too long ago in the premicrocomputer era, people spent many hours just to calculate the inverse matrix of a 5×5 matrix!

An extension of the simple system of linear equations, by changing the parameters and exogenous variables, will lead to economic policy models, which essentially work like the comparative static analysis we have studied in Chapter 4. The vision of Wassily

Leontief of using vectors and matrices to represent input and output relations among the interdependent industries, not only gives the elements of vectors and matrices "meat and bones" (they are no longer a mere collection of dry numbers), but also is a great contribution to economic analysis and planning. Thus, this chapter concludes by introducing the essence of the input–output analysis. We will show that input–output analysis is also a simple extension of the method of solving a system of simultaneous equations, and can be handled by Excel functions easily.

10.1 Vectors

10.1.1 *Why vectors and matrices?*

In Chapter 3 and 4, we have seen a market model with three equations and three endogenous variables (D, S, and P) for a single commodity. In the real world, there are many commodities in the markets. Each commodity has its own demand and supply equations and equilibrium condition. Thus for n commodities, we have $3 \times n = 3n$ equations. If n is large, the number of equations will be so large that it is inefficient to write each of the 3n equations. This motivates us to use vectors and matrices to convert a linear equation system into a compact matrix form.

To illustrate the procedures, we will use the example of a simple market model in Chapter 3. Let

$$D = 18 - 2P, \quad S = -6 + 6P, \quad \text{and} \quad D = S. \tag{10.1}$$

To convert this system into matrix form, we first rearrange the equations such that the terms with endogenous variables D, S, and P are collected on the left-hand side and the constants on the right-hand side as follows:

$$\boxed{\text{D} \quad \text{S} \quad \text{P} \quad \text{constant}}$$

$$\begin{aligned} 1D + 0S + 2P &= 18 \\ 0D + 1S - 6P &= -6 \\ 1D - 1S + 0P &= 0. \end{aligned} \tag{10.2}$$

Next, collect the coefficients of the variables and write them into a row and column form; collect variables D, S, and P into a column vector; and collect the three constants into a column vector. Now we have the system of linear equations in matrix form:

$$\underset{\substack{\text{Matrix}\\ A}}{\overset{\substack{\text{coefficients of variables}\\ D \quad\ S \quad\ P}}{\begin{bmatrix} 1 & 0 & 2 \\ 0 & 1 & -6 \\ 1 & -1 & 0 \end{bmatrix}}} \underset{\substack{\text{vector}\\ x}}{\begin{bmatrix} D \\ S \\ P \end{bmatrix}} = \underset{\substack{\text{vector}\\ c}}{\overset{\text{constant}}{\begin{bmatrix} 18 \\ -6 \\ 0 \end{bmatrix}}} . \tag{10.3}$$

Here, the first group of numbers in the brackets is called a **coefficient matrix A**, the second column of variables is called a **variable vector x**, and the third column of numbers is called a **constant vector c**. Thus, the system of linear equations may be simplified by writing it as

$$\mathrm{Ax} = \mathrm{c}. \tag{10.4}$$

The process is based on scalar products of matrices, which will be explained in Sec. 10.2. The study of vectors and matrices is a mathematical field called **Matrix Algebra**, a part of Linear Algebra. In this chapter, we assume that the readers have studied **Linear Algebra** either in high school or in college mathematics courses, which will be helpful in learning matrix algebra. Thus, what we would like to present here is a review of some important basic properties of vectors and matrices.

Vectors

Let $a_1, \ldots, a_n$ be a set of n real numbers. An ordered n numbers in brackets or parentheses,

$$a = \begin{bmatrix} a_1 \\ a_2 \\ \vdots \\ a_n \end{bmatrix} \qquad a' = a^T = (a_1\, a_2 \cdots a_n), \tag{10.5}$$

is called a **vector of order n** or an **n-vector** with **elements** (or **component** or **coordinate**), $a_1, a_2, \ldots, a_n$. Written vertically, it is a **column vector**, horizontally, it is a **row vector**. For clarity, elements may be separated by a comma. A column vector can be changed to a row vector, and vice versa, by the **transpose** operator, as in (10.5). The mathematical convention is to write a vector as a column vector, usually denoted by a lower case letter and in boldface. In this book, we also write a vector in column form, but do not use the boldface font. In the above example, a is a column vector, and its transpose is denoted by "a'" or "a^T", which is a row vector. The transpose of a row vector a' is a column vector, by $(a')' = a$. To save space, when it is necessary to write the contents of a column vector, we write $a = (a_1, a_2, \cdots, a_n)'$. On a spreadsheet, a vector is a range of cells either in a single column or in a single row.

A vector may be represented by its ith element and can be written as $a = [a_i]$. When only one of the elements in a vector takes the value of one and the rest of elements take the value of zero, then the vector is called **a unit vector**. If all the elements take the value of one, then it is called **the sum vector**.

Geometrically, a vector may be represented by an arrow from origin to a point in space. The first element is on the X-axis and the second element is on the Y-axis, and the two elements together determine the position of the point in a two-dimensional space. Similarly, an n-vector is a point in an n-dimensional space.

The two column vectors a and b are defined in Table 10.1, C4:C8 and D4:D8, respectively. They may be entered by typing in the numbers, or may be entered as rounded

Table 10.1 Vector operations

	A	B	C	D	E	F	G	H	I	J	K	L	M	N
1	**Vector Operations**													
2	Elements		Col vectors				Row vectors							
3	#		a	b				=ROUND(RAND()*10,0)						
4	1		0	2										
5	2		1	5		a'=	aT	0	1	7	1	8		
6	3		7	8		b'=	bT	2	5	8	0	7		vectors
7	4		1	0										
8	5		8	7										
9														
10	(10.5)			(10.7)		(10.7)		(10.8)		(10.10)				
11	Col vectors			Addition		Subtraction				Scalar Product				
12	a	b		a+b		a-b		ka		a'b		a'a	b'b	(a-b)'(a-b)
13	ai	bi		ai+bi		ai-bi		3ai		aibi		ai^2	bi^2	(ai-bi)^2
14	0	2		2		-2		0		0		0	4	4
15	1	5		6		-4		3		5		1	25	16
16	7	8		15		-1		21		56		49	64	1
17	1	0		1		1		3		0		1	0	1
18	8	7		15		1		24		56		64	49	1
19														
20	**Three methods of finding a scalar product**													
21	**a. Sum method**											↓	↓	↓
22	(10.10) J22: =sum(J14:J18)									117		115	142	23
23	**b. Unnamed matrix multiplication method**													
24	(10.10) J24: =MMULT(H5:L5,D4:D8) →									117		115	142	23
25	**Named matrix multiplication method**													
26	(10.10) J26: =MMULT(aT,b)					a'b=b'a →				117		115	142	23
27	**c. SUMPRODUCT method**													
28	(10.10a) J28: =sumproduct(a,b) →									117		115	142	23
29	(10.10a) J29: =sumproduct(aT,bT) →									117		115	142	23

random numbers and changed to values.[1] It is suggested that the numbers be enclosed by borders. Excel has a special transpose command. To take the **transpose of** a, select C4:C8 and enter ˆC. Move the pointer to a blank space, say, H5, and enter

$$\text{<"Home"><Clipboard, Paste!><Transpose>.} \tag{10.6}$$

The transpose of vector a is a row vector a′ in H5:L5. Similarly, the transpose[2] of column b is a row vector b′ in H6:L6. Name all four vectors as a, b, aT, and bT. Note that we cannot name H5:L5 "a′", since Excel considers "a′" to be a. Thus, instead of using "a′", we should

[1]To **change to values**, select vectors a and b, and ˆC <"Home"><Clipboard, Past!><Paste Values>.

[2]Since two vectors are adjacent, we may take their transpose simultaneously. Selecting C4:D8 in Table 10.1, enter ˆC, and moving pointer to H5, enter (10.6), we have H5:L6. This is the same as the matrix transpose.

name it aT. Similarly, we name b' as bT. In addition, since it is hard to type superscript and subscript in an Excel spreadsheet, we should type a_i as ai, b_i as bi, etc.

10.1.2 *Vector operations*

As vectors and matrices are new algebraic entities, like the real number system, we have to define the vector and matrix operations. Let two column n-vectors be $a = (a_1, a_2, \ldots, a_n)'$ and $b = (b_1, b_2, \ldots, b_n)'$.

a. Equality: a and b are equal, that is, $a = b$, if and only if $a_i = b_i$ for all $i = 1, 2, \ldots, n$. That is, each and every corresponding element of the two vectors is equal.

b. Addition and subtraction: **Addition** of two vectors is defined as

$$a + b \equiv [a_i + b_i] = \begin{bmatrix} a_1 + b_1 \\ \vdots \\ a_n + b_n \end{bmatrix}. \tag{10.7}$$

Each and every corresponding element is added together in the resulting vector. Similarly, in the case of **subtraction**, $a - b$, we will take each element in the first vector a_i minus the corresponding element in the second vector b_i to obtain the resulting vector.

c. Scalar multiplication: Let k be a scalar (a real number). Then, **scalar multiplication** is defined as

$$ka \equiv [ka_i]. \tag{10.8}$$

That is, every element in a vector is multiplied by k in the resulting vector.

Implementation of vector operations

To implement the operations of $a + b$ and $a - b$ of (10.7), enter

$$\begin{array}{ll} \text{D14: A14+B14} & \text{for } a_1 + b_1, \\ \text{F14: A14−B14} & \text{for } a_1 - b_1, \end{array} \tag{10.9}$$

and copy to D15:D18 and F15:F18, respectively. Note that if the operation is on the rows that have the same names as a and b (say, in a blank space like cell N4 in row 4), we might simply enter $a + b$ in N4. Then Excel would remember the positions of the corresponding elements in a and b, and would show 2 in N4. Copying N4 down to N5:N8, we would have the addition. Otherwise, we would have to enter the cell address in D14 and F14 as shown in (10.9).

The scalar multiplication (10.8) and scalar product (10.10) are implemented as follows:

H14 :	=3*A14	$3 * a_1$
J14 :	=A14*B14	$a_1 * b_1$
L14 :	=A14^2	a_1^2
M14 :	=B14^2	b_1^2
N14 :	=(A14–B14)^2	$(a_1 - b_1)^2$

We enter the desired vector operation on the first element of the vector. Then we copy it down to rows 15–18. Row 13 shows that the new vector below each entry is obtained by operating on the individual elements of two vectors. For example, the column label $a_i + b_i$ in D13 implies that each element in vector D14:D18 is an addition of a_i in vector a and b_i in vector b, $i = 1, 2, \ldots, 5$.

10.1.3 *Scalar product*

The scalar product of a and b

The **scalar product** of the two vectors a and b is defined as

$$a'b \equiv a^T b = [a_i]'[b_i] = a_1 b_1 + a_2 b_2 + \cdots + a_n b_n = \sum_i a_i b_i = b'a. \tag{10.10}$$

This is called a **row–column multiplication**. Each element in a is multiplied by the corresponding element in b, and the result is summed up to a scalar, so it is called a **scalar product**. Equivalently, a'b may also be written as a·b or b·a (where both a and b are defined as column vectors) and called a **dot product**. The Excel command for the scalar product for vectors a and b, where both are n-vectors (having the same number of elements), is[3]

$$=\text{SUMPRODUCT}(a, b). \tag{10.10a}$$

When we use this command, we must type a and b both in column vectors or both in row vectors; one cannot be a column vector and the other a row vector. The scalar product will play a crucial role in the definition of matrix multiplication below.

Numerical examples

There are three methods of finding a scalar product in Excel, as shown in rows 20–29. For quick reference, we have shown the equation number and the Excel cell formula in the bordered cells of column J in the corresponding rows of the bordered cells.

[3]If a, b, and c are same n-vectors, the SUMPRODUCT(a, b, c) will return $\sum_i a_i b_i c_i$ for $i = 1$ to n. It can include up to 30 n-vectors (arrays).

a. The **sum method**: In J22, L22, M22, and N22 we sum the ranges J14:J18, L14:M18, M14:M18, and N14:N18, respectively. They are the scalar products $a'b, a'a, b'b$, and $(a-b)'(a-b)$, respectively, as shown in row 12.

b. The **matrix multiplication method**. This method uses the Excel function =MMULT(·). This is a row–column multiplication. Hence, the first vector a is a row vector and the second vector b should be a column vector. If the ranges are not named, enter the Excel function below:

J24:	=MMULT(H5:L5, D4:D8)	$=a'b$
L24:	=MMULT(H5:L5, C4:C8)	$=a'a$
M24:	=MMULT(H6:L6, D4:D8)	$=b'b$
N24:	=MMULT(H5:L5−H6:L6, C4:C8−D4:D8)	$=(a'-b')(a-b)$.

They are the (row range)∗(column range). The N24 in the last column uses (range–range) format before multiplication. If the ranges are named, then it is much easier and less confusing to use the named ranges, as shown below:

J26: =MMULT(aT, b)
L26: =MMULT(aT, a)
M26: =MMULT(bT, b)
N26: =MMULT(aT − bT, a − b).

This method is better in the sense that it shows the contents of the operation clearly.

c. The **sumproduct method** In this case, enter in the following cells:

J28: =SUMPRODUCT(a, b) or = SUMPRODUCT(b, a)
L28: =SUMPPRODUCT(a, a)
M28: =SUMPRODUCT(b, b)
N28: =SUMPRODUCT(a − b, a − b).

Among these three methods, the sumproduct method is probably the easiest method.

Example 10.1 A cost function. The cost function is defined as $C = wL + rK$, where C, L, and K are cost, labor, and capital, and w and r are wage rate and price of capital, respectively. All variables are positive numbers. Let us assume $w = 2, r = 1.5$, and $C = 30$. Write the cost function in vector form. Then draw the cost line, find its intercepts and slope, and indicate them in the chart.

Answer: We can write $C = (w, r)(L, K)' = (2, 1.5)(L, K)' = 2L + 1.5K = 30$.

Solving for K in terms of L, we have $K = 20 - 4L/3$. This cost function solved explicitly for K is drawn for various L in Fig. 12.1 in Chapter 12. The slope of the cost function is, from Section 3.3.3, $-4/3$. □

Another useful application of the sumproduct function is to find various sums of squares. In Table 7.4, let us name column B6:B23 as C_, and C6:C23 as X, also name the mean of

C in B26 as mC, the mean of X in C26 as mX, the coefficient a in I9 (or I4) of Table 7.3 as a, and b in H9 (or H4) of Table 7.3 as b. Then, we have

$$\begin{aligned}\text{SST} &= \text{sumproduct}(\text{C\_}-\text{mC}, \text{C\_}-\text{mC}) = 23{,}466{,}162,\\ \text{SSE} &= \text{sumproduct}(\text{C\_}-\text{a}-\text{b}*\text{X}, \text{C\_}-\text{a}-\text{b}*\text{X}) = 9649,\\ \text{SSR} &= \text{sumproduct}(\text{a}+\text{b}*\text{X}-\text{mC}, \text{a}+\text{b}*\text{X}-\text{mC}) = 23{,}456{,}513,\end{aligned}$$

where $\widehat{C} = a + bX$. Thus, we can calculate the sums squares without using a calculation table like columns F6:H23 in Table 7.4.

10.1.4 *Application to price indexes*

The sumproduct function can be applied for calculation of various price indexes, as shown in the following examples.

Example 10.2 The value of a commodity basket. Let p_{it} be the price of commodity i at time t and $i = 1, \ldots, 10$. Let q_{it} be the quantity of commodity i purchased at time t. Let p_{it} be a random number from 1 to 10 and q_{it} be from 1 to 100. Write the prices of 10 commodities in column vector form p^t and their quantities in column vector form q^t. Find the value of the basket of 10 commodities, $v = p^{t\prime}q^t$.
Answer: Since t can be arbitrary, we take t = 0 in Table 10.2. p^0 and q^0 for the price and quantity vectors for 10 commodities at time t = 0 (base period) are shown in columns B and C. The value of the basket of 10 commodities is the sum of the column of $p^{0\prime}q^0$, and is 1926 in F12. The scalar product can be calculated by the sum method or the sumproduct method. □

Table 10.2 Calculation of price indexes, TC, La, Pa, and Fi

	A	B	C	D	E	F	G	H	I
1	Com#	p0	q0	p1	q1	p0q0	p1q1	p0q1	p1q0
2	1	2	90	7	91	180	637	182	630
3	2	9	51	3	44	459	132	396	153
4	3	2	49	6	43	98	258	86	294
5	4	9	42	10	72	378	720	648	420
6	5	0	8	6	38	0	228	0	48
7	6	6	65	7	98	390	686	588	455
8	7	2	70	8	55	140	440	110	560
9	8	2	43	8	24	86	192	48	344
10	9	3	17	8	16	51	128	48	136
11	10	6	24	7	12	144	84	72	168
12	sum					1926	3505	2178	3208
13	Total Cost inde		181.98			181.98			
14	Laspeyres		166.56			166.56			
15	Paasche		160.93					160.93	
16	Fisher					163.72			

Example 10.3 Price indexes. Let t = 0 as the base period and t = 1 as the current period. Using p^t and q^t, the two vectors defined in Example 10.2, we have the following definitions of price index numbers:

Total cost index[4] $$\boxed{TC = \frac{p^{1\prime}q^1}{p^{0\prime}q^0}} \tag{10.11}$$

Laspeyres price index $$\boxed{L_a = \frac{p^{1\prime}q^0}{p^{0\prime}q^0}} \tag{10.12}$$

(use q^0 as the weight; Example: the Consumer Price Index)

Paasche price index $$\boxed{P_a = \frac{p^{1\prime}q^1}{p^{0\prime}q^1}} \tag{10.13}$$

(use q^1 as the weight; Example: the GNP deflator)

Fisher price index $$\boxed{F_i = (L_a * P_a)^{1/2} = \sqrt{\frac{p^{1\prime}q^0}{p^{0\prime}q^0}\frac{p^{1\prime}q^1}{p^{0\prime}q^1}}} \tag{10.14}$$

The individual price ratio of commodity i between period 0 and period 1 is p_{i1}/p_{i0}, for $i = 1, 2, \ldots, 10$. Thus, L_a is the ratio of prices at time t and at time 0, both prices being weighted by quantities at base period t = 0, i.e. vector q^0. P_a is the price ratio of prices at time 1 and at time 0, both prices being weighted by quantities at period t = 1, i.e. vector q^1.

The prices and quantities of 10 commodities for t = 0 and t = 1 are given in columns B–E in Table 10.2. They are generated by random numbers 0 to 10 and 0 to 100 for price and quantity, respectively. Naming the four columns p0, q0, p1, and q1 as their column titles, and using the sumproduct method or the sum method introduced in Sec. 10.1.3, we can calculate the above four price indexes as shown in rows 12 to 16 of Table 10.2.

---

[4]The formula of the total cost index is also called the **value index**, and is the same as the **Standard and Poor's (S&P) 500 Composite Index** of stock prices. This index comprises 500 firms in industry, utilities, transportation, and finance, and is defined simply as the market value of the 500 stocks at the period divided by the market value of the 500 stock prices at the base period (1941–1943 = 100). See Lee *et al.* (2000), Chapter 19.

For example, the sumproduct method for calculating L_a is, after naming,

$$L_a = \text{SUMPRODUCT}(p1, q0)/\text{SUMPRODUCT}(p0, q0) = 166.56,$$

which is shown in C14. When we use the sum method, the same result of $L_a = 166.56$ is shown in F14, which is the ratio of the sum of the entries of the p1q0 column (3208 in I12) divided by the sum of the entries of the p0q0 column (1926 in F12)). $L_a = 166.56$ means, over all, the prices increase 66.56% in period 1 compared with the prices in the base period 0, according to Laspeyres price index formula.

From Table 10.2, we can also see that different price index formulas yield different results. When we use the Total Cost index formula, the prices increase by 81.98%; If the Fisher price index formula is used, the prices increase by 63.72%; and with the Paasche price index formula, the prices increase by 60.93% in period 1 as compared with period 0. □

Example 10.4 The sum vector. The Dow Jones Industrial Average (DJIA) index comprises only 30 representative firms in industry, and was originally simply the unweighted index of the 30 stock prices.[5] If 1 is the sum vector, then the DJIA is defined simply as $DJ_t = 1'p^t/d_t$, where p^t is the closing stock prices of the 30 representative companies in the United States, and d_t is the **Dow divisor**, which originally was the number of the companies, but the current divisor has made many adjustments for stock splits, company spinoffs, dividends, etc. For our purpose, however, we simply define it as $d_t = 1'p^0$, that is, the sum of the original stock prices of the 30 companies. Thus, we have

$$\boxed{DJ_t = \frac{1'p^t}{1'p^0},} \tag{10.15}$$

which is the ratio of the sum of prices of 30 stocks at time t and the sum of prices of the same stocks at time 0. Generate 30 stock prices for two periods, p0 and p1, by random variables ranging from 50 to 150, and construct the DJ index.

Answer: Simply extend p0 and q0 columns in Table 10.2 from 10 to 30, and use either the sum method or the sumproduct method.

DJIA is a very simple price index, and yet, despite many criticism, it is the most popular and most used index among many indexes. It shows that the construction of a price index need not be very complicated. □

10.2 Matrices

A matrix is a collection of vectors. Therefore, its properties are extensions of those of the vectors that we explained in the previous section.

[5]See Chapter 19, Lee *et al.* (2000); Chapter 18, Lind *et al.* (2002). Readers are encouraged to consult Wikipedia for more information.

Table 10.3 Matrix addition and scalar multiplication

	A	B	C	D	E	F	G	H	I	J	K	L
1	Equation number						(8)			(9)		
2	A			B			A+B			2A		
3	1	4		0	3		1	7		2	8	
4	2	5		1	4		3	9		4	10	
5	3	6		2	5		5	11		6	12	
6												

10.2.1 *Definitions*

A **matrix** is a rectangular array of real numbers, arranged between parentheses or brackets. It is the same as a range on a spreadsheet. In this book, it will be enclosed by borders, as shown in Table 10.3. The numbers are called the **elements of the matrix**. The subscripts of a_{ij} are position numbers showing that a_{ij} is the element at the ith row and the jth column. The **size** or the **order** of a matrix is expressed by

$$\text{(the number of rows)} \times \text{(the number of columns)}.$$

For example, all the matrices in Table 10.3 are (3×2) matrices.

A **square matrix** is a matrix $A = [a_{ij}]_{mxn}$, where $m = n$. If $m \neq n$, then A is **a rectangular matrix**. For the square matrix, the **principal (or main) diagonal** of a matrix comprises the diagonal elements from the northwest corner to the southeast corner. The diagonal elements from the northeast corner to the southwest corner are called the **secondary diagonal**.

Relation between Matrices and Vectors

A matrix may be regarded as a collection of row vectors or column vectors. If we denote the ith row of matrix A by subscript i, A_i, or the jth column of A by superscript j, A^j, then matrix A may be written as an equivalent form as

$$A = [a_{ij}]_{2\times 3} = \begin{bmatrix} a_{11} & a_{12} & a_{13} \\ a_{21} & a_{22} & a_{23} \end{bmatrix} = \begin{bmatrix} A_1 \\ A_2 \end{bmatrix} = [A^1, A^2, A^3]. \tag{10.16}$$

If, in a square matrix, $a_{ij} = a_{ji}$, so that it is symmetric along the diagonals, it is called **a symmetric matrix**. If a square matrix has the principal diagonal elements, which are one, and all the off-diagonal elements are zero, then the matrix is called an **identity matrix**, I_{nxn}. If all the elements are zero in a matrix, then it is a **zero matrix**, $0 = [0]$.

10.2.2 *Matrix operations*

Like vectors, matrices are new algebraic entities and have their own operation system. Let $A = [a_{ij}]_{mxn}$ and $B = [b_{ij}]_{mxn}$. The following definitions apply.

Matrix operations

Equality: $A = B$, if and only if $a_{ij} = b_{ij}$ for all i and j. That is, every element of A must be equal to the corresponding element in B. For example, given

$$\begin{bmatrix} a_{11} & a_{12} \\ a_{21} & a_{22} \end{bmatrix} = \begin{bmatrix} 1 & 2 \\ 3 & 4 \end{bmatrix},$$

then $a_{11} = 1, a_{12} = 2, a_{21} = 3$, and $a_{22} = 4$.

Addition and subtraction: $A+B = [a_{ij}]+[b_{ij}] = [a_{ij}+b_{ij}]$. Each and every corresponding element is added to form a new matrix. For example, given

$$A = \begin{bmatrix} 9 & 1 \\ 5 & 4 \end{bmatrix}, \quad B = \begin{bmatrix} 1 & 3 \\ 7 & 6 \end{bmatrix}, \tag{10.17}$$

then we have

$$A + B = \begin{bmatrix} 9+1 & 1+3 \\ 5+7 & 4+6 \end{bmatrix} = \begin{bmatrix} 10 & 4 \\ 12 & 10 \end{bmatrix}.$$

Similarly, for subtraction, we have $A - B = [a_{ij}] - [b_{ij}] = [a_{ij} - b_{ij}]$. Note that to perform matrix addition and subtraction, two matrices must have the same size.

Scalar multiplication: $kA = k[a_{ij}] = [ka_{ij}]$. That is, every element of A is multiplied by the scalar k. For example, for $k = 3$, and A in (10.17), we have

$$3\begin{bmatrix} 9 & 1 \\ 5 & 4 \end{bmatrix} = \begin{bmatrix} 27 & 3 \\ 15 & 12 \end{bmatrix}.$$

Excel implementation

Excel does not provide functions for matrix addition, subtraction, and scalar multiplication. However, they are easy to implement on spreadsheets.

Two matrices A and B are given in Table 10.3. To add A to B, in cell G3, enter =A3+D3, and then copy G3 down to G3:H5. To multiply 2 to A, in J3, enter =2 * A3, and then copy it down to J3:K5. This method takes advantage of relative references.

The range copy command

Another method is to use the **range copy command**. After entering A and B, select the output range G3:H5 first. The range will be shaded except for G3, in which the pointer resides. In G3, enter = and select matrix A, which is A3:B5, and enter +. Select matrix B, which is D3:E5, and then enter ctrl+shift+enter. The addition will be filled in range G3:H5 as shown, and each cell in the range will show {=A3:B5 + D3:E5}, enclosed by large braces.

Similarly, to use the **range copy command** to obtain matrix 2A, select the output range J3:K5 first, and in J3, enter =2*A3:B5, which will show automatically when the pointer selects the range A3:B5. Applying the range copy command, every cell in J3:K5 will show {=2*A3:B5}, enclosed by large braces.

10.2.3 *Matrix multiplication (the row–column multiplication)*

We rewrite the system of three Eqs. (10.1)–(10.3) in matrix form and obtain Ax = c in (10.4). This is based on the rule of scalar product, as we will explain below. We first define the row vectors $A_i, i = 1, 2, 3$, of A as follows:

$$\begin{matrix} A_1 \\ A_2 \\ A_3 \end{matrix} \begin{bmatrix} 1 & 0 & 2 \\ 0 & 1 & -6 \\ 1 & -1 & 0 \end{bmatrix}_{3\times 3} \begin{bmatrix} D \\ S \\ P \end{bmatrix}_{3\times 1} = \begin{bmatrix} 18 \\ -6 \\ 0 \end{bmatrix}_{3\times 1}. \tag{10.18}$$

Then, we may write, by the rules of the scalar multiplication and scalar product and by the definition of equality of vectors,

$$\begin{bmatrix} A_1 \\ A_2 \\ A_3 \end{bmatrix}_{3\times 3} x_{3x1} = \begin{bmatrix} A_1x \\ A_2x \\ A_3x \end{bmatrix}_{3\times 1} = \begin{bmatrix} c_1 \\ c_2 \\ c_3 \end{bmatrix}_{3\times 1}. \tag{10.19}$$

Note that every row vector in A is multiplied by the column vector x_{3x1}, and we obtain a scalar product. More generally, we may define the matrix multiplication as follows.

When two matrices A and B are multiplied, the product is denoted AB, where A is called the **lead matrix (premultiplier)**, and B is called the **lag matrix (postmultiplier)**.

The lead matrix $A = [a_{ij}]_{mxp}$ is **conformable** with the lag matrix $B = [b_{ij}]_{qxn}$ if the number of columns in A is equal to the number of rows in B, that is, p = q. The **product** AB = C matrix is defined only when A is conformable with B (Fig. 10.1). From the nature of row–column multiplication, we can see that the conformable requirement is necessary. Since a row of A, say A_i, will be multiplied by a column of B, say B^j, the number of elements in both vectors must be equal, so that we can do one-to-one correspondence multiplication in a scalar product of two vectors (see Secs. 10.1.2 and 10.1.3) and obtain the result as a scalar (real number).

Furthermore, since A has m rows and B has n columns, the product AB = C can only have m rows (like A) and n columns (like B). Thus, the size of AB = C is m × n. The relations are illustrated in (10.19) and can be seen directly in (10.20) and (10.21).

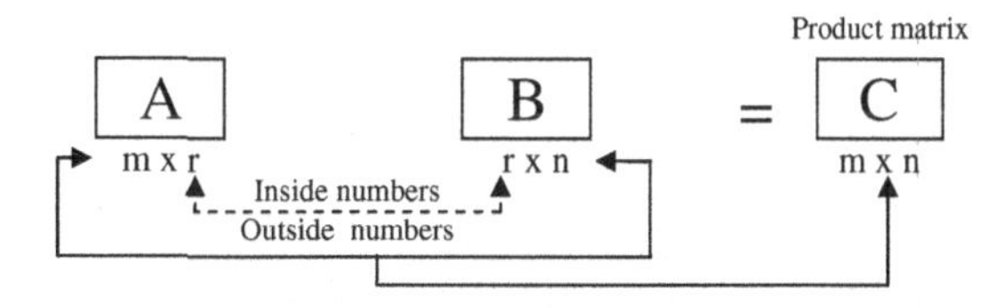

Fig. 10.1 Conformability and the product of matrix multiplication

Let $A_{m\times r} = [a_{ij}]$ and $B_{r\times n} = [b_{ij}]$. Then the **product AB** $= C_{m\times n} = [c_{ij}]$, which is defined as

$$c_{ij} \equiv A_i B^j = \sum_{k=1}^{r} a_{ik} b_{kj}, \tag{10.20}$$

where A_i is the **ith row of A** and B^j is the **jth column of B**, $i = 1, 2, \ldots, m$, and $j = 1, 2, \ldots n$. Note that k ranges from 1 to r, and the k in the second subscript of A and the k in the first subscript of B reflect the requirement that the number of columns in A must be the same as the number of rows of B. On the other hand, the number of rows (m) of A and the number of columns (n) of B determine the size of C as $(m \times n)$, as shown in the illustration in Fig. 10.1.

Example 10.5 An illustration of matrix multiplication. Let $A = [a_{ij}]_{3\times 2}$ and $B = [b_{ij}]_{2\times 2}$. Then A has 2 columns and B has 2 rows, so A and B are conformable in matrix multiplication. The product $AB = C$ is, by the definition of equality of matrices,

$$\begin{matrix} A_1 \\ A_2 \\ A_3 \end{matrix}\overset{A}{\begin{bmatrix} a_{11} & a_{22} \\ a_{21} & a_{22} \\ a_{31} & a_{32} \end{bmatrix}}_{3x2} \overset{B^1 \quad B^2}{\begin{bmatrix} b_{11} & b_{12} \\ b_{21} & b_{22} \end{bmatrix}}_{2x2} = \overset{C = [A_i B^j]}{\begin{bmatrix} A_1B^1 & A_1B^2 \\ A_2B^1 & A_2B^2 \\ A_3B^1 & A_3B^2 \end{bmatrix}}_{3x2} \equiv \overset{C = [c_{ij}]}{\begin{bmatrix} c_{11} & c_{12} \\ c_{21} & c_{22} \\ c_{31} & c_{32} \end{bmatrix}}_{3x2} \tag{10.21}$$

For this special case, c_{ij} in (10.21) is

$$c_{ij} = A_i B^j = (a_{i1}, a_{i2})(b_{1j}, b_{2j})' = a_{i1}b_{1j} + a_{i2}b_{2j} \tag{10.22}$$

The following example shows the numerical operation of the formula. □

Example 10.6 Matrix multiplication. Let the matrices A and B be given as follows:

$$\overset{\text{Lead Matric}}{A = \begin{bmatrix} 1 & 2 \\ 1 & 0 \\ 1 & 2 \end{bmatrix}_{(3\times 2)}} \qquad \overset{\text{Lag Matrix}}{B = \begin{bmatrix} 3 & 0 \\ 3 & 4 \end{bmatrix}_{(2\times 2)}} \tag{10.23}$$

Then,

$$AB = \begin{bmatrix} (1)(3) + (2)(3) & (1)(0) + (2)(4) \\ (1)(3) + (0)(3) & (1)(0) + (0)(4) \\ (1)(3) + (2)(3) & (1)(0) + (2)(4) \end{bmatrix} = \begin{bmatrix} 9 & 8 \\ 3 & 0 \\ 9 & 8 \end{bmatrix} \equiv C.$$

In actual calculation, it may be easier to do "row–column multiplications" as follows: We systematically fix the columns of B, say, B^j, and multiply the rows of A, say $A_i, i = 1, 2, 3$, to column B^j, and place the scalar product in the corresponding element in $C^j, j = 1, 2$, in $C = AB$. □

10.2.4 *Excel function for matrix multiplication*

The Excel function for **matrix multiplication** is

$$\boxed{\text{=MMULT(array1, array2)=MMULT(A,B),}} \tag{10.24}$$

where we use the more familiar matrices A and B to represent array1 and array2. A is the lead matrix and B is the lag matrix, and A and B must be conformable.

Example 10.7 Matrix multiplication. The Excel function (10.24) is applied to the matrices A and B given in Example 10.6 above for comparison. We proceed as follows:

Step 1. Enter matrices A and B as shown in Table 10.4. It is suggested that the labels A and B are entered in A2 and D2 for clarity, and the sizes of the matrices be designated at the lower right-hand corner of the matrices. This will enable us to check whether the matrices are conformable, and at the same time, we will know the size of the product AB, which in this case is 3×2.

Step 2. Select and shade a 3×2 matrix in range G3:H5 to prepare for the product AB.

Step 3. In G3, enter the Excel function

=MMULT(A3:B5,E3:E4).

Step 4. Apply the range copy command (see Chapter 6), namely, <control+shift+enter>. Now we have the product in range G3:H5. Matrix transpose and matrix inverse in rows 8–11 will be explained in Secs. 10.3.1 and 10.3.2. □

Table 10.4 Matrix multiplication, transpose, and inverse

	A	B	C	D	E	F	G	H	I	J	K	L	M
1	Matrix multiplication												
2	A			B			AB						
3	1	2		3	0		9	8					
4	1	0		3	4		3	0					
5	1	2				2×2	9	8					
6		3×2						3×2					
7													
8	Matrix transpose							matrix inverse					
9	A'				B'			invB			invB*B = I		
10	1	1	1		3	3		0.33	0.00		1	0	
11	2	0	2		0	4		-0.25	0.25		0	1	

The product matrix constructed in this way is "alive" in the sense that if the contents of A or B matrix are changed, the results of AB will change automatically. Thus, A and B can be used as a **template** and can be "reused".

Example 10.8 Matrix AB and matrix BA may not be the same. In the real number system, we have ab = ba. In matrices, this relation may not hold. By checking the sizes of matrices, we know that AB is not equal to BA if A and B or B and A are not conformable, and even if they are conformable, we still may not have AB = BA. For example, let

$$\mathrm{A} = \begin{bmatrix} 0 & 1 \\ 1 & 0 \end{bmatrix}, \quad \mathrm{B} = \begin{bmatrix} 2 & 4 \\ 1 & 3 \end{bmatrix}. \tag{10.25}$$

Then, we can show that AB $\neq$ BA, as

$$\mathrm{AB} = \begin{bmatrix} 1 & 3 \\ 2 & 4 \end{bmatrix}, \quad \mathrm{BA} = \begin{bmatrix} 4 & 2 \\ 3 & 1 \end{bmatrix}.$$

□

Example 10.9 Multiplication of A by an identity matrix is A. Given $\mathrm{A}_{2\times 3}$ as

$$\mathrm{A} = \begin{bmatrix} 1 & 2 & 3 \\ 4 & 5 & 6 \end{bmatrix}, \quad \text{then IA} = \text{AI} = \text{A}.$$

In this case, since A is 2×3, to meet the conformable requirement in matrix multiplication, the lead matrix I must be 2×2 and the lag matrix I must be 3×3. We may write $\mathrm{I}_{2\times 2}\mathrm{A} = \mathrm{AI}_{3\times 3} = \mathrm{A}_{2\times 3}$ for clarity. □

Example 10.10 a′b and ab′ are different. If a and b are vectors, using Excel function, we may verify that a′b is a scalar (scalar product) and ab′ is a matrix. For example, let

$$\mathrm{a} = \begin{bmatrix} 1 \\ 2 \end{bmatrix}, \quad \mathrm{b} = \begin{bmatrix} 2 \\ 3 \end{bmatrix}, \qquad \text{then, a}'\text{b} = \text{b}'\text{a} = 8. \tag{10.26}$$

On the other hand,

$$\mathrm{ab}' = \begin{bmatrix} 1 \\ 2 \end{bmatrix} \begin{bmatrix} 2 & 3 \end{bmatrix} = \begin{bmatrix} 2 & 3 \\ 4 & 6 \end{bmatrix}, \quad \text{but}$$

$$\mathrm{ba}' = \begin{bmatrix} 2 \\ 3 \end{bmatrix} \begin{bmatrix} 1 & 2 \end{bmatrix} = \begin{bmatrix} 2 & 4 \\ 3 & 6 \end{bmatrix}.$$

□

10.2.5 *A system of parametric equations*

The method of matrix multiplication can be applied to cases in which matrices A and B consist of parameters or variables. Thus, in the example in Sec. 10.1.1, (10.3) can be rewritten in scalar product form as shown in (10.18) and (10.19). Unlike other microcomputer software for quantitative methods, such as *Mathematica* or *Derive*, Excel cannot rewrite a parametric system of simultaneous equations automatically into a matrix form. Manual conversion is required. The following are some examples.

Example 10.11 Lag matrix is a variable column vector. Let A and x be

$$A = \begin{bmatrix} 3 & 0.5 \\ -1 & 4 \\ -0.7 & 9 \end{bmatrix}_{(3\times 2)} \quad \text{and} \quad x = \begin{pmatrix} x_1 \\ x_2 \end{pmatrix}_{(2\times 1)} .$$

Then,

$$Ax = \begin{bmatrix} 3x_1 + 0.5x_2 \\ -x_1 + 4x_2 \\ -0.7x_1 + 9x_2 \end{bmatrix}_{(3\times 1)} .$$

In this case, the size of Ax is (3×1), which is a column vector, and each element of Ax is the scalar product of a row of A and the column x. □

Example 10.12 Matrix expression of a set of linear equations. Let a system of two linear equations be given as

$$\begin{cases} -5p_1 + \ \ p_2 = -12 \\ \ \ \ p_1 - 3p_2 = -16 \end{cases} .$$

Write this equation system in matrix form.

Answer: Using the three steps in Sec. 10.1.1, we rewrite the equations in matrix form as

$$\begin{matrix} \begin{bmatrix} -5 & 1 \\ 1 & -3 \end{bmatrix} & \begin{bmatrix} p_1 \\ p_2 \end{bmatrix} & = & \begin{bmatrix} -12 \\ -16 \end{bmatrix} \\ A & x & = & c \end{matrix} \qquad (10.27)$$

□

Example 10.13 A parametric model of a linear equation system. Let a system of three linear equations be

$$\begin{cases} Y = C + I_o + G_o \\ C = a + b(Y - T) \\ T = d + tY \end{cases} . \qquad (10.28)$$

Write this equation system in matrix form.

Answer: In this example, we have three equations in three variables. The steps of writing the equations into matrix form are as follows:

Step 1. Determine the endogenous variables, exogenous variables, and the coefficients. In this case, the endogenous variables are Y, C, and T, the exogenous variables are I_0 and G_0, and the coefficients are a, b, d, and t.

Step 2. Rearrange each equation in the order of Y, C, T on the left-hand side and the constant on the right-hand side as follows :

$$\begin{array}{cccccc} & & \text{LHS} & & & \text{RHS} \\ & \text{Y} & \text{C} & \text{T} & & \text{constant} \\ \left\{\begin{array}{l} \\ \\ \\ \end{array}\right. & \begin{array}{r} 1\text{Y} \\ -b\text{Y} \\ -t\text{Y} \end{array} & \begin{array}{r} -1\text{C} \\ +1\text{C} \\ +0\text{C} \end{array} & \begin{array}{r} +0\text{T} \\ +b\text{T} \\ +1\text{T} \end{array} & \begin{array}{c} = \\ = \\ = \end{array} & \begin{array}{r} I_o + G_o \\ a \\ d \end{array} \end{array} \qquad (10.29)$$

Step 3. Write the parametric coefficient matrix A, the endogenous variable vector x, and the constant vector c as follows:

$$\begin{array}{ccccc} \begin{array}{ccc} \text{Y} & \text{C} & \text{T} \end{array} & & & & \text{constant} \\ \begin{bmatrix} 1 & -1 & 0 \\ -b & 1 & b \\ -t & 0 & 1 \end{bmatrix} & \begin{bmatrix} \text{Y} \\ \text{C} \\ \text{T} \end{bmatrix} & = & & \begin{bmatrix} I_o \ +G_o \\ a \\ d \end{bmatrix} \\ \text{A} & \text{x} & = & & \text{c} \end{array} \qquad (10.30)$$

□

Example 10.14 The reverse process. From (10.30), A is a parametric coefficient matrix and x is a variable vector, as shown below

$$A = \begin{bmatrix} 1 & -1 & 0 \\ -b & 1 & b \\ -t & 0 & 1 \end{bmatrix}_{(3\times 3)}, \quad x = \begin{bmatrix} \text{Y} \\ \text{C} \\ \text{T} \end{bmatrix}_{(3\times 1)}.$$

Show that the product Ax is equal to the left-hand side of the three equations in (10.29). Answer: We can use row–column multiplication to find Ax as follows:

$$Ax = \begin{bmatrix} Y - C \\ -bY + C + bT \\ -tY + T \end{bmatrix}_{(3\times 1)}, \qquad (10.31)$$

which is the left-hand side of (10.29). □

10.3 Transpose and Inverse Matrices

Two important operations involving a matrix are taking the transpose and inverting the matrix.

10.3.1 *The transpose of a matrix*

Let a matrix be $A = [a_{ij}]$. The **transpose** of A, denoted as A' or A^T, is a matrix in which the rows of A' are the columns of A, or, equivalently, the columns of A' are the rows of A. For example, if A is given as in the matrix below, the rows of A, A_1, and A_2 become the column of A', that is $(A')^1$ and $(A')^2$, where superscripts indicate the column numbers. Equivalently, the columns of A, namely, A^1, A^2, and A^3 become the rows of A', namely, $(A')_1, (A')_2$, and $(A')_3$, where subscripts indicates the row numbers.

$$A = \begin{bmatrix} a_{11} & a_{12} & a_{13} \\ a_{21} & a_{22} & a_{23} \end{bmatrix}, \quad A' = \begin{bmatrix} a_{11} & a_{21} \\ a_{12} & a_{22} \\ a_{13} & a_{23} \end{bmatrix}$$

The Excel command

Excel does not have a function for matrix transpose. It belongs to one of the copy/paste commands. The command for transpose has already been introduced in (10.6) of Sec. 10.1.1.

Example 10.15 Matrix transpose. In Table 10.4, to take the transpose of matrix A, select the range of A, A3:B5, and <^C>. Moving the pointer to A10, and applying the command sequence in (10.6), we have A' in A10:C11. Similarly, for B, we can find B' in E10:F11. □

Let A be a square matrix. Using the definition of matrix transpose, we see that A is a **symmetric matrix** if and only if $A = A'$. In this case, the elements of square matrix A are symmetric along the main diagonal. In Table 10.4, we can see B is not a symmetric matrix, because B' is not equal to B. However, if $b_{21} = 3$ in D4 is changed to 0, then B is a symmetric matrix.

10.3.2 *The inverse of a matrix*

Given any two square matrices A and B, such that

$$AB = BA = I, \tag{10.32}$$

then B is called the **inverse matrix** of A and is written as $B = A^{-1}$. Similarly, A is the inverse matrix of B and is written as $A = B^{-1}$. Thus, (10.32) can be written as

$$AA^{-1} = A^{-1}A = I, \quad \text{or} \quad BB^{-1} = B^{-1}B = I.$$

The Excel command

The Excel function for the inverse matrix is

=MINVERSE(array)=MINVERSE(A).

(10.33)

Example 10.16 Inverse Matrix. In Table 10.4, B is a (2×2) square matrix. We may use the following steps to find its inverse matrix B^{-1}:

Step 1. Move the pointer to a blank cell, say H10, and select a range which has the same size as B, H10:I11, and shade the range (or, you may copy B down to H10:I11).

Step 2. In cell H10, enter Excel command,

=MINVERSE(D3:E4).

Step 3. Apply the range copy command. The matrix B^{-1} will be shown in H10:I11. In this case, if B^{-1} exists, then B is called a **nonsingular matrix**. We can use matrix multiplication to show that $BB^{-1} = B^{-1}B = I$, as in K10:L11. □

If B^{-1} does not exist, an error sign #NUM! will appear. The inverse matrix constructed in this way is "alive" in the sense that if any element in matrix B is changed, the result of B^{-1} will change automatically. Thus, the original 2×2 matrix B and its inverse B^{-1} can be "reused" as a **template** for finding any 2×2 inverse matrix.

The concept of inverse matrix resembles the inverse in the real number system. If a is a real number, its inverse number is $1/a$, which may be written as a^{-1}. Hence, $aa^{-1} = a^{-1}a = 1$. The difference is that while a^{-1} always exists for any real number a except for $a = 0$, the inverse A^{-1} of matrix A may not exist in the matrix system.

Example 10.17 The case of no corresponding inverse matrix. In Table 10.4, if we change the element b_{11} in D3 from 3 to 0 in matrix B, then the entries of B^{-1} will show #NUM! in H10:I11. In this case, B^{-1} does not exist, and B is called **a singular matrix**. □

Example 10.18 Excel for Matrix inversion and multiplication. Table 10.5 presents a 4×4 matrix A in the range A2:D5. We first find its inverse matrix using Excel command MINVERSE(A2:D5) and enter the result of invA or A^{-1} in A7:D10. Then, we multiply A^{-1} by A and enter it in A12:D15, and also multiply A by A^{-1} and enter it in A17:D20.

Notice that we have not quite achieved the identity matrix in either case. This is due to rounding errors. Examination of the numbers will show we have come very close. To show the identity matrix, we need to reproduce $A^{-1}A$ in other blank range, say F12:I15. However, note that we cannot copy A12:D15 directly to F12:I15, as A12:D15 was constructed using the range copy command. Therefore, we go to cell F12, enter =A12, and then copy A12:D15 to F12:I15. Change F12:I15 entries into integers by using the "decreasing decimal" command, and the identity matrix shows in F12:I15. Do the same for AA^{-1} in F17: I20. □

10.4 Solving a System of Simultaneous Linear Equations

Let a system of simultaneous linear equations be given as $Ax = c$, as shown in (10.4). Multiplying both sides by A^{-1} on the left side,[6] we have $A^{-1}Ax = A^{-1}c$. By the property

[6]You cannot multiply A^{-1} on the right since x or c as a premultiplier is not conformable with A^{-1} as a postmultiplier.

Table 10.5 Matrix inversion and multiplication

	A	B	C	D	E	F	G	H	I
1	Matrix A								
2	12	3	21	11					
3	2	11	5	45					
4	16	21	3	51					
5	8	10	0	9					
6	InvA $=A^{-1}$								
7	0.0126	-0.1479	0.1585	-0.1739					
8	-0.0067	0.1277	-0.1660	0.3100					
9	0.0433	0.0717	-0.0897	0.0964					
10	-0.0037	-0.0104	0.0435	-0.0788					
11	$A^{-1}A$					$A^{-1}A$			
12	1	-4.4409E-16	-1.11022E-16	-1.11022E-15		1	0	0	0
13	-4.441E-16	1	-1.11022E-16	-8.88178E-16		0	1	0	0
14	0	0	1	-1.11022E-16		0	0	1	0
15	1.1102E-16	0	0	1		0	0	0	1
16	AA^{-1}					AA^{-1}			
17	1	-3.1919E-16	1.66533E-16	0		1	0	0	0
18	-2.776E-17	1	0	4.44089E-16		0	1	0	0
19	0	-4.4409E-16	1	0		0	0	1	0
20	-6.939E-18	-3.0531E-16	-1.66533E-16	1		0	0	0	1

of an inverse matrix that $A^{-1}A = I$, we have $Ix = x = A^{-1}c$. This is the **inverse matrix method** of solving for the variable vector x, denoted as $x^{\#}$,

$$\boxed{x^{\#} = A^{-1}c,} \tag{10.34}$$

in which A^{-1} must exist in order to have the solution of the system of numerical simultaneous linear equations. This result indicates that, we may solve the system of numerical simultaneous linear equations by first finding the inverse matrix of A, and then multiplying it to the constant vector c.

Using the demand and supply market model in Sec. 10.1.1, the inverse matrix method of solving the model may be arranged in Table 10.6. It is suggested that the original system of equations be entered in the table (rows 3 to 5) to avoid errors in rearranging the variables.

Solution procedure

The solution procedure has the following five steps:

Step 1. Rearrange the equations. Move the terms with endogenous variables D, S, P to the LHS of the equality and move the constant terms and the exogenous variables (in this model, there is no exogenous variable) to the RHS of the equality, as in G3:K5.

Step 2. Write the system of equations in **matrix form**, the coefficient matrix A as in B9:D11, and the constant vector c as in F9:F11. Write the variable names in B8:D8, and the equation numbers in A9:A11 (not shown) for easy verification.

Table 10.6 A design for solution: A single market model

	A	B	C	D	E	F	G	H	I	J	K
1	**A Design for Solution: Ax = c.**										
2	Step 1 Rearrange variales						D	S	P		
3			1	D = 18-2P			D		+2P	=	18
4			2	S = -6+6P				S	+6P	=	-6
5			3	D = S			D	+S		=	0
6	Step 2 Write in matrix form										
7		A				c					
8		D	S	P							
9		1	0	2		18					
10		0	1	-6		-6					
11		1	-1	0		0					
12				(3x3)		(3x1)			The inverseN-arrangement		
13	Step 3 Inverse matrix and the solution						Step4				
14		invA				x# = invA*c					
15		0.75	0.25	0.25		12	D#				
16		0.75	0.25	-0.75		12	S#				
17		0.13	-0.13	-0.13		3	P#				
18				(3x3)		(3x1)					
19	Notes	B14\VERSE(B8:C10)}									
20		F14 (B14:D16,F8:F10)}									
21	Step 4 Interpretation										

Step 3. Find the inverse matrix, invA, by entering Excel command

=MINVERSE(B9:D11),

in B15, and the result is in B15:D17.

Step 4. Multiply c **by inv**A. First select the output range F15:F17. In F15, enter

=MMULT(B15:D17, F9:F11),

and apply the range copy command. We have the **solution vector** $x^{\#}$ in F15:F17. The process and method of arranging the matrices look like the upside down N, we call this **the inverse N arrangement**.

Step 5. Identify the equilibrium variables as in G15:G17. The equilibrium demand $D^{\#}$, the equilibrium supply $S^{\#}$, and the equilibrium price $P^{\#}$ are given as a solution vector (12, 12, 3) in step 4, which has the same values as obtained by the graphical method in Chapter 3.

Note that the order of variables in the solution vector $x^{\#}$ is the same as the order of variables arranged in the coefficient matrix A in Step 2. The choice of the variable order is arbitrary, but once the order is chosen, it should be used consistently during the solving process.

There is a strong resemblance with (10.34) in the real number system. If a and c are real numbers, then $ax = c$ can be solved for x as

$$x^{\#} = c/a = c(1/a) = a^{-1}c,$$

for $a \neq 0$. In the matrix system, the condition is that matrix A is nonsingular.

Example 10.19 From Example 10.12, $\mathbf{Ax} = \mathbf{c}$, the two equations in matrix form in (10.27), is reproduced in (10.35):

$$\begin{bmatrix} -5 & 1 \\ 1 & -3 \end{bmatrix} \begin{bmatrix} p_1 \\ p_2 \end{bmatrix} = \begin{bmatrix} -12 \\ -16 \end{bmatrix}. \tag{10.35}$$

To make Table 10.7 more general, we name the coefficients and constants as in the first two rows, the parameter table, of Table 10.7, and enter the named coefficients and constants in A and c of the inverse N arrangement (hence, if any parameter changes in row 2, A and/or c also change. Change the parameters and see how the solution x# changes). Now we solve for the variable vector $x^{\#} = A^{-1}c = (p_1^{\#}, p_2^{\#})'$. The matrix operations and the result are presented in Table 10.7.

We have the solutions $p_1^{\#} = 3.7$ and $p_2^{\#} = 6.6$. As we have mentioned before, the (2×2) coefficient matrix A and (2×1) constant vector c can be reused as a **template**. Enter any other numbers in A and/or c, and the new solution for variable vector x will show instantly in the table. □

10.5 Macroeconomic Policy Models

One of the advantages of using Excel is that we can find numerical solutions instantly on the spreadsheets, without going through tedious manual calculations, especially matrix multiplication and matrix inversion. We can apply these techniques in macroeconomic analysis.

Table 10.7 Solving a two-equation system

Parameter table:

a	b	c	d	e	f
-5	1	1	-3	-12	-16

The inverse N arrangement:

p1	p2		c	
-5	1		-12	
1	-3		-16	
↓		↗	↓	
invA			X#=(invA)*c	
-0.214	-0.071		3.714	p1#
-0.071	-0.357		6.571	p2#

10.5.1 *The model and the markets*

The model

Equations (10.36)–(10.43) present a large 8-equation Keynesian IS-LM model, which is an extension of the Keynesian national income models we have discussed in Chapters 3 and 4. The model is as follows:

Goods market	$Y = C + I + G_0 + X$	Equil. cond. of goods market	(10.36)
	$C = a + 0.8(Y - T)$	Consumption function	(10.37)
	$T = -25 + 0.25Y$	Tax function	(10.38)
	$I = 65 - 20R$	Investment function	(10.39)
	$X = 60 - 0.1Y$	Net exports function	(10.40)
Money market	$1.5L = 5Y - 50R$	Money demand function	(10.41)
	$M = M_0$	Money supply function	(10.42)
	$L = M$	Equil. cond. of money market	(10.43)

There are eight endogenous variables: Y, C, I, X, T, L, M, and R, they denote national income, consumption, investment, net exports (i.e. exports minus imports), tax revenue, money demand, money supply, and interest rate in percent (= 100r%), respectively. There are two exogenous variables: money supply M_0 and government expenditures G_0. There is one parameter, a, which is the subsistence level of consumption. The eight endogenous variables will be determined from the eight equations in the model.

The markets

In the above model, Equations (10.36)–(10.40) describe the **goods market**, and (10.36) is the **equilibrium condition of the goods market**. The last three equations describe **the money market**, and (10.43) is the **equilibrium condition of the money market**, showing that the demand of money equals the supply of money. Given the model, the government may change the value of an exogenous variable M_0, or G_0, or parameter a, to achieve a macroeconomic policy objective.

In Table 10.8, we demonstrate the convenience and ease of using the Excel inverse matrix method to solve the eight endogenous variables in the model, which is almost impossible to do manually. In Step 1, the eight equations are rearranged in rows 2–9 by moving all the endogenous variables to the left-hand side of the equality sign, and the constants, exogenous variables, and parameter a to the right-hand side of the equality sign. In Step 2, enter the coefficient matrix in C14:J21. The arrangement of Table 10.8 is similar to that of Table 10.6.

10.5.2 *The policy instruments and the policy impact matrix*

The policy instruments

A change in parameter a is called the **consumption policy** (encouraging or discouraging consumption through increasing or decreasing the value of basic consumption level a).

Table 10.8 A macroeconomic policy model

	A	B	C	D	E	F	G	H	I	J	K	L	M	N	O	P	Q
1	Step 1 Rearrange the equations																
2		1	Y-C-I-X=G0														
3		2	C-0.8Y+0.8T=a														
4		3	T-0.25Y=-25														
5		4	I+20R=65											policy scenario			
6		5	X+0.1Y=60											Cons	Mon	Fis	
7		6	L+50R-5Y=0										defau	Case1	Case2	Case3	Parameter
8		7	M=M0									a	15	30	15	15	table
9		8	L-M=0									M0	1500	1500	3000	1500	
10												G0	94	94	94	188	
11	Step 2 Write in matrix form																
12			A									c					
13			Y	C	I	X	T	L	M	R				Cons	Mon	Fis	
14		1	1	-1	-1	-1	0	0	0	0			94	94	94	188	
15		2	-0.8	1	0	0	0.8	0	0	0			15	30	15	15	
16		3	-0.25	0	0	0	1	0	0	0			-25	-25	-25	-25	
17		4	0	0	1	0	0	0	0	20			65	65	65	65	
18		5	0.1	0	0	1	0	0	0	0			60	60	60	60	
19		6	-5	0	0	0	0	1	0	50			0	0	0	0	
20		7	0	0	0	0	0	0	1	0			1500	1500	3000	1500	
21		8	0	0	0	0	0	1	-1	0			0	0	0	0	
22																	
23	Step 3 Inverse matrix and the solution												Step 4x# = invA*C				Inverse N
24			InvA										defau	Case1	Case2	Case3	arrangement
25			0.4	0.4	-0.3	0.4	0.4	-0.2	0.2	0.2		Y#	342	348	582	379	
26			0.24	1.2	-1	0.2	0.2	-0.1	0.1	0.1		C#	240	259	384	263	
27			-0.8	-0.8	0.6	0.2	-0.8	-0.1	0.1	0.1		I#	-18	-30	102	-93	
28			-0.04	-0	0	-0	1	0	-0	-0		X#	26	25	2	22	
29			0.1	0.1	0.9	0.1	0.1	-0	0	0		T#	60	62	120	70	
30			0	0	0	0	0	0	1	1		L#	1500	1500	3000	1500	
31			0	0	0	0	0	0	1	0		M#	1500	1500	3000	1500	
32			0.04	0	-0	0	0	0	-0	-0		R#	4	5	-2	8	
33																	
34	Step 4 Interpretation											Step 5	Impact matrix				
35														Cons	Mon	Fis	
36												Y#		6.0	240.0	37.6	
37												C#		18.6	144.0	22.6	
38												I#		-12.0	120.0	-75.2	
39												X#		-0.6	-24.0	-3.8	
40												T#		1.5	60.0	9.4	Comparative
41												L#		0.0	1500.0	0.0	Static analysis
42												M#		0.0	1500.0	0.0	
43												R#		0.6	-6.0	3.8	

A change in M_0 is called **monetary policy** (increasing the amount of M_0 is the **easy monetary policy** and decreasing the amount of M_0 is the **tight monetary policy**). A change in G_0 is called **fiscal policy** (expansion or contraction of government spending by increasing or decreasing the amount of G_0). Thus, in this model, we have three **policy**

instruments, a, M_0, and G_0, and we call them case 1, case 2, and case 3, as shown in M5:P10.

The default values of the three instrument variables are given as follows: a = 15, M_0 = 1500, and G_0 = 94. These are entered in M8:M10. For illustration, we consider the following cases:

Case 1. Only parameter a is doubled (the consumption policy), see N8 in row 8.
Case 2. Only M_0 is doubled (the monetary policy), see O9 in row 9.
Case 3. Only G_0 is doubled (the fiscal policy), see P10 in row 10.

In each case, the value of an instrument variable is doubled from the original default level, and the new sets of values for a, M_0, and G_0 are entered in M8:P10. Corresponding to the policy scenarios, we have different constant vectors, which incorporate the policy changes. These constant vectors are shown in N14:P21. Other things being equal, the policy changes are indicated by bordered cells.

The inverse matrix and the solution matrix

For a large coefficient matrix, A, like this example, it is easier to find the range of the inverse matrix by simply copying the coefficient matrix to the range of the inverse matrix, C25:J32. Then, as in Step 3, select the range of the inverse matrix and find the inverse matrix, invA. In Step 4, multiply invA to the expanded constant matrix $C_{8\times4}$, and we have the 8×4 solution matrix $X^{\#}$ instantly in M25:P32, for the default case and the three policy cases. For reference, in L25:L32, we add the names of the endogenous variables next to each row of the solution matrix.

The impact matrix

We now compare the effects of changing policies, i.e. Cases 1–3, with the default solutions. The effects are measured by taking the difference between the new policy solutions and the default solutions for all endogenous variables. The **impact matrix** is shown in N36:P43. In cell N36, enter

$$\text{N36:} \ = \text{N25}-\$\text{M25},$$

and copy it to N36:P43, and we have the impact matrix.

10.5.3 *Policy implications*

For the default case, the equilibrium national income[7] is 342 units (same below), and the equilibrium[8] consumption is 240. The average propensity to consume (C/Y) is 0.70, as contrasted with the marginal propensity to consume, which is 0.8 in (10.37). The equilibrium

[7]The measurement may be in billions or trillions of dollars.
[8]As familiar from Chapters 3 and 4, they are in equilibrium since the solutions satisfy the two equilibrium conditions in the model. All other variables are also equilibrium variables.

investment is −18, the equilibrium net exports is 26, the equilibrium tax revenue is 60, and the money supply and demand are maintained at 1500. In its ratio to the national income, the investment ratio is 12%, the average tax rate is 20%, and the money multiplier (M/Y) is about 4. The equilibrium interest rate is 4%, quite consistent with the Federal prime rate of the mid-2000s.

For case 1, from N36:N43 of the impact matrix, we can see that, other things being equal, the consumption policy of doubling parameter a, from 15 to 30, will increase national income, consumption, tax revenue, and interest rate. On the other hand, the policy will reduce the equilibrium investment and net exports. The qualitative (direction) and quantitative (magnitude) changes in all endogenous variables are shown clearly in N36:N43.

For case 2, from O36:O43 of the impact matrix, we can see that, other things being equal, the monetary policy of doubling the money supply may greatly stimulate the increase in equilibrium national income, consumption, investment, and tax revenue. It reduces net exports. The equilibrium interest rate may become negative, −2%. This is an unrealistic scenario and can be ruled out.

Lastly, for case 3, other things being equal, range P36:P43 shows that the fiscal policy of doubling government expenditure will increase the equilibrium national income, consumption, tax revenue, and interest rate. The interest rate, in fact, is 8%, an unusually high rate. On the other hand, the policy will reduce the equilibrium investment and net exports. The net exports become negative, which means the imports exceed the exports. An interpretation is that the increase in government expenditures may result in overspending (consumption) in public goods, such that it will crowd out private investment and raise government purchases of foreign goods (imports).

Based on our model, we may observe that during a time of recession, if the government wants to stimulate the increase in national income and consumption, monetary policy is quite effective, but is not feasible, and fiscal policy appears to be better than consumption policy, which increases national income 37.6 units, as compared with 6 units for consumption policy. Consumption policy is still better than the default policy (of doing nothing). While this is a very simple prototype policy model, it catches the essence of the workings of the government economic policymaking procedures.

Table 10.8 is only a demonstration of the use of Excel commands in solving a system of simultaneous linear equations. For real-world applications, we need to build a much larger macroeconomic model with many endogenous variables and with more realistic assumptions in economic conditions. This is a topic related to the advanced econometrics and beyond the scope of this book.

10.6 Input–Output Models

Another useful application of matrix algebra is **Leontief Input–Output Analysis** between industries in economic planning. Suppose we have an interdependent economy in which each commodity is used as an intermediate good to produce the same kind of good and/or other commodities, as well as for consumers' final consumption. For example, industrial goods like GM's trucks can be used in GM's factories while producing other trucks, the trucks can also be used in agriculture firms in transporting and in producing food, and we can use the trucks directly as our transportation equipment for final consumption. These input–output interrelationships among industries can be expressed by an input-coefficient matrix, a gross-output vector, and a final consumption vector. They can be arranged together to form an input–output model. Our discussions and illustrations of Excel commands in the previous sections can be used to find the solutions of equilibrium (gross) output vector for economic planning.

10.6.1 *The input-coefficient matrix and basic assumptions*

We now give economic meaning to the elements of an input-coefficient matrix.

Input coefficients

The general input relationships among industries are denoted by an element a_{ij} in an input-coefficient matrix A. It is the value of the ith good used as input to produce one dollar's worth of the jth good. Note that in input–output models, all goods are measured in terms of value in dollars. For example, if $a_{21} = 0.25$, then \$0.25 of the second good is used as an input to produce \$1.00 worth of the first good. The element a_{ij} is called an **input-coefficient** (or more precisely, an **input–output coefficient**).

More specifically, if x_{ij} denotes the total value, say \$100 (million or billion, same below), of the ith good used in the production of the jth good, and the total output of commodity j, called the **gross output** of the jth good, is $x_j = \$400$, then

$$a_{ij} \equiv x_{ij}/x_j = \$100/\$400 = 0.25. \tag{10.44}$$

Since we can write $0.25 = \$0.25/\$1.00$, we may interpret 0.25 as the 0.25 "dollar's worth" of commodity i used to produce one "dollar's worth"[9] of commodity j.

Industrial relations

In a simple economy with two industries, say, agriculture and manufacturing, the input-coefficient matrix may be arranged systematically as a (2 × 2) matrix A:

$$A = \begin{array}{c} \\ \text{Agr} \\ \text{Mfg} \end{array} \begin{array}{c} \begin{array}{cc} \text{Agr} & \text{Mfg} \end{array} \\ \begin{bmatrix} a_{11} & a_{12} \\ a_{21} & a_{22} \end{bmatrix} \end{array} = \begin{array}{c} \begin{array}{cc} \text{Agr} & \text{Mfg} \end{array} \\ \begin{bmatrix} 0.30 & 0.11 \\ 0.25 & 0.35 \end{bmatrix} \end{array}. \tag{10.45}$$

Rows of A show the **source industries**, and columns of A show the **destination industries**. We may express this as $i \to j$ (from i to j).

[9]If the price of an orange is $p = \$0.50$ per orange, then $1/p = 1/0.50 = 2$ means one dollar can buy two oranges. Hence, "one dollar's worth of orange" is two oranges. If $x_j = p_j q_j$, then $x_j = q_j/(1/p_j)$, thus x_j may also be measured in the unit of dollar's worth of commodity j. We do not pursue this problem here.

Two basic assumptions

For simplicity, we make the following assumptions:

A1: Each industry uses a **fixed input–output ratio**.
A2: Each industry produces only **one homogeneous good**.

The first assumption means that the amount of the ith good (input) needed to produce one dollar's worth of the jth good (output) is fixed, i.e. coefficient a_{ij} for all i and j is a constant in the input-coefficient matrix A. Since a_{ij} is constant, (10.44) shows that an increase in total input x_{ij} by k times implies that the total output x_j must also increase by k times. This relation must hold for all i and j. Hence assumption A1 implies constant return to scale in production. The fixed input-coefficient matrix also implies that the production technology remains constant for every industry. Therefore, an input–output model is essentially a short-run model.

The second assumption A2 means that one industry produces only one good (the assumption of **no joint production**). If Dell produces microcomputers as well as software, then we divide Dell into two companies, at least in theory, and treat the hardware production plant and the software production plant as two separate firms.

10.6.2 *The structure of the economy*

The input-coefficient matrix A yields both column (input) and row (output) interpretations.

On the columns

The column shows the **cost structure** of a simple economy consisting of an agricultural industry (industry 1) and a manufacturing industry (industry 2). The first column of (10.45) shows that the agricultural industry (Agr) used $a_{11} = 0.30$ dollar of agricultural goods and $a_{21} = 0.25$ dollar of manufacturing goods (Mfg) in the production of \$1 worth of agricultural goods. They sum up to less than \$1. The rest (\$1 – \$0.30 – \$0.25 = \$0.45) represents the amount of primary inputs (i.e. labor, land, etc.) used in the production of \$1 worth of agricultural goods. The decomposition of one dollar's worth of each output is shown as a pie chart consider it as a one dollar coin) in Fig. 10.2. The second column can be interpreted similarly.

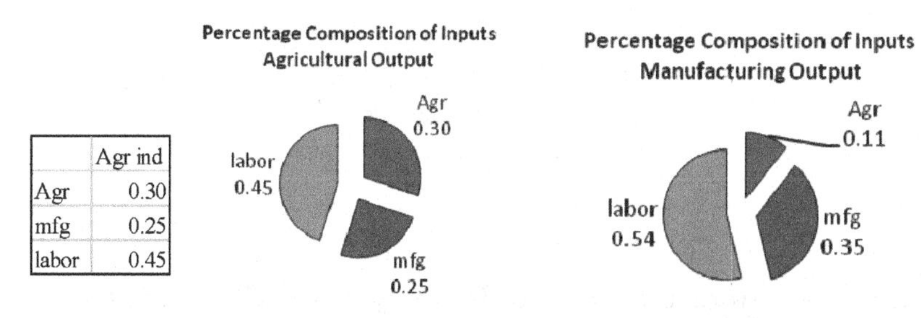

Fig. 10.2 The cost structure of two industries

The pie chart

The pie chart is drawn by using Excel charts, as follows. Construct a simple table for the first column of the input-coefficient matrix of (10.45), as shown on the left-hand side of Fig. 10.2. Then, enter

<"Insert"><Charts, Pie><Exploded Pie>. (Exploded pie is drawn.)

Then <"Chart Tools,Design"><Chart Layouts, Quick Layout!><Layout 1>.

The data labels are added. Clicking the data labels and moving outside the pie chart, we have Fig. 10.2.

Similarly, we can construct the table and pie chart for the manufacturing industry.

On the rows

Each row of the input-coefficient matrix A shows the **distribution structure** of an output to each industry. For example, in the first row of (10.45), $a_{11} = 0.30$ dollar of agricultural output goes to produce one dollar's worth of agricultural output, and $a_{12} = 0.11$ dollar of agricultural output goes to produce one dollar's worth of manufacturing output.

Let x_j be the gross output produced by the jth industry. Then $a_{ij}x_j$ is the total amount of the ith good used to produce x_j dollars of the jth commodity. That is the ith commodity demanded by the jth industry. In the above example, if $a_{21} = 0.25$, and if \$1000 worth of the first commodity x_1 is produced, then the amount of the second commodity used to produce x_1 is $a_{21}x_1 = 0.25 \times 1000 = 250$ dollars. This amount represents the demand for the second commodity by the first industry.

The household sector and equilibrium conditions

Let d_j represent the amount of the jth commodity demanded by households for final consumption. Then the gross output x_j produced by the jth industry must be distributed either to industries as intermediate inputs or to households as final consumption goods. For the two-industry case, we have the following conditions:

$$a_{11}x_1 + a_{12}x_2 + d_1 = x_1 \tag{10.46}$$

$$a_{21}x_1 + a_{22}x_2 + d_2 = x_2 \tag{10.47}$$

Industrial demand | Households demand | Total supply

Total demand

Example 10.20 A numeric model and the system of equations. Given the input-coefficient matrix A in (10.45), and if $d = (3, 10)'$, then we can write the input–output model (10.46) and (10.47) as follows:

$$0.30x_1 + 0.11x_2 + 3 = x_1, \quad (10.48)$$

$$0.25x_1 + 0.35x_2 + 10 = x_2. \quad (10.49)$$

These two equations show **the equilibrium condition** that

$$\text{Aggregate demand (AD)} = \text{Aggregate supply (AS)},$$

in each commodity market, which is equivalent to the equilibrium condition in the demand and supply market model. □

Specifically, the first Eq. (10.46) or (10.48) shows the equilibrium condition of the first commodity market, that is, the left-hand side shows the total demand and the right-hand side shows the total supply of the first commodity. Similarly, the second Eq. (10.47) or (10.49) shows the equilibrium condition of the second commodity market, that is, the left-hand side shows the total demand and the right-hand side shows the total supply of the second commodity.

Each equation in the input–output model shows that "demand equals supply", the equilibrium condition in each market. Unlike the usual market model of demand and supply, this model does not specify the demand and the supply of each commodity as a function of price, as we might expect from the usual market model,[10] or as a function of income, as in the Keynesian national income model.[11] We may say that the model in (10.46) and (10.47) is a **reduced form** of the usual market model, in which only the equilibrium condition between demand and supply of commodities in the model is postulated.

10.6.3 *Solving input–output models*

After setting up equilibrium conditions (10.46) and (10.47), we face the problem of solving the system of simultaneous equations. Since a_{ij}'s are constants, we have two equations in two variables, x_1 and x_2, for two given final demands d_1 and d_2.

To solve this system of equations, we follow the steps below.

Step 1. Arrange the variables. Moving the first two terms with variables in the left-hand side of each equation to the right-hand side and rewriting the equations, we have

[10] In the usual market model, as shown in Chapter 3, we have three equations in three variables. The demand and supply are explained by p. In the input–output model, we only have the demand for each sector, and demand and supply are not explained by other variables. It only states that $D = S$.

[11] In the national income model, as shown in Chapter 3, we have four equations in four variables. The input–output model only shows $AS = AD = C + I_0 + G_0$ without explaining that C is a function of Y.

$$(1 - a_{11})x_1 - a_{12}x_2 = d_1 \tag{10.50}$$
$$-a_{21}x_1 + (1 - a_{22})x_2 = d_2 \tag{10.51}$$

$$\underbrace{\phantom{(1 - a_{11})x_1 - a_{12}x_2}}_{\text{Net supply of output to households}} \quad \underbrace{}_{\text{Net demand of households}}.$$

Step 2. Matrix form. Rewrite (10.50) and (10.51) in matrix form:

$$\begin{bmatrix} 1 - a_{11} & -a_{12} \\ -a_{21} & 1 - a_{22} \end{bmatrix} \begin{bmatrix} x_1 \\ x_2 \end{bmatrix} = \begin{bmatrix} d_1 \\ d_2 \end{bmatrix}, \tag{10.52}$$

or,

$$\boxed{(I - A)x = d,} \tag{10.53}$$

where $x = (x_1, x_2)'$ is the variable vector of gross output, and $d = (d_1, d_2)'$ is the constant vector of final demand.

Step 3. Find the solution. Using the inverse matrix method, we have

$$x^{\#} = (I - A)^{-1}d. \tag{10.54}$$

Note the inverse matrix in (10.54) is $(I - A)^{-1}$, i.e. the inverse matrix of the technology matrix $(I-A)$, not the input-coefficient matrix A. The solution $x^{\#}$ is called the **equilibrium gross output**.[12]

The fundamental equation of input–output models

Matrix $(I - A)$ is called the **technology (or input–output) matrix**, since each column shows the input–output structure of each industry. Inputs are shown with nonpositive signs (some a_{ij} can be zero), and outputs are shown with nonnegative signs. In general, the diagonal elements in $(I - A)$ are nonnegative, showing the "net output" of producing one dollar's worth of the jth output. Here the value 1 means one dollar's worth of output. The element a_{jj} is the amount of the jth commodity used as an input to produce one dollar's worth of the jth commodity (i.e. electricity is used to produce electricity, and coal is used to produce coal). $(1 - a_{jj})$ is the "net" amount of the jth commodity produced by the jth industry that will be distributed among other industries and consumers' final demand. The off-diagonal elements of $(I - A)$ are nonpositive, showing the amount of input used in each industry. Equation (10.53) is called the **Fundamental Equation of Input–Output Models**.

10.6.4 *Implementation on Excel spreadsheets*

The input-coefficient matrix A in (10.45) and the final demand vector d in (10.46) and (10.47) are known values. In Excel, we set up a calculation table like Table 10.9.

---

[12] x_j is called **gross output** of the jth good since $(1 - a_{jj})x_j$ is the **net output** of the jth good.

Table 10.9 The basic scheme of input–output models

	A	B	C	D	E	
1	A			I		Input Coefficient matrix
2	0.30	0.11		1.00	0.00	
3	0.25	0.35		0.00	1.00	
4	0.55	0.46			2x2	
5	I-A			d		The inverse N arrangement
6	0.70	-0.11		3.00	d1	
7	-0.25	0.65		4.00	d2	
8		2x2		2x1		
9	inv(I-A)			x#=inv(I-A)*d		
10	1.52	0.26		5.59	x1#	
11	0.58	1.64		8.30	x2#	
12		2x2		2x1		
13	Manpower planning					Manpoer planning
14	A0			total labor req		
15	0.45	0.54		7.00		
16		1x2				
17	Labor allocation					
18	2.52	4.48				
19	L1=a01(x1#)	L2=a02(x2#)				

Step 1. We enter the (2×2) matrix A in A2:B3 and enter the (2×2) identity matrix I in D2:E3.

We will enter the technology matrix $(\mathrm{I} - \mathrm{A})$ below input-coefficient matrix A. In A6, enter =D2−A2, copy A6 to A6:B7, and we have $(\mathrm{I} - \mathrm{A})$ in range A6:B7. The final demand vector $\mathrm{d} = (3, 4)'$ is entered in D6:D7.

Step 2. Use Excel function =MINVERSE to enter the inverse matrix inv(I−A) in A10:B11.

Step 3. Multiply inv(I−A) to the final demand vector d to obtain the $\mathrm{x}^{\#} = \mathrm{inv}(\mathrm{I}-\mathrm{A}) * \mathrm{d} = (5.59,\ 8.30)'$ in D10:D11. This is the equilibrium gross-output vector. The equilibrium gross output for the first (agricultural) good is $\mathrm{x}_1^{\#} = \$5.59$ and the equilibrium gross output for the second (manufacturing) good is $\mathrm{x}_2^{\#} = \$8.30$.

10.6.5 *The basic problem of input–output models*

The basic mathematical problem of input–output models is to find a unique, nonnegative solution vector $\mathrm{x}^{\#} = (\mathrm{x}_1^{\#}, \mathrm{x}_2^{\#})'$ for Eq. (10.53), given the nonnegative final demands $\mathrm{d} = (\mathrm{d}_1, \mathrm{d}_2)' \geq 0$. The model (10.53) is called **viable (or productive)** if, for any final demand vector $\mathrm{d} \geq 0$, (10.53) has a nonnegative solution $\mathrm{x}^{\#} \geq 0$.

Proposition 10.1. Solow's condition. *If every column sum of input-coefficient matrix A is less than one, i.e.*

$$\sum_{\mathrm{i}} \mathrm{a}_{\mathrm{ij}} < 1 \quad \textit{for all } \mathrm{j}, \tag{10.55}$$

then the input–output model (10.52) *or* (10.53) *is viable.*

Without getting into too much mathematics, we can verify the proposition by using the following example.

Example 10.21 Check Solow's condition. This proposition is satisfied for the coefficient matrix A in Table 10.9. The first column sum is 0.55 and the second column sum is 0.46, both of which are less than one, as shown in A4:B4. The solutions x_1 and x_2 in D10:D11 are both positive values. Therefore, the input–output model in Table 10.9 is viable. □

10.6.6 *Manpower planning*

Another important application of the above input–output model is in manpower planning for producing equilibrium outputs. Let $A_0 = (a_{01}, a_{02})$ be a row vector of the amount of primary input (labor) used in the production of $1.00 worth of the jth commodity. The a_{0j} is called the **labor coefficient of industry j**. Then

$$a_{0j}x_j^{\#}, \quad \text{for } j = 1, 2,$$

is the amount of labor cost required in producing the jth equilibrium commodity $x_j^{\#}$. The total amount of labor required to produce all the required equilibrium gross outputs is

$$L^{\#} = A_0x^{\#} = a_{01}x_1^{\#} + a_{02}x_2^{\#}, \tag{10.56}$$

where $A_0 = (a_{01}, a_{02})$ is the labor input-coefficient (row) vector and $x^{\#}$ is the equilibrium gross output (column) vector. $L^{\#}$ is called the **equilibrium total labor requirement**. It is the total amount of labor needed to produce the equilibrium gross outputs $x_1^{\#}$ and $x_2^{\#}$. If initial **labor endowment** in the economy is denoted by L, then

if $L^{\#} = L$, there is a **full employment** at the equilibrium,
if $L^{\#} < L$, there is an **unemployment** at the equilibrium, and
if $L^{\#} > L$, there is a **labor shortage** at the equilibrium.

Thus the input–output model may also be used for **manpower planning** in an economy.

Example 10.22 Required labor inputs and manpower planning. Continuing from Example 10.20, the labor input coefficient for the agriculture good is $a_{01} = 1 - 0.30 - 0.25 = 1 - 0.55 = 0.45$, and the labor input coefficient for the manufacturing good is $a_{02} = 1 - 0.11 - 0.35 = 1 - 0.46 = 0.54$. The labor input-coefficient vector is $A_0 = (a_{01}, a_{02}) = (0.45, 0.54)$, which is shown in A15:B15. Multiply vector A_0 by the equilibrium gross-output vector $x^{\#} = (5.59, 8.30)'$, we have the equilibrium total labor requirement $L^{\#} = A_0x^{\#} = 7.00$. We have the **equilibrium labor allocation for the agricultural industry** $L_1 = a_{01}x_1^{\#} = 0.45(5.59) = 2.52$ in cell A18, and the **equilibrium labor allocation for the manufacturing industry** $L_2 = a_{02}x_2^{\#} = 0.54(8.30) = 4.48$ in cell B18.

If current endowment of labor L is 6.50, there will be a shortage of labor in producing the equilibrium outputs $x^{\#}$, and then wages will tend to rise and the equilibrium outputs may not be achieved. If labor endowment L is 8, there will be a surplus of labor, and then wages will tend to fall. Only if labor endowment L is 7, the same as the equilibrium labor requirement $L^{\#}$, can the equilibrium outputs be produced under full employment of labor. □

10.7 Summary

The main idea of this chapter revolves around the basic matrix equation $Ax = c$ for a linear equation system, and the method of solution using Excel commands, arranged systematically in Table 10.6.

Since A is a matrix and x and c are vectors, the first part of this chapter explains the rules of mathematical operations of vectors (Sec. 10.1) and matrices (Sec. 10.2), with the analogy between the matrix system and the familiar real number system. It is important to realize that a matrix is a collection of vectors, either a collection of row vectors or column vectors. This will help understand the procedure of matrix multiplication. There is no analogy with vector or matrix transpose in the real number system. However, there is a similarity between the inverse matrix and the inverse in real number system (Sec. 10.3), and the solution to the matrix equation can be understood easily with the analogy (Sec. 10.4).

As shown in Fig. 10.3, the study of scalar product in vectors leads to its application in index numbers (Sec. 10.2), and the study of matrix multiplication and inversion leads to the method of solving the variable vector $x = A^{-1}c$, its applications in macroeconomic policy models (Sec. 10.5), and the input–output models (Sec. 10.6). We use these economic topics here to demonstrate the power of Excel functions and spreadsheets in matrix algebra. The readers are encouraged to pursue more advanced topics in each field.

Finally, as we have noted in this chapter, one of the features of Excel matrix functions is that they are "alive" in the sense that they can be reused as templates.

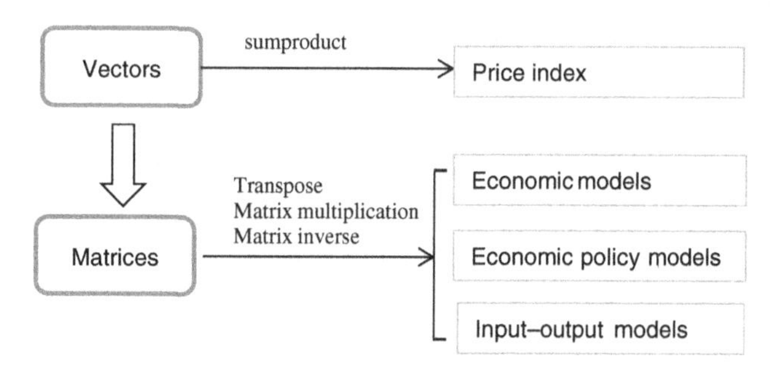

Fig. 10.3 Summary of the chapter

Review of Basic Equations and Functions

Vector operations

Let $a = [a_i]_{nx1}$, $b = [b_i]_{nx1}$

(a) $a = b$ iff $[a_i] = [b_i]$

(10.7) (b) $a+b = [a_i + b_i]$

(10.8) (c) $ka = [ka_i]$

(10.10) Scalar product

$$a'b \equiv a^T b = [a_i]'[b_i] = \sum_i a_i b_i = \text{sumproduct}(a,b)$$

(10.11) Total cost index $TC = \frac{p^{1\prime}q^1}{p^{0\prime}q^0}$

(10.12) Laspeyres price index $L_a = \frac{p^{1\prime}q^0}{p^{0\prime}q^0}$

(10.13) Paasche price index $P_a = \frac{p^{1\prime}q^1}{p^{0\prime}q^1}$

(10.14) Fisher price index

$$F_i = (L_a * P_a)^{1/2} = \sqrt{\frac{p^{1\prime}q^0}{p^{0\prime}q^0}\frac{p^{1\prime}q^1}{p^{0\prime}q^1}}$$

(10.15) Dow Jones industrial average index

$$DJ_t = \frac{1'p^t}{1'p^0}$$

Matrix operations

Let $A = [a_{ij}]$, $B = [b_{ij}]$

(a) $A = B$ iff $[a_{ij}] = [b_{ij}]$

(10.17) (b) $A + B = [a_{ij} + b_{ij}]$

(c) $kA = [ka_{ij}]$

(10.21) Matrix multiplication

$$\begin{matrix} A_1 \\ A_2 \\ A_3 \end{matrix}\begin{bmatrix} a_{11} & a_{22} \\ a_{21} & a_{22} \\ a_{31} & a_{32} \end{bmatrix}_{3x2} \begin{bmatrix} b_{11} & b_{12} \\ b_{21} & b_{22} \end{bmatrix}_{2x2} = \begin{bmatrix} A_1B^1 & A_1B^2 \\ A_2B^1 & A_2B^2 \\ A_3B^1 & A_3B^2 \end{bmatrix}_{3x2}$$

$$\equiv \begin{bmatrix} c_{11} & c_{12} \\ c_{21} & c_{22} \\ c_{31} & c_{32} \end{bmatrix}_{3x2}$$

(10.22) $c_{ij} \equiv A_i B^j = \sum_{k=1}^{r} a_{ik} b_{kj}$,

(10.34) Solution

If $Ax = c$, A is non-singular, then

$$x^\# = A^{-1}c$$

(10.53) Given A and d, the fundamental equation of input-output models is

$$(I - A)x = d$$

(10.54) Equilibrium gross output

$$x^\# = (I - A)^{-1}d$$

(10.55) Solow's condition

$$\sum_i a_{ij} < 1 \text{ for all } j$$

(10.56) Equilibrium labor requirement

$$L^\# = A_0\, x^\#$$

Key Terms: Economics and Business

Key Terms: Excel

Homework Chapter 10

Vectors and Matrices

10-1 Matrix addition and subtraction Consider the matrices

$$\mathrm{A} = \begin{pmatrix} 15 & 8 & 12 \\ -7 & 0 & 1 \\ 2 & 3 & -4 \end{pmatrix}, \quad \mathrm{B} = \begin{pmatrix} -10 & 0 & 2 \\ 6 & -0.67 & -1 \\ 5 & 5 & 4 \end{pmatrix}.$$

(a) Show that $\mathrm{A}+\mathrm{B}=\mathrm{B}+\mathrm{A}$ (Hint: Show that the RHS is the same as the LHS)
(b) What is $\mathrm{A}-\mathrm{B}$?
(c) Calculate $2\mathrm{A}-3\mathrm{B}$
(d) Calculate $4(\mathrm{A}-\mathrm{B})\mathrm{A}$

10-2 The transpose of a matrix Let the matrices be

$$A = \begin{pmatrix} 4 & 6 \\ 1 & 2 \end{pmatrix}, \quad B = \begin{pmatrix} -3 & 5 & 0 & 1 \\ 7 & 2 & -4 & 3 \end{pmatrix}.$$

(a) Derive the transpose, A' and B'.
(b) Prove the three properties of taking the transpose of a matrix: $(A')' = A, (A+B)' = A' + B'$, and $(AB)' = B'A'$.

(Hint: Before taking the transpose of AB, the function of underlining AB must be deleted by changing the entry of the cells to values. Note that $A + B$ in this case is not defined).

10-3 The matrix inverse Let the matrices A and B be the same as those in HW10-1. Show that

(a) The inverse of an inverse matrix is the matrix itself: $\text{inv}(\text{inv}(A)) = A$ (i.e., $(A^{-1})^{-1} = A$)
(b) The transpose of an inverse matrix is the inverse of the transpose of the matrix: $(\text{inv}(A))' = \text{inv}(A')$ (i.e., $(A^{-1})' = (A')^{-1}$).
(c) The inverse of a product of matrices is the product of the inverse of the two matrices in reversed order: $\text{inv}(AB) = \text{inv}(B) * \text{inv}(A)$ (i.e., $(AB)^{-1} = B^{-1}A^{-1}$)

10-4 Simulation Let A be a 5×10 matrix generated randomly from 0 to 1, and let B be a 5×5 matrix from 1 to 10. All with one decimal place. Answer the same questions as for HW10-2.

10-5 Simulation Let A be a 5×5 matrix generated randomly from 0 to 1, and let B be a 5×5 matrix from 1 to 10. All with one decimal place. Answer the same questions as for HW10-3.

10-6 System of equations Solve the following equations.

In each of the following cases, reproduce the format of Table 10.6 and find the solution (do not use Table 10.6 as a template).

(a) $x/2 + 4y = 16$
$3x - 10y = 40$

(b) $10x + y - 2z = 30$
$x - y = 1$
$3x - 4y + z = 15$

10-7 Systems of equations Using the results in HW10-6 as templates, solve the following equations:

(a) $q = 30 + 5p$
$q = 100 - 2p/3$

(b) $4p_1 + p_2 = 3$
$5p_1 - p_2 + 2p_3 = 1$
$p_1 - 5p_2 + 7p_3 = 2.$

(c) Use the matrices obtained in (b) as templates, and solve the equations in (a).
(d) In (a), if the two equations show demand and supply relations, which one is the demand equation, and which one is the supply equation? Why?

10-8 Random number simulation Let A be a 10×10 matrix of random numbers from 0 to 10, and let c be a 10×1 vector of random numbers from 100 to 200. Find the solution x in $\mathrm{Ax} = \mathrm{c}$. Enter <F9> several times and watch how the numbers dance. All numbers should have two decimal places.

Price indexes

10-9 Price indexes Construct a table, which is the same as Table 10.2, where pt and qt are random numbers such that p0 ranges from 0 to 10, q0 from 100 to 200, p1 from 5 to 20, q1 from 100 to 150.

Find the four price indexes (Total cost, Laspeyres, Paasche, and Fisher) between the two periods.

10-10 The Edgeworth Price Index Instead of choosing a base period quantity or a current period quantity as weights, Edgeworth suggested using the average quantity of these two periods:

$$\mathrm{E_t} = \frac{\sum_{i=1}^{n} p_{it}(q_{i0} + q_{it})}{\sum_{i=1}^{n} p_{i0}(q_{i0} + q_{it})},$$

which is called **the Edgeworth price index**.

(a) Write $\mathrm{E_t}$ in scalar product form.
(b) Using the table in HW10-9, find $\mathrm{E_t}$ by using Excel sumproduct formula.
(Hint: The numerator will be sumproduct(pt, (q0 + q1)) if pt, q0, and q1 are named.)

10-11 The Standard and Poor's (S&P)[13] 500 Composite Index This index comprises 500 firms in industry, utilities, transportation, and finance, and is defined simply as the market value of 500 stock at tth period divided by the market value of the 500 stock at the base period:

$$\mathrm{SP_t} = \frac{p^{t\prime} q^t}{p^{0\prime} q^0}.$$

This formula is the same as the total cost index in Example 10.3. Extending the table generated in HW10-9 to 500 prices and quantity, calculate $\mathrm{SP_t}$. (Hint: Use a horizontal split box to divide the screen).

10-12 Price indexes Using random numbers, generate the following price and quantities for 100 commodities, which are numbered from 1 to 100, with one digit after the decimal point. The base year is 2005.

(a) 2005, p05 = random numbers from 100 to 150
q05 = random numbers from 100 to 150.
(b) 2006, p06 = random numbers from 100 to 200
q06 = random numbers from 100 to 200.
(c) 2007, p07 = random numbers from 100 to 250
q07 = random numbers from 100 to 250.

[13]See Chapter 19, Lee, Lee, and Lee (2000); Chapter 18, Lind, Marchal, Mason (2002).

(d) Write the Laspeyres price index, Paasche price index, and Fisher's ideal price index for t = 06, and 07. Calculate these three indexes for 06 and 07.
(e) Assume that the commodities consist of stocks. Using the formula for the S&P 500 Index and the simple DJIA, calculate these two indexes for 06 and 07.
(f) What conclusion can you draw from the Fisher index for 2006 and 2007?
(g) Comparing the Laspeyres, Paasche, and Fisher price indexes for the years 2006 and 2007, can you find any relationship among them in terms of magnitude?
(h) What is the use of the Laspeyres price index and Paasche price index? Where are they applied in our society? From your calculation, can you see what the pros and cons are of the calculating procedure of these price indexes?

Macroeconomic models

10-13 The simple IS/LM macroeconomic model The Simple IS/LM macroeconomic model is given as

$$\begin{aligned} Y &= C + I + G, \\ C &= 200 + 0.8Y, \\ I &= 1000 - 2000r, \\ M &= -1000r + 0.1Y, \end{aligned}$$

where Y is national income, M is real money supply, r is interest rate (expressed in fraction), and G is government expenditure. The endogenous variables are Y, C, I, and r. The levels of government purchases and of money supply are exogenous, and are given as $G = 2000$ and $M = 1500$.

(a) Find the equilibrium values $Y^{\#}$, $C^{\#}$, $I^{\#}$, and $r^{\#}$.
(b) In the above problem, reduce the four-equation model to a two-equation model (show how you reduce the four equations):

$$\begin{aligned} 0.2Y + 2000r &= 1200 + G; \\ -1000r + 0.1Y &= M. \end{aligned}$$

(c) Solve for the actual values of Y and r, assuming that $G = 3000$ and $M = 2000$.
(d) If money supply decreases by 100, what are the effects on $Y^{\#}$ and $r^{\#}$?
(e) If government expenditure increases by 100, what is the effects on $Y^{\#}$ and $r^{\#}$?

10-14 Multiple constant vectors Let the Keynesian model be

$$\begin{aligned} Y &= C + I + G, \\ C &= 200 + 0.8Y, \\ I &= 1000 - 2000r, \end{aligned}$$

where the endogenous variables include national income (Y), consumption (C), and investment (I). The exogenous variables include government spending G and the interest rate (r).

Set up this model with a 3×3 coefficient matrix. Find the equilibrium Y, C, and I by using the following assumptions:

$$\text{(a) Let } G = 100, r = 0.05; \quad \text{(b) Let } G = 100, r = 0.06;$$
$$\text{(c) Let } G = 150, r = 0.05; \quad \text{(c) Let } G = 150, r = 0.06.$$

Find the solution for each of the four cases.
(Hints: Despite the question, you should find the solutions of all four cases simultaneously by using the same format as Table 10.8. The four cases should be entered in columns M–P.)

10-15 (Klein 1997, Section 4.4) The profit functions of two airplane manufacturers, the European firm Airbus and the American firm Boeing, are

$$A = -B/2 + F,$$
$$B = -A/2 + G,$$

where A is the profit of Airbus, B is the profit of Boeing, F is a subsidy from the governments of Europe to Airbus, and G is a subsidy from the United States government to Boeing.

(a) Express these two equations as a matrix system with A and B as the endogenous variables and F and G as the exogenous variables.
(b) Determine the solution if each government provides a subsidy of $100 million.
(c) What happens to the profit of each firm if the European governments increase their subsidy to $200 million, while the United States government keeps its subsidy equal to $100 million?

(Hints: In this case, since only two different scenarios are involved, you need not name F and G. Reproduce Table 10.5 without the naming box).

10-16 The IS/LM model (Hoy *et al.* 2001, p. 306) The simple IS/LM macroeconomic model is given as

1.	$Y = C + I + X$	Equilibrium condition of the commodity market
2.	$C = 210 + 0.8Y$	Consumption function
3.	$I = 20 - R$	Investment function
4.	$200 - 0.3Y + 0.5E = X$	Export function
5.	$480 = -R + 0.5Y$	Equilibrium condition of the money market
6.	$E = 180$	Exchange rate (the price of foreign currency).

where X is net exports, and E is the indirect exchange rate (e.g., Yen/dollar), and R is the interest rate in percentage. The endogenous variables are Y, C, I, X, R, and E.

(a) Using a 6×6 coefficient matrix, find the equilibrium values of the endogenous variables. Make sure the equilibrium values are clearly designated.
(b) Why are they called "equilibrium" values?
(c) As an economic model, is this model realistic? Explain your answer.

(Hints: $(Y\#, C\#, I\#, X\#, R\#, E\#) = (1000, 1010, 0.00, -10, 20, 180)$)

10-17 The Keynesian IS/LM/BP model (Hoy, op. sit.) Among the six questions in HW10-16, if equation 5 is written in three equations, we have:

5a.	$M = 480,$	Supply function of money,
5b.	$L = -R + 0.5Y$	Demand function of money,
5c.	$M = L$	Equilibrium condition of the money market.

Furthermore, Eq. (6) may also be expanded as

6a.	$A = R - 10$	Capital asset balance,
6b.	$B = X + A$	Balance of payment composition,
6c.	$B = 0$	Equilibrium condition in the foreign trade sector.

Thus, we have three markets and a total of 10 equations, 1–4, 5a–5c, and 6a–6c, in 10 variables: $Y, I, C, X, R, E, M, L, A, B$. Using a 10×10 coefficient matrix, find the equilibrium values of the 10 endogenous variables. Are they realistic? why?
(Hints : $(Y^{\#}, C^{\#}, I^{\#}, X^{\#}, R^{\#}, E^{\#}, M^{\#}, L^{\#}, A^{\#}, B^{\#}) = (1000, 1010, 0.00, -10, 20, 180, 480, 480, 10, 0)$)

10-18 The reduced form of the IS/LM/BP model In HW10-16 and HW10-17 above, 1–4, 5a–5c, and 6a–6c, we have three equilibrium conditions: one each for the commodity market (1), the money market (5c), and the foreign sector (6c). Reduce the 10 equations into three equilibrium conditions for the three markets.

1.	$0.5Y + R - 0.5E = 430.$	Commodity market
2.	$0.5Y - R = 480.$	Money market
3.	$0.3Y - 0.5E = 190.$	Foreign exchange market

(a) What are the endogenous variables?
(b) Using a 3×3 coefficient matrix, find the equilibrium values of the endogenous variables.
(c) Find the equilibrium values of other endogenous variables. Are they the same as those in HW10-17?
(d) The right-hand side of the second equation represents money supply $M_0 = 480$. If money supply increases to 500, what are the effects on equilibrium values? Does this make sense? Why? Show your calculation.

10-19 The policy model In the model of Table 10.8, let $a = 15$. Reduce this eight-equation model to a three-equation model of Y, T, and R (Hint: After substitution, the three equations will be (10.36), (10.38), and (10.43)).

(a) Write the model in matrix form and find the equilibrium values for $Y^{\#}, T^{\#}$, and $R^{\#}$.
(b) Eliminate T in the three-equation model to obtain a two-equation model of Y and R.
(c) Find the equilibrium income and interest rate. Are your $Y^{\#}$ and $R^{\#}$ different from the three-equation model in (a)?

10-20 A simple policy model II A simple IS/LM macroeconomic model is given as follows

$$
\begin{aligned}
&Y = C + I + X + G_0 && (1)\\
&C = 300 + 0.8(Y - T) && (2)\\
&0.5Y + 0.7X = 400 && (3)\\
&I + 2000r_0 = 2000 && (4)\\
&L = -1000r_0 + 0.3Y && (5)\\
&M = 6000 + 0.5Y && (6)\\
&L = M, && (7)
\end{aligned}
$$

where M is money supply, L is money demand, X is net exports, T is tax, Y, C, I, G_0, r_0 are national income, consumption, investment, government expenditure, and interest rate, respectively. The values of r_0 and G_0 will be given below.

(a) How many variables are in this model? What are they? How many are endogenous variables? Why are they called endogenous?
(b) How many are exogenous variables? What are they? Why are they called exogenous?
(c) Set up this model with a 7×7 coefficient matrix. Find the equilibrium value of Y, C, I, X, T, L, and M for the following three cases (purely for grading purposes, **do not change the order of the variables and the sequence of the equations**). A change in the value of G_0 is called the **fiscal (F) policy** of the government, and that of r_0 is called the **monetary (M) policy** of the government. Thus, the government has three policy choices in this model.
The default values of parameters are given as follows: $r_0 = 0.05$ and $G_0 = 5000$. Suppose the economy is in the process of recovery and the government attempts to stimulate the economy by encouraging consumption and by tax reduction. We consider the following three cases:

Case 1. Reduce r_0 to 0.04 (the monetary (M) policy);
Case 2. Increase G_0 to 5500 (the fiscal (F) policy);
Case 3. Reduce r_0 to 0.04 and increase G_0 to 5500 (using both monetary and fiscal (mixed) policy).

The arrangement of your answer should be similar to that of Table 10.8.
(d) From your results, can $X^{\#}$ be negative? Why? What is the meaning of negative $X^{\#}$ if it is?
(e) Why are these variables, $Y^{\#}$, etc. called equilibrium values? Please be specific in indicating what is meant by "equilibrium". How many markets are there in the model? What are they?
(f) We now consider the impact of monetary and fiscal policies. The impact is measured by subtracting the default value of the original equilibrium values from the values resulting from the new policy. Enter the effects in the last Step 5 table as in Table 10.8.
(g) What is the impact of the monetary policy on all the equilibrium values? Which variable increases and which decreases?
(h) Throughout all the policy options, L and M always remain the same, either 520 or 550. Why?
(i) Since the economy is in recession, the Bush Administration is trying to stimulate the economy by encouraging consumption C (not necessarily national income) and reducing taxes T. According to this model, which economic policy is most effective for this purpose? Note that you have

four choices: the M policy, the F policy, the mixed policy, or no change at all. Is your answer realistic? Justify your answer.

10-21 A simple policy model III A simple IS/LM macroeconomic model is given as

$$
\begin{aligned}
&\mathrm{Y} = \mathrm{C} + \mathrm{I} + \mathrm{X} + \mathrm{G_0} && (1)\\
&\mathrm{C} = \mathrm{a} + 0.8\mathrm{Y} && (2)\\
&\mathrm{I} + \mathrm{R} = 20 && (3)\\
&\mathrm{X} = 200 - 0.3\mathrm{Y} + 0.5\mathrm{E} && (4)\\
&\mathrm{L} = 0.5\mathrm{Y} - \mathrm{R} && (5)\\
&\mathrm{M} = \mathrm{M_0} && (6)\\
&\mathrm{L} = \mathrm{M} && (7)\\
&0.3\mathrm{Y} - (\mathrm{R} - 10) = 0.5\mathrm{E} + 200. && (8)
\end{aligned}
$$

The default values of parameters are given as follows: $\mathrm{a} = 210, \mathrm{M_0} = 520$, and $\mathrm{G_0} = 0$. Suppose the economy is in the process of recovery and the government attempt to stimulate the economy by three policy tools. We consider the following three cases:

Case 1. increase $\mathrm{G_0}$ to 5 (the fiscal (F) policy);
Case 2. increase $\mathrm{M_0}$ to 550 (the monetary (M) policy);
Case 3. increase the basic consumption to 212 (the consumption (C) policy).

Answer the same questions (a) to (i) in HW10-20.

10-22 A policy model Consider the following Keynesian macroeconomic model:

$$
\begin{aligned}
&\mathrm{Y} = \mathrm{C} + \mathrm{I} + \mathrm{G_0} + \mathrm{X} && (1)\\
&\mathrm{C} = \mathrm{a} + 0.9(\mathrm{Y} - \mathrm{T}) && (2)\\
&\mathrm{T} = \mathrm{T_0} && (3)\\
&\mathrm{I} = 65 - \mathrm{R} && (4)\\
&\mathrm{X} = 60 - 0.1\mathrm{Y} && (5)\\
&\mathrm{L} = 5\mathrm{Y} - 1000\mathrm{R} && (6)\\
&\mathrm{M} = \mathrm{M_0} && (7)\\
&\mathrm{L} = \mathrm{M}, && (8)
\end{aligned}
$$

where T is tax revenue, R is the interest rate in percent, X is net exports, and a is a parameter. Other variables have the usual meaning.

The default values of parameters are given as follows: $\mathrm{a} = 15, \mathrm{M_0} = 1500, \mathrm{G_0} = 94$, and $\mathrm{T_0} = 20$. We consider the following five cases:

Case 1. no change in government policies (the default policy);
Case 2. "a" doubled (the consumption policy);
Case 3. $\mathrm{M_0}$ doubled (the monetary policy);
Case 4. $\mathrm{G_0}$ doubled (the fiscal policy);
Case 5. $\mathrm{T_0}$ halved (the tax policy);

Answer the same questions as in HW10-20.

The input–output models

10-23 A 2×2 input–output model Let the input-coefficient matrix and the final demand vector be given as follows.

$$A = \begin{bmatrix} 0.2 & 0.5 \\ 0.3 & 0.3 \end{bmatrix}, \quad d = \begin{bmatrix} 3 \\ 10 \end{bmatrix}.$$

(a) Find the equilibrium gross outputs.
(b) Let the input coefficient of the first industry be $a_{01} = 0.5$, and that of the second industry be $a_{02} = 0.2$. Assume that the economy has labor endowment of 10 units. Find the equilibrium labor requirement for the two industries. What is the total equilibrium labor input requirement?
(c) In order to achieve the production of equilibrium gross output, does the economy have sufficient labor resource? How do you know? What are the consequences?

10-24 A 3×3 input–output model Let the 3 × 3 input matrix be

$$\mathbf{A} = \begin{bmatrix} 0.2 & 0.3 & 0.2 \\ 0.4 & 0.1 & 0.2 \\ 0.1 & 0.3 & 0.2 \end{bmatrix}$$

$d = (10, 5, 6)'$, and $A_0 = (0.3, 0.3, 0.4)$. Answer the same questions as HW10-23.

10-25 The structure of the US economy (Miller and Blair, 1985, p. 425) Table HW10-25 shows seven sectors of the 1977 US National input–output Table. The column titles 1, 2, ..., 7 correspond to the row titles of the first column.

(a) Give an economic interpretation of the fourth row (Manufacturing);
(b) Give an economic interpretation of the last row (labor);

Table HW10-25 1977 US National input–output table

	A	B	C	D	E	F	G	H	I	J
1		Sector	1	2	3	4	5	6	7	d
2	1	Agriculture	0.2463	0.0004	0.0035	0.0470	0.0014	0.0044	0.0007	27405
3	2	Mining	0.0021	0.0713	0.0091	0.0573	0.0007	0.0183	0.0049	-29295
4	3	Construction	0.0107	0.0375	0.0011	0.0064	0.0141	0.0297	0.0202	206809
5	4	Manufacturing	0.2020	0.0952	0.3720	0.3819	0.0601	0.0862	0.0134	578483
6	5	Trade & Trans	0.0576	0.0221	0.1096	0.0651	0.0603	0.0253	0.0092	323867
7	6	Services	0.1007	0.1265	0.0792	0.0750	0.1772	0.1749	0.0253	648710
8	7	Other	0.0027	0.0037	0.0032	0.0077	0.0079	0.0088	0.0020	220323
9		Labor	0.125	0.319	0.225	0.192	0.375	0.369	0.413	

(c) Give an economic interpretation of column 1 (agriculture);
(d) Give an economic interpretation of the last column (J).
(e) Find the equilibrium output x*;
(f) Find the equilibrium total labor requirements L*;
(g) Find the equilibrium labor requirement for each sector.
(h) Draw a cost structure pie chart for the agricultural and manufacturing industries.

10-26 A simulation model We simulate Table HW10-25 as follows. In the table, add two rows below "labor": row 10 is called profit, and row 11 is the column "sum" of C2:I10. In C2, enter =rand() and copy it down to C2:I10 (including labor and profits). The column sums are all larger than 1. Copy the whole table, that is, A1:J11 to a table below it, say, in A13:J23, and call the copied table the New Table. In the C2 position of the New Table, say C14, enter C2/C$11, and copy C14 to the whole New Table, including the labor and profit rows, but excluding column d. This will simulate the input-coefficient matrix, including the labor and profit coefficients (called the value added sector) rows. In column J, J14:J20 (the final demand sector) enter random numbers from 200 to 1000. This will simulate the household final demand. Lastly, change the whole New Table to values. This will give a table simulating that in Table HW10-25.

Answer the same questions as those in HW10-25.

Extra homework problems

10-27 Uses of CPI We have indicated in Example 10.3 that the consumer price index (CPI) is based on the Laspeyres index L_a. For simplicity, we take this index to be CPI. The following definitions are useful.

(a) **Real income** (or **deflated income, or income in constant dollars**) is defined as (nominal income*100)/CPI.
(b) **The purchasing power of money** (or **purchasing power of the dollar**) is defined as the inverse of CPI.
(c) **The inflation rate** is defined as the percentage change (or growth rate) of CPI over time.

We use the consumption and income data of Table 5.7. In Table 5.7, in D4, enter "CPI", in D5, enter 100.00, and in D6, enter =round(rand()*5, 2) + D5 + 0.1, assuming a minimum inflation rate per year of 0.1. Copying D6 down to D22, we have a time series of the CPI. Using this set of CPI, find the real income, real consumption, purchasing power of money, and inflation rate from 1980 to 1997 in the USA. Draw a composite chart of CPI, inflation rate, and purchasing power of money, and give economic interpretations to your findings.

Excel matrix functions

10-28 Templates The coefficient matrix A in Table 10.6 is a 3×3 matrix, but that of Table 10.7 is a 2×2 matrix. Using Table 10.6 as a template, find a way to derive the result of Table 10.7 (that is, without constructing Table 10.7).
(Hints: Enter A of Table 10.7 in B9:C10 of Table 10.6, enter 0 to B11:C11 and D9:D10, and 1 to D11. Rewrite column c, and we have x# of Table 10.7).

Part IV

Optimization

Chapter 11

Production and Utility Functions — 3D Graphics

Chapter Outline

Objectives of this Chapter

In Chapters 3, 4, and 10, the equations and models we have been dealing with are linear. Many nonlinear equations arise in economics and business applications. This chapter introduces nonlinear production functions, with references to nonlinear utility functions. The most popular nonlinear production function probably is the Cobb–Douglas (CD) form of production function. Its functional form is simple and symmetric, and yet it provides all the interesting and important features of a production function. We first discuss in detail the properties of the partial productivity curves, the total productivity surfaces, and isoquants, using the powerful Excel charts. This will be followed by discussions of average and marginal productivities of labor and capital, which then relate to the marginal rate of technical substitution (MRTS) and factor market equilibrium.

After a detailed discussion of the general properties of the CD production function, we introduce the constant elasticity of substitution (CES) production function. With the help of powerful Excel charts, for the first time in the literature, we are able to graph five different cases of the CES production function accurately with colors. The same method can be applied to graphing other functional forms, including one- and two-variable profit functions and saddle points.

Appendix 11A explains the four methods of constructing the basic sensitivity table, two of which have already been covered in the previous chapters. Appendix 11B shows the shape of the production surface under different returns to scale. For the first time in the literature, we also present accurately the production function under different returns to scale in 3D and 2D, using the Excel charts.

Readers may concentrate only on the properties of the CD production function in Secs. 11.1, 11.2, and the topics on profit maximization in Sec. 11.5. Those who are more technically oriented are encouraged to explore the other sections and the appendixes.

11.1 The Cobb–Douglas Production Function

11.1.1 *An overview of production functions*

In the process of production, a firm combines labor, capital (like machinery), and other resources (like land and raw materials) as inputs to produce a certain output Q. This relation is usually expressed as follows(Fig. 11.1):

Labor, capital, and other resources are called the **factors of production**. For the time being, if we omit other resources for simplicity, then the input and output relation can be summarized into a production function as

$$Q = F(L, K). \tag{11.1}$$

Traditionally, labor is called the **primary factor of production** and capital is called the **secondary factor of production**.[1] The general assumption here is that the function F is a continuous and differentiable function. By writing the production function like (11.1), we are assuming that the production function represents the maximum output that can be produced using various combinations of labor and capital.

11.1.2 *The Cobb–Douglas production function*

There are many types of production function. The most common one is called the **Cobb–Douglas production function** in honor of its proponents, Professor Charles Cobb (a mathematician) and Professor Paul Douglas (an economist). It is written as

$$\boxed{Q = AL^{a}K^{b},} \tag{11.2}$$

Fig. 11.1 Production relations — input and output

[1]The readers may notice that the classical theory of production ignores the interindustrial inputs and outputs as discussed in Chapter 10. In this case, the secondary factors, capital, may include intermediate outputs used as inputs.

where L and K are variables and A is a parameter representing the **technological factor**, which determines the extra output obtainable even if labor and capital remain constant. The parameters a and b represent the **labor share** and **the capital share of output**, respectively. As such, they are usually assumed to be positive fractions. If the sum is unity, $a + b = 1$, we say the production function has **constant returns to scale (CRS)**, if $a + b > 1$, it has **increasing returns to scale (IRS)**, and if $a + b < 1$, it has **decreasing returns to scale (DRS)** (see Appendix 11B for explanations). The labor share of output, a, usually ranges from 60% to 70% of GDP, and the capital share of output, b, usually ranges from 30% to 40% of GDP (see Solow, 1957). Thus, the empirical form of the CD production function will generally be

$$Q = AL^{0.6}K^{0.4}, \tag{11.3}$$

which is a CD production function in CRS. It is implemented in Table 11.1. Parameters are given names and values as shown in E4:F6. Cell F4 is named "aa" (double a) for labor share a, since Excel cannot distinguish between the capital letter A and the lower case a, and we would like to use capital A as the technology factor in F6. For symmetry, we also name cell F5 "bb" (double b) for capital share b. Equation (11.3) indicates that, if $a = 0.6, b = 0.4$, and $A = 2$, then for the values of labor at 4 units and capital at 8 units, output Q is given as 10.6 units in E16, as can be calculated directly from Eq. (11.3). For a different labor and capital combination, we will obtain a different output in the table.

Table 11.1 The Cobb–Douglas production surface

	A	B	C	D	E	F	G	H
1	Properties of the Cobb–Douglas production function							
2		$Q = AN^aK^b$						
3					Parameter Table			
4			Labor share		aa	0.6		
5			Capital share		bb	0.4		
6			tech factor		A	2		
7								
8			C14:	=A*L^aa*K^bb				
9								
10	Sensitivity Table							
11	**The Total Factor Productivity Surface**							
12	Q		Capital					
13		0	0	4	8	12	16	20
14	Labor	0	0.0	0.0	0.0	0.0	0.0	0.0
15		2	0.0	5.3	7.0	8.2	9.2	10.0
16		4	0.0	8.0	10.6	12.4	13.9	15.2
17		6	0.0	10.2	13.5	15.8	17.8	19.4
18		8	0.0	12.1	16.0	18.8	21.1	23.1
19		10	0.0	13.9	18.3	21.5	24.1	26.4
20		12	0.0	15.5	20.4	24.0	26.9	29.4
21		14	0.0	17.0	22.4	26.3	29.5	32.3
22		16	0.0	18.4	24.3	28.5	32.0	35.0
23		18	0.0	19.7	26.0	30.6	34.3	37.5

The output sensitivity table

Instead of calculating output each time, Table 11.1 presents the relationships between some possible levels of inputs and the resulting outputs systematically. This table is the same as the sensitivity table in Chapter 8 showing the relations among interest rates, terms, and payments. It is called the **total factor productivity (TFP) table** or the **output sensitivity table**. It shows that, when both levels of labor and capital are given, the maximum output from using these levels of input is also determined. Thus the name "total factor productivity" (the largest amount of output that can be produced from using all production factors). In this example, labor and capital are variables and they can have different combinations of values, but aa, bb, and A are constant parameters and their values are given in cells F4:F6.

11.1.3 *Creating an output sensitivity table*

We now explain how to use Excel to construct the output sensitivity (or TFP) table for the CD production function. In Table 11.1, let the quantity of labor range from $0, 2, \ldots, 18$ units in B14:B23, and name the range "L". Let the quantity of capital range from $0, 4, \ldots, 20$ units in C13:H13, and name the range "K".

In C14, enter the production function in Excel format as

$$=A*L\hat{\ }aa*K\hat{\ }bb. \tag{11.4}$$

Copying C14 down to C14:H23, we have the output sensitivity table (for other methods of entry, see Appendix 11A).

The interpretation of this table is already given above, namely, the number 10.6 in E16 means that 4 units of labor and 8 units of capital are combined to produce the maximum amount of 10.6 units of output (If **free disposability** in production is allowed, the same amounts of L and K can produce less than 10.6 units, that is, the output can be anywhere between 0 and 10.6 units). Other entries can be interpreted similarly. Note that when the quantity of labor (or capital) changes, say, from $0, 1, \ldots, 10$, then the outputs in the sensitivity table will also change automatically, without any need to retype the table.

11.1.4 *The partial productivity curves*

Mathematically, given the production function (11.1), a **labor productivity function (or curve)** when K is fixed at certain level is denoted as

$$Q = f(L|K) = g(L) = AL^{a}K_0^{b}, \tag{11.4a}$$

where K is held constant; thus, it is a function of L only. Similarly, a **capital productivity function (or curve)** when L is fixed at certain level is written as

$$Q = f(K|L) = h(K) = AL_0^{a}K^{b}, \tag{11.4b}$$

which is a function of K only.

For example, if capital is fixed at 8 units, and labor changes from 0 to 18 units, as shown in E14:E23 of Table 11.1, we obtain the **labor productivity curve** at a given 8 units of capital. It shows the output levels for various amounts of labor with K = 8 units. Similarly, when labor is fixed at 4 units, and capital changes from 0 to 20 units, as shown in Table 11.1, C16:H16, we have the **capital productivity curve** at a given 4 units of labor. It shows the output levels for various amounts of capital with L = 4 units. Both curves show the maximum possible outputs when one factor changes while another factor remains at a fixed level. Hence, these are called **partial productivity curves**, as compared with total productivity surface, in which both factors can change at the same time.

Labor productivity curves

Figure 11.2(a) draws six labor productivity curves, each corresponding to a fixed level of capital input in C13:H13, taken from six ranges C14:C23, D14:D23, ..., H14:H23 in Table 11.1. These 2D curves are drawn using the Excel commands in Chapter 3.

The amount of fixed capital is shown in the legend of Fig. 11.2(a). The position of each curve depends on the given amount of capital associated with labor. In general, the more capital associated with labor, the more productive the labor becomes. In other words, the larger the amount of fixed capital, the higher the labor productivity curves. But when capital is fixed, as labor increases, the output increases at a decreasing rate, making the labor productivity curve concave to the labor axis. (This is the same as moving along the curve versus the shifting of the curve in demand and supply analysis in Chapters 3 and 4) So, given a level of labor, say L = 4 units, the productivity level depends on how much capital is available to work with. If capital is fixed at 8 units, the productivity level will be 10.6 units, which can be read from Table 11.1 and also from Fig. 11.2(a). The lowest curve

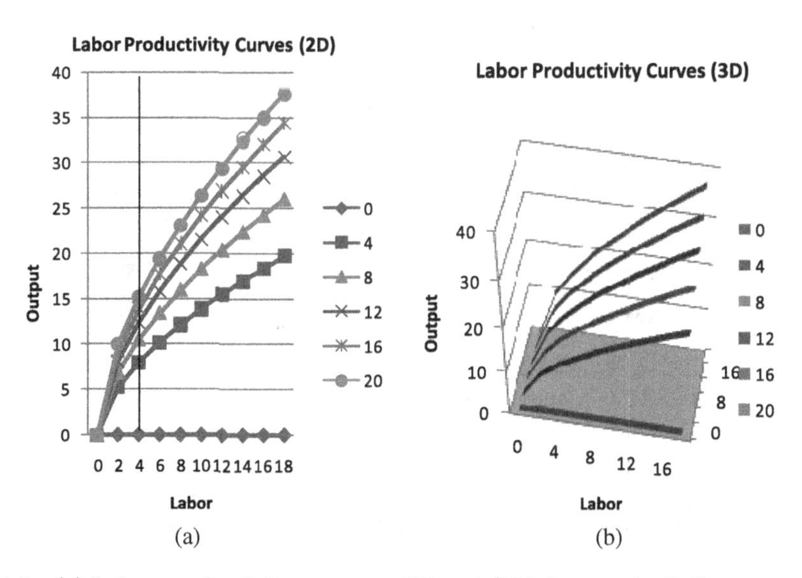

Fig. 11.2 (a) Labor productivity curves — 2D and (b) labor productivity curves — 3D

shows that when capital input is zero, there is no output at any level of labor, indicating that capital is an **essential input** for this production: No capital, no labor productivity at any level of labor input.

Figure 11.2(b) draws the labor productivity curve in 3D. To draw a 3D chart, change the chart type[2] in Fig. 11.2(a). Copy 2D to the right, select the chart area,[3] and then <RM> <Change Chart Type···> and select <3D Line>. Rename the chart as shown.

Capital productivity curves

A capital productivity curve shows the change of output when capital input changes with a given amount of labor. In Fig.11.3(a), each capital productivity curve represents a range of output, C14:H14, C15:H15,..., C23:H23 of Table 11.1. We reproduce (copy) Fig. 11.2(a) below the figure and rename it Fig. 11.3(a). Instead of starting from scratch, we may switch the row and column to derive the 10 capital productivity curves. Select Fig. 11.2(a), enter <RM>, and

<Select Data><Switch Row/Columns><OK>.

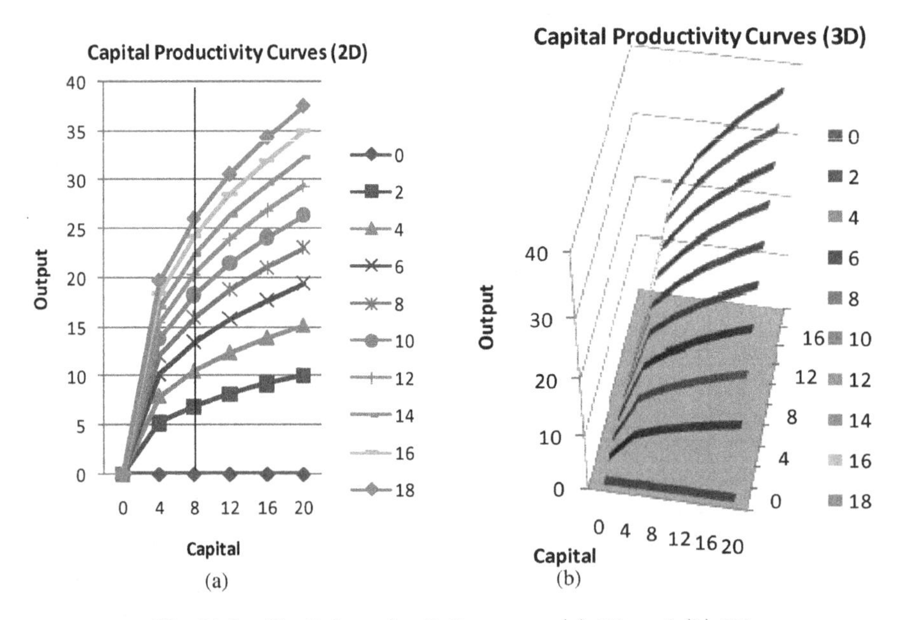

(a)
(b)

Fig. 11.3 Capital productivity curves (a) 2D and (b) 3D

[2]All the diagrams in this chapter are based on Table 11.1. Hence, different figures can be drawn by simply changing the chart type. To preserve the original chart, always make a copy of the chart and change the chart type of the copied chart.

[3]The chart may also be drawn from the Table by entering <"Insert"> <Charts, Line> < 3D Line>.

The chart in the window changes.[4] There are 10 productivity curves corresponding to the 10 ranges of output in rows 14–23 of Table 11.1. Changing the title and axis labels, we have the capital productivity curves as shown in Fig. 11.3(a). Similarly, we can construct the 3D chart, Fig. 11.3(b), from Fig. 11.3(a). They have a similar interpretation to that of the labor productivity curves.

It is a remarkable feature of Excel that the partial productivity curves can be drawn so easily by only switching of columns and rows.

11.1.5 *The total productivity surface*

Combining the above two productivity curves, we have the **total productivity surface** in 3D or the **production surface** in 2D. Figure 11.4 is drawn by copying any of the four charts in Figs. 11.2 and 11.3 to a blank area and using <RM> to change the chart type to "Surface". It may also be drawn by selecting Table 11.1, including the labels, and choosing the "surface" chart type. To compare with the curves in Figs. 11.2 and 11.3, we need the wireframe 3D. Click the chart and under <"Chart Tools, Design">, click the down arrow of the group name <Chart Styles!>. When six lines of the various styles are shown, select the first chart with grids on the second line (Style 9). We have Fig. 11.4.

Identifying the axes

One of the problems in drawing 3D charts is the identification of the axes. The labor axis can be identified by having $0, 2, \ldots, 18$ units, and the capital axis by having $0, 4, \ldots, 20$ units. But, if both axes have the same units, then they are hard to identify. In this case, select the chart (Fig. 11.4). The range of the chart is indicated in Table 11.1 as B13:H23, enclosed by blue borders. Click the small blue square box at the right lower corner (a corner of H23) of the borders, move up the corner to L = 10, and see the change in the axis of

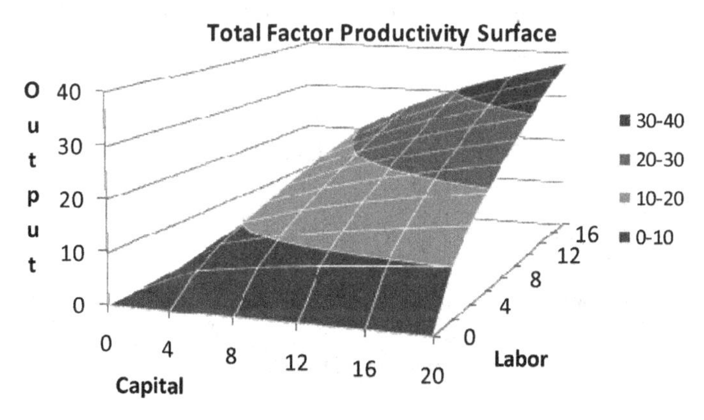

Fig. 11.4 Production surface

[4]This can also be down by clicking <"Insert"><Data, Switch Row/Columns>.

Fig. 11.4. One of the axes should have a range from 0 to 10, instead of 0 to 18. We then know that the axis which shrinks to range 0 to 10 is the axis of labor. After the labor axis is identified, the other axis must be capital. Select the chart again and move the chart range back to L = 18. You may also use the same method to identify the capital axis. Add or correct axis titles if necessary.

Click anywhere in the 3D chart, select <RM>, and a dialog box will appear. Select <3D Rotation ···>. Experiment by rotating X, Y, Z axes to different angles to see different surfaces, and have fun. In particular, move the origin of the TFP surface to face you, and identify the labor and capital productivity curves derived in Figs. 11.2 and 11.3.

11.1.6 *Isoquant maps*

When we cut the cake like Fig. 11.4 horizontally along the output (Z) axis, and project the contour on the (x, y) plane (capital and labor axes), we have an **isoquant** (iso means the same, quant means quantity) for a given level of output, as shown in Fig. 11.5. To draw Fig. 11.5, copy the 3D chart of Fig. 11.4 to a blank space, and select the chart. Click <RM>, and choose <Change Chart Type ···> in the menu. Under "Surface", choose the "Contour" chart (the third one), and <OK>. We have Fig. 11.5. The axes of this chart are the factors of production and the chart is also called the **factor** (or **input**) **space**.

Definition

Formally, an **isoquant** shows the combinations of all L and K which yield the same output level Q_0. In notation,

$$\{(L, K) | Q_0 = F(L, K)\}. \tag{11.4c}$$

Three isoquants are drawn in Fig. 11.5. Excel determines the number of isoquants automatically. The legend shows that the current range of the output is 0–40 at an interval

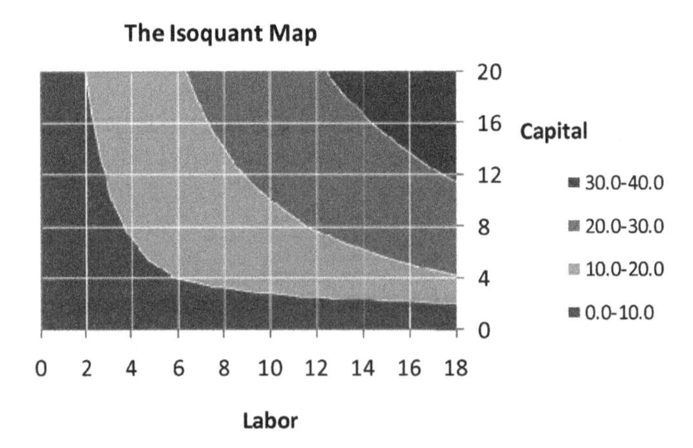

Fig. 11.5 The isoquant map

of 10 units. The scale will change according to the size of the chart when the chart is expanded (try to expand the chart).

Note that the convention in Economics is to put labor (the primary factor of production) on the horizontal axis and capital on the vertical axis. If labor is not on the horizontal axis, the axes may be interchanged by selecting the chart, and[5]

<RM><Select Data···><Switch Row/Column><OK>.

Then, the axes in the chart are interchanged: that is, the labor (or capital) axis in Fig. 11.5 changes to the capital (or labor) axis. The axis labels have to be changed manually. You may also use the method of identifying the axes mentioned above.

A dense isoquant map

If the production surface is continuous, then there are an infinite number of isoquants filling the input space without intersections. Thus, we may want to fill the factor space with as many isoquants as possible. For this purpose, we have to go back to the 3D chart. Copy the 3D chart Fig. 11.4 (not the isoquant map[6] Fig. 11.5) to a blank space and double click the vertical Z-axis. Select <Format Axis> from the dialog box. The <Format Axis> dialog box appears (see Fig. 3.11(b)). Select <Format Axis, Axis Options>. In the "Axis Options" dialog box, change the Major unit from "Auto" to @Fixed 1.0. Also, select <Format Axis, Number> and in the "Number" window, change the "Decimal places" to zero (just to avoid crowding) and select <OK>. We then have Fig. 11.6. If the legend is too crowded, either enlarge the chart to give more space to the legend, or change the legend font size, or press <RM><Format Axis, Axis Options>. In the "Axis Options" dialog box, change

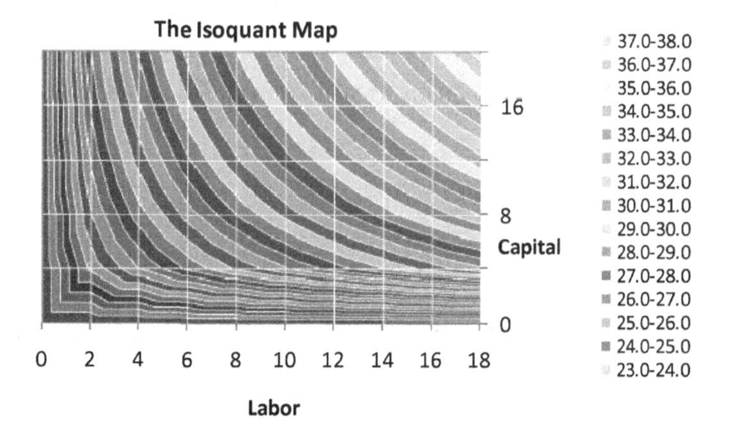

Fig. 11.6 Dense isoquant map

[5]You may also use the ribbon. Click <"Chart Tools," "Design"><Data, Switch Row/Column>.

[6]Unlike Excel 2003 or earlier, the scale of the legend in the 2D chart can be changed only by going back to the 3D chart. In the 3D chart, change the scaling of the third (Z) axis, and return to the 2D chart.

"Maximum" from "Auto" to "30". After changing the vertical axis, convert the 3D chart, Fig. 11.4, back to get the isoquant map Fig. 11.6.

Compared with Fig. 11.5, Fig. 11.6 shows the isoquants of output levels from 23 to 38, with intervals of one unit. Reduce the font of the legend or increase the height of the chart, to make sure that the isoquants are drawn from 23 to 38 by detecting the change of colors from one level of output to another level.

The isoquant map in Fig. 11.6 is in color.[7] It can be changed to black and white wireframe by selecting the chart and <RM>, and[8] selecting <Change Chart Type ..., Surface> to <Wireframe Contour>, as shown in Fig. 11.7. This is the black and white isoquant map that we often see in textbooks.

The corresponding isoquant map is also shown roughly in Table 11.1 by connecting the cells with the same or close numbers. Three isoquants are drawn manually in Table 11.1.

11.1.7 *The derivation of isoquants by solving production functions*

Instead of using Excel to draw the isoquant map automatically, the general practice in economics and business is to solve the production function for K in terms of L, and find the value of K for each given output Q at a certain level. Thus, from (11.2), we have $K^b = Q/AL^a$. Using Rules of Exponents, raising both sides by 1/b, we have

$$K = (Q/AL^a)^{(1/b)}. \qquad (11.5)$$

To draw (11.5), we copy the whole sheet of Table 11.1 to a new sheet. In the new sheet, rename it as Table 11.2. Referring row 16 (the shaded row) of Table 11.1 as a guide line, let

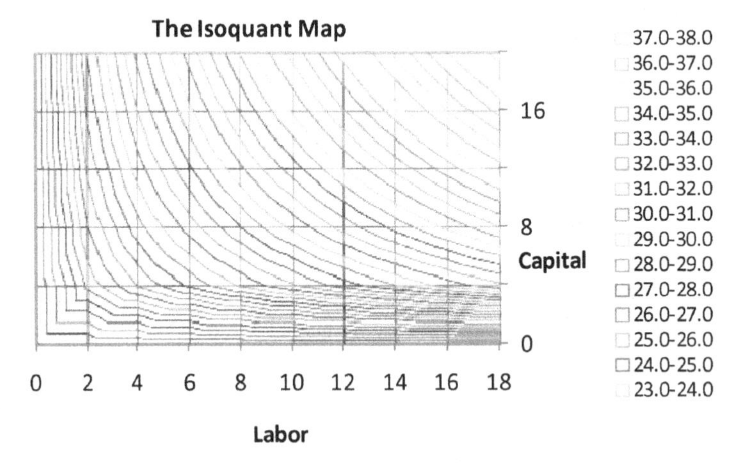

Fig. 11.7 "Wireframe contour"

[7] The colors come automatically and the colors are not as bright as those in Excel 2003. It appears that Excel does not allow changing the color of 3D manually.

[8] You may also go through the ribbon by clicking <"Chart Tools", "Design"><Type, Change Chart Type> <Change Chart Type, Surface><Wireframe Contour>.

Table 11.2. Isoquant sensitivity table

	A	B	C	D	E	F	G	H
8		Q	=A*L^aa*K^bb					
9		C14	K =(Q/(A*L^aa))^(1/bb)					
10								
11	Isoquants Sensitivity Table							
12	K		Output Q					
13			8	10.6	12.4	15	20	26
14		0						
15	Labor	2	11.3	22.9	33.8	54.5		
16		4	4.0	8.1	12.0	19.3	39.5	76.2
17		6	2.2	4.4	6.5	10.5	21.5	41.5
18		8	1.4	2.9	4.2	6.8	14.0	26.9
19		10	1.0	2.0	3.0	4.9	10.0	19.3
20		12	0.8	1.6	2.3	3.7	7.6	14.7
21		14	0.6	1.2	1.8	2.9	6.0	11.6
22		16	0.5	1.0	1.5	2.4	4.9	9.5
23		18	0.4	0.8	1.3	2.0	4.1	8.0

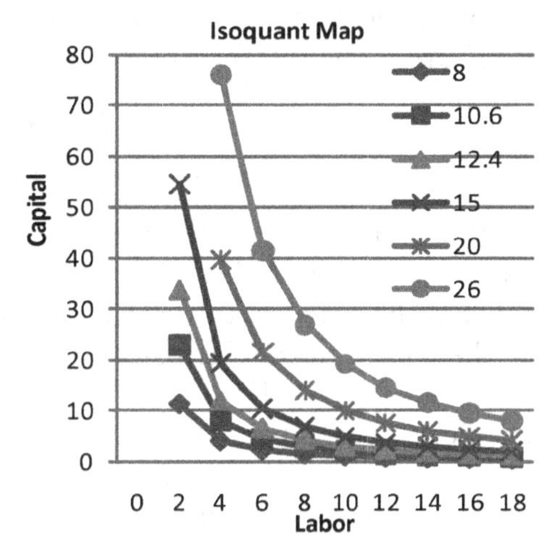

Fig. 11.8 Isoquant map

Q take values of 8, 10.6, 12.4, 15, 20, and 26, which are entered in place of K in row 13 of Table 11.2. Name C13:H13 Q (no need to erase the name K in Table 11.1). Enter Eq. (11.5) in C14 of Table 11.2 as

$$=(Q/(A^*L\hat{\ }aa))\hat{\ }(1/bb). \tag{11.6}$$

Equation (11.6) is then copied[9] to C14:H23. The range C14:H14 of Table 11.2 will give #Divid/0!, because we cannot divide something by zero. The values in cells G15 and H15 are too large to be contained in Fig. 11.8. Thus, both are deleted to give the curves in Fig. 11.8. Note that the main entries of the table are the value of K, not Q, as we were reminded in A12.

The isoquant at each output level shown in C13:H13 in Table 11.2 and also in Fig. 11.8 can be located in the more complicated (densy) isoquant map in Fig. 11.6. Copy Fig. 11.8 and place it nearby and draw a single isoquant, say for Q = 26, by deleting all other isoquants. This can be done by clicking each isoquant and pressing the <delete> key. We have constructed a single isoquant curve for Q = 26 like the one in Fig. 11.11 in Sec. 11.2.3.

11.2 Average and Marginal Productivities of a Factor

In addition to partial and total productivities, we can define average and marginal productivities from the production function.

[9]If parameters A, aa, and bb are not defined, enter the numbers 2, 0.6, and 0.4 in their places.

11.2.1 *Average productivities*

The **average productivity of labor** (APL) (or capital, APK) is defined as the ratio of output divided by labor (or capital). For the CD production function (11.2), the APL and APK are, using the rules of exponents, respectively,

$$\begin{aligned} Q/L &\equiv \mathrm{APL} = \mathrm{AL}^{a-1}\mathrm{K}^{b}; \\ Q/K &\equiv \mathrm{APK} = \mathrm{AL}^{a}\mathrm{K}^{b-1}. \end{aligned} \tag{11.7}$$

A family of APL for a given level of capital, similar to Table 11.1, is given in Table 11.3. The construction of this table may need some explanation. Since the frame of the sensitivity table is the same as Table 11.1, we copy Table 11.1 to a new sheet, delete the upper part of the copied table, and place the "Table for APL" in row 3 as shown in Table 11.3. We first name row C5:H5 as KK, and column B6:B15 as LL, instead of K or L, just to avoid conflict[10] with Table 11.1 in the same workbook. Instead of using the names for A, aa, and bb, we enter the given values directly as shown in the equation for Q/L in (11.7) in B2. Entering this formula in C6, and copying it down to C6:H15, we have the sensitivity table for the APL. The table is charted in Fig. 11.9 in 3D curves.

The chart shows clearly that an APL is a downward sloping curve which is convex to the labor axis, showing that, as labor increases, the APL decreases. The level of APL increases as capital increases, but the shape of the APL curve remains the same.

Table 11.3 Average productivity of labor

	A	B	C	D	E	F	G	H
1	(11.7)	Q/L	=AL^(aa-1)K^bb					
2	C6	=2*LL^(0.6-1)*KK^0.4						
3	Table for APL							
4			Capital					
5			0	4	8	12	16	20
6	Labor	0						
7		2	0.0	2.6	3.5	4.1	4.6	5.0
8		4	0.0	2.0	2.6	3.1	3.5	3.8
9		6	0.0	1.7	2.2	2.6	3.0	3.2
10		8	0.0	1.5	2.0	2.4	2.6	2.9
11		10	0.0	1.4	1.8	2.2	2.4	2.6
12		12	0.0	1.3	1.7	2.0	2.2	2.5
13		14	0.0	1.2	1.6	1.9	2.1	2.3
14		16	0.0	1.1	1.5	1.8	2.0	2.2
15		18	0.0	1.1	1.4	1.7	1.9	2.1

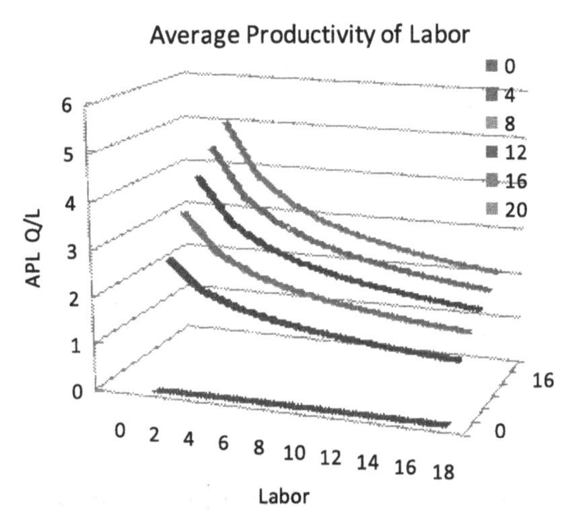

Fig. 11.9 Average productivity of labor

[10]Another possible way of assigning a unique name and still retaining the variable name is K0, K01, K02, K_, K_1, capital1, capital2, etc.

11.2.2 *Marginal productivities*

The marginal productivity of labor, MPL (or capital, MPK) is the effect of a small change in labor input on the output, holding capital at a constant level. Mathematically, it is the small change of the production function when labor (or capital) changes one unit holding capital (or labor) constant. For the CD production function, they are

$$\begin{aligned} \left.\frac{\Delta Q}{\Delta L}\right|_{\text{Kconst}} &\equiv \text{MPL} = \text{AaL}^{\text{a}-1}\text{K}^{\text{b}} = \text{aQ/L}, \\ \left.\frac{\Delta Q}{\Delta K}\right|_{\text{Lconst}} &\equiv \text{MPK} = \text{AbL}^{\text{a}}\text{K}^{\text{b}-1} = \text{bQ/K}. \end{aligned} \tag{11.8}$$

In this particular form of production function (11.2), average and marginal productivity (MP) functions are related by the right-hand side of (11.8). Since a and b are fractions, we see that MPL is a fraction of APL, and MPK is also a fraction of APK. Thus, in the diagram, the MPL (or MPK) must be lower (with smaller values) than the APL (or APK). This can be seen from the smaller range of the vertical axis in Fig. 11.10, as compared with that in Fig. 11.9, although the shapes and positions of the curves look the same. The APL and the MPL at K = 8 are compared in Fig. 11.13. The MPL is consistently lower than the APL for all levels of output.

Table 11.4 is constructed by copying Table 11.3, pasting it below, and making some adjustments. The ranges of capital and labor are renamed KKK and LLL to avoid conflict. The formula in C6 is from the first equation in (11.8), and is reproduced in E1 of Table 11.4.

Table 11.4 Marginal productivity of labor

	A	B	C	D	E	F	G	H	I
1	(11.8)		ΔQ/ΔL		=aAL^(a-1)K^bb				
2	C6	=2*0.6*LLL^(0.6-1)*KKK^0.4							
3	Table for MPL								
4			Capital						
5			0	4	8	12	16	20	w
6	Labor	0							1.2
7		2	0.0	1.6	2.1	2.5	2.8	3.0	1.2
8		4	0.0	1.2	1.6	1.9	2.1	2.3	1.2
9		6	0.0	1.0	1.3	1.6	1.8	1.9	1.2
10		8	0.0	0.9	1.2	1.4	1.6	1.7	1.2
11		10	0.0	0.8	1.1	1.3	1.4	1.6	1.2
12		12	0.0	0.8	1.0	1.2	1.3	1.5	1.2
13		14	0.0	0.7	1.0	1.1	1.3	1.4	1.2
14		16	0.0	0.7	0.9	1.1	1.2	1.3	1.2
15		18	0.0	0.7	0.9	1.0	1.1	1.3	1.2

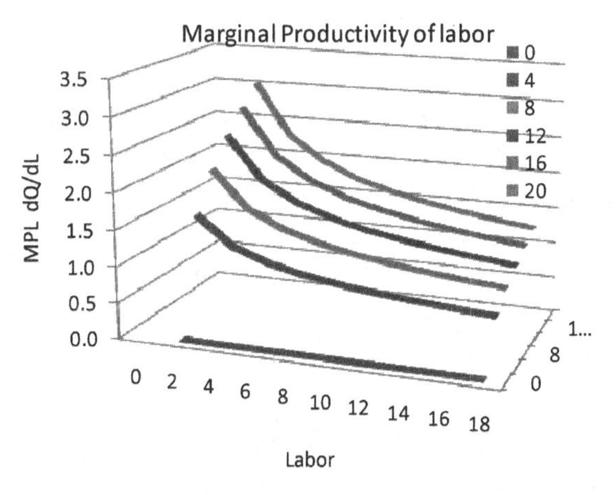

Fig. 11.10 Marginal productivity of labor

After the table is set up, instead of redoing the chart, copy Fig. 11.9 to the next of Table 11.4, and rename it Fig. 11.10. When the copied Fig. 11.10 is clicked, the data area of Table 11.3 will be enclosed by borders. Drag the blue borders in Table 11.3 down to fit into Table 11.4 (as explained with Table 3.1 in Chapter 3). The chart changes automatically from APL curves to MPL curves. Revising the chart and axis titles, we have Fig. 11.10.

The law of diminishing returns

The chart shows that as labor input increases, the MPL decreases steadily. In economics, it is said that the **law of diminishing returns** (or more precisely, the **law of diminishing marginal product**) holds.[11] Figure 11.10 shows clearly that when labor continue to increase one unit, MPL decreases accordingly.

As with APL, the level of MPL depends on the level of capital that is held constant. The larger the level of capital, the higher the MPL, but the shape of the MPL remains the same.[12] This may not be the case for other forms of production functions.

11.2.3 *The law of diminishing marginal rate of technical substitution in production*

In the previous sections, we have discussed the four productivity relations between output and labor or capital. We now study the relations between labor and capital given a level of output.

In economics, we generally assume that MP of a factor is positive. An extra one unit increase in labor will increase output by MPL. If labor increases by ΔL units, output will increase by MPL*ΔL units. This is the partial increment of output for ΔL. Similarly, if capital increases by ΔK units, output will increase by MPK*ΔK units. This is the partial increment of output for ΔK. Thus, if both labor and capital increase by ΔL units and ΔK units, then the total increment of output, ΔQ, will be

$$\Delta Q = (MPL)\Delta L + (MPK)\Delta K. \quad (11.9)$$

Example 11.1 Numerical illustration. The intuitive interpretation of (11.9) is as follows. If MPL = 34 units per man hour, holding capital input constant, and ΔL = "change in labor input" = 2 man hours, then, MPL*ΔL = "change in output due to the change in labor by ΔL" = (34)(2) = 68 units.

If MPK = 96 units per machine hour, and ΔK = change in capital = 1 machine hour, then MPK*ΔK = "change in output due to the change in capital by ΔK" = (96)(1) = 96 units.

[11] In mathematics, this means that the second-order partial derivative of output with respect to labor is negative.

[12] In mathematics, this means that the cross second-order partial derivatives of output with respect to labor and capital are positive.

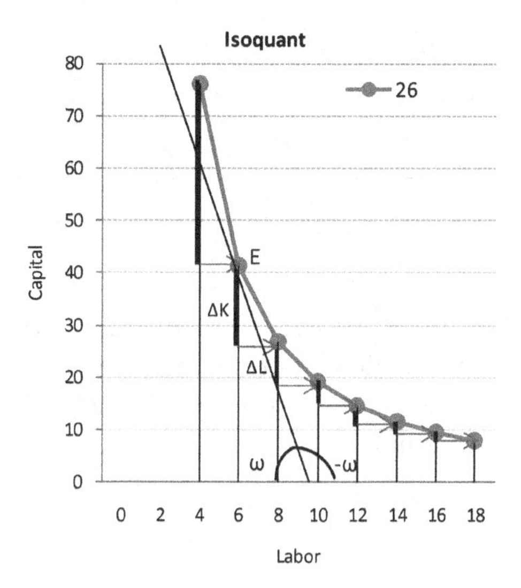

Fig. 11.11 The law of diminishing marginal rate of technical substitution

Hence, a firm can produce extra output $\Delta Q = (34)(2) + (96)(1) = 164$ units from the extra inputs of L and K. □

The slope of an isoquant

Suppose that production occurs along an isoquant, say at Q = 26 in Fig. 11.11, before and after the changes in labor and capital. This means that $\Delta Q = 0$ along the isoquant, and from (11.9), we may solve ΔK for ΔL and write[13]

$$\frac{\Delta K}{\Delta L} = -\frac{\text{MPL}}{\text{MPK}} \tag{11.10}$$

which is negative because both MPL and MPK are positive. $\Delta K/\Delta L$ is the slope of the isoquant, say at point E. If we denote the ratio of the marginal productivities[14] on the right-hand side of (11.10) by ω, then the slope is shown as $-\omega$. It shows that when labor increases by ΔL, the output will increase by MPL*ΔL, and then capital must decrease by ΔK, so that the output will decrease by MPK*ΔK to offset the increase in output due to the increase of labor. Therefore, the total output level will not change and stay on the same isoquant. Thus, the relation between the change of labor and the change of capital is negative along an isoquant.

[13]Mathematically, we may also solve ΔL for ΔK and write $\Delta L/\Delta K$. In this case, K is taken as the independent variable and L the dependent variable. In general practice, economists prefer to think of labor as the primary factor of production (not to speak of Marx's labor theory of value) and take labor as the independent variable.
[14]ω has a special meaning, as defined in Sec. 11.3.1. Here we take it as a usual Greek letter like θ.

The marginal rate of technical substitution

Since it is easier to make economic interpretations with positive number, economists take the absolute value of the slope in (11.10), to have

$$\boxed{\left|\frac{\Delta K}{\Delta L}\right| = -\frac{\Delta K}{\Delta L} = \frac{MPL}{MPK} \equiv MRTS.} \tag{11.11}$$

Equation (11.11) is called the **marginal rate of technical substitution** (MRTS) between factors. Diagrammatically, as in the demand curve analysis we studied in Chapter 3, MRTS is the supplementary angle of the slope at point E, namely, tan ω, see Fig. 11.11. In the current example, from (11.8), we have

$$\left|\frac{\Delta K}{\Delta L}\right| = -\frac{\Delta K}{\Delta L} = \frac{MPL}{MPK} = \frac{aK}{bL}.$$

The MRTS depends on the constant a/b multiplied by the capital–labor ratio K/L.

The MRTS measures the degree of substitutability between labor and capital at the same output level. For example, when we increase the use of labor by one unit, the MRTS determines how many units of capital can be substituted (i.e. decreases the use of capital) by using this extra unit of labor, so that we can maintain the same level of output. Strictly speaking, it is the **MRTS of labor for capital**, since we use labor to substitute for capital.

We may also increase capital to substitute for labor by decreasing labor one unit at a time. In this case, we have **MRTS of capital for labor**. In a continuous analysis, MRTS of labor for capital and of capital for labor will be the same, and may be called simply **MRTS between capital and labor**.[15]

The law of diminishing MRTS

Figure 11.11 also shows that when we continue to increase the use of labor one unit at a time, the amount of capital that the extra unit of labor can replace becomes smaller and smaller, as shown by the shorter dark vertical line segments in Fig. 11.11. This is a general rule in production, and it is called the **law of diminishing marginal rate of technical substitution** (MRTS). This law holds whenever the isoquant is convex to the origin.

11.2.4 *Marginal productivity and perfect competition*

Equilibrium conditions in factor markets

According to the theory of production, if perfect competition prevails in the factor markets, to maximize profits, a firm will produce output until the MP of a factor equals the price of using that factor. If w is the price of a unit of labor (**real wage rate**) and r is the price

[15]Mathematically, one is the derivative from the left, and the other is the derivative from the right. If the isoquant is a smooth continuous curve, in the limit, the left and right derivatives should be the same.

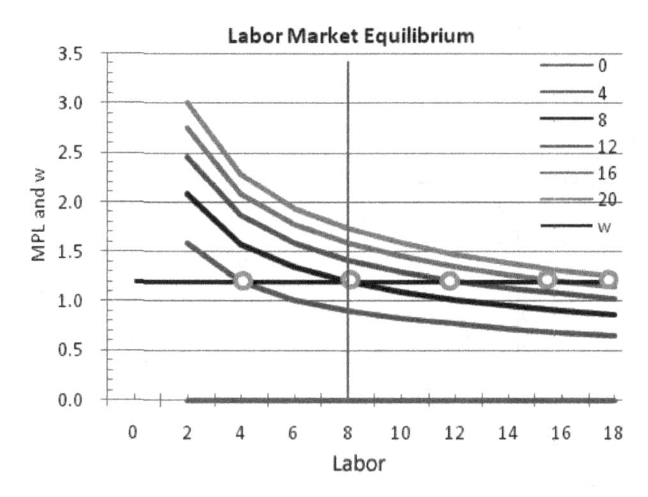

Fig. 11.12 Factor market equilibrium

of a unit of capital (**real rental rate**), then the firm will hire labor and capital until the following equalities hold:

$$\boxed{\mathrm{MPL} = \mathrm{w}, \quad \mathrm{MPK} = \mathrm{r}} \tag{11.12}$$

These are called the **equilibrium conditions in factor markets**. In Table 11.4, wage rate w = 1.2 is entered in column I, and the chart is drawn as Fig. 11.12. Five equilibrium points, marked in circles, are shown corresponding to the MPL curves using different amounts of capital, 4, 8, 12, 16, and 20 units. When the capital is given as, say, K = 8 units, then the labor market reaches an equilibrium point at L = 8 units, as shown by the intersection of the dark w line with the dark MPL curve at K = 8 units. This condition may also be seen in the shaded cells of Table 11.4.

Demand function for factors

Mathematically, the two equilibrium conditions in (11.12) consist of two nonlinear equations with two variables, L and K. Like the system of linear equations, under certain mathematical conditions,[16] they can be solved for the factor prices w and r, given parameters A, a, and b in (11.8). Writing the solutions in functional form, we have

$$\boxed{\mathrm{L} = \mathrm{L}(\mathrm{w}, \mathrm{r}), \quad \mathrm{K} = \mathrm{K}(\mathrm{w}, \mathrm{r})} \tag{11.13}$$

They are called the **demand functions for labor and capital**.

[16]That is, it should satisfy the conditions of the Implicit Function Theorem.

Table 11.5 Three labor productivity curves at k = 8

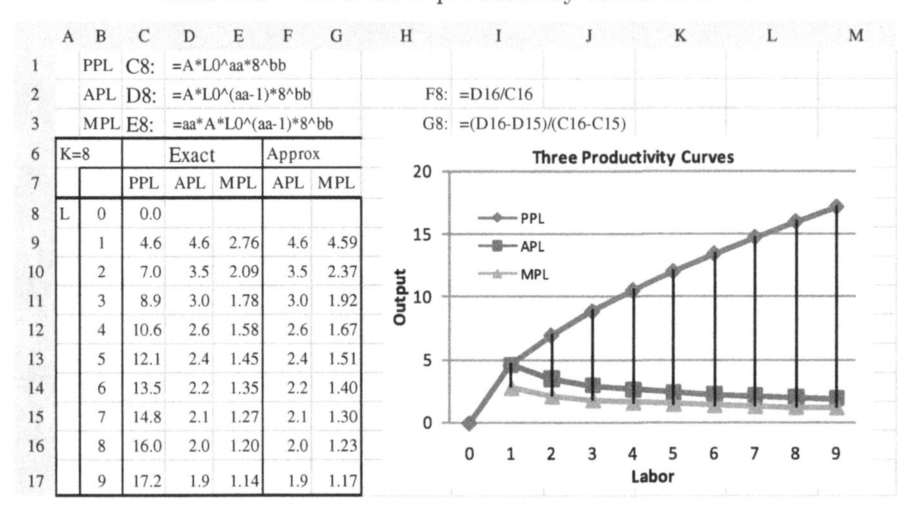

	A	B	C	D	E	F	G	H	I
1		PPL	C8:	=A*L0^aa*8^bb					
2		APL	D8:	=A*L0^(aa-1)*8^bb				F8:	=D16/C16
3		MPL	E8:	=aa*A*L0^(aa-1)*8^bb				G8:	=(D16-D15)/(C16-C15)
6	K=8			Exact		Approx			
7			PPL	APL	MPL	APL	MPL		
8	L	0	0.0						
9		1	4.6	4.6	2.76	4.6	4.59		
10		2	7.0	3.5	2.09	3.5	2.37		
11		3	8.9	3.0	1.78	3.0	1.92		
12		4	10.6	2.6	1.58	2.6	1.67		
13		5	12.1	2.4	1.45	2.4	1.51		
14		6	13.5	2.2	1.35	2.2	1.40		
15		7	14.8	2.1	1.27	2.1	1.30		
16		8	16.0	2.0	1.20	2.0	1.23		
17		9	17.2	1.9	1.14	1.9	1.17		

Fig. 11.13 Three productivity curves

11.2.5 *The relations among the three productivity curves*

The partial, average, and MP curves of labor at K = 8 units are grouped together in Table 11.5 and Fig. 11.13 for comparisons. The range B8:B17 is named L0, and the three parameters are named as in Table 11.1. Since MPL measures the change in output when labor increases one unit, the interval of labor is taken as an increase of one unit at a time from 0 to 9 units. The formulas for ranges C8:C17, D8:D17, and E8:E17 are shown in rows 1–3, respectively, in the table. C8:C17 calculates partial productivity of labor (PPL), which is the output that labor produces at K = 8, using the corresponding units of labor in B8:B17. D8:D17 shows APL based on Eq. (11.7), and E8:E17 has MPL calculated from (11.8). They are "exact" by definitions. We have deleted D8:E8 to avoid dividing by zero.

Approximation of APL and MPL

On the other hand, F8:F17 calculates APL by taking PPL in C8:C17 and dividing it by the corresponding units of labor used in B8:B17. G8:G17 calculates MPL by taking the difference of the consecutive PPLs, e.g. C9 − C8, C10 − C9, C11 − C10, etc., to find the increment of output when labor increases one unit at a time. These are the methods of calculating APL and MPL without using calculus-like formulas (11.8). The results in F9:G17 are the approximations for D9:E17. Note that we have deleted F8:G8 to avoid dividing by zero.

Comparison of the three productivities

The three continuous productivity curves in columns C, D, and E are illustrated in Fig. 11.13. Note that the MPL lies below the APL, and both curves have negative slopes. The **law of diminishing returns** can also manifest itself in the partial productivity curve as the **law of diminishing marginal product**. When labor continues to increase one unit, as shown by the vertical lines, the marginal product of labor also increases (MPL > 0), but the amount of additional output decreases as the use of labor increases each time. In other words, from the partial productivity curve, we can see that the output increases at a decreasing rate as the use of labor continues to increase. This is the general law of production.

We can also see the law mathematically. In (11.8), the MPL and MPK are both positive. However, under the CRS, we have $a + b = 1$ and $a - 1 = -b$. Hence, $MPL = Aa(K/L)^b$. Other things being equal, an increase in L will decrease the MPL. This proves that the MPL curve is downward sloping, and the **law of diminishing marginal product of labor** holds. Similarly, we can prove that MPK is also downward sloping.

11.3 The CES Production Function

A more general form of the production function is called the **Constant Elasticity of Substitution (CES) production function**. It has a complicated functional form:

$$\boxed{Q = A(aL^s + bK^s)^{\frac{1}{s}},} \quad A > 0, \quad 0 < a < 1, \quad b = 1 - a, \quad s < 1, \qquad (11.14)$$

where, as in the case of the CD production function, A is a positive constant, a and b are fractions, and s is a constant less than one. It can be shown that (11.14) is also a CRS production function.

11.3.1 *The elasticity of substitution*

To explain the name of the function, we first define the important concept of the elasticity of substitution (ES) between two factors.

In general, the ES between two factors of production, L and K, is defined as the percentage change of K/L, the **factor ratio**, divided by the percentage change of MPL/MPK, the ratio of the marginal productivities of the two factors. If we denote $K/L = k$ and $MPL/MPK = \alpha$, then the **elasticity of substitution** can be defined as

$$\boxed{ES = \frac{\%\Delta \text{ of } (K/L)}{\%\Delta \text{ of } (MPL/MPK)} = \frac{\%\Delta \text{ of } k}{\%\Delta \text{ of } \alpha}.} \qquad (11.15)$$

In the Theory of Economic Growth, elasticity of substitution is also denoted as σ.

ES under perfect competition in factor markets

Under perfect competition in factor markets, equilibrium conditions (11.12) hold. Hence, we may write $MPL/MPK = w/r \equiv \omega$, which is called the **factor price ratio**, or the

wage–rental ratio. Under this condition, the elasticity of substitution can be written as

$$\mathrm{ES} = \frac{\%\Delta \text{ of } (\mathrm{K/L})}{\%\Delta \text{ of } (\mathrm{w/r})} = \frac{\%\Delta \text{ of k}}{\%\Delta \text{ of } \omega} = \sigma. \tag{11.16}$$

Like MRTS, ES measures the **substitutability** of labor and capital when the factor price ratio changes, but unlike MRTS, which depends on the unit of measuring labor and capital, it is a pure number, like price elasticity of demand. In fact, its concept is similar to the price elasticity of demand, which is the ratio of the percentage change of quantity and the percentage change of the price, and measures the sensitivity of price change. However, it is different from the price elasticity of demand in that it is the ratio of the percentage change of the capital–labor ratio over the percentage change of the factor price ratio.

In terms of Fig. 11.11, the capital–labor ratio is the slope of a ray (not shown) though the origin and point E. Suppose point E at L = 6 changes to E′ at L = 8. Both ω and k will decrease. On the other hand, if E changes to E′ at L = 4, both ω and k will increase. Thus, ES is generally positive. This is the case if the law of diminishing MRTS holds, and the isoquants are convex to the origin. The ES measures the degree of sensitivity of such change, that is, the percentage change of the capital–labor ratio when the factor price ratio changes one percent.

It can be shown that, the CES production function (11.14) has the ES

$$\mathrm{ES} \equiv \frac{1}{1-\mathrm{s}}, \tag{11.17}$$

which must be positive. Parameter $\mathrm{s}(= (\mathrm{ES}-1)/\mathrm{ES})$ is also called the **substitution factor**. In this particular production function, since s is a constant, the ES is also a constant. Thus, the production function (11.14) is called the constant-elasticity-of-substitution (CES) production function.

Example 11.2 The ES of the Cobb–Douglas production function. Dividing the two equations in (11.8), we have $\omega = \mathrm{MPL/MPK} = (\mathrm{aQ/L})/(\mathrm{bQ/L}) = (\mathrm{a/b})\mathrm{k}$. Taking the difference both sides, we have, letting $\Delta\omega = \omega_1 - \omega_0, \Delta \mathrm{k} = \mathrm{k}_1 - \mathrm{k}_0$,

$$\Delta\omega = (\mathrm{a/b})\Delta\mathrm{k}. \tag{11.18}$$

Since $(\Delta\omega/\omega)$ is the $\%\Delta$ of ω and $(\Delta\mathrm{k/k})$ is the $\%\Delta$ of k, dividing both sides by ω, and substituting into (11.16),

$$\mathrm{ES} = \frac{\%\Delta \text{ of k}}{\%\Delta \text{ of } \omega} = \frac{\omega}{\mathrm{k}}\frac{\mathrm{b}}{\mathrm{a}} = 1. \tag{11.19}$$

since $\omega = (\mathrm{a/b})\mathrm{k}$. Thus, the ES of a CD production function is unitary and it does not depend on returns to scale. □

11.3.2 *Properties of a CES production function*

In (11.17), since ES is positive, s can range only from $-\infty$ to 1. Depending on the value of s, we have five cases of ES. They are shown in Table 11.6 and illustrated in Fig. 11.14

Table 11.6 CES production function and ES

	A	B	C	D
1	Case	Prod function	s	=1/(1-s)
2				ES
3	e	s → inf, ES → 0 (Leontief pf)	-5.0	0.17
4	d		-4.5	0.18
5	d		-4.0	0.20
6	d		-3.5	0.22
7	d		-3.0	0.25
8	d	s <0, 0 < ES < 1	-2.5	0.29
9	d	CES (Asy to axes)	-2.0	0.33
10	d		-1.5	0.40
11	d		-1.0	0.50
12	d		-0.5	0.67
13	c	s → 0, ES = 1 (CD)	0.01	1.01
14	b	0 < s < 1, ES > 1 (CES)	0.50	2.00
15	a	s →1, ES → inf (Perfect ES)	0.90	10.00
16	Note:	→ means go to		

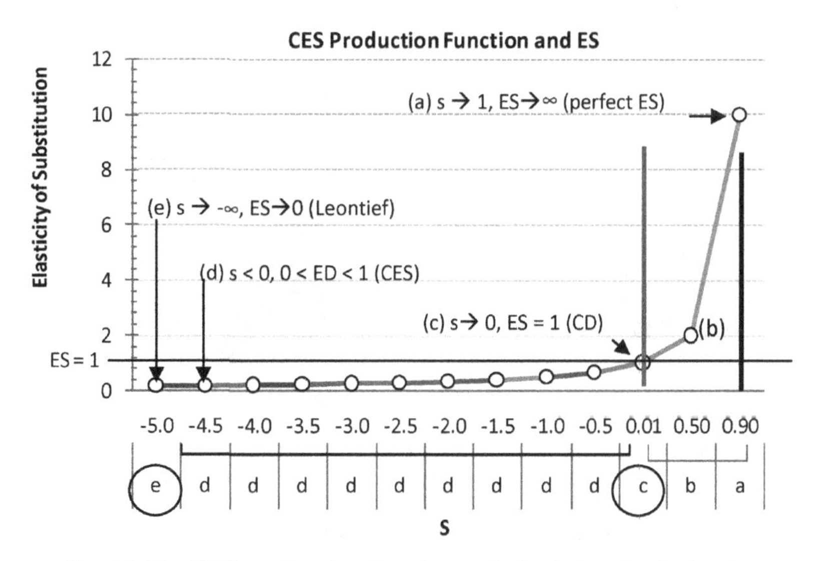

Fig. 11.14 CES production function and elasticity of substitution

(case a to e). Note that, in drawing Fig. 11.14, the range for the chart in Table 11.6 should be A2:A15 and C2:D15. Based on Fig. 11.14, we may take advantage of the graphic ability of Excel and draw 3D and 2D charts for each of the five cases. They are shown in the 10 charts of Fig. 11.15. The methods of drawing Figs. 15(a)–15(e) are rather complicated and will be explained in Sec. 11.3.3. Here, we first discuss the algebraic properties of the CES production function bases on the ES.

a. If s → 1, then ES → infinity. Hence, in Fig. 11.15(a), we enter s = 0.9 to indicate that s approaches 1. For s = 0.9, ES = 10, as shown in row 15 of Table 11.6. In this case, the production surface is a plane and the isoquants become a family of straight lines, as shown in 2D of Fig. 11.15(a).

b. If $0 < s < 1$, then $ES > 1$. In particular, we enter $s = 0.5$, then $ES = 2$. See row 14 of Table 11.6. The substitution between labor and capital is elastic, thus the bending of

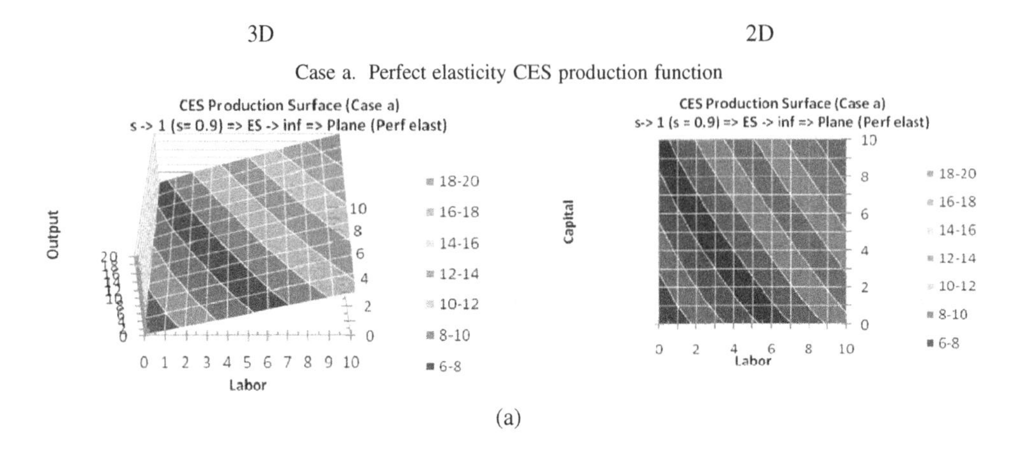

(a)

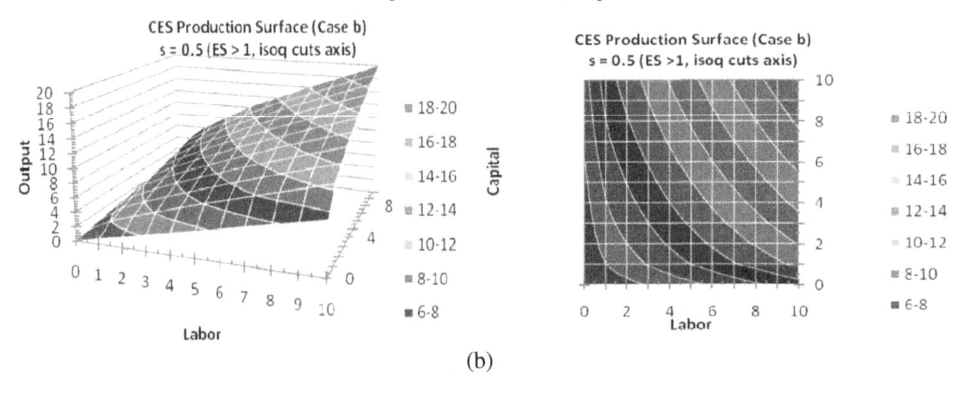

(b)

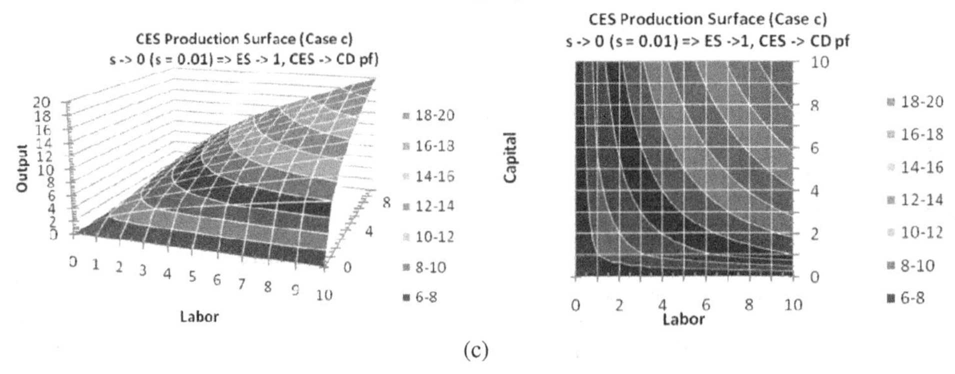

(c)

Fig. 11.15 The CES production — 3D and 2D

3D 2D

Case d. isoquants asymmetric to both axes

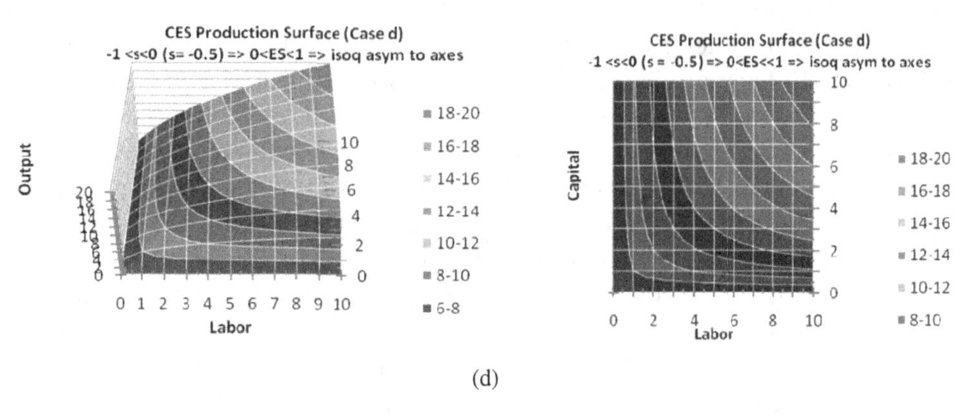

(d)

Case e. The Leontief production function

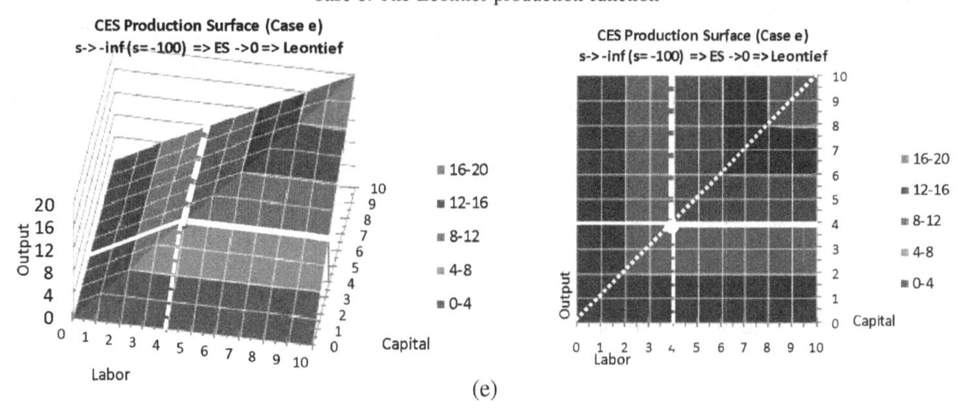

(e)

Fig. 11.15 (*Continued*)

the isoquants is less severe, and as such, the isoquants cut the L and K axes. See 2D of Fig. 11.15(b).

c. If $s \to 0$, then, $ES \to 1$. Note that if we enter $s = 0$ in (11.17), then, $ES = 1$, but for $s = 0$, the exponent of the CES production function (11.14) is divided by zero; hence, the CES production function is not defined. To avoid this problem, we approximate $s = 0$ by using $s = 0.01$, then $ES = 1.01$, as shown in row 13 of Table 11.6. In this case, we have a CD production surface like Fig. 11.15(c). The isoquants are rectangular hyperbola, with the labor and capital axes as their asymptotic lines. Mathematically, we write[17]

$$\boxed{\lim_{s \to 0} Q = AL^{a}K^{b}.} \tag{11.20}$$

[17]For a proof, see Chiang and Wainwright (2005).

Thus, the CD production function is a particular case of the CES production function when s approaches 0.

d. If $s < 0$, then $0 < ES < 1$. In this case, we may take many values of s from 0 to $-\infty$. Table 11.6 shows 10 values from -0.5 to -5 at an equal interval of 0.5. All values of ES are less than 1. They are inelastic, and are illustrated in Fig. 11.14. For example, at $s = -0.5, ES = 0.67$ and is a positive fraction. See row 12 of Table 11.6 and Fig. 11.14. In this case, since ES is inelastic, the production surface bends more severely along the center ridge line and so do the isoquants. Hence, each tail of the isoquant is asymptotic to a line inside the factor space and converges asymptotically to a line parallel to the axis for each axis. See the 2D chart of Fig. 11.15(d).
e. If $s \to -\infty$, then $ES \to 0$. In Table 11.6, we approximate infinity by using $s = -5$. Then, at this value of s, $ES = 0.17$. This will make the illustration easier in Fig. 11.14. Since the results for $s = -100$ and $s = -5$ are almost the same, we use $s = -100$ to draw the production surface for this case in Fig. 11.15(e). The production surface has a clear ridge line, and looks like a pyramid (see the 3D chart of Fig. 11.5(e)). The isoquants are rectangular kinked lines. Labor and capital have a fixed proportion, and efficient production can occur only along the ridge line (see below). We call this type of production function the **fixed-coefficient production function**, or the **Leontief production function**. Mathematically, we write[18]

$$\boxed{\lim_{s \to \infty} Q = A \min(L, K).} \tag{11.21}$$

This shows that the production takes place at the minimum of the two factors, and only along the ridge line through the origin. See Sec. 11.3.4 for more discussion.

11.3.3 *The shape of the CES production surface*

In the previous section, we have seen that, as s changes, the shape of the CES production surface changes dramatically. The data table for the 10 charts in Figs. 11.15(a)–11.15(e) is given in Table 11.7. Setting up Table 11.7 in a new worksheet, name the range C8:M8 "K" and the range B9:B19 "L". In addition, we name each of the four cell in range D1:D4 aa (for A), a, b, and s. We also construct sub-table H1:M4 for guidance in changing s. The values for s in the range I2:M2 are the suggested values based on the five cases described in Sec. 11.3.2 above, except case e. In C9, enter Eq. (11.14) directly as

```
=aa*(a*L^s+b*K^s)^(1/s),
```
(11.22)

and copy it to C9:M19. In cell F4, we enter the ES equation (11.17). The current table illustrates Case a, that is, $s = 0.9$ (in D4), or $ES = 1/(1 - s) = 10$ (in F4), which changes with s in D4. We change s in D4 by taking the five values from I2 to M2 each time and drawing the 10 charts in Fig. 11.15.

---

[18]See Footnote 19.

Table 11.7 Five cases of the CES production function

	A	B	C	D	E	F	G	H	I	J	K	L	M
1			aa	2				Case	a	b	c	d	e
2			a	0.6				s =	0.9	0.5	0.01	-0.5	-100
3			b	0.4		ES		ES =	10	2	1.01	0.67	0.01
4			s	0.9	=>>	10		Name	PE	CES	CD	CES	Leo
5													
6			C9:	Q=A(aL^s+bK^s)^(1/s)					F4:	ES=1/(1-s)			
7			K										
8			0	1	2	3	4	5	6	7	8	9	10
9	L	0	0.0	0.7	1.4	2.2	2.9	3.6	4.3	5.1	5.8	6.5	7.2
10		1	1.1	2.0	2.8	3.5	4.3	5.1	5.8	6.5	7.3	8.0	8.8
11		2	2.3	3.2	4.0	4.8	5.6	6.3	7.1	7.9	8.6	9.4	10.1
12		3	3.4	4.4	5.2	6.0	6.8	7.6	8.4	9.1	9.9	10.6	11.4
13		4	4.5	5.5	6.4	7.2	8.0	8.8	9.6	10.4	11.1	11.9	12.7
14		5	5.7	6.7	7.5	8.4	9.2	10.0	10.8	11.6	12.4	13.1	13.9
15		6	6.8	7.8	8.7	9.6	10.4	11.2	12.0	12.8	13.6	14.4	15.2
16		7	7.9	9.0	9.9	10.7	11.6	12.4	13.2	14.0	14.8	15.6	16.4
17		8	9.1	10.1	11.0	11.9	12.7	13.6	14.4	15.2	16.0	16.8	17.6
18		9	10.2	11.3	12.2	13.1	13.9	14.7	15.6	16.4	17.2	18.0	18.8
19		10	11.3	12.4	13.3	14.2	15.1	15.9	16.8	17.6	18.4	19.2	20.0

As in Chapter 4 on comparative static analysis, the five cases are obtained using the Excel picture copy command after entering the suggested value of s in each case in Table 11.7.

Using the CES production function, we may also find partial productivity curves of labor and capital (PPL, PPK), average productivity curves of labor and capital (APL, APK), and marginal productivity curves of labor and capital (MPL, MPK). These topics are included in the Homework of this chapter.

11.3.4 *The fixed-coefficient production function*

It has been shown that the fixed-coefficient production function is the limit of the CES production function when the substitution factor s approaches minus infinity. This function, like the CD production function, is another important functional form of production function. It can be written more generally as

$$Q = A\min(aL, bK) = A\min\left(\frac{L}{a_L}, \frac{K}{a_K}\right), \tag{11.23}$$

where a and b are constants. Let $a = 1/a_L$ and $b = 1/a_K$, where $a_L = L/Q$ and $a_K = K/Q$. L/Q and K/Q are labor and capital **input coefficients** similar to the labor input coefficient defined in Chapter 10. Then, we have the right-hand side of (11.23). These input coefficients are fixed constants, hence the name of the fixed-coefficient production function. The formula (11.23) means that production is determined by the minimum of the two arguments: if $aL < bK$, then $Q = AaL = AL/a_L$ and the production depends only on labor; if $aL > bK$, then $Q = AbK = AK/a_K$, and the production depends only on capital. The production takes place along the line where $aL = bK$, that is, along the line of $K = (a/b)L = (a_L/a_K)L$, which is a straight line through the origin called the **ridge line**. See Fig. 11.15(e). Any other points that are not on the ridge line can only produce the same output at the point along the ridge line.

Illustration of production under a limitation factor

This is illustrated in the 2D graph of Fig. 11.15(e), which shows $a = b = 1$. In this case, we still assume $A = 2$. Suppose capital is the scarce factor, called the **limitation factor**, at $K = 4$. (It is the white horizontal line in the 2D graph of Fig. 11.15(e).) Since a_L and a_K are constant, 1, 2, 3, up to 4 units of labor can be combined with a fixed amount of capital in 1, 2, 3, up to 4 units, to produce output levels of 2, 4, 6, or 8 units. This is shown by the 45-degree line up to the intersection of the crossing point, and the increase in output can be seen in the 3D chart. But after $L = 4$ units, any increase in labor is useless, since there are only 4 units of capital, which can be combined, at most, only with 4 units of labor, to produce $2 \times 4 = 8$ units of output. Hence, the pyramid in the 3D chart is cut from the top at $K = 4$. The labor productivity curve at $K = 4$ consists of the dotted white ridge line up to $L = 4$ and the horizontal solid white line from $L = 5$ to $L = 10$. Any amount of labor more than 4 units will not increase the output and will be discarded, if the production is efficient. Given $K = 4$, the maximum output can only occur at point $(L, K) = (4, 4)$, located on the ridge line. A similar argument holds if labor is the limitation factor.

Thus, if either labor or capital is a limitation factor, then it will be fully employed, while the other excessive factor is a redundant factor of production and will be wasted (or discarded). Production occurs only along the ridge line.

11.4 Applications to Utility Functions

We may extend the interpretations of the production function in the previous sections to the theory of the **utility function**. A utility function is customarily written as $U = U(x, y)$, where the U on the left-hand side is the dependent variable and the U on the right-hand side is a functional notation similar to $F(x, y)$ or $G(x, y)$. Independent variables are usually denoted as the consumption of commodity x and of commodity y. Utility is measured by "util", which is an index (a fictitious unit of measurement).[19] We write the CD form of utility function and the CES form of utility function as

$$\boxed{U = Ax^a y^b \quad \text{and} \quad U = A(ax^s + by^s)^{\frac{1}{s}},} \tag{11.24}$$

where $A > 0$, $0 < a < 1$, $b = 1 - a > 0$, $s < 1$, as in the case of the production function.

[19] We are not going into the discussion of measurability of utility. For simplicity, we simply assume that utility is measurable.

With the utility functions defined, we may follow the production theory to explain the utility theory of consumption. By replacing L with x and K with y in (11.4a) and (11.4b), we have the **partial utility functions** (or curves):

$$U = Ax^a y_0^b, \quad U = Ax_0^a y^b.$$

The **average utility functions** U/x and U/y are, similar to (11.7),

$$U/x \equiv AU_x = Ax^{a-1}y^b \quad \text{and}$$
$$U/y \equiv AU_y = Ax^a y^{b-1},$$

and the **marginal utility functions**, denoted MU_x or MU_y, are similar to (11.8):

$$\left.\frac{\Delta U}{\Delta x}\right|_{Kconst} = MU_x = Aax^{a-1}y^b = aU/x$$
$$\left.\frac{\Delta U}{\Delta y}\right|_{Lconst} = MU_y = Abx^a y^{b-1} = bU/y.$$

MUx is the extra utility obtainable from consuming extra units of goods x.

We also have the **utility surface** and **utility indifference map** and **indifference curves**,

$$\{(x, y)|U_0 = U(x, y)\},$$

in place of the production surface and the isoquant map and isoquants (see Sections 11.1.6 and 11.1.7). The MRTS, (11.11), is called the **marginal rate of substitution between goods x and y:**

$$\boxed{\left|\frac{\Delta y}{\Delta x}\right| = -\frac{\Delta y}{\Delta x} = \frac{MU_x}{MU_y} = \frac{ay}{bx} = MRS.}$$

The word "technical" is dropped since it represents the subjective evaluation of x and y, not the technical relation between labor and capital as in production theory. Thus, the law of diminishing MP is replaced by the **law of diminishing marginal utility**, and the law of diminishing rate of technical substitution is replaced by the **law of diminishing marginal rate of substitution**, as explained in the next chapter (see Sec. 12.5.4).

The factor prices in Sec. 11.2.4 should be replaced by **commodity prices** p_x and p_y, and the equilibrium conditions of the factor market for profit maximization should be replaced by the **equilibrium conditions of commodity market** for utility maximization, in which the marginal utility of commodity x should be equal to the price of commodity x, etc.

$$MU_x = p_x, \quad MU_y = p_y.$$

Thus, we see the theory of utility function parallels that of the production function.

11.5 Profit Maximization and Saddle Points in 3D

To take advantage of the power of Excel graphics, the format of Table 11.7 may now be applied to generate other types of 3D surfaces.

11.5.1 *Profit maximization: One-variable case*

In Chapter 2, we defined the profit of a firm as the total revenue (TR) minus the total cost (TC). The basic assumption in microeconomics is that a firm maximizes its profits, given the TR curve and the TC curve, both of which are functions of output level. Our use of 3D diagrams makes the solution of the profit maximization problem especially easy and intuitive.

Definitions

The mathematical definitions of maximum and minimum values are intuitive and not difficult to understand. A function f(x) has **a maximum value** (or **minimum value**) at $x^{\#}$ if

$$f(x^{\#}) \geq f(x) \text{ (or } \leq \text{ for minimum)}, \tag{11.25}$$

for all values of x in the neighborhood of $x^{\#}$. Both the maximum value and the minimum value are called **extreme values** (or **extrema**) of f(x), and $x^{\#}$ is an **extreme point** of f(x). In Fig. 11.16, for example, the maximum point is at $x = 10$, and the maximum value of f(x) is 1100. Example 11.3 shows how to find these values.

The above definition does not exclude the case in which the extreme value is not unique and there is a flat area at $x^{\#}$. In this case, there are infinite extreme values. If the strict inequality holds in the above definitions, we say f(x) has a **strict maximum** at $x^{\#}(f(x^{\#}) > f(x))$ or **strict minimum** at $x^{\#}(f(x^{\#}) < f(x))$. In this case, the maximum or minimum point is unique. In general, we are interested in a strict maximum or a strict minimum unless otherwise specified.

An illustration

Graphically, the maximum value in the definition (11.25) is achieved if the profit curve touches a horizontal line parallel to the x-axis, and the profit curve is also concave to the horizontal axis, as shown by the profit function in Fig. 11.16. Example 11.3 explains how to draw the curves and obtain the maximum profit.

Example 11.3 Profit maximization: The one-variable case. Let the cost function be

$$TC = x^3 - 8x^2 + 50x + 100,$$

and the demand function be $x = 200 - p$, where x is a certain output level and p is the price per unit of the output. Let x range from 0 to 17 units. Find the profit-maximizing level of output and the maximum profit. Illustrate the functions and the maximum profit diagrammatically.

Answer: We first construct the profit function and then draw the chart and find the solution (or approximate solution).The first step is to find the TR function, TR = px, which is a function of x. From the demand function, x = 200 − p, we can solve for p = 200 − x. Substituting it into TR to eliminate p, we have

$$TR = px = (200 - x)x.$$

By definition, the profit function is TR minus TC, $\pi \equiv TR - TC$. The problem is similar to the one we have encountered in Sec. 2.2 of Chapter 2. See Table 2.3. We have come a long way, and have learned since then how to name the variable x and draw the chart. Let the variable x range from 0 to 17, and name the variable x, and then enter TR, TC, and profit exactly as they are given above. The table should be similar to Table 2.3 and is omitted here. Drawing the chart from the table, we have Fig. 11.16. The profit maximizing output is at $x^{\#} = 10$ at f, the maximum profit is $\pi^{\#} = 1100$, ef, the TR is $TR^{\#} = 1900$ at a, and the TC is $TC^{\#} = 800$ at b. Note that ab = ef. □

11.5.2 *Profit maximization: Two-variable case*

The two-variable case is much more interesting. For the two-variable case (or multivariable case), the definitions of maximum and minimum are a direct extension of (11.25). Formally, let $z = f(x, y)$. Then f has a **maximum value (or minimum value)** at $(x^{\#}, y^{\#})$, if

$$\boxed{f(x^{\#}, y^{\#}) \geq f(x, y) (\text{or} \leq \text{for minimum}),} \tag{11.26}$$

for all (x, y) in the neighborhood of $(x^{\#}, y^{\#})$.

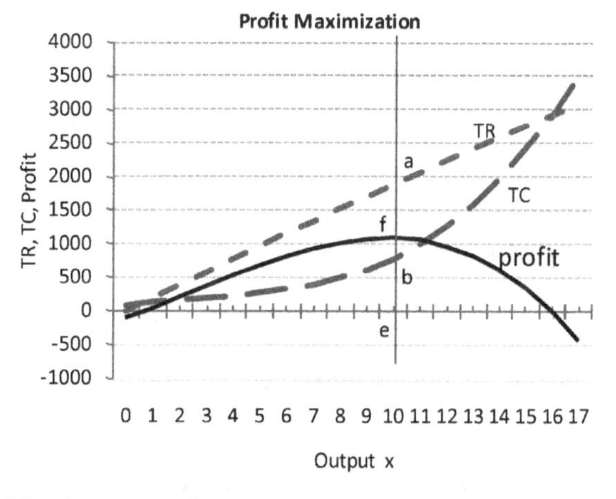

Fig. 11.16 profit maximization — one-variable case

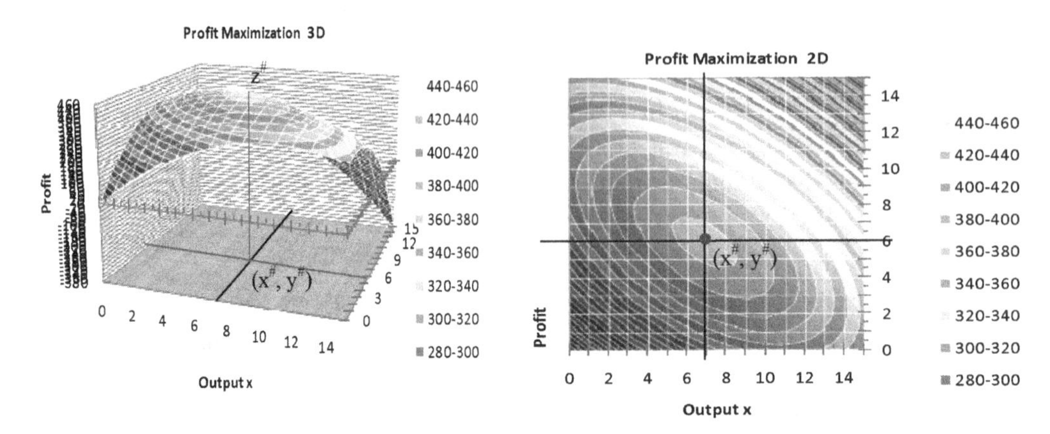

Fig. 11.17 Profit maximization — A two-variable case

This condition means that f(x, y) is maximized (or minimized) at a point $(x^{\#}, y^{\#})$, if the value of the function f(x, y) at point $(x^{\#}, y^{\#})$ is larger (or smaller) than the value of the function f(x, y) at any point in the neighborhood of $(x^{\#}, y^{\#})$. When strict inequality holds, then we have a **strict maximum** (or **minimum**). Generally, we are interested in a strict maximum (or minimum).

An illustration

Definition (11.26) is illustrated in Fig. 11.17, as explained in Example 11.4. If a point $(x^{\#}, y^{\#})$ yields a maximum (or a minimum) value of a function $z = f(x, y)$, then at $(x^{\#}, y^{\#}), z^{\#} = f(x^{\#}, y^{\#})$ is at a stationary value and cannot increase or decrease. Intuitively, from the diagram, a maximum (or a minimum) point, say $z^{\#}$, of the function must touch a plane which is parallel to the (x, y) space, and if, in the neighborhood of $(x^{\#}, y^{\#})$, the function is concave (or convex), then condition (11.26) will be satisfied, and $(x^{\#}, y^{\#})$ is a maximum (or minimum) point.

Example 11.4 Profit maximization. The two-variable case Let the TR and TC functions for two commodities x and y be given as[20]

$$TR = 65x + 75y, \quad TC = 4xy + 3x^2 + 4y^2.$$

Then the total profit function is

$$\pi = 65x + 75y - 4xy - 3x^2 - 4y^2. \tag{11.27}$$

Find the optimal output levels and the maximum profit.

[20]The interested reader may refer to Chapter 11 of Chiang and Wainwright (2005), or any other textbook on mathematical economics, for the mathematical method of solving the unconstrained optimization problem.

Table 11.8 Profit maximization: Two-variable cases

	A	B	C	D	E	F	G	H	I	J	K	…	Q
1		Variable		x	y	xy	x^2	y^2					
2		Coefficient		e	f	g	h	i		max			
3		Value		65	75	4	3	4		446			
4			B7:	=ex+fy-gxy-hx^2-iy^2									
5	x	y											
6		0	1	2	3	4	5	6	7	8	9	…	15
7	0	0	71	134	189	236	275	306	329	344	351	…	225
…	…	…											
13	6	282	329	368	399	422	437	444	443	434	417		147
14	7	308	351	386	413	432	443	446	441	428	407		113
15	8	328	367	398	421	436	443	442	433	416	391		73
…	…	…											
22	15	300	311	314	309	296	275	246	209	164	111		-375

Answer: Instead of solving this profit maximization problem directly, we can formulate it into a general form by taking advantage of the Excel naming method. Rewriting the coefficients in (11.27) into parametric form as

$$\pi = \mathrm{ex} + \mathrm{fy} - \mathrm{gxy} - \mathrm{hx}^2 - \mathrm{iy}^2, \tag{11.28}$$

we name and define the coefficients e, f, g, h, and i, as shown in range D2:H2 in Table 11.8. Name the range B6:Q6 "y" and the range A7:A22 "x". The parameter vector is $(\mathrm{e, f, g, h, i}) = (65, 75, 4, 3, 4)$, which is entered in D3:H3 in Table 11.8. In B7, enter the right-hand side of Eq. (11.28) and copy B7 to B7:Q22. We have the sensitivity table as shown. To find the maximum profit, in cell J3, enter =max(B7:Q22). We have max = 446, which is located at cell H14, at which the optimal output levels are $\mathrm{x}^\# = 7$ and $\mathrm{y}^\# = 6$. This solves the problem. □

Illustration

The results in Table 11.8 are illustrated in Fig. 11.17 in 3D and in 2D versions. The horizontal line and the vertical lines are added manually. The maximum profit is between 440 and 460 units in the graphs. The range is located at the top of the legend. The profit surface on the 3D chart has a unique maximum profit at $\pi^\#$ (denoted as $\mathrm{z}^\#$ in Fig. 11.17). The function

$$\pi = \{(\mathrm{x}, \mathrm{y}) | \pi^0 = \mathrm{f}(\mathrm{x}, \mathrm{y})\},$$

is the locus of the circles, which are called the **isoprofit curves**. The isoprofit curves are shown on the 2D chart as opaque circles with the maximum profit at the center area of the circles. Make sure that the horizontal axis shows output x and the vertical axis shows output y.

The maximum profit 446 units indeed satisfies the condition of maximum stipulated in (11.26), as H14 in Table 11.8 is surrounded by values lower than 446. In fact, 446 is the maximum profit globally. It is the maximum point for all values of x and y in the range of Table 11.8. The readers may try other values for the parameters and see the changes in the shape of the profit surface and the isoprofit curves, and also see the changes of the point of maximum profit.

11.5.3 *The nature of an optimal point*

Note that the profit surface in Fig. 11.17 (3D) shows that the maximum profit $z^{\#}$ of the profit function is not only maximized along the X-axis, but also along the Y-axis at the maximum profit. This is called **the first-order condition** or **the necessary condition of maximization** in calculus. Figure 11.17 (3D) also shows that the point is also maximized along any other directions in the neighborhood of the maximum profit. Thus, we may say that in the neighborhood of a point, if the point is maximized not only along the direction of the X- or Y-axis but also along any other straight lines passing through the point $(x^{\#}, y^{\#})$, then we can say that that point $(x^{\#}, y^{\#})$ is a maximum point. This is called **the sufficient condition of maximization** in calculus. Furthermore, we may also state that **a necessary and sufficient condition** of a point being a maximum (or minimum) is that the profit function is maximized along the X-axis and the Y-axis, and the profit surface is concave (or convex) at that point.

The profit function $\pi(x, y)$ in (11.28) is a general **quadratic function**. As mentioned at the end of the previous section, we may experiment with different values of the parameters and observe the changes in the profit surface and the isoprofit curves. In particular, when $e = f = g = 0$, and $h = 5, i = -5$, we have a saddle point at the origin, $(x, y) = (0, 0)$. This will be discussed in the following section.

Example 11.5 A two-product firm under perfect competition. A firm produces two outputs x and y, with prices $p_1 = 2$ and $p_2 = 5$, respectively. Here we assume that the commodity prices are constant, that is, the firm is small compared to the industry as a whole and the firm has no influence on the market prices of its products. Thus, the firm is said to be under **perfect competition in the commodity markets**. The firm's revenue from each commodity is p_1x and p_2y. Assume $p_1 = 2$ and $p_2 = 5$; then the firm's TR function is

$$TR = p_1x + p_2y = 2x + 5y.$$

Suppose the firm's TC function of producing both commodities is given by

$$TC = x^2 - 2xy + 3y^2.$$

Here, we also assume that the firm is under **perfect competition in the factor markets**. Now, the problem is to find the profit maximizing output levels $x^{\#}$ and $y^{\#}$, and the maximum profit.

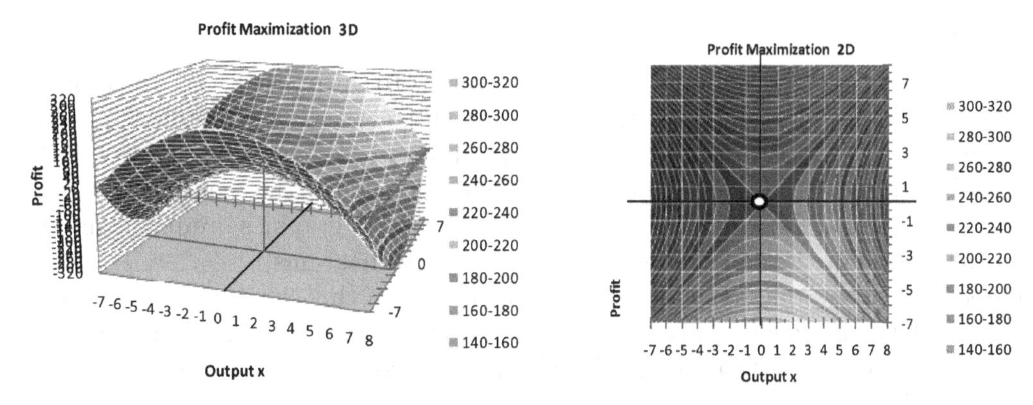

Fig. 11.18 A saddle point

Answer: The profit function is $\pi = \mathrm{TR} - \mathrm{TC}$. Hence, the problem is to find x and y such that

$$\max_{x,y} \pi = 2x + 5y - 2xy - x^2 - 3y^2.$$

This problem can be solved by using Table 11.8 and Fig. 11.17 as **templates**. To do this, copy Table 11.8 and Fig. 11.17 to a new spreadsheet. Enter the new coefficients in the parameter table D2:H3 in Table 11.8. Compared with the general equation (11.28), the major difference is that the coefficient of the cross term xy is +2 in this example, while the coefficient of xy in (11.28) is $-g$. Hence, enter $-g = 2$, or $g = -2$ in F3 of Table 11.8. The new parameter vector will be $(e, f, g, h, i) = (2, 5, -2, 1, 3)$. The table and the charts will change automatically. The maximum profit is found to be $\pi^{\#} = 7$ at $x^{\#} = 3$, $y^{\#} = 2$. □

11.5.4 *Saddle points and other surfaces*

Using Table 11.8 as a template, let the parameter vector be $(e, f, g, h, i) = (0, 0, 0, 5, -5)$, and let x and y range from -7 to 8 by changing the entries of A7:A22 and B6:Q6. Without any other adjustment, we have a function with a saddle point at the origin $(x, y) = (0, 0)$, as shown in Fig. 11.18. (it indeed looks like a horse's saddle). It is a **saddle point** in the sense that along the line $x = 0$, the function has a minimum at the origin, but along the line $y = 0$, the function has a maximum at the origin. Furthermore, along the 45-degree diagonal line from the southwest corner to the northeast corner and also along the diagonal line from the northwest corner to the southeast corner, the values of the profit function are zero. This can be verified from the table, and also the 3D and 2D charts, by moving the chart range of x or y in the table.

Other values of the parameters, such as $(e, f, g, h, i) = (0, 0, 0, -5, 5), (0, 0, 0, 5, 5)$, and $(0, 0, 0, -5, -5)$ may be entered to see the changes in the shape of the chart.

11.6 Summary

In general, we have four kinds of productivities: Total productivity surface, partial productivity curves (Sec. 11.1), average productivity curves, and marginal productivity curves (Sec. 11.2). The relations of these productivities are illustrated in Fig. 11.19. This chart applies to any production function, and with changes in the names of the variables, it can also be applied to utility functions (Sec. 11.4). Excel graphics are particularly useful in drawing these functions or curves either in 3D or 2D diagrams.

Three kinds of production functions are introduced in this chapter: The Cobb-Douglas (CD) (Sec. 11.1), the constant elasticity of substitution (CES) (Sec. 11.3), and the Leontief fixed-coefficient (Sec. 11.3) production functions. All of them can be illustrated in 3D and 2D graphics. The introduction to the CES production function leads to the concept of the elasticity of substitution (ES), which is fairly complicated. This chapter tries to

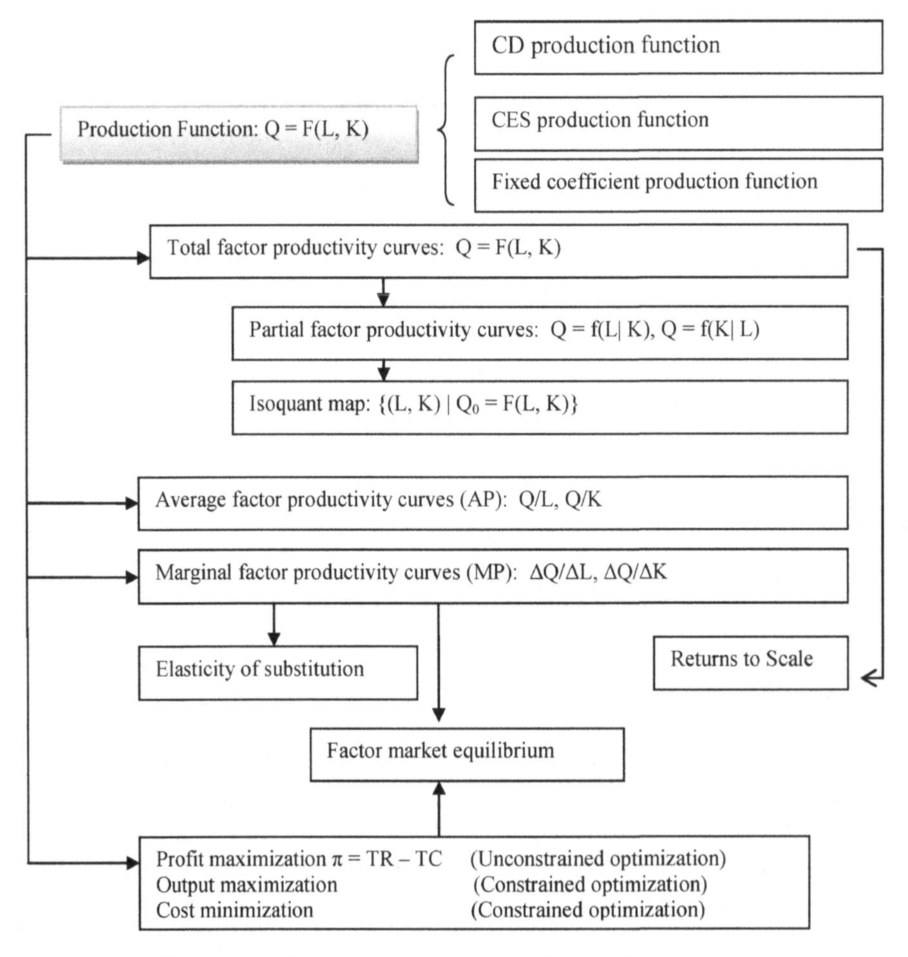

Fig. 11.19 Chapter summary of the theory of production

make it as easy as possible, and emphasizes its analogy with the price elasticity of demand.

The power of Excel graphics, combined with the naming method, can be extended to other types of functions without much difficulty. In Sec. 11.5, we introduce the method of solving profit maximization problems with one and two variables, using a general form of quadratic equation. We have shown that, by changing the value of parameters, the 3D graphics can change to saddle point and other graphs instantly. In fact, the possibilities of graphing are only limited by one's imagination.

In the next chapter, we will make good use of the knowledge of production and utility functions that we have introduced in this chapter, to study some topics on constrained maximization and the methods of solution.

Appendix 11A. Four Methods of Setting up a Sensitivity Table

Up to now, we have seen the importance of setting up a sensitivity table for economic and business analysis. In fact, the table is the basis for drawing a chart. There are four ways to construct a two-input sensitivity table. We use the CD production function (11.2) as an example.

(a) The **mixed references method**. In C14 of Table 11.1, enter =A*$B14^aa*C$13^bb, and copy to C14:H23. See Chapter 6.
(b) The **range-copy command (RCC) method**. In C14, enter =A*B14:B23^aa*C13:H13^bb, where the range B14:B23 represent labor and the range C13:H13 represents capital. Selecting the range and applying the RCC, we obtain the table. See Chapter 6.
(c) The **naming method**. Name the range B14:B23 "L" (or labor) and C13:H13 "K" (or capital). In C14, enter = A*L^aa*K^bb. Then either apply RCC as (b) above, or copy C14 directly to C14:H23 without using RCC.
(d) The **data table method**. This method requires using two-input cells to represent the variables, in this case, labor and capital. Since we do not use this method, the interested readers may read other books on Excel.

The naming method (c) without RCC is probably the easiest, most flexible, and intuitive, as we have used it in the text.

Appendix 11B. Returns to Scale and Shape of the Production Function

Let m be a factor of multiplicity. Multiplying m to factors of production, we have

$$A(mL)^a(mK)^b = m^{a+b}AL^aK^b = m^{a+b}Q.$$

Thus, if $a + b = 1$, doubling the scale of the factors of production will also double the output. Hence, we say that there are constant returns to scale (CRS) in production. On the other hand, if a + b > (or <) 1, then doubling the scale of the factors will more (or less) than double the output. Hence, we say that there are increasing (or decreasing) returns to scale.

To view the effects of returns to scale on the shape of the production surface, we use a method similar to the comparative static analysis in Chapter 4. Table 11B.1 is the same as that in Table 11.1, except that a range C3:E6 is added, and that, for symmetry, labor and capital now range from 0, 1, ..., 9. The equation in C14 is the same as that in Table 11.1, which is extended to C14:L23.

For CRS in C4:C5, we take aa = bb = 1/2; for IRS in D4:D5, we take aa = bb = 4/3; and for DRS in E4:E5, we take aa = bb = 1/3. These are suggested values. The current values are in B4:B5. Enter the suggested values of CRS in B4:B5, and draw the 3D and 2D charts. Use the picture copy command to copy the charts as pictures. Continue to enter the suggested values of IRS and DRS into B4:B5 one by one. We have the six pictures as in Figs. 11B.1(a)–11B.1(c). All the 3D charts are rotated to have angles X = 30° and Y = 0°.

Examining the charts reveals some interesting patterns.

(a) CRS: The rays through the origin are straight lines, and the spacing of the isoquants between the consecutive output units are the same.
(b) IRS: The rays through the origin bend upward, and the spacing of the isoquants between the consecutive output units gets narrower as the units increase.
(c) DRS: The rays through the origin bend downward, and the spacing of the isoquants between the consecutive output units gets wider as the units increase.

Table 11B.1 Returns to scale and the shape of the production surface

	A	B	C	D	E	F	...	L
1	Returns to Scale and the Production Function							
2		$Q = AN^aK^b$						
3	Parameter		CRS	IRS	DRS			
4	aa	0.5	0.5	1.33	0.33			
5	bb	0.5	0.5	1.33	0.33			
6	A	2	1.0	2.7	0.7	aa+bb		
7								
8			C14:	=A*L^aa*K^bb				
9								
10	Sensitivity Table							
11	**The Total Factor Productivity Surface**							
12	Q		Capital					
13			0	1	2	3	...	9
14	Labor	0	0.0	0.0	0.0	0.0		0.0
...		...						
23		9	0.0	6.0	8.5	10.4		18.0

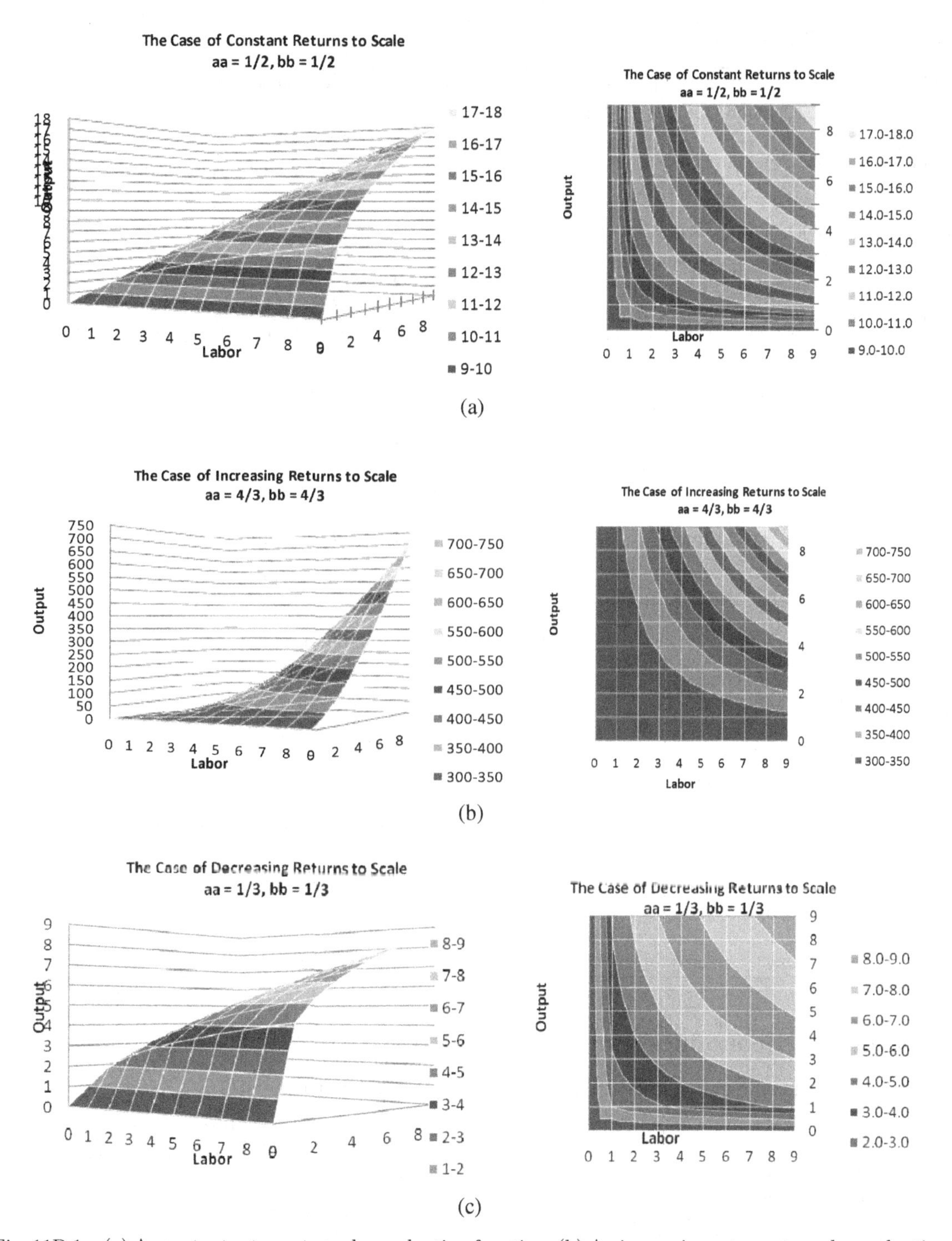

Fig. 11B.1 (a) A constant returns to scale production function. (b) An increasing returns to scale production function. (c) A decreasing returns to scale production function

Review of Basic Equations and Formulas

(11.1) $Q = F(L,K)$ General production function

(11.2) $Q = AL^aK^b$ Cobb-Douglas prod function

$a + b <=> 1$

(11.4a) $Q = f(L|K) = g(L)$

(11.4b) $Q = f(K|L) = h(K)$

(11.4c) $\{(L, K)| Q_0 = F(L, K)\}$ isoquant

(11.5) $K = (Q/AL^a)^{(1/b)}$ isoquant

(11.7) $Q/L \equiv APL = AL^{a-1} K^b = APL$

(11.7) $Q/K \equiv APK = AL^a K^{b-1} = APK$

(11.8)

$$\left.\frac{\Delta Q}{\Delta L}\right|_{Kconst} \equiv MPL = AaL^{a-1}K^b = aQ/L$$

$$\left.\frac{\Delta Q}{\Delta K}\right|_{Lconst} \equiv MPK = AbL^aK^{b-1} = bQ/K$$

(11.9) $\Delta Q = MPL\ \Delta L + MPK\ \Delta K$

(11.11) $\left|\frac{\Delta K}{\Delta L}\right| = -\frac{\Delta K}{\Delta L} = \frac{MPK}{MPL} \equiv MRTS$

(11.12) Market equil. cond.

$MPL = w$, $MPK = r$

Factor demand functions

(11.13) $L = L(w, r)$, $K = K(w, r)$

CES production function

(11.14) $Q = A(aL^s + bK^s)^{\frac{1}{s}}$

$A, a, b, > 0, s \leq 1$

(11.15) Elasticity of substitution

$$ES = \frac{\%\Delta\ of\ (K/L)}{\%\Delta\ of\ (MPL/MPK)} = \frac{\%\Delta\ of\ k}{\%\Delta\ of\ \alpha}$$

Elasticity of substitution (ES) under perfect competition in the factor market

(11.16)

$$ES = \frac{\%\Delta\ of\ (K/L)}{\%\Delta\ of\ (w/r)} = \frac{\%\Delta\ of\ k}{\%\Delta\ of\ \omega} = \sigma$$

ES of CES production function

(11.17) $ES \equiv \frac{1}{1-s}$

(11.20) $\lim_{s\to 0} Q = AL^aK^b$

(11.21) $\lim_{t\to\infty} Q = A\min(L, K)$

Fixed coefficient production function

(11.23)

$Q = A\min(aL, bK) = A\min\left(\frac{L}{a_L}, \frac{K}{a_K}\right)$

(11.24) $U = Ax^ay^b$

(11.24) $U = A(ax^s + by^s)^{\frac{1}{s}}$

(11.25)Maximum (or Minimum)

$f(x^\#) \geq f(x)$ (or $\leq$) for all x in the nhd of $x^\#$

(11.26) Maximum (or Minimum)

$f(x^\#, y^\#) \geq f(x, y)$ (or $\leq$)

for all (x, y) in the nhd of $(x^\#, y^\#)$

Key Terms: Economics and Business

Key Terms: Excel

Homework Chapter 11

Production functions and productivity curves

11-1 Estimation of the Cobb-Douglas production function Table HW11-1 shows part of the original data used by Cobb and Douglas (1928).[21] Q is index of manufactures output per year, L is index of average number of workers employed, and K is index of fixed capital.

(a) Draw the scatter diagrams between Q and L, and Q and K, with L and K being independent variables.
(b) Using the rule of logarithm (see Example 8.12, Section 8.5.3), reduce the Cobb-Douglas production function (11.2) to

$$\ln \mathrm{Q} = \mathrm{c} + \mathrm{a} \ln \mathrm{L} + \mathrm{b} \ln \mathrm{K}$$

where c = ln A. This equation is called a **log-linear equation**.
(c) Estimate the above log-linear equation by taking ln Q, ln L, and ln K as variables. Write it in usual regression form, with the t statistic and the prob-value of the t distribution (two-tailed test), F statistic and the prob-value of the F distribution, below the Excel Regression Table. What is the value of R^2?
(d) Evaluate the significance of the coefficients at the 5% significance level.
(e) What conclusions can you make from the above analysis?

[21] The original Cobb and Douglas (1928) estimated the production function from 1899 to 1922 (1899 = 100). We took their data from 1918 to 1922 and converted their data to index (1918 = 100). Their original estimated results are a = 0.75, b = 0.25, and A = 1.10.

Table HW11-1 Estimation of a production function

	A	B	C	D
1	Year	Q	L	K
2	1918	100	100	100
3	1919	98	97	106
4	1920	104	97	111
5	1921	80	74	114
6	1922	108	81	118

11-2 The Cobb–Douglas production The Cobb–Douglas production function is given as (11.3), $Q = 2L^{0.6}K^{0.4}$, where L changes from 0, 1, to 15, and K from 0, 2, to 30. Each chart must be labeled clearly.

(a) Draw the partial productivity curve of capital when labor is fixed at 11.
(b) From Table 11.5, draw the exact and approximate average productivity curve of capital when labor is fixed at 11. Do both curves overlap?
(c) From Table 11.5, draw the exact and approximate marginal productivity curve of capital when capital is fixed at 11. Do both curves overlap?
(d) Describe the relationship between the marginal productivity curve and the average productivity curve. Which one is larger? Do they have the same shape and position?
(e) Draw the total factor productivity surface of the production function.
(f) Draw the isoquant map of the function.

(Hints: (a) Just answer the question that L = 11. Do not construct the whole sensitivity tables. (b) and (c), the exact average and marginal productivities of capital are given by equations (11.7) and (11.8). Approximate of MPK = $(Q_i - Q_{i-1})/(K_i - K_{i-1})$).

11-3 The Cobb–Douglas production function Let the CD production function be given as

$$Q = AL\hat{}(0.7)K\hat{}(0.3).$$

Let A = 3. L changes from 0, 1, 2, to 15, and K from 0, 2, to 30.

(a) Draw the 3D production surface and label the three axes.
(b) Draw the 2D input space (isoquants) with an output scale of 1 unit. The 3D and 2D charts must have wireframe.
(c) Draw the 2D and 3D APL curves for each given unit of capital in one chart.
(d) Draw the 2D and 3D MPL curves for each given unit of capital in one chart.
(e) Arrange the above four charts in one page in Sheet2 and print it out.
(f) Do the same questions as (c), (d), and (e) for the MPK. Arrange the two charts in one page and print it out.
(g) Find the MRTS (11.11) at L = 10 along the isoquant of Q = 30? What is the economic interpretation?
(h) What is the ES between factors?

11-4 The Cobb–Douglas production function Continuing HW11-3 from/and using charts, answer the following questions:

(a) What is the returns to scale of the production function?
(b) Does the law of diminishing returns hold?
(c) Does the law of diminishing marginal rate of substitution hold?

Prove or derive your assertion and give an economic interpretation of each of the above three questions.

11-5 Production surfaces Reproduce the production surface of Fig. 11.4 with attractive colors of your choice. Rotate the 3D chart once from above (the origin facing you directly), once from the other side (with labor, not capital, on the left-hand side), and once with the opening facing you (you see the floor of the chart). Print out the above three 3D charts.

11-6 Isoquants Using the CD production function (11.3), derive the isoquant formula as in (11.5) by solving labor in terms of capital (that is, labor L now is the dependent variable). Draw the isoquant map, which is similar to Fig. 11.8 except that now the capital is on the horizontal axis.

11-7 The marginal rate of technical substitution (MRTS) The slope (11.10) is derived from the total increment equation of (11.9). Since (11.9) is symmetric with respect to ΔL and ΔK, we may write (11.10) as $\Delta L/\Delta K$. Using the isoquant derived for $Q = 26$ in HW11-6, interpret the MRTS in this case. Does the law of diminishing MRTS still hold in this case? Explain the law and give an economic interpretation.

11-8 The CES production function Show that the CES production function (11.14) is a constant-return-to-scale (CRS) production function by using the equation method and the charting method as expounded in Appendix 11B. Explain the economic meaning of CRS.

11-9 CES production function Given a CES production function (11.14), let $L = 0, 1, \ldots, 18$, $K = 0, 1, \ldots, 10$, $a = 0.6$, $b = 0.4$, $A = 2$, $s = -0.5$.

(a) Draw the labor productivity curves for all K in the sensitivity table.
(b) Draw the 3D production surface.
(c) Draw the 2D isoquant map if output intervals are unity.
(d) Draw the average productivity curves for all K (APK) in the sensitivity table.
(e) What is the elasticity of substitution for this CES production function?

(Hints: (d): No need to simplify the function. Simply enter = Q/L, where Q is in (11.14). (e): Use (11.17)).

11-10 Scale of returns Determine the scale of returns for the following functions. Show your findings step by step. Set up a sensitivity table for L and K being 0, 1,...,10. Draw a 3D surface and a 2D wired color isoquant map for each case, make sure that the labor axis is on the left hand side facing you.

(a) $Q = L^4 - 2L^2K^2 + K^4$

(b) $Q = \dfrac{aL^2 + bLk + cK^2}{dL + eK}$, $a = b = c = d = e = 1$

(c) $Q = bL^aK^{1-a} + cK$, $(a, b, c) = (2, 0.7, 2)$
(d) $Q = aL\log(\frac{K}{L}) + bL$, $(a, b) = (2, 2)$
(e) $Q = \sqrt{L} + \sqrt{K}$
(f) $Q = 10L + 100K - 3LK - 4L^2 - 3K^2$

11-11 Big Mac operations Let the Big Mac production function be given as

$$Q = 1000L^{0.6}K^{0.4}.$$

Let L change from 0, 100 to 1500, and K from 0, 4 to 40. Make sure that labor is on the horizontal axis.

(a) Draw the 3D production surface and label the three axes.
(b) Draw the 2D input space (isoquants) with an output scale of 10,000 units. Make sure labor is on the horizontal axis.
(c) Draw the labor productivity curve for given $K = 8$ and 20.
(d) The marginal productivity of labor (MPL) for this model is given as

$$MPL = 1000(0.6)L^{-0.4}K^{0.4}.$$

Draw the MPL curves for $K = 8$ and $K = 20$.
(e) The average productivity of labor (APL) for this model is given as

$$APL = 1000L^{-0.4}K^{0.4}.$$

Draw the APL curves for $K = 8$ and $K = 20$.
(f) Arrange the above five charts in one page and print it out.
(g) Show the cell formula in the first cell of each table, or each calculation, above the table or column.

Utility functions

11-12 Utility function Let the utility function be given as

$$U = \text{sqrt}(x) + \text{sqrt}(y).$$

All the charts and axes of the charts must be labeled clearly.

(a) Draw the utility surface for $x = 0, 1, \ldots, 9$ and $y = 0, 1, \ldots, 10$. Put commodity x in the column and commodity y in the row. Utility is measured up to one decimal place.
(b) Draw the utility surface in 3D.
(c) Draw the indifference curves (the same as the isoquants) for the two commodities. Let the interval of utility in the legend be 0.2 util and the maximum utility be 5 util.

11-13 Logarithmic utility function Let the utility function be

$$U(x,\ y) = \frac{1}{2}\ln(x) + \frac{1}{2}\ln(y),$$

where x is the number of commodity X and y is the number of commodity Y consumed each month. Let x and y range from 0, 1, ..., 10.

(a) Set up the utility sensitivity table x as the column input labels (labeling the rows) and y as the row input labels (labeling the columns). Entries of the table should have two decimal places.
(b) Draw the utility curves of x when commodity y is fixed at $y = 2, 4, 6$, and 8 (4 curves in one chart).
(c) The marginal and average utility curves for X are given as

$$MU_x = 1/(2x)$$
$$AU_x = (\ln(x) + \ln(y))/(2x).$$

Find the MU_x and AU_x curves at $y = 5$ for x ranges from 0, 1 to 10.
(d) Draw the utility surface in 3D. Make sure to label the chart and the axes.
(e) Draw the indifference curves for the two commodities. Let the interval of the utility in the legend be 0.1 util.

11-14 GPA indifference curves Suppose that from past experience you have figured out that your grade point average, GPA, from your two courses, College Mathematics and Western Literature, takes the form

$$GPA = \frac{2}{3}(\sqrt{M} + \sqrt{2L}),$$

where M is the number of hours per day spent studying for Mathematics and L is the number of hours per day spent studying for Literature. Let the hours spent on L range from 0, 1, to 15 for the rows (the X-axis) and M range from 0, 1, to 12 for the columns (the Y-axis).

(a) Draw the 3D diagram for the GPA surface with wireframe.
(b) Draw the GPA indifference curves for both courses. Change the interval of the grade in the legend to 0.2. Give an interpretation of the GPA indifferent curves.
(c) Draw the partial GPA productivity curves for hours spent on Mathematics given $L = 0, 2, 12$. What is the interpretation of these curves?
(d) Do the same as (c) above for the average GPA productivity curves.
(e) The marginal productivity of GPA for mathematics is give as

$$MPGPA_M = (1/3)M^{-1/2}.$$

Do the same as (c) above for these marginal GPA productivity curves.
(f) The marginal productivity of GPA for Literature is give as

$$MPGPA_L = (2/3)(2L)^{-1/2}.$$

Find the marginal rate of substitution (MRS) between the two courses. Does the law of diminishing marginal rate of technical substitution (MRTS) hold? Why? Let M = 3, show MRTS for L = 1, 2, ..., 15. What is the interpretation?

11-15 A utility function The utility a person derives from consuming apple A and banana B is described by the utility function

$$U(A, B) = (A - 2)^{0.4}(B - 3)^{0.6}.$$

In the following questions, we first set up the utility sensitivity table for A = 2, 2.5, ..., 8 and B = 4, 4.5, ..., 10. Put A in the column (row labels) and B in the row (column labels). Entries of the tables should have one decimal place. All the chart titles and axes of the charts must be labeled clearly.

(a) Draw the utility sensitivity table for A and B.
(b) The marginal utility function for A(MU_A) is given as

$$MU_A \equiv \frac{\Delta U}{\Delta A} = 0.4(A - 2)^{-0.6}(B - 3)^{0.6}.$$

Draw the marginal utility curve for A = 1, 1.5, ..., 8 at B = 7 and 10. (You may attach it to the utility sensitivity table but should label the columns clearly.)
(c) The average utility function for A(AU_A) is

$$AU_A = \frac{U}{A} \equiv \frac{(A - 2)^{0.4}(B - 3)^{0.6}}{A}.$$

Draw the average utility curve for A = 1, 1.5, ..., 8 at B = 7 and 10. (You may attach it to the utility sensitivity table but should label the columns clearly.)
(d) What is the shape of the MU_A curve (sloping upward, downward, or flat)? What is the economic meaning of the shape of the curve? Does the shape make sense? Why?
(e) Draw the utility surface in 3D. Make sure that the variable A is on the horizontal axis.
(f) Draw the indifference curves for the two commodities. Let the scale of the legend be 0.5 util.

Profit maximization

11-16 Profit maximization of a firm Let the CD production function be given as

$$Q = AL^{0.5}K^{0.4}.$$

Let A = 3. L changes from 20, 21,..., 35 and K changes from 10, 12,..., 40. The output price is 6, labor is paid at an hourly rate of w = 4.5, and the rental price of capital is r = 3. Find the firm's maximum profit level. What are the profit-maximizing levels of labor and capital?

(Hints: Set up the cost function, and follow Example 11.4)

11-17 The profit function of an oligopolistic firm The profit of firm A is given as

$$\pi = 72x - 4x^2 - 4xy,$$

where x is the output of firm A and y is the output of firm B. Let x and y range from 1, 2, ..., 10.

(a) Set up the profit table for given x and y.
(b) Draw the profit surface in 3D. Use a text box to denote the isoprofit curve for $\pi = 100$.
(c) Draw the isoprofit curves in 2D. Use a text box to show the isoprofit curve for $\pi = 100$.
(d) Draw the profit curve for output x when y is given at $y = 5$. Use a dotted line to denote the profit of firm A when it produces at $x = 5$.

11-18 A two-product firm under monopolistic competition Suppose the firm in Example 11.5 is under monopolistic competition, that is, the commodity prices depend on outputs the firm produces. Thus, the demand function for each commodity is given by

$$x = 20 - 3p_1 + p_2,$$
$$y = 30 + p_1 - 3p_2.$$

The cost function is the same as in Example 11.5.

(a) The total revenue is $TR = p_1x + p_2y$. Using the elimination method, solve the above two equations in terms of p_1 and p_2. Substituting into TR, we have TR in terms of x and y.
(b) Show that the profit function is

$$\pi = (90/8)x + (110/8)y + (7/4)xy - (27/8)y^2 - (11/8)x^2.$$

(c) Let $x = 0, 0.1, \ldots, 1.5$ and $y = 0, 0.5, \ldots, 5$. Find the maximum profit and optimal outputs $x^{\#}$ and $y^{\#}$.

(Hints: (c) $(x^{\#}, y^{\#}, \pi^{\#}) = (0.6, 2.0, 18.602)$)

11-19 Discriminatory pricing policy A telephone company has two demands for its service.

The demand on weekdays is $x = 100 - 0.2p_1$.
The demand during the night and on holidays is $y = 20 - 0.1p_2$.

Since x and y are essentially the same commodity, the cost depends on the total level of production: $C = 300 + 100(x + y)$. Let $x = 0, 10, \ldots, 100$ and $y = 0, 1, \ldots, 10$.

(a) Find the profit maximizing output levels.
(b) Find the profit maximizing prices.
(c) Find the price elasticity of demand at equilibrium in each market.
(d) From the results in (c), give an economic interpretation of this discriminatory pricing policy for the two markets.

(Hints: First solve for p_1 and p_2 and substitute into the TR function. $\pi = 400x - 5x^2 + 100y - 10y^2 - 300$. We have $x^{\#} = 40, y^{\#} = 5$).

11-20 A profit function Let a profit function be

$$\pi = 50x + 80y - 3xy - 2x^2 - 4y^2,$$

where x and y are outputs ranging from 0, 1, ..., to 15. Find the maximum profit from the sensitivity table, from the colored 3D graph (in which the legend ranges from −100 to 500 and the major unit (interval) is 50), and also from the 2D contour map (in which the legend ranges from −100 to 500 and the major unit is 10). What is the maximum profit? What are the profit-maximizing levels of outputs?
(Hints: Use =max(range) to find the maximum profit in the sensitivity table. $(x^{\#}, y^{\#}, \pi^{\#}) = (7, 7, 469)$)

Saddle points and other functions

11-21 Construction of a 3D table Let the function be

$$f(x, y) = x^4 - 2x^2y^2 + y^4,$$

where x and y range from −5, −4, ..., to 5. Using the following three methods (see Appendix 11A), draw the colored 3D surface and the 2D contour map. Please copy the cell formula of the first cell (the cell at the northwest corner) and place it above each table.

(a) The mixed reference method.
(b) The RCC method.
(c) The naming method.

11-22 Graph of a function Let the function be

$$f(x, y) - ax^2 + by^2,$$

where x and y range from −5, −4, ..., to 5. Using the naming method by naming a and b cells, find the colored 3D and 2D contour graphs for the following four cases: (a) $a = 1, b = 1$; (b) $a = 1, b = -1$; (c) $a = -1, b = 1$; (d) $a = -1, b = -1$. Find the maximum, minimum, or saddle point in each graph and use an empty circle to show its location. Copy all the picture graphs in a new sheet. You should have eight different graphs on one page.

11-23 Graph of a function Let the function be

$$f(x, y) = xy - \ln(x^2 + y^2),$$

where x and y range from −5, −4.5, ..., to 5. Find the colored 3D and 2D contour graphs. There are two saddle points: (1, 1) and (−1, −1). Color the cells of saddle points in the table with yellow, and use an empty circle to show the location of the saddle points in the 3D and 2D graphs.

11-24 Saddle point (full and half) Let the function be

$$f(x, y) = x^2 - y^2,$$

where x and y range from $-5, -4, \ldots,$ to 5.

(a) Draw the colored 3D surface and 2D contour map. The saddle point is at (0, 0). Using a white circle to denote the saddle point, paint the cell (0, 0) in yellow in the table. The chart title should be "Fig. HW11-24(a) A Saddle Point, x = −5 to 5, y = −5 to 5", and "Fig. HW11-24(b). A Contour Map with a Saddle Point, x = −5 to 5, y = −5 to 5". The range statement in the title should be placed on the second line.
(b) Make sure the X-axis and the Y-axis are correctly labeled.
(c) The range of the legend for both graphs should be from −25 to 25, with the major unit of scale (interval of the vertical axis) being 5.
(d) Picture copy Fig. HW11-25(a) and Fig. HW11-25(b) to a new sheet, called "saddle", and place them side by side.
(e) Use and revise the original Fig. HW11-25(a) and Fig. HW11-25(b) to construct the following pairs of 3D and 2D graphs. Use a white circle to denote the point (0, 0) in each graph.

$$\begin{aligned} &(a)\ \ x = -5 \text{ to } 0, \quad y = -5 \text{ to } 5,\\ &(b)\ \ x = -5 \text{ to } 5, \quad y = -5 \text{ to } 0,\\ &(c)\ \ x = -5 \text{ to } 0, \quad y = -5 \text{ to } 0. \end{aligned}$$

(f) Place the above six graphs systematically below the pictures of Fig. HW11-25(a) and Fig. HW11-25(b) in the "saddle" sheet. Thus, you have a total of eight graphs, in addition to two original Excel graphs, in one page.
(g) Explain the shape of the lines through (0, 0) along the x-axis and the y-axis, and explain how the saddle point is formed. Explain the relationship between the 3D graph and the 2D contour map. Use letters such as a, b, ... to show that they correspond to each other in shape and colors.

11-25 Saddle point (full and half) Repeat the questions in HW11-24, this time for the cases

$$\begin{aligned} &(a)\ \ x = 0 \text{ to } 5, \quad y = -5 \text{ to } 5,\\ &(b)\ \ x = -5 \text{ to } 5, \quad y = 0 \text{ to } 5,\\ &(c)\ \ x = 0 \text{ to } 5, \quad y = 0 \text{ to } 5. \end{aligned}$$

Rotate the graph so that the saddle point (0, 0) faces you. Explain how 3D and 2D graphs correspond to each other.

11-26 Saddle point (full and half) Let the function be

$$f(x, y) = x^4 - 4x^2y^2 + y^4,$$

where x and y range from $-5, -4.5, \ldots,$ to 5. The saddle point is at (0, 0). Repeat the questions in HW11-24. The vertical range of the 3D and 2D graphs should be −1,500 to 1,000, and the major unit (the interval of the vertical axis) should be 200.

Chapter 12

Constrained Optimization in the Theories of Production and Consumption — Using Excel Solver

Chapter Outline

Objectives of this Chapter

In Chapter 11, we discussed production and utility functions and their functional properties, such as their slopes, their shapes, etc. This chapter goes a step further and discusses how these functions can be applied to analyze rational behavior of two basic economic entities, the representative firm and the representative household.

The first four sections of the chapter discuss the theory of production: maximizing output under a cost constraint, and minimizing cost under an output constraint. We solve the constrained output maximization problem by using the graphic method first and then using the Excel Solver method. The last two sections are devoted to the theory of consumption: maximizing utility under a budget constraint, and minimizing budget subject to a given utility function. In the latter, we solve both problems by using the Excel Solver first, and then using the graphic method to illustrate the results.

Lastly, the decomposition of price change into the substitution effect and the income effect is illustrated, and compensating variation of income is calculated. The procedure of graphic illustration and the circular flow of the economy between the firm and the household and their economic behavior are illustrated diagrammatically in the summary of this chapter.

12.1 The Theory of Production

12.1.1 *The cost constraint*

The production function depicts the technological relation between inputs and output. How does a firm decide how much to produce using labor and capital? The firm cannot produce an infinitely large amount of output, since its production activities are constrained by its ability to acquire resources, among other things. The firm's **resource constraint** is expressed by the **cost equation**

$$\boxed{C = wL + rK,} \tag{12.1}$$

where C is the cost (or expenditure) of the firm, namely, the total amount of the firm's budget to acquire labor and capital; w is the wage rate; and r is the price of capital (rental or interest rate accrued to capital). We assume perfect competition in the factor markets. This implies that w and r are constant.

To draw the cost function, let $w = 2$ and $r = 1.5$. Note that we have chosen the units of labor and capital so that w is higher than r. The values of w and r depend on their different units of measurement, not on the same units of measurement; that is, labor may be measured per hour, per day, or per month, etc., and capital may be measured per machine, per ten machines, etc. Let the firm's cost of operation be $C = 30$ (such as, 30,000, 300,0000 dollars, etc.). We have entered this information in the parametric table in Table 11.1, which is reproduced as Table 12.1 (ignore the curves in Table 12.1 for the time being). In Table 12.1, enter a new box G4:I6. Create the name in range I4:I6 as the left column: wage, rent, and cost. For clarity, it is always advisable to enter the equations of the model like those in G2 and G3.

The constraint is written in such a way that all the variables are on one side, and it can be solved with respect to either L or K. If the constraint is solved for K, with L being the independent variable on the right-hand side of the equation, then we have

$$K = (C/r) - (w/r)L. \tag{12.2}$$

In the current example, (12.2) can be written as

$$K = (30/1.5) - (2/1.5)L = 20 + 1.3L.$$

This constraint should be attached to the right-hand side of the sensitivity table, as labor is listed in the column range B14:B23 in Table 12.1. On the other hand, if the constraint is solved for L, with K being the independent variable on the right-hand side of the equation, then we have

$$L = (C/w) - (r/w)K. \tag{12.3}$$

The constraint should be attached to the bottom of the sensitivity table, as capital is listed in the row range in C13;H13 in Table 12.1.

For convenience, we take L as the independent variable as in (12.2). Thus, in I14, enter Excel formula for (12.2) as specified in D9 — as usual, C should be entered as C_ — and

Table 12.1 Production surface and the cost

	A	B	C	D	E	F	G	H	I
1	Properties of the Cobb–Douglas production function								
2		$Q = AL^aK^b$					C = wL + rK		
3				Parameters			K = C/r - (w/r)L		
4		Labor share		aa	0.6		Wage	ww	2
5		Capital share		bb	0.4		Rent	rr	1.5
6		Scale factor		A	2		Cost	C	30
7									
8			C14:	=A*L^aa*K^bb					
9			I14:	=C_/rr-(ww/rr)*L					
11	Production Surface and Cost Constraint								
12				Capital, K					Cost
13			0	4	8	12	16	20	K
14	Labo	0	0.0	0.0	0.0	0.0	0.0	0.0	20.0
15	Labor	2	0.0	5.3	7.0	8.2	9.2	10.0	17.3
16		4	0.0	8.0	10.6	12.4	13.9	15.2	14.7
17		6	0.0	10.2	13.5	15.8	17.8	19.4	12.0
18		8	0.0	12.1	16.0	18.8	21.1	23.1	9.3
19		10	0.0	13.9	18.3	21.5	24.1	26.4	6.7
20		12	0.0	15.5	20.4	24.0	26.9	29.4	4.0
21		14	0.0	17.0	22.4	26.3	29.5	32.3	1.3
22		16	0.0	18.4	24.3	28.5	32.0	35.0	-1.3
23		18	0.0	19.7	26.0	30.6	34.3	37.5	-4.0

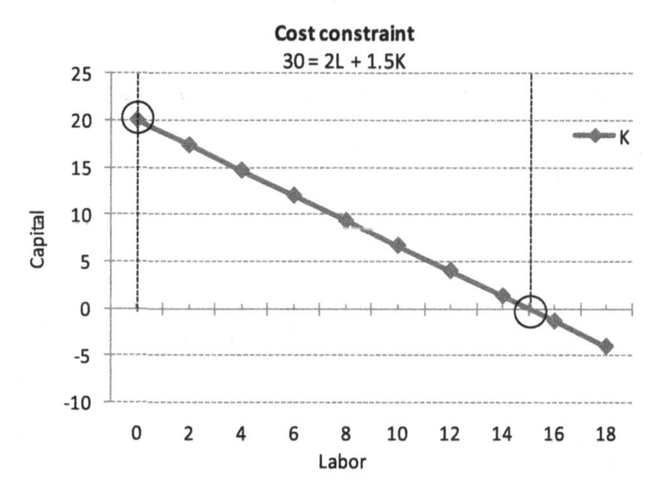

Fig. 12.1 Cost constraint

then copy I14 down to I23. We then have the values of K for each corresponding L. Draw the **cost line** as in Fig. 12.1. The L-intercept on the horizontal axis of the line is 15 (that is, $(L, K) = (15, 0)$), which falls between $L = 14$ and $L = 16$. At the point $L = 0$, the K-intercept on the vertical axis is 20 (that is, $(L, K) = (0, 20)$), which is located in cells I14.

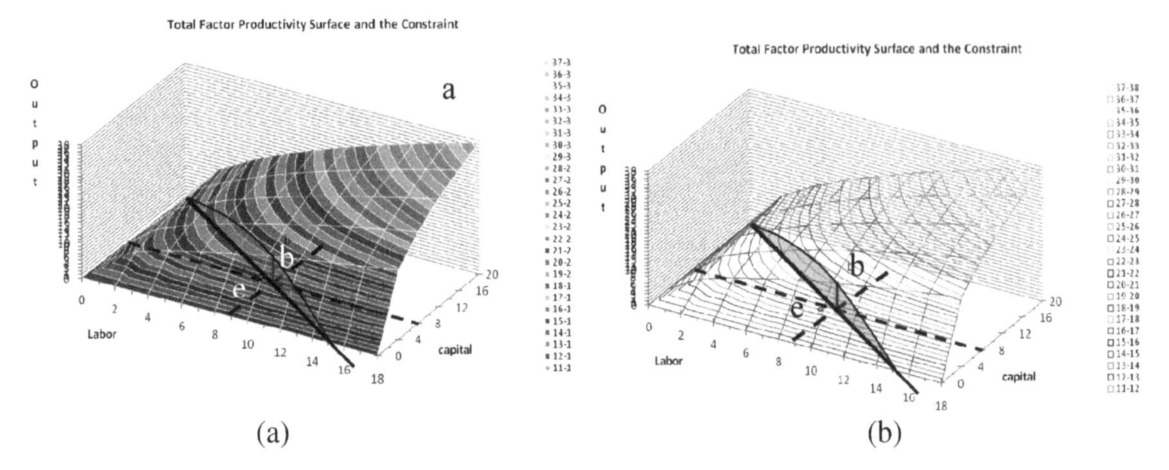

Fig. 12.2 The cost constraint (a) 3D and (b) wireframe 3D

This cost constraint is also imposed on the output table of Table 12.1 as a line connecting cells from C21 to H14 in Table 12.1. Also, see Fig. 12.2(a).

Note that the entries in C14:H23 of the table are output levels produced by the combinations of labor in B14:B23 and capital in C13:H13. However, the entries in I14:I23 are capital requirements to combine with labor in B14:B23 along the same cost line, $C = 30$. We put them together just for convenience of data entry.

12.1.2 *The production surface and the cost constraint*

The firm chooses a combination of labor and capital so as to maximize output, subject to the cost constraint. In standard mathematical notation, the problem is

$$\max_{L,K} Q = AL^a K^b, \quad \text{Objective function} \tag{12.4}$$

$$\text{s.t. } C = wL + rK \quad \text{Constraint,} \tag{12.5}$$

where we want to maximize output Q, in the Cobb–Douglas production function, with respect to labor L and capital K, subject to (s.t.) the linear cost constraint C at given wage rate w and price of capital r. The first equation (12.4) is called the **objective function**, and the variables to be changed to find the maximum of the objective function are generally indicated below the "max" sign. The second equation (12.5) is called the **constraint**, in this case, the **cost constraint**. If there are values $L = L^\#$ and $K = K^\#$, such that (12.4) and (12.5) are satisfied, then the pair $(L^\#, K^\#)$ is called the **optimal solution** to this maximization problem, $L^\#$ is called the **optimal labor**, and $K^\#$ is called the **optimal capital**. $(L^\#, K^\#)$ is called the **optimal allocation of resources**.

The general method of solution is to use the Lagrangian method.[1] Here, we use the diagrammatic method.

For our example, we use the Cobb–Douglas production function introduced in the previous chapter, $Q = 2L^{0.6}K^{0.4}$, and the cost constraint illustrated in Fig. 12.1. Therefore, the model in (12.4) and (12.5) can be written as

$$\max_{N,K} Q = 2L^{0.6}K^{0.4}, \tag{12.6}$$

$$\text{s.t. } 30 = 2L + 1.5K. \tag{12.7}$$

The production surface and the cost constraint

We now combine the production surface (11.3) of the previous chapter and the cost constraint in Fig. 12.1, and draw Fig. 12.2. The cost constraint depends only on capital and labor. It is drawn in Fig. 12.1, and in the (L, K) space of Fig. 12.2. Since it does not depend on the output (the vertical axis), the cost constraint is a vertical plane in 3D, cutting across the production surface along the solid line in the (L, K) space. Unlike the profit maximization problem in Fig. 11.17, where we find the maximum point of the whole surface, the current problem is to find the maximum point of the surface along this vertical plane formed by constraint (12.7). In Fig. 12.2(a), it appears that the maximum point is at point "e", which has a maximum height "eb" on the production surface, along the solid straight line of the cost constraint connecting the intercepts $(L, K) = (0, 20)$ and $(L, K) = (15, 0)$. As will be shown below and from the diagram, the optimal values of labor and capital are at $(L^{\#}, K^{\#}) = (9, 8)$, and the constrained maximum output is $Q^{\#} = 17$. Figure 12.2(b) shows the constrained maximum output in the wireframe 3D.

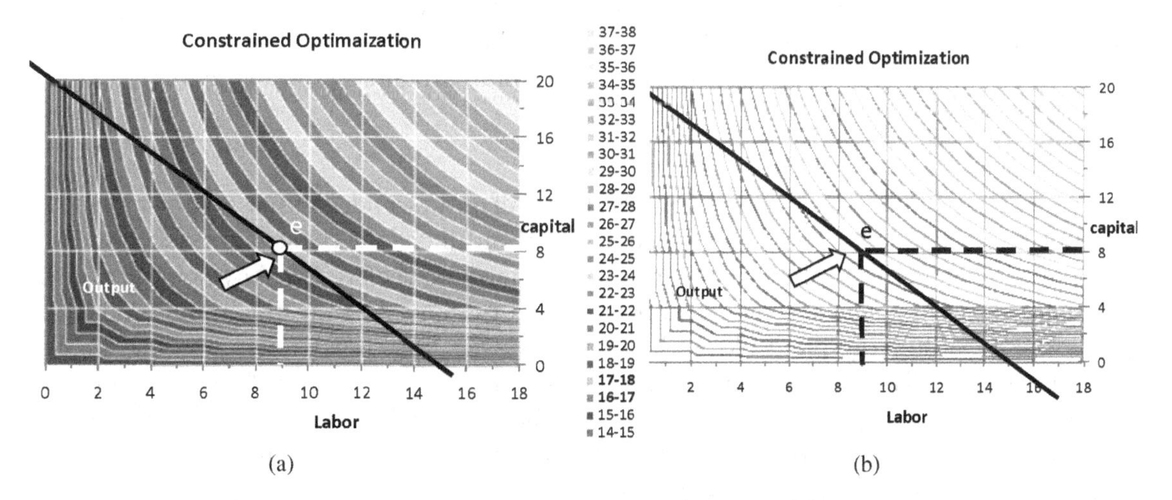

Fig. 12.3 Isoquants and the cost constraint (a) 2D and (b) wireframe 2D

[1]See any textbook on Mathematical Economics, for example, Chiang and Wainwright (2005).

The isoquant map

The general practice in Economics is to draw the cost constraint on an **isoquant map**, instead of a production surface, as shown in Fig. 12.3. The cost constraint is drawn manually based on Fig. 12.1. The maximum output under the cost constraint is found at the tangent point of the cost line with the highest level of isoquant.

From the sequence of the colors in the legend of Fig. 12.3(a), we may roughly find the tangent point of the cost line with the highest isoquant at Q = 17. If the tangent point is hard to pinpoint, you may enlarge the chart to find it, or find it directly from the intersection of L = 9 row and K = 8 column in Table 12.1. We will denote the point of tangency as "e" in Figs. 12.3(a), and 12.3(b), from which we remove the color for clarity. The output at this point is called the **optimal output**. The direction of the arrow shows that the producer "moves up" along the isoquants until the highest isoquant, which is tangent with the cost constraint. At this output level, approximately 17 units (to be exact, it is 17.17 units), the producer used 9 units of labor and 8 units of capital. Thus, the **optimal labor** is 9 units and the **optimal capital** is 8 units. The triplet

$$(\mathrm{L}^{\#}, \mathrm{K}^{\#}, \mathrm{Q}^{\#}) = (9, 8, 17), \qquad (12.8)$$

denotes the **solution of the constrained optimization of production.**

The isoquant map in the sensitivity (or data) table

The isoquant map of Fig. 12.3 can be traced roughly in the sensitivity table of Table 12.1. The cost constraint connects the two intercepts from the middle of cells C21 and C22 to H14. The selected isoquants are connected among cells with the same or close levels of output. The cost constraint touches the isoquant between Q = 16 (E18) and Q = 18.3 (E19), implying that the optimal level of output should be around 17, the optimal labor between 8 to 10 (i.e., $\mathrm{L}^{\#} = 9$), and the optimal capital around 8 (i.e., $\mathrm{K}^{\#} = 8$).

Proof that point e is the optimal output level

How do we know that point e gives the highest level of output under the cost constraint? Two conditions make point e the maximum possible output under the constraint: the isoquant is tangent at point e, and the isoquants are strictly convex to the origin.[2] Any point below the cost line will yield a smaller amount of output, and the firm does not use up its production budget at C = 30. Any point along the cost line other than point e will produce smaller output, although the firm uses up its production budget at C = 30. If the isoquants are convex to the origin, this can be seen by the fact that, when a point moves upward or downward along the cost constraint from point e, the isoquants will intersect with the cost constraint and give lower output levels than the isoquant at point e. Lastly, although any

[2]As we have seen in Fig. 11.11, and will see later, the isoquant is strictly convex to the origin if the law of diminishing marginal rate of technical substitution is satisfied.

point in the northeast part of point e will give a higher output level, it is out of the reach of the firm, since it will cost more than C = 30 (the firm's budget) to produce the higher output level. This shows that point e gives the maximum output level under the given cost constraint C = 30.

The condition that the curve must be strictly convex at point e is important. If the isoquants are not strictly convex to the origin, say, if they are concave to the origin in the neighborhood of e, then the isoquant which is tangent to the cost line at e will be the **minimum point** in the neighborhood. The producer will be better off by moving away from e. Thus, in this case, e is the unstable optimal point. You may illustrate this by drawing a concave isoquant, which touches the constraint from below at e.

Some notes on graphing

The cost constraint, the dotted lines, and the vertical line in Figs. 12.2 and 12.3 are added manually. In Fig. 12.2(b), the area between the curve and the constraint is painted blue with 50% transparency. The letters e and b are in the text boxes. The entries in the legend are in single space and stretched to contain as many intervals as possible. Figure 12.3(b) shows the wireframe 2D chart of Fig. 12.3(a).

12.1.3 *The nature of optimization*

In Sec. 11.2.3, we have shown that the slope of the isoquant is $\Delta K/\Delta L = -MPL/MPK$ (see (11.10)). If L and K also change along the cost constraint, taking the difference $\Delta L = L_1 - L_0$, and $\Delta K = K_1 - K_0$ and holding the cost C constant, we have

$$0 = w\Delta L + r\Delta K.$$

We can solve the $\Delta K/\Delta L$ as

$$\frac{\Delta K}{\Delta L} = -\frac{w}{r}, \tag{12.9}$$

which is the **slope of the cost constraint** in Fig. 12.1.

Since the condition of optimization is that the cost constraint should touch the highest isoquant at point e, the slope of the isoquant must be equal to the slope of the cost constraint. Thus, when the output is maximized at point e, we have the following **slope condition**:

$$\boxed{\frac{MPL}{MPK} = \frac{w}{r}, \quad \text{or} \quad \frac{MPL}{w} = \frac{MPK}{r}.} \tag{12.10}$$

Economic interpretation of the slope condition

The second part of (12.10) depicts the **law of equal marginal product per dollar**. This means that, at point e, the marginal product of labor from hiring one extra dollar's worth of labor is equal to the marginal product of capital from using one extra dollar's worth of capital, and therefore, the firm will not change the labor and capital combination again. The firm has reached the optimal output level under the given cost constraint.

More specifically, since w is wage per hour (or per worker) 1/w is the amount of labor a firm can buy by spending \$1.00 on labor. For example, if wage rate is w = \$6.00 per hour, then 1/w = 1/6 = 0.17 is the amount of labor \$1.00 can buy, namely, 10 min of work. On the other hand, MPL is the amount of output an extra one unit of labor can produce. For example, extra one unit of labor can produce 100 units of output. Hence, MPL/w = (1/6)(100) = 17 is the units of output the firm can produce if it spends one extra dollar on hiring labor.

Similarly, MPK/r is the units of output the firm can produce if it spends one extra dollar on buying capital. Hence, the second part of (12.10) depicts the **law of equal marginal product per dollar**. It states that, at equilibrium, the firm will allocate resources such that the total marginal product producible by spending one dollar on labor is the same as that producible by spending one dollar on capital. In other words, the firm will employ labor and capital until each dollar the firm spends on both resources yields the same marginal product.

Example 12.1 Comparative static analysis. The cost constraint in (12.7) is given as 2L + 1.5K = 30, and it is listed in column I of Table 12.1. In I4 of Table 12.1, change wage rate from 2 to 3, and find the new optimal labor, capital, and output. Observe the changes in the position of the optimal point e. Denote the new equilibrium point by e′ in a textbox. Does the new optimal point make sense? Why, or why not.

Answer: When w changes to 3, the new cost constraint has the Y-intercept at (0, 20), the same as the old constraint, but its X-intercept is at (10, 0), instead of original (15, 0). Hence the new constraint is steeper than the old constraint. From the isoquant map, we can find that the new optimal values are $(L^{\#}, K^{\#}, Q^{\#}) = (6, 8, 13.5)$. Compared with the original optimal value $(L^{\#}, K^{\#}, Q^{\#}) = (9, 8, 17)$, optimal use of labor decreased, but no change in capital, resulting in reduced output. This make sense since as the wage rate increases, the firm uses less labor, and so output decreases. □

12.1.4 *Factor demand functions*

In Microeconomics, the slope condition (12.10) and the cost constraint

$$C = wL + rK$$

must hold for a given cost C. The condition (12.10) and the cost constraint constitute a system of two simultaneous equations in the two variables, L and K. In Mathematics, the law of equal marginal product per dollar and the cost constraint together are called the **necessary condition for constrained maximization in production**.[3] The system of

[3]The **sufficient condition for maximization** (for the two variables with one constraint case) is that if the necessary condition and the law of diminishing marginal rate of technical substitution hold at a point $(L^{\#}, K^{\#})$, then $(L^{\#}, K^{\#})$ is the maximum point. See Chiang and Wainwright (2005), Chapter 12.

these two equations can be solved[4] explicitly for each variable as a function of parameters w, r, and C:

$$\boxed{\begin{aligned} L &= L(w, r, C), \\ K &= K(w, r, C). \end{aligned}}$$

These two functions are called the **demand function for labor** and the **demand function for capital** of the firm respectively. The difference between these demand functions and the previous factor demand functions (11.13) is that, in the current constrained optimization case, the factor demand functions also depend on cost C.

Example 12.2 Derivation of factor demand functions. In (11.8), we have MPL = aQ/L and MPK = bQ/K. Hence, substituting them into (12.10) and multiplying out, we have

$$bwL - arK = 0.$$

Combining this with the cost constraint and writing it in matrix form, we have

$$\begin{bmatrix} bw & -ar \\ w & r \end{bmatrix} \begin{bmatrix} L \\ K \end{bmatrix} = \begin{bmatrix} 0 \\ C \end{bmatrix}.$$

For w = 2, r = 1.5, C = 30, a = 0.6, and b = 0.4, in the Cobb–Douglas production function, we can solve the matrix equation (see Chapter 10) and obtain $L^{\#} = 9$ and $K^{\#} = 8$. Substituting them into the objective function (12.6), we have $Q^{\#} = 17.1716$. Thus, the optimal solution is indeed given by (12.8). □

12.2 Using Excel Solver in Constrained Optimization

While the diagrammatic method of solution is intuitive and clear, the solution of constrained maximization is only an approximation. The problem of using Excel 2007 is that the 3D and 2D color scheme of Excel 2007 is not as good as that of Excel 2003, and the difference between the colors of different intervals is obscure and hardly distinguishable. It is very difficult to find the optimal level of output from the 3D production surface in Fig. 12.2, or the 2D isoquant map in Fig. 12.3, using Excel 2007. Thus, we need to find an exact solution to optimization problems such as (12.6) and (12.7).

To find the exact solution, if it exists, we may use the Excel Solver, which is an add-in listed in <"Data"><Analysis, Solver>.

---

[4]Under certain mathematical assumptions, they can be solved theoretically by using the Implicit Function Theorem. See, for example, Chiang and Wainwright (2005), Chapters 8 and 12.

12.2.1 *Installing the Solver*

If the Solver button is missing under the <"Data"> tab, then it should be added by using the following procedure:

<"Microsoft Office"><Excel Option><Add-Ins><Solver Add-in>
<Go···><Add-Ins, Add-Ins available: xSolver Add-in><OK><Yes>

<Yes> is to confirm that you want to install the Solver. The Solver is then installed.

Set up the table for the Solver

Table 12.2 is the suggested form of table setup for the constrained optimization problem of (12.6) and (12.7). In this table, we have followed the usual format of writing the constrained optimization problem. The objective function is in row 8, and the constraint is in row 11. The variables, named L and K, which Excel calls "changing cells", are entered in B8:C8. Since we do not know their optimal values, arbitrary initial values of 1 are entered in B8:C8. The parameters of the production function, named aa, bb, and A, are listed in D2:D4, and the factor prices, named w and r, are listed in B11:C11. The equations are entered as follows:

D8 : =A*L^aa*K^bb, "Target Cell" (Objective function),
D11 : =w*L+r*K, "Cell Reference",
F11 : 30, "Constraint".

12.2.2 *Solver procedure*

To invoke the Excel Solver, the following procedures are helpful:

Step 1. Click <"Data"><Analysis, Solver>. The "Solver Parameters" dialog box, Fig. 12.4, appears. Inside the window, in the "Set Target Cell:" box, enter D8; for "Equal

Table 12.2 The Solver setup for constrained optimization

	A	B	C	D	E	F	
1			Named parameters				
2			aa	0.6			
3			bb	0.4			
4			A	2			
5		(varaibles)					
6		Changing cells		Target cell			
7	max	L	K	AL^aK^b	=	Q	Objective function
8	(L, K)	1	1	2	?	Q#	(12.6)
9							
10	constr	w	r	wL+rK	=	C	Constraint
11		2	1.5	3.5	?	30	(12.7)
12				Cell ref		Constr	
13							
14	(L, K)	9	8	17.17163	?	Q#	

To": select Max. In the "By Changing Cells:" box, enter B8:C8. The $ sign is added automatically.

Step 2. To add the cost constraint, click <Add>, and the "Add Constraint" dialog box, Fig. 12.5, appears. In the "Cell Reference:" box, enter D11 as the cost constraint, w*L+r_*K; and for "Constraint", F11, the fixed cost budget of 30 . They are related by "equality", in the middle of the dialog box. Excel will add the $ sign and = sign before $F$11 automatically, as in Fig. 12.5. Since there is no other constraint to <Add> at this time, click <OK> to close the "Add Constraint" window to go back to "Solver Parameters" window, which now should appear as Fig. 12.4.

Step 3. Click the upper-right button <Solve> in the "Solver Parameters" dialog box Fig. 12.4. After a few seconds, the "Solver Results" dialog box, as in Fig. 12.6, appears,

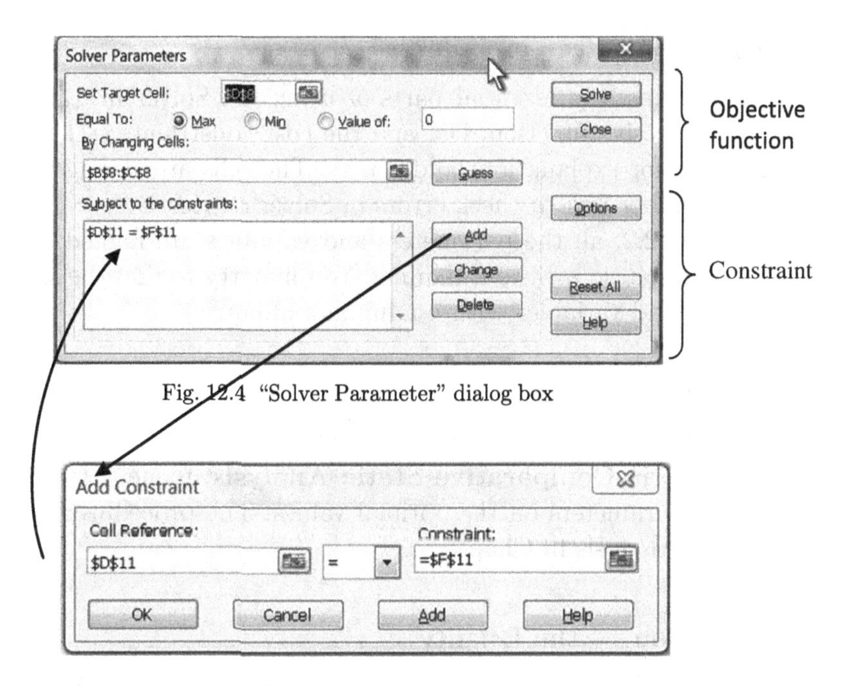

Fig. 12.4 "Solver Parameter" dialog box

Fig. 12.5 "Add Constraint" dialog box

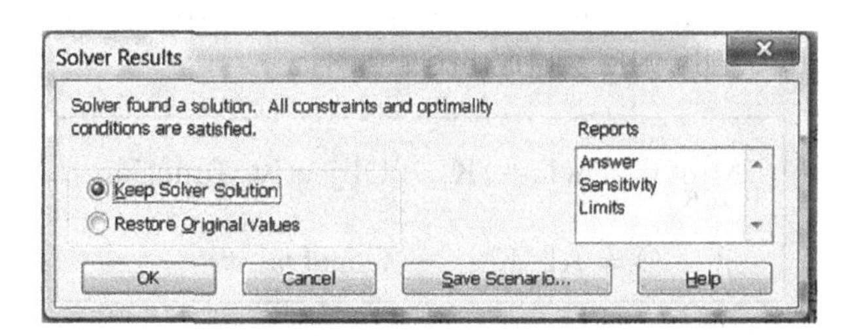

Fig. 12.6 "Solver Results" dialog box

indicating either "Solver found a solution", or "error". In this example, the solution is found and the assumed initial solution in row 8 of Table 12.2 should change to the optimal solution as shown in row 14. From row 14, we can read the optimal solution,

$$(L^{\#}, K^{\#}, Q^{\#}) = (9, 8, 17.171),$$

as illustrated as point b in Figs. 12.2(a) and 12.2(b), and also as point e in Figs. 12.3(a) and 12.3(b). The right-hand side of Fig. 12.6, the "Solver Results" dialog box, indicates that we may also generate one to three reports: Answer, Sensitivity, and Limits.[5] The solution is the same as that solved analytically in Example 12.2.

Note that at the optimal solution, the cost constraint must be satisfied. This means that we should have D11 = F11 = 30 in Table 12.2 when the optimal solution is found.

12.2.3 *Summary of the Solver procedures*

From the above three steps, the essential parts of using the Solver in Table 12.2 are the variables B8:C8, the objective function, D8, and the cost constraint, D11 and F11. Other entries in Table 12.2 are for explanations and clarity. The cells are arranged in such a way that it is easier to remember and to check errors in Solver entries.

Note, that in Table 12.2, all the parameters and variables are named. This is just for clarity. We may find the solution without naming. You may try to find the solution without naming the parameters and variables with a minimum of entries.

Comparative static analysis

After the solution is found, as in Table 12.2, one of the parameters, aa, bb, A, w, or r, may be changed to perform **Comparative Static Analysis**, namely, to see the effect of change in one of these parameters on the optimal values. The procedures are the same as for Comparative Static Analysis in Chapter 4.

12.3 Cost Minimization — the Duality

The duality to output maximization is cost minimization. Given the production technology represented by the Cobb–Douglas production function, the firm would like to choose levels of labor and capital combination such that the cost of production for a given output level is minimized. Mathematically, the problem is

$$\min_{L,K} \ C = wL + rK, \quad \text{Objective function} \tag{12.11}$$

$$\text{s.t.} \quad Q = AL^{a}K^{b}. \quad \text{Constraint} \tag{12.12}$$

[5]For details, see Cornell (2006).

Using the previous numerical example, the problem can be written as

$$\underset{\mathrm{L,K}}{\mathrm{Min}}\ \mathrm{C} = 2\mathrm{L} + 1.5\mathrm{K}, \tag{12.13}$$

$$\text{s.t. } 17 = 2\mathrm{L}^{0.6}\mathrm{K}^{0.4}, \tag{12.14}$$

where Q can be any given output level. In order to show the duality between output maximization and cost minimization, Q = 17 is taken from the previous example. In this case, there is only one isoquant, which is given at Q = 17, and many possible isocost lines. We call (12.12) the **production constraint**, but, more accurately, it is a **technology constraint** showing how the given output can be produced under the Cobb–Douglas type production technology.

12.3.1 *Sensitivity table*

To find the minimum cost, we use Table 12.1 as a template, instead of doing everything from the beginning. Copy Table 12.1 to a new worksheet[6] (call it Outputmax), and rename the original table Table 12.3 as a cost minimization working table (we work on the original

Table 12.3 Cost minimization table

	A	B	C	D	E	F	G	H	I
1	Cost minimization								
2		Q	L^aK^b= A						
3			Parameter Table						
4	Labor share		aa	0.6	Wage		ww	2	
5	Capital share		bb	0.4	Rent		rr	1.5	
6	Scale factor		A	2	Cost		C	30	
7									
8			C14	=ww*L+rr*K					
9			I14:	=(17/(A*L^aa)^(1/bb)					
10	Sensitivity Table								
11	**The Cost Surface**								
12	C		Capital, K						Q=17
13			0	4	8	12	16	20	K
14	Labor	0	0	6	12	18	24	30	
15	L	2	4	10	16	22	28	34	
16		4	8	14	20	26	32	38	26.3
17		6	12	18	24	30	36	42	14.3
18	→	8	16	22	28	34	40	46	9.3 ←
19		10	20	26	32	38	44	50	6.7
20		12	24	30	36	42	48	54	5.1
21		14	28	34	40	46	52	58	4.0
22		16	32	38	44	50	56	62	3.3
23		18	36	42	48	54	60	66	2.8

[6]You may also create a new workbook. This is recommended to avoid the conflict in using the same range names. If we use the old workbook, which contains Tables 12.2, then aa, bb, and A in Table 12.2 may be renamed aaa, bbb, a01, a011, etc.

table to pressure the original range names). Make sure that all the names (aa, bb, etc.) are still valid by selecting each cell to see whether the cell is named correctly. Delete the entries of old data in C14:H23. Then, at C14, enter the objective function (12.11):

C14: =ww*L+rr*K,

and copy C14 to C14:H23. This gives the **isocost surface** as shown in Fig. 12.7(a). The isocost map is its 2D counterpart and is shown in Fig. 12.7(b).

12.3.2 *The nonlinear constraint*

In the minimization problem, the constraint (12.12) is the Cobb–Douglas production function, which is a nonlinear function. Solving the production function for capital K (since Table 12.3 uses L as the independent variable), we have

$$K = \left(\frac{Q}{AL^a}\right)^{1/b}. \tag{12.15}$$

To implement this in the table, erase the old constraint entries in I14:I23 of Table 12.3. Then, in I14, enter[7] (12.15) as

=(17/(A*L^aa))^(1/bb),

and copy I14 to I14:I23. We obtain the corresponding values of K to different values of L in B14:B23 at given Q = 17. This completes the cost minimization table. Since I14 is divided by L = 0 and I15 has large value of K = 74.47, without loss of generality, we erase the values in these two cells.

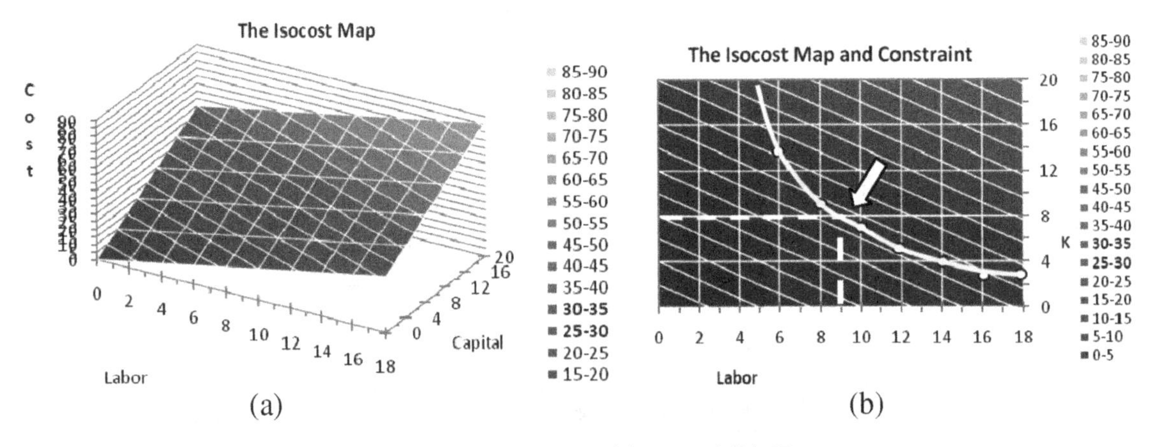

Fig. 12.7 Isocost Map (a) 3D and (b) 2D

[7]You may also use the range copy command. However, in this simple case, and for the possible change in the formula, the regular copy command may be preferable.

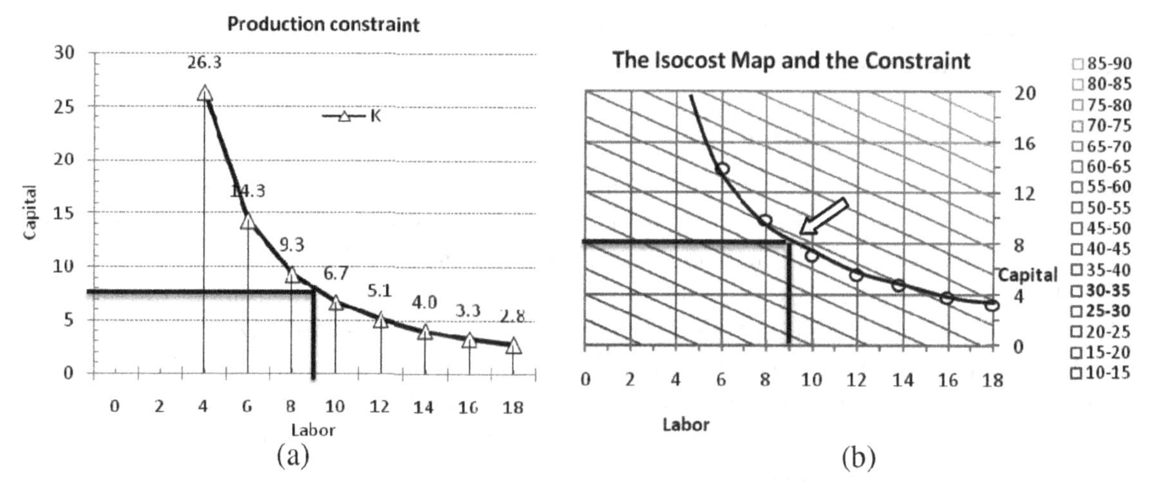

Fig. 12.8 (a) The production constraint. (b) Isocost map and production constraint

12.3.3 *Graphic methods of solution*

After constructing Table 12.3, we now draw the charts. The **cost surface** is drawn in Fig. 12.7(a). Note that, for each level of labor (or capital), cost increases linearly as capital (or labor) increases. When the isocost lines are projected on the factor space, we obtain the isocost map, as shown in Fig. 12.7(b).

The constraint, Q = 17, is plotted in Fig. 12.8(a). Since the output constraint is a nonlinear curve, it is difficult to impose it on the isocost map. We need to choose at least three points to draw a curve. Here, we circle all the points in the range of labor from 6 to 16 using Table 12.3, and connect them linearly, as shown Figs. 12.7(b) and 12.8(b). The minimum cost for a given production function, Q = 17, is obtained by the tangent point of the cost line and the nonlinear output constraint, as shown by the downward arrow in both figures, indicating that cost is minimized at this point (the lowest cost line that can produce Q = 17).

Optimal allocation of resources under cost minimization

Thus, from the isocost map, we see that the optimal allocation of the factors is 9 units of labor and 8 units of capital, and the minimum cost is, by inspecting the change of the shade of the color of the isocost map, $30, the same as the cost constraint in output maximization. Thus, the equilibrium values are the same as the case of the output maximization. The equilibrium triplet is

$$(\mathrm{L}^{\#}, \mathrm{K}^{\#}, \mathrm{C}^{\#}) = (9, 8, 30), \tag{12.16}$$

except that now C, instead of Q, is the variable in the objection function. This correspondence between the constrained output maximization problem (the original problem in

Sec. 12.2) and the constrained cost minimization problem (in Sec. 12.3) is called the **duality in production theory**.[8]

12.3.4 *The Solver solution*

Table 12.4 presents the shorter version, using the Solver table, for the cost minimization problem. Since all the entries are similar to the original problem of Table 12.2, we may copy Table 12.2 to another sheet and rename the original table, Table 12.4. In Table 12.4, the variables are the same B8:C8 as in the original problem. Exchange the roles of the objective function and the constraint: D8:F8 is now the constraint, A = 2 and Q = 17.17, and D11 is the target cell of the objection function. Then, without renaming the parameters and variables, invoke the three-step procedure of the Solver in Sec. 12.2, we have the dual solution of (12.16) in Table 12.4, in which row 8 will be replaced by the equilibrium solution in row 14 ($L^{\#} = 9$ and $K^{\#} = 8$), and row 11 will be replaced by the equilibrium solution in row 15 ($C^{\#} = 29.997 = 30$). Note that, if the constraint in F8 is entered as 17 without decimals, instead of 17.17, then the equilibrium cost will be $C^{\#} = 29.7002$ at $L^{\#} = 8.9$ and $K^{\#} = 7.91$.

Note also that at the optimal solution, the constraint D8 = F8 must be satisfied, and "=" should hold in E8, as in row 14. Similarly, "=" should also hold in E11, showing the minimum cost has been achieved.

12.4 Simultaneous Presentation of an Objective Function and a Constraint

In the previous maximization problem or minimization problem, the entries of the objective function and constraint in the data tables of Tables 12.1 and 12.3 are quite different in nature. The entries of the objective function are either output (Table 12.1) or cost (Table 12.3), but the entries of the constraint are capital. In this section, we will make all the entries the levels of capital by solving both functions in terms of capital with respect to labor. This method does not require separate imposition of the constraint on the isoquant map or isocost map.

Table 12.4 Cost minimization

	A	B	C	D	E	F	
6		Changing cells		Cell ref		Constr	
7	constr	L	K	AL^aK^b	=	Q	Constraint
8	(L, K)	1	1	2	?	17.17	(12.13)
9							
10	min	w	r	wL+rK	=	C	Objective function
11		2	1.5	3.5	?	C#	(12.14)
12				Target cell			
13							
14	(L, K)	9	8	17.17	?	17.17	
15		2	1.5	29.9972	?	C#	

[8]If C ≠ 30, the equilibrium points in the original problem and in the dual problem will not match.

Solving **constraints** (12.1) and (12.12) for K in terms of L, we have (12.2) and (12.15), which are reproduced below for convenience:

$$K = \frac{C}{r} - \frac{w}{r}L, \tag{12.2}$$

$$K = \left(\frac{Q}{AL^a}\right)^{1/b}. \tag{12.15}$$

Given C and Q, both equations give values of K when the value of L is chosen. In the cost minimization problem, copy Table 12.3 to a new sheet and rename the original table, Table 12.5. Make sure that the names of parameters and variables still hold in the new sheet. Since the parameter tables are the same, we only need to change the cost surface of Table 12.3 to the isocost sensitivity table. Table 12.5 shows rows 11–23 of the new table. C13:H13 is now the range of cost from $20, 25, \ldots, 45$, and is named CC (which overlaps with the previous name K in Table 12.3). Labor (L) is the same as in Table 12.3, except that it now ranges from $4, 5, \ldots, 13$. Then, in C14, enter formula (12.2) for K as

C14: =(CC/rr)−(ww/rr)*L,

and copy C14 down to C14:H23. The range I14:I23 is the same as that in Table 12.3, except that now its values (Q's are rounded to whole numbers) are changed to reflect the new values of labor. The entries of the data part, C14:I23, now show the values of K for given C in (12.2) and for given Q in (12.15).

Unlike Figs. 12.3(b) and 12.8(b), in which the constraint and the object function are drawn separately, we now draw both at the same time in a chart. Selecting the range B13:I23 of Table 12.5, we draw Fig. 12.9. The equilibrium point a, which is indicated by

Table 12.5 Isocost map and output constraint

	A	B	C	D	E	F	G	H	I
11	Isocost Map and Output Constraint								
12	K		Cost C						Output Q
13			20	25	30	35	40	45	Q=17
14	Lbr	4	8	11	15	18	21	25	26
15	L	5	7	10	13	17	20	23	19
16		6	5	9	12	15	19	22	14
17		7	4	7	11	14	17	21	11
18		8	3	6	9	13	16	19	9
19		9	1	5	8	11	15	18	8
20		10	0	3	7	10	13	17	7
21		11	-1	2	5	9	12	15	6
22		12	-3	1	4	7	11	14	5
23		13	-4	-1	3	6	9	13	4

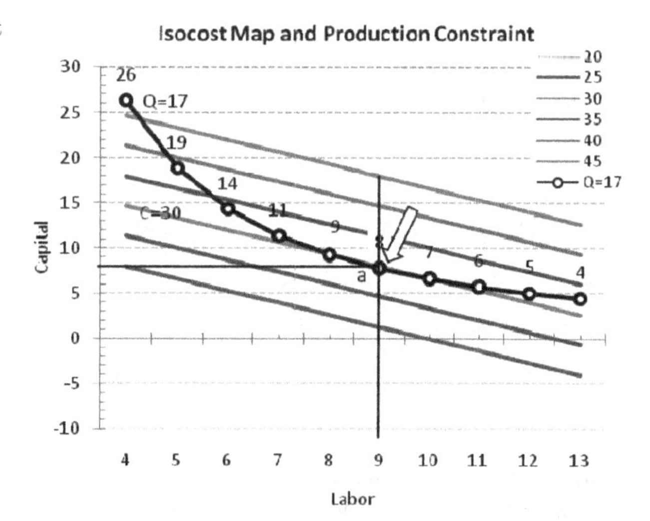

Fig. 12.9 Isocost map and production constraint

the downward arrow, is the point at which the output constraint is tangent to the isocost curve, at C = 30.

Naturally, this result can also be read from Table 12.5. The isocost line for C = 30 is given in E14:E23. At L = 9 in row 19, the value of K on the isocost line is 8, and the value of K along the output constraint is also 8, at which the isoquant curve touches the isocost line.[9] The chart shows that this cost level is the lowest along the output constraint.

Note that it does not make sense to draw the 3D surface for Table 12.5, as the entries of the table are values of another variable, K, given some selected levels of cost. Note also that to construct Table 12.5, we have to learn the appropriate range of C and L, using the method of trial and error. In general, this method is not practical. In our case, we can chose the range of C and L in Table 12.5, since we already know their values either from the solution of the output maximization problem (Table 12.2) or from the Solver (Table 12.4): the cost is given as 30 and the optimal labor is 9. We have tried to construct the isocost line for C = 30 so that the C = 30 line is at the middle of the isocost map. Since we presume C = 30 is known, the method suggested in this section can be used as an illustration of the nature of minimization, but it cannot be used in solving the minimization problem.

Example 12.3 Output maximization. Using Table 12.5, assuming the same Cobb–Douglas production function and the values of parameters, construct the output maximization problem using the formulas in (12.2) and (12.15), with labor ranging continuously from $4, 5, \ldots, 13$, and output ranging from $5, 9, \ldots, 25$, and with C = 30. Show the numbers in the table with one decimal place.
Answer: In this case, Table 12.5 should have output 5, 9, ..., 25 in range C13:H13, and the range should be named QQ. In C14, enter formula (12.15) for K as =(QQ/(A*L^aa))^(1/bb), and in I14, enter formula (12.2) for K as = (C_/rr) – (ww/rr)*L. Copy them down to row 23. Fig. 12.9 will be changed accordingly. Make necessary revisions of the table titles, etc. □

In general, the problem of finding the optimal allocation of resources is the center of the theory of production in Microeconomics, and is an important part of the Static Analysis in Economics.

12.5 Utility Maximization and Comparative Static Analysis

12.5.1 *Utility maximization*

Assume that there are only two commodities, X and Y. Suppose there is a utility function $U = U(x, y)$, where x is the quantity of commodity X consumed and y is the quantity of

[9]From Table 12.5, and also Fig. 12.9, it appears that at L = 8, K(on isocost) = K(on output constraint) = 9, and also at L = 10, K(on isocost) = K(on output constraint) = 7, hence the output constraint overlaps with the isocost line C = 30. This will not be the case if we add one or two decimals to the numbers.

commodity Y consumed. Let the consumer's **budget constraint** be M = px + qy, where $M is the total budget (or expenditure), p is the unit price of X, and q is the unit price of Y. Then, the rational behavior of a consumer[10] is

$$\max_{x,\,y} U = U(x, y), \tag{12.17}$$

$$\text{s.t. } M = px + qy, \tag{12.18}$$

where (12.17) corresponds to Eq. (12.4) when the Cobb–Douglas production function is replaced by the utility function and (12.18) corresponds to (12.5) when the cost function is replaced by the budget constraint. The unit of utility is called **util**, which is a convenient fictitious concept.

In the theory of consumption (or demand), a consumer allocates within his/her individual budget $M between two commodities to maximize his/her utility function, U(x, y), which specifies the utility that the consumer can derive by consuming these two commodities. The prices (p, q) of the two commodities are given to the consumer, since the consumer faces perfect competition in the commodity markets and so the consumer is a price taker. Mathematically, it is assumed that a rational consumer will find the optimal consumption bundle $(x^{\#}, y^{\#})$, such that it is the solution of (12.17) and (12.18).

The budget constraint assumes that the consumer spends all budget $M between these two commodities, no more and no less. This also means no robbing and no cheating as the consumers have to pay prices for consumption. It is also assumed implicitly that what is spent is also what is consumed.

12.5.2 *The Excel Solver*

The similarity between the utility maximization problem (12.17) and (12.18) and the output maximization problem (12.4) and (12.5) should be noted. To apply the spreadsheet program, we need to specify the form of the utility function. Like the production function, the utility function may be of the Cobb–Douglas type, the CES type, the quadratic type, etc. To contrast our approach with that of Carfill (2000), we adopt his utility function as

$$U(x, y) = x^2y.$$

In Sec. 12.1, we first derived the output surface and the isoquant map, and then found the **optimal allocation of resources**, which maximizes the output. The problem with this approach is that, in order to construct the output surface, we have to decide the ranges

[10]Sometimes, economists use capital letters X and Y to represent the commodities, and small letters x and y to show the quantities of X and Y that are consumed or purchased. However, we do not make a strict distinction between them. Note also that in static analysis, when a good is purchased, it is assumed that it is consumed, no hoarding or storing.

of the values of labor and capital so that the optimal output can be shown neatly from the isoquants. Depending on the range of the variables, the optimal output may not exist in certain ranges, and we have to use trial and error to find the appropriate range. To avoid this problem, it is more convenient to use the **Solver** to find the optimal values first, and then find the appropriate ranges of values for the variables to be included in the optimal points for visual presentation. Thus, in this section, we reverse the procedures in the previous sections and start by using the Solver.

Utility maximization via Solver

Let the utility maximization problem be

$$\text{Max} \quad U(x, y) = x^2 y, \tag{12.19}$$

$$\text{s.t.} \quad 2x + y = 300, \tag{12.20}$$

where we assume that $p = \$2$, $q = \$1$, and $M = \$300$. The solution table is shown in Table 12.6.

The structure of the table is similar to that of Table 12.2. Enter the initial values $x = 10$, $y = 10$ in row 8 (not shown). Enter $p = 2$, $q = 1$, and $M = 300$ in row 11 (not shown). Invoking the Solver, as explained in Sec. 12.2.2 above, we have the optimal solution table as shown in rows 13–18. They are obtained by copying the optimal values of the original table (not shown) in range A7:G11, pasting them as values in rows 13–18, and denoting this copied table with the old table's number. We denote the original table, rows 7–11, as the "New table" for use in comparative static analysis below. From the old table, we see

Table 12.6 A Solver scheme for utility maximization

	A	B	C	D	E	F	G	
1		parameters		default	parameters		default	
2		a	2	2	p	4	2	
3		b	1	1	q	1	1	
4		aa	1	1	M	300	300	
5								
6	New	table			objective			
7		max	x	y	=aa*x^a*y^b	=	U	Objective function
8	q1	(x,y)	50.00	100.00	250000.00	?	U*	(12.19)
9								
10		Constr	p	q	px+qy	=	M	Constraint
11	p1		4	1	299.999999	?	300	(12.20)
12								
13	Old	table			objective			
14		max	x	y	=aa*x^a*y^b	=	U	Objective function
15	q0	(x,y)	100.00	100.00	1000000.01	?	U*	(12.19)
16								
17		Constr	p	q	px+qy	=	M	Constraint
18	p0		2	1	300.000001	?	300	(12.20)

that the optimal solution is

$$(x^{\#}, y^{\#}, U^{\#}) = (100, 100, 1000000). \tag{12.21}$$

Comparative static analysis

To perform the comparative static analysis when price p of x changes from 2 to 4, we change $p \doteq 2$ to $p = 4$ in row 11 in Table 12.6. The new budget constraint M′ is in E11:G11:

$$M' = 4x + y = 300. \tag{12.22}$$

Using $x = 100, y = 100$ as the initial values in C8:D8, the Solver will find the new optimal solution,

$$(x^*, y^*, U^*) = (50, 100, 250000), \tag{12.23}$$

as shown in row 8. Thus, by comparing row 8 with row 15, we see that a doubling of price p from 2 to 4 cuts the optimal consumption of x by half, from 100 to 50, but has no effect on the consumption of Y, and reduces the utility from 1,000,000 util to a mere 250,000 util in this example.[11]

In column A, p^0 and q^0 are the original price and quantity, and p^1 and q^1 are the changed price and quantity at maximum utility.

12.5.3 *The sensitivity table and graphics*

Based on the above two optimal solutions derived from the Solver, we may now construct the sensitivity table of utility, with a range of x and y from 0 to 150 and with an interval of 10. This is shown in Table 12.7. Since the values (index numbers) of the utility function are large, as shown in Table 12.6, for our convenience in drawing the figures, we have divided the utility function (12.19) by 1000. That is, we write the utility function as

$$U(x, y) = x^2y/1000. \tag{12.24}$$

Naming A4:A19 as x and B3:Q3 as y, entering (12.24) in B4, and copying down to B4:Q19, we have the utility sensitivity table. The utility surface and the indifference map are shown in Figs. 12.10(a) and 12.10(b), respectively.

Illustration of the optimal point in the indifference map

In order to show the level of utility at the optimal point, the indifference map of Fig. 12.10(b) is drawn with a maximum utility level of 1500 and an interval of 100.

[11]In column A, p^0 and q^0 are the original price and quantity, and p^1 and q^1 are the changed price and quantity for comparative static analysis. From these data, we can calculate total cost index, Laspeyres, Paasche, Fisher price indexes. See Example 10.3.

Table 12.7 Sensitivity table for utility maximization

	A	B	C	D	E	F	G	H	I	J	K	L	M	N	O	P	Q	R	S
1				U=x^2y/1000															y'=300-4x
2	x	y																y=300-2x	
3		0	10	20	30	40	50	60	70	80	90	100	110	120	130	140	150	y	y'
4	0	0	0	0	0	0	0	0	0	0	0	0	0	0	0	0	0	300	300
5	10	0	1	2	3	4	5	6	7	8	9	10	11	12	13	14	15	280	260
6	20	0	4	8	12	16	20	24	28	32	36	40	44	48	52	56	60	260	220
7	30	0	9	18	27	36	45	54	63	72	81	90	99	108	117	126	135	240	180
8	40	0	16	32	48	64	80	96	112	128	144	160	176	192	208	224	240	220	140
9	50	0	25	50	75	100	125	150	175	200	225	250	275	300	325	350	375	200	100
10	60	0	36	72	108	144	180	216	252	288	324	360	396	432	468	504	540	180	60
11	70	0	49	98	147	196	245	294	343	392	441	490	539	588	637	686	735	160	20
12	80	0	64	128	192	256	320	384	448	512	576	640	704	768	832	896	960	140	-20
13	90	0	81	162	243	324	405	486	567	648	729	810	891	972	1053	1134	1215	120	-60
14	100	0	100	200	300	400	500	600	700	800	900	1000	1100	1200	1300	1400	1500	100	-100
15	110	0	121	242	363	484	605	726	847	968	1089	1210	1331	1452	1573	1694	1815	80	-140
16	120	0	144	288	432	576	720	864	1008	1152	1296	1440	1584	1728	1872	2016	2160	60	-180
17	130	0	169	338	507	676	845	1014	1183	1352	1521	1690	1859	2028	2197	2366	2535	40	-220
18	140	0	196	392	588	784	980	1176	1372	1568	1764	1960	2156	2352	2548	2744	2940	20	-260
19	150	0	225	450	675	900	1125	1350	1575	1800	2025	2250	2475	2700	2925	3150	3375	0	-300

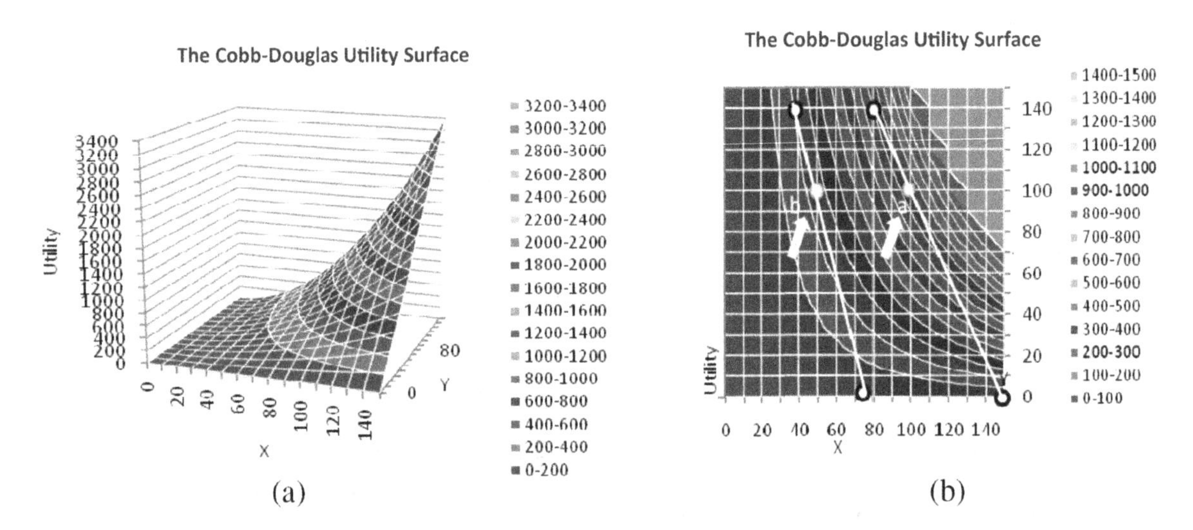

Fig. 12.10 A Solver scheme for utility maximization

In Table 12.7, the budget constraints are shown in column R, y = 300 − 2x, for (12.20) and column S, y′ = 300 − 4x, for (12.22). In the table, x is entered in the column, hence x is taken as the independent variable.[12] To draw the budget constraint on the indifference

[12]We may solve the budget constraint as an explicit function of y, x = (300−y)/2. Then, y is the independent variable. In this case, the budget constraint should be entered in row 20.

map, we note that, for the original budget constraint (12.20),

$$x = 80 \text{ implies } y = 140 \text{ (draw a circle in Fig. 12.10(b))},$$
$$x = 150 \text{ implies } y = 0 \text{ (draw another circle in Fig. 12.10(b))}.$$

Connecting the intercepts (the two circles), the original budget constraint, $M = 2x + y = 300$, is drawn as a solid white line in Fig. 12.10(b). According to the Solver, Table 12.6, the budget line is tangent to an indifference curve of 1,000,000 util at point a, which is located at $(x^{\#}, y^{\#}) = (100, 100)$, and is denoted by a white circle in Fig. 12.10(b). In the legend, the indifference curve is located between 900,000 (util/1000) = 900 and 1,100,000 (util/1000) = 1100. These ranges are marked by bold face in the legend.

Similarly, after the price of x has changed from 2 to 4, the new budget line is $M' = 4x + y = 300$. In Table 12.7,

$$x = 40 \text{ implies } y = 140 \text{ (draw a circle in Fig. 12.10(b))},$$
$$x = 75 \text{ implies } y = 0 \text{ (draw another circle in Fig. 12.10(b))}.$$

The new budget line, $M' = 4x + y = 300$, is shown as the lower solid white line in Fig. 12.10(b). According to the Solver, Table 12.6, it is tangent to an indifferent curve of 250,000 util at point b, also denoted by a white circle, at $(x^{\#}, y^{\#}) = (50, 100)$. In the legend, the indifference curve is located between 200,000 util/1000 = 200 and 300,000 util/1000 = 300. The range is also marked by boldface in the legend.

Illustration of the optimal point in the sensitivity table

To make the sensitivity table more interesting, the two indifference curves, which are tangent to the two constraints, are drawn in Table 12.7. The solid line shows the original constraint, y, and the indifference curve is the connecting cells that are painted in yellow. The maximum utility cell under budget constraint is attained at L14, at which U = 1,000,000 util/1000 = 1000. The new budget constraint, y′, is shown by the dotted line, and the indifference curve is also the connecting cells that are painted in yellow. The maximum utility cell under the new budget constraint is attained at L9, at which U = 250,000 util/1000 = 250. The relations of the indifference curves in the table, in the utility surface, and in the indifference map should be understood clearly.

12.5.4 *The law of diminishing marginal rate of substitution*

Total increment in utility

As with the theory of production, given a utility function, $U(x, y)$, from consuming good x and good y, we define the marginal utility of good x, MU_x, as the extra utility that the consumer enjoys by consuming one extra unit of good x. If x increases by a small amount, Δx, then $(MU_x)\Delta x$ is the partial increment of utility enjoyed by consuming Δx units of x. Similarly, $(MU_y)\Delta y$ is the partial increment of utility enjoyed by consuming Δy extra units of y. For the two commodities x and y together, the total increment of utility, ΔU, that the consumer can derive from consuming Δx and Δy, is

$$\boxed{\Delta U = (MU_x)\Delta x + (MU_y)\Delta y.} \tag{12.25}$$

Example 12.4 Let the utility function be $U = U(A, B)$, where A is the amount of apples and B is the amount of bananas that a consumer consumes. Let marginal utilities be $MU_A = 35$ util and $MU_B = 10$ util. Find the total increment of utility, ΔU, if the consumer consumes an extra 2 units of apples and an extra 3 units of banana.
Answer: The total increment of consumer's utility, ΔU, is

$$\Delta U = (MU_A)\Delta A + (MU_B)\Delta B = (35)(2) + (10)(3) = 100.$$

The total increment of utility from consuming an extra 2 units of A and an extra 3 units B is 100 util. The relation is illustrated in Fig. 12.11. □

Marginal rate of substitution

If the changes in consumptions of x and y take place along the indifference curve, then there is no change in total utility level, and $\Delta U = 0$. In this case, we can solve for Δy in terms of Δx in (12.25), find $\Delta y/\Delta x$, and then take its absolute value. We define the result as the **marginal rate of substitution** between x and y:

$$\Delta y/\Delta x = -MU_x/MU_y,$$

$$\boxed{\left|\frac{\Delta y}{\Delta x}\right| = -\frac{\Delta y}{\Delta x} = \frac{MU_x}{MU_y} = MRS,} \tag{12.26}$$

which is the negative of the slope $\Delta y/\Delta x$ of the indifference curve, and the MRS is equal to the ratio of MU_x and MU_y. Since MU_x and MU_y are generally assumed to be positives, the MRS is also positive. As in the case of production theory, the **law of diminishing**

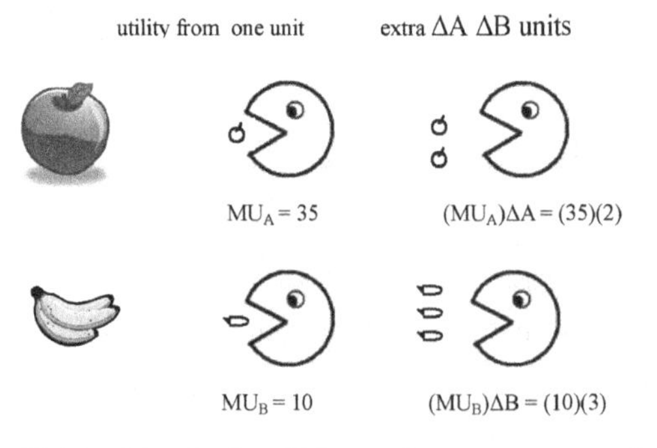

Fig. 12.11 Marginal utilities and extra consumptions

marginal rate of substitution also holds in utility theory: The extra amount of good y becomes smaller and smaller when we continue to use one unit of good x to replace good y and still remain on the same indifference curve. This is the case if the indifference curve is convex to the origin. Equation (12.26) shows that as we consume more x and less y, MU_x decreases and MU_y increases along the indifference curve. Hence, the value of MRS decreases. This can be seen clearly from the consumption of apples and bananas in Example 12.4. □

12.5.5 *The law of equal marginal utility per dollar*

The slope of the budget constraint

If changes in x and y occur along a budget constraint, holding budget M constant, i.e. $\Delta M = 0$ for the changes in $\Delta x = x_2 - x_1$ and $\Delta y = y_2 - y_1$, we have

$$\Delta M = p\Delta x + q\Delta y = 0. \tag{12.27}$$

The slope of the budget constraint (12.18) is, holding M constant,

$$\frac{\Delta y}{\Delta x} = -\frac{p}{q}.$$

The law of equal marginal utility per dollar and the marginal utility of money

Since Fig. 12.10(b) shows that at the maximum point a, the slope of the budget constraint must be the same as the slope of the indifference curve, from the above equation and (12.26), we have

$$\frac{MU_x}{MU_y} = \frac{p}{q} \tag{12.28}$$

or

$$\boxed{\frac{MU_x}{p} = \frac{MU_y}{q} = \lambda.} \tag{12.29}$$

Equation (12.29) has a well-known economic interpretation. MU_x is the marginal utility of consuming one extra unit of x. Since p is the price of x, 1/p is the amount of commodity x that \$1.00 can buy. For example, if one apple costs a quarter, then $1/0.25 = 4$ is the quantity of apples \$1.00 can buy. That is, 1/p is the **purchasing power of money** with respect to x. Hence, $MU_x/p = (MU_x)(1/p) = \lambda$ is the marginal utility derived from purchasing \$1.00 worth of x.

Similarly, $MU_y/q = (MU_y)(1/p) = \lambda$ is the marginal utility derived from purchasing \$1.00 worth of y. Thus, Eq. (12.29) states that, when the consumer's utility is maximized (a necessary condition), the marginal utility obtainable from spending \$1.00 on x is equal to the marginal utility obtainable from spending \$1.00 on y. Thus, (12.9) is called **the law of equal marginal utility per dollar.**

Note that unless $p = q = 1$, the MU of x is generally not equal to the MU of y : $MU_x \neq MU_y$. This shows that the term "per dollar" is crucial in this law.

The marginal utility of money

From (12.29), multiplying the same quantity to the numerator and denominator, we can write

$$\lambda = \frac{MU_x \Delta x}{p\Delta x} = \frac{MU_y \Delta y}{q\Delta y} = \frac{MU_x \Delta x + MU_y \Delta y}{p\Delta x + q\Delta y} = \frac{\Delta U}{\Delta M}, \tag{12.30}$$

where we apply (12.25) to the numerator and (12.27) for $\Delta M \neq 0$ to the denominator. $\Delta U/\Delta M = \lambda$ is called the **marginal utility of money**. It is the extra utility derived from spending one extra dollar. Thus, combining this with (12.29), we can state that at the optimal point of utility maximization, the extra utility derived from spending one dollar on x and one dollar on y must be equal to the extra utility derived directly from spending one dollar (the marginal utility of money).

12.5.6 *Derivation of demand functions*

Like Sec. 12.1.4 in the factor demand functions of production, the slope condition (12.28), or equivalently, the law of equal marginal utility per dollar (12.29), and the budget constraint (12.18) constitute a system of two simultaneous equations in two variables x and y. In Mathematics, they are the **necessary condition of constrained maximization in consumption**. Under certain mathematical assumptions, these two equations can be solved explicitly for the variables x and y as a function of parameters p, q, and M, as

$$x = f(p, q, M), \tag{12.31}$$

$$y = g(p, q, M). \tag{12.32}$$

They are called the consumer's **demand functions** for good x and for good y. The derivation of the demand functions above is the theoretical foundation of the demand function, $D = a - bp$, which we are already familiar with from Chapter 3. The difference is that the demand function in Chapter 3 is simplified as a linear function in its own price p only, and that another good price q and the consumer budget M are held constant and are included in its constant term a.

Example 12.5 The optimal quantity demanded. From utility function (12.19), we find $MU_x = 2xy$ and $MU_y = x^2$. Hence, from (12.28) or (12.29), we have

$$px - 2qy = 0.$$

Combined with budget constraint, $px + qy = M$, we can solve $x^{\#} = 2M/3p$ and $y^{\#} = M/3q$. For $p = 2$, $q = 1$, and $M = 300$, we have the optimal quantity demanded $x^{\#} = 100$ units $= y^{\#}$, which is the same as the optimal solution we obtained from the graphic method and from the Solver in (12.21). □

Example 12.6 Law of equal marginal utility per dollar. At the optimal point, we have $MU_x = 2x^{\#}y^{\#} = 2(100)(100) = 20{,}000$ util, and $MU_y = (x^{\#})\hat{}2 = (100)^2 = 10{,}000$ util. Since $p = 2$ and $q = 1$, the law of equal marginal utility per dollar in (12.29) holds for $20{,}000/2 = 10{,}000/1 = 10{,}000$ util $= \lambda$, which is the marginal utility of money.

The marginal utility of money can be derived by the Solver in Table 12.6. In Table 12.6, enter 301 in G11 (as the budget increases by $1.00), and enter $p = 2$ in C11. Other parameters are the same as shown in the New table, rows 7–11. Invoke <solve> in the Solver. Cell C8 will show 100 units and cell E8 will show the utility to be 1,010,033.37 util, as the optimal values of $x^{\#}$ and $y^{\#}$ also change slightly (small decimals). The change in utility, ΔU, can be calculated by comparing the maximum utility in the old (original) problem, as shown in E15 as 1,000,000.01 util, so $\Delta U = 1{,}010{,}033.37 - 1{,}000{,}000.01 = 10{,}033.36$ util. Since (12.29) gives 10,000 util, there is an error of 33.36 util, or about 0.33% of the original utility. This is due to the discrete approximation of marginal utility as the MU of money is defined in continuous terms, that is, when ΔM approaches 0. □

12.6 Decomposition of the Price Change

In Sec. 12.5.2, we discussed a case in which the price of good X, p, is doubled from $2 to $4, while the price of good Y, $q = 1$, remains constant. Since q does not change, the intercept of the Y-axis for both budget constraints (12.20) and (12.22) should be the same, $300/q = 300$, for both budget lines. Thus, the two budget lines in Fig. 12.10(b) should have the same Y-intercept at point (0, 300). This is not shown in Fig. 12.10(b), since the range of Y is taken only from 0 to 150.

12.6.1 *The indifference map*

To show the Y-intercept, copy Table 12.7 to a new sheet to preserve the original table, and then work on the original table.[13] In that table, Table 12.7, change the range of Y from 0 to 600, with step 50, by entering 50 in C3 and using fill-handle to copy B3:C3 up to N3. You can ignore columns O, P, and Q. The utility levels in the range from B4 to N19 change accordingly. To adjust the indifference map, activate Fig. 12.10(b) by clicking its chart area. The data part of the chart in new Table 12.7 will be enclosed by blue borders. Click and drag the blue square box at the bottom of the right-hand side border of the table up to

[13]We may work on the table in the new sheet also. The advantage of working on the old table is that we can use the originally named ranges of the old table. Note that you use the old Table 12.7 as a template to generate a new sensitivity table. This is another advantage of using a spreadsheet.

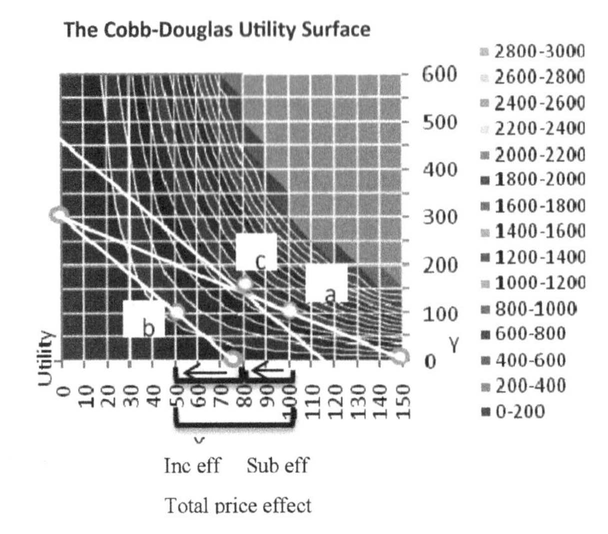

Fig. 12.12 Decomposition of total price effect

N19. Figure 12.10(b) now changes automatically to Fig. 12.12. Make necessary revisions of labeling as in Fig. 12.12.

The optimal point a before the price change is still at $(x^{\#}, y^{\#}, U^{\#*}) = (100, 100, 1000000)$, and the optimal point b after the price change is at $(x^{\#}, y^{\#}, U^{\#}) = (50, 100, 250000)$.

12.6.2 *Substitution and income effects*

If the price of X is doubled, the consumption of X falls to half, from $x^{\#} = 100$ to $x^{\#} = 50$ (see Table 12.6). Assuming other things are equal, that is, nominal income (or budget) and the price of Y do not change, this **total price effect** can be decomposed into two parts. One is the **substitution effect**, the decrease in the demand for X due to the increase in relative price when X is replaced by good Y in order to maintain the same utility level (indifference curve). The other is the **income effect**, the decrease in the demand for X due to the increase in its price when the consumer's real income (or purchasing power) becomes smaller.[14]

To find the substitution effect, consider a fictitious situation in which the consumer at the new optimal point b is given an extra income (or budget) ΔM, where Δ denotes the change. This extra income is to compensate for the consumer's loss in purchasing power due to the increase in price of X, such that the consumer can remain at the same indifference

[14]For a mathematical treatment of this topic, see Chiang and Wainwright (2005), pp. 375–380. For an elementary explanation without diagrams, see Samuelson and Nordhaus (1995), pp. 78–79 and with reversed diagrams, see Mankiw (1998), pp. 459–460.

curve as before the price change. The adjustment in income ΔM will shift the new budget line upwards until it is tangent to the original indifference curve at point c. This can be done by copying the new budget line so that the copied budget line is parallel to the new budget line and touches the old indifference curve (which passes through point a) at point c. Thus, the total price effect can be read from the figure as 50, which is divided into the **substitution effect** of 20 and the **income effect** of 30, as shown by the braces in the lower part of Fig. 12.12.

Thus, if the price of X doubles, and if the consumer is given the compensation income ΔM, the consumer will substitute good Y for X by reducing the consumption of X from 100 to 80, and increase the consumption of Y in order to remain at the same indifference curve. However, when the compensated income is taken away, the consumer reduces consumption of both X and Y: reduces consumption of X from 80 to 50, and also reduces the consumption of Y back to the original level. In general, Y may reduce more or less compared with the original level of Y, depending on the shapes of the indifference curves.

12.6.3 *Another illustration*

While Fig. 12.12 presents a bird's eye view of the substitution and income effects with color, too many indifference curves in the figure may be confusing. Since we already know, through the Solver, the maximum utility levels before and after the price change of X, we may draw only two indifference curves, as we did in Sec. 12.4 (see Table 12.5), and make simultaneous presentation of an objective function and the constraint.

From the adjusted utility function (12.24), we can solve for y in terms of x and U to obtain

$$y = 1000U/x^2. \tag{12.33}$$

To avoid confusion, we choose two values of U. Let U = 250 or 1000. In Table 12.8, we draw only two indifference curves in columns B and C and two constraints in columns D and E. Based on (12.33), column B shows the value of y for U = 250 and column C shows the value of y for U = 1,000. Column D is the value of $y = 300 - 2x$ from the original constraint, $M = 300 = 2x + y$; and column E is the value of $y = 300 - 4x$ from new constraint, $M' = 300 = 4x + y$, when p is doubled from 2 to 4. They are the same as columns R and S in Table 12.7.

The results in columns A–E of Table 12.8 are illustrated in Fig. 12.13 with much clarity on the effect of total price change, where we have reproduced points a, b, and c of Fig. 12.12. Both Table 12.8 and Fig. 12.13 show that, when x = 100, the Y-coordinate a of the indifference curve at U = 1000 is the same as that of the original constraint (y) (see the shaded part of row 16); and when x = 50, the Y-coordinate b of the indifference curve at U = 250 is the same as that of the new constraint (y′) (see the shaded part of row 11). Thus, the constraints are tangent to the indifference curves. Figure 12.13 presents a clear view of the substitution and income effects. Note that, unlike in Fig. 12.12, in order

Table 12.8 The substitution and income effects

	A	B	C	D	E	F
1	The substitution and income effects					
2					y'=300-4x	
3		U=1000*U/x^2		y=300-2x	y=476.2-4x	
4	x	U		constraints		
5		250	1000	y	y'	M(CV)
6	0			300	300	476.2
7	10			280	260	436.2
8	20	625.0		260	220	396.2
9	30	277.8		240	180	356.2
10	40	156.3	625.0	220	140	316.2
11	50	100.0	400.0	200	100	276.2
12	60	69.4	277.8	180	60	236.2
13	70	51.0	204.1	160	20	196.2
14	80	39.1	156.3	140	-20	156.2
15	90	30.9	123.5	120	-60	116.2
16	100	25.0	100.0	100	-100	76.2
17	110	20.7	82.6	80	-140	36.2
18	120	17.4	69.4	60	-180	-3.8
19	130	14.8	59.2	40		-43.8
20	140	12.8	51.0	20		-83.8
21	150	11.1	44.4	0		-123.8

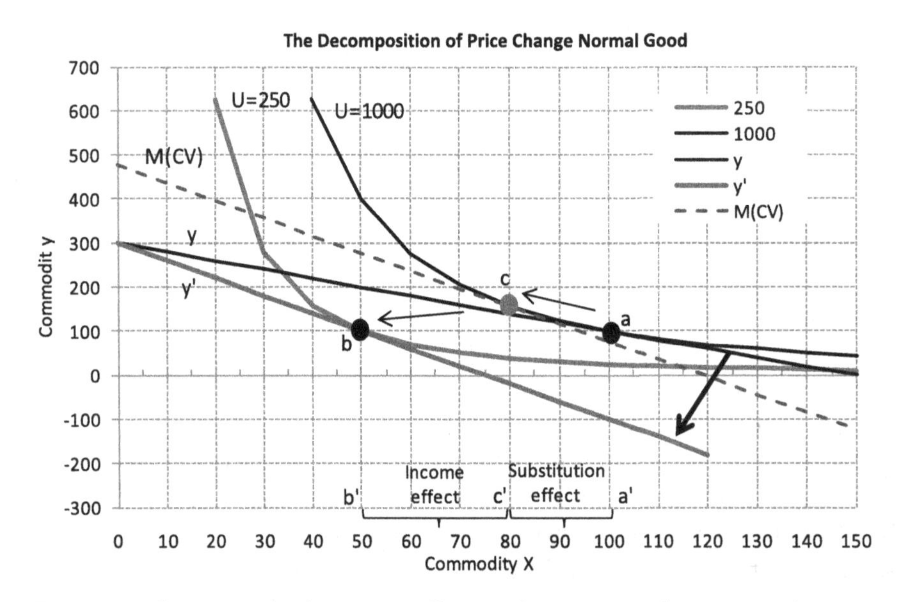

Fig. 12.13 Income and substitution effects and compensated variation of income

to draw the indifference curves through points a and b, we have to first know the levels of maximum utility.

12.6.4 *Compensating variation of income*

The duality problem

In the previous section, we stated that point c is obtained by compensating for the consumer's loss (or decrease) of purchasing power of budget (or income) due to the doubling of the price of X from 2 to 4, so that the consumer can remain at the original indifference curve at U = 1,000,000. The amount of compensating income (or budget) is called the **compensating variation (CV) of income**. How to measure this CV? We first introduce the **duality of the utility maximization**.

From the construction of Fig. 12.13, since the utility level is given at U = 1,000,000 and the budget (or expenditure) is unknown, the budget must be as small as possible for a given utility function. Thus, we have the duality to the maximization problem formulated in (12.19) and (12.20), similar to the duality in production (12.11) and (12.12):

$$\min_{x,y} \; M = px + qy, \quad \text{Objective function} \tag{12.34}$$

$$\text{s.t. } U(x,y) = x^2y/1000, \quad \text{Constraint} \tag{12.35}$$

where p = 4 and q = 1. As in the case of production, this problem can be solved by the Excel Solver. To use the Solver, copy Table 12.6 to a new sheet and change the original table for the dual problem as we did in Table 12.4, with the format and solution shown

Table 12.9 Derivation of the compensating variation of income

	A	B	C	D	E	F	G	
1		parameters		default	parameters		default	
2		a	2	2	p	4	2	
3		b	1	1	q	1	1	
4		aa	1	1	M	M#	300	
5								
6	Conptable							
7		Constr	x	y	=aa*x^a*y^b	=	U	Constraint
8	q1	(x,y)	79.37	158.74	1000000.00	?	1000000	(12.34)
9								
10		Min	p	q	px+qy	=	M	Objective function
11	p1		4	1	476.2203156	?	M#	(12.35)
12					Target cell			

in Table 12.9. The objective function is now in E11 and the constraint is in E8 and G8, with the same variables in C8:D8. Table 12.9 shows the result. The constrained minimum solution is

$$(x^{\#}, y^{\#}, M^{\#}) = (79.37, 158.74, 476.22).$$

Since the original budget (or income or expenditure) is 300, the CV of income is 476.22 – 300 = 176.22.

Illustration in table and figure

Column F of Table 12.8 shows the budget line with CV of income, M(CV) = 476.2 – 4x, which is drawn as a dotted line in Fig. 12.13. M(CV) is tangent to the indifference curve of U = 1000 at point c = (79.37, 158.74). Thus, from Tables 12.6 and 12.9, we can now state in numerical terms that, under the current model, when the price of X, p, is doubled from 2 to 4, the consumer reduces the consumption of X from 100 to 50, the **total price effect**, without changing the consumption level of Y.

However, the **substitution effect** indicates that, due to the doubling of p, the consumer will reduce the consumption of X by 20.63 (=100 – 79.37) and increase the consumption of Y by 58.74 (=158.74 – 100), if the consumer's loss of income (or purchasing power), 176.22, due to the doubling of p, is compensated.

The **income effect** states that, without the compensating income, the consumer will reduce both consumption of X and consumption of Y further: the consumption of X will decrease by 29.37(=79.37 – 50), and the consumption of Y will decrease by 58.749 (= 158.74 – 100).

Normal and inferior goods and the total price effect

In general, it is known that when the price of a good increases, the consumer decreases the quantity demanded for that good and substitutes it by another good, if the consumer stays at the same indifference curve. Thus, the **substitution effect** is always negative.

On the other hand, the effect of income change is not obvious. If the commodity is a **normal good**, then a decrease in income will decrease the demand for that commodity, as we have seen in Fig. 12.13. Thus, income effect due to price increase is negative. This means that an increase in p decreases income and then decreases the demand for x. Since the total price effect is the sum of the substitution effect and the income effect, the total price effect of increase in price p is negative, that is, both effects result in a decrease in the demand for x.

However, if the commodity is an **inferior good**, like a black-and-white TV (as compared with a color TV) or a used car (as compared with a new car), then income effect is negative, that is, when income increases, demand for the inferior good decreases, or when income decreases, demand for the inferior good increases. In this case, the income effect due to increase in price is positive. This means that an increase in p decreases income and increases the demand for x. In this case, the total price effect of an increase in price may be negative or

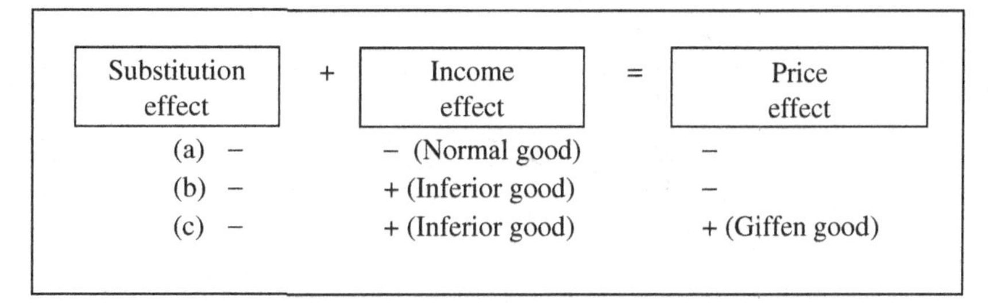

Fig. 12.14 The price effect.

positive, depending on the comparative magnitude of the substitution effect and the income effect. If the positive income effect is so large that it overrides the negative substitution effect, then the total price effect is positive, that is, increase in the price of x will increase the demand for x. In this case, we call x the **Giffen good**, and we have a **positively sloped demand curve**. The above relations are illustrated in Fig. 12.14.

12.7 Summary: Circular Flow of the Economy

The first part of this chapter is illustrated in Fig. 12.15. We start from the illustration of the nature of the problem by first drawing the cost constraint, which is imposed on the 3D output surface and the 2D isoquant map. The optimization condition that the slope of the isoquant and the slope of the cost constraint must be the same leads to the law of equal marginal productivity per dollar. Along with the cost constraint, we derive factor demand functions (Sec. 12.1). We then introduce the Excel Solver systematically, and use it to verify the visual solution obtained from the graphic method (Sec. 12.2).

The Solver is used to find the solution to the dual problem of minimizing cost subject to the output (or technology) constraint (Sec. 12.3). We then illustrate the minimization problem diagrammatically in 2D charts (Sec. 12.4).

The last two sections are devoted to the theory of consumption, the basic scheme of maximizing utility under a budget constraint, and applications of its results are illustrated in the flow chart of Fig. 12.16. In this part, we reverse the procedure by first finding the solution from the Solver, and using the results to illustrate the maximization problem diagrammatically. Similarly to production theory, the slope condition that the slopes of the indifference curve and the budget constraint also leads to the law of equal marginal utility per dollar. Along with the constraint, the commodity demand functions can be derived. We also derive the marginal utility of money (Sec. 12.5.5). Lastly, the decomposition of total price effect into the substitution effect and the income effect is illustrated. The derivation of CV of income gives rise to the problem of minimizing the budget (expenditure) subject to the utility function (Sec. 12.6). Using the Solver, we derive the numerical value of the CV of income.

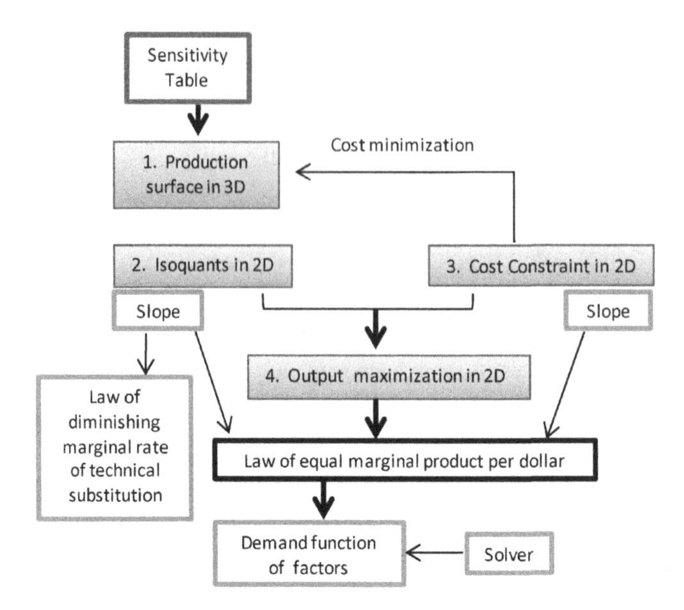

Fig. 12.15 The basic scheme of the theory of production

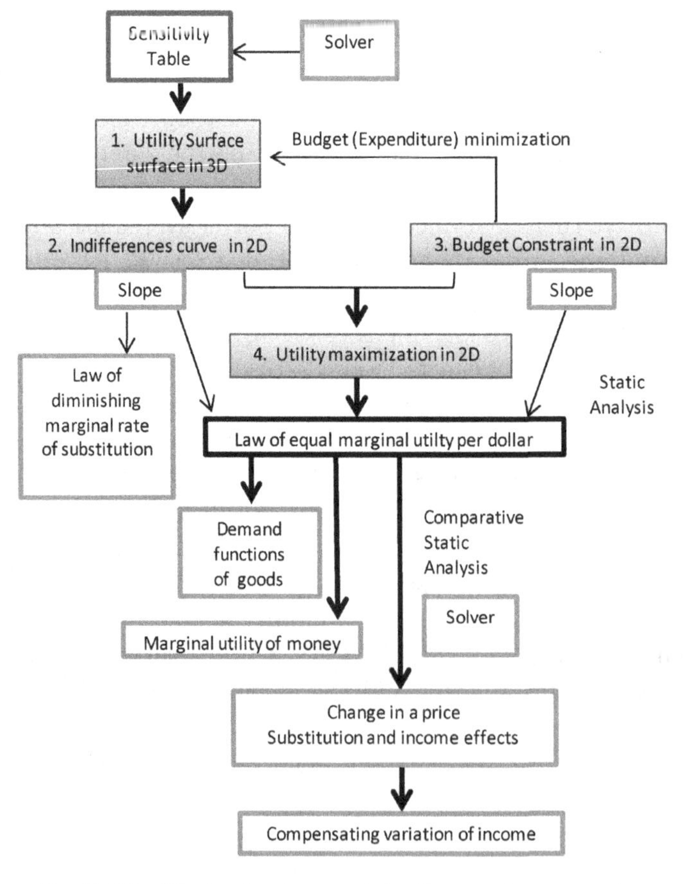

Fig. 12.16 The basic scheme of the theory of consumption

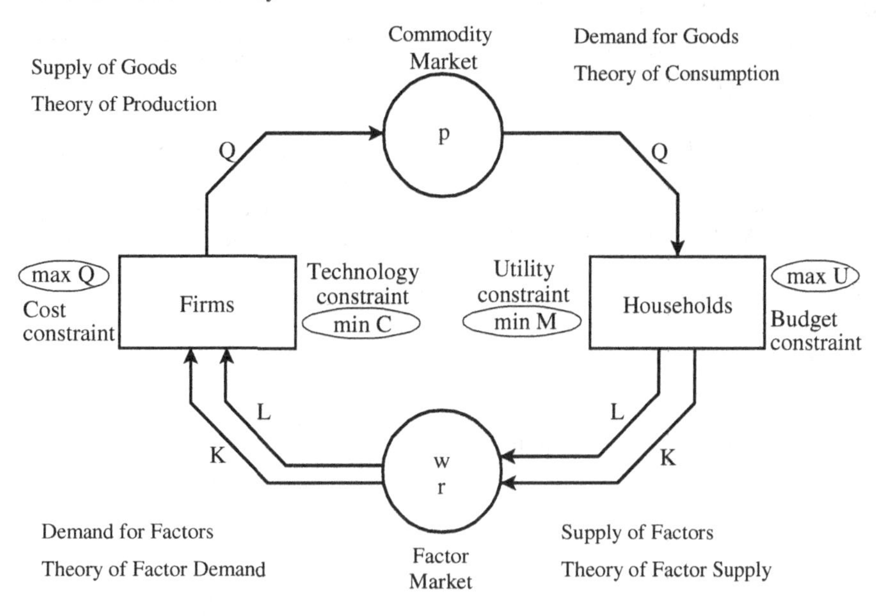

Fig. 12.17 The circular flow of the economy and microeconomics

In both theories of production and consumption, to find the solution of an optimization problem, all we need are steps 2 and 4 in the flow charts of Figs. 12.15 and 12.16. After setting up the sensitivity table and attaching the linear constraint to the right-hand side of the table, we may find the intercepts of the constraint and impose the constraint directly on the isoquants, as in Fig. 12.3, without drawing the cost constraint chart like Fig. 12.1 or the 3D output surface as in Fig. 12.2. This will simplify the charting procedure.

In general, we have seen the similarity between production theory and consumption theory. Within each theory, we also find the similarity between output maximization and cost minimization within the production theory, and between utility maximization and budget minimization within consumption theory. The relationships between firms and households, and their rational behavior in finding the solution to maximization or minimization problems, and related markets and economic theories are illustrated in the circular flow chart of Fig. 12.17. What we have discussed in this chapter is the essential part of microeconomic theory using Excel spreadsheets, graphics, and the Solver command.

Review of Basic Equations and Formulas

Output maximization (12.4) $\max_{L,K} Q = AL^a K^b$ (12.5) s.t. $C = wL + rK$	Utility maximization (12.17) $\max_{x,\,y} \; U = U(x, y)$ (12.18) s.t. $M = px + qy$
Total increment of output (12.25) $\Delta Q = (MPL)\Delta L + (MPK)\Delta K$	Total increment of utility (12.25) $\Delta U = (MU_x)\Delta x + (MU_y)\Delta y$
Slope condition (12.10) $\frac{MPL}{MPK} = \frac{w}{r}$,	Slope condition (12.28) $\frac{MU_x}{MU_y} = \frac{p}{q}$
Law of equal marginal product per dollar (12.10) $\frac{MPL}{w} = \frac{MPK}{r}$	Law of equal marginal utility per dollar Marginal utility of money (12.29) $\frac{MU_x}{p} = \frac{MU_y}{q} = \frac{\Delta U}{\Delta M}$
Factor demand function $L = L(w, r, C)$ $K = K(w, r, C)$	Commodity demand function (12.31) $x = f(p, q, M)$ (12.32) $y = g(p, q, M)$
Cost minimization (12.11) $\underset{L,K}{Min} \; C = wL + rK$ (12.12) $s.t. \;\; Q = AL^a K^b$	Budget minimization (12.34) $\underset{x,y}{Min} \; M = px - qy$ (12.35) $s.t. \;\; U = U(x, y)$
Simultaneous presentation (12.2) $K = (C/r) - (w/r)L$ (12.15) $K = (\frac{Q}{AL^a})^{1/b}$	Total price effect = Substitution effect + Income effect

Key Terms: Economics and Business

Key terms: Excel

Homework Chapter 12

12-1 From (12.10), we may write

$$\frac{\mathrm{w}}{\mathrm{MPL}} = \frac{\mathrm{r}}{\mathrm{MPK}}.$$

Give an economic interpretation of this equality.
(Hints: MPL is the amount of output one extra unit of labor can produce. Hence, 1/MPL is the amount of labor required to produce one extra unit of output. It is the marginal input coefficient of labor, and w/MPL is the marginal real cost of labor in producing one extra unit of output).

12-2 Example 12.1 In Example 12.1, using the Excel Solver, show that the new optimal values are indeed $(\mathrm{L}^{\#}, \mathrm{K}^{\#}, \mathrm{Q}^{\#}) = (6, 8, 13.5)$.

12-3 Example 12.2 In Example 12.2, the equations are given as $\mathrm{bwL} - \mathrm{arK} = 0$ and $\mathrm{wL} + \mathrm{rK} = \mathrm{C}$. The matrix method in Chapter 10 requires that the values of the parameters are given. (a) Using the substitution method, derive the parametric form of the demand functions for labor and capital. (b) For the value of parameters given in Example 12.2, verify that the values of the optimal factor demands obtained from the diagrammatic method is indeed the same as the values obtained from the demand functions.
(Hints: (a) From the first equation, solve for L as $\mathrm{L} = \mathrm{arK}/\mathrm{bW}$, substituting into the cost constraint, and noting that $\mathrm{a} + \mathrm{b} = 1$, we have the factor demand functions $\mathrm{K}^{\#} = \mathrm{bC}/\mathrm{r}$ and $\mathrm{L}^{\#} = \mathrm{aC}/\mathrm{w}$. (b) For the values of parameters given in the Example, we have $(\mathrm{L}^{\#}, \mathrm{K}^{\#}) = (9, 8)$).

12-4 Equivalent variation in income In Fig. 12.13, we have determined the extra amount of income necessary to compensate the loss to a consumer due to a price increase, so that the consumer can remain at the original utility level.

Conversely, we may also determine the maximum amount of income the consumer is willing to give up to avoid the change in the price of X. This amount of income is called the **equivalent variation** (EV) (see Cahill, 2000). In terms of Fig. 12.13, this is the amount of income that is parallel to the original constraint y and tangent to the indifference curve at $\mathrm{U} = 250$. Using a calculation table for Excel Solver similar to Table 12.9, find EV and illustrate it as in Fig. 12.13.
(Hints: $\mathrm{M(EV)} = 188.99$ at $\mathrm{x}^{\#} = \mathrm{y}^{\#} = 63$. Since $\mathrm{M}' = 300$, $\mathrm{EV} = 189 - 300 = -111$).

The Excel Solver for unconstrained optimization

12-5 The Solver for profit maximization of a firm Using the Excel solver, verify the solution of HW11-16, where the problem was to maximize $\pi = 3 * 6 * \mathrm{L}^{0.5}\mathrm{K}^{0.3} - 4.5\mathrm{L} - 3\mathrm{K}$.
(Hints: From Table 12.2, ignore or delete the constraint in rows 10 and 11, and also leave "Subject to the Constraint" blank in Fig. 12.4. $(\mathrm{L}^{\#}, \mathrm{K}^{\#}, \pi^{\#}) = (27, 25, 49.17)$)

12-6 The Solver for the profit function of an oligopolistic firm Using the Excel solver, verify the solution of HW11–17, that is, maximize $\pi = 52\mathrm{x} - 4\mathrm{x} - 4\mathrm{xy}$ for given $\mathrm{y} = 5$.
(see Hints of HW12-5. $(\mathrm{x}^{\#}, \pi^{\#}) = (6.5, 169)$)

12-7 The Solver for unconstrained maximization (A profit function) Using the Excel solver, verify the solution of HW11-20: maximize $\pi = 50\mathrm{x} + 80\mathrm{y} - 3\mathrm{xy} - 2\mathrm{x}^2 - 4\mathrm{y}^2$. (See Hints of HW12-5. $(\mathrm{x}^{\#}, \mathrm{y}^{\#}, \pi^{\#}) = (7.0, 7.4, 469.6)$)

Constrained optimization

12-8 Big Mac operations Continuing from HW11-11, suppose Joe is planning to open a Big Mac franchise. He is confronted with the following information on factors of production.

Labor cost $\mathrm{w} = \$6.00$ per hour,

Equipment cost $\mathrm{r} = \$200$ per unit,

Total operation budget $\mathrm{C} = \$10{,}000$.

From the literature, Joe knows the production function is of the form $Q = 1000L^{0.6}K^{0.4}$ and wants to maximize his big Mac production given his operation budget.

The problem is

$$\max_{x,y} \ Q = 1000L^{0.6}K^{0.4}.$$
$$\text{s.t. } wL + rK = 10,000.$$

(a) Construct the sensitivity table for $L = 0, 100, \ldots, 1500$, $K = 0, 4, \ldots, 40$.
(b) Using the naming method, name w and r, and take $w = 6$ and $r = 200$ as default.
(c) Solve the cost function in terms of K, attach the constraint table to the sensitivity table, and draw the constraint in a separate chart.
(d) Using the isoquant map, impose the cost function on it.
(e) In order to minimize his operation cost, approximately how many hours of labor must Joe hire and how much must he spend on equipment? Taking the initial values as (50, 50) or (100, 100) in the solver.
Show that if the initial values are taken as (1, 1) or (10, 10), the solution does not converge. Paste values of your results below the original calculation table.
(f) How many units of Big Mac does Joe have to produce to maximize his profit? Use an empty circle to mark the optimal point and a callout to indicate the optimal point along the indifference curve.
(g) From the results in (e), find the optimal output level in the sensitivity table and denote the cell with the yellow pen.
(h) Draw an isoquant passing through the optimal output cell in (g) and the cost constraint through the optimal cell.
(i) How do you know from the chart that the optimal point is really optimal? Give a proof or diagrammatic argument that it is really the optimal point.
(j) Justify your answer from the sensitivity table. Explain.
(k) If the wage rate increases to $10.00 per hour, what are the new optimal labor and capital? Use Excel Solver to find the answer, and use an additional heavy line and an arrow to show your answer on the indifference map.
(l) What is the new output level of Big Mac? Use an empty circle to identify the new equilibrium point and enclose the corresponding cell in the sensitivity table with borders.
(m) How does the answer to (h) differ from the one in (e)? What is the economic implication of the differences.

(Hints: (e) $(L^{\#}, K^{\#}, Q^{\#}) = (1000, 20, 209128)$; (k) Take the initial values as $(L, K) = (125, 125) \cdot (L^{\#}, K^{\#}) = (600, 20)$, (l) $Q^{\#} = 153, 922$)

12-9 Utility maximization Let the **utility function** be given as

$$U = \text{sqrt}(x) + \text{sqrt}(y).$$

Let the price of $X = 2.25$, the price of $Y = 1$, and income $= \$9$. Thus the budget constraint is

$$2.25x + y = 9.$$

All the charts and axes of the charts must be labeled clearly.

(a) Draw the utility surface for x = 0, 1, ..., 9 and y = 0, 1, ..., 10. Put commodity x in the column and commodity y in the row. Utility is measured up to one digit after the decimal point.
(b) Attach the constraint to the last column of the table.
(c) Draw the utility surface in 3D.
(d) Draw the indifference curves for the two commodities. Let the scale of the legend be 0.2 util and the maximum utility be 5 util.
(e) Draw the budget constraint in a separate chart.
(f) Using drawing tools, impose the budget line on the indifference curves.
(g) Approximately what is the value of the optimal consumption levels for X and Y? Use a callout to indicate the optimal point along the indifference curve.
(h) Using the yellow background color, find the indifference curve corresponding to the maximum level of utility in the utility table.
(i) Using the Excel Solver, verify your solution from the diagram.

(Hints: (i) $x^* = 1.2$, $y^* = 6.2$, $U^* = 3.6$.)

12-10 Utility maximization In HW12-9, let the price of x be 1 and that of y be 2, with income being \$300. Using the table and charts already give in HW12-9, find the optimal allocation of the commodities x and y, and the maximum utility. You may use the diagram in HW12-9, except now that x ranges from $0, 50, \ldots, 400$, y ranges from $0, 10, \ldots, 100$.

(Hints: $x^* = 200$, $y^* = 50$, $U^* = 21.2$.)

12-11 GPA maximization (Klein, 2001, p. 341) Continuing from HW11-14, suppose it is late in the semester and you have two exams left: College Mathematics (M) and Western Literature (L). You must decide how to allocate your working time during the study period. After eating, sleeping, exercising, and maintaining some human contact, you have 12 hours each day in which to study for your exams. Thus, your time constraint equation is

$$\mathrm{M} + \mathrm{L} \doteq 12.$$

You have figured out that your grade point average, GPA, from your two courses takes the form

$$\mathrm{GPA} = \frac{2}{3}(\sqrt{\mathrm{M}} + \sqrt{2\mathrm{L}}),$$

where M is the number of hours per day spent studying for College Math and L is the number of hours per day spent studying for Western Lit.

Let the hours spent on L range from 0 to 15 for the rows (the X-axis) and let M range from 0 to 12 for the columns (the Y-axis).

(a) Draw the 3D diagram for the GPA surface.
(b) Draw the GPA indifference curves for both courses. Change the major unit scale of the legend to 0.2.
(c) Draw the time constraint equation as a line on the GPA indifference map.

(d) What is the optimal number of hours per day spent studying for each course? If you follow this strategy, what will your GPA be?
(e) Using the Excel Solver, verify your optimal solution. (Hints: $M^* = 4, L^* = 8, GPA^* = 4.0$.)

12-12 A utility function Continuing from HW11-15, the utility you derive from consuming apples A and bananas B is described by the utility function

$$U(A, B) = (A - 2)^{0.4}(B - 3)^{0.6}.$$

In the following questions, we first set up the utility sensitivity table for $A = 2, 2.5, \ldots, 9$ and $B = 4, 4.5, \ldots, 10$. Put A in the column (row titles) and Y in the row (column titles). Entries of the tables should have one decimal place. All the chart titles and axes of the charts must be labeled clearly.

(a) Construct the utility sensitivity table for A and B. (you may copy the sensitivity table from HW11-15).
(b) If the price of A is \$2 and that of B is \$3, and the budget is \$24, then the constraint is

$$2A + 3B = 24.$$

Attach the constraint table to the utility table and draw the constraint in a separate chart.
(c) Draw the utility surface in 3D. Make sure that the variable A is on the horizontal axis.
(d) Draw the indifference curves for the two commodities. Let the scale of the legend be 0.5 util.
(e) Using a drawing tool, impose the budget line on the indifference curves.
Approximately what is the value of the optimal levels for A and B? Use an empty circle to mark the optimal point and a callout to indicate the optimal point along the indifference curve. What is the optimal level of utility?
(f) How do you know that the optimal point is really optimal? Give the proof or diagrammatic argument that it is really an optimal point.
(g) If the price of A doubles to \$4.00, what are the new optimal consumption levels of A and B? What is the new utility level? Use an additional line and an arrow to show your answer on the indifference map.
(h) Using the Excel Solver, verify your optimal solution derived from the graphs.

(Hints: (f) $(A^{\#}, B^{\#}, U^{\#}) = (4.2, 5.2, 2.2))$; (j) $(A^{\#}, B^{\#}, U^{\#}) = (7.2, 5.6, 3.4))$

12-13 Leisure and income (Toumanoff and Nourzad, 1994, 142–143) A person can spend her time working, earning money income (M) at a wage rate of $w = \$5/\text{hour}$. Any time not spent working is leisure (L), and her income is only from working. There are 24 hours in a day, so the time constraint is $T = 24$. Thus, if she works $T - L$ hours, her wage rate will be $M/(T - L) = 5$. Suppose she has a utility function $U = 10\,ML^2$. She maximizes her utility with respect to money income and leisure. All the charts and the axes of the charts must be labeled clearly.

(a) Set up a sensitivity table of her utility for $M = 0, 5, 10, \ldots, 70$ and $L = 0, 2, 4, \ldots, 24$.
(b) Draw her utility surface in 3D.

(c) Draw the labor-leisure indifference curves in 2D. Let the scale of the legend be 20,000 util and maximum utility be 320,000 util.
(d) Draw the time constraint in a separate chart.
(e) Using drawing tools, draw the time constraint on the indifference map, and find the optimal combination of income and leisure. Use a text box and an arrow to indicate the optimal point along the indifference curve.
(Hints: $L^* = 16, M^* = 40, U^* = 102,400$. Thus, the budget line should be tangent to the indifference curve at about the 100,000 level of utility.)
(f) Set up the maximization formally and verify the above result (item e) by using the Solver. Should the solution from the Solver and the optimal point in the diagram in e be the same?
(g) From the chart, estimate the maximum utility of the consumer.
(h) If the wage increases to \$6.00, will she work more, or less, or the same? How much will her income and the utility level be? Using the same indifference map derived above, clearly indicate your answers by an arrow. ($L^* = 16, M^* = 48, U^* = 122{,}880$.)
(i) Verify the above (item h) by using the Solver.

12-14 Nash equilibrium of duopolistic firms (a three-part question) Two duopolistic firms are competing in a market, producing the same product and facing the same demand function:

$$p = 100 - 0.5(x + y),$$

where x is the amount of product produced by Firm A, and y by Firm B. The duopolistic relationship is illustrated as follows.

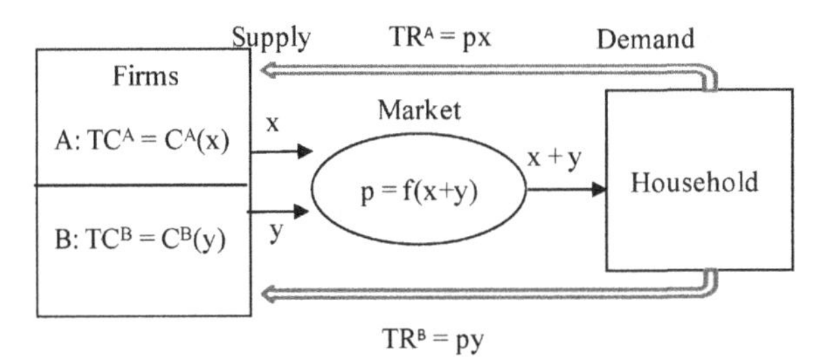

It shows that the two firms, which have different cost functions, produce the same output and charge the same price to consumers.

Part I. Firm A

The total revenue for Firm A is

$$TR^A = px = (100 - 0.5(x + y))x.$$

The cost for Firm A to produce x is linear, $TC^A = 5x$. Thus firm A's profit function is

$$\pi^A = TR^A - TC^A = (95 - 0.5y)x - 0.5x^2 \tag{1}$$

Table HW12-14a Sensitivity table for duopoly firm A

	A	B	C	D	E	F	G	H
1			Y ⟶				y=(95-x)/0.5	yy=50-0.25x
2			5	10	...	100	y (rfA)	yy (rfB)
3	X	20	1650					
4	↓	25						
5		30						
19		100						

Firm A then maximizes its profits subject to Firm B's given output y. The necessary condition of Firm A's profit maximization given Firm B's output y is given by[15]

$$x = 95 - 0.5y, \tag{2}$$

which depends on Firm B's output, and is called the **reaction function** of Firm A (rfA).

(a) Draw the isoprofit curves for Firm A in 3D and 2D, using the sensitivity table, Table HW12-14a. Note that the ranges of x and y are different. The 2D chart title should be "Fig. HW12-14a Isoprofit curves for Firm A".
The isoprofit curves should have an interval of 200, and the minimum profit levels (the value of the z-axis in the chart) should be nonnegative. Let Firm A's output x be on the horizontal axis (you may have to switch the row/column). Label the axes as output x and output y.

(b) In Table HW12-14a, enter the reaction function (2) of Firm A on the right-hand side of Table HW12-14a as a function of x. Draw the line in a separate chart. Label the axes, and give the chart title.

(c) Impose the reaction function on A's isoprofit chart by first marking the intercepts with white circles in A's isoprofit chart.

(d) What is the relation between the reaction function and the isoprofit curves for Firm A?

Part II. Firm B

Now, duopolist B has a total revenue function

$$TR^B = py = (100 - 0.5(x + y))y$$

and a nonlinear cost function $TC^B = 0.5y^2$. Thus, Firm B's profit function is

$$\pi^B = TR^B - TC^B = (100 - 0.5x)y - y^2. \tag{3}$$

The necessary condition of Firms B's profit maximization given Firm A's output is given by

$$y = 50 - 0.25x, \tag{4}$$

[15]The reaction function is derived by taking the first-order derivative of the profit function (1). Note that, for (1), the second-order condition of the profit maximization is also satisfied.

which depends on Firm A's output, and is called Firm B's **reaction function**.

(e) Using the same format as Table HW12-14a, set up the sensitivity table for Firm B, call it Table HW12.14b, and then draw the isoprofit curve (3) of Firm B. Call it Fig. HW12-14b. The isoprofit curves should have an interval of 200, and the minimum profit levels (the value of the z-axis of the chart) should be nonnegative. Let Firm A's output x be on the horizontal axis (you may have to switch the row/column). Label the axes as output x and output y.

(f) In Table HW12-14a, enter the reaction function (4) at the right hand side of the table, as shown (since the independent variable x is entered as a column), and draw (4) in a separate chart. Label the axes and give the chart its title.

(g) Impose the reaction function on B's isoprofit chart by first marking the intercepts with white dots in Fig. HW12-14b.

(h) What is the relation between the reaction function and the isoprofit curves for Firm B?

Part III. Market

(i) The market equilibrium is obtained by the intersection of the two reaction functions. The intersection can be found by solving Eqs. (2) and (4). Using the graphic method, draw the two reaction functions in a chart, and called it Fig. HW12-14c. Using the matrix method in Excel (not the algebraic method) find the market equilibrium output for A and B.

(j) Denote the pair of outputs $(x^{\#}, y^{\#})$ at this intersection as N. It is called the **Nash equilibrium**. Mark point N in Figs. HW12-14a and HW12-14b. Do Firms A and B have the same level of profit? What can you say about the economic property of N? In what sense is it an equilibrium point? Find the Nash equilibrium point N in Table HW12-14a and HW12-14b, and find the isoprofit curve connecting N in each table.

(k) Draw the 3D diagrams of the isoprofit surface for Firms A and B. The x-axis should be on your left and the y-axis should be on your right, and the hollow area should face away from you. Place the diagram on a new blank sheet, and place them side by side.

(Hints: You have Tables HW12-14a and b, Fig, HW12-14a, b, and c in this question, (i) $(x^{\#}, y^{\#}) = (80, 30)$; (j) No. $(\pi^{A\#}, \pi^{B\#}) = (3200, 900)$)

12-15 Duopoly II In HW12-14, denote the firm A's profit function[16] as

$$\pi^A = 72x - 4x^2 - 4xy.$$

Then Firm A's reaction function (RFA) can be shown as

$$x = 9 - (1/2)y.$$

[16]This profit function is derived by assuming that the output price is $p = 80 - 4(x + y)$, and Firm A has a total revenue function $TR^A = px = (80 - 4(x + y))x$, and cost function $TC^A = 8x$.

If there is another firm, Firm B, which has a profit function[17]

$$\pi^B = 72y - 4y^2 - 4xy,$$

then Firm B's reaction function (RFB) can be shown as

$$y = 9 - (1/2)x.$$

Following the same questions in (a)–(k) in HW12-14, draw the profit indifference curves for A and B and find the market equilibrium outputs for both firms.
(Hints: $x^{\#} = 6$ and $y^{\#} = 6, \pi^{A\#} = 144$ and $\pi^{B\#} = 144$.)

[17]This profit function is derived by assuming the output price is $p = 80 - 4(x + y)$, and Firm A has a total revenue function, $TR^B = py = (80 - 4(x + y))y$ for firm B, and cost function $TC^B = 8y$.

Part V

Research Methods and Presentation

Chapter 13

Research Methods — Excel Data Analysis

Chapter Outline

Objectives of this Chapter

In this chapter, we take a leave from economic theory and statistical analysis and work on several very useful data mining tools for handling large data sets. They include sorting, filtering, subtotaling a data set, and creating a PivotTable for data. There is no economic or business theory behind them. Rather, we should know the economic theory and statistical analysis, and probably a lot of common sense, to use these research tools. Unlike other textbooks which use data in CD, we first use the IF functions to generate the data, and then apply the data mining tools to extract information. In this sense, this chapter is self-contained: we do not need to go outside to find the data. However, external data are very important in statistical analysis. In the next chapter, we will discuss how to find and download some national and international data from the government and private sources.

13.1 Some Basic Definitions

13.1.1 *List and Database*

In Excel 2003, a **list** is a data set in a worksheet, and a **database** is a data set in an external file. Thus, what we have called a table in the previous chapters is in fact a list. To add more confusion, in Excel 2007, Excel adds a new feature or mode called "Table". We generally create a table as we have done before, and then designate the table as the "Excel Table" to take advantages of various Excel Table features. In this chapter, we refer to tables as either tables or lists, and when the Excel Table mode is activated, we call them Excel Tables.

Like a list or database, a table should not have a blank row or column as Excel determines the range of the table automatically. A table or list consists of **fields**, which are always in columns, and **records**, which are always in rows. A record is a collection of fields. The relationship between fields and records is shown in the following scheme.

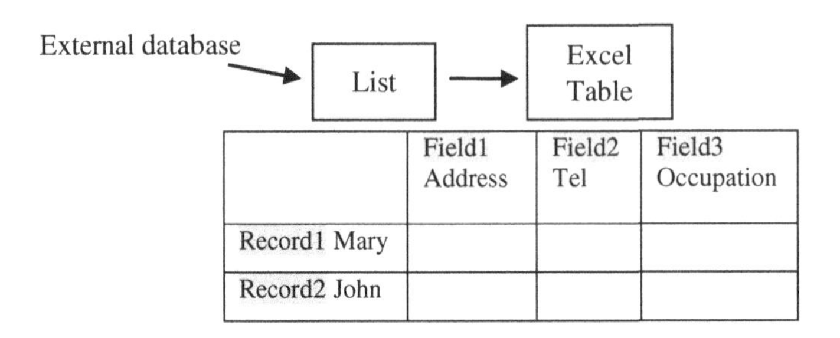

In manipulating data, if the list or table is **free standing**, that is, each of four sides of list or table has a blank row and column separating the list from other entries of the worksheet, or a side is adjacent to row or column header bar, then Excel will detect the range of the list or table automatically. Otherwise, the range of the list or table must be selected or named before any data operation. In this lecture note, we assume that the tables are free standing.

The method of extracting useful information from data is based on statistical analysis. The method of organizing data for statistical analysis is called **data organization**, **data analysis**, or **data mining**. As we will see in this chapter, there are four basic tools in Excel for data mining: sorting, filtering, subtotaling, and using the pivot table.

13.1.2 *Labels and values*

It is useful to give some definitions of different types of data in a worksheet.

A **label** is an ordinary text of a cell entry, which begins with an alphabet, a label indicator (′, ˆ, ″), a space, or symbols other than @, \$, +, −, (), #, and /. The row and column headers in the above table are labels; so is "Sum" or "Total", which are used to identify a value or a series of values. **Text** is a type of label. The label entry in a cell can contain about 32,000 characters.

A **string** is a series of labels. If a string is included in a formula, it must be enclosed in quotation marks ("). For example, "You did a good job" is a string.

A **value** is a cell entry, which begins with a digit or with symbols @, \$, +, −, #, and /. A value may be a result of computation from a formula. **Numbers** are a type of value, since they do not change unless new numbers are entered.

A **formula** is an entry of numbers or cell addresses using algebraic operations or mathematical equations or formulas. For example, the formula for calculating profit =TR−TC is=100−80 or =A3−B3, and the net profit after 8% tax is =(A3−B3)−(A3−B3)*0.08. A **function** is a built-in mathematical formula like=AVERAGE(...), =LINEST(...), =FV(...), etc. You may not know the **mathematical equations** for these functions, but you can calculate them by using Excel functions.

13.2 The IF Function

The IF function is one of the most useful functions of spreadsheets. It tests the contents of a cell or cells and makes decisions whether the contents are true or false. The syntax of the IF function is given as

$$\boxed{\text{=IF(logical\_test,value\_if\_true,value\_if\_false),}} \tag{13.1}$$

or we can simply write IF(test, true, false), which implies that if the test statement is true, then the formula returns the contents of "true"; otherwise, the contents of "false" will be returned. For more explanation, press <F1> for HELP and find the explanation of the "IF" function.

13.2.1 *Examples of the IF function*

The following example gives a basic idea of how to use the IF function. Suppose an instructor has the results of two tests in a class of 10 students. There are several ways to calculate the grade average, as explained below.

The basic data are in Test1 and Test2 columns of Table 13.1. They are random integers from 0 to 100 as entered in A3:B12. Enter the following formulas in A3 and B3:

A3: =RANDBETWEEN(0, 100)
B3: =RANDBETWEEN(0, 100).

Table 13.1 Applications of the IF function

	A	B	C	D	E	F	G	H	I
1	Examples of the IF function								
2	Test1	Test2	Compare	T1	T2	Or	And	And/formu	Comments on col.G
3	62	17	work harder	3	F	pass	fail	no 'sum'	'Avg not calculated'
4	63	7	work harder	3	F	pass	fail	no 'sum'	'Avg not calculated'
5	64	78	improved	2	C	pass	pass	142	71
6	66	24	work harder	3	F	pass	fail	no 'sum'	'Avg not calculated'
7	69	54	work harder	3	F	pass	fail	no 'sum'	'Avg not calculated'
8	33	41	improved	3	F	fail	fail	no 'sum'	'Avg not calculated'
9	2	60	improved	3	F	fail	fail	no 'sum'	'Avg not calculated'
10	19	83	improved	1	B	pass	fail	no 'sum'	'Avg not calculated'
11	71	96	improved	1	A	pass	pass	167	83.5
12	11	28	improved	3	F	fail	fail	no 'sum'	'Avg not calculated'

The formula=RANDBETWEEN(a, b) gives integer random numbers from a and b. Name range A3:A12 Test1 and B3:B12 Test2. Formulas for other columns in Table 13.1 are explained below.

13.2.2 *Nested IF functions*

Column C. If Test2 is **no less than** Test1, enter the comments "improved". Otherwise, enter "work harder". This is a string entry:

C3: =IF(Test2>=Test1,"improved","work harder").

Column D. **A simple imbedded function**: If Test2 is **greater than or equal to** 80, enter 1 (for level 1); if greater than or equal to 60 but less than 80, enter 2 (level 2); otherwise, enter 3 (level 3). This is a value entry.

D3: =IF(Test2>=80,1,IF(Test2>=60,2,3)).

Note that, since the first IF function already takes care of Test2 greater than 80, the range of the second IF function is 60–80, not 60–100. The latter IF is subject to all previous IFs.

Column E. **A multiple imbedded function**: If Test2 is greater than or equal to 90, enter A; if greater than or equal to 80, enter B; if greater than or equal to 70, enter C; if greater than or equal to 60, enter D; otherwise, enter F. This is a **multiple nested formula**. Note that no space is allowed in the entries, and that the number of left-side parentheses must be the same as the number of right-side parentheses.

E3: =IF(Test2>=90,"A",IF(Test2>=80,"B",IF(Test2>=70,
"C",IF(Test2>=60,"D","F")))).

13.2.3 *Logical IF functions*

Column F. **Use of OR:** If Test1 is greater than or equal to 60, or Test2 is greater than or equal to 60, enter "pass;" otherwise enter "fail". In this case, a student only needs to pass one of the two tests to receive the semester grade. Here, the OR formula is the statement =OR(x, y) = {x or y}, as shown by the shaded part of the two circles in Fig. 13.1. Enter

F3: =IF(OR(Test1>=60,Test2>=60),"pass","fail").

Column G. **Use of AND:** If Test1 is greater than or equal to 60, **and** Test2 is greater than or equal to 60, enter "pass;" otherwise enter "fail". In this case, a student must pass both

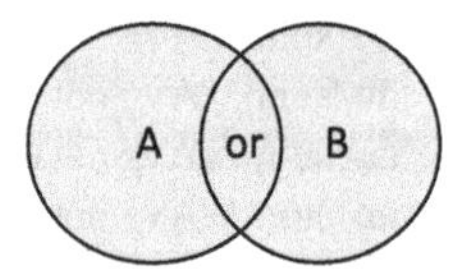

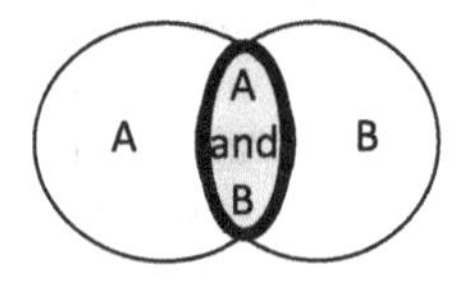

Fig. 13.1 Illustration of "and" and "or"

tests to receive a semester grade. Here the formula is =AND(x, y) = {x and y}, as shown by the two overlapping circles in Table 13.1. Enter

G3: =IF(AND(Test1>=60, Test2>=60), "pass", "fail").

Column H. More generally, the return can be **a function**: If Test1 is greater than 60, **and** Test2 is also greater than 60, then return the sum of Test1 and Test2, rounded to integer. The sum is then used as the semester total; otherwise enter the string "no sum", and no semester grade is given. Enter

H3: =IF(AND(Test1>60, Test2>60),round(Test1+Test2, 0), "no sum").

Note that both tests must be greater than 60.

Column I. The return can be the value of **a formula**, as in Column H: If cell G3 contains "pass", return the average of Tests 1 and 2, rounded to one decimal place; otherwise, enter the string 'Avg not calculated' with quotation marks. Note that if double quotation marks are used, as in ""The sum is ..."", the error window will appear. Similarly, the formula round((test1+test2)/2,1) or round(average(A4,B4),1) can be used, but not round(average(test1,test2),1), which gives the average of the sum of the column of Test1 and the sum of the column of Test2. Enter in

I3: =IF(G3="pass",round((A3+B3)/2,1), "'Avg not calculated'")

After filling in row 3 of Table 13.1, copy A3:I3 to A4:I12. Enter <F9> several times to see how the IF functions work.

A note on drawing Fig. 13.1

LHS: Circles A and B are drawn by using <"Insert"><Illustrations, Shapes!>, selecting "Oval", and holding the shift key to make an exact circle. The arc of one circle, say A, will be covered by the other circle, B. Select A and copy A to a blank space, calling it AA. Make A "no fill" by <"Drawing Tools, Format"><Shape Styles, Shape Fill><No Fill>; then AA is transparent. Move AA back to overlap A. Use a transparent textbox to fill in all the letters and words in the shapes.
RHS: The middle shaded part is drawn manually by the "Oval" shape.

13.2.4 *Generating a large set of data*

The IF function is useful in creating a large set of seemingly meaningful artificial data, as the following example shows. Suppose a company has 100 salespersons in four regions: North, East, West, and South (NEWS). The sales record for each odd month from 2006 to 2008 is given. We would like to construct a random number assignment for years, odd months, and regions. The final table is shown in Table 13.3. We take the following steps to generate the data.

Step 1. In Table 13.3, enter the column labels in row 2 and the formulas in Table 13.2 in row 3. (See Step 3 in Section 13.3.1 below.)

Step 2. Complete the table by copying row 3 to A3:L102.

Step 3. Copy the whole table to Sheet2 (to preserve the original table).

Step 4. Rename Sheet1 "original" and Sheet2 "copy". In the original sheet, change Table 13.3 to values. Since the data table is large, the best way is to select the whole worksheet and change the whole sheet to values.

Step 5. Delete the random columns, that is, columns B, C, and D in Table 13.3. The new table should look like Table 13.4. Note that the table is **free standing**, in the sense that no extraneous numbers or labels are adjacent to the table. Hence, Excel will recognize the range of the table automatically.

Table 13.2 The formula table

Col.Title	Cell	Formula	Explanation
Year	B3	=RANDBETWEEN(1,3)	Three years 2006-2008
Month	C3	=RANDBETWEEN(1,6)	Odd monthsJan, Mar, May, Jul, Sep, Nov
Region	D3	=RANDBETWEEN(1,4)	Four regionsN,E,W,S
Year	E3	=IF(B3=1,2006,IF(B3=2,2007,2008))	Assignment of year
Month	F3	=IF(C3=1,"Jan",IF(C3=2,"Mar",IF(C3=3,"May",IF(C3=4,"Jul",IF(C3=5,"Sep","Nov")))))	
			Assignment of month
Region	G3	=IF(D3=1,"North",IF(D3=2,"South",IF(D3=3,"East","West")))	
			Assignment of region
TC	H3	=ROUND(RAND()*(80-40)+40,2)	TC, 40 to 80
TR	I3	=ROUND(RAND()*(100-50)+50,2)	TR, 50 to 100
Profits	J3	=TR-TC	Profits
YTD-S	K3	=RANDBETWEEN(0,10)	Year to day sick leave, 0 to 10 days
YTD-V	L3	=RANDBETWEEN(0,30)	Year to day vacation leave, 0 to 30 days

Table 13.3 Construction of random database

	A	B	C	D	E	F	G	H	I	J	K	L
1	**Generating the Sales Record of the GiGo Company**											
2	Name	Year	Month	Region	Year	Month	Region	TC	TR	Profits	YTD-S	YTD-V
3	1	3	2	4	2008	Mar	South	42.96	88.26	45.30	1	28
...		2	5	1	2007	Sep	North	57.85	58.32	0.469	3	2
102	100	1	5	4	2006	Sep	South	57.48	79.48	22.01	4	4

Table 13.4 The database of the sales record

	A	B	C	D	E	F	G	H	I
1	Name	Year	Month	Region	TC	TR	Profits	YTD-S	YTD-V
2	1	2007	Sep	East	65.98	74.27	8.29	6	19
...	...	2008	Nov	North	55.85	61.03	5.18	2	0
101	100	2007	Mar	South	45.56	50.82	5.26	3	4

13.3 Excel Tables

13.3.1 *Constructing an Excel Table*

Instead of building Table 13.3 manually, we may also build it by using the **Excel Table** command. We start from a new sheet, Sheet2. Since the table should be free standing, we first delete the title of Table 13.3 in A1, and then proceed with the following steps.

Step 1. The frame of the table. The frame of the table should be entered first. Enter the Name of salesperson from 1 to 100 in A3:A102 as before, and enter column labels A2:L2 as shown in row 2 of Table 13.3. This will give the range of the list A2:L102.

Step 2. Invoke the Excel Table. Click anywhere in the table and enter

<“Insert”><Tables, Table>(or, simply clicking<^T>)<OK>. (13.2)

The “Create Table” dialog box, Fig. 13.2, appears.

Enter the list range A2:L102 and click “My table has headers”, and <OK>. To encourage the use of the left hand, we recommend using the <^T> command.

The range of A2:L102 of Table 13.3 will be colored with branded rows (with colored strips). The down arrows, like those in row 10 of Table 13.5 appear on the right-hand side of the column labels. Since columns B, C, D and E, F, G have the same column labels, the Table command automatically changes the entries of E2, F2, and G2 to Year2, Month3, and Region4. If any cell in the table is clicked, click the new tab

<“Table Tools, Design”>.

Step 3. Enter the formulas listed in Table 13.2 into row 3 of Table 13.5. For example, in B3, enter =RANDBETWEEN(1, 3). When the <Return> key is pressed, the formula in B3 will be automatically copied down to B4:B102. Since the cell formulas for cells C3 and D3 are similar, you only need to copy B3 to C3:D3. Then, in C3, change (1,3) in the formula to (1,6). When the <Return> key is pressed, the corrected formula in C3 will be copied automatically to C4:C102. Similarly, in D3, we have cell formula (1,4). When the <Return> key is pressed, the corrected formula in D3 will be copied automatically to D4:D102.

Similarly, enter the cell formula in E3, and when the <Return> key is pressed, E3 will be copied automatically to the whole column of the Table. Since F3 and G3 have similar

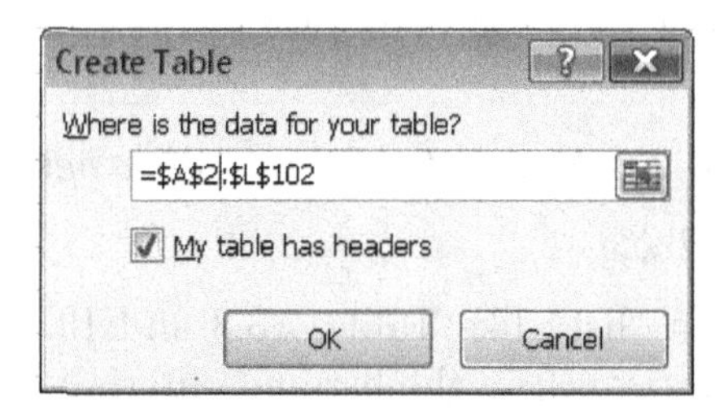

Fig. 13.2 Create Table dialog box

Table 13.5 Excel Table format.

	A	B	C	D	E	F	G	H	I
1	Nan	Yea	Mor	Regi	TC	TR	Profi	YTD-S	YTD-V
2	1	2006	July	North	56.33	56.12	-0.21	8	14
3	2	2007	Sep	East	41.87	53.97	12.10	5	3
100	99	2006	Sep	South	43.82	50.3	6.48	9	1
101	100	2008	Nov	South	63.68	66.27	2.59	9	20
102		100	100	100	41.36	99.64	13.34	517	74.89
103		Count	Count	Count	Min	Max	Avg	Sum	Var

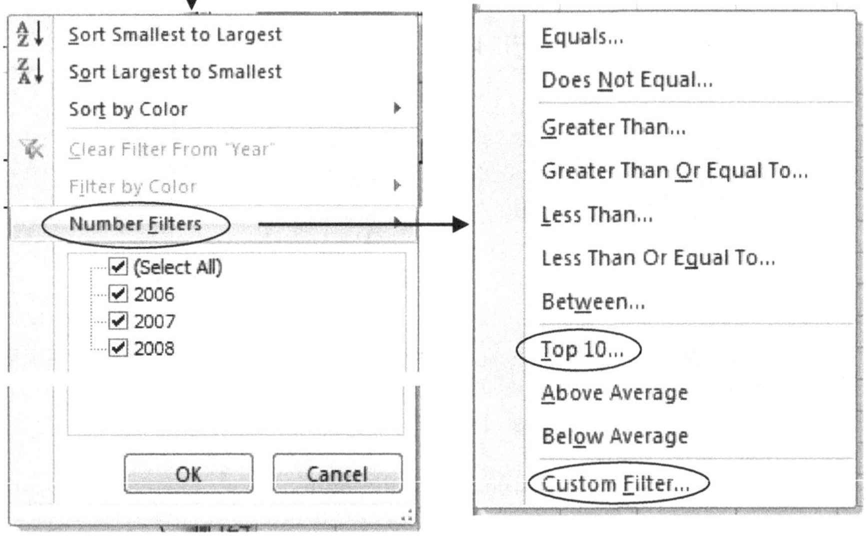

Fig. 13.3 <"Data", Sort & Filter><Filter><Year!>

Fig. 13.4 <Number filters>

IF functions, copy E3 to F3:G3 and enter <F2> to edit the formula, etc. Continue the process up to L3. This completes the construction of the Excel Table.

Steps 4, 5, and 6 These steps are similar to Steps 3, 4, and 5 of Sec. 13.2.4. After deleting columns B, C, and D of the random numbers in Table 13.3 and changing Year2, Month3, and Region4 back to Year, Month, and Region, we obtain an Excel Table, as shown in rows 1 to 101 of Table 13.5, that is similar to Table 13.4, except that the table has branded rows and down arrows in row 1. For the time being, we ignore rows 102 and 103 in Table 13.5.

13.3.2 *Changing a list to an Excel Table and vice versa*

Changing a list to an Excel Table

If we construct the table as a "list" like Tables 13.3 and 13.4, it still can be converted to an "Excel Table" by clicking any cell in the list and entering <^T>. The branded rows and down arrows appear as in Table 13.5.

De-activating the Excel Table mode

To cancel the Excel Table mode, and convert an Excel Table to a simple range, that is, a list, click anywhere inside the Excel Table, and select

<"Table Tools, Design"><Tools><Convert to Range>. (13.3)

Click <Yes> to the dialog box "Do you want to convert the table to a normal range?" The branded rows stay the same but the down arrows disappear. All the Excel Table commands are disabled, and clicking the table will not show the Table tab of (13.3) in the menu bar. (Question: How do we get rid of the color of the table?)

13.3.3 *Copying and moving within the Excel Table*

In terms of row and column selection, the Excel Table becomes a mini spreadsheet itself: When the pointer moves to the leftmost side of the row (the first cell of the row), say A3 in Table 13.5, or to the header row of each column, say C1 in Table 13.5, or to the first cell in the upper northwest corner of the Table, say A1 in Table 13.5, the pointer changes to a horizontal arrow, down arrow, or a slanted arrow, respectively.[1] (You may also enter ˆA to select all of the Table.) When the arrow is pressed, the row or the column within the table, or whole table, will be selected, and can be copied or moved. This is convenient way to select rows or columns within Excel Table if the table is very large.

13.3.4 *Hiding rows*

The rows in a list or an Excel Table can be hidden by selecting

<"Home"><Cells, Format!><Visibility, Hide & Unhide!><Hide Rows>. (13.4)

The method of hiding rows will be given in Filtering in Sec. 13.5.

13.3.5 *Methods of data analysis*

Using Tables 13.4 and 13.5, in either the list mode or the Excel Table mode, we introduce the following three methods of "data mining". These will be explained in the following sections.

Method Basic commands

a. Sorting: by month, by year, by region, in the ascending order of profits. The list commands are

<"Data"><Sort & Filter, Sort!>.

[1]When the pointer moves to, say, cell B1 (or A3) in Table 13.5, an arrow appears above "Year" (or, to the left of "2"), and the "Year" column (or row 3) within the table will be selected. If the pointer moves farther up (or left) to column heading B (or row heading 3), another arrow appears, and the whole column B (or row 3) of the worksheet will be selected. Please note the differences. Move the table from the column (or row) heading bar and try again.

Without clicking <Sort>, you may also click the AZ↓ button for sorting from A to Z, or the ZA↓ button in the same group for sorting from Z to A.

b. Filtering: by month, by year, by region, in the ascending order of profits. The list commands are

<"Data"><Sort & Filter><Filter>.

c. PivotTable: by year, month, and region simultaneously. The list commands are

<"Insert"><Tables, PivotTable!>.

These commands are based on the list. If the Excel Table format is used, they are imbedded in the Table commands, as we will see in the next sections. Note that if the result is not what you have expected, you can always use the "undo" button to undo the operation.

In the following data analysis, we will take advantage of the new Excel Table format. In case the table is not formatted as Excel Table, like Table 13.4, it can be formatted as Excel Table easily by pressing <^T>.

13.4 Sorting

Sorting is a rearrangement of the order of rows or columns in either an ascending or a descending order of values or labels. If the list is not detached and the sorting is for the entire list, you need to select the table range first.

13.4.1 *Single column sorting*

Sorting by rows

If you click the down key next to Year (or Month, or Region) in Table 13.5, a menu window like Fig. 13.3 appears. If the contents of the column are numeric, the first two commands will show <Sort Smallest to Largest> and <Sort Largest to Smallest>. However, if the column contents are alphabetic, then the first two rows will show <Sort A to Z> and <Sort Z to A>. In the current case, after you click <Sort smallest to Largest>, the whole list will be sorted by year: 2006, 2007, and 2008. A slim upward arrow will be added next to the down arrow in B1 of Table 13.5, indicating the currently active column.

Example 13.1 Sort the whole table by Month, and then by Region. Sort them by clicking <Sort Z to A>. Note that the current sorting will override the previous sorting. □

Sorting by custom list

In Example 13.1, "month" will be sorted alphabetically (Jul comes before Mar), not in the order of months. In this case, we may **sort with a custom list**. Moving the pointer to anywhere in the Month column, click

<"Data"><Sort & Filter, Sort!>.

A Sort dialog box Fig. 13.6 comes up. Click the down arrow <Order, A to Z!> and select <Custom List...>, and a Custom Lists dialog box appears. Select the sequence of <Jan, Feb,...> from the Custom Lists dialog box,[2] and then click <OK> to go back to the Sort dialog box. Click <OK> again, and the months will be sorted in the correct order.[3]

13.4.2 *Multicolumn sorting*

You may sort the data by as many levels as you need by clicking the "Add Level" button on the upper left corner of the first row of Fig. 13.6. In general, however, only three levels are used.

Example 13.2 We want to sort Table 13.5 first by "Year", then by "Month", and then by "Region". This can be done by clicking <"Data"><Sort & Filter, Sort!><"Sort, Add Level

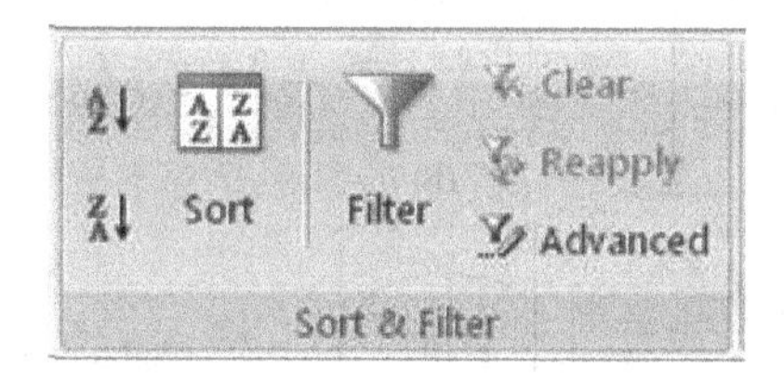

Fig. 13.5 <"Data"><Sort & Filter>

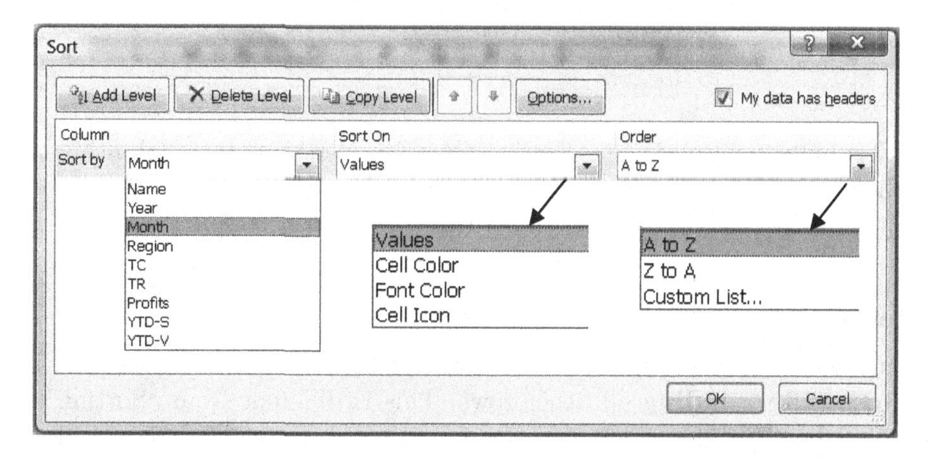

Fig. 13.6 <"Data"><Sort & Filter><Sort>

[2]If the custom list does not include abbreviated names of months, you have to create one. See Appendix 2B of Chapter 2 for creating items in the custom list.

[3]Another method is clicking the down arrow of the month column (C1) in Table 13.5, and clicking <Sort by Color><Custom Sort><Sort, Order!><Custom Lists>. Select the desired month sequence and press <OK><OK>.

Table 13.6 <"Data", Sort & Filter><Filter>

B111 | f_x =SUBTOTAL(103,B11:B110) (c)

	A	B	C	D	E	F	G	H	I
1		All data			Filtered data		(Jan 2006, South and West)		
2			Profit		Profit	code #		% of total	
3		Count()	100		5	103		5.00	
4		Max()	54.5		45.2	106		82.85	
5		Min()	-28.3		15.4	105		-54.31	
6		Average()	13.3		31.1	101		232.72	
7		Var()	311.7		143.5	110		46.05	
8		Stdev()	17.7		12.0	107		67.86	
9									
10	Nam	Year	Mo	Regi	T	T	Pro	YTD	YTD
64	50	2006	Jan	South	55.28	70.66	15.38	4	30
70	91	2006	Jan	South	49.58	89.09	39.51	10	30
97	16	2006	Jan	West	45.03	90.21	45.18	0	28
100	32	2006	Jan	West	52.6	76.11	23.51	3	23
109	69	2006	Jan	West	45.3	76.98	31.68	9	14
111	Total	5	5	5	49.56	80.61	31.05	5.2	25.0
112	code	103	103	103	101	101	101	101	101
113	name	Count	Count	Count	Avg	Avg	Avg	Avg	Avg
114	Formula for the profit column				=SUBTOTAL(101,G11:G110)				

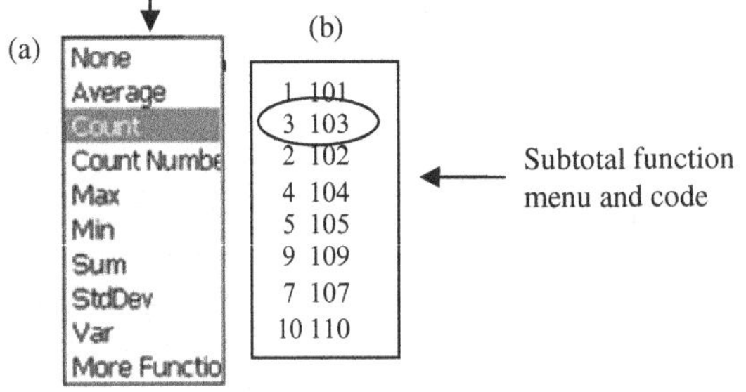

(a)

None
Average
Count
Count Numbe
Max
Min
Sum
StdDev
Var
More Functio

(b)

1	101
3	103
2	102
4	104
5	105
9	109
7	107
10	110

Subtotal function menu and code

Column", Sort by!> two times, and selecting "Year" under "Sort by", "Month" under "Then by", and then "Region" under the second "Then by", <OK> . The data will be sorted by year, and within the year, by month, and within the month, region. □

13.5 Filtering

As the name shows, filtering is similar to sorting, but it sorts and presents the rows by certain criteria (filter), **hiding** all other rows. This is different from "Sorting", as sorting does not hide other rows. Excel has two filtering method, simple and advanced. The simple method uses the header row of the table, the advanced method sets up a separate criterion and can filter the subset to a place outside the table.

13.5.1 *Simple filtering*

Suppose that we want to find how many South and West regions are included in the January 2006 data. We filter 2006 first, and then ask how many rows of January are included in the

2006 data. This will give all the data for January 2006. We then ask, within this restricted date, how many are in the South or West regions. What is the number of firms in the South or West in January 2006? What are the maximum, minimum, average, variance, and standard deviation of their profits? What are their statistics compared with those of the total firms? We also ask what the average TC, TR, Profit, YTD-S, and YTD-V are for these two regions in January 2006. The following steps may be taken.

a. **Level 1**: When the down arrow in the Year column (like B10 of Table 13.6) is clicked, a dialog box, Fig. 13.3, appears. Click <Select All> to deselect all the entries. (The default is <Select All>; it is a toggle key.) Then click 2006 and press <OK>. This will "filter" out the rows of year 2006 only, and other rows will be hidden. The down arrow will be reduced and a filter logo comes out next to the down arrow, as in B10 of Table 13.6, indicating that the Year column has been filtered. At the same time, the number of the filtered rows appears next to the "Ready" indicator in the status bar at the bottom of the worksheet, as in "34 of 100 records found".
b. **Level 2**: Next, click the down arrow of Month, and click <Select All> to deselect all entries and then select "Jan" <OK>. The rows of Jan among the year 2006 rows that will be filtered out, and a filter logo will be added to C10, indicating that the Month column has been filtered.
c. **Level 3**: Next, click the down arrow of Region, deselect all, and then select South and West, <OK>. The rows of South and West among the Jan 2006 rows will be filtered out, and a filter logo will appear in D10, indicating the Region column has been filtered. Rows 10 to 109 of Table 13.6 show the results. Only five rows (from different rows, see the row headings with blue numbers) out of 100 rows satisfy the above three levels of criteria. We will call this final table the **filtered table**.

Copying and moving a filtered Excel Table

As with any range, clicking anywhere in the Excel Table, you can use <^A> to select the filtered table, and <^C><^V> to copy only the filtered table to somewhere else.

13.5.2 *The total row*

Showing the total row

We may find the descriptive statistics of the five rows of the filtered table in Table 13.6. Some common descriptive statistics can be found in the Total Row at the end of the filtered table. To add the Total Row, click

<"Table Tools, Design"><Table Style Option, Total Row>. (13.5)

The Total Row is attached to the end of the table in row 111 of Table 13.6, with the word "Total" shown in A111 automatically.

Subtotal functions

If a cell in Total Row, say, B111, is clicked, a down arrow appears. Click the down arrow to reveal a menu box (a) in Table 13.6. The menu box shows some of the subtotal functions,[4] which have the format of

$$\boxed{\text{=SUBTOTAL(function\_num,ref1,ref2,...)},} \qquad (13.6)$$

where the "function_num" is the code for calculation type. The code is listed on the right-hand side (b) of the menu box in Table 13.6. In cell B111 of Table 13.6, we enter the subtotal function "Count", ("count" the rows of the filtered table), the code of which is 103 (circled), and 5 will be shown in B111, indicating that there are five rows in the filtered table. The formula of this subtotal function is shown in the formula bar (c) of Table 13.6. Note that the range of the data, namely, the number of rows, is automatically set to the whole range of the Table, B11:B110.

Copy B111 to B111:D111. The three cells will enter the number of the rows, 5, as expected. Similarly, click E111 and the down arrow, and selecting "Average", the average of TC for the five rows appears. The range of data is entered as E11:E110 automatically. Copying E111 to E111:I111, we have the averages of the five columns of the filtered table.

The subtotal function code

If the "function_num" in (13.6) is larger than 100 (see column F of Table 13.6), then the subtotal function will skip counting the contents of manually hidden rows. On the other hand, if the code is less than 100, the subtotal function will count the contents of every row, including the manually hidden ones, in the Excel Table. Note that the following codes are missing in the menu box (a): 6 and 106, for PRODUCT; 8 and 108, for STDEVP; 11 and 111, for VARP.

Manual entry of subtotal functions

Rows 1–8 of Table 13.6 show the difference between the total functions (in C3:C8) and the subtotal functions (E3:E8) for the profit column. They compare six descriptive statistics of total profit data (100 rows) in range C3:C8, and filtered profit data (5 rows) in E3:E8. Since the code numbers of subtotal functions are hard to remember, we have listed them in F3:F8 of Table 13.6 for reference, corresponding to the names of the functions in B3:B8.

---

[4]The name "subtotal" here is a generic name that includes not only the sum of numeric fields, but also Count, Average, Max, Min, Product, Count Names, Stdev, Stdevp, Var, and Varp. Thus, the "Subtotal" function is a convenient generic name for the descriptive statistics we studied in Chapter 5. Ref1, ref2, in the subtotal function (13.6) are ranges or references for subtotaling, and can range from ref1 to ref254.

We enter the functions manually for rows 3 to 8. To calculate the profit statistics for all 100 rows, enter the regular function (the **"total" function** in contrast to the "subtotal" function) in C3 of Table 13.6:

C3: =COUNT(G11:G110),

where the range G11:G110 of the profit column is entered automatically by typing = COUNT(, and then moving the pointer to the upper part of the column heading "profit", G10, and clicking the left mouse once when a down arrow appears.[5] Closing the right parenthesis, and pressing the return key, the number of total rows, 100, appears in C3. Similarly, we enter the other formulas in C4:C8.

To calculate the profit statistics for January 2006 in the South and West regions only, enter

E3: =SUBTOTAL(102,G11:G110), (13.7)

in E3 using the following steps: Type "=SUBTOTAL(", and a subtotal menu box similar to the submenu (a) of Table 13.6 appears. Scrolling down the submenu box and clicking the term "102 - Count" ("2 - Count" also works) twice, you will enter 102 of (13.7).[6] Enter the comma. To enter the range, move the pointer to the upper part of cell G10 until a black down arrow appears. If you click the down arrow once, the range G11:G110 (if you press <LM> twice, you are in error) will be selected and entered automatically in formula (13.7). Without entering the closing parenthesis, press the return key to complete the entry of (13.7). Similarly, we enter the other formulas in range E4:E8 of Table 13.6 for other subtotal functions. Columns C3:C8 and E3:E8 show clearly the difference between the "total" function (for the whole Excel Table) and subtotal function (for the filtered table).[7]

To compare the **subtotal** with the **total**, column H3:H8 calculates the ratio of filtered data over total data for each statistical function. In H3, enter = E3*100/C3, etc. H3 shows that only 5% of the total salespersons were in the South and West regions during January 2006, and their average profit is 232.72%, or 2.3 times larger than the average profits earned by the 100 salespersons. The standard deviation of these five salespersons is only 67% (in H8) of the total standard deviation, indicating that their profits dispersion is much smaller.

These manually calculated subtotal averages are the same as those calculated without using the Total Row. For example, the average profit calculated manually by the Subtotal

---

[5]If <LM> is clicked twice, the range will be selected for G10:G111, which includes column heading G10 and the Total Row G111.

[6]We have found that, in Excel 2007, the code for COUNT in the SUBTOTAL function is 102, but that in Total Row is 103.

[7]For the filtered rows (different from manually hidden rows), there is no difference between codes larger than 100 and codes less than 100. Thus, 102 in (13.7) can be 2 also.

function in E6 is the same as that obtained from the Total Row in G111; both are enclosed with borders and connected with an arrow to show the correspondence.

In general, if we are interested only in descriptive statistics, then Total Row may be the most efficient and useful in the sense that you do not need to enter or memorize a formula code and the range of the data.

Removing the filters

To go back to the original regional data of 2006, click <Select all> in Fig. 13.3 in the Region column. To go back to the original three-year data, click <Select all> in Fig. 13.3 in the Year column. Alternatively, if you are impatient, you can return to the 100 original rows by clicking the Filter logo under <"Data"> in Figs. 13.5 and 13.6. The Filter logo is a toggle key.

Number filters

Example 13.3 Top or Bottom 10. To find the top 10 profit makers, click the "Profit" down arrow in G1 of Table 13.5. Figure 13.3 appears. Choose "Number Filters", and a dialog box, Fig. 13.4, appears. Choose "top 10..." And the "Top 10 AutoFilter" dialog box appears. The down arrow of "Top" contains Top and Bottom. Note that 10 can be changed to any number, and the down arrow of "Items" has either "Items" or "percent". Enter "Top", "10", and "Items", and press <OK>. The top 10 items will be selected. □

Example 13.4 Top and Bottom 5. In YTD-V, we want to show the data of those who took less than 5 days and those who took more than 25 days of vacation within a three-year period. In this case, open <Custom Filter...> in Fig. 13.4, which is opened under the <Number Filters> dialog box of Fig. 13.4. When the Custom AutoFilter dialog box opens, click the downward arrow to select "is less than or equal" to "5" or "is greater than or equal" to "25" and press <OK>. Copy the whole list below the sorted list. □

13.5.3 *Advanced filtering*

The Advanced Filtering works from the list in a non-Excel Table mode. We would like to find out (a) if a person takes the YTD-V for over 15 days, and the YTD-S for over 5 days in a year, whether the excessive vacation days and sickness days will affect the profit earning of the person, and (b) whether there is a difference between the North and South regions.

Criteria region

The Advanced Filtering requires setting up a **criteria region**, as shown in A1:D3 in Table 13.7. The criteria region can be placed anywhere, except inside the table, on the worksheet. It should enter the **column headings** of the list (see the arrow). The second and the third rows are **criteria**. According to spreadsheet convention, if the criteria are placed in a row, then the criteria are related under **AND relations.** In this example, row

Table 13.7 Advanced filtering and the criteria region

	A	B	C	D	E	F	G	H	I
1	Region	YTD-V	YTD-S	Profit					
2	North	>=15	>=5	>=30	OR				
3	South	>=15	>=5	>=30					
4		AND							
5	Name	Year	Month	Region	TC	TR	Profit	YTD-S	YTD-V
91	98	2007	Nov	North	54.57	91.36	36.79	7	26
105	91	2006	Jan	South	49.58	89.09	39.51	10	30

2 states these criteria: North and YTD-V is ≥ 15 days and YTD-S is ≥ 5 days, and Profit is ≥ 30 for each heading. If the criteria are placed in a column, then the criteria are related under **OR relations**. In this example, we are asking to see North or South, in other words, row 2 or row 3.

After setting up the criteria region, we invoke the "Advanced Filtering" command. Clicking any cell inside the table, enter

<"Data"><Sort & Filter><Advanced>.

Then, an <Advanced Filter> dialog box (Fig. 13.7) appears. In the box, select "Filter the list, in-place" (that is, display in a filtered list, the default). Define the "List range", A5:I105. (This will be entered by Excel automatically if the Excel Table format is defined and the list is free standing. The $ sign will be added by Excel.) Define (if this is the first time) the "Criteria range" as A1:D3 (the $ sign will be added by Excel) as shown in Table 13.7, and click <OK>. The results are shown on the last three rows of Table 13.7. Make sure that the "AND" criteria are satisfied simultaneously in row 91 and row 105, and the "OR" criteria by two rows, one for North and one for South.

Fig. 13.7 Advanced filter dialog box

The results show that the effect of large numbers of days of sickness and vacation may not exert a drop in profit. For example, Salesperson #98 had 7 days of YTD sickness and 26 days of vacation, and yet he/she earned 36.79 units of profit, possibly due to the large TR over TC. May be this salesperson is more efficient than others in the Northern region. Furthermore, it appears that there is no difference between salespersons in the North and South. In fact, salesperson #91 in the Southern region appears to be more efficient, with higher numbers of sickness and vacations days, but also has a higher profit than #98.

Clearing an Advanced Filter

To clear the Advanced Filter and return to the original table, simply click

<"Data"><Sort & Filter><Clear>.

Computed Criteria

A more powerful filtering is called "**computed criteria**". In this case, we filter the profit rate, namely, the ratio of profit over TR, being greater than 50%, as shown by the formula in C2 of Table 13.8. Enter this formula in A2, and give a name in A1 (or you may leave it blank). For computed criteria, do not use the column headings. The formula in cell C2 contains cell addresses G4 and F4. They refer to the first row of the data table and no $ sign is added, so that Excel can go to G5 and F5, G6 to F6, and so on. Any cell address like G4 and F4 in the formula should be a relative reference (and outside the list should be an absolute reference) so that Excel can calculate the same formula (in C2) by moving downward. Because of the inequality, A2 will show either true or false.

In the Advanced Filter dialog box, Fig. 13.7, we choose the "Criteria range" as A1:A2 and <OK>. The filtering result shows that there are three salespersons whose profit ratio is larger than 50%. To verify this manually, column K calculate the profit ratios (note that column J must be blank to keep the table free standing). It shows that three persons indeed had profit rate greater than 50%.

Table 13.8 Computed criteria

G104 ▾ fx =SUBTOTAL(101,G4:G103)

	A	B	C	D	E	F	G	H	I	J	K
1	Profit/TR										Profit
2	FALSE		=G4*100/F4>50								rate
3	Name	Year	Month	Region	TC	TR	Profit	YTD-S	YTD-V		
6	22	2007	May	South	42.03	91.87	49.84	6	1		54.3
10	42	2006	July	South	44.6	99.13	54.53	4	3		55.0
95	16	2006	Jan	West	45.03	90.21	45.18	0	28		50.1
104	Total	3	3	3	43.9	93.7	49.85	3.3	10.7		
105		count	count	count	avg	avg	avg	avg	avg		

13.6 Subtotaling a list

Subtotaling a list means calculating subtotals at every change of the value of a sequence. For example, we calculate the number of salespersons and the average profit at the change of years from 2006 to 2007, 2007 to 2008, and at the end of 2008. Since the change of the value is crucial, the list must be sorted by a field first, say, by year, then the subtotal can be calculated for each change in the field. Subtotaling does not work with the Excel Table format. Thus, starting from Table 13.5, change the Excel Table to a list by converting to range by using (13.3).

Examples

To explain the working of subtotaling, we use the example that we want to find out how many sales persons are included in each of the three years, from 2006 to 2008, and the average of their TC, TR, Profit, YTD-S, and YTD-V for each year.

13.6.1 *Adding a subtotal*

We start from Table 13.3 or 13.4, where the Excel Table mode is removed — if not, use (13.3) to convert to range. Clicking any cell in the table, enter

<"Data"><Outline, Subtotal>.

The subtotal dialog box Fig. 13.8 appears. In "At each change in:" select Year, in "Use function:" select Count, and in "Add subtotal to" click YTD-V, to place the number in the last column. Accept the default "Replace current subtotal" and "Summary below data". Entering <OK>, we have a table similar to Table 13.9.

There are 100 rows in Table 13.9. To save space, we have hidden some rows and show only the essential part of the table. The table shows 34 entries in 2006 (circled in row 38),

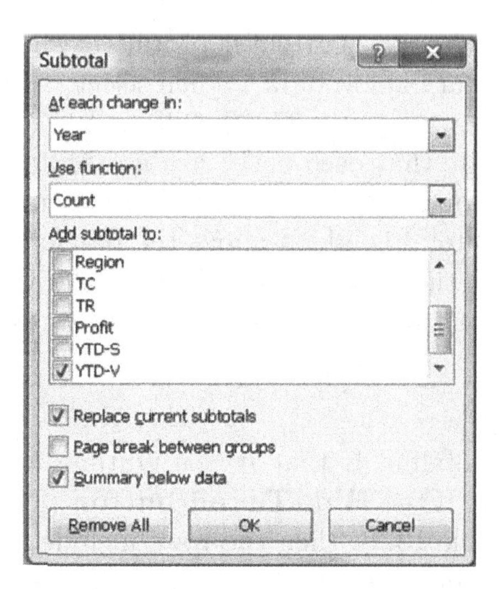

Fig. 13.8 Subtotal dialog box

Table 13.9 Adding two subtotals for each year

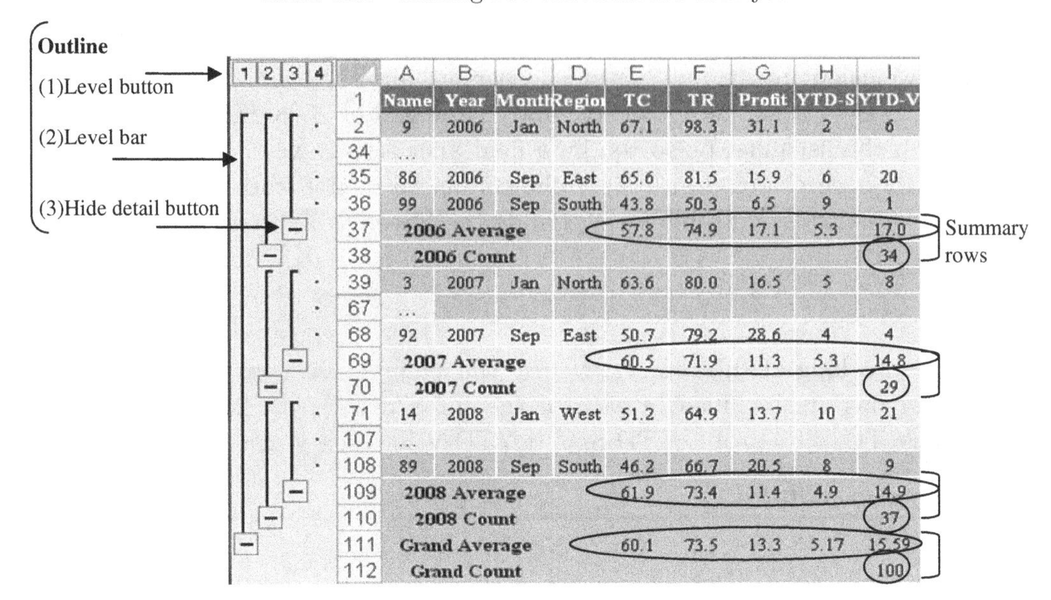

	A	B	C	D	E	F	G	H	I
1	Name	Year	Month	Region	TC	TR	Profit	YTD-S	YTD-V
2	9	2006	Jan	North	67.1	98.3	31.1	2	6
34	...								
35	86	2006	Sep	East	65.6	81.5	15.9	6	20
36	99	2006	Sep	South	43.8	50.3	6.5	9	1
37	2006 Average				57.8	74.9	17.1	5.3	17.0
38	2006 Count								34
39	3	2007	Jan	North	63.6	80.0	16.5	5	8
67	...								
68	92	2007	Sep	East	50.7	79.2	28.6	4	4
69	2007 Average				60.5	71.9	11.3	5.3	14.8
70	2007 Count								29
71	14	2008	Jan	West	51.2	64.9	13.7	10	21
107	...								
108	89	2008	Sep	South	46.2	66.7	20.5	8	9
109	2008 Average				61.9	73.4	11.4	4.9	14.9
110	2008 Count								37
111	Grand Average				60.1	73.5	13.3	5.17	15.59
112	Grand Count								100

29 entries in 2007 (row 70), and 37 in 2008 (row 110). See the circled numbers. The total number (Grand Count) is 100, as shown in row 112.

Adding the second subtotal

To find the averages of TC, TR, Profit, YTD-S, and YTD-V for each year, invoke the Subtotal dialog box Fig. 13.8 again. In "At each change in:" still select Year; in "Use function:" click the down arrow and select Average; in "Add subtotal to:" click the down arrow and select TC, TR, Profit, YTD-S, and YTD-V to place the average under each change of the year. Deselect "Replace current subtotal" to allow averages to be listed along with Count, and keep "Summary below data". Then select <OK>. We have added subtotal "Average" in the table, as shown in rows 37, 69, 109, and 111 (circled) in Table 13.9. These entries are "alive" in the sense that each entry has a subtotal function. For example, E37 in Table 13.10 is generated by a subtotal function shown in the formula box. Note that the latest addition of the subtotal is added above the previous subtotal in Table 13.9. The Average row is added above the Count row, and that the Grand Average in row 111 is added above the Ground Count in row 112 and attached to the end of the table.

13.6.2 *Outline*

One of the features of Subtotaling is that it has Outline features that you can hide or show in detail. In Table 13.9, if the **Hide Detail button** (3) is clicked (the minus button indicated by an arrow in Table 13.9), then the rows included in the **level bar** (2) of that button will be reduced and hidden. Similarly, if the **Show Detail button** (4), which is the

Table 13.10 Summary rows of subtotals

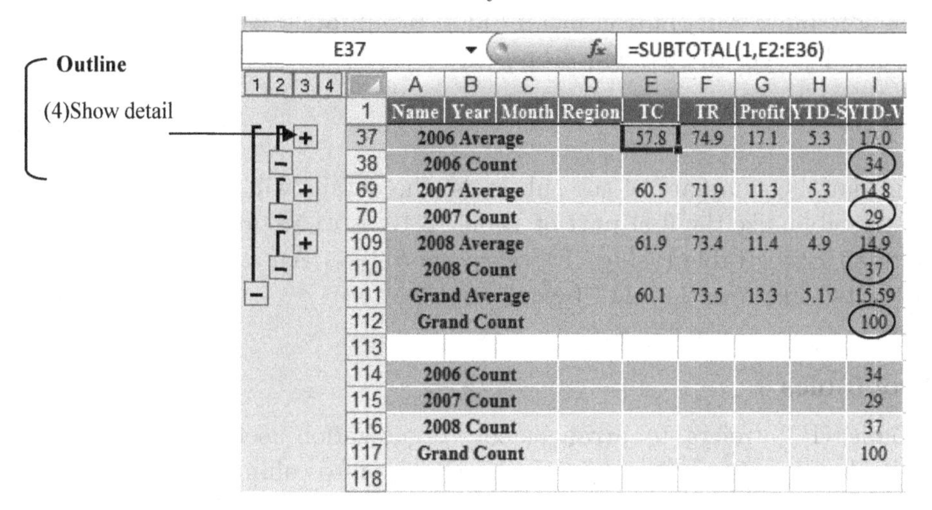

E37 =SUBTOTAL(1,E2:E36)

	A	B	C	D	E	F	G	H	I
1	Name	Year	Month	Region	TC	TR	Profit	YTD-S	YTD-V
37	2006 Average				57.8	74.9	17.1	5.3	17.0
38	2006 Count								34
69	2007 Average				60.5	71.9	11.3	5.3	14.8
70	2007 Count								29
109	2008 Average				61.9	73.4	11.4	4.9	14.9
110	2008 Count								37
111	Grand Average				60.1	73.5	13.3	5.17	15.59
112	Grand Count								100
113									
114	2006 Count								34
115	2007 Count								29
116	2008 Count								37
117	Grand Count								100
118									

plus button as shown in Table 13.10, is clicked, the rows associated with that button in the level bar will be expanded.

The plus and minus signs control the degree of detail locally. On the other hand, the **Level button** (1) (see Table 13.9) controls hide or shows the details globally. Table 13.10 shows the result of clicking the level 3 box in Table 13.9. All data in three changes of the year are suppressed at once, as shown by the three plus signs. If the level 2 box is clicked, only the results of Count will be shown, and if the level 1 box is clicked, only the Grand Average and Grand Count will be shown. To go back to the original table with 100 rows, click the level 4 box.

Copying the subtotal table

Note the discontinuity in the sequence of row headings. If A1:I109 in Table 13.10 is copied, you will get the full table of 100 rows, instead of the summary table as currently shown in Table 13.10. To extract only the summary table, say the Count rows, select rows 37, 68, 107, and 109 simultaneously (by holding the control key) and copy them down to rows 111 to 114 as shown in the bottom of Table 13.10. This may be facilitated by converting the filtered (list) table to an Excel Table (using ^T), selecting the desired rows, and copying and pasting.

To exit outline and subtotal

If you click a cell in the list or table, the Outline portion of Subtotal table can be deleted by selecting

<“Data”><Outline, Ungroup!><Clear Outline>. (13.8)

The subtotals can be removed from the list by clicking <"Data"><Outline, Subtotal> and clicking the <Remove All> button in Fig. 13.8. If Subtotals are removed, the Outline part on the left-hand side of the list will disappear.

Printing

If you print out the summary of the subtotals, the details will also be printed out. If only the summary table, like the first part of Table 13.10, is to be printed, we can either convert the list into an Excel Table (using ^T), select the table (^A), and print it out, or the table can be extracted like rows 114–117 before printing.

13.7 PivotTables

A PivotTable (PT) refers to arrangement, organization, selection, and calculation of complicated data in matrix form by "pivoting" rows into columns and columns into rows,[8] without creating new rows or column headings. It is probably the most powerful tools of spreadsheet data analysis. In this section, we want to show some very useful features of PivotTable operations and how to extract information from a PT.

13.7.1 *Constructing a PivotTable*

Excel places PivotTable under

<"Insert"><Tables, PivotTable><PivotTable>. (13.9)

We continue to assume that the list is given as Table 13.4, and is free standing. In this example, we would like to construct a region/month two-dimensional PT Report. Although we start from a list, we can convert a list to an Excel Table to take advantage of Excel Table commands, and can then convert back to a range (list), if necessary.

Click any cell in the list or table and follow the sequence of (13.9). The "Create PivotTable" dialog box (Fig. 13.9) appears.

Inside the dialog box, select "Table/Range". The box is filled automatically if the list is free standing. The default location for the PT is "New Worksheet". However, since we generally use PT for extensive data analysis, it is recommended that the PT be placed in the same worksheet, either overlapping with the list or next to the list. In this example, in Fig. 13.10(a), we place it at K4, the Northeast corner of the current list. In the "Location": box at the bottom of Fig. 13.9, Excel will enter the current sheet name with "!" from the current sheet tab "PT" in Fig. 13.9 and the cell address with $ signs.

Click <OK> and a dialog box (Fig. 13.10(a)) will appear, along with the PivotTable Field List (Fig. 13.10(c)).

[8]In Chapter 10, we called "pivoting" the "transposing" of column or row vectors. The term pivot table is a generic name, while PivotTable is a trademark of the Microsoft Corporation.

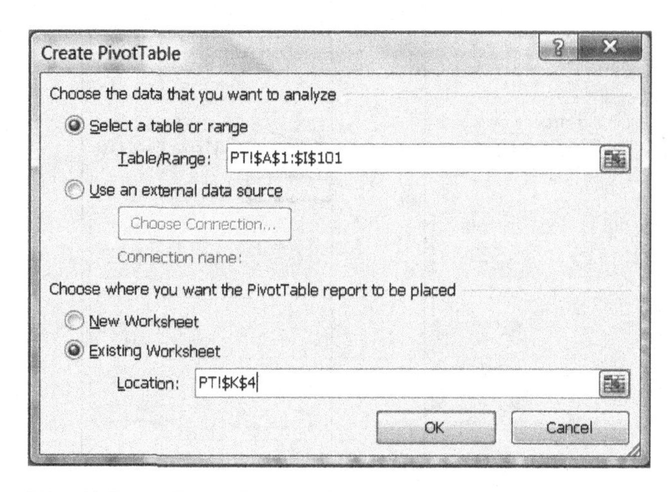

Fig. 13.9 <"Insert"><Tables, PivotTable><PivotTable>

Using Excel 2003 format

In the new Excel 2007 PivotTable format, Figs. 13.10(a) and 13.10(b) appear. We found that this format is not as intuitive or easy as the old Excel 2003 PivotTable format, although the new format has added some features. The best way is to use both formats. Thus, we install the Excel 2003 format by following procedure: Place the pointer in Fig. 13.10(a), and click

<"PivotTable Tools, Options"><PivotTable, Options!><Options>,

<PivotTable Options, Display>

<Display, xClassic PivotTable layout (enables dragging of the fields in the grid)>,

where x means that we check mark the box. Unless we deselect the box again, the classic PivotTable will stay in the future. When <OK> is pressed, the dialog boxes of the **PivotTable frame**, Figs. 13.10(b), and **PivotTable Field List**, 13.10(c), appear at the same time. If only PT frame (Fig. 13.10(b)) appears, clicking anywhere inside the PT frame (Fig. 13.10(b)), and selecting <RM> will cause a dialog box to appear. Then click <Show Field List> and the PT Field List (Fig. 13.10(c)) will appear.

13.7.2 *PivotTable reports*

The **PivotTable Field List** (Fig. 13.10(c)) consists of two parts. The first part is the **Field Section**, which constitutes all the field names of the original list; the second part, or **Area Section**, constitutes the four areas or boxes in the **PivotTable frame**, Fig. 13.10(b). It consists of four boxes: (1)″ the <Drop Row Fields Here> box; (2)″ the <Drop Column Fields Here> box; (3)″ the <Drop Data Item Here> box; and (4)″ the <Drop Page Fields Here> box. These four boxes are also reproduced in the lower part of the <PivotTable Field List>, Fig. 13.10(c), as four boxes (with confusing names): (1)′ <Row Labels>, (2)′ <Column

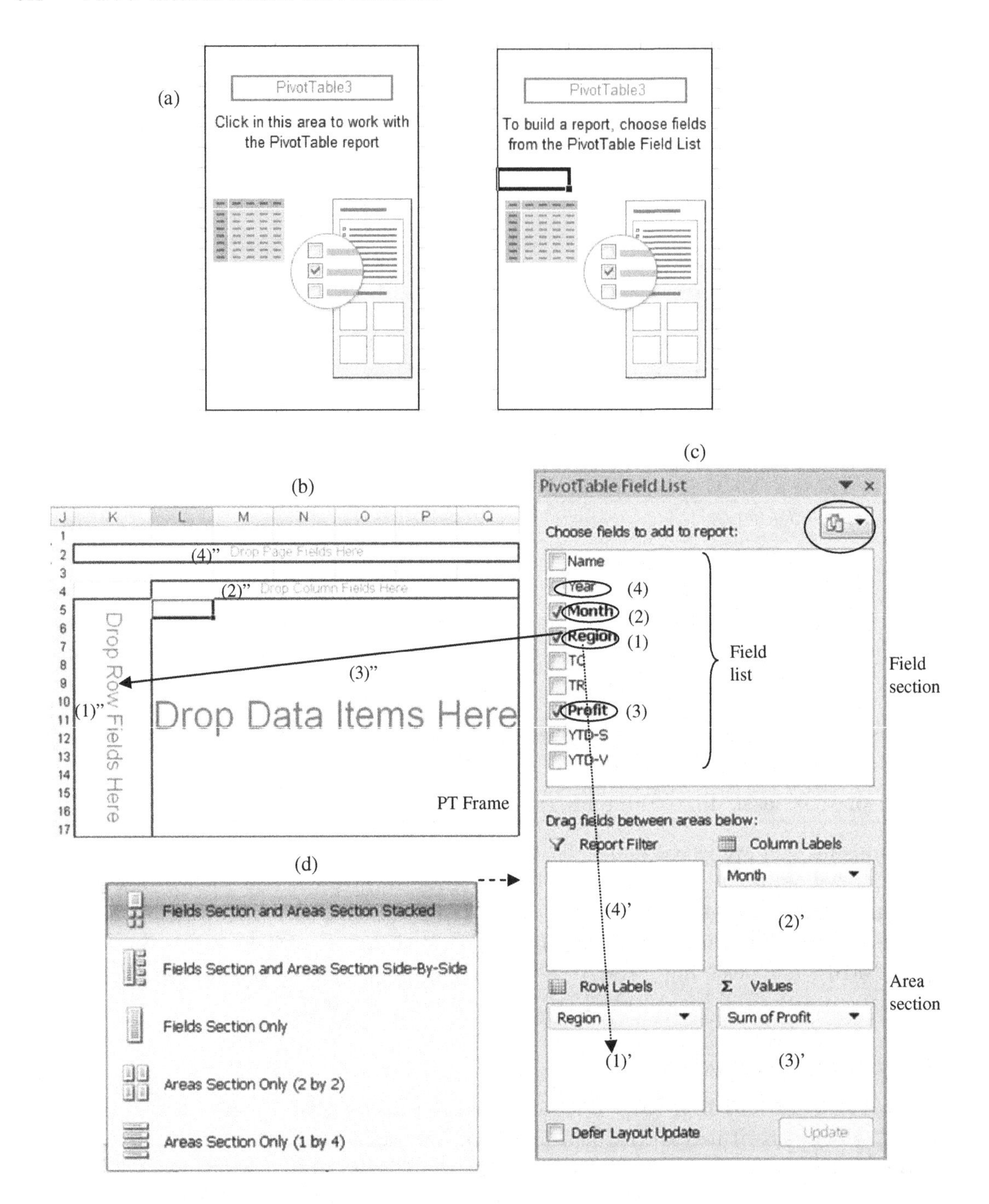

Fig. 13.10 (a) Structure of PivotTable — Excel 2007. (b) PivotTable frame. (c) The PivotTable field and (d) Arrangement of fields and area sections

Labels>, (3)′ <Values>, and (4)′ <Report Filter>. If these four boxes are not shown under the Field List, clicking the down arrow on the upper right corner of the <PivotTable Field List> (circled) will show the dialog box, Fig. 13.10(d). Clicking the first item (default) <Fields Section and Area Section Stacked> will then cause the lower part of Fig. 13.10(c) to appear. If the pointer clicks outside the PT frame, the PT Field List disappears, and if the pointer clicks inside the frame, the List reappears.

Two methods of constructing a two-axis PivotTable report

We now construct the **PivotTable**, or **PivotTable Report**, by moving the field names to the boxes. There are two ways to do so. The **Excel 2003 method** is shown by the solid line connecting Figs. 13.10(b) and 13.10(c). This method is to click and drag the field names, say, (2) "Month", in the box of <PivotTable Field List> to the (2)″ <Drop Column Fields Here> box; then click and drag (1) "Region" to the (1)″ <Drop Row Fields Here> box; and then click and drag (3) "Profit" to the (3)″ <Drop Data Item Here> box in Fig. 13.10(b). The three items in the PT Field List are checkmarked and we have Table 13.11. This is a **two-axis table**. Note that if the <PivotTable Field List> is in the way covering data, the Field List can be dragged by the heading row and moved to an open space.

The second method is the new **Excel 2007 method** as shown by the dotted lines in Fig. 13.10(c). When a field in the Field section of Field List (c) is clicked and dragged to a box in the Area section, say, (1) "Region" to (1)′ "Row Labels" box, the corresponding box (1)″ in Fig. 13.10(b) will be automatically filled, giving the same **two-axis table** as Fig. 13.11.

The Excel 2003 method is intuitive and straightforward, but the Excel 2007 method probably makes it easier to change the table arrangement and manage the external data, if needed. We will mainly follow the Excel 2007 method after the initial opening of the PT. (The 2007 procedures may also be found in the corresponding Excel 2003 method.) We use **PivotTable** and **PivotTable Report** interchangeably.

13.7.3 *PivotTable operations*

Transposing a PivotTable

The PT, Table 13.11, can be seen as a matrix. It can be transposed (or "pivoted") to change columns to rows and rows to columns. To do so, click the PT and the PT Field List Fig. 13.10(c) will appear. Then, by clicking and dragging (1) "Region" to the Column Labels box (2)′ and clicking and dragging (2) "Month" to the (1)′ Row Labels box, we transpose the PT, as with rows 4–9 of Table 13.12. The reader should verify that the data are rearranged but not changed (except the missing months).

Returning to Table 13.11, click the table to invoke the PT Field List, Fig. 13.10(c). We may remove "Month" in box (2)′ by clicking the down arrow next to "Month" in the Area section and clicking <Remove Field>. Table 13.11 collapses to two columns, K and L (not shown). Column K would show the four regions and column L would show

Table 13.11 PivotTable report

J	K	L	M	N	O	P	Q	R
1								
2	Drop Page Fields Here							
3								
4	Sum of Prof	Mon						
5	Region	July	Jan	Mar	May	Sep	Nov	Grand Total
6	East	-2.7	38.18	0.95	29.22	185.36	45.44	296.45
7	North	55.94	89.26	43.98	36.72	48.55	45	319.45
8	South	61.53	54.89	26.79	114.19	36.76	-34.18	259.98
9	West	-1.05	223.41	78.49	58.18	83.97	15.45	458.45
10	Grand Total	113.72	405.74	150.21	238.31	354.64	71.71	1334.33

the regional profit total (namely, column R) for three years. This is a **one-axis table of columns**.

Move (2) "Month" from the Field section in Fig. 13.10(c) back to the "Column Labels" box (2)′ in the Area section of Fig. 13.10(c). Then click the down arrow next to (1)′ "Region" and click <Remove Field>. The table collapses to three rows (not shown). Row 4 would show the column field, row 5 would show the months, and row 6 shows the monthly profit total for three years. We have a **one-axis table of rows**.

Hiding or unhiding columns or rows

Table 13.11 has two down arrows, which have the same function as filtering. Clicking the down arrow will show a menu box similar to Fig. 13.3, and rows and columns can be added or deleted. In Table 13.12, three months, Jan, Mar, and May, are filtered, and the filter logo appears on the right-hand side of K5. The filter mode can be cleared by clicking the down arrow and selecting <Clear Filter From "Month">.

Table 13.12 Transpose of PivotTable and the three-axis table

J	K	L	M	N	O	P
1						
2	Year	(All)				
3						
4	Sum of Profit	Region				
5	Month	East	North	South	West	Grand Total
6	Jan	38.18	89.26	54.89	223.4	405.74
7	Mar	0.95	43.98	26.79	78.49	150.21
8	May	29.22	36.72	114.2	58.18	238.31
9	Grand Total	68.35	170	195.9	360.1	794.26

(All)
2006
2007
2008
Select Multiple Items
OK Cancel

Fig. 13.11 Dialog box of Page Fields

A three-axis PivotTable

By activating the PT and clicking and moving (4) "Year", in the Field section of Fig. 13.10(c) to the "Report Filter" (that is, Page Fields) box (4)′ in the Area section of PT Field List, we create K2:L2 of Table 13.12. This is a third axis, in the sense that, if the down arrow in L2 is clicked, a Filtering dialog box of Page Fields, Fig. 13.11, appears (circled) on the right-hand side of the PT. It lists all the years in the data table. It functions like a third axis. When only <2006> is selected from the dialog box, the profit data for 2006 will be shown in the PT Report; when only <2007> is selected, the profit data for 2007 will be shown in the PT Report. It is like turning the "page" of a data book. If both 2006 and 2007 are selected, but 2008 is deselected, then the sum of total profits for 2006 and 2007 will be shown in the PT Report. Thus, we have a **three-axis PivotTable**.

13.7.4 *More PivotTable features*

Moving a column or row inside the PT

Moving a column or a row in the PT is similar to moving a column or row on a spreadsheet. To select a column or row title, click the side of the column or row, and drag it to a new place. This is the same procedure as moving spreadsheet rows and columns.

In Table 13.11, we want to move "North" in row 7 up to row 6. In this case, click K7, and move the pointer to a border of K7. A four-headed arrow will show at the tip of the pointer, as shown (circled) in part (a) of Table 13.13. (A black downward arrow may appear first. Keep trying until the four-headed arrow appears.) Move the pointer upward and the border pointed by the four-headed arrow changes to a black opaque wide band with a short vertical line at both ends of the line, as shown inside the circle in part (b). A text box below the arrow indicates the range of the band. Move the band up until it lies between row 5 and row 6 (see part (c)), just above the "East" cell, and release the LM. The "North" row is now moved above the "East" row. Similarly, in Table 13.11, we may click cell L5 and move "July" to the cell between "May" in O5 and "Sep" in P5.

Table 13.13 Moving a row in a PivotTable

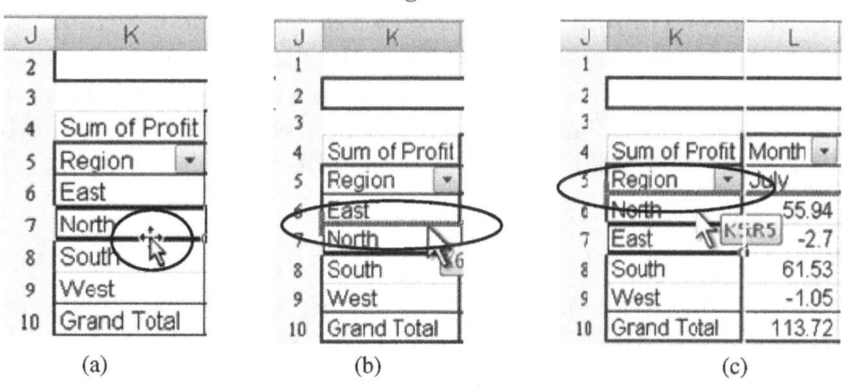

(a) (b) (c)

Table 13.14 Double-layer row labels

J	K	L	M	N	O	P	Q	R	S
4	Sum of P		Montl ▾						
5	Regio ▾	Year ▾	Jan	Mar	May	July	Sep	Nov	Grand Tota
6	⊟North	2006	92.08	38.96		22.69		27	180.73
7		2007	-2.82		13.78	43.77	-16.67	36.51	74.57
8		2008		5.02	22.94	-10.52	65.22	-18.51	64.15
9	North Total		89.26	43.98	36.72	55.94	48.55	45	319.45
10	⊟East	2006		10.56	23.43	5.36	46.25	-6.79	78.81
11		2007	38.18	-8.8		-8.06	52.92	37.85	112.09
12		2008		-0.81	5.79		86.19	14.38	105.55
13	East Total		38.18	0.95	29.22	-2.7	185.36	45.44	296.45
14	⊞West		223.41	78.49	58.18	-1.05	83.97	15.45	458.45
15	⊞South		54.89	26.79	114.19	61.53	36.76	-34.18	259.98
16	Grand Total		405.74	150.21	238.31	113.72	354.64	71.71	1334.33

Double-layer axes

From the three-axis table of Table 13.12, click and drag "Year" in the "Report Filter" box (4)′ in the Area section (Fig. 13.10(c)) to the space below "Region" in the Row Labels box (1)′. We now have the **two-layer row labels** of Table 13.14, in which "Year" is entered on the right of "Region". In this case, the table is arranged so that Region shows the total profit of each Year within the Region. Note the – and + signs before the region names. This is similar to the Subtotaling of Table 13.9: – is "Hide detail button", and + is the "Show detail button". In Table 13.14, North is shown in detail, and can be reduced (– is circled), but the details of the West (circled) and South regions are hidden.

If we click the down arrow of "Year" in the Row Labels box (1)′ of the Area section of Fig. 13.10(c) to open a dialog box, and click <Move Up> to move "Year" above the "Region" in the same box, the "Year" in Table 13.14 moves to the left side of "Region". That is, the position of K6:K8 and that of L6:L8 are exchanged. The table is then arranged by Year first, showing regional profits within each of the three years.

Similarly, clicking the down arrow of "Year" and clicking "Move to Column Labels" will cause "Year" to move to the space below "Month" in the Column Labels box (2)′. This gives us **two-layer column labels**, as shown in rows 5 and 6 of Table 13.15. To save space, we have reduced the May data and hidden the data in columns U and V. The table is arranged by Month first showing total profit for each of three years within each month. Moving "Year" above "Month" in the Column Labels box (2)′, we have a table arranged by year first, showing total profit for each of six months within the year (not shown).

Multiple data fields and field settings

Going back to Table 13.11, for simplicity, delete South from the row labels and delete all column labels except Jan. Move TC and TR from the Field Section to the Values box (3)′ in Table 13.10(c). Click the down arrow of <Sum of TC> in Table 13.10(c). When the menu

Table 13.15 Double-layer column labels

	J	K	L	M	N	O	P	Q	R	S	T	W	X
4		Sum of Prof	Mont	Yea									
5			⊟Jan			Jan Total	⊟Mar			Mar Total	⊞May	⊞No	Grand Total
6		Region	2006	2007	2008		2006	2007	2008				
7		North	92.1	-2.8		89.3	39.0		5.0	44.0	36.7	45.0	319.5
8		East		38.2		38.2	10.6	-8.8	-0.8	0.9	29.2	45.4	296.5
9		West	100.4	23.1	99.9	223.4	22.5	5.9	50.1	78.5	58.2	15.5	458.5
10		South	54.9			54.9	-10.1		36.9	26.8	114.2	-34.2	260.0
11		Grand Total	247.3	58.5	99.9	405.7	61.9	-2.9	91.2	150.2	238.3	71.7	1334.3

Table 13.16 PivotTable Report

	K	L	M	N	O	P	Q
1							
2	Year	(All)					
3							
4		Month	Data				
5		Jan			Total Count of TC	Total Sum of TR	Total Average of Profit
6	Region	Count of TC	Sum of TR	Average of Profit			
7	North	8	599.6	11.2	8	599.6	11.2
8	East	1	86.7	38.2	1	86.7	38.2
9	West	8	623.6	27.9	8	623.6	27.9
10	Grand Total	17	1309.8	20.6	17	1309.8	20.6

window appears, click

<Value Field Settings. . .> (to open a dialog box)
<Summarize value field by, Count><OK>.

The subtotaling function changes to "Count of TC". Similarly, click and drag TR and Profits to the "Values" box (3)′, and change "Sum of Profit" to "Average of Profit". The PT now changes to Table 13.16. The table shows Count of TC, Sum of TR, and Average of Profit for the three regions in January, and also the corresponding Grand Total.

13.7.5 *PivotCharts*

A PivotChart is similar to an ordinary chart, except that it recognizes the PivotTable and automatically creates a chart. For example, let us return to Table 13.11, in which July data is moved after May. Click anywhere in the PT, and select

<"PivotTables Tools,Options"><Tools, PivotChart>
<Insert Chart, Column, Clustered Column (the first chart)><OK>. (13.10)

Then, Fig. 13.11 will be created, along with a dialog window <PivotChart Filter Pane>, Fig. 13.12. Figure 13.11 shows the total profit of the six odd months in four regions. In Fig. 13.12, the down arrow next to the Region and Month will lead to another Filtering dialog box like Fig. 13.3, and you can add or delete some row fields or column fields from the chart, just as with Filtering, without redoing the chart. The filtering logo will appear next to the down arrow.

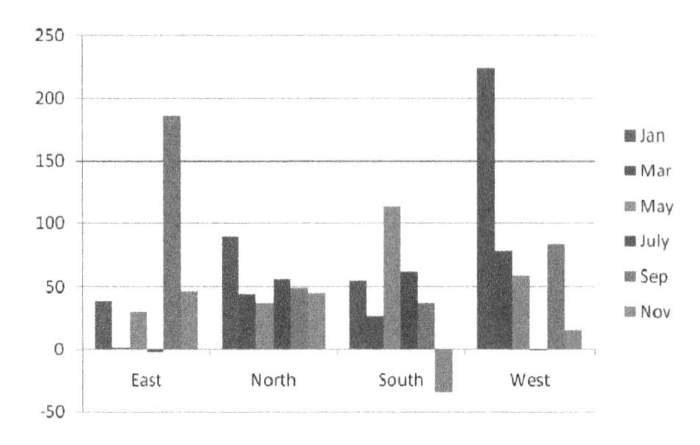

Fig. 13.11 PivotChart for Table 13.9

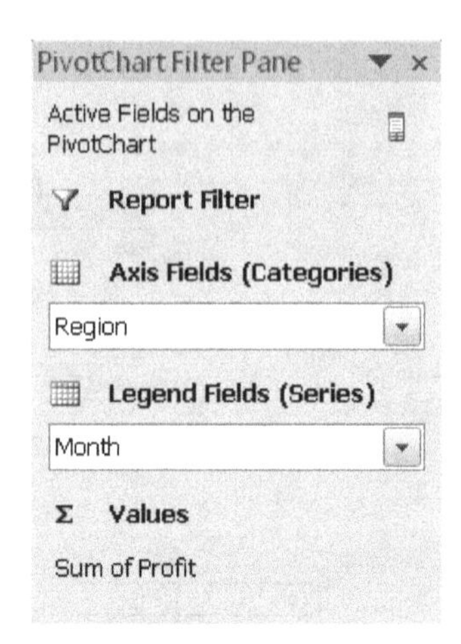

Fig. 13.12 PivotChart filter pane

Note that the row labels of "Region" are taken to be the X-axis (Axis Fields (Categories)), and the column labels of "Month" are taken to be the Y-axis (Legend Fields (Series)). When the PivotChart is clicked, <"PivotChart Tools"> is added to the ribbon. It has four tabs: Design, Layout, Format, and Analyze. The PivotChart can be edited like regular charts introduced in Chapter 3. The readers should consult Excel reference books for further studies.

13.8 Examples of Using Excel Data Analysis

Based on the database of the sales record of Table 13.4, we ask the following questions using the data analysis methods introduced in this chapter. Most of the questions can be answered in several ways. We choose only one of them to answer the question. If "PT Field List" is not shown when the PT is selected, click <RM> <Show Field List>.

13.8.1 *Examples of sorting and filtering*

Sorting

1. Given the data, who made the most profits? Least profits?
Answer: In this case, change Table 13.4 to an Excel Table (^T) to have Table 13.5. Clicking the down arrow next to the Profit G1 in Table 13.5 (we denote this by <Profit!> below), and clicking <Sort Largest to Smallest> when Fig. 13.3 appears, we have the solution: Salesperson #42 of the South region earned the most profits, 54.53 (thousand or million dollars), in July 2006 (not shown).

Similarly, clicking <Sort Smallest to Largest> in Fig. 13.3, we have the solution that Salesperson #75 of the North region earned the largest negative profit the least profit: −28.32 in November 2008 (not shown).

Filtering

2. Who are the five best profit earners? Five worst profit earners? Compare their performances.

Answer: The question is asking "who", hence, we should use the simple filtering method to find the top and the bottom performers and compare the average performance of each group. (Group comparison can be done by comparing averages, standard deviations, or other measures of concentration or diversion). Changing the list to Excel Table, click the down arrow next to Profit (<Profit!>). In the dialog box (Fig. 13.3), select <Sort Smallest to Largest>. We select Top five rows to get the second table below, and then select Bottom five rows to get the first table. Add the "Total" row to each table and find the averages.

Name	Year	Month	Region	TC	TR	Profit	YTD-S	YTD-V
91	2006	Jan	South	49.58	89.09	39.51	10	30
24	2008	Sep	North	55.12	99.16	44.04	2	30
16	2006	Jan	West	45.03	90.21	45.18	0	28
22	2007	May	South	42.03	91.87	49.84	6	1
42	2006	July	South	44.6	99.13	54.53	4	3
Total	**5**			**47.3**	**93.9**	**46.62**	**4.4**	**18.4**
				Avg	Avg	Avg	Avg	Avg

Name	Year	Month	Region	TC	TR	Profit	YTD-S	YTD-V
75	2008	Nov	North	79.33	51.01	-28.32	7	8
48	2006	Nov	South	78.69	57.99	-20.70	5	0
30	2007	Nov	South	72.61	52.42	-20.19	4	10
47	2008	May	East	71.28	53.85	-17.43	9	16
43	2007	Sep	North	67.19	50.52	-16.67	5	24
Total	**5**			**73.8**	**53.2**	**-20.662**	**6**	**11.6**

Comparing the average performance of the two groups, we find that the main difference is that the first group has much lower TC and much higher TR than the second, and that there seems not to be much difference between YTD-S and YTD-V. The performance of the high profit group is due to its much higher TR and has little to do with sickness or vacations taken by salespeople.

3. In 2008, who earned the highest profit? The least profit?

Answer: Since the question is asking about the single year 2008, the best way to answer this question is to filter out the 2008 data, and then sort the 2008 profit data by <Sort Largest to Smallest> (see Fig. 13.3). The results are that, in 2008, salesperson #24 earned the highest profit, and #75 earned the lowest profit.

Nam	Yea	Montl	Regio	TC	TR	Profi	YTD-S	YTD-V
24	2008	Sep	North	55.12	99.16	44.04	2	30
75	2008	Nov	North	79.33	51.01	-28.32	7	8

To show only the two relevant rows out of 37 rows in 2008, click the down arrow next to Profit, deselect <Select All> in the dialog box (Fig. 13.3), reselect 44.04 and −28.32, and select <OK>.

4. How many people earned positive profits? Earned zero profits? How many earned profits from 20 to 40?

Answer: Under <Profit!> click <Number Filters> (Fig. 13.3) and <Greater Than...> (Fig. 13.4), and enter 0. The answer is, using the count subtotal function, 74 salespersons earned positive profits. Under <Profit!>, no salesperson earned 0 profits. Under <Profit!>, for profits greater than or equal to 20 and less than or equal to 40, there are 35 salespersons. The table should be similar to the one in Example 2.

13.8.2 *Examples of using PivotTable*

A one-axis PivotTable

5. Which region had the largest year-to-day vacation? Is it related to the warmer weather in the southern region?

Answer: Using the PT, we see that the North region has the highest year-to-day vacation. The number of vacation days does not appear to be related to the weather. We have added the Count of YTD-V column to make sure that all 100 rows are included in the calculation. To add two statistics of a same field, click and drag YTD-V twice to the Values box (3)′ in the Field List (Fig. 13.10(c)). Then, if you click the down arrow and click <Value Field Settings...><Count><OK>, the Sum of the YTD-V column changes to the Count of YTD-V column.

	Values	
Row Labe	**Sum of YTD-V**	**Count of YTD-V2**
East	309	23
North	454	28
South	291	20
West	505	29
Grand Total	**1559**	**100**

6. In which month did the firm have the largest profits during the three years?
This is similar to Question 5. Creating a new PT, entering "Month" in the row label box (1)′, and entering Sum of Profit in the "Values" box (3)′, we have the answer: January.

A two-axis PivotTable

7. In which year did the firm have the largest profit? Smallest profit? In which month AND year did the sales of the firm achieve the highest profit?

Answer: The first part is simply asking which year without asking month, but second part needs to have monthly data for each year. The new PT should be as follows:

Sum of Profit	Column Labels			
Row Labels	**2006**	**2007**	**2008**	**Grand Total**
Jan	247.34	58.48	99.92	405.74
Mar	61.92	-2.89	91.18	150.21
May	79.87	75.05	83.39	238.31
July	82.58	41.66	-10.52	113.72
Sep	103.24	86.98	164.42	354.64
Nov	7.85	69.62	-5.76	71.71
Grand Total	**582.8**	**328.9**	**422.63**	**1334.33**
	247.34	-10.5		

Q11

From the table, we see that 2006 had the largest profit and 2007 had the smallest profit. For a large data set, to find the month and year of the largest and smallest profits, we use =MAX() and =MIN() function at the bottom of the table, and find the corresponding numbers in the table. The answer is indicated by the bordered cell: the maximum profit is 247.34 in Jan 2006, and minimum profit is -10.52 in July 2008.

8. Which region earned the largest profits in November?
Answer: The PT and the answer are similar to those for Question 7, with Month on the row labels, but Region on the column labels, and Sum of Profit in values. Our data shows North.

9. Can you verify that the Northern region is colder, which may be why more salespersons get sick, resulting in reduced profits? Show the table and your verification.
Answer: From the table below, the YTD-S in the North region is the second lowest, but the profit is the second highest. Apparently, there is no relation between the weather in the North region and the YTD-S and Profit.

	Values		
Row Labe	**Sum of Profit**	**Sum of YTD-S**	**Sum of YTD-V**
East	296.45	102	309
North	319.45	122	454
South	259.98	132	291
West	458.45	161	505
Grand Tot	**1334.33**	**517**	**1559**

Q9 Q 10

10. Can you say that the South region has better weather, and thus more people may take vacations, reducing profits?

Answer: From the PT of the previous question, the South region has the lowest YTD-V, but the Sum of Profit for the three years is the lowest among the regions. More people in the South worked throughout the three years, yet the profit is the lowest. Apparently, weather is not a factor in determining the profit of this region.

11. Can you say that the month of November, close to the end of the year, may be the reason that profits increase in November?

Answer: From the Grand Total of months in the PT for Question 7, November had exceptionally low profits over three years. Apparently, seasonality does not affect the November sales of the good the firm is selling.

12. Can you say that the East region is more populated, and this may be the reason it has a higher volume of sales, especially in November?

Answer: Volume of sales is measured by TR. We need to know the distribution of data by Region and Month. Hence, the PT is as follows:

Sum of TR	Column Labels					
Row Labels	East	North	South	West	Grand Total	
Jan	86.67	599.56	159.75	623.61	1469.59	
Mar	202.29	161.78	229.52	453.52	1047.11	
May	298.06	228.61	333.84	378.68	1239.19	
July	197.21	377.21	174.65	76.35	825.42	
Sep	626.57	269.11	234.19	527.65	1657.52	
Nov	294.66	367.22	311.18	134.64	1107.7	443.08
Grand Total	**1705.46**	**2003.5**	**1443.13**	**2194.45**	**7346.53**	=avg(Nov)

In three years, the East region had the second smallest total volume of sales, and the East region's sales in November are well below the average of the four regions. Thus, the argument does not seem to hold.

A three-axis PivotTable

13. Which region earned the largest profits in November 2008? November 2007?

Answer: There are three variables, Region, Month, and Year. They are arranged as follows. Using filtering, we conceal months other than November in the table.

Sum of Profit	Column Labels				
Row Labels	East	North	South	West	Grand Total
Nov	**45.44**	**45**	**-34.18**	**15.45**	**71.71**
2006	-6.79	27	-12.36		7.85
2007	37.85	36.51	-20.19	15.45	69.62
2008	14.38	-18.51	-1.63		-5.76
Grand Total	**45.44**	**45**	**-34.18**	**15.45**	**71.71**

Clearly, The East region has the largest profit in November 2008 and 2007.

14. Which region and month have zero vacations throughout the three years? In 2006, 2007, and 2008, separately?

Answer: The first part of question requires knowing the distribution of YTD-V for three years by Region and Month. This is shown by the Grand Totals. The second part needs knowledge of the distribution of YTD-V for each year among Region, Month, and Year.

Sum of YTD-V	Column Labels				
Row Labels	**East**	**North**	**South**	**West**	**Grand**
2006	**139**	**155**	**106**	**177**	**577**
Jan		95	60	65	220
Mar	9	8	8	29	54
May	24		14	14	52
July	33	37	3		73
Sep	44		1	69	114
Nov	29	15	20		64
2007	**82**	**169**	**41**	**137**	**429**
Jan	16	37		21	74
Mar	24			17	41
May		11	1	26	38
July	5	48	10	21	84
Sep	15	24	20	19	78
Nov	22	49	10	33	114
2008	**88**	**130**	**144**	**191**	**553**
Jan				60	60
Mar	6	17	42	36	101
May	43	30	39	73	185
July		4			4
Sep	23	55	28	22	128
Nov	16	24	35		75
Grand Total	**309**	**454**	**291**	**505**	**1559**

From the PT, no Region or Month shows zero vacations in three years (see the circled numbers).

But when the data is disaggregated by Year, then it shows that, in 2006, Jan in the East, May and Sep in the North, and July and Nov in the West have zero vacation days. In 2007, May in the East, March in the North, and January and March in the South show no vacation days. In 2008, January and July in the East, January in the North, January and July in the South, and July and November in the West have zero vacation days. The distribution of zero vacation days among region and month in a year appears to have no specific pattern.

13.8.3 *Relations among sorting, filtering, and PivotTable*

Comparing the PivotTable and filtering

15. Using the filtering method, show the results in Question 13.

Answer: In the PivotTable, Table 13.5, if you click <Year!><@2008>, the table first filters 2008 out, then filters Nov out by <Month!><@Nov> from 2008 data. It then sorts the Nov 2008 data by Region through <Region!>, in alphabetical order. The filtered table is shown below. Column J shows the total profits of the East, North, and South in Nov 2008. They match exactly with the Nov 2008 row in the PT of Question 13.

	A	B	C	D	E	F	G	H	I	J
1	Nam	Yea	Mont	Regio	T	T	Prof	YTD-!	YTD-'	
48	55	2008	Nov	East	60.13	75.43	15.30	1	16	
58	27	2008	Nov	East	69.98	69.06	-0.92	3	0	14.38
70	36	2008	Nov	North	56.67	66.48	9.81	1	16	
81	75	2008	Nov	North	79.33	51.01	-28.32	7	8	-18.51
86	100	2008	Nov	South	63.68	66.27	2.59	9	20	
101	64	2008	Nov	South	54.65	50.43	-4.22	6	15	-1.63
102	**Total**	**6**					**6**		**75**	

The PivotTable and sorting

16. Using the sorting method, show the results for Question 13.

Answer: Sort by Year first, then by month. The sorted table shows the same six "Nov 2008" data that the filtered table in Question 15 does. Extract this part below the sorted table and sort Region alphabetically, we have exactly the same table as the filtered table in Question 15 except that other data are not filtered out.

The PivotTable and subtotaling

17. Using the subtotaling method, show the results in Question 13.

Answer: We first change the list to the Excel Table, and sort first by Region, then by Month, and then by Year. Then we convert the Excel Table to a list and clear filtering. Find subtotal by Year, then by Month, then by Region. We have five levels of Outline. Click level 2, and then expand 2008 Nov Total, we have the following subtotaling table, which contains the same information as the table in Question 15.

1 2 3 4 5	A	B	C	D	E	F	G	H	I
1	Name	Year	Month	Region	TC	TR	Profit	YTD-S	YTD-V
72		**2006 Total**					582.80		
131		**2007 Total**					328.90		
137			**Jan Total**				99.92		
140			**July Total**				-10.52		
153			**Mar Total**				91.18		
172			**May Total**				83.39		
173	55	2008	Nov	East	60.13	75.43	15.30	1	16
174	27	2008	Nov	East	69.98	69.06	-0.92	3	0
175				**East Total**			14.38		
176	36	2008	Nov	North	56.67	66.48	9.81	1	16
177	75	2008	Nov	North	79.33	51.01	-28.32	7	8
178				**North Total**			-18.51		
179	100	2008	Nov	South	63.68	66.27	2.59	9	20
180	64	2008	Nov	South	54.65	50.43	-4.22	6	15
181				**South Total**			-1.63		
182			**Nov Total**				-5.76		
199			**Sep Total**				164.42		
200		**2008 Total**					422.63		
201		**Grand Total**					1334.33		

13.9 Summary

This chapter deals with organizing and managing a large data set. The data set is created by using the IF function. This method enables us to use a new data set without typing, scanning, or downloading the data set (Sec. 13.2). The data set is arranged in a special mode called the Excel Table, which facilitates the tasks of selecting, copying, moving, and subtotaling the rows and columns (Sec. 13.3). Whether in the Excel Table mode or not, the whole table can be rearranged either by alphabetical or numerical order. If duplicates are present, we can sort them in multiple levels using the sort dialog box. The sorting command is most useful if we are looking for largest or smallest items or ranking the data set (Sec. 13.4). If knowledge of all the data is not required, we can filter out only those data we are looking for without bothering with the other data. We can then apply the data analysis commands to the filtered data (Sec. 13.5). On the other hand, if some statistical calculation of each segment or category of the whole data set is needed, then we can use the subtotaling method to find the subtotal statistics of each segment or category (Sec. 13.6).

However, many features of the filtering and subtotaling methods can be found in the PivotTable (Sec. 13.7). It can rearrange and calculate a data set instantly in many flexible ways with a few mouse clicks: adding, deleting, changing the columns and rows, using the subtotal commands to calculate basic statistics of a subset of data, changing the PivotTable to a one-, two-, or three-dimensional table format, drawing charts directly from the PivotTable, and many other operations. It can be said that one of the advantages of using the Excel spreadsheets is the availability of PivotTable. The last section (Sec. 13.8) gives some examples of the use of sorting, filtering, subtotaling, and PivotTabling. Many examples can be answered in several ways. The readers should choose the most appropriate method.

Key Terms: Excel

Homework Chapter 13

13-1 Subtotaling time series Table HW13-1 generates the long-run real GDP per capita of eight countries. The data on 1911 and 1992 are taken from Maddison (1995). The long-run growth rates from 1911 to 1992 for each country are calculated as in row 2 of Table HW13-1. Using the 1911 data as given levels of the initial income, and the growth rates in row 2, we generate the data for the other years. Enter the "Year" and "No" columns and the labels in rows 1 and 2 as shown in the table.

Then, after entering initial income in row 5, we use the following formula in C6:

$$=\mathrm{pc}(0)^{*}\exp(g^{*}t) + (b - a)x + a,$$

Table HW13-1 Simulation of long-run real GDP per capita

C6: =C$5*EXP($B6*C$2)+RAND()*1000-500.

	A	B	C	D	E	F	G	H	I	J
1	**Long-run Real GDP Per Capita of Eight Countries**									
2	Gr rate g		0.021	0.021	0.024	0.015	0.018	0.03	0.0318	0.033
3	Initial per capital income									
4	Year	No	France	Germany	Italy	UK	USA	Korea	Taiwan	Japan
5	1911	0	3219	3602	2407	4815	5052	898	882	1304
6	1912	1	3623	3543	2756	5306	5508	1344	629.85	1298
7	1913	2	3573	3643	2429	4972	4921	755.2	1344.9	1679
86	...	...	...	...	...	...	...	...	...	...
87	1992	81	17959	19351	16229	15738	21558	10010	11590	19425

where pc(0) is the initial per capita income, g is the growth rate, t is time (year), and x is the random number. Let a = −500 and b = 500. Then, the equation means that starting from the initial per capita income pc(0), it grows at growth rate g over t periods, plus or minus a random number between a and b, starting from a. The cell formula in C6 is shown in Table 13-1.

Copy C6 to J86 to create the data set. Note that the numbers in the data set should be very close to the actual numbers in row 87 of Table HW13-1.

Using the table, answer the following questions:

(a) Find the average and standard deviation (stdev) for real GDP per capita for

Prewar period,	1911–1940,
Transitional period,	1941–1950,
Postwar period,	1951–1990.

Print out the summary table for the three periods.

(b) Find the average and standard deviation (stdev) for real GDP per capita for every ten-year period.

$$1911\text{–}1920, 1921\text{–}1920, \ldots, 1981\text{–}1990.$$

Print out the summary table of the 10 periods.

(c) The coefficient of variation is defined as CV = stdev*100/average. Extract the average and stdev for each period and find the CV for each period in the above two questions, using the following format in the next sheet:

Prewar period	average	stdev	CV,
Transitional period	average	stdev	CV,
Postwar period	average	stdev	CV.

(d) What is the economic interpretation of your results in question (a)? What are the pros and cons of dividing data like those in (a) and those in (b)?

(Hints: To answer question (a), you have to create a column, say call it "period3", and designate a letter to each period, say the prewar period a, the transition period b, etc. Then sort by the letters in ascending order of the "period" column. Do the same for the 10 periods in question (b). For an application of the method, see paper by Hsiao and Hsiao (2003)).

13-2 Can you answer questions (a) to (d) of HW13-1 by using the sorting method? If you can, using the method to answer the questions.

13-3 Can you answer questions (a) to (d) of HW13-1 by using filtering method? If you can, using the method to answer the questions.

13-4 Can you answer questions (a) to (d) of HW13-1 by using the PivotTable method? If you can, using the method to answer the questions.

13-5 The data created in Table HW13-1 are simulated data of long-run real GDP per capita. Convert the data into continuous annual growth rates of real GDP per capita of the eight countries. Using the subtotalling method, answer the same questions (a) to (d) in HW13-1 in terms of continuous growth rate (see (8.18)).

13-6 Using the PivotTable method, answer the questions in HW13-5.

13-7 Political affiliation A pollster surveyed 200 people who are affiliated with one of the three major political parties. The table is given to you as follows.

Gender: Male/Female,

Party: Republican/Democrats/Reform,

Race: African/Asian/White/Others.

Generate the random number assignment for gender, party, and race, using the method shown in the following table.

Income: ranges from 10,000 to 100,000 (with a unit of 1000).

Education: from 12 to 17 years, the "17 years" denoting some graduate work.

Family size: from 1 to 5, "5" denotes a family with more than 4 persons.

Sample	Gen	Party	Race	Gen	Party	Race	Y	Edu	FamilySize
1	2	2	3	M	Rep	Wh	42	16	3
2	1	3	1	F	Ref	As	43	15	4
...									
500	1	3	2	F	Ref	Af	52	16	4

Construct the following PivotTables (you must use PivotTables except for Table 5 below).

Your answer to each question must be based on the table. You should copy the table under the question and paint the cell which contains your answer in yellow to highlight your answer.

Table 1. The numbers of persons in each gender (in column) and party (in row) categories.
(a) How many females and males are in the sample?
(b) How many are Democrats, Republicans, and Reformers?
(c) How many males who are also Republicans?

Table 2. A three-dimensional table showing the race of each gender.
How many African Americans, Asian Americans, Whites, and Others are female Republicans?

Table 3. A three-dimensional table showing the gender of each race.
How many African Americans are male and Democrats?

Table 4. Construct one table (and only one table), which can answer the following questions:
(a) Can you say that women are generally richer (have larger incomes) than men?
(b) Can you say that whites are generally richer than the other races?

(c) Can you say African Americans generally have larger families?
(d) Can you say Asian Americans are generally better educated?

Table 5. Using subtotals, answer the question in Table 2. Present the subtotals in reduced form.

13-8 A multinational company A multinational company has 10,000 branches all over the world. A random sample of 500 branches was taken for three years, between 2005 and 2007, at the beginning month of each quarter of the year (Jan, Apr, Jul, Oct), from four regions: Asia, Africa, Europe, and Latin America. The choice of regions is also random.

(a) It is found that total cost ranges from $200 to 300 million, total revenue ranges from $250 to 350 million, and inventory ranges from $10 to 20 million. Branches are numbered from 1 to 500, and all other data are distributed randomly. The table should have the following column labels:

Branch, Year, Quarter, Region, TC, TR, Profits, Inventory.

(b) Copy (do not type) **the cell formula** of each of the first cell entries of each column on the cell above the column label, **before** you change the table to values (this will show how you create the data).
(c) Using **PivotTables** answer each of the following questions. (You should have EIGHT PivotTables, one for each question, two summary tables of subtotals, and one autofilter table). **Color the cell(s) which contain(s) your answer, and WRITE your answer in words AGAIN below the PivotTable.** To save paper, print out the first page of the table only (making sure the cell formulae are included).

Table 1. In which year did the company have the largest inventory?
Table 2. In which quarter of the year did the company have the largest profits?
Table 3. Which region earned the largest profits in the fourth quarter?
Table 4. Which region earned negative profits in the fourth quarter of 2007?
Table 5. Can you verify that Asia is prosperous, which may be the reason that the company earns more in Asia during the last quarter of the year?
Table 6. Can you say that the Latin America region had higher costs and inventories on average?
Table 7. Which region and quarter did not have any positive profits during the three years?
Table 8. How many branches are there in each region?
Table 9. Show that Table 8 can also be derived by the subtotal method, using the summary table. Copy the summary table, detached from the original data, and name it Table 13-8. Please make sure that the regions are clearly designated.
Table 10. Using the filtering method, find the 10 most profitable branches, arranged in decreasing order. Also, find the 10 least profitable branches.

13-9 Extra credit questions Find any three similar questions from HW13-8 and give the answer to your own questions. You may use any method of Excel Data Analysis.

13-10 World Development Report[9] Section 14.5.2 lists some useful websites where you can find domestic or international data. Visit the World Bank website (http://worldbank.org) and download the following 10 world development indicators for all countries in the world (they are also available either in Table 1 or under Selected World Development Indicators, inside the World Development Report published every year. The contents may also differ from year to year). In the following, the name of the data is followed by possible classification of the data:

Label part

y	Per capita GNP, classify into 3 regions: low, middle, and high
y2	Per capita GNP: classify into 7 regions: Low, Middle, Lower Middle, Upper Middle, High, OECD, and Non-OECD
Geo	5 regions: SSA= Sub-Saharan Africa; As=Asia; Eu&CA=Europe and Central Asia ME&NA= Middle East and North Africa; Am=Americas
Geo2	Subregions of the above 5 regions: Sub-Saharan Africa (2 subregions): East and Southern Africa, West Africa Asia (2 subregions): East Asia and Pacific, South Asia Europe and Central Asia (2 subregions): Eastern Europe and Central Asia, Rest of Europe Middle East and North Africa (2 subregions): Middle East, North Africa Americas (2 subregions): North America, South America

Data part

Pop	Population (millions)
Area	Area (Thousands of square kilometers)
PC	GDP per capita (listed separately under the table of Economic Activity).
PCg	GDP per capita average annual growth rate (percent), under the table of Economic Activity.
Inf	Average annual rate of inflation (percent)
LEx	Life expectancy at birth (years)
Lit	Adult literacy rate (percent), Total

Questions

(a) Using the sorting method, find which country has highest GNP per capita, lowest GNP per capita. how does it compared with that of USA?

(b) Using the sorting method, find which country has the smallest population and what is the country's area. Which country has the largest population, and what is its area? .Calculate the population density of each country (expressed in the number of people per km^2).

---

[9]The 2010 data can be downloaded from http://siteresources.worldbank.org/INTWDR2010/Resources/5287678-1226014527953/Statistical-Annex.pdf. The "permanent address of this page" is http://go.worldbank.org/FTD88BBDV0 (as of July 1, 2010). To import, see Section 14.5 The Text Import Wizard. Another useful data source is IMF, see Section 14.5.2. If the above data sources are not available, the readers may construct their own questions of the similar nature (see below) for applications of Excel data analysis.

(c) Using the sorting method, find how many high per capita income countries are in the geographic region of Americas and of Asia. What is the name of the countries?
(d) Using the subtotal method, calculate the number of countries, the total population, and the average life expectancy at birth for each category of per capita income y in Sheet2. Enter the world total. Calculate the percentage of countries and population in world total for each of three country categories.
(e) From the above data, what conclusion can you make?

13-11 World Development Report II Using the PivotTable and charts, answer the following questions.

(a) What are the basic economic characteristics of low, middle, high-income countries?
(b) What are the basic economic characteristics (GDP per capita, population, literacy rate, etc.) of world's ten regions (Geo2)?
(c) It is said that the Sub-Saharan region is the economically most depressed area. Can you show this from your data?
(d) It is said that East Asia and Pacific region is the economically most dynamic area. Can you show this from your data?
(e) In terms of per capita GDP, are the Asian NIEs and ASEAN-4 converging to OECD countries?

(Hints: (a) You may use the following table and illustrate diagrammatically. Your choice of variables may vary. In this case, the Pivot Table is in A1:E7. Rows 8 to 10 are added to show the percentage distribution of population, area, and the number of countries.

	A	B	C	D	E
1		y			
2	Data	Lo	Md	Hi	Grand Total
3	Average of PC	338.7	2337.5	19534.5	4776.0
4	Average of PCg	0.1	0.3	2.4	0.7
5	Sum of Pop	2934.4	1318.6	795.9	5048.9
6	Sum of Area	33609.0	57263.0	31086.0	121958.0
7	Count of Country	40	65	22	127
8	%Pop	58.1	26.1	15.8	100.0
9	%area	27.6	47.0	25.5	100.0
10	%number of country	31.5	51.2	17.3	100.0

(e) NIEs = Newly Industrializing Countries, including Hong Kong, Korea, Singapore, and Taiwan. Taiwan's data is missing in the dataset. Thus, NIEs here consists of only the first three. ASEAN-4 = Indonesia, Malaysia, Philippines, and Thailand. You have to create category for NIEs and ASEAN-4 to answer this question.)

13-12 A catch up game Based on the data from Table HW13-1 (or the World Development Report), answer the following questions.

(a) Are the Asian Newly Industrializing Economies (NIEs), Korea, and Taiwan catching up with Japan in terms of per capita GDP?

(b) Are the level of GDP per capita of Japan also catching up with that of the USA?
(c) How many years will it take for these countries to catch up if the current rates of per capita GDP growth are maintained in the future? Using the time profile method and the formula method to answer this quation.

Chapter 14

Research Presentation — Sharing Excel Tables and Charts

Chapter Outline

Objectives of this Chapter

The first part of this chapter examines the use of flow charts to summarize the data structures of economics, business, and statistics that readers have learned in previous chapters. We show how to draw the organization chart manually (Sec. 14.2). After understanding the method of drawing a flowchart, readers practice using a built-in organization chart in the homework section. The homework section also presents a summary diagram for different economic and statistical models covered in this book. The reader will better understand the structure or relationship expounded taught in the previous chapters by drawing the organization charts and diagrams.

Since the major purpose of data analysis using Excel is to get results, which will be explained and interpreted, it is important for the tables and the charts in Excel to be transferred into Word for writing reports or articles, or into PowerPoint for slide presentations. Therefore, we introduce the methods of sharing Excel Data (namely, tables and charts) with Microsoft Word (Sec. 14.3) and with Microsoft PowerPoint (Sec. 14.4). Lastly, since most of the data imported from websites are in text form, we show how to parse the text data into an Excel table. Some useful websites on general, national, regional statistical data are given at the end of the chapter (Sec. 14.5).

14.1 The Fields in Economics

Broadly speaking, economic analysis consists of the analysis of **static models** and **dynamic models**. This classification is based on whether time plays a meaningful and significant role in the model. Time is not explicitly considered in static models, but is in dynamic models. The study of the effects of changes in a parameter or an exogenous variable on the equilibrium values is called **comparative analysis**. Thus, we have **comparative static analysis** and **comparative dynamic analysis**. The relationship is shown in the first part of Fig. 14.2.

Among economic models, some models pose **nonoptimization problems**, such as finding the equilibrium values of a single market model or a national income model, or finding the convergence of a business cycle model to equilibrium values. On the other hand, some economic models study **optimization problems**, such as the maximization of profits, outputs, or utility, or the minimization of cost or loss in static or dynamic models. In many cases, the nonoptimization problems can be derived from the optimization problems. For example, the demand and supply functions of a single market model may be derived from the equilibrium condition (the first-order condition) of utility maximization or cost minimization problems, as we saw in Chapter 12. These relations are shown by directional arrows in the second part of Fig. 14.2.

Noting that economics is a field of social science that deals with individual persons and groups of people or institutions, the above structure of economic analysis reflects the following definition of economics. "**Economics** is the study of how people and society choose to employ scarce resources that could have alternative uses in order to produce various commodities and to distribute them for consumption, now or in the future, among various persons and groups in society". The important point here is "how" people do it. Samuelson and Nordhouse (1989, p. 4) give some other popular definitions of economics as follows:

- "Economics is the science of choice. It studies how people choose to use *scarce* or *limited* productive resources (land, labor, equipment, technical knowledge) to produce various commodities (such as wheat, beef, overcoats, concerts, roads, missiles) and distribute these goods to various members of society for their consumption."
- "Economics is the study of how human beings go about the business of organizing consumption and production activities."
- "Economics analyzes movements in the overall economy — trends in prices, output, and unemployment. Once such phenomena are understood, economics helps develop the policies by which governments can affect the overall economy."

Although our book does not specifically deal with economics, the methods of analysis and the examples given in the chapters of this book include many important aspects of economics and business.

14.2 Creation of an Excel Flowchart

We are now going to study how to create a flowchart, Fig. 14.2, which is also shown in Preface of this book. Most of the drawing tools were explained in Sec. 1.4 of Chapter 1, so the explanation is fairly simple. We first make **square grids** for drawing.

(1) Select the whole worksheet by clicking the "select-all" button or enter <^A>.

(2) **Making plotting paper** To make plotting papers with square cells, drag any column separator in the column heading bar to the left until "Width: 2.14 (20 pixels)" (depending on the height of the row, the default of the row height is 15.00 (20 pixels)) is displayed, as shown in the upper figure of Fig. 1.4(b). If necessary, you may also adjust any row separator in the row-heading bar so that the grids are more or less square shaped.

(3) **Adding a shape** Click <"Insert"><Illustrations, Shapes!><Stars and Banners, Curved Down Ribbon> as in the first row of Fig. 14.2, and place it in an appropriate place on the spreadsheet window. There are three yellow diamonds around the banner, which can be used to modify the shape of banner. Try to reshape the banner as you like it best.

(4) **Text** Select the banner and start writing. A flashing vertical line appears inside the banner. Write "Economic and Business Analysis" as shown in rows 4–7 in Fig. 14.2, add boldface, and center it. You may also use a text box to enter the title. The font size used here is 9 points.

(5) **Rectangles** To add rectangles, click <"Insert"><Illustration, Shapes><Rectangles, Rectangle> and draw a rectangle. Using <"Drawing Tools, Format"><Shape Styles, Shape Fill!><White Background 1>, change the background to white. Similarly, choosing <Shape Styles, Shape Outline!><Weight!>, we change the weight of the borders to 1 point. Click the square box and write "Static Analysis" as shown. Color the background with light yellow.

(6) **Making the squares uniform** To make the squares uniform, hold the control key and **copy** the first square to make the second, third, and fourth squares. Enter the text in each box as shown in Fig. 14.2. You may change the size or font of the three boxes by selecting them simultaneously (click each of the three boxes while holding down the shift key).

(7) **Connections** To make the connections among the squares, click <"Insert"> <Illustration, Shapes><Basic Shapes, right Brace>). The pointer changes to a thin plus sign. Click anywhere in a blank space on the spreadsheet to make a brace. A **brace** (a) appears in Fig. 14.1. There are two yellow diamonds in the upper and middle parts of the brace (you may have to make the brace wide enough to reveal the yellow

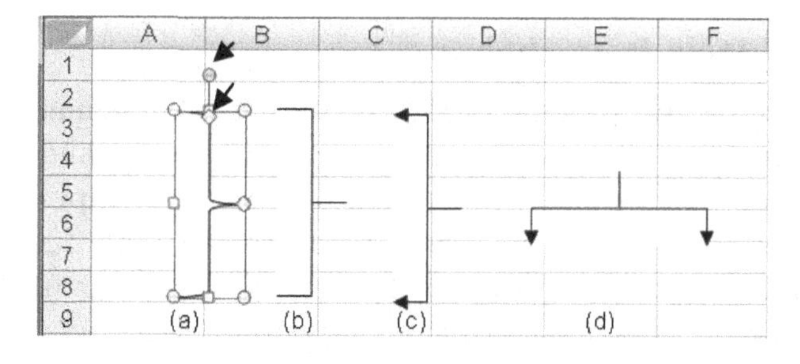

Fig. 14.1 Making a connection

diamond). Click the upper diamond and drag it to the upper edge of the brace. The round edge changes to a **square edge** as shown in (b) and (c). Click <"Drawing Tools, Format"><Shape Styles, Shape Outline!><Arrows!> and choose the **double arrow** (Arrow Styles 7), and you have a double arrow, as shown in (c).

(8) Finally, selecting the square bracket (c) and moving the pointer close to the green circle on top of the square brace, a black half-circle arrow appears. When the pointer is pressed, four black arrows will surround the green circle. Holding down the left mouse and **rotating** the pointer, we create a **horizontal connector** like (d) in Fig. 14.1.

(9) **Adjustment** We copy **connecter** (d) to a place under the "Static Analysis" box, and reproduce this connector as many times as needed, as in Fig. 14.2. Complete the rest of Fig. 14.2 by using the same method as above.

Note that to save time, all the boxes can be **group-selected** by clicking each box while holding the shift key. After the group selection, you may center the contents of the boxes, color them, change the font to 8 points, Times New Roman and move up or down to adjust the location of the boxes simultaneously. Similarly, you may also select multiple connectors by clicking each connector while holding the shift key. You may ignore chapter entries.

(10) **The grid lines** We also like to show the **gridlines**, instead of a white background. For this purpose, select the whole spreadsheet, <^A>, and click

$$\begin{aligned}&<\text{“Office”}><\text{Print, Print Preview}><\text{Print, Page Setup}>\\&<\text{Page Setup, Sheet}><\text{Print, @Gridlines}><\text{OK}>.\end{aligned} \tag{14.1}$$

Gridlines appear in the print preview mode. Click <Preview, Close Print Preview> to return to the current spreadsheet.

14.3 Sharing Excel Data with MS Word

The results of Excel tables or charts from data mining are usually transferred to Microsoft Word (or other word processors) for extensive analysis, explanation, or interpretation. This section introduces the methods of transferring the Excel results to Microsoft Word. There are at least three methods of imbedding Excel tables and charts to Microsoft Word

14.3.1 *Paste Excel table from clipboard*

We construct a 4×4 table like part (a) of Fig. 14.3 in Excel. The numbers can be arbitrary random numbers. Enter ^C to copy the table to the clipboard and ^V to insert it from the clipboard into Word. Right after ^V is pressed in Word, a clipboard logo, the "**Paste Options**", appears at the lower right corner of the copied table in Word (circled), as in part (a) of Fig. 14.3. If the pointer moves inside the table, a four-arrow-headed cross sign in a small square box also appears, as shown at the upper left corner of the table in part (a) of Fig. 14.3. Clicking the "Paste Options" sign, a small menu appears on the right side of the

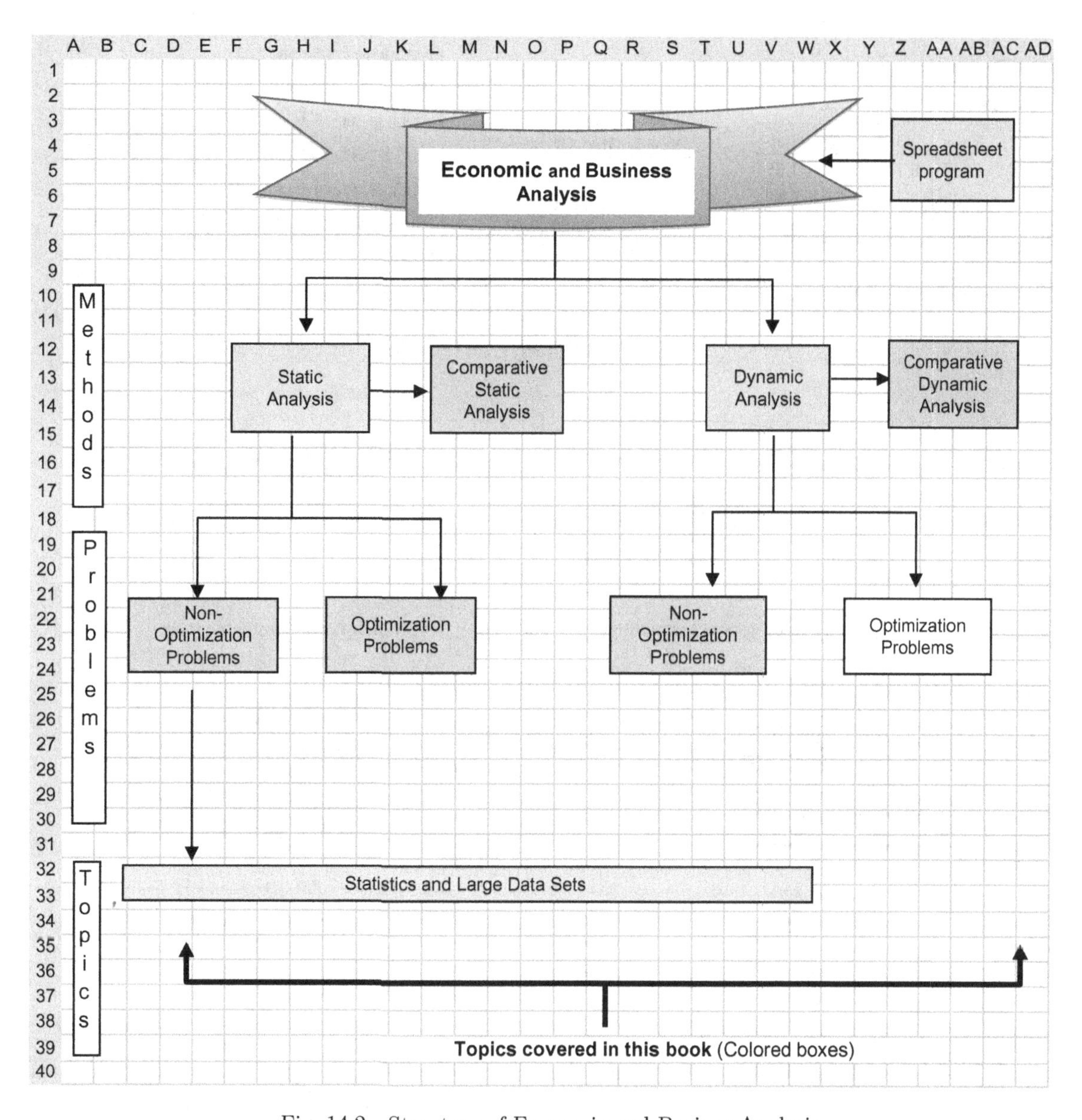

Fig. 14.2 Structure of Economic and Busines Analysis

table, as in part (b) of Fig. 14.3. There are six ways to paste the table from the clipboard, as explained below.

(a) Keep source formatting. This is the default. Select and shade the bullet eye. Click the <Return> key. The table is pasted as HTML (see Sec. 14.5.2 for this acronym) in Word, preserving the original formatting in the Excel table. When the four-headed cross sign in the left upper corner (circled) is clicked, the whole table will be selected, as in

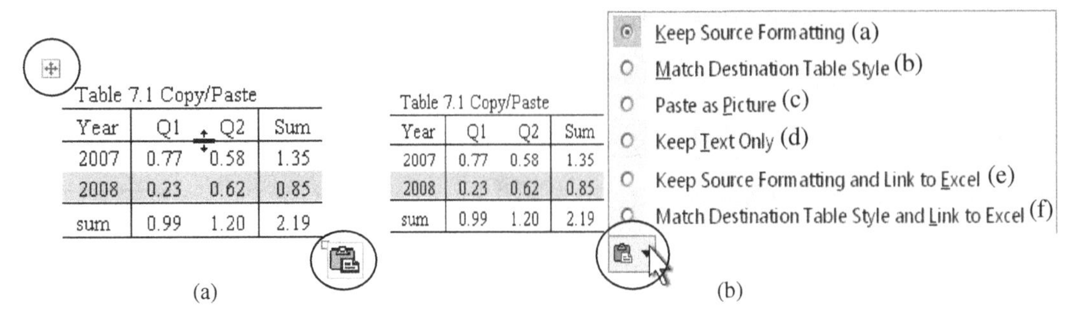

Fig. 14.3 <Copy> in Excel, <Paste> in MS Word

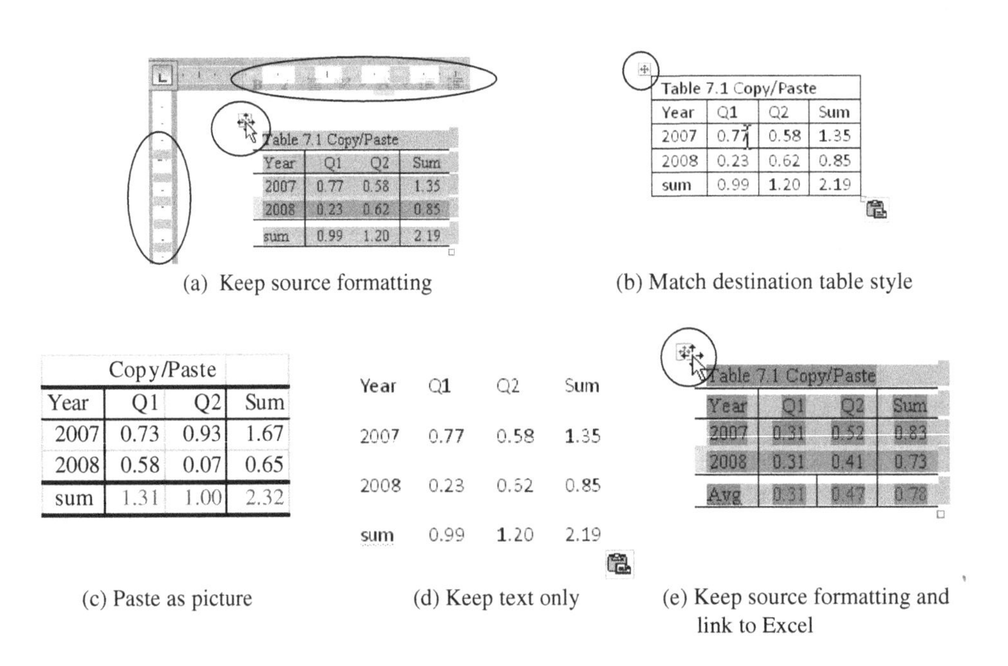

Fig. 14.4 <Copy> in Excel, and <Paste> in MS Word

part (a) of Fig. 14.4. At the same time, the “Table Tools” tab appears in the title bar, and two subtabs, “Design” and “Layout”, appear in the menu bar for editing the table in Word. The gridlines of the cells in the original chart disappear (even after you invoke (14.1)). You can add borders or grids by selecting <“Table Tools, Design”><Table Styles, borders><All Borders>. You may change the border color. The cell size can be adjusted by <“Table Tools, Layout”><Cell Size!,Table Row Height>, and clicking the up or down arrow.

The boxes and texts may be misaligned and may need editing. By selecting the table, and clicking the borders of a row or a column, the row height and column width can be adjusted. A long table may be broken between the pages. However, the contents of the table may be edited directly. To avoid breaking up of the HTML table in Word, we may imbed the table into the **Drawing Canvas**, which is at the bottom of the <"Insert"> <Illustrations, Shapes!> drop-down box. However, the HTML table cannot be imbedded directly into the Drawing Canvas. To do so, make a text box in the canvas and then paste the table into the text box. In this case, the grid of the table cannot be preserved.

The horizontal and vertical tab bars appear in the horizontal and vertical heading bars when the table is selected (make sure the <View Ruler> button is on), as in the upper and left parts of part (a) of Fig. 14.4 (circled). They correspond to the vertical and horizontal borderlines of the cells of the table. These separators can be selected and moved to size the column width or row height of the cells. The problem with this HTML table is that it is hard to adjust the sizes of cells, and a long table may be cut into two parts between two pages. The whole table can be erased by selecting the small square box, clicking <RM>, and selecting <Cut>.

(b) Match destination table style. This is shown in (b) of Fig. 14.4. It is similar to (a), but the cells are bordered. This table is the same as one created from the Word's table command.[1]

(c) Paste as picture. See part (c) of Fig. 14.4. When the table is selected, <"Picture Tools, Format"> appears in the menu bar for editing the table.

(d) Keep text only. See part (d) of Fig. 14.4. In this case, the table format disappears, the contents of the table are copied as texts, and the numbers are separated by tab characters.

(e) Keep source formatting and link to Excel. See part (e) of Fig. 14.4. In this case, when the table is selected in Word, the contents of the table will be shaded, and if the small square box (circled) is clicked, the whole table is selected with shades. This is the same as clicking <"Home"><Clipboard, Paste!><Paste Special...><@Paste Link> under the HTML format. When the original table in Excel has changed, say, the last row changes from "sum" to "average" in Excel, the corresponding table in Word will be changed automatically, or by selecting the table and entering <F9> as shown in the last row of part (e) of Fig. 14.4. To edit or break the link, click anywhere on the table in Word, and go to

<"Office"><Prepare><Edit Link to Files><Break Link>,

or click the table and RM and select <Update Link> or <Linked Worksheet Object><Links...>. A dialog box appears for editing. Note that for the linking to Excel file to be effective, the original Excel file must be open.

(f) Match destination table style and link to Excel. The result is the same as (b) with an Excel link like item (e) above. Note that, as with (e), for the linking to Excel file to be effective, the original Excel file must be open.

[1]In Word, a table can be created by using <"Insert"><Tables, Table!><Excel Spreadsheet>.

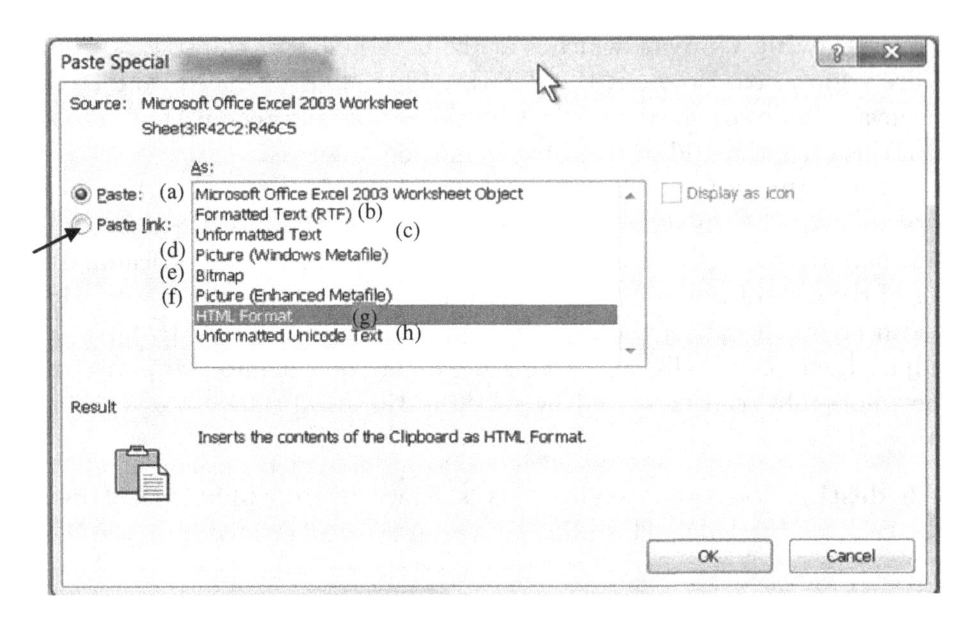

Fig. 14.5 Contents of the "Paste Special" command

14.3.2 *Using the paste special command*

If Microsoft Word is open in a compatible mode (that is, in Excel 97-2003 format), then the above clipboard method may distort the table when it is pasted in Word. In this case, we may use Excel's **paste special** command. Suppose we still have part (a) of Table 14.3 on the Excel spreadsheet. Select the table in Excel and enter <^C>. Keeping the Excel file open, move the pointer to Word, and in Word, click

<"Home"><Clipboard, Paste!><Paste Special>

A dialog box Fig. 14.5 appears. There are eight methods of pasting, numbered from (a) to (h) in Fig. 14.5. They can be grouped in four categories, most overlapping with the items in the "Paste Options" dialog box in Fig. 14.3.

(a) *Paste as Microsoft Excel 2003 Worksheet Object*

Click item (a) and <OK>. The grid lines are light, just intended to show the positions of the entries. Click the chart twice and the whole Excel program and entire Excel workbook will appear. You may edit the chart as if it is in the Excel program, and the Excel ribbon appears for you to use. Click any place outside the chart to return to Word and the Word ribbon. This is very convenient, since you can edit the chart in Excel. However, the file tends to be large.

(b)(g) *Paste as formatted text*

Paste either as formatted text (RTF) (b) or in HTLM Format (g) Both results are the same as part (a) of Fig. 14.4. The HTLM format (g) is a newer format than the RTF format (b), and has less distortion.

(c)(h) *Paste unformatted text*

Paste as unformatted text (c) or unformatted Unicode text (h). Both results are similar to part (d) of Fig. 14.4. The Unicode text can accommodate nonregular ANSI characters.

(d)(e)(f) *Paste as a picture*

Paste as Picture (Windows Metafile) (d), Bitmap (e), or Picture (Enhanced Metafile) (f). The three pasting results are similar to part (c) of Fig. 14.4. The pictures can be edited by using <"Picture Tools, Format"> tab. Generally, Bitmap picture will give better image, but the file is larger. The tables and charts in this book use (f) to transfer from Excel to Word.

Paste Linking

All the above pasting features can be linked to the original Excel table by clicking the Paste Link button on the upper left side of Fig. 14.5.

14.3.3 *Pasting the chart as a picture format*

For copying and pasting an Excel chart, Excel 2007 also comes with an older picture copy command. In Excel, select an Excel chart and click <"Home"><Clipboard, Paste!><As Picture!><Copy as Picture><Copy Picture, Appearance, @As shown on screen><OK>. Then, open Word, and select ^V to paste it in Word or in the graphic canvas[2] in Word. The chart or table is copied as a picture and can only be edited as a picture. If the original Excel worksheet has embedded grid lines (see below), the grid lines will be shown in the picture.

14.4 Sharing Excel Data with MS PowerPoint

We now use the computer as a slide projector to display the presentation on the computer monitor or on the screen. This is a slide show or slide presentation.

14.4.1 *Making slides*

In an Excel spreadsheet, we will construct the table and chart of the Total Productivity Surface, Table 11.1, and Labor Productivity Curve, Fig. 11.2(b). (If they are not available, construct a simple 3 × 2 data table similar to Fig. 14.3(a), with random numbers for Q1 and Q2 from 2007 to 2009 without sums. Deleting the year label, draw a column chart.) The table and chart are reproduced in A11:R23 in parts (a) and (b) of Fig. 14.7. We will call the table Part (a) and the chart Part (b). These are originally entered in Excel. Our purpose is to copy them from Excel to PowerPoint to make a slide show.

[2]Graphic canvas is useful if items in Shapes, like a circle or an arrow, are added to a table or a chart. The canvas keeps all shapes together when the table or chart is moved.

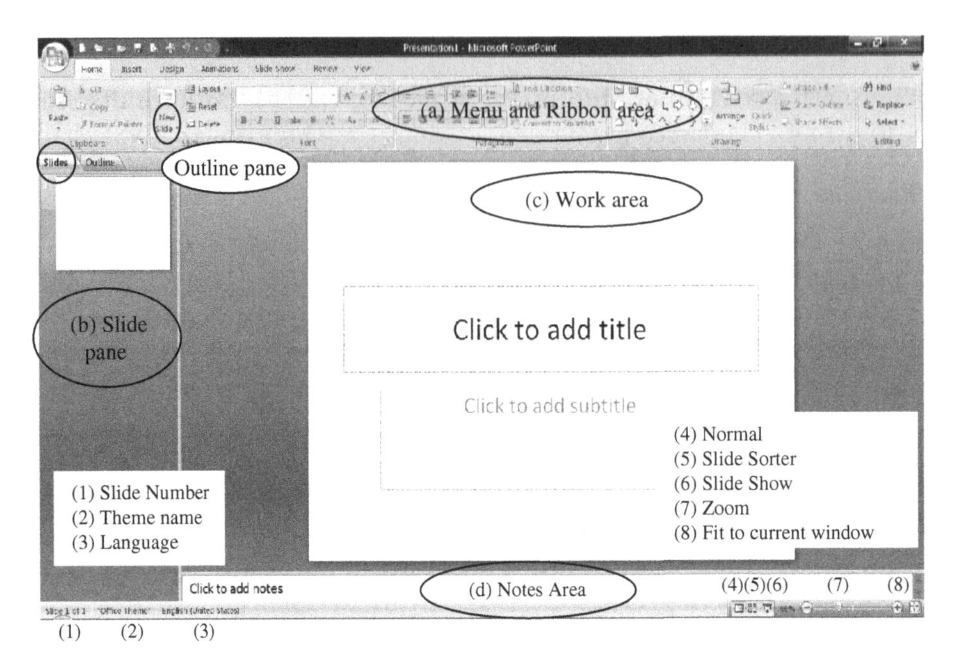

Fig. 14.6 PowerPoint window

(1) Open the PowerPoint (PP) program. When we start the PowerPoint, the **PowerPoint** window Fig. 14.6 appears. It consists of four parts: (a) The menu and ribbon area, (b) the slide pane, which lists all the current slides, (c) the work area, for editing the slides, and (d) the notes area for the slide, for keeping notes about each slide.

In the "Click to add title" box (c), enter "Transferring Tables and Charts from Excel Worksheets". In the "Click to add subtitle" box (c) of Fig. 14.6, enter "Different Methods". We have the first slide in the slide pane (b) of Fig. 14.7. If you like, you may change the letter style of the titles by clicking <"Insert"><Text, WordArt> and choosing a style for the letters.

(2) Open a new slide. After entering the title page as slide 1, the next slide is created by copying the table and chart from Excel and pasting them to the work area (c) of PP. This new slide is different from the title page slide. To add a **new slide** in the work area, click in PP.

<"Home"><Slides, New Slide!><Title and Content>.

You may choose any one of the nine pre-structured slides. A new blank slide appears in the PP window, and a small blank slide is added below slide 1 in the slide pane (b). It has number 2 on the left, and we will call it slide 2. In the work area of slide 2, the "Click to add title" box, and "Click to add subtitle" appear again.

(3) Copy the table and chart in Excel. Open the Excel spreadsheet, select table (a) and chart (b) in range A11:R23 of the Excel spreadsheet, and enter ^C.

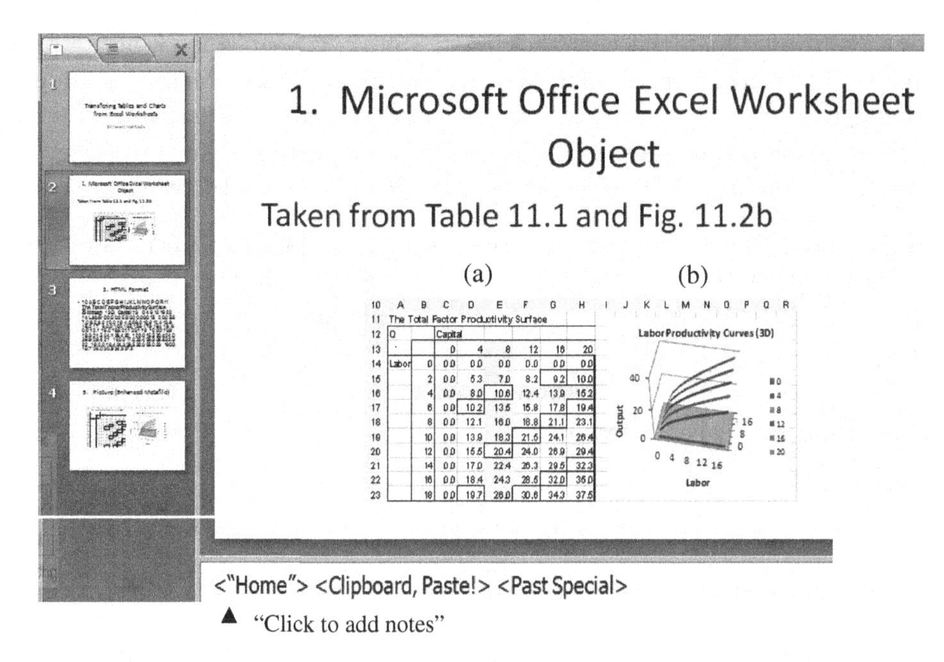

Fig. 14.7 Paste special in PowerPoint program

(4) Paste Special in PP. Moving the pointer to the lower part of slide 2, click

<"Home"><Clipboard, Paste!><Paste Special>.

A "Paste Special" dialog box, similar to Fig. 14.5, appears. From this dialog box, click the first item,

<Paste Special, @Paste As: Microsoft Office Excel Worksheet Object><OK>.

The table and chart are embedded in the PP work area. The result is the same as part (a) of "Microsoft Excel 2003 Worksheet Object" in Sec. 14.3.2. Click the chart twice; the whole Excel spreadsheet will open in PP and you can edit, using the Excel spreadsheet commands. To close the Excel spreadsheet in PP and return to the PP window, click anywhere outside of the table.

Note that the borders of slide 2 in the slide pane of Fig. 14.7 is shaded to show that slide 2 is the active slide in the PP work area.

(5) HTML Format. Open the new slide as in step 2, repeat steps 3 and 4 above, and select

<Paste Special, @paste As: HTML Format><OK>,

and we have slide 3 as shown in the slide pane of Fig. 14.7. The HTML will erase the table format and give unformatted distorted text, as shown by the work area of slide 3 in the slide pane of Fig. 14.7. It appears that this format is not much of use.

(6) Picture Format. Open the new slide as in step 2, repeat steps 3 and 4 above, and select Bitmap, Picture (Enhanced Metafile) or Picture (Window Metafile) under <Paste Special> dialog box, and you can paste the table and chart from Excel to the work area of the PP window, as shown in slide 4 of Fig. 14.7. It is a picture, like (c) of Fig. 14.4, and can be edited by <"Picture Tools, Format"> when the picture in the work area is selected.

(7) Other Format. Repeat the same procedure, and the last two formats of the Special Paste, "Formatted Text (RTF)", and "Unformatted Text" will give pasted results similar to part (d) of Fig. 14.4. We have obtained slide 5 and slide 6. In both cases there are no table frames.

14.4.2 *Slide view and presentation*

Viewing the slides

To **show the slides** either on the computer screen or on the projector screen, click

<"Slide Show"><Start Slide Show, From Beginning> or <From Current Slide>.

The slide will fill up the computer screen. Use <Pageup> and <Pagedown> keys **to move the slides**, or press LM or the space bar, or <RM><Next> to go to the next slide. Get out of the slide show by pressing the <ESC> key or clicking the LM at the end. You may start the slide show anytime by clicking the "Slide Show" button (6) in Fig. 14.6. You may also click <Start Slide Show, **Custom Slide Show**> and select only the slides you want to present.

Presentation

While viewing a slide in the Slide Show mode, when RM is pressed a dialog box comes out, as in Fig. 14.8. You may then **go to any slide** by pressing <Go to Slide> and choosing the destination. If you want to make a point, you may **annotate or underline** the contents of the slide by selecting <Pointer Options> and a pen to draw a shape, say a circle, as shown in Fig. 14.8. The drawing can be **erased** by <Eraser> or <Erase All Ink on Slide>, which will show up in the dialog box on the right side of Fig. 14.8 when a pen or highlighter is used.

Setting up the slide show

You may control the slide show by using <"Slide Show"> <Set Up, Set Up Slide Show!> and choosing **show type** as <@Presented by a speaker (full screen)>. If you want to let the slides run unattended, you have to set up **automatic timing** for running each slide by clicking <"Slide Show"> <Set Up, Rehearse Timings!>. Then, the <Rehearsal> toobar appears at the upper left corner of the slideshow window. While the time is ticking, you press <Enter> or <Next> key to record the time interval of moving to the next slide. Then, <Esc><Yes>. The "Slide Sorter" window comes out. Click <Start Slide Show, From Beginning> to start the show.

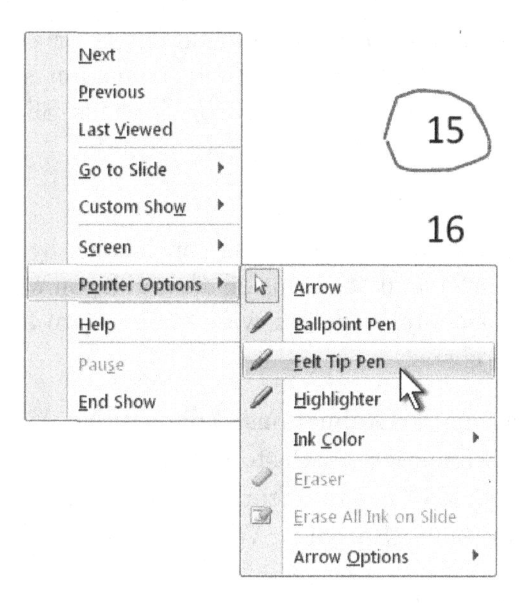

Fig. 14.8 Slide view management

14.4.3 *Slide management*

We can use the ribbons to edit slides, add themes, change colors and fonts, etc. At the lower right-hand side of the work area, in addition to (6) **Slide Show** (which is shorthand for <"View"><Presentation Views, Slide Show>), there are the **Normal** button (4) (shorthand for <"View"><Presentation Views, Normal>) to go back to Figs. 14.6 or 14.7, and the **Slide Sorter** button (5) (shorthand for <"View"><Presentation Views Slide Sorter>). You may also click and experiment with <Slide Master>,<Handout Master> and <Notes Master> under the <"View"><Presentation Views> category.

Moving, copying, hiding the slides

As with Excel, you can copy, move, or delete the slides in the slide pane (of Fig. 14.7, or part (b) of Fig. 14.6) or in the Slide Sorter (press (5) in Fig. 14.6 or <"View"><Presentation Views, Slide Sorter>). This can be done by **selecting a slide or a group of slides** (by pressing the control key and click the slides) and then entering the ^C ^V, ^X^V sequence, or using the <Delete> key, or clicking the RM and selecting an item from the dialog box. You can **hide or unhide a slide** either by clicking <"Slide Show"><Set Up><Hide Slide>, or by selecting a slide or a group of slides in the slide pane or the Slide Sorter and choosing <RM><Hide Slide>. The slide number in the slide pane or in the Slide Sorter will be enclosed by a square with a backward slash indicating that the slide is hidden. <Hide Slide> is a toggle key, and you can unhide the slide by clicking <RM><Hide Slide> again.

Slide transition

One of the unique features of PowerPoint is the "**Animations**" tab in the ribbon. You may customize the transition between slides by <"Animations"><Transition to This Slide,

Transition Scheme!> and choose the way the slide enters or exits. You may also choose a transition sound, what kind of sound, and the transition speed. You can apply the modification to all slides by clicking <Transition to This Slide, Apply to All> under <"Animations">.

Text animation

If, at the end of the presentation, you show the conclusions by putting them in order, as "Conclusion 1", "Conclusion 2", and "Conclusion 3", and if you want to have each conclusion **showing up separately**, then take the following steps. Open **a text box** then enter and shade "Conclusion 1", and click

<"Animations"><Animations, Animate: No Animation!>
<Custom Animation!><Custom Animation, Add Effect!>
<Entrance!><5 Fly In><Start!><On Click><Direction!>
<From Bottom><Speed!><Very Fast>.

The upper left corner of Conclusion 1 will show a number in a square, indicating its sequence in the animation. You may click <Play> at the bottom of the "Custom Animation" dialog box to see the effect.

Next, write "Conclusion 2" inside the **same** text box in which the animated "Conclusion 1" is located. When you enter "Conclusion 2" below "Conclusion 1" and press the return key, "Conclusion 2" will be automatically animated with the same animation scheme as "Conclusion 1". If you start a new text box and enter "Conclusion 3", no animation will be added. Also see the instruction inside Fig. 14.9 for more information. If you enter "Conclusion 3" in the same text box, but you do not want animation for Conclusion 3, click the animation number and press <**Remove**>. The animation will be removed from Conclusion 3, which will appear with the title when the title slide first appears, rather than appearing in a sequence of items on the title slide.

To test your animation scheme, click the <**Play**> button at the bottom of the "Custom Animation" dialog box. <On Click> does not work in Normal view ((4) in Fig. 14.6) mode. To show the actual effect, click <"Slide Show"><Start Slide Show, From Current Slide>. The nonanimated part appears first. Click LM or the space bar each time to bring up "Conclusions" one by one.

14.4.4 *Printing the slides*

There are several options for printing slides when you choose an option from the "Print What:" at the bottom of the <Office><Print!><Print> dialog box. There are four choices:

- <Slides> Prints the slides one by one;
- <Handouts> Prints 2, 3, 4, 6, or 9 slides per page;
- <Notes Pages> Prints the slide with speaker's notes;
- <Outline View> Prints the Outline pane of Fig. 14.6.

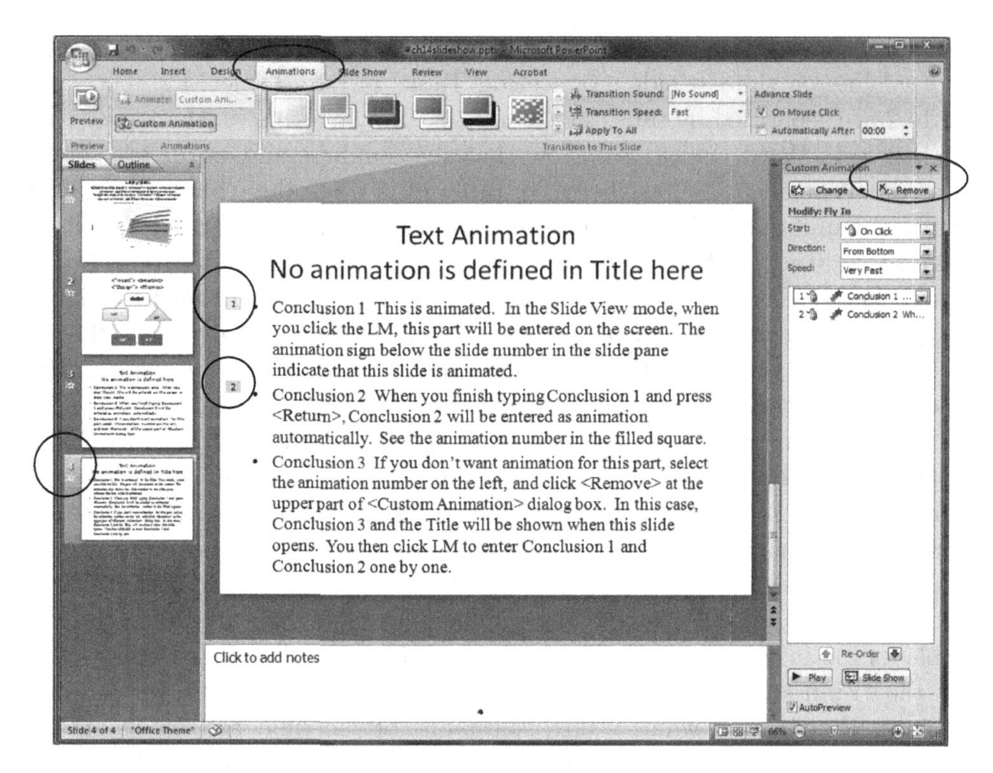

Fig. 14.9 Text animation

For example, if you select <Handouts>, "6 slides per page", and "horizontal order of the appearance of the slides", and <OK>, the slides will be printed out in that format.

14.5 The Text Import Wizard

When we download a dataset from a website, the saved data may be in a text file or an Excel file (and sometimes the data may be downloaded directly into an Excel worksheet). If it is a text file, the data should be converted into an Excel file in order to take advantage of the Excel data analysis we have expounded in the previous chapters. This section constructs simulated downloaded data in a text file of Microsoft Word, and then shows how to convert the text file into an Excel file. The process of separating text data into columns in table format in Excel is called **parsing**.

14.5.1 *Using the text import wizard*

On the Excel spreadsheet, create a 6×5 matrix of random numbers in range A1:E6, as shown in Table 14.1.

(1) **Opening Microsoft Word.** In Excel, <copy> A1:E6 and <paste> the range as an unformatted text in Word (see (c) of Fig. 14.5). This gives us Table 14.2, which is the

same as part (d) of Fig. 14.4. The construction of an artificial text file is shown by the dotted line on the left side, connecting Table 14.1 to Table 14.2. The numbers in Table 14.2 are separated by tabs. This can be seen by clicking <"Home"><Paragraph, Show/Hide> in Word, which will show that the numbers and words are separated by right arrows (namely, tabs), as shown in Table 14.2. Click the <Show/Hide> button to return to the unformatted text (without arrows), as shown in part (d) of Fig. 14.4.

(2) **A text separated by tabs or space.** This is the case with Table 14.2. If Table 14.2 is copied back to Excel, it will automatically **parse**, and you will get the original table, Table 14.1, in Excel. This relation is shown by the solid line on the right-hand side connecting Table 14.2 to Table 14.1. This means that if the numbers in Word are separated by tabs and arranged systematically, as in Table 14.2, then the numbers in Word will be parsed automatically into columns when they are copied to Excel.

(3) **Constructing a data set with comma separators.** In many cases, when the data are downloaded or scanned (by scanner) and pasted in Word, the numbers are separated by either spaces or commas. To simulate this situation, we will separate the numbers by commas, instead of tabs (right arrows). After revealing the show/hide markers, as in Table 14.2 in Word, we see that the numbers are now separated by right arrows (as in Table 14.2). In Word, enter ˆF (that is, <"Home"><Editing, Find!><Find. . . >) to get the "Find and Replace" dialog box, as shown in Fig. 14.10. In the dialog box, to fill in <"Find", Find what:> box, select one of the right arrows, as shown in the shaded right arrow (the first arrow) in Table 14.2, copy and paste it inside the "Find what" box in Fig. 14.10. In the <"Replace"><Replace with:> box, enter a comma (,) and one space (by pressing the space bar once). Then, click the <Replace All> button. Table 14.2 changes to Table 14.3. All the numbers are separated by the comma and space combination. The construction of the artificial text file from Table 14.1 to Table 14.3 is indicated by the dotted line on the left.

(4) **A text separated by commas or others.** Copy Table 14.3 in Word back to the Excel spreadsheet, and the table will look like Table 14.4, which is in text format, in the sense

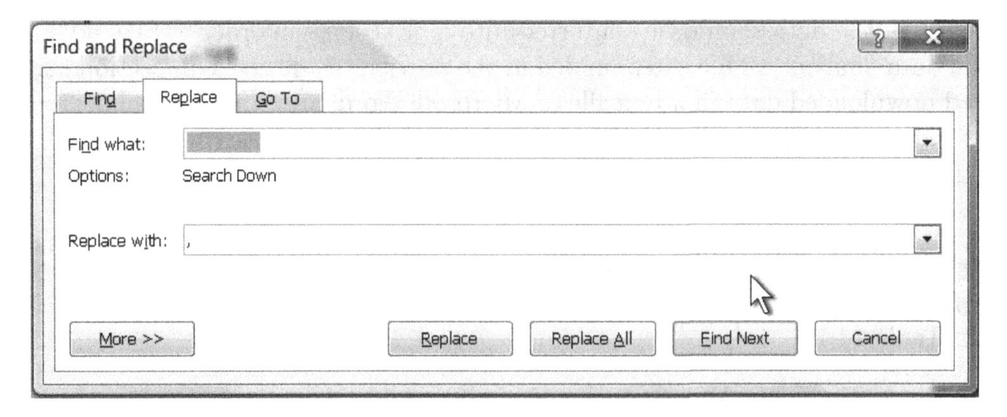

Fig. 14.10 Find and replace dialog box

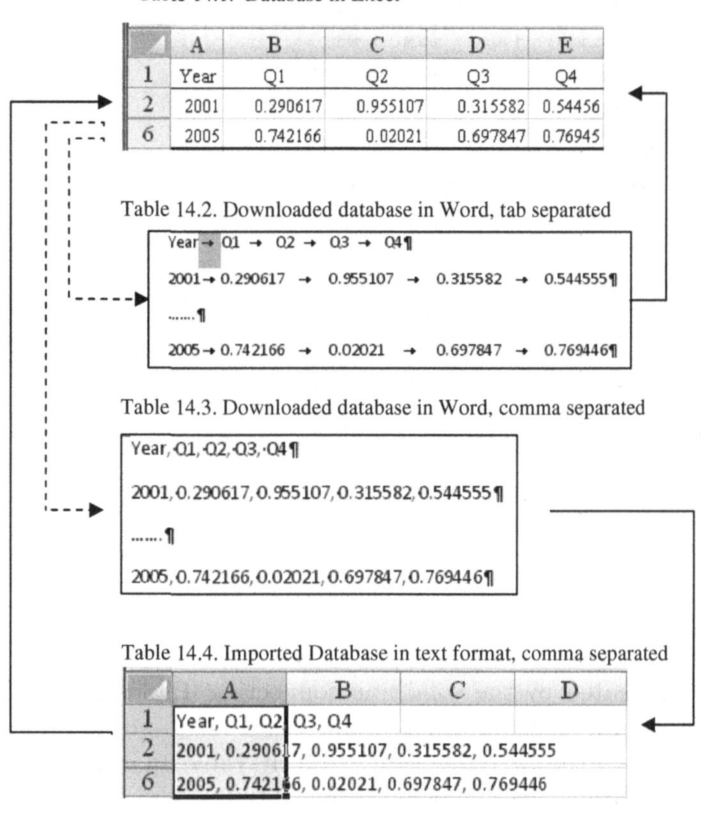

Table 14.1. Database in Excel

	A	B	C	D	E
1	Year	Q1	Q2	Q3	Q4
2	2001	0.290617	0.955107	0.315582	0.54456
6	2005	0.742166	0.02021	0.697847	0.76945

Table 14.2. Downloaded database in Word, tab separated

Year→Q1 → Q2 → Q3 → Q4¶
2001→0.290617 → 0.955107 → 0.315582 → 0.544555¶
.......¶
2005→0.742166 → 0.02021 → 0.697847 → 0.769446¶

Table 14.3. Downloaded database in Word, comma separated

Year,·Q1,·Q2,·Q3,·Q4¶
2001,0.290617,0.955107,0.315582,0.544555¶
.......¶
2005,0.742166,0.02021,0.697847,0.769446¶

Table 14.4. Imported Database in text format, comma separated

	A	B	C	D
1	Year, Q1, Q2, Q3, Q4			
2	2001, 0.290617, 0.955107, 0.315582, 0.544555			
6	2005, 0.742166, 0.02021, 0.697847, 0.769446			

that the entries are concentrated in column A and overflow to other neighboring cells. We now want to separate the numbers in the text format into appropriate columns as in Table 14.1. Select the first column A1:A6 in Table 14.4, and click

<"Data"><Data Tools, Text to Columns>.

A three-step "**Text to Column**" wizard dialog box appears.

Step 1. The "Convert Text to Columns Wizard-Step 1 of 3" dialog box is shown in Fig. 14.11. Since we have commas as separators, click <@Delimited>, and go to <Next>.

Step 2. In the "**Delimited**" field,[3] a "Step 2 of 3" dialog box, Fig. 14.12, comes out. Make sure that Comma is selected, and the numbers are separated correctly by vertical lines in the lower window.

Step 3. We then go to <Next>, that is, to the "Step 3 of 3" dialog box (not shown). The row title will be shaded. If you want to keep the original format of Table 14.3, you

[3] "Delimited" means having limits or boundaries added. A comma or space or a slash can be a "delimiter" to separate the data in a database.

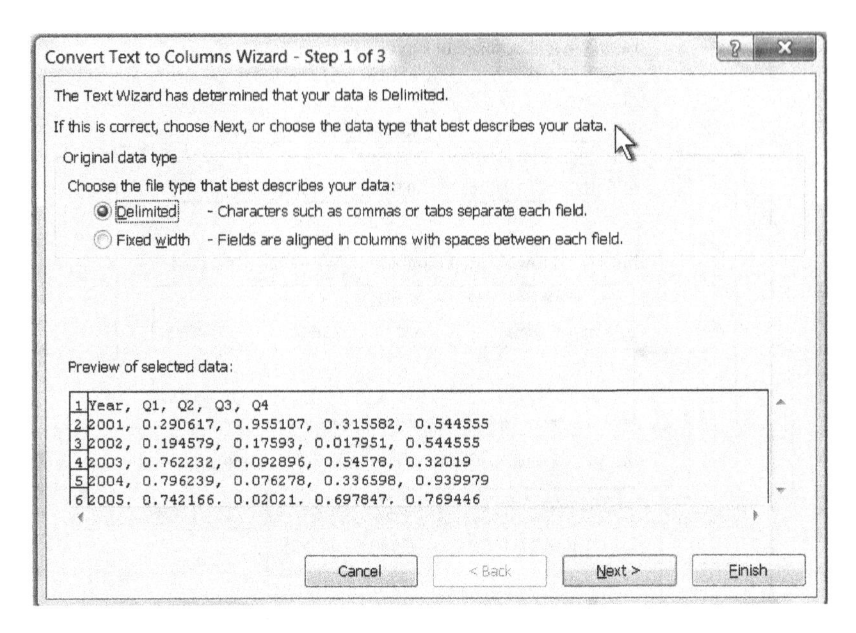

Fig. 14.11 Parsing, Step 1 of 3

may designate the "Destination:", say A11. Click <Finish>. Table 14.4 goes back to Table 14.1, as indicated by the solid line connecting Tables 14.4 and 14.1. The Word file, Table 14.3, is now changed to Excel file Table 14.1, and no comma is shown in the table. Thus, we open the text file Table 14.3 in Table 14.4, and then we **parse** it as Table 14.1. The relationship is shown by solid lines on the right-hand side, connecting Table 14.3 to Table 14.4, and by another solid line, connecting Table 14.4 to Table 14.1, on the left-hand side.

(5) If "**Fixed with**" is selected in Fig. 14.11, we will have Fig. 14.13. You can click a point along the ruler and add a **separating arrow** so that the numbers are separated correctly. You can move the arrow by dragging it to a correct position, or remove the arrow by dragging it outside of the data area. Click <Next> or <Finish> to return to Table 14.1. In this case, a comma will be added at the end of the number in each cell. To get rid of the commas, use "Find and Replace" in Fig. 14.10 to find the commas and replace nothing (just clicking the "Replace with" box without entering anything).

14.5.2 *Some useful websites*

Text Import Wizard is most useful when the data are downloaded from the websites through Internet. This section introduces some basic terminology for searching websites and some useful websites with resources for economics and country studies. The websites are current to July 2010. Since they may be revised or moved, readers may use a search engine like Google or Yahoo to find the exact location.

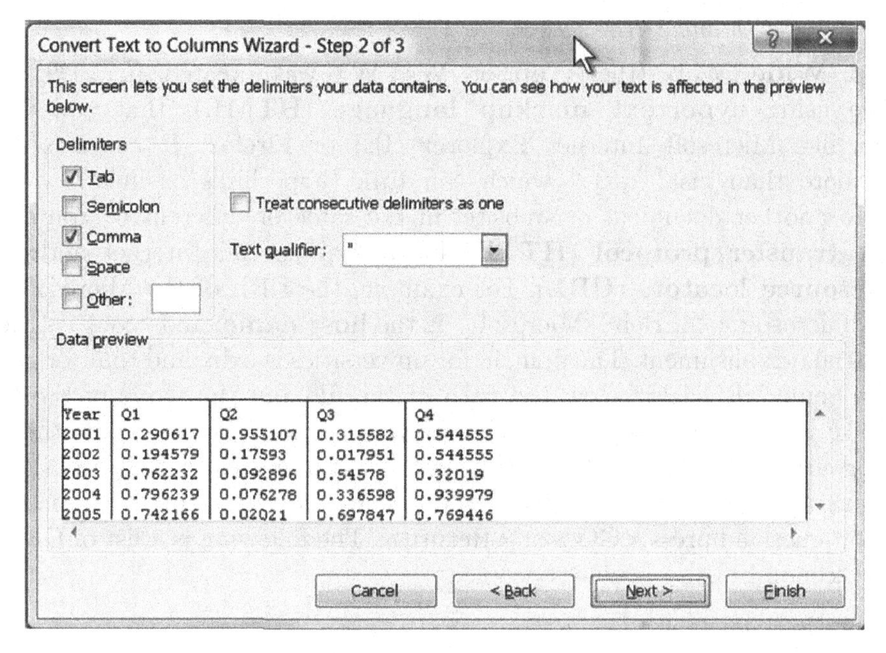

Fig. 14.12 Parsing, Step 2 of 3 — with delimited fields

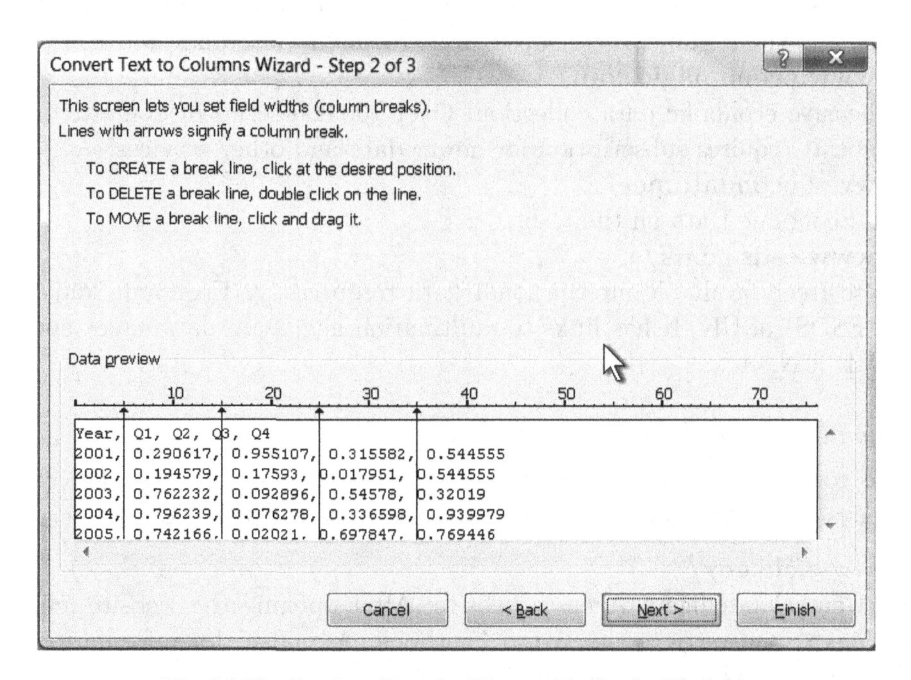

Fig. 14.13 Parsing, Step 2 of 3 — with fixed with fields

The World Wide Web and URL

The **World Wide Web** (the **Web** or **WWW**) was created in 1989, and is an encoded file using **hypertext markup language** (**HTML**), that can be read by a **browser**, like Microsoft Internet Explorer (IE) or Firefox. It contains **hypertext** (meaning "more than just" text)[4] which can **link** (hyperlinks or links) one document or website to another document or website, in the same or different computers, through **Hypertext transfer protocol** (**HTTP**) located at special Internet addresses called **uniform resource locators** (**URL**). For example, the URL of the Microsoft website is http://www.microsoft.com. Here "Microsoft" is the **host name**, and "com" is the **domain** (for commercial establishment. The domain for universities is .edu, and that for governments is gov, etc). Some URLs also show the path of the file and the name of the file and its extension, like html or htm, separated by a **forward slash** or a period at the end of the URL. If a forward slash/at the end of URL is present, it indicates that the website is a folder (not a file) name. When you want to visit the website, you type this URL in the **address box** of the browser and press <GO> or <Return>. The following is a list of URLs of useful websites for economists and business people.

Resources for Economists on the Internet

- **http://www.rfe.org/**
  This is a site sponsored by the American Economic Association. Probably the most comprehensive site for all aspects of economic research. (also known as Bill Coffe's page)
- **http://www.economagic.com/**
  Comprehensive economic data collection. Over 100,000 series of domestic and international data. It requires subscription for newer data and other services.
- **http://www.econdata.net/**
  Regional Economic Data on the web.
- **http://www.esds.ac.uk/**
  A guide to freely available international data resources by Economic and Social Data Service (ESDS) of UK. It has links to multination aggregate databanks, such as World Bank, IMF, UN, etc.

Country information and data

Most of the country or regional statistical data are available on line. In economic and business analysis, the following sites are useful. Enter the following URLs:

- **http://www.cia.gov/**
  The US Central Intelligent Agency website; After opening the website, click <library publications>, and retrieve the World Factbook. A "must" for information about any country in the world; Great maps and appendices;

---

[4]Unlike ordinary static "text", "hypertext" is **dynamic** in the sense that, when the text is clicked, a dialog box or window appears, a web page opens, a video clip runs, or some other action occurs.

- **http://www.worldbank.org/**
  The World Bank website; After entering the website, click <Data and Research> <Key Statistics> <World Development Indicators>. For example, you may download the data on GDP (current US$) and GDP Growth (annual %) for a particular country.

In the following URL, add **http://www.** at the beginning of the text.

- **National Bureau of Economic Research (NBER), nber.org**
  <"Data">
- **International Monetary Fund (IMF), imf.org**
  <"Data and Statistics><Global Data><IMF Financial Data>, <Exchange Rates Data>
- **Organization for Economic Cooperation and Development (OECD) (OECD), oecd.org**
  <Find, Statistics><Data warehouse and databases>
- **United Nation Statistics Division, unstats.un.org**
  Click <Economic Statistics> or <Demographic & Social Statistics> or <Environment and Energy Statistics>
- **Asian Development Bank, adb.org**
  <"Economics and Statistics"><"Database and Development Indicators">
  <"Key Indicators">
- **Japanese Government Statistics, stat.go.jp/English**
  <"Statistics>
- **Chinese Government Statistics, stats.gov.cn/English**
  <"Statistical Data>

The last two URLs are listed as examples of the data available in two specific countries. Most countries have their own statistics department. They are listed in the ESDS website mentioned above.

14.6 Summary

The first part of this chapter is an advanced version of applications of drawing tools introduced in Sec. 1.4 of Chapter 1. As Fig. 1.14 states, this class covers basic concepts and methods in economics, business, statistics, mathematics, and spreadsheet programs. Figure 14.2 illustrates the relations among these five topics. Mathematics is not explicitly labeled as such in the chart, but it is implied in the word "Analysis". Up to this point, we have not covered dynamic analysis or comparative dynamic analysis. They will be studied in the next chapter, as it is usually a more advanced topic in undergraduate teaching and in many cases, is skipped in curriculum planning. We list them in Fig. 14.2 to give a bird's eye view of the whole field of economic and business analysis. The methods of analysis are generic in the sense that they hold in almost all subfields of economics and business, including international economics, labor economics, health economics, resources economics, financial economics, marketing, etc.

We have emphasized in this book the importance of using spreadsheet programs to understand topics in economics and business. The results and conclusions derived from statistical analysis and mathematical methods need to be explained, interpreted, and analyzed. This is done by writing reports, papers, or articles, and the tables and charts may have to be shown in these documents. As explained in Sec. 14.3, there are many ways to do so. They can be in HTML format, Excel object, formatted or unformatted texts, and they may also link with the original tables or charts in Excel. Each of these methods has its own merits and demerits. For clarity of the shared tables and charts, and to keep the file reasonably small, we use <Copy> in Excel and <Paste Special><Picture (Enhance Metafile)> of Fig. 14.5(f) in Word.

Similarly, we may share Excel data with the PowerPoint program for presentation and illustration. As with sharing with Word, there are many ways to share data with PowerPoint, as explained in Sec. 14.4. A unique feature of PowerPoint presentation is that it can be animated in many ways. It is important, however, that the animation should not be used too often or too much, lest the audience should be distracted and fail to concentrate on the contents of your presentation!

Lastly, the reversed procedure of transferring Excel data to Word is transferring Word text to Excel tables. This is often the case if we download data from websites. Sec. 14.5 first set up a simulated downloaded text in Word, and then the text is transferred to Excel for parsing. For a well arranged text file, parsing is easy, but for a long concatenated file, rearrangement of the data before parsing, to let Excel know where to start and where to end, is required. This needs a certain skill and experience.

Key Terms: Economics and Business

Key Terms: Excel, Word, and PowerPoint

Homework Chapter 14

14-1 Circular flow of an economy and microeconomic theory Draw the "Circular Flow of Economy" Fig. 12.17 in Chapter 12.

14-2 Circular flow of an economy (National income) Using AutoShape drawing tools, draw the circular flow of economic activity as shown in Fig. HW14-2, consisting of business and firms on one side and the households and government on the other. In the upper inner circle, the households spend money on the final goods and services, which are part of National Income Spent (GNP), and the upper outer circle is part of the National Income Produced. The lower inner circle measures the National Income Earned, including wages, profits, rents, dividends, interest, etc. The three measures of National Income must be identical in equilibrium.

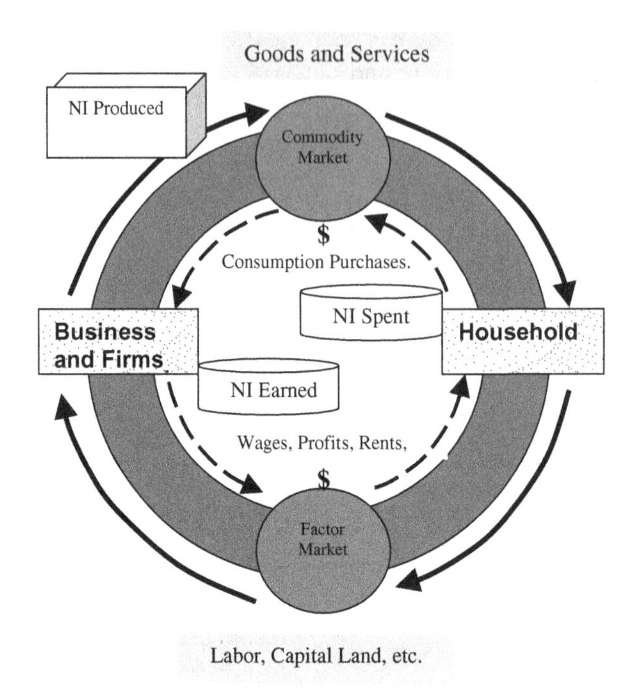

Fig. HW14-2 The circular flow of economic activity

14-3 Scope of economic theory In HW14-2, we have shown the circular flow of economic activity. How is it related to our previous chapters? (a) The macroeconomic models of Chapters 3 and 4 deal with National Income produced (Y) and National Income Spent (C and I). The equilibrium condition is shown in the commodity market. This is illustrated as the upper part of the circular flow. (b) The input–output model in Chapter 10 relates to the inter-industrial relations in the production sector, given the final demand sector. This is shown on the left-hand side of the circular flow involving Business and Firms. (c) The discussion of production functions and utility functions in Chapter 11, and the output maximization models and the utility maximization models in Chapter 12, are related to Business and Firms on the left-hand side and the households and government on the right-hand side of the diagram. (d) Finally, when we combine both the business and firms sector and the households and government sector, then we have a general equilibrium (GE) model. We have not discussed the GE model in this book. Based on the drawing of HW14-2, redraw Fig. HW14-3.

14-4 Organization chart Using the organization chart of Excel (<"Insert">, <Illustrations, SmartArt><Choose a SmartArt Graphic, Hierarchy> and then choose <Hierarchy>), draw an organization chart similar to the one below (Fig. HW14-4).

14-5 Probability Using the organization chart of Excel, draw the following probability chart in Square Shadows diagram style (Fig. HW14-5).

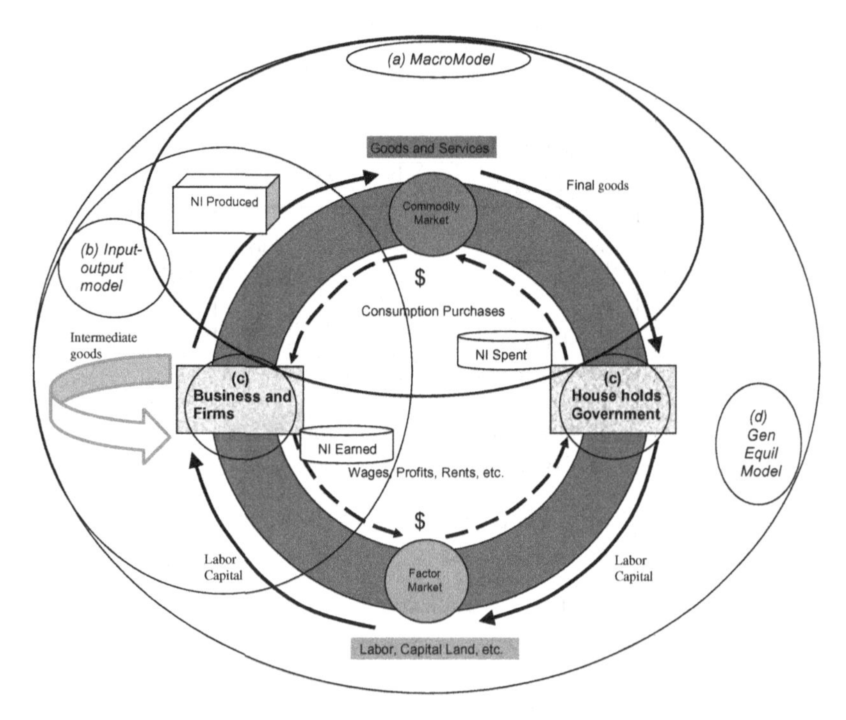

Fig. HW14-3 Scope of some economic theories

14-6 Probability distribution Using Excel, draw the following probability distribution chart (Fig. HW14-6).

14-7 M&A and BOP In international balance of payment (BOP), the exchange of goods and financial assets are recorded as the transfer of ownership (transfer of physical ownership or legal transfer). Hence, cross-border mergers and acquisitions (M&A) through the exchange of stocks are recorded statistically as the transfer of the ownership of the financial assets, namely, stock (not money).

If M&A occurs through stock exchanges, the purchasing firm A in country X gives its stocks to the shareholders of the selling firm B in country Y to compensate firm A's acquisition of firm B's stock. In BOP, the exchanges are recorded as X country's outward foreign direct investment (OFDI), and Y country's inward foreign direct investment (IFDI). The relations are shown in Fig. HW14-7.

On the other hand, after stock exchanges, the firm B's stockholders receive firm A's stock. In the BOP statistics, it is recorded as Y's investment to X. If the firm B stockholders' investment in firm A (which owns firm B now) is less than 10% of firm A's total stock value, then the investment is counted as country Y's outward foreign **portfolio** investment (OFPI) to country X (or country X's IFPI from Y), otherwise, if it exceeds 10%, then it is counted as country Y's outward foreign **direct** investment (OFDI) to X (or X's IFDI from Y) in BOP statistics. The relation is shown in the lower part of the chart. Reproduce the chart. (JETRO 2001, p. 16).

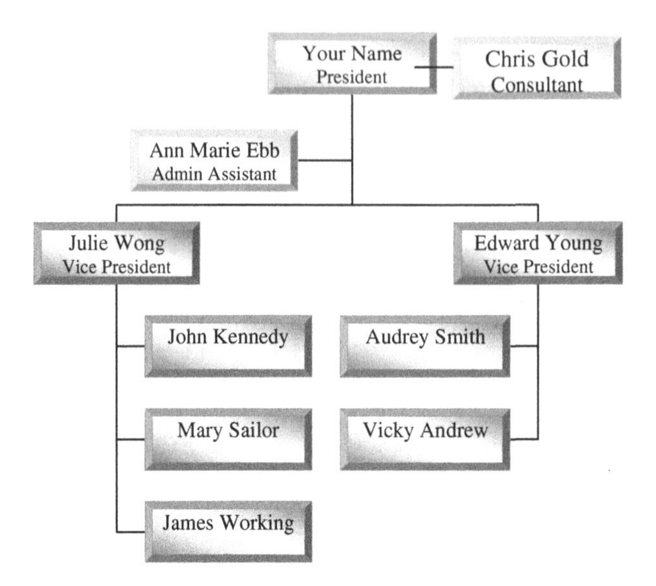

Fig. HW14-4 An organization chart of a company

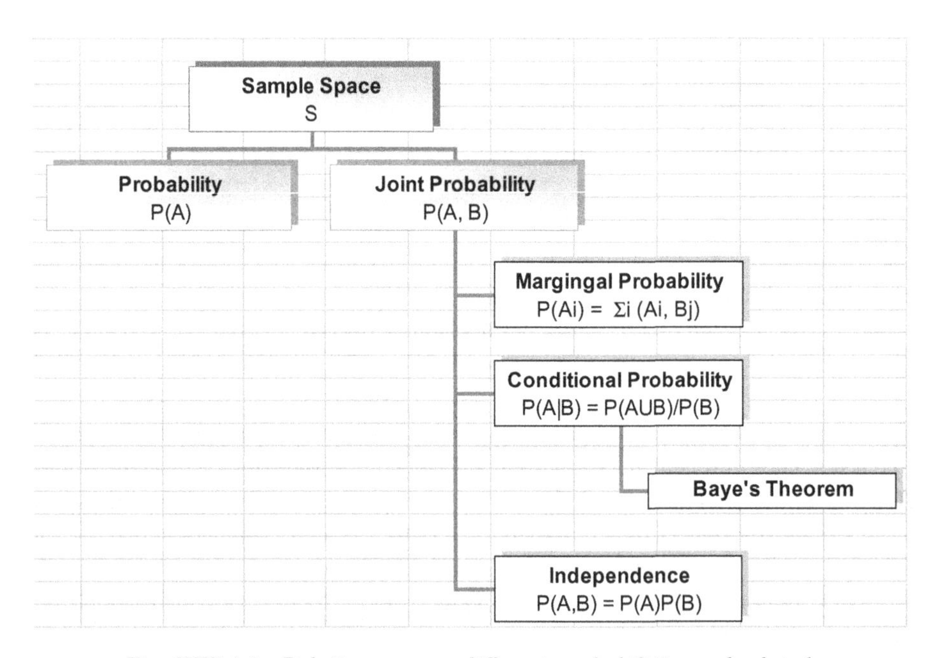

Fig. HW14-5 Relations among different probabilities and related

Pasting Excel Tables to Word

14-8 Copying from Excel to Word Construct a table consisting of the sales of the GiGo Company from 2004 to 2008 for each of four quarters. Let the sales be random variables from 50.00 to 100.00. The last row should show the quarterly average of the sales of five years, and the last

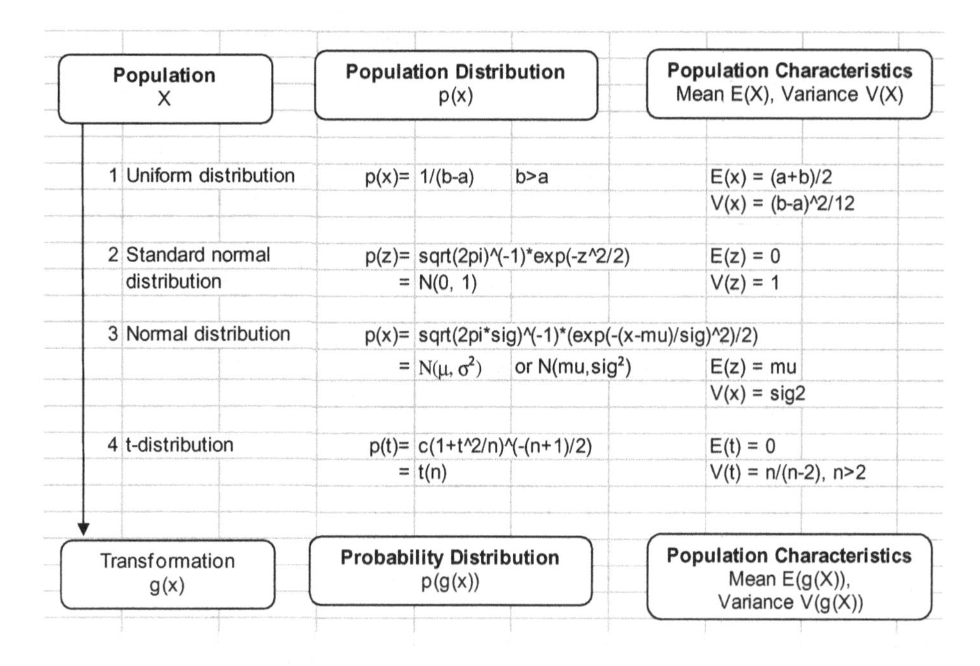

Fig. HW14-6 Some important probability distributions

column should show the sum of the sales of four quarters. Add borders to the data part. Paste this table from clipboard to Word in five ways, as shown in Fig. 14.4.

14-9 Paste special in Word Given the table in HW14-8, using the eight methods in the <Paste Special> command in Fig. 14.5, paste the table into Word. For each table of the <Paste Special> command in Word, give the title to each table, as shown by the title of each part in Fig. 14.4.

Pasting Excel Tables to PowerPoint

14-10 Presentation and animation Similarly to Fig. 14.5, there are eight methods of the special paste command in Excel. (a) Using the Excel table in HW14-8, construct eight PowerPoint slides using each of these eight methods. The title of each slide should be "Paste as xxx" where xxx is the name of the paste special. (b) Reproduce the "Microsoft Office Graphic Object" slide, and call the reproduced slide 9. Hide the original Graphic Object slide. (c) In slide 9, animate the slide by letting the slide enter from the bottom, with a transition sound of "applaud". (d) Let the four-quarter data enter the slide, one column at a time, on a click. (e) In the Notes area of slide 9, describe the process of programming parts (c) and (d). (f) Print out all the slides in two pages.

14-11 Animation Draw a column chart for the Excel table in HW14-8, and make a slide of the chart in the Microsoft Graphic Object. Animate the column entries by <Wipe, By Element in Series>. Write the sequence of animation programming in the Notes area and print out the slide and the notes.

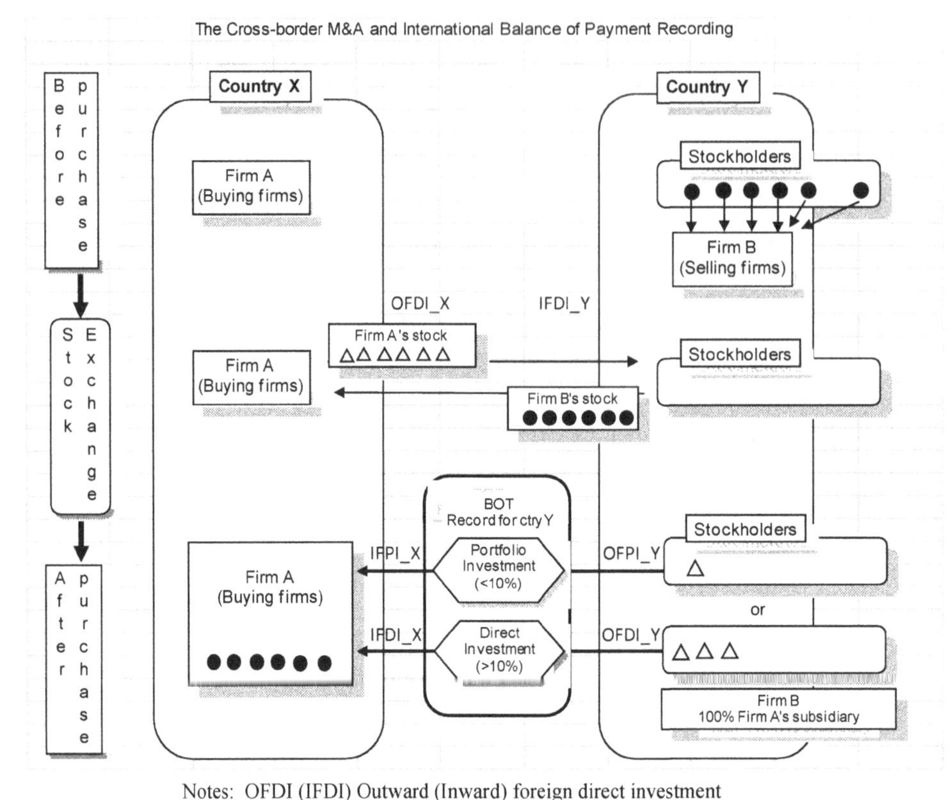

Notes: OFDI (IFDI) Outward (Inward) foreign direct investment
OFPI (IFPI) Outward (Inward) foreign portfolio investment

Fig. HW14-7 The cross-border M&A and International balance of payment

Text Import Wizard

14-12 Parsing Given the table in HW14-8, copy it as Table HW14-12 in Word with comma separators, copy the table back to Excel, and parse it to restore it to the original table.

14-13 Parsing data from PowerPoint In slide 3 of Fig. 14.7, we find that when an Excel table is pasted as HTML Format in a PowerPoint slide, the table format will be lost and the numbers will become a continuous sequence of numbers. Given the table in HW14-8, paste it as HTML Format in the PowerPoint slide. Copy this PowerPoint slide back to Excel, and parse it to have the original table.

(Hints: In the PowerPoint slide, rearrange the data so that each row starts from a year. The last row should start from average. Then, copy and paste the slide to Excel. If necessary, copy row by row. Use <Text to Columns> and click <Fixed width> in Step 1 of 3.)

Websites

14-14 The US economy and the Japanese economy Write a report of one or two pages comparing the US economy with the Japanese economy. What are the major differences between these two large economies in the world today? Explain briefly. You may download data and use charts to make your points. Specifically, what are the levels of real GDP per capita? How about their nominal GDP and nominal GDP per capita? Please try to write in your own words, and make sure the sources of data and texts you use are duly credited.

14-15 Some economic and business websites Visit CNNMoney.com or the Wall Street Journal website and print out today's closing stock prices for the shares of IBM (ibm), Microsoft (msft), Dell, and Home Depot. Please make sure that the date of the data is indicated. Go to the <Finance><Historical Prices> section of www.yahoo.com and graph the past 52-week daily closing prices for the Dow Jones Industrial Average Index (DJI). Do not download extraneous information.

14-16 Economic time series Print any two long annual data of economic time series (like GDP, government expenditure, tax revenue, the consumer price index, prime interest rates, etc.) for at least 20 years, for any two countries from a website like http://www.esds.ac.uk or www.Econommagic.com, www.econstats.com (for commodity prices), www.statistics.gov.uk/statbase/, etc. Draw charts to compare the economic performance of the two countries, and present them in PowerPoint. Print out the PowerPoint slides on one page.

Part VI

Dynamics and Comparative Dynamics

Chapter 15

An Introduction to Dynamic Analysis — Linked Cells

Chapter Outline

Objectives of this Chapter

In Chapters 3 and 4, we introduced static and comparative static analysis in microeconomics and macroeconomics. The variables in these chapters, such as price and quantity, income and consumption, do not change over time, and time is not explicitly considered. In reality, however, variables do change over time, as we encountered in the time series of income and consumption in regression analysis in Chapters 5–7. In economics and business, the study of effects of changes in variables over time is called **dynamic analysis**. This field studies how variables change over time, whether they converge toward equilibrium values (**stability**), and whether the equilibrium exists (**existence**) and is unique (**uniqueness**). Usually, the equilibrium values obtained in dynamic analysis are the same as those obtained in static analysis. In this way, dynamic analysis is closely related to static analysis. For example, in the market model of Chapter 3, we found equilibrium price and equilibrium quantity in static analysis. We now ask how the price and quantity change from a certain point (called the initial condition) to converge toward the equilibrium values. It is said that static analysis is like using a camera to take a picture of an object, and dynamic analysis is like using a camcorder to trace the movement of the object from one point to another. Both methods are important parts of economic and business analysis.

The basic tools used in dynamic analysis are **theory of difference equations** and **theory of differential equations**. The former is discrete analysis, the latter, continuous analysis. We encountered these two kinds of analysis in Chapter 8 when we studied compound interest and the force of interest, and discrete growth rates and continuous growth rates. Both are important tools in engineering and physics, and are closely related. The Excel spreadsheets are particularly suited for discrete analysis. Thus, in this chapter, we introduce the theory of difference equations. Our topic is modeling, solving, and illustrating the first-order linear constant coefficient difference equation. Although this is the simplest dynamic model, it contains many important methods, which are directly extendable to higher-order difference equations. Thus, this chapter should be a good introduction to more general theory of dynamic analysis in economics and business.

15.1 Introduction

In the lecture note on static analysis, we saw that the interaction between aggregate demand (AD) and aggregate supply (AS) leads to equilibrium between the two activities. However, if the time factor is introduced into the observation of the actual economy, AD and AS may not be in equilibrium at each point of time, or in a month, or in a year. The difference in AD and AS leads the economy to "fluctuate" and change, growing, declining, or staying at the same level. The study of fluctuation and change in economics is called **economic dynamics**, a field of economics, which studies business cycles and economic growth and development. This chapter introduces some basic concepts and methods of dynamic models.

15.1.1 *Difference equations*

In mathematics, the equations with time lags are called **difference equations**. The **order of a difference equation** is the largest lag in the model. In economic and business applications, the coefficients of Y(t) and its lagged terms Y(t − 1), Y(t − 2), etc., are usually constant. Thus, they take a form like

$$Y(t)+aY(t-1) = f(t), \tag{15.1}$$

$$Y(t)+aY(t-1)+bY(t-2)=f(t). \tag{15.2}$$

If the right-hand side term f(t) is zero, then the difference equation is called **homogeneous**. If it is a constant, or a function of time, then f(t) is called the **nonhomogeneous term** and the difference equation is called **nonhomogeneous**. The above two equations are **linear** since all the lagged variables do not have product terms or power terms. Thus, we call the first (or second) equation a **linear constant coefficient nonhomogeneous difference equation of the first (or the second) order**.

For the first-order equation, we prefer to write

$$\boxed{Y(t) = aY(t-1) + c,} \tag{15.3}$$

and call this equation the **standard form of the first-order difference equation**. In this chapter, for convenience, we use this standard form.[1]

15.1.2 *Backward difference operator*

In (15.1) and (15.2), t is the current period and we look backward. Let t lag one period and take t − 1 as the current period. Then, starting from t − 1 and looking backward, we have

$$Y(t-1) + aY(t-2) = f(t-1), \tag{15.4}$$

$$Y(t-1) + aY(t-2) + bY(t-3) = f(t-1). \tag{15.5}$$

Subtracting these two equations from (15.1) and (15.2), respectively, we have

$$\Delta Y(t) + a\Delta Y(t-1) = \Delta f(t), \tag{15.6}$$

$$\Delta Y(t) + a\Delta Y(t-1) + b\Delta Y(t-2) = \Delta f(t), \tag{15.7}$$

where[2]

$$\begin{aligned} \Delta Y(t) &= Y(t) - Y(t-1), \\ \Delta Y(t-1) &= Y(t-1) - Y(t-2), \\ \Delta Y(t-2) &= Y(t-2) - Y(t-3), \\ \Delta f(t) &= f(t) - f(t-1). \end{aligned} \tag{15.8}$$

As in Secs. 3.3.3 (Slope of demand curves), 4.1.2 (Comparative static analysis), 5.4.1 (Growth equations), and 11.2.3 (MRTS), $\Delta Y(t)$ denotes the change of variable $Y(t)$ from the previous value of $Y(t)$, namely, $Y(t-1)$. The difference here is that the previous value of $Y(t)$ refers to the value of $Y(t)$ at the previous time. In other words, $Y(t)$ is the function of time. Similarly, $\Delta Y(t-1)$ or $\Delta Y(t-2)$ denote changes of $Y(t-1)$ or $Y(t-2)$ from the previous values of $Y(t-2)$ or $Y(t-3)$, respectively. In the theory of difference equations, Δ is called the **backward difference operator**. This operator is used in many econometric studies. Expressions (15.4) and (15.6) are equivalent, in the sense that one form can be derived from the other. Equations (15.6) and (15.7) are called the (backward) **difference form**, and (15.1) and (15.2), the (backward) **shift form** (or **standard form**), of the difference equation. The difference form can be derived from the shift form directly by multiplying the backward difference operator Δ by both sides of (15.1) and (15.2).

Example 15.1 The first-order difference equation. The dynamic extension of the single market model is straightforward. We simply assume that the supplier's decision is based on the previous price formation, rather than the current price of the commodity Q. Thus, a dynamic single market model may be formulated as

---

[1]We found that if we use form (15.1), then $Y(t) = -aY(t-1) + c$. In this case, a in the iteration methods in Table 15.2 should be $(-a)$, which is inconvenient and unsightly. Thus, without loss of generality, we define the standard form as (15.3).

[2]We may derive the shift from the difference form (exercise).

$$D(t) = a - bP(t) \qquad a > 0, \quad b > 0 \qquad \text{Demand function,} \tag{15.9}$$
$$S(t) = -c + dP(t-1) \qquad c > 0, \quad d > 0 \qquad \text{Supply function,} \tag{15.10}$$
$$D(t) = S(t) \qquad \text{Equilibrium condition.} \tag{15.11}$$

Substituting the functions into the equilibrium condition, we have

$$P(t) = -(d/b)P(t-1) + (a+c)/b, \tag{15.12}$$

which is the constant coefficient linear difference equation of the first order given in (15.3), with

$$\text{coefficient of } Y(t-1) = -d/b, \quad \text{and}$$
$$\text{nonhomogeneous term } f(t) = (a+c)/b.$$

Using the backward difference operator, (15.12) can be written as

$$\Delta P(t) = -(d/b)\Delta P(t-1). \tag{15.13}$$

Example 15.2 The second-order difference equation. A dynamic model corresponding to the static macroeconomic model of Chapters 3 and 4 may be formulated as:

$$Y_t = C_t + I_t + G_t \qquad \text{Equilibrium condition,} \tag{15.14}$$
$$C_t = a + bY_{t-1} \qquad \text{Consumption function,} \tag{15.15}$$
$$I_t = I_0 + d(C_t - C_{t-1}) \qquad \text{Investment function,} \tag{15.16}$$
$$G_t = G_0 \qquad \text{Government expenditure.} \tag{15.17}$$

Like the static model, Eq. (15.14) is called the **equilibrium condition** since the left-hand side is aggregate supply ($AS = Y_t$) and the right-hand side shows the aggregate demand ($AD = C + I + G$). As in the static analysis, the aggregate market must be cleared for equilibrium to prevail at each period. (15.15) is the consumption function, which depends on the income of the previous year, denoted Y_{t-1}.

b is the marginal propensity to consume (MPC). It assumes that the consumer's decision about current consumption depends on the income the consumer has received in the previous year. (15.16) states that the firm's decision on current investment depends on the change in consumption expenditure, namely, the difference between current consumption and previous consumption. d is a parameter called the **accelerator**.[3] Equation (15.17) states that government expenditure stays constant over time. This model is called **dynamics** since the time factor enters into the model in a meaningful way and cannot be omitted, as it can be with the static model.

[3]So called since in Mechanics, if $D = f(t)$ is the distance D travels, which is a function of time t, then $\Delta D/\Delta t$ is the **velocity** of motion, and $\Delta^2 D/\Delta t^2 = \Delta(\Delta D/\Delta t)/\Delta t$ is the **acceleration** of motion. In the original formulation of the model, Hicks wrote $I_t = I_0 + v(C_{t-1} - C_{t-2})$, in which consumption is lagged twice, instead of $(C_t - C_{t-1})$, which is lagged once, and so v is called the **accelerator**.

Since the values of some variables depend on the values of variables in the previous year(s), this model is called a **dynamic macroeconomic model**. Since the equilibrium condition holds at each moment of time, as shown by (15.14), we have a **model of equilibrium dynamics**. In this model, there are six variables in four equations. We count the lagged and unlagged variables as the same variables. The endogenous variables are Y_t, C_t, I_t, and G_t, and the exogenous variables are I_0 and G_0. When C_t depends on current income Y_t and $C_t = C_{t-1}$, the model reduces to the static model of Chapters 3 and 4.

Note **the use of notations**. For convenience in writing, the lagged variables Y_{t-1} and C_{t-1} may also be written as $Y(t-1)$ and $C(t-1)$, or even Yt−1 and Ct−1 in Excel. They refer to the income and consumption of the previous year, and not the multiplication of $t-1$. In the above writing, Y_t, $Y(t)$, or Yt, is the value of current income. For simplicity, the subscript t may be omitted. If this is the case, we write the lagged variables as Y_{-1} and C_{-1}.

As in the static model, we can eliminate the last three equations by substituting them into the equilibrium condition and obtain a difference equation in Y_t,

$$Y(t) - b(1+d)Y(t-1) + bdY(t-2) = a + I_0 + G_0, \tag{15.18}$$

where the right-hand side is a constant. This is the second-order difference equation.

By lagging (15.18) one period and deducting it from (15.18), in terms of the backward difference operator, (15.18) is equivalent to

$$\Delta Y(t) - b(1+d)\Delta Y(t-1) + bd\Delta Y(t-2) = 0.$$

□

15.2 The First-Order Difference Equation

In this chapter, we study the simplest form of the difference equation, the standard form of the first-order equation (15.3) with the nonhomogeneous term being constant: $f(t) = c$:

$$Y(t) = aY(t-1) + c. \tag{15.3}$$

The following examples show that we have already encountered this form of the first-order difference equation in Chapters 8 and 9.

15.2.1 *An example: Simple interest*

In Chapter 8, we defined simple interest as

$$F(t) = P(1+it) = P + iPt, \tag{15.19}$$

where i is the constant interest rate per period, and P is the constant principal at period 0. $F(t)$ is the future value of principal P at time t with simple interest i, since the interest for the previous year is $F(t-1) = P + iP(t-1)$. Subtracting from $F(t)$, we have

$$F(t) - F(t-1) = iP \equiv c. \tag{15.20}$$

Since the interest rate and principal are given, iP is a constant, denoted here as "c". Thus, the **simple interest** in every period is a constant amount of money c, which is a fixed percentage of the principal (iP), paid in return for the use of money during that period.

The direct method of finding the value at each period

To trace the movement, or **dynamic path**, of F(t), rewrite (15.20) as

$$\boxed{F(t) = F(t-1) + c,} \tag{15.21}$$

which is a constant coefficient nonhomogeneous first-order difference equation. The difference between expression (15.21) and the generic expression (15.3) is cosmetic, in that function Y is now replaced by function F. Note that in (15.21), t cannot start from 0, since $F(-1)$ is not defined. We have to start from $t = 1$, and then, given the value of F(0) at $t = 1$, we can trace the movement of F(t) from $t = 1$ to $t = 2, \ldots$ up to some period s, like $F(0), F(1), \ldots, F(s)$. The starting value F(0) is called the **initial value** or the **initial condition** of F(t). The process is shown in columns B and C of Table 15.1. Clearly, the current dependent function F(t) in column C is a function of the previous independent function in column B. The previous value $F(t-1)$ is substituted into the current value F(t) in each period t. This relation is shown by the arrows between columns B and C in Table 15.1. Column C is the change of the value of F(t) in the consecutive period. This process is repeated up to the sth period.

The iteration method of finding the value at each period

On the other hand, we may repeat the process of substitution by replacing the value of the current period by the value of the period before that, and then repeating the process as often as needed, back to the initial period of $t = 0$. After substitution, we eliminate the intermediate value of each period and keep adding the extra current value to the initial value, and we obtain the cumulative value at time t from the starting value F(0). This is shown in column D of Table 15.1, for t up to s. Since s is arbitrary, we may write s as t

Table 15.1 Iteration method of $F(t) = F(t-1) + c$

	A	B	C	D	E	F
1	Time	Leading	Change in	Change from	Change in	Change from
2	t	term	each period	initial period	each period	initial period
3			(15.21)	(15.22)	(15.23)	(15.24)
4	0	F(0)		=F(0) + 0c	100.0	100.0
5	1	F(1)	=F(0)+c	=F(0) + 1c	105.0	105.0
6	2	F(2)	=F(1)+c	=F(0) + 2c	110.0	110.0
7	3	F(3)	=F(2)+c	=F(0) + 3c	115.0	115.0
	...	...	...	...	...	...
14	s	F(s)	=F(s-1)+c	=F(0) + sc	=F(s-1)+c	150.0

to have

$$\boxed{F(t) = F(0) + ct.} \tag{15.22}$$

Hence, if the value of F(0) is given, then all other values of F(t), t = 1, 2, ... will also be given.

Example 15.3 Numerical example of simple interest. To illustrate (15.21) and (5.22) numerically by a table and chart, let i = 0.05 and P = $100. Then iP = c = 5. Thus, in each period, a fixed amount of $5.00 is paid for the use of $100. The initial value in this interest rate model is called the **principal** of the loan, and was given as P = F(0) = 100 as in (8.1) of Chapter 8. In terms of the difference equation, we can write

$$F(t) = F(t-1) + 5, \tag{15.23}$$

or, equivalently,

$$F(t) = F(0) + 5t. \tag{15.24}$$

Equations (15.23) and (15.24) are implemented in columns E and F of Table 15.1 as follows. Enter F(0) = 100 in E4 and F4 as the initial condition. In E5, enter

E5: =E4+5 for t = 1 in (15.23).

Copying E5 to E14, we have the future value of P = 100 from t = 1 to 10, in each cell in E4:E14. Here, we have combined columns B and C by adding c = 5 to the value in the previous period to obtain the value of the current period. This is illustrated in Fig. 15.1. For example, in the chart, at t = 8, F(8) is the sum of F(7) in the previous period plus c = 5, giving the current value, which is F(8) = 140.

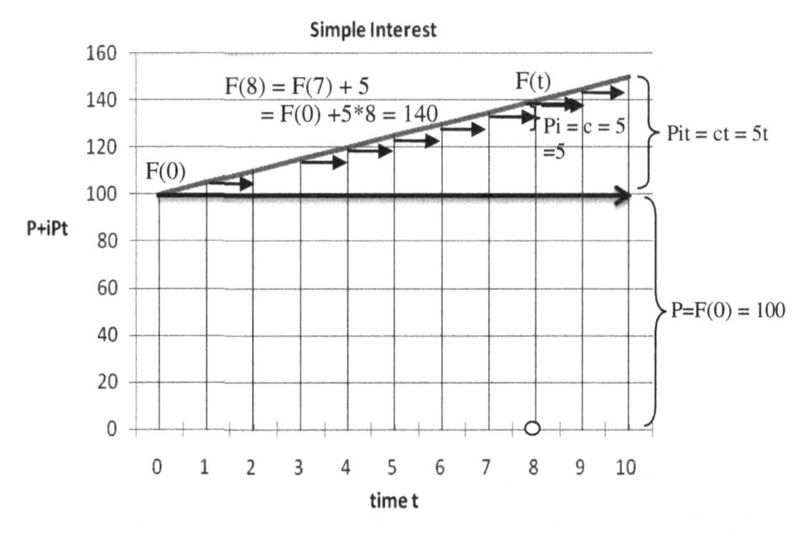

Fig. 15.1 The structure of a first-order difference equation (1)

Column F shows formula (15.24). In F4, enter $P = F(0) = 100$, and enter in F5

F5: =$F$4+5*A5.

Copying F5 down to F14, we have (15.24) at each t. In Fig. 15.1, at $t = 8$, we have $F(8) = F(0) + ct = 100 + 5*8 = 140$, as shown on the right side of the chart with right braces.

The constant interest at each period, $c = Pi$, or the increment of the total amount at each period over the previous period, is $F(t) - F(t-1) = 5$. This is illustrated by the vertical distance from the arrow-head to the F(t) line. In mathematical terms, F(t) in (15.23) gives an **arithmetic series**. □

Thus, we have shown numerically and diagrammatically that the difference equation (15.23) and the cumulative form (15.24) (the right-hand side of which depends only on time t and the initial condition) are equivalent. As will be defined below, (15.24) is called the **definite solution** of the difference equation (15.23).

15.2.2 *The future value problem without annuity*

Unlike simple interest, if interest is compounded each period, the interest accrued in the current period is based on the amount of money in previous period. Following the practice in Chapter 8, denoting the period by n and interest for each period by r, as in (8.5), we have

$$F(n) - F(n-1) = \Delta F(n) = rF(n-1). \tag{15.25}$$

Rewriting, we have

$$F(n) = F(n-1) + rF(n-1) = (1+r)F(n-1) = aF(n-1),$$

where $a = (1 + r)$. The above expression shows that the ratio of the consecutive values is constant. F(n) generates a **geometric series**: The ratio of the values of the consecutive terms is constant a.

15.2.3 *The future value problem with annuity*

The compound interest problem (15.25) is a particular case of a future value problem without annuity. If, in each period, a fixed amount of money c is paid or received, using the conventional notation of $t(= n)$, (15.25) can be written as

$$F(t) - F(t-1) = \Delta F(t) = rF(t-1) + c. \tag{15.26}$$

Thus, in each period, in addition to compound interest $rF(t - 1)$ from the last period, an amount of \$c is added. We rewrite (15.26) as

$$\boxed{F(t) = aF(t-1) + c,} \tag{15.27}$$

where $a = (1+r)$, or, for that matter, a can be any constant. Note that when $a = 1$, (15.27) reduces to the simple interest model (15.21), and when $c = 0$, $a \neq 1$, it reduces to the compound interest model (15.25).

Table 15.2 Iteration method of $F(t) = bF(t-1) + a$

	A	B	C	D	E	
1	Time	Leading	Change in	Change from	Change from	
2		term	each period	initial period	initial period (Def solution)	
3	t		(15.27)	(15.28) =A(t)+B(t)	(15.29)	
4	0	F(0)	F(0)	F(0)	F(0)	
5	1	F(1)	=a*F(0)+c	=a*F(0) +c	=a*F(0) +c(1-a)^1/(1-a)	
6	2	F(2)	=a*F(1)+c	=a^2*F(0) +ac +c	=a^2*F(0) +c(1-a)^2/(1-a)	
7	3	F(3)	=a*F(2)+c	=a^3*F(0) + (a^2 +a + 1)c	=a^3*F(0) +c(1-a)^3/(1-a)	
...	...	...	...	...		
14	s	F(s)	=a*F(s-1)+c	=a^s*F(0)	=a^s*F(0)	= A(t)
15				+ (a^(s-1) +a^(s-2) +... +1)c	+c(1-a^s)/(1-a)	+ B(t)

The iteration method of solution

The basic iteration scheme of (15.27), which is similar to Table 15.1, but much more complicated, is shown on an Excel spreadsheet in Table 15.2. Column B shows the leading term, and column C enters formula (15.27), and shows the amount of money in each period: the interest compounded for one period from the last period, plus the annuity c. Column D is obtained by substituting the previous leading function $F(t-1)$ for the independent term $F(t-1)$ of the current leading function $F(t)$, to calculate the cumulative value at time t $(F(t))$ from the starting value $F(0)$. The last row of column D shows the following formula:

$$\boxed{F^{\#\#}(t) = a^t F(0) + (a^{t-1} + a^{t-2} + \cdots + 1)c} \tag{15.28}$$

where, since s is arbitrary, we have replaced it by t, and denoted this special form by adding ##, as explained below. The terms in the parentheses are geometric series, similar to those obtained in (9.4). Using the result from Appendix 9A, (15.28) can be written as the right-hand side of (15.29) for $a \neq 1$:

$$F^{\#\#}(t) = a^t F(0) + c\frac{1-a^t}{1-a} \quad (a \neq 1) \tag{15.29}$$

$$= F(0) + ct \quad (a = 1) \tag{15.30}$$

The second case (15.30) is obtained from (15.22) or (15.28) (Column D of Table 15.1) when $a = 1$.

Example 15.4 Numerical example of compound interest. For $c = 1$ (an annuity of \$1 at each period), $r = 0.05$, the compound interest formula for each period is

$$F(t) = 1.05F(t-1) + 1, \tag{15.31}$$

which is (15.27) with $a = 1.05$ and $c = 1$. Then, for a given principal $F(0) = 100$, the future value of \$100 at time t with an annuity of \$1 is, from (15.28),

$$F^{\#\#}(t) = 1.05^t(100) + (1.05^{t-1} + 1.05^{t-2} + \cdots + 1)^*1; \qquad (15.32)$$

or, from (15.29),

$$F^{\#\#}(t) = 1.05^{t*}100 + ((1.05)^t - 1)/0.05, \qquad (15.33)$$

the second term of which is the same formula obtained in (9.4) and (9A.2) of Chapter 9.

The numerical difference equation (15.31) is calculated in Table 15.3. Column B contains the leading function F(t) for each t. We enter the initial value, 100, in row 4, as shown. Column C traces the change in each period as shown in (15.31), namely, enter

C5: =(1 + 0.05)*C4 + 1 for t = 1 in (15.31).

For convenience, column F presents (15.28) in two parts. D5 and E5 in Table 15.3 show the first part, A, and second part, B, respectively, as follows:

D5: =(1.05)^A5*$D$4	The first part (A) in (15.28);
E4: =0	The second part (B) for t = 0;
E5: =(1+0.05)^(A5−1)*1+E4	The second part[4] for t = 1;

The sum of the first and the second part in F5: =D5+E5 gives the future value (15.28) at t = 1. Copying D5:F5 to D6:F14, we have (15.28) for all t.

Table 15.3 Iteration method of $F(t) = bF(t-1) + a$

	A	B	C	D	E	F	G	H
1	Time	Leading	Change in	Iteration method@			ΔF(t)=	Change from
2	t	function F(t)	each period (Diff equ)	Comp sol A(t)=a^t*F(0)	Partial sol B(t)=a^(t-1)*c+B(t-1)	Def sol F(t)=A+B	F(t)-F(t- 1)	initial period Def sol
3			(15.31)	(15.28) A(t)	(15.28) B(t)	(15.28)	(15.8)	(15.33)
4	0	F(0)	100.0	100.0	0.0	100.0		100.0
5	1	F(1)	106.0	105.0	1.0	106.0	6.0	106.0
6	2	F(2)	112.3	110.3	2.1	112.3	6.3	112.3
7	3	F(3)	118.9	115.8	3.2	118.9	6.6	118.9
13	9	F(9)	166.2	155.1	11.0	166.2	8.9	166.2
14	10	F(10)	175.5	162.9	12.6	175.5	9.3	175.5
15	Note: @ B(0)=0. Also see (15.32) and Footnote 4.							

[4]To derive this formula, from the second part B(t) of (15.28) or the formula in column E of Table 15.3, since $B(t) = (a^{t-1} + a^{t-2} + \ldots + 1)c = \sum_i a^{t-1}c, 1 \le i \le t, B(0) = 0$, by a direct substitution, we have $B(t) = a^{t-1}c + B(t-1)$.

Column G shows the differences between the present and previous values of consecutive periods. Unlike the simple interest rate, which increases at a constant rate (see Fig. 15.1), the compound interest rate increases over time. Column D will be explained in (15.38).

Column H contains (15.29),

H5: =(1.05)^A5*$H$4+((1.05)^A5−1)/0.05,

which is the same as column C and F, as expected. □

15.3 The Standard Method of Solving Difference Equations

15.3.1 *Definition of solutions*

Definite solution

In general, F(t) in (15.21) and (15.27), or Y(t) in (15.3), is an unknown function of the independent variable t. When this function is expressed as a function of t alone, as in (15.29) or (15.30), for a given initial value F(0), then F(t) (which will be denoted $F^{\#\#}$ in later sections) is called a **definite solution** of the corresponding difference equation. When F(0) is given, F(1) will be determined uniquely, as shown in Tables 15.1 or 15.2. The subsequent functions F(2), F(3), etc. can also be determined uniquely, thus the name of the solution.

Example 15.5 Solutions of the simple interest problem. (a) What do you call Eq. (15.21) in terms of its order, coefficient, and homogeneity? (b) Using the substitution method, show that (15.22) is indeed a solution of difference equation (15.21) where $a = 1$.
Answer:

(a) Equation (15.21) is called a constant coefficient $(a = 1)$ first-order difference equation with a constant homogenous term.
(b) Since $F(t-1) = F(0) + c(t-1)$, substituting into (15.21), we can show that RHS $\equiv$ LHS.

□

15.3.2 *The standard method of deriving the general solution*

We have applied the iteration method to derive the definite solutions (15.29) and (15.30). For a higher-order difference equation, the iteration method becomes very cumbersome and not easy to apply. A systematic method of solution is introduced below. Let a first-order constant coefficient difference equation be given as (15.27).

Step 1. Particular solution. From the standard format (15.27),

$$F(t) = aF(t-1) + c. \tag{15.34}$$

We know that any **solution** of this difference equation must satisfy both sides of the equation identically. This means that, if $F_p(t)$ is a solution, we must have

$$F_p(t) - aF_p(t-1) \equiv c. \tag{15.35}$$

Since the right-hand side is constant, we may conjecture that the solution denoted $F_p(t)$ must also be constant. Otherwise, the left-hand side will be a function of t and never be identical to the right-hand side. Hence, let $F_p(t) = F_p(t-1) = k$. Substituting into (15.35), we have $k - ak = c$, or, if $a \neq 1$,

$$\boxed{F_p(t) = k = c/(1-a)} \tag{15.36}$$

Clearly, by substitution, (15.36) satisfies (15.34) identically. Hence, (15.36) is a solution of (15.34). This solution is called the **particular solution** of the difference equation (15.34) and is denoted $F_p(t)$.

Step 2. Complementary solutions. The truncated equation of (15.34),

$$F(t) - aF(t-1) = 0, \tag{15.37}$$

is called a **homogeneous equation** corresponding to (15.34). (In contrast, (15.34) is called a **nonhomogeneous equation**, and is the same as (15.37) when $c = 0$). Since (15.37) is constructed from (15.34), it is also called the **complementary equation** of (15.34). Using the iteration method, with the initial condition $F(0)$, and letting $c = 0$ in columns D and E of Table 15.2, we have a solution of (15.37):

$$F(t) = a^t F(0).$$

Since $F(0)$ is arbitrary, writing it as A, we have the **general solution of the homogeneous difference equation** (15.37):

$$\boxed{F_c(t) = Aa^t} \tag{15.38}$$

Equation (15.38) is also called **the complementary solution** of (15.34) and is denoted $F_c(t)$.

Step 3. General solution. It can be shown by substitution that the **general solution** of the difference equation (15.34) is the sum of the complementary solution and the particular solution:

$$F^{\#}(t) = F_c(t) + F_p(t), \tag{15.39}$$

or

$$\boxed{F^{\#}(t) = Aa^t + \frac{c}{1-a}} \tag{15.40}$$

which consists of one unknown constant A.

Step 4. The definite solution. To find the **definite solution** of the difference equation (15.34), let $t = 0$ in the general solution (15.40). Then,

$$F^{\#}(0) = A + F_p(0) = A + \frac{c}{1-a},$$

or solving for A,

$$A = F_0 - F_p = F^{\#}(0) - \frac{c}{1-a}. \tag{15.41}$$

Substituting A into the general solution (15.40), and writing $F^{\#}(0)$ as F_0 and $F_p(t)$ as F_p, we have the same definite solution (15.29) obtained from the iteration method.

$$\boxed{\begin{aligned} F^{\#\#}(t) &= (F_0 - F_p)a^t + F_p && (15.42)\\ &= a^t F_0 + c\frac{1-a^t}{1-a} \quad (a \neq 1) && (15.43)\\ &= F_0 + ct \qquad (a = 1) && (15.44)\end{aligned}}$$

A special case

In the iteration method in Sec. 15.2.1, we have already shown that, when $a = 1$, the particular solution is (15.30). How is this result related to the standard method of solution we have just expounded? For $a = 1$, $F_p(t) = k$ does not work in (15.35), as we will have a contradiction, $k - k = c \neq 0$. In this case, we try $F_p(t) = kt$, hoping that t will cancel out. Substituting into (15.34), we have $kt - k(t-1) = c$, or $k = c$. Hence, we have $F_p(t) = ct$.

The iteration method for $F(t) = F(t-1)$, $a = 1$, and $c = 0$, in Table 15.1, gives $F(t) = F(t-1) = F(0)$ from the last row of Table 15.1. Since this is the solution of the homogeneous equation (15.37) corresponding to the original difference equation (15.34), the complementary solution $F_c(t)$ for (15.34), with $a = 1$, is $F_c(t) = F(0) = A$, for some arbitrary constant A. Hence, the general solution is

$$F^{\#}(t) = F_c(t) + F_p(t) = A + ct. \tag{15.45}$$

Since $F^{\#}(0) = A$, the definite solution is $F^{\#\#}(t) = F_c(t) + F_p(t) = F^{\#}(0) + ct$, which is (15.44) for $a = 1$.

A flow chart

The above four steps are illustrated in a flow chart, Fig. 15.2. We first start from the original nonhomogeneous equation to find the particular solution, and then construct a homogeneous equation to find the complementary solution. We sum these two solutions to obtain the general solution. Lastly, given the initial condition, we can determine the constant to derive the definite solution.

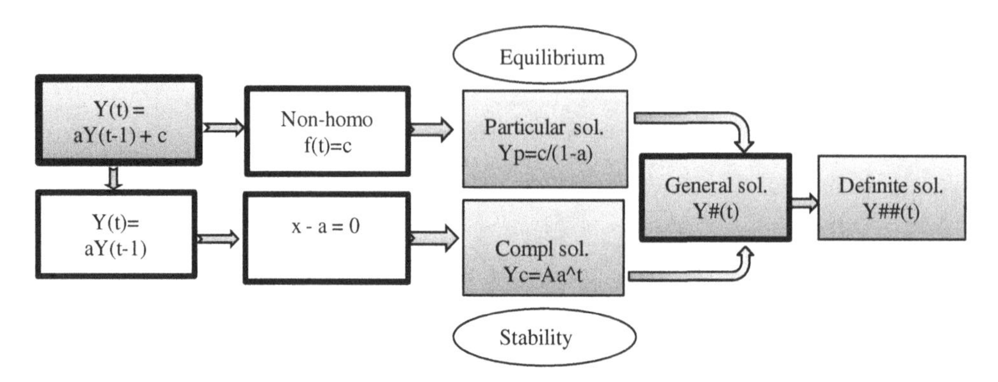

Fig. 15.2 The solution process of a first-order difference equation

Example 15.6 Compound interest with annuity. For $a = 1 + r$, and $r \neq 0$, the difference equation (15.27) is $F(t) = (1 + r)F(t - 1) + c$. The definite solution (15.29) can be written as follows, given $F(0) = P$:

$$F(t) = (1 + r)^t P + c\frac{(1 + r)^t - 1}{r}.$$

Thus, given the amount of principal P as the initial condition F(0), and given annuity c, the future value at time t is given by the above formula. The first term, the complementary solution, is the compound interest of the principal at time t, which starts counting from $t = 0$, as shown in (8.5). The second term, the particular solution, is the future value of the annuity at time t; this starts counting from $t = 1$ if the annuity is paid at the end of the previous period, as shown in (9.4). It should be noted that the interest on the principal is added to the future value of the annuity. When $c = PMT = 1$, the right-hand side is FVIFA in (9.5). □

Example 15.7 A future value problem with annuity. We reformulate the example in Example 15.4 in accordance with the standard method of solution. For the interest rate per period $r = 0.05$ and the principal $P = F(0) = \$100.00$, with an annuity of $c = \$1.00$, we have, from (15.31),

$$F(t) = 1.05F(t - 1) + 1. \tag{15.46}$$

The particular solution is, from (15.36), $F_p(t) = c/(1 - a) = -1/0.05$. The complementary solution is, from (15.38), $F_c(t) = (1.05)^t A$. The general solution (15.40) is the sum of both solutions:

$$F^{\#}(t) = A(1.05)^t - 1/0.05. \tag{15.47}$$

To find the definite solution, let $F(0) = 100$ at $t = 0$. Then $F^{\#}(0) = 100 = A - 1/0.05$, or $A = 100 + (1/0.05)$. Substituting into the general solution (15.47), we have the definite solution

$$\begin{aligned} F^{\#\#}(t) &= \left(100 + \frac{1}{0.05}\right) 1.05^t - \frac{1}{0.05} \\ &= 120(1.05)^t - 20 \qquad (15.48) \\ &= (100)1.05^t + \frac{1.05^t - 1}{0.05}. \qquad (15.49) \end{aligned}$$

This result is the same as (15.33) by the iteration method. □

Example 15.8 Investment multiplier. We want to study the effect of government spending of \$0.4 trillion on public construction, by assuming that the MPC is $b = 0.6$ and the current equilibrium income is \$4.5 trillion. In each period, consumers spend 60% of the previous extra income. Hence, the current extra consumption is $C(t) = bC(t-1)$, a geometric series. This is the same form as (15.25), and the iteration method yields the equivalent form $C(t) = Ab^t$, as the general solution (15.38) of the homogeneous difference equation (15.37). The initial government spending is $A = C(0) = 0.4$. Note that, since b is a fraction, $C(t)$ converges toward 0 when t goes to infinity. □

15.4 Equilibrium and Stability

We have seen in (15.39) that the general solution $F^{\#}(t)$ of a nonhomogeneous constant coefficient difference equation is the sum of a particular solution $F_p(t)$ and the complementary solution $F_c(t)$. A particular solution $F_p(t)$ of the nonhomogeneous equation (15.34) is also called the **equilibrium value** of the difference equation. It is a **stationary equilibrium** if $F_p(t)$ is a constant, and a **moving equilibrium** if $F_p(t)$ depends on time. If the time path of the general solution $F^{\#}(t)$, (15.40), of a nonhomogeneous equation **converges** toward the equilibrium value $F_p(t)$, then the dynamic path of the nonhomogeneous equation (15.34) is called **stable**, or the **model is stable**; otherwise, if the path **diverges** from the equilibrium value, the **path is unstable**, or the **model is unstable**. The condition for stability of the nonhomogeneous equation is called the **stability condition**. A convergent or divergent time path may be **oscillating**, **monotonic**, or **constant**.

Dynamic Analysis

The study of the existence, uniqueness, and stability of the equilibrium values in a dynamic model is called **dynamic analysis**. For example, it is well known that Samuelson's model (15.18) is stable if the product of MPC b and accelerator d is less than one, that is, $bd < 1$. As in static analysis, after we find the law of motion of the equilibrium path of the endogenous variables, we may study the effects of changes in the parameters of the model on the equilibrium path of the endogenous variables. This type of study is called **comparative dynamic analysis**.

15.4.1 *The dynamic path of the complementary solution*

The convergence of the dynamic path depends on the sign and magnitude of the coefficient a and the constant A in complementary solution $F_c(t) = Aa^t$ in (15.38). We first discuss the **shape of** $\mathbf{a^t}$, assuming $A = 1$. Theoretically, parameter a can take any values from $-\infty$ to $+\infty$. There are seven important categories that make the path of the complementary solution different. The seven ranges or intervals of a, designated as (a), ..., (g), are shown in Fig. 15.3. For each interval, we pick a number to represent the numbers in the interval to construct Table 15.4. The representative numbers are listed in row 2 of Table 15.4. To enter data in Table 15.4, we name A3:A8 "t" and B2:H2 "a". Entering = a^t in cell B3 and copying B3 to B3:H8, we have the table. We select A1:D8 to draw Fig. 15.4(a), and A1:A8 and F1:H8 (including rows 1 and 2) to draw Fig. 15.4(b). The paths are explained below.

Cases (a), (b), and (c)

From Figs. 15.3, 15.4(a), and columns B, C, and D of Table 15.4, we see that if coefficient a is negative ($a < 0$), the path oscillates. In particular, if $a < -1$ (say $a = -1.5$ as in (a)), the path oscillates and diverges to infinity; if $a = -1$ (case (b)), the path oscillates indefinitely

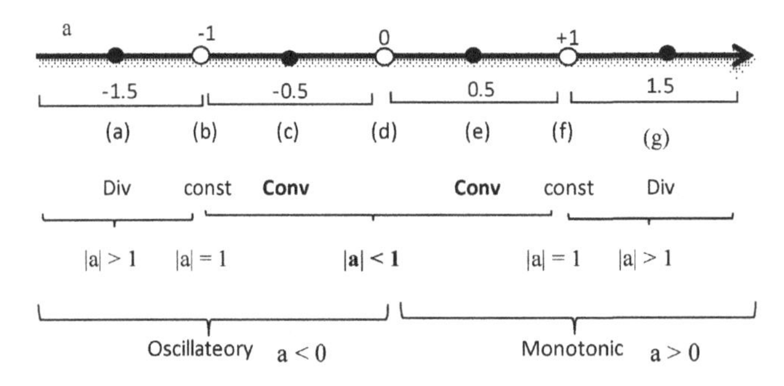

Fig. 15.3 Behavior of the complementary solution

Table 15.4 Behavior of the complementary solution $F_c(t) = a^t$

	A	B	C	D	E	F	G	H
1		(a)	(b)	(c)	(d)	(e)	(f)	(g)
2		-1.5	-1.0	-0.5	0.0	0.5	1.0	1.5
3	0	1.0	1.0	1.0		1.0	1.0	1.0
4	1	-1.5	-1.0	-0.5	0.0	0.5	1.0	1.5
5	2	2.3	1.0	0.3	0.0	0.3	1.0	2.3
6	3	-3.4	-1.0	-0.1	0.0	0.1	1.0	3.4
7	4	5.1	1.0	0.1	0.0	0.1	1.0	5.1
8	5	-7.6	-1.0	0.0	0.0	0.0	1.0	7.6

Unstable Stable Unstable

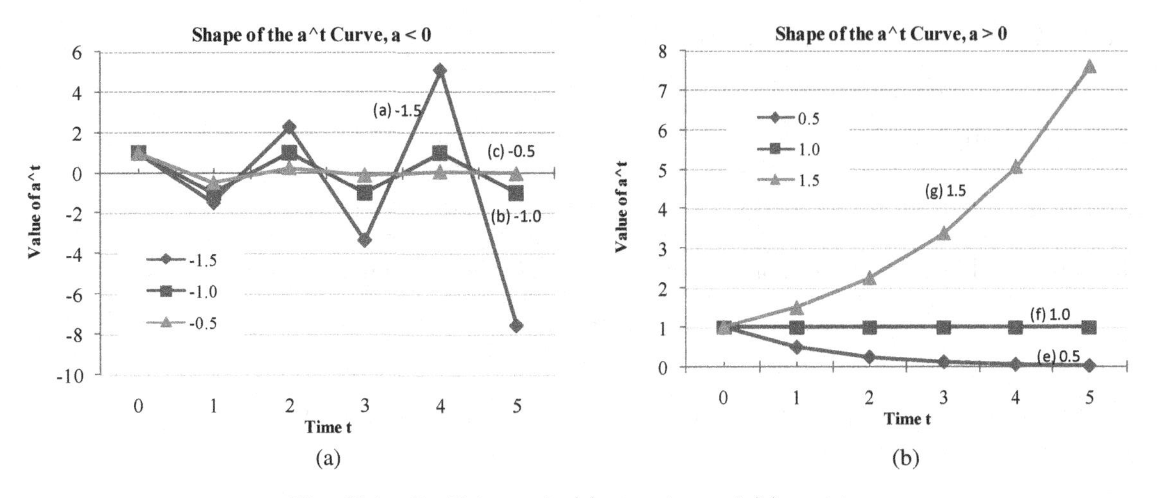

(a) (b)

Fig. 15.4 Coefficient a is (a) negative and (b) positive

with constant amplitude, and if $-1 < a < 0$ (say $a = -0.5$ in case (c)), the path oscillates and converges toward zero.

Cases (e), (f), and (g)

From Figs. 15.3, 15.4(b), and columns F, G, and H of Table 15.4, we see that, if coefficient a is positive ($a > 0$), the path is monotonic. In particular, if $0 < a < 1$ (say, $a = 0.5$ in Case (e)), the path converges monotonically toward zero; if $b = 1$ (Case (f)), the path is constant without fluctuation, and if $a > 1$ (say $a = 1.5$ as in case (g)), the path diverges monotonically to positive infinity.

The stability condition

From Figs. 15.3 and 15.4, we see that Case (c) ($-1 < a < 0$) and Case (e) ($0 < a < 1$) will render convergence of the general solution ($F^{\#}(t)$ to $F_p(t)$). Hence, the stability condition is that the absolute value of a must be less than 1:

$$\boxed{|a| < 1, \quad \text{for } a \neq 0} \tag{15.50}$$

If $a = 0$, then $F_c(t) = 0$, and $F^{\#}(t) = c$ from (15.40). The path is constant.

15.4.2 *The effects of the value of A*

If the constant A is positive, it only increases the magnitude or scale of the path (the scale effect). There is no change in the basic shape and trend of the path (the reflection effects). If A is negative, there will be a scale effect as well as reflection effects: positive coordinates, say (a, b), become negative coordinates, (−a, −b), and the time path is the reflection of the graph of $|A|a^t$ across the horizontal axis. This can be seen from both charts in Fig. 15.4 by flipping down the original chart along the horizontal axis.

15.5 A Dynamic Market Model

15.5.1 *The solutions*

In Example 15.1, we have seen that the first-order difference equation for the market model is

$$\boxed{P(t) = -(d/b)P(t-1)+(a+c)/b.} \tag{15.51}$$

By the standard method of solution in Sec. 15.3.2, the particular solution (15.36) is

$$P_p(t) = \frac{(a+c)/b}{1-(-d/b)} = \frac{a+c}{b+d}, \tag{15.52}$$

which is the static equilibrium value of price $P^{\#}$ that we derived in Chapter 4. The complementary solution (15.38) is

$$P_c(t) = A(-d/b)^t, \tag{15.53}$$

where A is an arbitrary constant. The general solution is, from (15.39),

$$\boxed{\begin{aligned} P^{\#}(t) &= P_c(t)+P_p(t) \\ &= A\left(-\frac{d}{b}\right)^t + \frac{a+c}{b+d}. \end{aligned}} \tag{15.54}$$

To determine constant A, let $t = 0$. Then, $P^{\#}(0) = A+P_p(t)$. Solving for A and substituting into $P^{\#}(t)$, we have the definite solution

$$P^{\#\#}(t) = (P^{\#}(0) - P_p(t))\left(-\frac{d}{b}\right)^t + \frac{a+c}{b+d}. \tag{15.54}$$

Note that we call either $P^{\#}(t)$ or $P^{\#\#}(t)$ the equilibrium price path, since the model maintains the equilibrium condition that $D(t) = S(t)$ at each time period.

15.5.2 *The stability condition*

From Fig. 15.3 and (15.50), the stability condition for this market model is that the absolute value of the term in parentheses is less than one:

$$\left|-\frac{d}{b}\right| = \left|\frac{d}{-b}\right| < 1, \tag{15.55}$$

which implies that

$$|d| < |-b|, \quad \text{or} \quad d < |b|. \tag{15.56}$$

That is, the stability condition is that the absolute value of d is less than the absolute value of −b. Hence, if both b and d are positive, the stability condition is that the slope d of the supply curve must be less than the slope b of the demand curve. Intuitively, this makes sense, since in this model, the dynamic factor is the fluctuation of the supply function, which depends on the coefficient d. Thus, the model is stable if the slope coefficient d is "subdued", or restricted, by the slope b of the demand curve.

15.6 An Excel Solution Table of First-Order Difference Equations

15.6.1 *A numerical example*

A numerical counterpart of the Cobweb model (15.9) to (15.11) is given as follows:

$$D(t) = 25 - 4P(t), \tag{15.57}$$

$$S(t) = 2 + 3.6P(t-1), \tag{15.58}$$

$$D(t) = S(t), \tag{15.59}$$

where we assume default parameter values a = 25, b = 4, c = −2, and d = 3.6. Here, the supply curve has a positive vertical intercept so as to make the price changes inside the positive range (see the parameter table in Table 15.6). From the static model, we find that the equilibrium price is $P^{\#} = (a + c_{-})/(b + d) = 3.026$, and the equilibrium quantity is $Q^{\#} = (ad - bc_{-})/(b + d) = 12.895$.

15.6.2 *Standard form and its solutions*

In terms of this numerical model, the difference equation (15.51) can be written as

$$\boxed{P(t) = -0.900P(t-1) + 5.750} \tag{15.60}$$

The complementary solution (15.53) is $P_c(t) = A(-0.9)^t$, and the particular solution (15.52) is $P_p(t) = 3.026$, which is the same as the equilibrium price found in the static model. The general solution (15.54) is

$$P^{\#}(t) = A(-0.9)^t + 3.026. \tag{15.61}$$

Given P(0) = 1 when t = 0, we have A = −2.026. The definite solution (15.55) is

$$P^{\#\#}(t) = -2.026(-0.9)^t + 3.026. \tag{15.62}$$

If P(0) = 0, then, from (15.62),

$$0 = A + 3.026, \quad \text{or} \quad A = -3.026. \tag{15.63}$$

15.6.3 *Using the Excel table to solve the dynamic market model*

We now standardize the above solution procedure on the Excel spreadsheet. We may either start from the original system of Eqs. (15.58)–(15.60), or from the generic first-order equation (15.61). We first start from the system of equations, which are implemented in Table 15.5. To avoid confusion with the coefficients of the generic first order difference

Table 15.5 Solution of the first-order difference equation

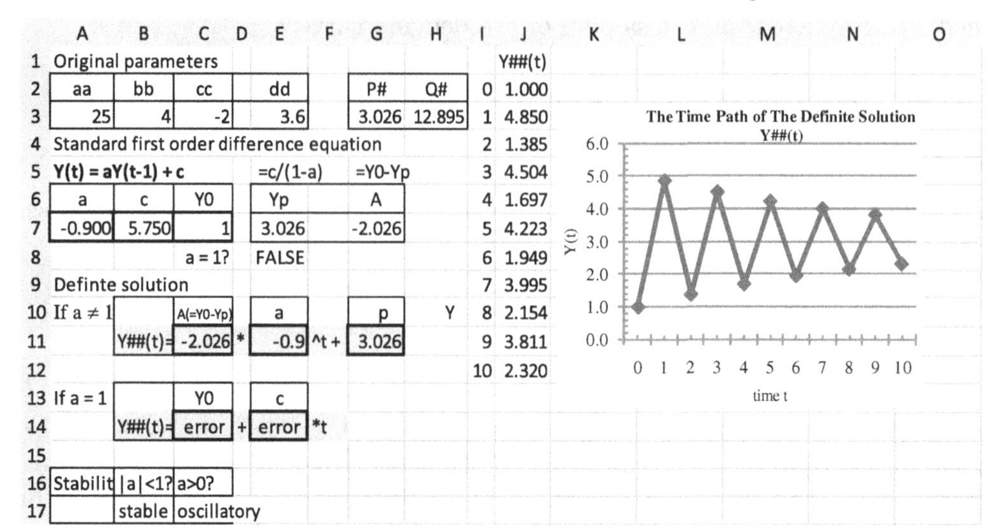

	A	B	C	D	E	F	G	H	I	J
1	Original parameters									Y##(t)
2	aa	bb	cc		dd		P#	Q#	0	1.000
3	25	4	-2		3.6		3.026	12.895	1	4.850
4	Standard first order difference equation								2	1.385
5	Y(t) = aY(t-1) + c				=c/(1-a)		=Y0-Yp		3	4.504
6	a	c	Y0		Yp		A		4	1.697
7	-0.900	5.750	1		3.026		-2.026		5	4.223
8			a = 1?		FALSE				6	1.949
9	Definte solution								7	3.995
10	If a ≠ 1		A(=Y0-Yp)		a		p	Y	8	2.154
11		Y##(t)=	-2.026	*	-0.9	^t +	3.026		9	3.811
12									10	2.320
13	If a = 1		Y0		c					
14		Y##(t)=	error	+	error	*t				
15										
16	Stabilit	|a|<1?	a>0?							
17		stable	oscillatory							

equation, we define aa = 25, bb = 4, cc = 2, and dd = 3.6, and name the four cells in A3:E3 using the values in row 2 above these cells. The equilibrium price and quantity are entered in G3 as = (aa+cc)/(bb+dd), which is $P^{\#} = 3.026$; and in H3 as =(aa*dd − bb*cc)/(bb+dd), which is $Q^{\#} = 12.895$.

Next, we find the generic form of the first-order difference equation (15.36), which is reproduced below:

$$Y(t) = aY(t-1) + c. \tag{15.3}$$

The coefficients a and c in generic equation (15.3) are equivalent to, from (15.51), $-(d/b)$ and $(a+c)/b$, respectively. Thus, in Table 15.5, fill in A7 and B7 as follows:

$$\text{A7: } =-(\text{dd/bb}) \quad \text{B7: } =(\text{aa} + \text{cc})/\text{bb}.$$

This will give the coefficient a of $Y(t-1)$ and the constant term c in (15.3) as $a = -0.900$ and $c = 5.750$, as shown in (15.61). Conversely, given the system of equations, (15.61) can be read from A7 and B7.

15.6.4 *Programming the definite solution*

Our problem now is, given $Y(0) = P(0) = 1$, to find the definite solution of difference equation (15.61). Name A7:E7 as the values in row 6. Enter in E7 and G7:

E7: $=c\_/(1-a)$ the particular solution (15.36), and

G7: $=Y_0 - Y_p$ the constant A in (15.41).

Then, the solution is given in row 10 to row 14. If $a \neq 1$, in row 11, fill in the blank space as follows:

C11: =G7 $\quad$ = A, that is, = Y0 − Yp,

E11: =a $\quad$ the base of the complementary solution,

G11: =Yp $\quad$ copy the particular solution Yp.

The definite solution (15.63) is shown in row 11.

In Table 15.5, row 8 checks whether a = 1 or not. In E8, we enter the "true or false" function =a=1. If a = 1, E8 will show "TRUE", and the definite solution will show in row 14; if $a \neq 1$, it will show "FALSE", and the definite solution will be shown in row 11. In C14 and E14, enter

C14: =IF(NOT(a = 1), "error", Y0) and

E14: =IF(NOT(a = 1), "error", c_).

Thus, if $a \neq 1$, C14 and E14 will show "error" blocking the access. Hence, for the current example in the table, $a \neq 1$, we have "error" in row 14.

Stability

Row 17 checks the stability of the dynamic path. From (15.50), enter

B17: =IF(abs(a)<1, "stable", "unstable") and

C17: =IF(a>0, "monotonic", "oscillatory").

B17 shows whether the dynamic path is stable (convergence to equilibrium) or unstable (divergence from equilibrium), and C17 shows whether the path is monotonically or oscillatorily converging or diverging. In the current example in the table, the path is stable and oscillatory.

The time path of the difference equation

For convenience, we have drawn the chart for the definite solution on the right-hand side of the calculation table. Naming I2:I12 "t", in J2, enter $-(Y_0 - Y_p) * a\hat{}t + Yp$. Then, if the definite solution changes, the time path also changes accordingly, providing the visual presentation of the time path of the variable.

Solutions of general dynamic models of the first order difference equation

Since all linear first-order difference equations with constant nonhomogeneous term can be rewritten to standard form (15.3), we may copy Table 15.5 to another sheet, say Sheet2, erase the first three rows, and let the solution table start from row 4, using the standard equation listed in row 5. Thus, a, c, and Y0 are now the parameters the value of which should be filled to obtain the definite solution (row 11 or row 14) of the difference equation (15.3).

Example 15.9 A market model. In Chapters 3 and 4, we had the following market model

$$D = 18 - 2P, \quad S = -6 + 6P, \quad D = S.$$

(a) Assuming the supplier's supply decision depends on the price of the previous period, rewrite the model in dynamic form.
(b) Derive the standard first-order difference equation.
(c) What is the particular solution of this dynamic model?
(d) Is the model (or equilibrium price) stable? Show your answer. If so, will the equilibrium quantity also be stable?
(e) Let $Y(0) = 5$. Find the definite solution of this model.
(f) Does the dynamic path of price converge to the particular solution? What is the value of the particular solution?
(g) If the path is stable, change some of the four parameters in row 2 to make it unstable; if the model is unstable, change the parameters to make it stable. Why does your choice of the parameter work?
(h) After you change the value of a parameter, is the path oscillatory or monotonic? If it is oscillatory, change the parameter to make it monotonic, and vice versa.
(i) What will happen to the solution if $a = 1$? Under what condition can coefficient a in A7 of Table 15.5 be 1? Find a value of parameters which will make $a = 1$.
(j) Does it make economic sense to have $a = 1$. Explain.
(k) With $Y(0) = 100$, what are the general and definite solutions of this model after you find $a = 1$ in (i)?
(l) Is this model (in (k)) stable? Can it be stable? Can it be oscillatory? Why?

Answer:

(a) The dynamic model will be $S(t) = -6 + 6P(t - 1)$, and other equations stay the same.
(b) The standard first-order difference equation for this model is, from (15.51) or from row 7 of the calculation Table 15.5, $P(t) = -3P(t - 1) + 12$.
(c) The particular solution is $P_p = 3$.
(d) No. Since $a = -3$, it is unstable and oscillatory (also, read row 17 of Table 15.5). Since $D^{\#}(t) = Q^{\#}(t) = 18 - 2P^{\#}(t)$, the stability condition of $Q^{\#}$ is the same as that of $P^{\#}$. Thus, if $P^{\#}$ is unstable, so is $Q^{\#}$.
(e) From row 11 of Table 15.5, $P^{\#\#}(t) = 2(-3)^t + 3$.
(f) The definite solution $P^{\#\#}(t)$ will not converges to the particular solution since $a = -3$. The value of the particular solution is 3.
(g) From (d), the price path is unstable. Since $a = -dd/bb$, we may change either dd or bb such that $dd < bb$, to make the model stable. We may take $dd = 1$.
(h) Since, if $dd = 1$, $a = -0.5 < 0$, the model is oscillatory and stable. To make it monotonic, $a = -dd/bb$ must be positive. If $dd = -1$, (that is, the supply curve of the static model has a positive intercept with the quantity axis), then the model is monotonic and stable.
(i) If $a = 1$, the particular solution will be divided by 0. Since $a = -dd/bb$, we may change dd or bb or both to make $a = 1$. Let $dd = -2$. Then, $a = 1$.
(j) If $a = 1$, the economic meaning is that the response of the buyer of the good to the price change (bb) is exactly the same as the response of the supplier of the same good to the

price change (dd). This probably will happen very rarely, since buyers and suppliers make decisions independently. Thus, it does not make economic sense.

(k) The general solution in this case is $P^{\#}(t) = A + 12t$, and the definite solution is, for $Y(0) = 100$, from row 14 of Table 15.5, $P^{\#\#}(t) = 100 + 12t$.

(l) No. The model in unstable: $P^{\#}(t)$ explodes as t increases to infinity. No, since the dynamic part, ct, enters in the solution linearly, and so it can never be stable and oscillatory. □

15.7 The Phase Diagram and the Time Paths — The Cobweb Model

Using the numerical model presented in (15.61), we may draw the charts of the dynamic paths in Excel easily. There are two kinds of charts. One is the usual **time path**, drawing the price P(t) or quantity Q(t) path against time. The other is called the **phase diagram**, which is the dynamic path of a nontime variable, say P(t), against the other variable, say Q(t). Thus, time is hidden behind the variables. Instead of treating dynamic analysis and comparative dynamic analysis separately, we set up a table like the one we discussed in Chapter 4. This is a good time to review Chapters 3 and 4.

The parameter table

The first part of Table 15.6 is the usual parameter table, which is already familiar from Table 4.1 of Chapter 4. We start from a new workbook. In row 6, enter the values of the parameters and name them a, b, c, and d as in row 5. In row 10, we enter the suggested values of b against a fixed value of d at d = 3.6. These values of b are for the discussion of the stability and the pattern of the dynamic path of the price. As in Table 4.1, in G11, we enter =(a*d−b*c)/(b+d)>0 to check the nonnegativity of the equilibrium output $Q^{\#}$, and in G12, we enter =b>d to check the stability of the price path. We have preserved the conditional formatting in the parameter table.

Note that, in Table 15.6, like Table 15.5, we have taken the intercept of the supply curve as c = −2, in order to have a positive intercept with a vertical axis. This is needed to keep the price path of Fig. 15.5 inside the first (positive) quadrant.

The Data Table (1)

We first look at the market model (15.58)–(15.60) more carefully. When the initial price P(0) = 0 is given, S(1) is given from (15.59). Then, D(1) is given from the equilibrium condition (15.60). Given D(1), P(1) will be determined from the demand equation (15.58) as P(1) =(a−D(1))/b. When P(1) is given, then we repeat the above process to obtain P(2), and so on.

This process is implemented in The Data Table (1) of Table 15.6. Enter

B16:	0	The initial value P(0) = 0,
D16:	=−c_+d*B16	S(1) from (15.59),
C17:	=D16	Equilibrium condition S(1)=D(1) (15.60),
B17:	=(a−C17)/b	New price P(1) from D(1) (15.58).

Table 15.6 Drawing for Cobweb Model

	A	B	C	D	E	F	G
1	**Dynamic and Comaparative Dynamic Analysis**						
2			**of a Single Market Model (Cobweb Model)**				
3	**A. The Paraneter Table**						
4		Parameters				Equi Values	
5		a	b	c	d	P#	Q#
6	Curr	25	4	-2	3.6	3.026	12.895
7	Deflt	25	4	-2	3.6	3	12
8	Chg	0	0	0	0	0.0	0.9
9	Chg	chg a	chg b	chg c	chg d	chg P#	chg Q#
10	Sugg	b = 4, 3.2, 3.6, -2, -3.6, -5					
11	Non-negativity condition: Q# = (ad-bc)/(b+d) >0						TRUE
12	Stability condition: b>d						TRUE
13							
14	**B. Data Table (1)**						
15		P	Dt=St	St+1			
16	0	0		2			
17	1	5.75	2	22.7			
18	2	0.575	22.7	4.07			
19	3	5.2325	4.07	20.837			
20	4	1.0408	20.837	5.7467			
21	5	1.8133	5.7467	19.328			
22							
23	**C. Data Table (2)**						
24	time	P	Dt=St	St+1	Cobweb		
25	0	0		2			
26		0		2	2		
27	1	5.75	2	22.7	2		
28		5.75	2	22.7	22.7		
29	2	0.575	22.7	4.07	22.7		
30		0.575	22.7	4.07	4.07		
31	3	5.2325	4.07	20.837	4.07		
32		5.2325	4.07	20.837	20.837		
33	4	1.0408	20.837	5.7467	20.837		
34		1.0408	20.837	5.7467	5.7467		
35	5	4.8133	5.7467	19.328	5.7467		
36		4.8133	5.7467	19.328	19.328		

After deriving P(1), we can repeat the process in D17:

D17: =−c_+d*B17 Starting the new process.

Copying B17:D17 to B18:D21, we have the Data Table (1). To draw the chart for this table, select B15:D21, then choose <"Insert"><Charts, Scatter!><Scatter with Straight Lines and Markers (4th chart)>, and we have the demand and supply curves as shown on the right-hand side of The Data Table (1). There are markers on the curves, but they are not connected between the curves.

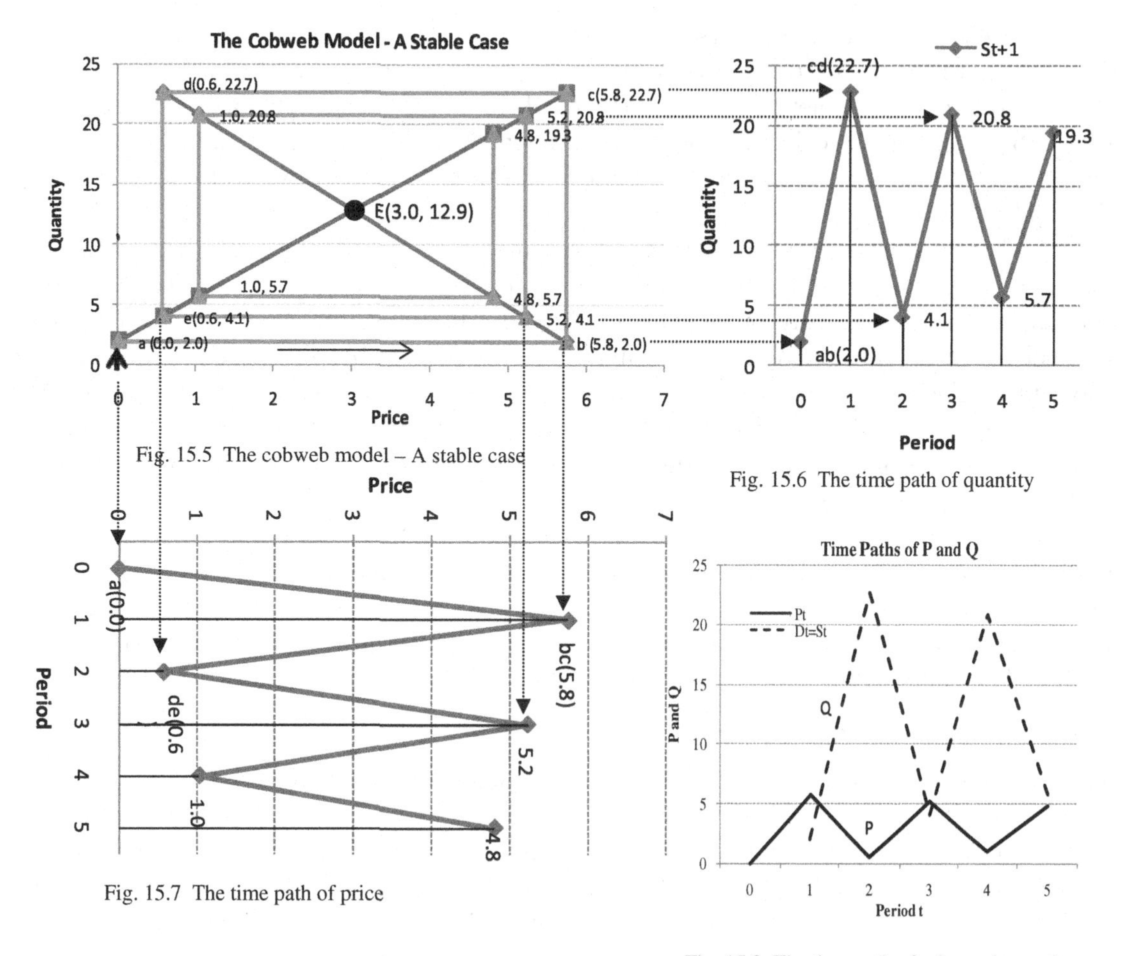

Fig. 15.5 The cobweb model – A stable case

Fig. 15.6 The time path of quantity

Fig. 15.7 The time path of price

Fig. 15.8 The time path of price and quantity

The Data Table (2)

To connect the markers, we revise the table. Copying the range A14:D21 of The Data Table (1) down to A23, rename the copied table Data Table (2). The original 6 rows, namely, rows 16–21, are now in 6 new rows, from rows 25 to 30. Inserting a blank row below each of the rows in new table, we have the Data Table (2) as shown in A25:D36, except that rows 26, 28, 30, 32, 34, and 36 are currently blank (not shown).

In the first blank row, in cells B26 and D26, enter

B26:	0	Duplication of the initial value;
D26:	=−c_+d*B26	Repeating S(1) as in row 25.

Copying B27:D27 to B27:D36, we have duplicated Data Table (1) row by row. At this point, the chart still shows the demand and supply curves only.

Construction of the cobweb charts

To connect the markers, enter =D25 in E26, to repeat D25, copy E26 down to E36. The column, E26:E36, will then connect all the markers. We can show this as follows: Selecting B24:E36 (including the column labels), and drawing the scatter diagram with "straight lines and markers" (the fourth box in the Scatter drop-down window), we will have a chart like Fig. 15.5.

To edit the chart, right click (<RM>) the demand curve and select <Add Data Labels>, which will add the Y coordinates. Do the same for the supply curve. To add the X coordinates, right click any one of the labels and <RM>, select <Format Data Labels>, and select <@X Value> in the <"Format Data Labels, Label Options"> dialog box. In the same dialog box, <"Format Data Labels, Number"><Number><Decimal places: [1]> and <Close>. The X-values will be added to the left of the Y-values in the labels, and will have one decimal place. Click any one of the labels again, all the labels will be selected, and change the font size to 8 for all the labels, and <enter>.

Lastly, double click one of the labels to select that label only, and edit the label by adding the parentheses and letters as shown in Fig. 15.5, like a(0.0, 2.0), b(5.8, 2.0), etc. Do this for the first several labels to trace the changes in points. Finally, we add arrows for the direction of movement, and a circle to show the equilibrium point, we have the chart as shown. The equilibrium point should be at E(3.0, 12.9) from F6 and G6 of Table 15.6.

Interpretation of the Cobweb model

We now trace the movement of the price path. Starting from the initial value of price at 0, the quantity supplied will be 2 units at a(0.0, 2.0) as denoted by a heavy arrow. Since these 2 units are supplied with no price (or because the price is extremely low), competition among consumers bids the price up to 5.8 at b(5.8, 2.0) (follow the long arrow to the right), at which the commodity market will be cleared. With a price as high as 5.8, the supplier will supply the commodity up to 22.7 units in the next period at c(5.8, 22.7). The market will be flooded by the commodity, which can be cleared only when the price drops to 0.6 at point d(0.6, 22.7). But then, at price 0.6, the supplier loses money and reduces the supply in the next period to 4.1 at point e(0.6, 4.1). This completes the up and down cycle of price and quantity in two periods. The process is then repeated until the equilibrium point at E(3.0, 12.9) is reached, as shown in F6 and G6 of Table 15.6.

In general, in this model, the supplier moves the quantity, and the consumer determines the price to clear the market, and both suppliers and consumers over react to the price change. The reaction may be subdued, or exaggerated. In Fig. 15.5, the magnitude of the changes in price and quantity is subdued, and decreases in each round. Eventually, it converges to the equilibrium values ($P^{\#}$, $Q^{\#}$), which is the particular solution $P_p(t) = 3.026$ of (15.62) of the difference equation (15.61). This model is also called the **Cobweb model** due to the appearance of the chart.

The time paths of quantity and price

We may extract the time path of quantity and price from the cobweb model. Figure 15.6 shows the time path of quantity Q(t). It is the line chart (with markers) drawn from ranges

A15:A21 for time period, and D15:D21 for supply (=demand). We have added the data labels so that you can trace the corresponding change in quantity from period 1 to 5 between Figs. 15.5 and 15.6. Note that we have made the height of the quantity axis in both charts the same for easy comparison. Clearly, the amplitude of the change decreases over time, and the time path of Q(t) converges with the equilibrium quantity $Q^{\#}(t) = 12.9$.

Figure 15.7 shows the time path of price. It is the line chart drawn from ranges A15:A21 for time period and B15:B21 for price. The chart is converted to picture first and then transposed horizontally to match Fig. 15.5. Following the data labels, we can trace the change in price over the periods between Figs. 15.5 and 15.7. Like the time path of quantity, the amplitude of the price change decreases over time and converges with equilibrium price $P^{\#} = P_p(t) = 3.0$.

Figure 15.8 combines Figs. 15.6 and 15.7 in one chart. It is the drawing of range A15:C21 of Table 15.6. Note the different amplitudes of fluctuation between price and quantity. The pattern of fluctuations may be clearer if the time period is extended from 5 to, say, 20.

The unstable case, the monotone case, and other cases

Row 10 of Table 15.6 suggests five other values of b in decreasing order. These values should be compared with the value of d, which is, for convenience, set at d = 3.6. When b = 3.2, b < d, and the price is unstable, the price and quantity explode. Figure 15.9 illustrates this case for P(0) = 1 (entering 1 in B16 and B25:B26). If the time paths for price and quantity were drawn, the amplitude of the fluctuations would appear as in Figs. 15.6 and 15.7, except that the fluctuation of the paths would become larger and larger, and the paths would be explosive.

When b = 3.6 = d, we have repetition of price and quantity at the same amplitude each period, as shown in Fig. 15.10. The time paths of price and quantity have the same amplitudes every period, namely, the lines connecting the peaks and troughs of the changes in price or quantity are parallel to each other in Fig. 15.8. In this case, both consumers and

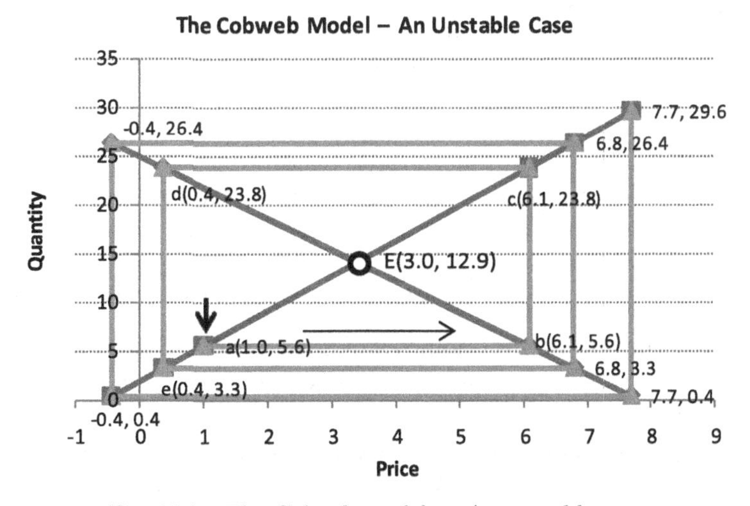

Fig. 15.9 The Cobweb model — An unstable case

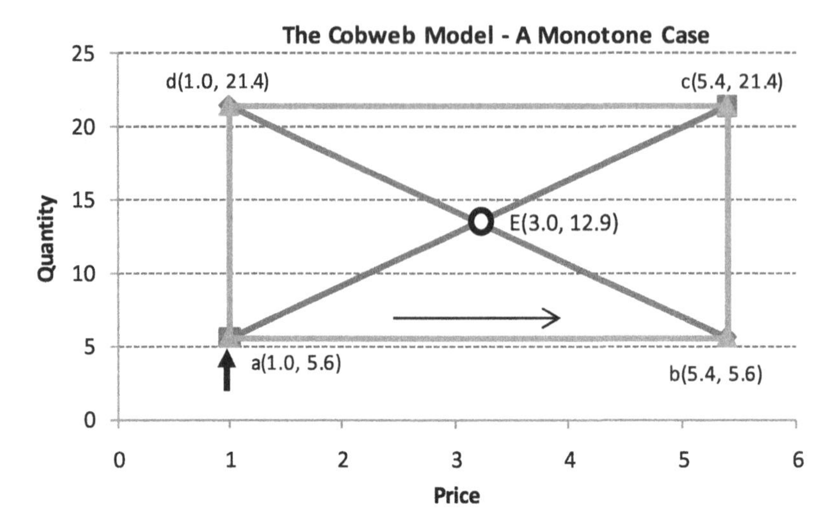

Fig. 15.10 The Cobweb model — A monotone case

suppliers do not revise their expectations of changes in price and quantity, and the market repeats itself period after period with the same fluctuations.

If b is allowed to be negative, that is, if the demand curve is positively sloped, then the demand curve cuts the supply curve from below if $|b| > d$, from above if $|b| < d$, and parallel to the supply curve if $|b| = d$. The three cases of the cobweb model, Figs. 15.5, 15.9, and 15.10, will still hold, and the cobweb model will show that the paths of price and quantity are convergent, divergent, or monotone, respectively.

15.8 Applications to Theory of Economic Growth

There are two important applications of the first-order difference equations. One is called the Harrod's growth model and the other is the Solow's model of neoclassical economic growth. These two models are the basis of dynamic analysis in macroeconomics and were developed and extended fully during the 1960s.

15.8.1 *Harrod's growth model*

Harrod presented a model of economic growth based on the investment and saving relation as follows:

$$S(t) = S_0 + sY(t), \quad I(t) = I_0 + d(Y(t) - Y(t-1)), \quad S(t) = I(t). \tag{15.64}$$

The first equation is the saving function. s is a fraction called the **marginal propensity to save** (MPS), $0 \leq s \leq 1$. The second equation is the investment function, which depends on the difference between the past and present national income. If current income exceeds the previous year's income, investors will think the trend will continue and increase current investment; if less, then investors will decrease investment. The parameter d is a positive number and called the **accelerator** (or **velocity**). The last equation is the **equilibrium condition** that at each time period, saving must equal investment. This is equivalent to the equilibrium condition AD = AS that we saw in Chapters 3 and 4.

Substituting the first two equations into the equilibrium condition, we have

$$\boxed{Y(t) = \left(\frac{d}{d-s}\right) Y(t-1) + \left(\frac{S_0 - I_0}{d-s}\right).} \tag{15.65}$$

Thus, given the initial condition $Y(0) = Y_0$, the general solution is, from (15.40),

$$Y^{\#}(t) = A\left(\frac{d}{d-s}\right)^t + \left(\frac{I_0 - S_0}{s}\right). \tag{15.66}$$

which is called the **equilibrium path of national income**. Since, in general, we have d > s, the Harrod model is generally unstable. Assuming that the fixed saving (S_0) and the fixed investment (I_0) are the same in an economy, Harrod defined the warranted rate of growth as $g_w = (Y^{\#}(t) - Y^{\#}(t-1))/Y^{\#}(t-1)$. Thus, substituting $Y^{\#}(t)$ in the model, the warranted rate of growth along the equilibrium path is

$$\boxed{g_w = \left(\frac{s}{d-s}\right)}$$

Example 15.10 A numerical example. Let s = 0.1, d = 4, $S_0 = 4500$, $I_0 = 4000$, and $Y(0) = 6000$.

(a) Find the general and definite solutions.
(b) Draw the dynamic path of the national income.
(c) If $S_0 = I_0$, what is the Harrod's warranted rate of growth?

Answer:

(a) Either from direct calculation or from the calculation table, Table 15.5, since from (15.48), $a = d/(d-s) = 1.026$, $c = (S_0 - I_0)/(d-s) = 128.205, 1 - a = -0.026, F_p = c/(1-a) = -5,000$. Hence, we have $Y^{\#}(t) = A(1.026)t - 5,000$, and

$$Y^{\#\#}(t) = 11000(1.026)^t - 5000.$$

(b) The time path of the income can be seen from the attached chart of Table 15.5 as a monotonic slightly upward bending (almost a straight line) unstable path.
(c) From the formula for g_w, for $S_0 = I_0$, the warranted rate of growth is 2.6% per year.

□

15.8.2 *The neoclassical model of economic growth*

Unlike the Harrod model, the **neoclassical model of economic growth** specifies the functional form of national income, but is still based on the same equilibrium condition that saving equals investment. The model is as follows.

$$I(t) = K(t) - K(t-1) + dK(t-1) \quad 0<d<1, \qquad \text{Investment function} \qquad (15.67)$$

$$S(t) = sY(t-1), \quad 0 < s < 1, \qquad \text{Saving function} \qquad (15.68)$$

$$I(t) = S(t) \qquad \text{Equilibrium condition} \qquad (15.69)$$

$$Y(t) = AL(t)^a K(t)^b, \quad a + b = 1, \qquad \text{Production function} \qquad (15.70)$$

$$L(t) = (1+n)L(t-1), \quad 0 < n < 1, \qquad \text{Labor equation} \qquad (15.71)$$

where d is the rate of capital depreciation, a positive number, and s is the saving rate, a positive fraction. We have five variables in five equations. The variables are (I(t), K(t), S(t), Y(t), L(t)). This model is also called the **Solow model**. The first equation shows that the current investment is the difference between the increment of capital between the current and previous periods and the amount of depreciation of the previous capital stock. Thus, I(t) here is defined as **gross investment**. The saving function states that current saving depends on the amount of saving of the previous income. Equation (15.69) states that the economy is in equilibrium if gross investment is the same as gross saving. Equation (15.70) shows that real income (that is, the output) has the form of the Cobb–Douglas production function we studied in Chapter 11. Since the production function introduces a new variable, labor, L(t), labor is explained by (15.71), which merely states that in this economy labor grows at a constant **labor growth rate** n.

Derivation of the fundamental equation of the growth model

As in this chapter and others, we want to reduce the five growth equations into a single equation. Substituting the first four equations into the equilibrium condition (15.69), we have

$$K(t) - (1-d)K(t-1) = sY(t-1) = sAL(t-1)^a K(t-1)^b. \qquad (15.72)$$

We thus have two variables, K(t) and L(t), in two Eqs. (15.72) and (15.71). To eliminate L(t), we divide both sides by L(t), using (15.71), noting that the RHS is

$$\frac{Y(t-1)}{L(t)} = \frac{AL(t-1)^a K(t-1)^b}{L(t)} = \frac{AL(t-1)^{a-1} K(t-1)^b}{1+n} = \frac{Ak(t-1)^b}{1+n}, \qquad (15.73)$$

where we define $K(t)/L(t) \equiv k(t)$ (making sure that K(t) and k(t) are different). After dividing (15.72) by L(t) and using k(t), we have

$$k(t) - \left(\frac{1-d}{1+n}\right) k(t-1) = \left(\frac{sA}{1+n}\right) k(t-1)^b. \qquad (15.74)$$

In this model, k(t) is called the **equilibrium**[5] **capital–labor ratio** since under this model, k(t) satisfies the equilibrium condition (15.69) for all periods. Thus, the system of five variables and five equations is now reduced to one equation in one variable k(t). This is a first-order nonlinear constant coefficient difference equation. Subtracting both sides by k(t − 1), we have

$$\boxed{k(t) - k(t-1) = \left(\frac{sA}{1+n}\right) k(t-1)^b - \left(\frac{d+n}{1+n}\right) k(t-1)} \tag{15.75}$$

which is called the difference equation version of the **fundamental equation of the neoclassical!growth model**.

Balanced equilibrium capital–labor ratio

As in Example 15.2, we define the **balanced equilibrium capital–labor ratio**, or simply, the **balanced growth path**, as the condition when the capital–labor ratio is the same at every period, k(t) = k(t − 1), that is, when the economy converges to a static condition. Let $k(t) = k(t-1) = k^{\#}$. Then, from (15.75)

$$k^{\#} = \left(\frac{sA}{d+n}\right)^{1/(1-b)}, \tag{15.76}$$

which is a constant. Since $y(t) = Ak(t)^b$, an increase in the balanced equilibrium capital–labor ratio $k^{\#}$ will increase the per capita output, which is a measure of improvement in social welfare. Since $1/(1-b) > 1$, $k^{\#}$ can be increased by raising the levels of marginal (and average) propensity to save s and the scale of production A, and also by decreasing the rates of depreciation and growth rate of labor.

Example 15.11 A numerical model. Let a = 0.7, b = 0.3, d = 0.05, s = 0.1, n = 0.02, and A = 2, in the above neoclassical model.

(a) Rewrite the model using these parameters (to get the feeling of the model) and find the fundamental equation of the neoclassical growth model.
(b) What is the value of the balanced equilibrium capital–labor ratio $k^{\#}$? How do you derive this ratio from the fundamental equation? What is the economic significant of this ratio?
(c) Set up a data table and draw a chart illustrating the balanced growth path $k^{\#}$.
(d) Is the balanced growth path globally stable? Why?
(e) Under what conditions do we have an unstably balanced growth path in the economy? A growth path with no balanced growth equilibrium?

[5]It is "equilibrium" since, along the dynamic path of k(t), the equilibrium condition (15.69) is satisfied at each t.

(f) How will the value of the balanced growth path change if, other thing being equal, the saving rate changes from 0.1 to 0.12? What is the elasticity of change of the balanced growth due to change in the saving ratio?

Answer:

(a) Simply substitute the values of the parameters into the model. The coefficients of the fundamental equation (15.75) are named in row 2 of Table 15.7 (enter a = 1 – b in B2. We only change b in C2). We define the first term on the right-hand side of the fundamental equation as y and the second term as z: that is, $y = C_0k(t-1)\hat{}b$ and $z = C_{00}k(t-1)$, where the coefficients C0 and C00 are defined in B4 and B5 of Table 15.7. For this model, C0 is calculated in D4 and C00 in D5. Thus, the fundamental equation for this model is

$$k(t) - k(t-1) = 0.196k(t-1)^b - 0.069k(t-1).$$

(b) From D7, the value of the balanced equilibrium capital–labor ratio is $k^\# = 4.481$. This is shown in Fig. 15.11. It can be derived from the fundamental equation by letting $k(t) = k(t-1) = k^\#$. See the text for its economic significance.

(c) In Table 15.7, let k(t – 1) range from 0, 0.5 to 10. Enter y = C0*k(t – 1)^b in C11. In D11, enter z = C00*k(t – 1). Selecting range B10:D31, draw the Solow model in Fig. 15.11. Note that the slope of line z is C00, which is not a 45° line. The balanced growth path is located at the value of $k^\#$, at which curve y and line z intersect.

(d) From Fig. 15.11, if the equilibrium capital–labor ratio is less than the balanced growth path, $k(t) < k^\#$ (k(t) is on the left of $k^\#$), and then $y > z$, hence, $\Delta k(t) = k(t) - k(t-1) > 0$. This means that the equilibrium capital–labor ratio k(t) will increase over time and converge to $k^\#$. If $k(t) > k^\#$ (k(t) is on the right of $k^\#$), then $\Delta k(t) < 0$. This means that the equilibrium capital–labor ratio k(t) will decrease over time and

Table 15.7 Derivation of the neoclassical growth model

	A	B	C	D	E	F	G
1		a	b	d	s	n	AA
2		0.7	0.3	0.05	0.1	0.02	2
3	(15.75)	k(t)-k(t-1) = C0k(t-1)^b - C00k(t-1)					
4	C0 =	sA/(1+n)		0.196			
5	C00=	(d+n)/(1+n)		0.069			
6		(1-d)/(1+n)		0.931			
7	k# =	(s*A/(d+n))^(1/(1-b))		4.481			
8							
9		k(t-1)					
10			y	z			
11		0.0	0.000	0.000			
12		0.5	0.159	0.034			
30		9.5	0.385	0.652			
31		10.0	0.391	0.686			

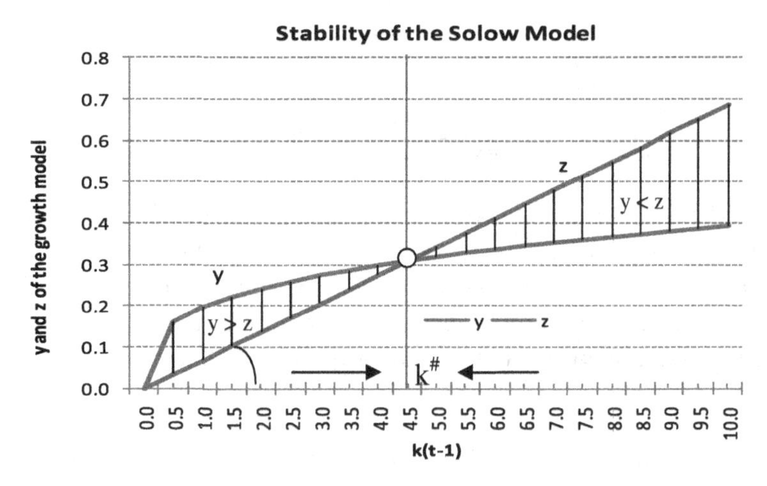

Fig. 15.11 Stability of the Solow model

converge to $k^{\#}$. Hence, the balanced growth path $k^{\#}$ is **stable**. Since the production function is concave, the curve y is also concave, and function z is a line, hence, if y and z intersect at $k^{\#} \neq 0$, the balanced growth path $k^{\#}$ is unique and stable for all $k(t) > 0$. It is **globally stable**.

(e) The two functions y and z intersect at 0, which means $k^{\#} = 0$ and is unstable. There is no economic meaning. The other possibility is that b is equal or greater than 1, in which case z is a line or a convex curve which will not intersect with line y for all $k(t) > 0$.

(f) When the saving rate changes from 0.10 to 0.12, the value of the balanced growth $k^{\#}$ changes from 4.481 to 5.814. Thus, the % change in saving is 20% (=(0.12−0.1)*100/0.1) and that in balanced growth is 29.74%(=(5.814 − 4.481)*100/4.481). The elasticity of change is 1.487. That is, one percent increase in saving will increase the balanced capital–labor ratio by 1.487%. □

15.9 The Balanced Growth Path and the Golden Rule

From Eq. (15.75), along the balanced growth path, we have $k(t) = k(t-1)$. That is, from (15.71), rewriting,

$$\frac{K^{\#}(t)}{L^{\#}(t)} = \frac{K^{\#}(t-1)}{L^{\#}(t-1)}, \quad \text{or} \quad \frac{K^{\#}(t)}{K^{\#}(t-1)} = \frac{L^{\#}(t)}{L^{\#}(t-1)} = 1 + n.$$

Capital stock grows at the same growth rate n as labor. Similarly, from (15.70),

$$\boxed{\frac{Y^{\#}(t)}{Y^{\#}(t-1)} = (1+n)^a(1+n)^b = (1+n)} \tag{15.77}$$

Thus, output $Y^{\#}$ also grows at rate n. Similarly, saving and investment are also growing at rate n:

$$\frac{S^{\#}(t)}{S^{\#}(t-1)} = \frac{sY^{\#}(t)}{sY^{\#}(t-1)} = 1 + n = \frac{I^{\#}(t)}{I^{\#}(t-1)},$$

since $S^{\#}(t-1) = I^{\#}(t-1)$ under the equilibrium growth path. All five variables in the model grow at the same labor growth rate n. Hence, Solow's model is also called the **balanced growth model**, and the dynamic paths of these variables are also called the **balanced growth paths**. Since the economy is growing at a constant rate, it is also said to be in a **steady state** or to be a **steady state economy**.

Economic implications

A balanced growth economy is an ideal economy in which all the economic variables are growing at the same rate as the growth rate of labor. We may also assume that the labor growth rate is the same as the population growth rate. Thus, the growth of the economy is limited by the population growth rate. This result is seemingly plausible in the long run. In the short run, however, the level of balanced capital–labor ratio can be increased by increasing the saving ratio s and production scale (or technology) A. This is also plausible. However, it is interesting to note that while an increase in labor growth rate will reduce the balanced capital–labor ratio in the short run, in the long run, it will increase the rate of economic growth.

Another important implication of the steady state economy is that it can be used as a base for measuring a country's growth performance, and as such, it can be used to compare growth performance among countries. See Weil (2005, Chapter 3).

Example 15.12 Let $L(t) = 1.02L(t-1)$. Find the balanced growth paths for $K^{\#}(t), Y^{\#}(t)$, and $I^{\#}(t)$, with $L(0) = 30$, $K(0) = 200$, $Y(0) = 100$, $I(0) = 20$, $n = 0.07$, and $t = 0, 1, \ldots, 10$. Use the naming method and enter the

(a) Difference equation form: $X(t) = (1 + n)X(t-1)$
(b) Solution form, $X(t) = X(0)(1 + n)^t$.
(c) Is a balanced growth economy realistic? Why is it important in macroeconomic analysis?
(d) What is the output per worker in a balanced growth economy? How can a balanced growth economy raise output per worker?

Answer:

(a) In row 2 of Table 15.8, name the initial conditions and the balanced growth rate, n, and enter the initial condition of each variable in B6:E6. In B7, enter = (1 + nn)*B6, copy it down to B6:E16, and draw the chart on the right.
(b) The solution form is, by using the calculation table, Table 15.5, $L(t) = 30(1.02)^t$. Similarly for other variables. In B7, enter = B$6*(1+nn)^$A7, and copy B7 to B7:E16. The table should be exactly the same as the table obtained by using the difference equation form (a)
(c) See the subsection on "Economic implication" above.

Table 15.8 A balanced growth economy

	A	B	C	D	E
1	L0	K0	Y0	I0	nn
2	30	200	100	20	0.07
3					
4	time	Difference equation form			
5		L#(t)	K#(t)	Y#(t)	I#(t)=S#(t)
6	0	30.0	200.0	100.0	20.0
7	1	32.1	214.0	107.0	21.4
8	2	34.3	229.0	114.5	22.9
9	3	36.8	245.0	122.5	24.5
10	4	39.3	262.2	131.1	26.2
11	5	42.1	280.5	140.3	28.1
12	6	45.0	300.1	150.1	30.0
13	7	48.2	321.2	160.6	32.1
14	8	51.5	343.6	171.8	34.4
15	9	55.2	367.7	183.8	36.8
16	10	59.0	393.4	196.7	39.3

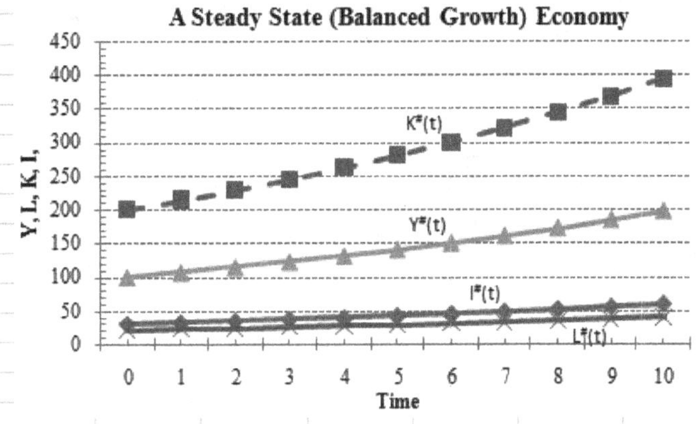

A Steady State (Balanced Growth) Economy

Fig. 15.12 A balanced Growth Economy

(d) The balanced growth output per worker is, from (15.73) and (15.76) after dividing the denominator and the numerator by $Y^{\#}(t-1)$,

$$Y^{\#}(t)/L^{\#}(t) = Ak^{\#}(t)^{b} = A\left(\frac{sA}{d+n}\right)^{b/(1-b)}. \tag{15.78}$$

Thus, a balanced growth output per worker can be raised by increasing saving rate s and the production scale (technology) A. □

15.9.1 *The Golden Rule of accumulation*

The balanced growth path is derived for given parameters, especially the saving ratio. We also see that per capita income (15.78) under the balanced growth path can be raised by increasing the saving ratio. Raising the per capita income will indirectly raise the per capita consumption, a better indicator of social welfare. The economic planner may want to choose a saving ratio such that the per capita consumption can be maximized along the balanced growth path.

In order to identify this saving rate, we first find the per capital consumption function. Since the equilibrium condition of macroeconomics is, from (15.69), $Y(t) = C(t) + I(t) = C(t) + S(t)$, but $S(t) = sY(t-1)$, we find the following:

$$C(t) = Y(t) - sY(t-1) = Y(t)\left(1 - s\frac{Y(t-1)}{Y(t)}\right).$$

Table 15.9 Derivation of the Golden Rule

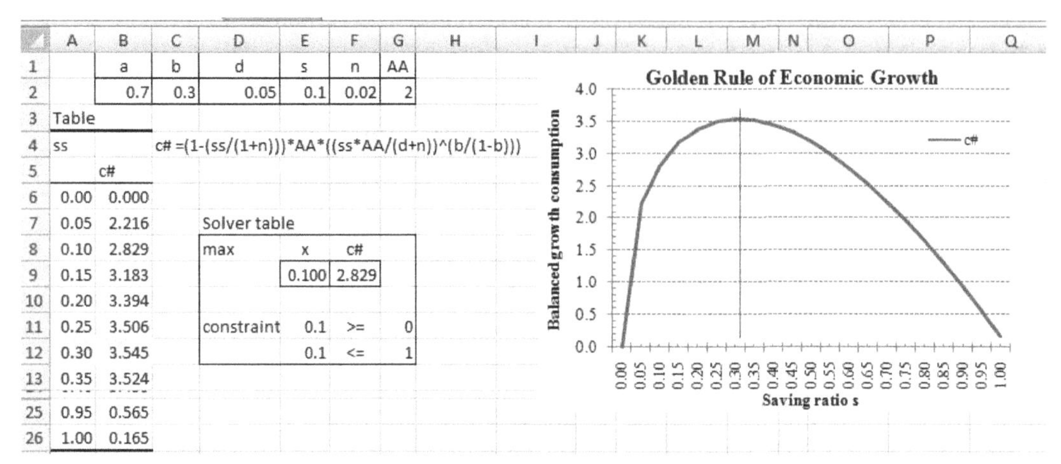

	A	B	C	D	E	F	G
1		a	b	d	s	n	AA
2		0.7	0.3	0.05	0.1	0.02	2
3	Table						
4	ss		c# =(1-(ss/(1+n)))*AA*((ss*AA/(d+n))^(b/(1-b)))				
5		c#					
6	0.00	0.000					
7	0.05	2.216		Solver table			
8	0.10	2.829		max	x	c#	
9	0.15	3.183			0.100	2.829	
10	0.20	3.394					
11	0.25	3.506		constraint	0.1	>=	0
12	0.30	3.545			0.1	<=	1
13	0.35	3.524					
25	0.95	0.565					
26	1.00	0.165					

Fig. 15.13 Derivation of the Golden Rule

Dividing both sides by L(t), writing per capita consumption as c = C/L, then, under the balanced growth path, substituting (15.77) and (15.78) into c, we have

$$\boxed{c^{\#} = \left(1 - \frac{s}{1+n}\right) A \left(\frac{sA}{d+n}\right)^{b/(1-b)}} \tag{15.79}$$

which is a function of s. The problem now is

$$\max c^{\#}, \quad \text{subject to } 0 \leq s \leq 1. \tag{15.80}$$

We first illustrate the balanced growth consumption function as a function of saving ratio. In Table 15.9, we name the parameters in row 2, and name the column A6:A26 as the range of the saving ratio ss (to avoid overlapping with named s in E2). In B6, enter formula (15.79), as shown in C4, and then copy B6 down to B26. The consumption function is drawn in Fig. 15.13.

From the table and chart, the per capita consumption under the balanced growth economy forms a concave function throughout the range of the saving rate, and the global unique maximum is achieved at s = 0.30. At this saving ratio, the country's welfare is maximized. Thus, the economy is in the best possible state, and the welfare of the people is also at the best possible state. The economy and society are in **Bliss**.

Using the Excel Solver to find the Golden Rule

We can use the Excel Solver to find the exact solution of the maximization problem (15.80). We follow the format introduced in Chapter 12. The Solver table is enclosed in D8:G12 of Table 15.9. Denote the saving rate x. The "changing cell" (variable) x is assigned to cell E9, which is named x. Enter the long formula in C4 using x, instead of ss of column A or s in E2, into the "target cell" F9. Enter the initial value x = 0.1 in E9. Enter =x in E11 and E12 (when =x is entered in E11 and E12, both cells will show 0.1), and 0 and 1 in G11 and G12 as the two constraints. Invoking the Excel Solver, the exact solution is found in E9 as $x^{\#}(= s^{\#}) = 0.306$ and in the target cell F9 as $c^{\#} = 3.546$, same as the values in A12:B12 in Table 15.9.

15.10 Summary

In this chapter, we first introduced various new terms and concepts of economic dynamics and difference equations. We found first that, without saying so, we had already encountered them in Chapters 8 and 9, as the dynamic problems of calculating future values of simple and compound interest, with or without annuity. Reformatting the interest problem in the terms of a simple first-order constant coefficient linear difference equation, we applied the iteration method to find the solution of the interest problems of difference equations (Sec. 15.2). The iteration method becomes very complicated and tedious for solving higher order difference equations. In Sec. 15.3, we introduced the standard method of solving linear difference equations: the procedure is to find the particular solution and the complimentary solution, and then find their sum, which is the general solution. From this, for a given initial condition, we find the definite solution. The relations among these solutions are illustrated in Fig. 15.2. This procedure is also valid for higher-order linear difference equations.

After learning the procedure, we discussed the equilibrium and stability of the dynamic path (Sec. 15.4). The equilibrium value is shown as the particular solution, and the stability of the dynamic path can be seen from the complementary solution (see Fig. 15.2). In this way, mathematical theory of difference equations is related to dynamic economic and business analysis. To illustrate these relations, we introduced the dynamic market model when the supply function depended on the previous price (Sec. 15.5). Using this market model, we derived the four solutions and then implemented the solution procedure on an Excel spreadsheet table. This was done in Sec. 15.6. Using this table, we can instantly find the definite solution of any first-order constant coefficient linear difference equation, and every step of finding the solution is shown on the spreadsheet.

This leads to the interesting interaction between demand and supply due to price changes, which is known as the **Cobweb Model** (Sec. 15.7). We have shown that, by a simple and ingenious modification of the data table for comparative static analysis, introduced in Chapter 4, the Cobweb model can be illustrated and explained diagrammatically in an Excel spreadsheet, showing both the phase diagram and the time path charts.

The last section (Sec. 15.8) illustrated how first-order difference equations can be applied to the theory of economic growth, by introducing **Harrod's model** and **Solow's model**. As with the fundamental equation of the input–output model of the Leontief model, in Chapter 10, we systematically introduced the neoclassical one-sector growth model and derived the **fundamental equation of the neoclassical growth model**. As in statics and comparative statics, all these models are based on the mini-general equilibrium theory of demand and supply, the foundation of economic and business analysis. We have defined the **balanced growth path** and the **Golden Rule of capital accumulation**, which leads the economy and society to a state of **Bliss**. Now that we have shown how the economy and business can achieve a state of Bliss, we have reached the time to end this book.

After introducing the first-order constant coefficient linear difference equation, we may go on to the second order and higher-order difference equations and nonlinear difference equations. However, if we proceed further, we will need to introduce complex numbers, trigonometric and sinusoidal functions, characteristic equations, characteristic roots, etc. Although they can be illustrated beautifully on spreadsheets, it will take another volume to explore them, as we have done elsewhere. We plan to introduce this material in a more advanced textbook.

Review of Basic Equaitons and Formulas

General form (15.3) $Y(t) = aY(t-1) + c$ (15.27) $F(t) = aF(t-1) + c$ Particular solution (15.36) $F_p(t) = k = \frac{c}{1-a}$ Complementary solution (15.38) $F_c(t) = Aa^t$ General solution (15.39) $F^{\#}(t) = F_c(t) + F_p(t)$ (15.40) $= Aa^t + \frac{c}{1-a}$ Definite solution (15.42) $F^{\#\#}(t) = (F_0 - F_p)a^t + F_p$ (15.43) $= a^t F_0 + c\frac{1-a^t}{1-a}$ $(a \neq 1)$ (15.44) $= F_0 + ct$ $(a = 1)$ Stability condition (15.50) $\|a\| < 1$, for $a \neq 0$ $F^{\#}(t) = c$, for $a = 0$ A Dynamic Market (Cobweb) model (15.9) – (15.11) (15.51) $P(t) = -(d/b)P(t-1) + (a+c)/b$.	Harrod's growth model (15.64) (15.65) $Y(t) = \left(\frac{d}{d-s}\right)Y(t-1) + \left(\frac{S_0 - I_0}{d-s}\right)$. Neoclassical growth model (15.67) – (15.71) Fundamental equation of the neoclassical growth model (15.75) $\Delta k(t) = \left(\frac{sA}{1+n}\right)k(t-1)^b - \left(\frac{d+n}{1+n}\right)k(t-1)$, Balanced equilibrium capital-labor ratio (15.76) $k^{\#} = \left(\frac{sA}{d+n}\right)^{1/(1-b)}$, Balanced growth path (15.77) $\frac{K^{\#}(t)}{K^{\#}(t-1)} = \frac{L^{\#}(t)}{L^{\#}(t-1)} = \frac{Y^{\#}(t)}{Y^{\#}(t-1)}$ $= \frac{S^{\#}(t)}{S^{\#}(t-1)} = \frac{I^{\#}(t)}{I^{\#}(t-1)} = 1+n$ Per capita consumption under balanced growth (15.79) $c^{\#} = (1-\frac{s}{1+n})A\left(\frac{sA}{d+n}\right)^{b/(1-b)}$ Golden rule (15.80) max $c^{\#}$, subject to $0 \leq s \leq 1$

Key terms: Economics and Business

Key terms: Mathematics

Homework Chapter 15

15-1 Forward difference equations Instead of backward difference, let $t - 1 = s$ in (15.1), or let $t - 2 = s$ in (15.2), then we have **forward difference form of difference equations**. Write these two backward difference equations, the dynamic market model (15.12), and the dynamic macroeconomic model (15.18) in the forward difference form of difference equations. Since t and s are dummy, change s to t in the new difference equations.

15-2 Solutions Show that, writing (15.3) as $F(t) = aF(t-1) + c$ and $a \neq 1$, the general solution (15.40) and the definite solution (15.43) are indeed solutions for this difference equation.
(Hints: Since F(t) is given by (15.3), we only need to show that the right-hand side of (15.3) is the same as the left-hand side of (15.3)).

15-3 Iterative method From (15.25), $F(n) - F(n-1) = rF(n-1)$, let $n = 1, 2, \ldots, 10$, and let $F(0) = F_0$ be given.

(a) Find the definite and general solutions using the iterative method, by constructing a table like Table 15.2. Column B should be the leading function, column C should be the iterative process of the difference equation, and column D should be the definite solution (change from the initial period).
(b) Let $r = 0.05$ and $F(0) = 1$. Construct a table like Table 15.3, namely, with column C showing the numerical change in F(t) for each period (the difference equation), and with columns D, E, and F showing the change in the definite solution (15.28).
(c) Is the table similar to Table 9.3? Explain if there are differences
(d) Column G should show the value of the first difference, as in column G of Table 15.3.
(e) Draw the curve in a way similar to Fig. 15.1. What does the F(t) curve look like?

15-4 Reconstruction of the Excel calculation table Using the market model given in Example 15.9, reconstruct the Excel calculation of Table 15.5. Make sure your solution is the same as those obtained directly from Table 15.5.

Interest rate problems

15-5 Future value with annuity If Mary deposits \$1000 per year in the bank at an interest rate of 6%, compounded annually for ten years, how much will she accumulate by the end of ten years? Set up the difference equation and solve the problem by using the standard method of solution.

15-6 Future value with annuity John saves \$100 per month in a bank at an interest rate of 6% per year compounded monthly. How much will he have after 5 years? 10 years? Set up the difference equation and solve the problem using the standard method of solution.

15-7 Amortization Nick borrows \$10,000 from a bank at an annual interest rate of 6% for ten years. If he plans to return the debt by equal installments at the end of each year, what will the amount of his first installment payment (PMT) be? Set up the difference equation with interest rate r for the general case. Then solve the problem by using the standard method of solution.
(Hints: The difference equation is $\Delta F(t) = 0.06F(t-1) - PMT$, or $F(t) = (1+r)F(t-1) - PMT$, for $F(0) = \$10{,}000$, where PMT is unknown. Set $F^{\#}(10) = 0$).

15-8 The solution table Let the standard first order linear difference equation be

$$Y(t) = aY(t-1) + c.$$

Find the definite solution of the following equations using the calculation table, Table 15.5. State the stability and the shape of the solution.

(a) $Y(t) + 4Y(t-1) + 5 = 0$, $Y(0) = 20$.
(b) $Y(t) - Y(t-1) - 2 = 0$, $Y(0) = 15$.
(c) $2Y(t) = -Y(t-1) + 5$, $Y(0) = 10$.
(d) $4Y(t-2) - 3Y(t-3) = 5$, $Y(0) = 30$.
(e) $L(t) = 1.05L(t-1)$, $L(0) = 10$.
(f) $K(t) = 1.02K(t-1)$, $K(0) = 20$.
(g) $\Delta L = 0.05L(t-1)$, $L(0) = 10$.
(h) $\Delta L(t) = 1.05$, $L(0) = 10$.
(i) $2K(t+4) + 2K(t+3) + 22 = 0$, $K(0) = 11$.
(Hints: (d)Let $t-2 = s$. See HW15-1; (i) Let $t+4 = s$)

15-9 A market of Fuji apples The market of Fuji apples in Denver can be described by the dynamic system of equations, (15.9)–(15.11) in Example 15.1, with the following parameters:

$$a = 85, \quad b = 0.7, \quad c = 20, \quad d = 0.5.$$

The quantity is measured in dozens and the price in dollars.

(a) Write this market model as a system of three equations.
(b) How many equations and variables are there in this model?

(c) Find the difference equation in terms of price.
(d) Solve the difference equation algebraically using the standard solution method, assuming that the initial condition is $P(0) = 100$.
(e) Verify your manual solution by using the solution table, Table 15.5.
(f) Is the system stable? What is the shape of the dynamic path?
(g) Find the equilibrium time path of quantity like Fig. 15.6.
(h) Draw the dynamic paths of price and quantity in one chart, like Fig. 15.8.
(i) Draw the cobweb chart like Fig. 15.5.
(j) Using the Cobweb chart, trace and interpret the economic meaning of the cobweb model.

15-10 The Cobweb models In the basic Cobweb model of (15.9)–(15.11) in Example 15.1, the following values of the parameters are estimated for demand and supply functions (say, by using the least squares regression):

(a) $a = 180$, $b = 0.4$, $c = 30$, $d = 0.2$, given $P(0) = 250$;
(b) $a = 180$, $b = 0.4$, $c = 30$, $d = 0.8$, given $P(0) = 250$;
(c) $a = 180$, $b = 0.4$, $c = 30$, $d = 0.4$, given $P(0) = 250$;
(d) $a = 180$, $b = 0.4$, $c = 30$, $d = -0.4$, given $P(0) = 250$;
(e) $a = 180$, $b = 0.4$, $c = 30$, $d = -0.8$, given $P(0) = 250$;
(f) $a = 180$, $b = 0.4$, $c = -30$, $d = 0.2$, given $P(0) = 250$.

Find the equilibrium price and quantity for each case. For (a), (b), and (c), trace the cobweb of price movement as shown in Fig. 15.5, and relate the time paths of quantity and price as in Figs. 15.5–15.8. (Hints: (a) $Y^{\#\#} = -100(-0.5)t + 350$, $P^{\#} = 350$, $Q^{\#} = 40$).

15-11 A dynamic market model Suppose that the demand and supply of the hog market are given as

$$D(t) = a - bP(t) + 50(1 + r)^t,$$
$$S(t) = -c + dP(t - 1),$$
$$S(t) = D(t).$$

Here D, S, and P, are quantity demanded, quantity supplied, and price. The supply depends on the last year's price the farmers received, and the demand is a function of this year's price and the steady demand of hogs from foreign countries over the years. Let the time range from 0 to 20, and the initial time is $P(0) = 20$.

(a) Using the naming method, draw the graph of the equilibrium quantity (S and P) and price, assuming that

$$a = 18, \quad b = 3, \quad c = 3, \quad d = 4, \quad r = 0.02.$$

Please make sure that axes and lines are labeled, and the values of the parameters are shown in the chart title.
(Hints: Write the equations on the spreadsheet, and use the following table format:

t	S(t)	D(t) = S(t)	P(t)
0			20
1	77	77	−2.7

where $P(t) = (a - D(t) + 50(1 + r)^t)/b$, for $t = 1, 2, \ldots, 20$, which is derived from inverting the demand function. The construction of the above table indicates that, starting from the second equation, when $P(0) = 20$ is given, $S(1) = -c + dP(0) = 77$ is determined, and from the third equation, we have $D(1) = S(1) = 77$. Thus, the D(t) column is a copy of the S(t) column. After D(1) is given, P(1) is determined at $t = 1$ as -2.7, and so on).

(b) Name the case when $b = 3$ "Case 1". Change b to the following cases and draw the chart for each case. Collect all five charts in one page.

Case 2. $b = 6$; Case 3. $b = 4$; Case 4. $b = -6$; Case 5. $b = -3$.

(c) Why are the paths of the graph called the dynamic equilibrium paths? In what sense is it equilibrium? In what sense is it dynamic?
(d) Which case is stable? Which case is unstable?
(e) Which case(s) does (do) not make economic sense? Explain. In reality, which case is most economically meaningful? Why?

National income models

15-12 A dynamic macroeconomic model In the macroeconomic model (15.14)–(15.17) of Example 15.2, we have parameters (a, b, and d) and four variables: endogenous variables (Y_t, C_t) and exogenous variables (I_t, G_t). More generally, we can set up a general model as follows: we retain the equilibrium condition (15.14), but newly define the behavior equations as follows:

$$C(t) = C_0 + aY(t) + bY(t-1),$$
$$I(t) = I_0 + cY(t) + dY(t-1), \quad \text{and}$$
$$G(t) = G_0 = 0.$$

(a) Derive step by step the difference equation for this model:

$$Y(t) = \left(\frac{b+d}{1-a-c}\right) Y(t-1) + \left(\frac{C_0 + I_0}{1-a-c}\right).$$

(b) The equilibrium income and consumption are defined as the case when the variables do not change over time; that is $Y(t) = Y(t-1) = Y^{\#}$. Find the values of equilibrium income and consumption in terms of parameters and exogenous variables.
(c) For the equilibrium income and consumption to be economically meaningful, what conditions should be imposed on the parameters and the exogenous variables?
(d) Modify the calculation table Table 15.5 such that, for any given set of parameters, exogenous variables, and the initial condition Y(0), the solution of this model will appear in rows 9–14 in the table. (Note: Change $P^{\#}$ to $Y^{\#}$, and $Q^{\#}$ to $C^{\#}$, and add C_0 and I_0 to the right of $C^{\#}$ in Table 15.5.)

(e) Let $C(t) = 600 + 0.5Y(t) + 0.2Y(t-1)$, and $I(t) = 100 + 2Y(t) - 2Y(t-1)$. Derive the difference equation for this model, starting from $Y(0) = 4000$. Find the equilibrium income and consumption. Do they make sense? Why?

(f) Let $C(t) = 1000 + 0.6Y(t) + 0.25Y(t-1)$, and $I(t) = 400 + 0.1Y(t-1)$. Derive the difference equation for this model, assuming $Y(0) = 4000$. Find the equilibrium income and consumption. Do they make sense? Why?

(g) Let $C(t) = 600 + 0.75Y(t-1)$, and $I(t) = 100 + 0.1Y(t-1)$. Derive the difference equation for this model, assuming $Y(0) = 10{,}000$. Find the equilibrium income and consumption. Do they make sense? Why?

15-13 Balanced growth and Golden Rule

(a) Reproduce Table 15.7, Fig. 15.11, on the balanced growth path.
(b) Reproduce Table 15.8, Fig. 15.12, on the balanced growth economy.
(c) Reproduce Table 15.9, Fig. 15.13, on the derivation of the Golden Rule.

15-14 Comparative dynamic analysis of balanced growth In Sec. 15.9.1, we found that the Golden Rule of capital accumulation for the neoclassical model specified in Table 15.9 with $s = 0.306$.

(a) Is this Golden Rule saving ratio realistic?
(b) What is the value of the balanced growth capital–labor ratio at this saving ratio?
(c) When the Golden Rule saving ratio is entered in Table 15.7, can you find the value of the balanced growth path in Fig. 15.11? Why?
(d) In the above question, the y curve and the z line do not intersect, do they intersect after all? If so, draw the two lines that show the intersection. If not, how can you make them intersect?
(e) Denote the Golden Rule balanced growth path $k^{\#\#}$. Will the original balanced growth path, $k^{\#}$ converge to the new path $k^{\#\#}$? Why?

(Hints (d) Since $k^{\#} \cong 22$, let the range of $k(t-1)$ in Table 15.7 be from 18.0 to 27.0). (e) Yes).

15-15 Comparative dynamic analysis In Table 15.9, insert a blank space in rows 3 and 4. As in Chapter 4, row 3 is the default row. Copy the numbers in C2:G2 to C3:G3. Row 4 is the "suggested values" row. (a) In C4, enter 0.4; (b) in D4, enter 0.10; (c) in F4, enter 0.04; (d) in G4, enter 4 and (e) also 10. Other things being equal, using the default chart Fig. 15.13 as a template, draw a new consumption curve for each of these five cases, and find the new Golden Rule saving ratio. Use Excel Solver to derive and verify the value of Golden Rule saving ratio obtained from the chart.

Postscript
Microsoft Excel 2007 and Excel 2010

As we have mentioned in the Preface, the contents of this book have been developed from 1985 to the present in a course entitled **Microcomputer Application in Economics**. The original intention was to use spreadsheets to explain, analyze, illustrate, interpret, and integrate major topics in economics and business. Over the years, we have used VisiCalc, SuperCalc, Lotus 123, Quattro Pro, Excel 98, and Excel 2003, all of them available to the students or in the University of Colorado computer laboratory at various times. When I decided to arrange my lecture notes and choose topics to publish as a book, Microsoft introduced Excel 2007, which uses "ribbon" format, and has quite a different appearance from all the previous versions. The spreadsheet program used in this book is Excel 2007. Recently, Microsoft introduced Excel 2010. Fortunately, so far as this book is concerned, the differences are very minor; the Excel commands and all the Excel formulas used in this book are still valid in Excel 2010. This section will show some major differences between the two versions to help readers make quick transition from Excel 2007 (hereafter, **Excel7**) to Excel 2010 (hereafter, **Excel10**). The numbering of new Excel10 figures and tables corresponding to those of Excel7 is indicated by -P (**P**ostscript) in the numbering of figures in this section. Readers will see from the comparison that, so far as the contents of this book are concerned, the changes are non-consequential, and they can adapt to the new version easily.

Page 5 Fig. 1.1 Worksheets and a workbook
Page 9 Fig. 1.3 The northwest corner of the workbook

The major difference appears in the new <"**File**"> tab in Excel10, which has replaced the <"Office"> tab (1) in Excel7. We compare the contents of the two tabs in Fig. 1.1-P. In Excel10, a new tab "**Info**" (shown with a hexagon) is created on the left column of the <"File">. When "File" is clicked, the window opens for <info> window of the currently opened sheet and file, providing information and choices of action on the file as shown in the middle part of Fig. 1.1-P.

All other subtabs are directly accessible from <"File">. For example, if you want to preview or print the file, Excel7 requires <"Office"> <Print!, Print Preview>, or <"Office"> <Print!, Print>, then the "Print Preview", or "Print" dialog box, appears. In Excel10, the procedure is simply <"File"> <Print>, both "Print Preview" and "Print" dialog box appear in the same window. It saves one mouse clicking. Other sub-tabs in

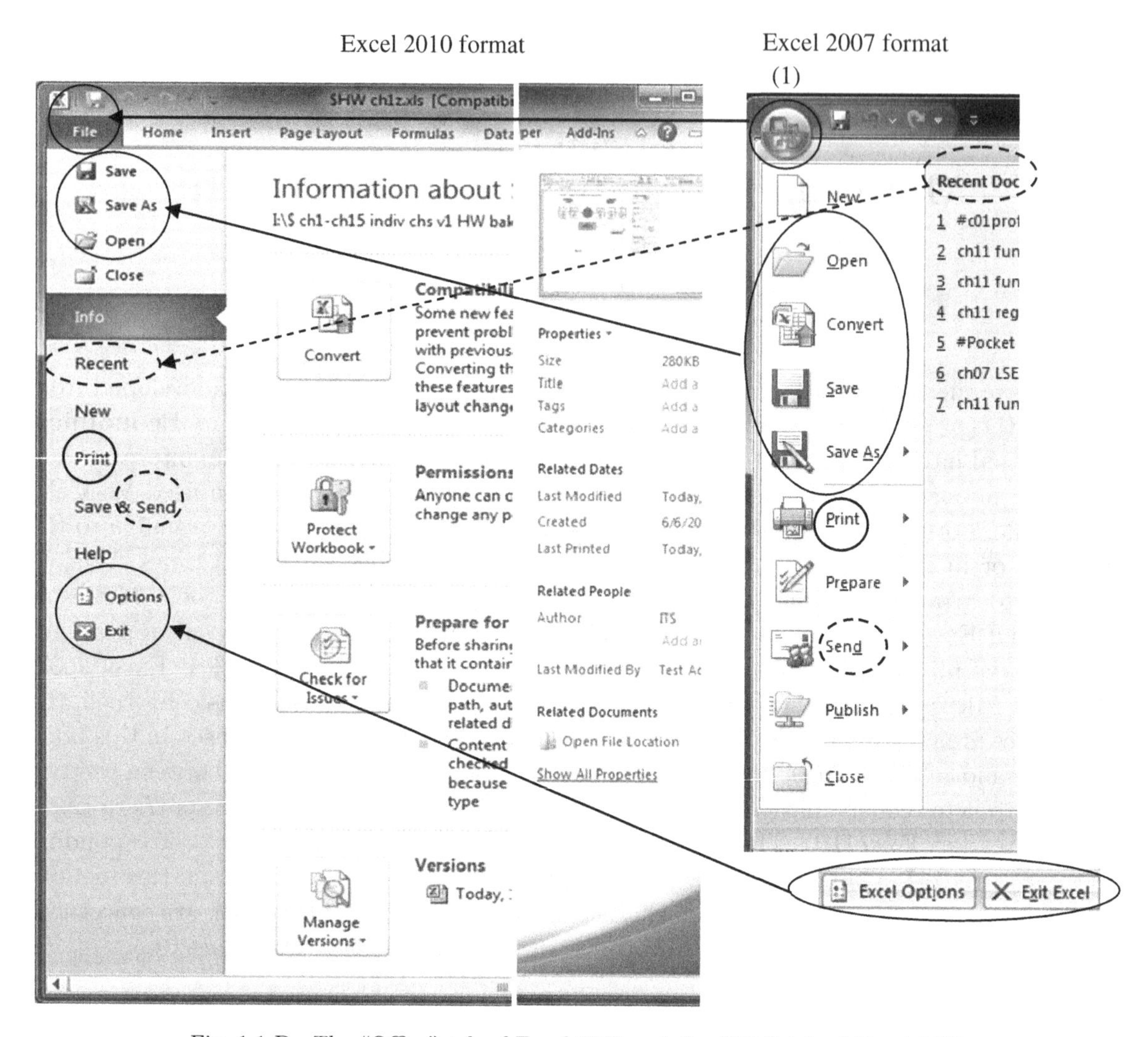

Fig. 1.1-P The “Office” tab of Excel 2007 and the “File” tab of Excel 2010

<“Office”> tab in Excel7 are also rearranged in Excel10, see the circles and the arrows connecting the circles from right to left in Fig. 1.1-P.

Page 27 Fig. 1.15 Print preview and page setup

As mentioned above, Excel10 has combined <“Office”><Print!><Print> and <“Office”><Print!><Print Preview> of Excel7 in one in <“File”><Print>. The old <Page Setup> tab is moved to the bottom of the <Print> dialog box. Clicking new <Page Setup> will lead to exactly the same dialog box as that in Fig. 1.15. Three commonly used commands in old <Page Setup>, namely, page orientation (a), scaling (b), and margin tab in Fig. 1.15, are shown directly in the new <Print> dialog box. See the three connecting lines from the right to the left in Fig. 1.15-P.

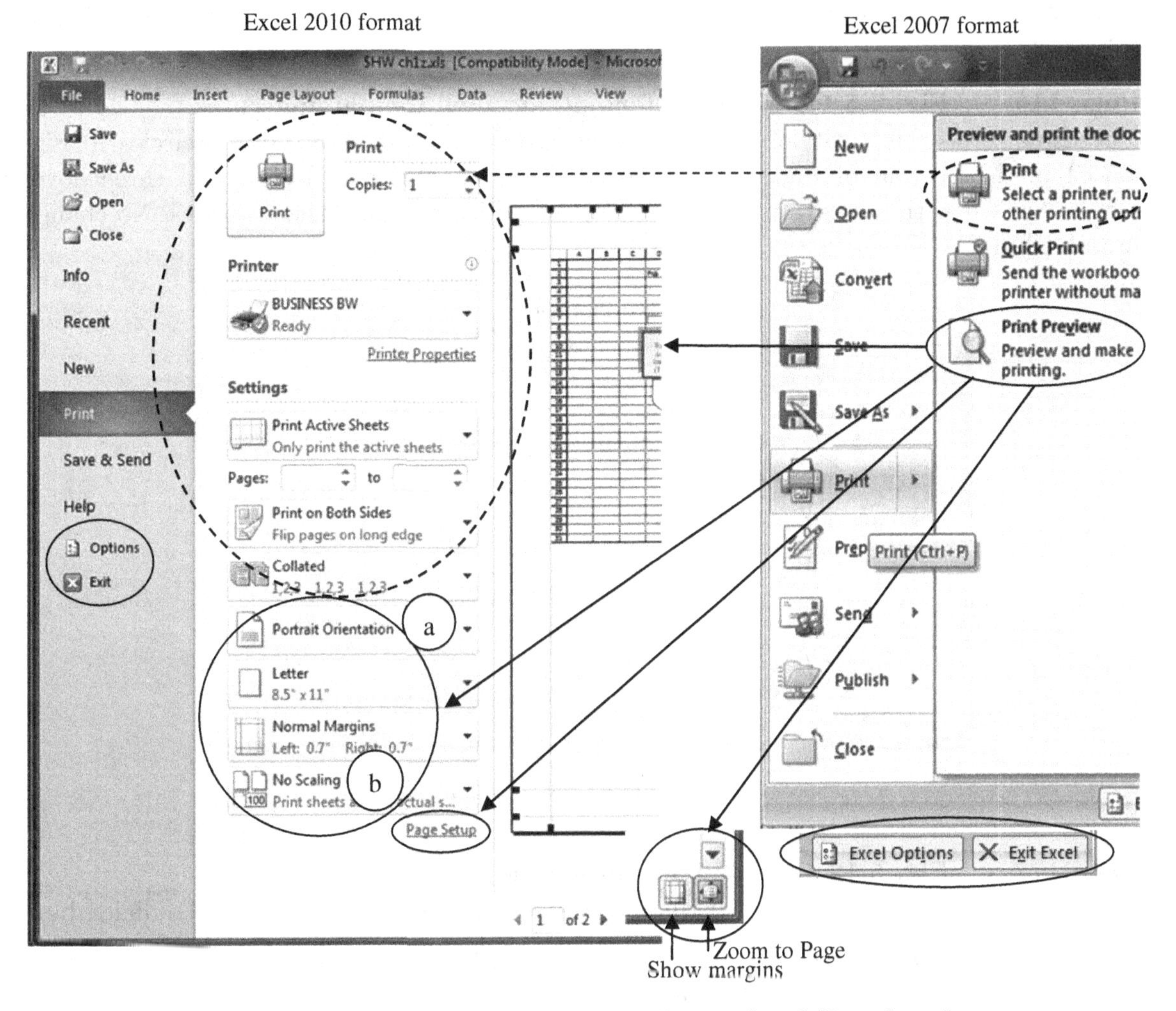

Fig. 1.15-P Location of "Print," "Print Preview," and "Page Setup"

In Excel10, after printing, you may return to the worksheet by clicking <"Home"> tab, or any tab; or exit the file by clicking <"File", Exit> or <"File", Close>.

Page 75 Fig. 3.10(b) Formatting dialog box

In Excel10, Microsoft restored the pre-Excel07 feature of clicking twice a chart element to invoke a dialog box to edit that element. For example, in Fig. 3.10, clicking the supply curve twice (the first click will select the curve), "Format Data Series" dialog box will appear, skipping the small dialog box of Fig. 3.10(a). This is an extra convenience of editing the chart elements.

In Excel10, Microsoft has added "**Glow and Soft Edge**" button on the left side of all the formatting dialog boxes. Since the dialog boxes are the same as those of Excel7, the above two new features, and all other new features of charting, do not affect the using of this book.

Page 115 Fig. 4.5 Conditional formatting-new formatting rules

The three boxes: "Icon Style", "Reverse Icon Order", "Show Icon Only" (circles in Fig. 4.5-P) in the bottom of Fig. 4.5 are moved closer to "Format Style", and three down arrows (shown with a hexagon in Fig. 4.5-P) are added to three "Icon" boxes. No change in procedures.

Excel 2010 format

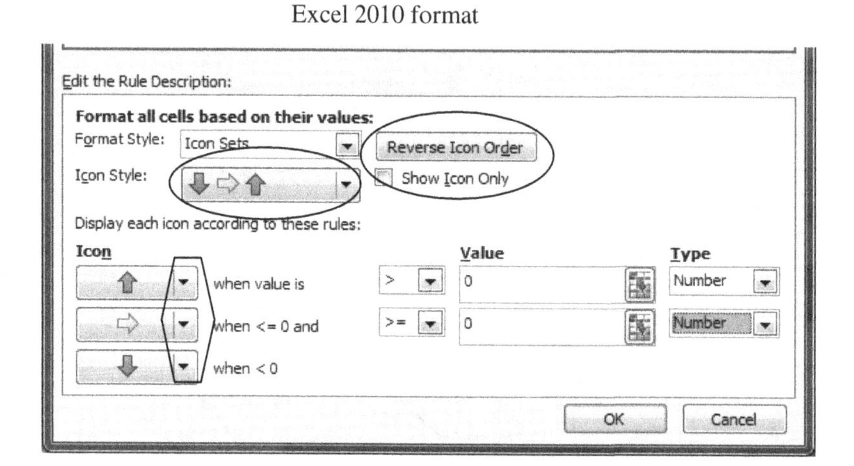

Fig. 4.5-P Conditional formatting

Page 177 Fig. 5.12 Format trendline dialog box

In Excel10, the "**Glow and Soft Edge**" tab has been added to the left of the dialog box.

Page 451 Fig. 12.4 "Solver Parameter" dialog box
Fig. 12.5 "Add Constraint" dialog box
Fig. 12.6 "Solver Results" dialog box

In Excel10, the tabs have been rearranged, as shown by the solid and dotted circles in Fig. 12.4 and Fig. 12.4-P. A new box for "Select a Solving Method" (enclosed in hexagon) has been added: GRG Nonlinear (Generalized Reduced Gradient method of solving nonlinear problems), Simplex LP (Simplex method of solving linear programing), and Evolutionary (method for solving nonlinear systems of equations). In this book, we only use the default "GRG Nonlinear" method.

The contents and arrangement of Fig. 12.5 "Add Constraint" dialog box, and Fig. 12.6 "Solver Results" dialog box in Excel10, are the same as those of Excel7, except that "Help" tabs are deleted from these two Excel10 dialog boxes (not shown).

Page 500 Table 13.6 <"Data", Sort & Filter><Filter>

In Excel 2010, our comment in Footnote 6 (page 503) about Count and CountA in (a) and (b) of Table 13.6 still holds. The error is not corrected.

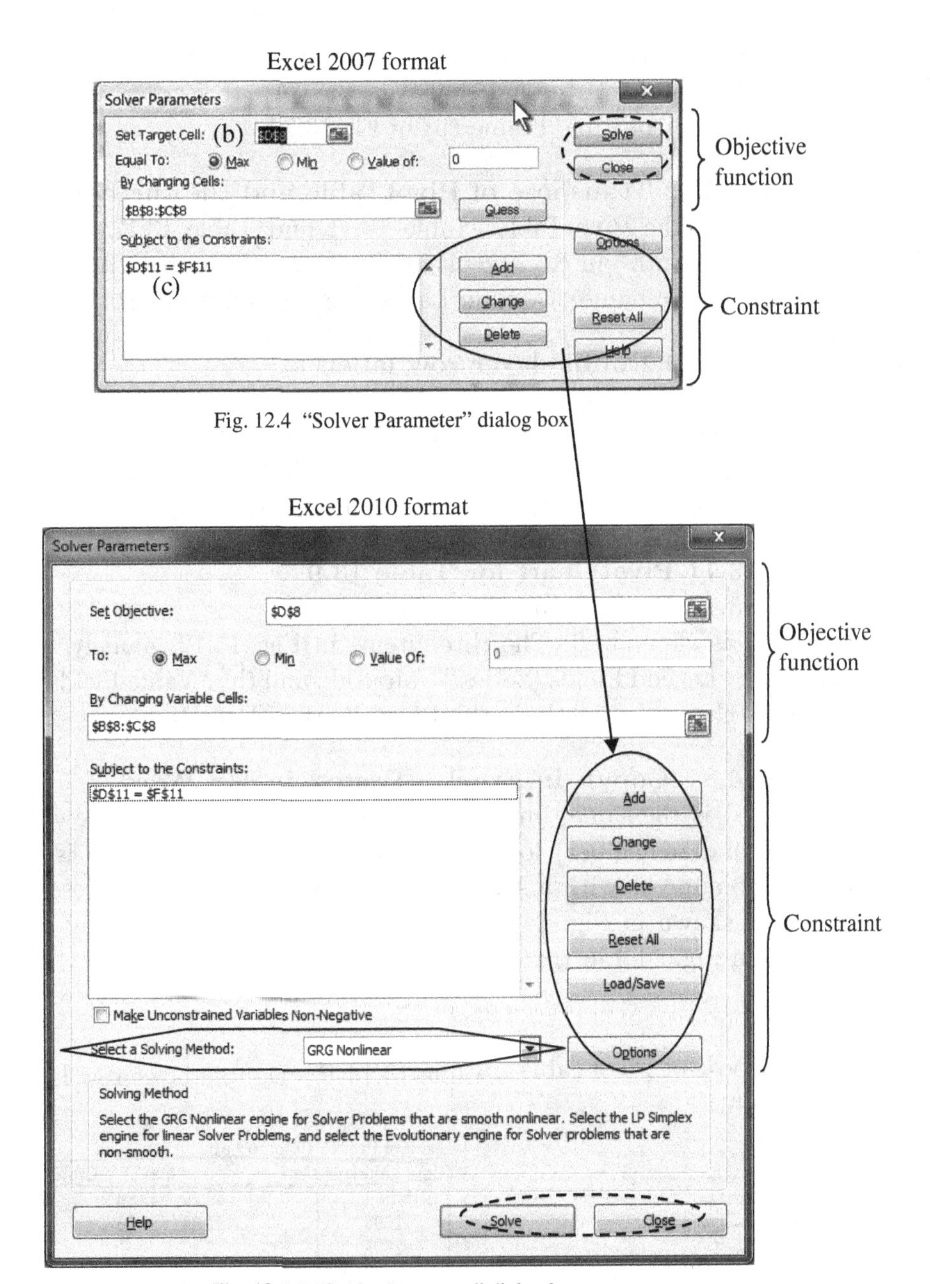

Fig. 12.4 "Solver Parameter" dialog box

Fig. 12.4-P "Solver Parameter" dialog box

Page 512 Fig. 13.10(a) Structure of PivotTable-Excel 2007

In Excel 2010, the same process of clicking <"Insert"> <Tables, PivotTable!> <PivotTable> will show the same "Create PivotTable" dialog box of Fig. 13.9. When <OK> is clicked, Excel 2010 will show the second box: "To build a report, ..." box of Fig. 13.10(a) (the first box: “Click in this area to ...” box is eliminated), along with PivotTable Field List (c) of Fig. 13.10. When one of the four boxes in the “Area” section of Fig. 13.10(c) is filled, Excel will show the PivotTable Frame (b) of Fig.13.10.

Page 514 Table 13.12 Transpose of PivotTable and the three-axis table

In Excel10, the format of the PivotTables Table 13.11 and Table 13.12 are the same as those of Excel7, except “Month” in K5 is replaced by its generic name “Row Labels”, and “Region” in L4 by its generic name “Column Labels.” See Table 13.12-P.

Page 516 Table 13.14 Double-layer row labels

In Excel10, the first row label “North” has a separate row which lists “North Total” for each month with bold face, followed by the second row labels of “Year”. The row of “North Total” in Excel7 is eliminated, as shown by Table 13.14-P. There is no change in Table13.15 “Double-layer column labels,” except the generic names mentioned in Table 13.12-P.

Page 518 Fig. 13.11 PivotChart for Table 13.9

Page 518 Fig. 13.12 PivotChart filter pane

In Excel10, Fig. 13.12 is eliminated. The three items in Fig. 13.12, namely, “Axis Fields (Categories)”, Region!, “Legend Fields (Series)”, Month!, and the “Value Fields”, are shown as buttons and can be changed directly in the PivotChart in Excel10.

Page 538 Fig. 14.3 <Copy> in Excel, <Paste> in MS Word

In Excel10, Fig. 14.3(a) is the same, but the 6 items in Fig. 14.3(b) are now replaced by 6 (hard to recognize and remember) logos, see Fig. 14.3-P. When a logo is clicked, the explanation of the logo appears with a short-hand capital letter in parentheses at the end of the explanation, as shown in Fig. 14.3-P. The alphabet at the beginning of each item corresponds to those in Fig. 14.3 in this book.

Table 13.12-P Transpose of PivotTable and the three-axis table

J	K	L	M
1			
2	Year	(All)	
3			
4	Sum of Profit	Column Lab	
5	Row Labels	East	North
6	Jan	38.18	89.26
7	Mar	0.95	43.98
8	May	29.22	36.72
9	Grand Total	68.35	169.96

Excel 2010

Table 13.14-P Double-layer row labels

J	K	L	M	R
4	Sum of Profit	Column Labe		
5	Row Labels	Jan	Mar	Grand Total
6	North	89.26	43.98	319.45
7	2006	92.08	38.96	180.73
8	2007	-2.82		74.57
9	2008		5.02	64.15
10	East	38.18	0.95	296.45
11	2006		10.56	78.81
12	2007	38.18	-8.8	112.09
13	2008		-0.81	105.55
14	South	54.89	26.79	259.98
15	West	223.41	78.49	458.45
16	Grand Total	405.74	150.21	1334.33

Excel 2010

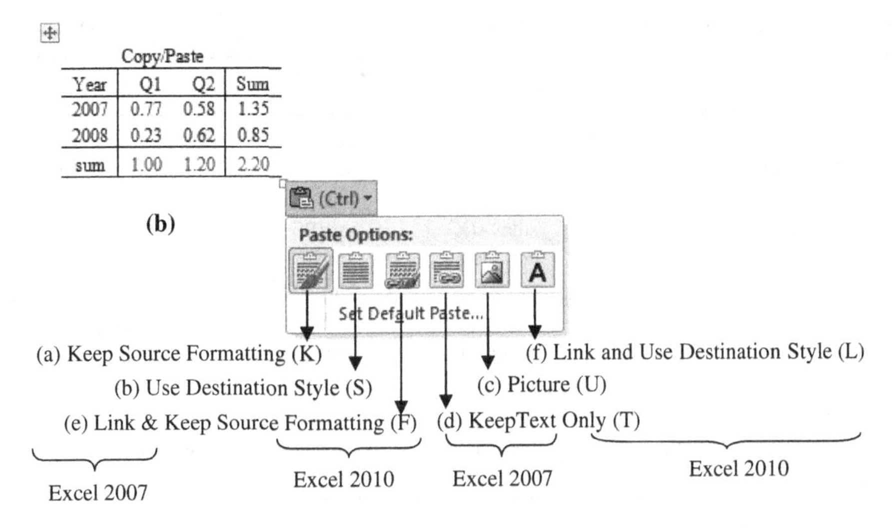

Fig. 14.3-P <Copy> in Excel, <Paste> in MS Word

Page 546 *Text animation in PowerPoint 2010*

In PowerPoint 2010, the <"Animations"> tab in PowerPoint 2007 has been separated into two tabs, <"Transitions"> and <"Animations">. Thus, in page 546, the Animation command sequence for PowerPoint 2010 may be revised as

<"Animations"><Animation, Fly In><Timing, Start!><On click>
<Animation, Effect Options!><Direction, from below>

The readers should be able to figure out how to modify the other PowerPoint 2007 command sequences for PowerPoint 2010.

Change in Excel functions

Some Excel statistical functions in Excel7 are revised and renamed in Excel10, but the old functions are still available as "**compatible functions**" in Excel10. Therefore, no revision of the functions used in the book is necessary. Nevertheless, we have listed revised functions in Table P1 for the convenience of the readers who want to use new functional form. Microsoft claims that the equivalent new formulas have been improved and more accurate, but at our level of exposition, it may only add more confusion.

In Table P1, sample and "population" variance and covariance functions in Excel10 are now separated clearly by adding .p and .s after var, but still retaining the confusing term "population." (in fact, all six functions also apply to sample data. The difference is whether the sum of mean square deviation (SSx) or the sum of the products of mean deviation (SSxy) is divided by n or n−1, as shown in the Note column of Table P1). In Excel10, the standard normal probability function (pdf) and the inverse of the standard normal cumulative distribution function (cdf) are now available, as shown in rows 8 and 9 of Table P1. The function names of the t distribution have changed to reflect right- or two-tailed

Table P1 Comparison of Excel formulas used in this book in Excel 2007 and Ecel 2010

#	Section page	equ#		Excel7 Formula	Book Notation	Excel10 Formula	Note*	Name
1	154	5.2.2	(5.12)	varp(X)	s_n^2	var.p(X)	SSx/n	"population" variance
2	156	5.2.3	(5.13)	stdevp(X)	s_n	stdev.p(X)	sqrt(SSx/n)	"population" standard deviation
3	158	5.2.4	(5.14)	var(X)	s^2	var.s(X)	SSx/(n-1)	(sample) variance
4	159	5.2.5	(5.16)	stdev(X)	s	stdev.s(X)	sqrt(SSx/(n-1))	(sample) standard devitation
5	161	5.3.1	(5.20)	covar(X,Y)	cov	covariance.p(X,Y)	SSxy/n	"population" covariance
6				n.a.		covariance.s(X,Y)	SSxy/(n-1)	(sample) covariance
7	211	6.7.2	(6.31)	normsdis(z)		norm.s.dist(z,1)	cumlative=1	standard normal cdf, P(Z<Z0)
8				n.a.	N(0,1)	norm.s.dist(z,0)	cumlative=0	standard normal pdf
9				n.a.		norm.s.inv(p)		inverse standard normal cdf
10	213	6.7.3	(6.34)	normdist(x,μ,σ^2,1)		norm.dist(x,μ,σ^2,1)	cumlative=1	(general) normal cdf, P(X<X0)
11	213	"	(6.34)	normdist(x,μ,σ^2,0)	N(μ,σ^2)	norm.dist(x,μ,σ^2,0)	cumlative=0	(general) normal pdf
12	215	6.8.1	(6.36)	norminv(p,μ,σ^2)		norm.inv(p,μ,σ^2)		inverse (general) normal cdf
13	253	7.5.1	(7.37)	tdist(t,df,1)		t.dist.rt(t,df)	tail=1	right-tailed t-dist, P(t<-t0 or t>t0)
14	253	"	(7.38)	tdist(t,df,2)		t.dist.2t(t,df)	tail=2	two-tailed t-dist., P(t<-t0 and t>t0)
15	256	7.5.3	(7.42)	tinv(p,df)		t.inv.2t(p,df)		inverse of two-tailed t-dist.
16	257	7.6.2	(7.46)	Fdist(F,df1,df2)		F.dist.rt(F,df1,df2)		right-tailed F-dist, P(F>F0)
17				n.a.		F.dist(F,df1,df2,1)	cumlative=1	F-cdf, P(F<F0)
18				n.a.	F(df1,df2)	F.dist(F,df1,df2,0)	cumlative=0	F-pdf
19	258	7.6.3	(7.48)	Finv(p,df1,df2)		F.inv.rt(p,df1,df2)		inverse of right-tailed F-dist.
				n.a.		F.inv(p,df1,df2)		inverse of F-cdf

Notes: cumulative=1= true, 0 = False. n.a. = not available in Excel7.

probability, as shown in rows 13 to 15. Three new F-functions are added to derive cdf, pdf, and the inverse of the cdf of the F distribution. The F distribution pdf (row 18) can be used to draw Fig. 7.5. Otherwise, since F statistic is already available from the Excel Regression Table, we don't need to use these three new functions.

Fig. P1(a) shows that, in Excel10, when "=norm" is entered in a cell, all the functions related to the normal distribution appears in the drop-down list. Each of the compatible functions from Excel7, the last four functions in this case, is denoted with a yellow triangle in the function logo on the left. The readers may double-click any function and enter the function in cell E2. Fig. P1(b) is the function list in Excel 2010 when "=t" is entered in a cell, and Fig. P1(c) is the corresponding list in Excel7. The corresponding equations are shown by arrows connecting the circles. When "=td" or "=ti" is entered in a cell in Excel10 (not shown), the drop-down list will show the compatible function with a yellow triangle.

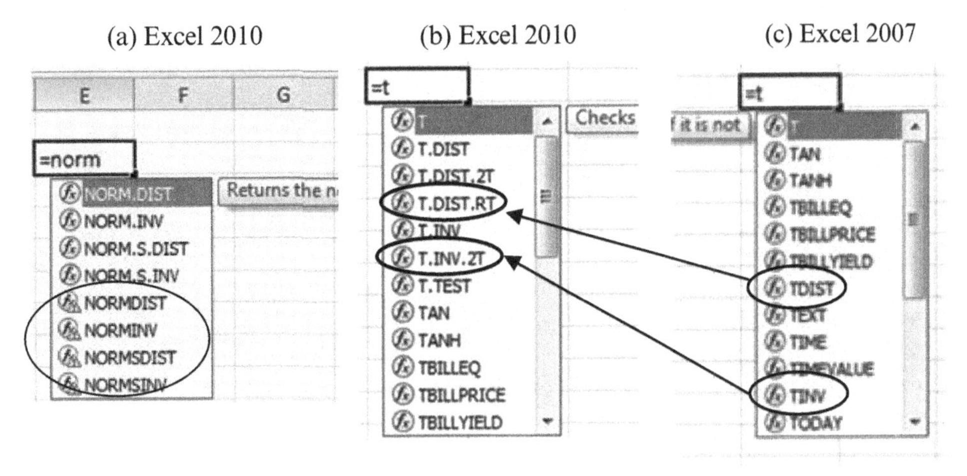

Fig. P1 Comparsion of the function list in Excel 2007 and Excel 2010.

Summary

So far as the purpose, contents, methods, procedures, and results of this book are concerned, changes in Excel 2010, as compared with Excel 2007, are rather cosmetic. As noted in this Postscript, if there is any change in Excel10, the readers of this book should be able to figure out easily the new location or names of the commands and formulas used in this book.

Finally, this book contains much more than Excel procedures. To those who do not use Excel much, or who have Excel-phobia, we recommend that they may sit back (from the computer) and read the book, skipping the part which expounds hands-on experiments of making tables and charts. They will miss the best opportunity to learn one of the most useful modern computer skills. However, they should still find in this book the concepts and exposition relating economic and business analysis with practical applications illuminating, backed by numerous helpful tables and charts, and enjoy the nice computer-drawn and accurate 2D and 3D charts, like the five cases of the CES production functions and three cases of returns to scale in Chapter 11, or 3D charts of production and utility functions in Chapter 12.

References

(Chapter number shows the chapter the reference is used in this book)

Cahill, Miles, and George Kosicki (2000), "Exploring Economic Models Using Excel," *Southern Economic Journal*, 66(3), pp. 770–792. Chapter 12.

Chiang, Alpha C., and Kevin Wainwright (2005), *Fundamental Methods of Mathematical Economics*, 4th ed. New York: McGraw-Hill. Chapters 11, 12.

Cobb, Charles W. and Paul H. Douglas (1928), "A Theory of Production," *American Economic Review*, Vol. 18, supplement. Chapter 11.

Cornell, Paul (2006), *Beginning Excel What-if Data Analysis Tools: Getting Started with Goal Seek, Data Table, Scenarios, and Solver*, Berkeley, CA: Apress, Chapter 12.

Dorfman, Robert, Paul A. Samuelson, and Robert M. Solow (1958), *Linear Programming and Economic Analysis*, NY: McGraw-Hill Book Co. Chapter 10.

Dougherty, Christopher (2007), *Introduction to Econometrics*, 3rd ed. NY: Oxford University Press. Chapter 7.

Fischer, Stanley, and Rudiger Dornbusch (1983), *Economics*, McGraw-Hill. Chapter 1.

Goldberg, Samuel (1958), *Introduction to Difference Equations, with illustrative examples from Economics, Psychology, and Sociology*. NY: Wiley. Chapter 15.

Gujarati, Damodar (1992), *Essentials of Econometrics*, NY: McGraw-Hill, 1992. Chapters 5, 7.

Gujarati, Damodar (1995), *Basic Econometrics*, 3rd ed. NY: McGraw-Hill. Chapter 7.

Gujarati, Damodar (2003), *Basic Econometrics*, 4th ed. NY: McGraw Hill. Chapter 7.

Harnett, Donald L. and James L. Murphy (1985), *Statistical Analysis for Business and Economics*, 3rd ed., MA: Addison-Wesley. Chapter 7.

Hess, Peter (2002), *Using Mathematics in Economic Analysis*, NJ: Prentice-Hall. Chapters 11, 12.

Hoel, Paul G. (1956), *Introduction to Mathematical Statistics*, John Wiley and Sons. Chapters 5, 6.

Hoy, Michael, John Livernois, Chris McKenna, Ray Rees, and Thanasis Sengos (2001), *Mathematics for Economics*, 2nd ed. MA: MIT Press. Chapter 10.

Hsiao, Frank S.T., and Mei-Chu W. Hsiao (2003), "Miracle Growth in the Twentieth Century-International Comparisons of East Asian Development," *World Development*, Vol. 31, No. 2. pp. 227–257. Chapter 13.

Hsiao, Frank S.T., and Mei-Chu W. Hsiao (2006), "FDI, Exports, and GDP in East and Southeast Asia — Panel Data versus Time-Series Causality Analyses," *Journal of Asian Economics*, Vol. 17, No. 6. pp. 1082–1106. Chapter 1.

JETRO Investment White Paper, 2001 (in Japanese), Tokyo: Japan External Trade Organization (JETRO). Chapter 14.

Kelejian, Harry H., and Wallace E. Oates (1989), *Introduction to Econometrics, Principles and Applications*, NY: Harper & Row, Publishers. Chapters 5, 7.

Klein, Michael W. (2001), *Mathematical Methods for Economics*, 2nd ed. Boston: Addison Wesley. Chapters 8, 11, 12.

Lee, Cheng F., John C. Lee, and Alice C. Lee (2000), *Statistics for Business and Financial Economics*, 2nd ed., Singapore: World Scientific. Chapter 10.

Lind, Douglas A., William G. Marchal, and Robert D. Mason (2002), *Statistical Techniques in Business and Economics*, 11th ed., NY: McGraw-Hill Irwin. Chapter 10.

Lott, William F., and Subhash C. Ray (1992), *Applied Econometrics: Problems with Data Sets*, NY: Dryden Press. Chapters 5, 7.

Maddison, Angus (1995), *Monitoring the World Economy, 1820–1992*, Development Centre, Paris: OECD. Chapter 13.

Mankiw, N. Gregory (1998), *Principles of Economics*, NY: The Dryden Press. Chapter 12.

Meier, Robert C., William T. Newell, and Harold L. Pazer (1969), *Simulation in Business and Economics*, Prentice-Hall, NJ: Englewood. Chapters 5, 6.

Miller, Ronald E., and Peter D. Blair (1985), *Input-Output Analysis, Foundations and Extensions*, Prentice-Hall. NJ: Englewood Cliffs. Chapter 10.

Mizrahi A. and Sullivan M. (1988), *Mathematics for Business and Social Sciences*, 4th ed., NY: John Wiley & Sons. Chapter 8.

Nishimura, Kazuo (1986), *Introduction to Micro Economics* (in Japanese), Tokyo: Iwanami Shoten. Chapter 11.

Norton, Peter (2000), *Introduction to Computers*, 4th edition, NY: Glencoe/McGraw-Hill. Chapter 14.

Ramanathan, Ramu (1995), *Introductory Econometrics with Applications*, 3rd ed., NY: The Dryden Press. Chapters 5, 7.

Root, Franklin R. (1990), *International Trade and Investment*, 6th ed., OH: South-Western Publishing Co. Chapter 2.

Salvatore, Dominick (1982), *Statistics and Econometrics*, Schaum's Outline Series, NY: McGraw-Hill, pp. 156–159. Chapters 5, 7.

Samuelson, Paul (1980), *Economics*, 11th ed. NY: McGraw-Hill. Chapter 1.

Samuelson, Paul A., and William D. Nordhaus (1995), *Microeconomics*, 15th ed. New York: McGraw-Hill. Chapter 12.

Santerre, Rexford E., and Stephen P. Neun (1996), *Health Economics, Theories, Insights, and Industry Studies,* Chicago: Irwin. Chapter 4

Solow, Robert M. (1957), "Technical Change and the Aggregate Production Function," The *Review of Economics and Statistics*, Vol. 39, pp. 312–320. Chapter 11.

Toumanoff, Peter, and Farrokh Nourzad (1994), *A Mathematical Approach to Economic Analysis*, Minneapolis/St. Paul: West Publishing Co. Chapter 11, 12.

Weil, David N. (2005), *Economic Growth*, NY: Addison-Wesley. Chapter 15.

World Bank (2009), *World Development Report, Reshaping Economic Geography.* Washington, D.C.: The World Bank Group. Chapter 13.

Index